Lake Pavin

Télesphore Sime-Ngando • Pierre Boivin
Emmanuel Chapron • Didier Jézéquel
Michel Meybeck
Editors

Lake Pavin

History, geology, biogeochemistry, and sedimentology of a deep meromictic maar lake

Springer

Editors
Télesphore Sime-Ngando
LMGE, Laboratoire Microorganismes: Génome et Environnement, UMR CNRS 6023
Université Clermont-Auvergne, Université Blaise Pascal
Aubière, France

Emmanuel Chapron
ISTO Institut des Sciences de la Terre d'Orléans, Observatoire des Sciences de l'Univers en région Centre
UMR 7327 (CNRS-Université d'Orléans-BRGM)
Orléans Cedex, France

GEODE UMR 5602 CNRS
Université de Toulouse 2 Jean Jaurès
Toulouse Cedex, France

Michel Meybeck
METIS UMR CNRS 7619
Université Pierre et Marie Curie
Paris, France

Pierre Boivin
LMV, Université Clermont-Auvergne - CNRS - IRD - OPGC
Aubière, France

Didier Jézéquel
IPGP (Institut de Physique du Globe de Paris)
Sorbonne Paris Cité, Univ. Paris Diderot
Paris, France

ISBN 978-3-319-39960-7 ISBN 978-3-319-39961-4 (eBook)
DOI 10.1007/978-3-319-39961-4

Library of Congress Control Number: 2016947055

Printed on acid-free paper

This Springer imprint is published by Springer Nature
The registered company is Springer International Publishing AG Switzerland

Preface

The volcanic mountain Lake Pavin (45°29′41′ N, 002°53′12″E) enjoys an increasing amount of interest both in terms of touristic activities, with more than 200,000 persons attracted each year by the beauty of the site, and in terms of scientific research with hundreds of references published in peer-reviewed international journals. Remarkably studies go back to the eighteenth century (Montlosier 1789, http://www.edition-originale.com/-MONTLOSIER-Essai-sur-la-theorie-des-volcans-d-Auvergne-Clermont-1802.html,35329). These publications resulted from the works of scientists from Europe but also from around the world (USA, China, Japan, etc.), attracted by the geomorphological and biological peculiarities of the deep (Z_{max} = 92 m) meromictic Lake Pavin. This maar lake offers a unique environment characterized by (i) a permanently anoxic monimolimnion from about 60 m downwards, (ii) a small surface area (44 ha) about equal to the drainage basin area (50 ha), and (iii) a substantially low human influence with no river inflow.

We propose the first multidisciplinary scientific volume centered on a maar lake (Lake Pavin) but importantly in comparison with other similar temperate lakes. The work groups different chapters in five main parts, reflecting the scientific research and disciplines conducted so far in Lake Pavin: (i) general limnology, history, and comparative legends (Chaps. 1, 2, and 3); (ii) origin, volcanology, and geological environment (Chaps. 4, 5, 6, 7, 8 and 9); (iii) geochemistry and biogeochemical cycles (Chaps. 10, 11 and 12); (iv) biology and microbial ecology (Chaps. 13, 14, 15, 16, 17, 18, 19, 20 and 21); and (v) sedimentology and paleolimnology (Chaps. 22 and 23). The biological part refers to the study of the biology of the lake, including pelagic and benthic communities. However, most of the chapters in this part mainly deal with microbial ecology, which has been extensively studied in Lake Pavin during the past decades. Major methodological, conceptual, and empirical advances in environmental sciences during the past decades indeed are known from the molecular study of uncultured microbes and viruses in natural environments.

Much more than a local interesting study site, we aim to provide an extensive rigorous limnological text centered on Lake Pavin, of value and interest for an international scientific audience. The syntheses of the main characteristics of Lake Pavin are, for the first time, set in a firmer footing comparative approach, encompassing regional, national, European, and international aquatic science contexts. In this book we synthesize heretofore very scattered knowledge on a unique site whose characteristics make it an ideal reference site to examine questions on freshwater lake origin, physics, geochemistry, biogeochemistry, sedimentology, biology, and variability in time and space. The targeted audience is researchers and advanced students, primarily in the fields of limnology, biogeochemistry, and aquatic ecology. Because the book is an outgrowth of an extensive teaching experience of the main authors, it can also serve as a useful reference for aquatic resource scientists, professionals, and engineers.

We thank all the contributing authors and coauthors of the book, some of whom also have reviewed submissions from colleagues. We are particularly indebted to non-anonymous reviewers who work hard for many months and provided invaluable comments. Thanks are owed to A. Cheronet (publishing editor for Environmental Sciences, Springer) and her staff for their precious assistance at different steps of the editing process. Finally, F. Rieu conducted the preliminary bibliographic research on Lake Pavin, on which the contents and organization of the book are based upon.

Aubière, France Télesphore Sime-Ngando

Contents

Contributors

G. Abril EPOC (Laboratoire Environnements et Paléoenvironnements Océaniques et Continentaux), UMR 5805, Université de Bordeaux, Pessac, France

P. Agrinier IPGP (Institut de Physique du Globe de Paris), Sorbonne Paris Cité, Univ. Paris Diderot, UMR 7154 CNRS, Paris, France

Patrick Albéric ISTO Institut des Sciences de la Terre d'Orléans, Observatoire des Sciences de l'Univers en région Centre, UMR 7327 (CNRS-Université d'Orléans-BRGM), Orléans Cedex 2, France

Christian Amblard LMGE, Laboratoire « Microorganismes: Génome & Environnement », UMR CNRS 6023, Université Clermont-Auvergne, Université Blaise Pascal, Aubière, France

Fabien Arnaud EDYTEM Environnement, Dynamiques et Territoire de la Montagne, UMR 5204 CNRS, Université de Savoie, Le Bourget du Lac Cedex, France

Eléonore Attard LMGE, Laboratoire « Microorganismes: Génome & Environnement », UMR CNRS 6023, Université Blaise Pascal, Aubière, France

Equipe Environnement et Microbiologie, UMR CNRS-IPREM 5254, Université de Pau et des Pays de l'Adour, Pau, France

Corinne Bardot LMGE, Laboratoire « Microorganismes: Génome & Environnement », UMR CNRS 6023, Université Clermont-Auvergne, Université Blaise Pascal, Aubière, France

Aude Beauger GEOLAB UMR 6042 CNRS-Université Blaise Pascal-Maison des Sciences de l'Homme, Clermont-Ferrand Cedex 1, France

Karim Benzerara Institut de Minéralogie, de Physique des Matériaux, et de Cosmochimie (IMPMC), Sorbonne Universités, UPMC Université Paris, CNRS UMR 7590, MNHN, IRD UMR 206, Paris, France

L. Bergonzini GEOPS (Géosciences Paris Sud), Université Paris-Saclay, Univ. Paris-Sud, Orsay Cedex, France

Sylvain Bernard LMCM Laboratoire de Minéralogie et Cosmochimie du Museum, UMR 7202 CNRS-Museum National d'Histoire Naturelle (MNHN) USM 0205, Paris, France

Yvan Bettarel LMGE, Laboratoire Microorganismes: Genome et Environnement, UMR CNRS 6023, Université Clermont Auvergne, Université Blaise Pascal, Aubière, France

Institut de Recherche pour le Développement (IRD) MARBEC Marine Biodiversity, Exploitation & Conservation UMR IRD (n° 248), Ifremer, Univ. Montpellier, CNRS (n° 9190), Montpellier, France

Corinne Biderre-Petit LMGE, Laboratoire Microorganismes: Génome et Environnement, UMR CNRS 6023, Université Clermont-Auvergne, Université Blaise Pascal, Aubière, France

Pierre Boivin LMV, Université Clermont-Auvergne - CNRS - IRD - OPGC, Aubière, France

Céline Bonhomme LEESU, UMR MA 102, École des Ponts ParisTech, Marne-la-Vallée, France

Guillaume Borrel LMGE, Laboratoire « Microorganismes: Génome & Environnement », UMR CNRS 6023, Université Clermont-Auvergne, Université Blaise Pascal, Aubière, France

Gilles Bourdier LMGE, Laboratoire Microorganismes: Génome et Environnement, UMR CNRS 6023, Université Clermont-Auvergne, Université Blaise Pascal, Aubière, France

Gisèle Bronner LMGE, Laboratoire Microorganismes: Génome et Environnement, UMR CNRS 6023, Université Clermont-Auvergne, Université Blaise Pascal, Aubière, France

Vincent Busigny IPGP (Institut de Physique du Globe de Paris), Sorbonne Paris Cité, Univ. Paris Diderot, UMR 7154 CNRS, Paris, France

Benoît Caron Laboratoire de Pétrologie, Géochimie et Minéralogie Magmatiques, Sorbonne Universités, UPMC Univ Paris 06, CNRS, Institut des Sciences de la Terre de Paris (iSTeP), Paris, France

Jean-François Carrias LMGE, Laboratoire Microorganismes: Génome et Environnement, UMR CNRS 6023, Université Clermont-Auvergne, Université Blaise Pascal, Aubière, France

Emmanuel Chapron Present Address: ISTO Institut des Sciences de la Terre d'Orléans, Observatoire des Sciences de l'Univers en région Centre, UMR 7327 (CNRS-Université d'Orléans-BRGM), Orléans Cedex 2, France

Present Affiliation: GEODE UMR 5602 CNRS, Université de Toulouse 2 Jean Jaurès, Toulouse Cedex 9, France

Marie Charpin LMGE, Laboratoire Microorganismes: Genome et Environnement, UMR CNRS 6023, Université Clermont-Auvergne, Université Blaise Pascal, Aubière, France

Léo Chassiot ISTO Institut des Sciences de la Terre d'Orléans, Observatoire des Sciences de l'Univers en région Centre, UMR 7327 (CNRS-Université d'Orléans-BRGM), Orléans Cedex 2, France

Jonathan Colombet LMGE, Laboratoire Microorganismes: Génome et Environnement (LMGE), UMR CNRS 6023, Université Clermont-Auvergne, Université Blaise Pascal, Aubière, France

Julie Cosmidis Department of Geological Sciences, University of Colorado, Boulder, CO, USA

Didier Debroas LMGE, Laboratoire Microorganismes: Génome et Environnement, Université Clermont-Auvergne, Université Blaise Pascal, Aubière, France

Anne-Lise Develle EDYTEM Environnement, Dynamiques et Territoire de la Montagne, UMR 5204 CNRS, Université de Savoie, Le Bourget du Lac Cedex, France

Isabelle Domaizon INRA, UMR 42 CARRTEL, Thonon les Bains, France

Eric Dugat-Bony INRA-AgroParisTech, UMR782 Génie et Microbiologie des Procédés Alimentaires, Thiverval-Grignon, France

François Enault LMGE, Laboratoire Microorganismes: Génome et Environnement, UMR CNRS 6023, Université Clermont-Auvergne, Université Blaise Pascal, Aubière, France

Gerard Fonty LMGE, Laboratoire « Microorganismes: Génome & Environnement », UMR CNRS 6023, Université Clermont-Auvergne, Université Blaise Pascal, Aubière, France

François Desmolles Departmental Federation for Fisheries and the Protection of Aquatic Environments, Lempdes, France

Mélanie Gerphagnon LMGE, Laboratoire Microorganismes: Génome et Environnement, UMR CNRS 6023, Université Clermont-Auvergne, Université Blaise Pascal, Aubière, France

Alain Gourgaud LMV, Université Clermont-Auvergne - CNRS - IRD - OPGC, Aubière, France

Vincent Grossi LGLTPE, Laboratoire de « Géologie de Lyon, Terre, Planètes, Environnement », UMR CNRS 5276, Université Claude Bernard, Villeurbanne, France

Mylène Hugoni LMGE, Laboratoire Microorganismes: Génome et Environnement, UMR CNRS 6023, Université Clermont-Auvergne, Université Blaise Pascal, Aubière, France

Alexis Jaillard Institut Polytechnique LaSalle Beauvais, BEAUVAIS cedex, France

Didier Jézéquel IPGP (Institut de Physique du Globe de Paris), Sorbonne Paris Cité, Univ. Paris Diderot, UMR 7154 CNRS, Paris, France

Marlène Jobard LMGE, Laboratoire Microorganismes: Génome et Environnement, UMR CNRS 6023, Université Clermont-Auvergne, Université Blaise Pascal, Aubière, France

Isabelle Jouan-Dufournel LMGE, Laboratoire Microorganismes: Génome et Environnement, UMR CNRS 6023, Université Clermont-Auvergne, Université Blaise Pascal, Aubière, France

Sylvaine Jouhannel LMV, Université Clermont-Auvergne - CNRS - IRD - OPGC, Aubière, France

Etienne Juvigné Laboratoire de Géomorphologie et de Géologie du Quaternaire, Sart-Tilman, Bât. 11, Université de Liège, Liège, Belgium

Benoit Keraval LMGE, Laboratoire « Microorganismes: Génome & Environnement », UMR CNRS 6023, Université Clermont-Auvergne, Université Blaise Pascal, Aubière, France

Philippe Labazuy LMV, Université Clermont-Auvergne - CNRS - IRD - OPGC, Aubière, France

Patrick Lajeunesse CEN Centre d'études nordiques, Département de Géographie, Université Laval, Québec, Canada

Delphine Latour LMGE, Laboratoire Microorganismes: Génome et Environnement, UMR CNRS 6023, Université Clermont-Auvergne, Université Blaise Pascal, BP 80026, Aubière, France

Pierre Lavina Institut Polytechnique LaSalle Beauvais, BEAUVAIS cedex, France

Oanez Lebeau Institut de Physique du Globe de Paris, Sorbonne Paris Cité, Université Paris Diderot, UMR 7154 CNRS, Paris, France

Alberic Leclerc Ecole Supérieure des Géomètres et Topographes, Le Mans, France

Grégoire Ledoux CEN Centre d'études nordiques, Département de Géographie, Université Laval, Québec, Canada

Emilie Lefevre LMGE, Laboratoire Microorganismes: Génome et Environnement, UMR CNRS 6023, Université Clermont-Auvergne, Université Blaise Pascal, Aubière, France

Present address: Department of Civil and Environmental Engineering, Duke University, Durham, NC, USA

Anne-Catherine Lehours LMGE, Laboratoire « Microorganismes: Génome & Environnement », UMR CNRS 6023, Université Clermont-Auvergne, Université Blaise Pascal, Aubière, France

Cécile Lepère LMGE, Laboratoire Microorganismes: Génome et Environnement, Université Clermont-Auvergne, Université Blaise Pascal, Clermont-Ferrand, France

CNRS, UMR 6023, LMGE, Aubière, France

Hervé Leyrit Institut Polytechnique LaSalle Beauvais, BEAUVAIS cedex, France

F. Lopes ECP (Ecole Centrale Paris), Châtenay-Malabry Cedex, France

Pascale Lutz Institut Polytechnique LaSalle Beauvais, BEAUVAIS cedex, France

Isabelle Mary LMGE, Laboratoire Microorganismes: Génome et Environnement, UMR CNRS 6023, Université Clermont-Auvergne, Université Blaise Pascal, Aubière, France

Michel Meybeck METIS, CNRS-Unversité Paris 6¸UMR 7619, Paris Cedex 05, France

Didier Miallier Laboratoire de Physique Corpusculaire, Clermont Université, Université Blaise Pascal, Clermont-Ferrand, France

Gil Michard IPGP (Institut de Physique du Globe de Paris), Sorbonne Paris Cité, Univ. Paris Diderot, UMR 7154 CNRS, Paris, France

Yannick Miras GEOLAB UMR 6042 CNRS-Université Blaise Pascal-Maison des Sciences de l'Homme, Clermont-Ferrand Cedex 1, France

Sébastien Monchy LMGE, Laboratoire Microorganismes: Génome et Environnement, UMR CNRS 6023, Université Clermont-Auvergne, Université Blaise Pascal, Aubière, France

Present address: INSU-CNRS, UMR 8187 LOG, Laboratoire d'Océanologie et de Géosciences, Université du Littoral Côte d'Opale, ULCO, Wimereux, France

Jean-Pierre Morel LMGE, Laboratoire « Microorganismes: Génome & Environnement », UMR CNRS 6023, Université Clermont-Auvergne, Université Blaise Pascal, Aubière, France

Nicole Morel-Desrosiers LMGE, Laboratoire « Microorganismes: Génome & Environnement », UMR CNRS 6023, Université Clermont-Auvergne, Université Blaise Pascal, Aubière, France

Stéphanie Palesse LMGE, Laboratoire Microorganismes: Genome et Environnement, UMR CNRS 6023, Université Clermont-Auvergne, Université Blaise Pascal, Aubière, France

Pierre Peyret Clermont-Université, Université d'Auvergne, EA CIDAM 4678, Clermont-Ferrand, France

Noah J. Planavsky Department of Geology and Geophysics, Yale University, New Haven, CT, USA

Michel Poulin Centre de Géosciences, Ecole des Mines ParisTech, Paris, France

Angia Sriram Pradeep Ram LMGE, Laboratoire Microorganismes: Genome et Environnement, UMR CNRS 6023, Université Clermont-Auvergne, Université Blaise Pascal, Aubière, France

Serena Rasconi LMGE, Laboratoire Microorganismes: Génome et Environnement, UMR CNRS 6023, Université Clermont-Auvergne, Université Blaise Pascal, Aubière, France

Present address: WasserCluster – Biological Station Lunz, Inter-university Centre for Aquatic Ecosystem Research, Lunz am See, Austria

Viviane Ravet LMGE, Laboratoire Microorganismes: Génome et Environnement, UMR CNRS 6023, Université Clermont-Auvergne, Université Blaise Pascal, Aubière, France

Simon Roux LMGE, Laboratoire Microorganismes: Génome et Environnement, UMR CNRS 6023, Université Clermont-Auvergne, Université Blaise Pascal, Aubière, France

Mohamed Saad LEESU, UMR MA 102, École des Ponts ParisTech, Marne-la-Vallée, France

Gérard Sarazin IPGP (Institut de Physique du Globe de Paris), Sorbonne Paris Cité, Univ. Paris Diderot, UMR 7154 CNRS, Paris, France

Markus J. Schwab GFZ, Helmholtz Zentrum Potsdam, Potsdam, Germany

Télesphore Sime-Ngando LMGE, Laboratoire Microorganismes: Génome et Environnement, UMR CNRS 6023, Université Clermont-Auvergne, Université Blaise Pascal, Aubière, France

Najwa Taïb LMGE, Laboratoire Microorganismes: Génome et Environnement, UMR CNRS 6023, Université Clermont-Auvergne, Université Blaise Pascal, Aubière, France

Bruno Tassin LEESU, UMR MA 102, École des Ponts ParisTech, Marne-la-Vallée, France

Pauline Thierry Laboratoire de Pétrologie, Géochimie et Minéralogie Magmatiques, Sorbonne Universités, UPMC Univ Paris 06, CNRS, Institut des Sciences de la Terre de Paris (iSTeP), Paris, France

Jean-Claude Thouret LMV, Université Clermont-Auvergne - CNRS - IRD - OPGC, Aubière, France

Benoît Villemant Laboratoire de Pétrologie, Géochimie et Minéralogie Magmatiques, Sorbonne Universités, UPMC Univ Paris 06, CNRS, Institut des Sciences de la Terre de Paris (iSTeP), Paris, France

Brigitte Vinçon-Leite LEESU, UMR MA 102, École des Ponts ParisTech, Marne-la-Vallée, France

Eric Viollier IPGP (Institut de Physique du Globe de Paris), Sorbonne Paris Cité, Univ. Paris Diderot, UMR 7154 CNRS, Paris, France

Renata Zocatelli ISTO Institut des Sciences de la Terre d'Orléans, Observatoire des Sciences de l'Univers en région Centre, UMR 7327 CNRS-Université d'Orléans-BRGM, Orléans Cedex 2, France

William Zylberman Institut Polytechnique LaSalle Beauvais, BEAUVAIS cedex, France

Part I

Limnology, History and Comparative Legends

Michel Meybeck

1 Pavin, the Birthplace of French Limnology (1770–2012), and Its Degassing Controversy (1986–2016)

Michel Meybeck

Abstract

Lake Pavin is located in the Auvergne Mountains, central France, at 1300 m a.s.l. This small lake (0.44 km^2), partially fills an explosive volcanic crater in the Cezallier, a young volcanic area south of the Mont Dore. Its deepest part, from 65 to 92 m, is permanently anoxic, a very rare limnic phenomenon termed *meromixis*. Pavin is the cradle of French limnology, having been first surveyed in 1770, and then regularly studied since 1880 by local botanists and zoologists from Clermont-Ferrand, along other pristine lakes of the Cezallier. Pavin scientists, such as Delebecque the founder of French limnology and his friend Martel the founder of speleology, visiting the area in 1892, always used up-to-date sampling techniques and methodologies, often borrowed from other disciplines. Meromixis at Pavin was described for the first time in the 1950s and after 1970, its deep waters attracted new teams of isotope geochemists, water chemists, microbiologists and sedimentologists, often from foreign origin. After the unexpected and deadly limnic explosion of Nyos Lake (Cameroun) in 1986, the possibility of Pavin degassing was investigated and concluded to a lack of risk under present conditions. Pavin exceptional history and corpus of legends, referring to its repeated misbehavior and latent fear, perceived locally and in the greater area (Chaps. 2 and 3), remained unknown to contemporary scientists until now. To allow for the re-interpretation of these complex sources, a sensory grid of maar-lakes degassing is proposed here, based on scientists' observations or reports at other maar- lakes very similar to Pavin, Nyos (1986) and Monoun (1984) in Cameroun, Albano (398 BC) and Monticchio (1770s–1820s) in Italy.

Keywords

Pavin Lake • Limnology history • Lake degassing • Maar-lake • Nyos lake • Degassing grid

"*This lake, located on the top of the Mont Dore Mountain is, by its shape and its details, one of the most beautiful and most singular lakes of our country and adds to the many beautiful monuments that nature has provided Auvergne with.*" (Depping 1811)

"*Crater lakes can be used as useful test tubes to elucidate transfer processes in aquatic systems… It is a giant rain-gauge to register atmospheric fallout.*" (Martin 1985)

"*Why a dramatic event [as the one that occurred at Nyos] could not have happened in the volcanic ranges of Massif Central?*" (Tazieff, 2 September 1986, La Montagne, daily newspaper)

1.1 Introduction

Pavin Lake in Auvergne province is different from all other French lakes. It combines a rare origin, a *maar-lake*, *i.e.* a lake filling a crater lake resulting from a volcanic explosion, with an exceptional mixing type, the *meromictic character*, *i.e.* its bottom waters do not mix with the surface waters. In addition Pavin Lake presents a lack of direct human impacts

M. Meybeck (✉)
METIS, CNRS-Université Paris 6, UMR 7619, Place Jussieu, 75252 Paris Cedex 05, France
e-mail: michel.meybeck@upmc.fr

T. Sime-Ngando et al. (eds.), *Lake Pavin*, DOI 10.1007/978-3-319-39961-4_1

and a long Human history including a rich corpus of legends characterized by latent lake fear. It is is also considered as the most beautiful lake of the Massif Central (see above, Depping 1811). For geochemists it is a hydrosystem with limited forcing from its watershed that could be considered as a test tube (see above, Martin 1985). These points make it an exceptional lake in many fields of Earth Sciences and of Humanities.

"Pavin", as its visitors have called it for centuries, has also been identified soon after the catastrophic lake degassing at Nyos (Cameroon), in August 1986, as a lake with a potential degassing risk, a matter still debated today (see above Tazieff, the French volcanology leader). Pavin has also attracted over the last 140 years scientists from many countries: in the last 30 years they have found very specific chemical and microbiological features that will be addressed in more than twenty chapters of this book, so that Pavin is now one of the most investigated lake in Europe.

The first three chapters of this book present Pavin from the point of views of three different actors: (i) the scientific community at Pavin, from the beginning of scientific exploration to the recent degassing controversy (Chap. 1), (ii) Pavin neighbors and visitors on which Pavin history is based (Chap. 2), (iii) the local population and their perception of Pavin through its multiple legends, stories and associated beliefs (Chap. 3). These points of view cover both scientific and non-scientific knowledge concerning this natural entity. In each chapter the introductive sections and the conclusions are presented to the lay reader in each of these fields, including synthetic tables or figures. The core of the chapters provides the detailed argumentation on which this analysis is based. A preliminary analysis about Pavin resulted in some new hypotheses and led to the conclusions that Pavin had probably degassed at the sixteenth century (Meybeck 2010), which could be related to the fear that Pavin inspired over centuries, to its exceptional name – the terrifying- and to its relation with intense local religious practices.

This first chapter briefly sets up the Pavin scene, in comparison to other lakes in Auvergne and to other maar-lakes in Europe, and then presents the works of scientists at Pavin, from 1770 to 1986. Finally, the controversy on degassing risk opened by Tazieff at Pavin, after the Lake Nyos catastrophic event in 1986, is presented. This concern was one of the focuses of the international workshop convened in 2009 at Besse (Jezequel et al. 2010a, b). At each period the study of Pavin by scientists is placed in its historical context: Pavin has often been at the forefront of French limnology, then of international limnology. The new findings at Pavin after 1986, in geology and volcanology, limnology, biology and microbiology, are only shortly mentioned here as they are fully developed in the many chapters of this volume.

1.2 Analysis of Pavin Actors, History and Perception Through an Interdisciplinary and Intercomparative Approach

History and legends are so entangled at Pavin that a specific interdisciplinary approach has been needed to collect, analyze and synthetize hundred sources on which this work is based. This analysis has also required, in these chapters, an inter-comparison with few other maar-lakes located in Cameroon (Nyos and Monoun), in Italy (Averno, Albano, Nemi and Monticchio) and in Germany (Eifel lakes). There is no temporal limit to our analysis: we are aiming to analyze the societal connexion to such dangerous places as Pavin and its companions over the Longue Durée. This extends over 2500 years for some lakes. These comparisons have permitted to decipher some of the Pavin attributes, well known since centuries, sometimes still ignored by Earth Scientists and Social Scientists.

There is no current standard approach to investigate the history of lake degassing, although we have been preceded by Italian volcanologists, geologists and geochemists working on Lake Albano (Funiciello et al. 2002, 2003, 2010) and Monticchio Lakes (Caracausi et al. 2009). We have also used many works on legends and degassing impacts to populations made at Lake Nyos by Evgenia Shanklin (1989, 2007). In order to make a significant progress we worked through a non-linear and iterative approach: (i) assembling a maximum of written sources, of religious, archeological, folkloric and iconographic elements, and re-attributing a first set of these sources to Pavin, (ii) gaining from the knowledge of other maar-lakes, particularly for the degassing description, (iii) re-interpreting a second set of Pavin sources with the degassing sensory descriptors (see Table 1.2, this chapter), (iii) re-interpreting legends, fantastic stories, local religious history and iconography (Chap. 3).

The main steps of this research are the following.

Step 1. Preliminary analysis (Meybeck 2010)

Our preliminary analysis led to the following conclusions and working hypotheses:

(i) Pavin had degassed during the sixteenth century,
(ii) The generated fear, well documented during at least the late XVIth and the seventeenth centuries, was important,
(iii) Pavin legends are complex. The commonly accepted legend in the twentieth century, the Sunken City legend, has been forged at the end of the nineteenth century,

(iv) Catastrophic events that occurred in Pavin area in the 1300s could be related to Pavin Lake but are not documented by historical sources,
(v) Pavin Lake *"misbehavior"* – a term referring to the maar-lake abnormalities perceived by locals, coined at Nyos by Shanklin (1989) – and the nearby Vassivière Christian pilgrimage could be linked.

Step 2. In-depth material collection

Works on Pavin and its area were collected first, then those on nearby Vassivière mountain and its pilgrimage and nearby Creux de Soucy cavity, the towns of Besse and Mont Dore (see Fig. 1.2c). The research has also been extended to major texts on Auvergne history and description (sixteenth to nineteenth centuries), to all Auvergne maps until the mid-1800s, to Auvergne tales and legends collected in the XIXth. The quest has also been extended to other recent scientific findings, myths and tales on feared lakes in France and in Switzerland, and on maar-lakes in Italy and Germany (developed in Chap. 3), in connexion with their limnological characteristics (Chap. 1). Major *Cosmographiae*, i.e. early global geography descriptions and geography works (sixteenth to nineteenth centuries) have also been considered. The regional outlook has been completed by the guidebooks on Besse and Mont Dore areas, on Auvergne (nineteenth to twenty-first centuries). Relevant articles on Pavin in the local newspapers and magazines (*La Montagne, La Galipote, Eruptions*), focused on 1986–1987 and 2009–2012 were also considered. The resulting corpus includes more than 300 references.

Step 3. Sensory description of lake degassing (Nyos Lake and Italian lakes)

Nyos publications and oral presentations of the French team members gathered at the international workshop held in 2009 at Besse were analyzed to find witness description of the post-event lake, soon after the peak degassing and of the damages to Humans and livestock's. In addition we used the historical descriptions of past degassing in Italian maar-lakes in the early 1800s found by volcanologists of the INVG (Istituto Nazionale di Geofisica e Vulcanologia, Palermo) for the Piccolo and Grande Monticchio Lakes (Basilicate) and by Rome Earth Scientists (Geological Sciences, University Roma Tre) for the degassing event that occurred in 398 BC in Lake Albano (Latium) (Chiodini et al. 1997; Caracausi et al. 2009; Funiciello et al. 2003, 2010; De Benedetti et al. 2008). The combination of these scientific assessments results in a set of sensory degassing descriptors (Chap. 1, Table 1.2).

Step 4. Re-analysis of Pavin textual material

The re-analysis of the Pavin textual material is done here with several assumptions: (i) Pavin is a gaz-containing maar-lake; as such its degassing is not impossible, as reported in other similar lakes, (ii) past degassing can be triggered by occasional event- slumps, rock fall, earthquake- or related to local variability of CO_2 emissions; it might have occurred any time, (iii) if it occurred with enough intensity to be noticed by local people, this non-scientific knowledge could have been reported, orally transmitted or expressed through local beliefs and customs, including religious, (iv) reporting, representation and explanation of these events are specific of their time and should therefore be interpreted in their historical context (Chap. 2), (v) this re-analysis can be done through a grid of sensory descriptors, established on other degassing lakes, Nyos, Monticchio, Albano Lakes mainly. The Pavin material has then been split into two categories: the perfectly contextualized sources, well dated, written by identified, educated people, which are discussed at Chap. 2 and all other sources as myths, legends, fantastic stories, miracles, religious iconography, which are discussed at Chap. 3. *[Original citations, in English, German and Italian, have been translated in French by the author].*

Step 5. Legends, beliefs and fears in Pavin and other maar-lakes

All traditional Pavin stories were analyzed: some of them can be considered as distorted descriptions of past Pavin state, others, as the Sunken City legend, are very recent (end of the nineteenth century). A fantastic story, published in 1632, so far not attributed to Pavin, and a legend featuring Pagan times, collected in the late nineteenth century, are re-attributed to Pavin. Similarities with legends, pagan and/or early Christian cults, from other mountain lakes and maar-lakes, in Italy and at Eifel, in Germany, are discussed in Chap. 3.

Step 6. Chronological presentation of Pavin state

A tentative chronology of Pavin state and of Pavin perception by different actors, local populations, scholars and scientists is proposed.

Several recursive loops from steps 2–5 have been necessary: most texts have been analyzed several times as new findings were made (Fig. 1.1).

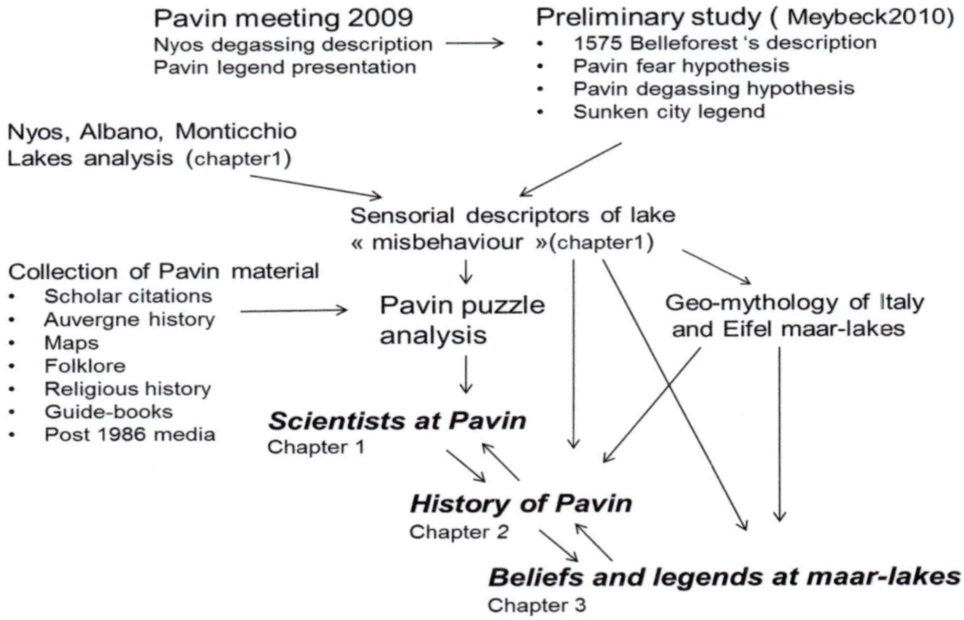

Fig. 1.1 Simplified methodology applied for the analysis of Pavin Lake history and degassing

1.3 Pavin, a Typical Maar-Lake Above any Contamination Source

1.3.1 Pavin General Features

Pavin lake is located in Auvergne, one of the few recent volcanic provinces in Europe (Fig. 1.2a). It is adjacent to the Mont Dore volcano massif (named *Mons Aureus* in the Antiquity period, then Mont d'Or) and belongs to the Western Cézallier volcanic district (Fig. 1.2b). It is located just at the hydrological divide between the Loire and the Dordogne watersheds, at an altitude of 1197 m. It is only topped by the Montchal volcano (also Montchalme) (1417 m), 0.5 km further south (Fig. 1.2c).

Pavin is a typical maar-lake, i.e. a lake located in an explosive crater depression. Maar-lakes are rare when considering the global distribution of lakes (Meybeck 1995a, b) and have in common several features: (i) the aerial lake basin is annular, (ii) they do not receive stream waters and their water balance is realized through direct rainfall or snowfall and groundwater inputs,(iii) many of them have no natural outlet, (iii) their detrital inputs are often limited to shore erosion, so that lake sedimentation is limited and/or controlled by internal sources. Pavin, with an area of 44 ha, 400 m in diameter, almost round in shape, with a present depth of 92 m, probably 4 m deeper before the excavation of natural aerial outlet in the late 1800th, is a sort of giant pluviometer, mostly fed by atmospheric inputs and few springs within the crater. It is frozen each year during 3–4 months and partially mixes twice a year (dimictic lake), when the water column is at 4 °C, but the mixing does not affect the whole water column, its bottom waters, termed *monimolimnion*, are therefore permanently anoxic, a characteristic of *meromictic* lakes. It has an aerial outlet flowing in a steep notch.

Until 1980 Pavin was in ultra-pristine conditions and oligotrophic, i.e. with very limited algal production. Today it is regarded as a natural laboratory for the study of the oxic-anoxic water interface, for the iron cycle during Precambrian period or for archeobacteria analysis (Jezequel et al. 2010a, b). Its classification as a unique European natural site has been proposed (Meybeck 2010).

The nearest town is Besse, some kilometers downstream, located since the twelfth century on the south side of the Couze Pavin River thalweg. Throughout the Middle Age until the mid-1800s the lake was isolated and largely avoided by local people, only coming in winter to take away some wood, due to their long-lasting fear of the *Lacus Pavens*, the terrifying lake (see Chap. 2). This reputation possibly originated since the Antiquity (see Chaps. 2 and 3). Until the mid 1800s Pavin was practically devoid of any fishes and there was no direct access by a carriage way, the lake was essentially observed from the rim until the mid-1800s (Lecoq 1835a, b). The first path to Pavin was traced in the mid-1800s by Henri Lecoq, an Auvergne naturalist who also brought the first boat and introduced chars to the lake (Chap. 2). The present circular pedestrian path around the lake has been built up in 1909 by the city of Besse and the access paved road in the 1960s. During centuries the pedestrian access to the nearby Vassivière mountain, where an important pilgrimage is located, was passing in the valley beneath the lake where the outlet meets the Couze Pavin river, probably on the left bank of the Couze Pavin (Fig. 1.2c).

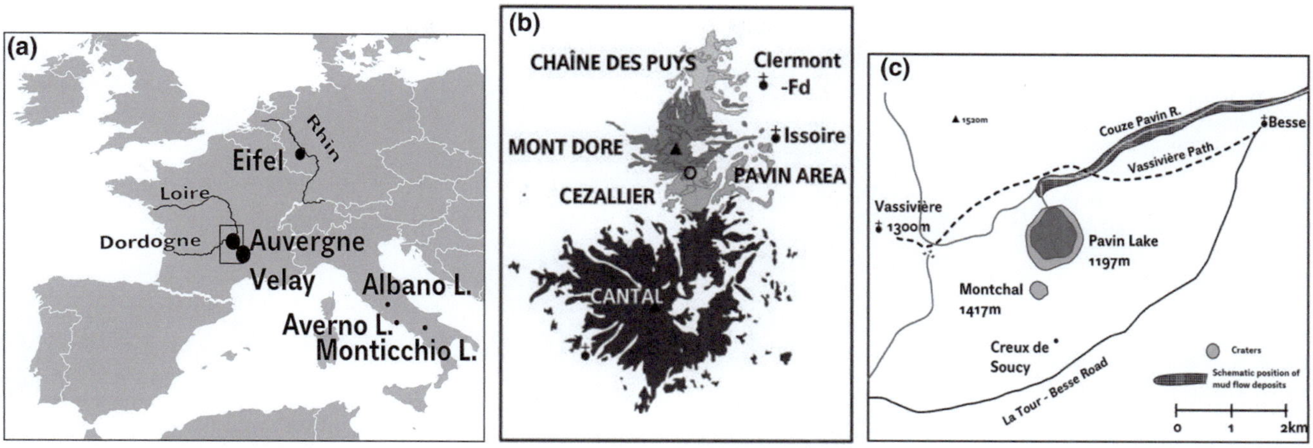

Fig. 1.2 (**a**) Position of European volcanic lake districts. (**b**) Auvergne volcanic districts. (**c**) Pavin area with Besse, Vassivière and Creux de Soucy, ancient roads and footpaths (early 1800s)

Until the mid-1950s the Pavin area was nearly void of permanent populations, excepted for few shacks and a chapel on top of the "Vassivière Mountain", at an altitude of 1300 m, the history of which could be connected to Pavin (Meybeck 2010 and Chaps. 2 and 3). From there, one can see the edge of the crater, but not the lake.

Pavin explosive crater has been formed on the side of Montchal volcano (Fig. 1.2a), slightly younger, characterized by a small crater and a scoriaceous lava flow. The whole Pavin area corresponds to a relatively recent volcanism (see Chap. 7), in which CO_2 degassing evidence has been found since the mid-nineteenth century (Martel 1894; Bakalowicz 1971; Lavina and del Rosso 2009; Gal and Gadalia 2011).

Many aspects of Pavin characteristics are very rare, some were noted since 1892: its water balance, its very special chemistry (richness in silica; extreme iron levels in anoxic deep waters), its deep groundwaters inputs, its thermic anomaly with slightly warmer deep waters. They are similar to some other volcanic lakes located in more active volcanic district as in Japan (Yoshimura 1937; Touchart and Ishiguro 1999). The meromictic nature of Pavin, now unique in France, is also commonly found in Japan (Yoshimura 1938), in Eifel (Germany) and in Latium (Italy), see further.

1.3.2 Pavin Compared to Other Lakes of the Cézallier Lake District

Pavin is surrounded by several lakes of major scientific and / or ecological importance. They are all located between 1050 and 1300 m, on the watershed divide between the Dordogne and Allier rivers (Fig. 1.3a). These lakes are in pristine condition with very limited Human pressures, excepted for pasture: the population density on their basin is less than 1 person per km^2. The diversity of these lakes, their origins and morphology is known since the survey of Delebecque (1898) (Table 1.1). Pavin, Chauvet, and La Godivelle d'en Haut are considered as maar-lakes, while Montcineyre results from the recent damming of a post-glacial depression by a volcanic cone. There are many post-glacial peat-bogs with residual lakes as Chambedaze, La Godivelle d'en Bas and Bourdouze, all with high biodiversity value with numerous relict species of the last glaciation, some well-known for their paleolimnological records of Holocene climate and vegetation, others protected at the national and European level (Meybeck 2010).

The *Creux de Soucy* Lake, located in a cavity within the Montchal lava flow, some 35 m deep, one kilometer South of Pavin, is a steep underground pond, 10 m deep, still under exploration by speleologists (Bakalowicz 1971; O. Pigeron, pers.com) (Figs. 1.2c and 1.3b). It is near 2 °C all year round and is so far unique in continental Europe. This cavity has been described in 1575 by a royal cosmographer, Belleforest, and has been thought since that time to feed Pavin with groundwater (See Sects. 2.3.3 and 2.3.5). The cavity has been visited for the first time in 1892 (Sect. 2.3.6) and its exploration still continues today: a vertical drowned corridor, found in the late 1960s, now extends the lake depth to 50 m.

La Godivelle d'en haut Lake has been intensively equipped in the 1960s for the study of lake evaporation in altitude (1300 m asl) and of water budget (snow and rain inputs, groundwater outputs) (Jacquet and Mandelbrot 1966). The Chambedaze Lake and peatland is famous for its palynological and tephra studies (Guenet and Reille 1988; Juvigné et al. 1993).

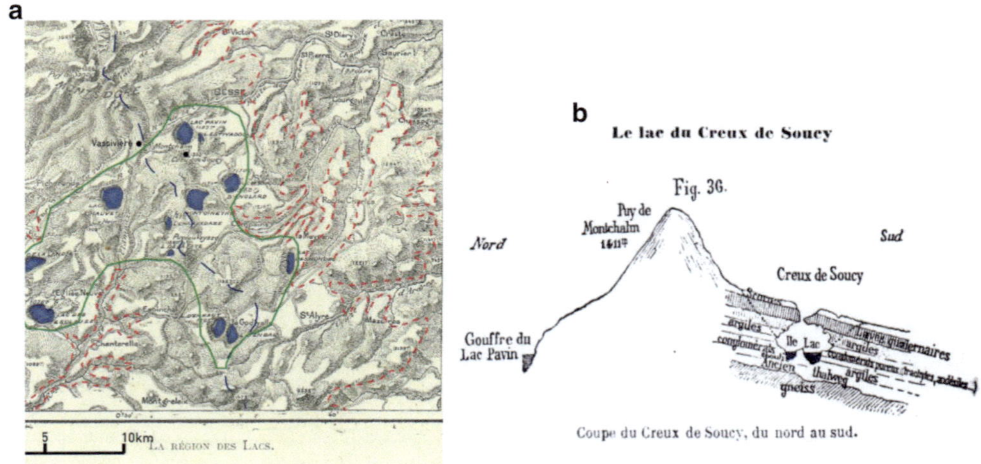

Fig. 1.3 The Cezallier lake district in Auvergne (Berthoule 1890) with the 1050 m altitude contour line in *red* and a possible delineation of the most pristine lakes in *green*. *Blue dotted line*: watershed divide between Dordogne and Allier Rivers (Meybeck 2010, based on Berthoule 1890). 3b North-south cross section Pavin-Montchal- Creux de Soucy by Gobin (1896) (BCU library, Clermont-Ferrand)

Table 1.1 Lmnological characteristics of Cezallier lakes, Pavin and other European maar-lakes in Central and South Italy and in Eifel

Lake	L.A.	ZM	V	Zm	Alt	C.A.	H.I.	Origin	Outlet
Cezallier									
Pavin	0.44	92	23.0	65	1197		Low	maar	aerial
Chauvet	0.54	64	17.3	N.A.	1162		Very low	maar	aerial
Godivelle Haut	0.15	44	2.74	N.A.	1239		Very Low	maar	subterranean
Servières	0.15	26		N.A.	1202		Very low	maar	
Montcineyre	0.4	18		N.A.	1182		Very low	volc. dam	subterranean
Chambedaze	0.02	(2)	(0.04)	N.A.	1180		Very low	peat bog	aerial
Bourdouze	0.15	4.5	(0.6)	N.A.	1168		Very low	peat bog	aerial
Godivelle Bas	0.16	(3)	(0.5)	N.A.	1200		Very low	peat bog	aerial
Central and South Italy									
Albano	5.8	167	450	70	293	3.68	Medium	maar	Tunnel (398 BC)
Nemi	1.7	32	30	15–20	316	10.3	Medium	maar	Tunnel (400 s BC)
Monticchio Grande	0.41	36	3.27	15	654	2.4	Medium	maar	Dug-out canal
Monticchio Piccolo	0.14	38	2.4	15	656	1.07	Medium	maar	Dug-out canal
Averno	0.55	34	6	7	<10		High	maar	Tunnel (500 s BC)
Eifel									
Pulver	0.38	74	14.3	N.A.	411	0.8	Low	maar	Subterranean
Weinfelder	0.17	51	4.3	N.A.	484	0.35	Low	maar	Subterranean
Ulmen	0.055	39	1.12	20	420	0.155	High	maar	Medieval tunnel
Laacher <1164	4.83	65	167	N.A.	289	12.2		maar	
Laacher> 1854	3.32	51	103	N.A.	275	12.2	Low	maar	Tunnel (1164)

L.A. lake area (km^2), *ZM* present maximum depth (m), *V* volume (Mm^3), *Zm* present monimolimnion depth (m), *C.A.* catchment area (km^2), *H.I.* human impacts, *N.A.* non applicable

The Cézallier lake district (Fig. 1.3a) has an exceptional level of naturalness index, as defined by Maichado (2004), (Meybeck 2010). The diversity of lake origin and morphology, water and sediment chemistry, hydrology, sedimentology, trophic status, fauna and flora, is exceptionnal as already noted by Berthoule (1890), Delebecque (1898), Bruyant and Eusebio (1904). However,this rare ensemble is not yet recognized as exceptional by geologists or naturalists (Sabouraud 2004; de Wever et al. 2006; Michel 2008) although some of its elements have been individually protected, mostly on ecological criteria (PNRVA 2004; Meybeck 2010).

1.3.3 Pavin Compared to Other European Maar-Lakes

Western continental Europe has three major volcanic lakes districts: Auvergne, Central Italy (Latium and Campania) and Eifel, between the Mosel River and Aachen (Fig. 1.2a) (Table 1.1). All Eifel lakes are termed "*maars*".

1.3.3.1 Eifel Lakes

The Eifel volcanic district, located in Germany, West of the Middle Rhine Valley, is famous for its deep round lakes which fill volcanic explosion craters (Buchel 1993). These depressions, termed locally maars, gave their name to other crater lakes of similar origins. Many of them are now completely filled with sediment, some of them remain filled with water, among them some famous ones: Laacher See, Pulver Maar, Ulmen Maar, Weinfelder Maar (Table 1.1). They are characterized by their round shape, circular drainage basin, absence of natural aerial outlet. This region, between 300 m and 500 m a.s.l., is exposed to Human pressures from villages and agriculture: many of these lakes have received a nutrient excess and have been eutrophied, with the exception of Laacher See, Pulver and Gemündener Maars (Scharf and Björk 1992).

The Eifel lakes have been first studied by August Thienemann, the father of German limnology (Thienemann 1914–1915), who found the permanent anoxia of Pulver Maar which was used later to define the lake meromicticity (Findenegg 1935). More recently they have been intensively studied for their chemistry and ecology (Scharf and Björk 1992). Although this synthesis was realized after the catastrophic Nyos event which occurred in 1986 (see further), no consideration was given to their CO_2 content in bottom waters and /or potential degassing risks. These studies were performed later (Aeschbach-Hertig et al. 1996). Unlike for the Italian maar-lakes no use of geomythology has yet been realized in this legend –rich region (see Chap. 3).

The Laachersee Lake is well known among earth scientists. It was formed 12,900 years ago by a sudden explosion which sent into the atmosphere about 6 and 20 km^3 of ashes hundreds of kilometers away within a few days. This ash layer is now used as a sedimentological tracer in many western European lakes (Bogaard et Schmincke 1995). The residual lake, now 51 m deep, has no natural aerial inlet or outlet. On its shore there is still evidence of a gentle degassing activity.

1.3.3.2 Italian Lakes

In Latium half a dozen lakes of volcanic origins are found (Margaritora 1992; Niessen et al. 1993; Chondrogianni et al. 1996; Elwood et al. 2009; Carapezza et al. 2008) (Table 1.1). Albano and Nemi Lakes are found in maars aged 38,000–40,000 years. Albano Lake, 3.5 km long by 2.3 km wide, is located in an important maar with twin underwater depressions forming a single lake. Albano waters are anoxic below 50 m. Lake Nemi is some 3 km SE of lake Albano, on the other side of Monte Cavo. In both lakes drainage basin is annular and their hydrological budget is therefore exceptional, with regards to most other lakes (Pourriot and Meybeck 1995) with a dominance of groundwaters inputs and outputs: there are neither stream inputs nor natural aerial outlet, a peculiarity also found in most other maar-lakes. Albano crater rim is today at 70 m above the water level. These two meromictic lakes have been drained by dug-out tunnels in Antiquity (see Sects. 3.5.3 and 3.7). Both lakes have been eutrophied.

The twin meromictic maar-lakes, Lago Piccolo di Monticchio (LPM) and Lago Grande di Monticchio (LGM), are located in the Mount Vulture, an isolated volcano, between Campania and Basilicate (Southern Italy), also known for its deep forest. These lakes are located in a remote environment and are separated by a narrow exundated isthmus of 50 m. Both are anoxic below 15 m today, characterized by a marked positive thermal gradient in deep waters (+1.1 °C/10 m) and contain important amount of CO_2, from mantelic origin (Chiodini et al. 1997; Schettler and Alberic 2008; Mancino et al. 2009; Caracausi et al. 2009).

Averno Lake, *i.e.* the lake without birds for the Greek colony in Italy in Antiquity, is a small maar lake located in the Phlegrean Fields, a very active volcanic district West of Naples. The lake is meromictic and a few years ago a massive fish kill has been reported (Caliro et al. 2008) suggesting a rollover, a partial mixing of deep anoxic waters. As in Pavin, microbiologists are studying its very peculiar bacteria communities (Bianchi et al. 2010). Antique tunnels and an outlet canal have also been dug-out on Averno shores (see Sect. 3.5.3). Averno has received sewage effluents in the past so that is present bottom anoxia can also be related to this pollution.

1.4 Pavin Scientific Exploration (1770–1985)

1.4.1 Chevalier's Expedition (1770)

Pavin scientific history begins in September 1770. Before that year Pavin is still thought to be bottomless and the origin of its waters, their linkage with the nearby Creux de Soucy cavity, are one of its many mysteries, as reported by local people (See 2.3.3). A high-rank royal civil engineer, Chevalier, is asked to make an expedition to disproof these "tales". He comes to Pavin on September 27 and 28, 1770, fully equipped with a sounding line and a lead weight, as those used by marine engineers, with a Réaumur thermometer, and with material to measure relative altitudes, but he has to make his own raft. His objectives are multiple: (i) to measure the depth of Pavin and of the nearby Creux de Soucy, (ii) to measure their relative altitudes, to check the Soucy-Pavin connection, (iii) to measure the temperatures of these lakes. All objectives are fulfilled: Pavin depth is measured at 288 ft (96 m), the Creux de Soucy lake level is determined to be above Pavin level, allowing a possible subterranean connection. Chevalier sends a bucket into the Creux de Soucy water: to his surprise the water is extremely cold for the summer season (5 °R or 6.25 °C). He also finds a set of springs, hidden in the vegetation, opposite the lake outlet, at the bottom of *"a volcanic cliff whose lava layer was recovered by ashes"*, one of the first report of the volcanic nature of Pavin crater. Chevalier has not made a thermal profile of Pavin, as De Saussure did in 1767 for Lake Geneva (Touchart 2003), but he is the first French scientist to have studied the major components of a lake system: its origin and basin rock composition, its water inputs, its depth and temperature. His measurements are still valid today, if one considers the lowering of the lake, from 96 m to 92 m, when the natural sill was dug out in the late 1800s.

Soon after Chevalier's expedition the first accurate maps of Pavin region were realized by Cassini de Thury, within the general mapping of France, and published between 1775 and 1815. They included four other crater lakes, within 20 km from Pavin, Chauvet, Godivelle d'en Haut, Servières – and two volcanic dams lakes – Chambon and Montcineyre. In these official maps, Pavin is still referenced in his original spelling, *Paven* (see Sect. 2.3.4). Few years after, the altitude of the lake is measured by L.F. Ramond, a mathematician topographer and prefect of Puy-de-Dôme: Pavin Lake water level is at 1203 m (today 1197 m) and the Montchal Volcano at 1411 m a.s.l.

This expedition is as soon widely celebrated by the Auvergne scholars (Legrand d'Aussy 1788; Delarbre 1795; 1805). For them it is the scientific proof of the lack of grounded truth of "Pavin stories" (see Sect. 2.3.5): Pavin has a definite depth, it is fed with groundwaters, and it can be connected with the Creux de Soucy. Legrand d'Aussy (1788) is particularly fascinated by Pavin and would like to know the *"shape and verticalness of the slope, composition of the water, where do these water come from, what is the temperature profile with depth, what is the water pressure at 48 toises, the density of water at different depth, the light penetration in lake bottom, etc"*. This visionary program will only be started 100 years later and is still the main research objectives of dozens of scientists that are studying Pavin today.

1.4.2 Lecoq, the Great Auvergne Naturalist, Normalizes Pavin... with Fishes (1847–1871)

Henri Lecoq (1802–1871), a botanist, geologist and hydrologist is attracted by Pavin since his first visit in 1831, during which the marvelous Pavin stories were still narrated by local people (See 2.3.6 and 3.3). Lecoq will spend his life to refute these tales one by one. In 1847 he manages to have the first boat to be assembled at the lake and makes soundings which confirm those of Chevalier (95 m), a decade later he introduces fish to the lake.

This operation, one of the first ever in France, is realized with his colleague Barnabé Rico, who is charged by Lecoq of the new and innovative piscicultures at Clermont, then Besse. This required building a pathway up to the lake, then bringing materials, boats and juveniles to Besse from the Clermont pisciculture facility, some 50 km away. In January 1859 the first fish juveniles are introduced: "92,000 trouts (*Salmo trutta*), 20,000 common salmons (*Salmo salar*), 18 Heusch salmons (*Hucho hucho*), 8000 chars (*Salvelinus Umbla*), some coregons (*Coregonus Fera*): 120018 salmonidae; 130 Cyprinidae and 200 adults crayfish (*Astacus fluviatilis*), in total" (Rico 1876; Berthoule 1890).

Before these introductions only gudgeons were present at Pavin. They were very peculiar by their size and their speckles (Rico 1876) and *"unique by their spotted and flecked colours"* (Bruyant and Eusebio 1904) but no attempt is made to further study them. In another isolated maar-lake in Mount Vulture (Italy), similar fish speciation had been observed for a small fish, *Cyprinus Vulturius* Tenore, although the reason for this speciation, as isolation or specific water chemistry, was not discussed by Gussone and Tenore (1838).

The pisciculture facility, soon constructed at Besse by Lecoq and Rico, is followed by another one in the nearby crater lake Chauvet (Berthoule 1890). At Pavin, it turns to be a great scientific, commercial and touristic success. Three years after the introduction, the first salmons (as big as 1100 g and 58 cm), trouts (1700 g for 54 cm, aged 38 months) and chars (up to 750 g for 43 cm), are caught and crayfish population is now developed. Some specimens of extraordinary sizes are reported: one 18.5 kg eel (1873), two salmons (8 kg and 12 kg, 80 cm, caught in 1874 in the same net) and one of the Heusch salmons, a 105 cm female weighting 14.5 kg with its eggs, from the juveniles introduced at only 9 cm some 15 years before. The rapid growth of these species probably benefitted from the lack of fishes before the introduction. The average catch for the first 10 years is 157 kg / year, i.e 3.7 kg/ha/year. Fishes are sent to the best hotels in Mont Dore, Clermont and Besse.

This successful salmon introduction is first challenged by Paris scientists in charge of the national fish introduction plan, then gradually recognized by them, but Rico remained hurt by the deny of his successful work until his last day. During the Lecoq-Rico period (1859–1873), the lake outlet is leveled and lowered; a little stone cabin is built by the lake. It will be used by Clermont scientists who are working on Pavin in the late 1870s and will attract tourists until the 1950s (See Fig. 2.7).

Lecoq has not realized important limnological studies at Pavin, except the description of a fresh water sponge *Euspongilla lacustris* (Lecoq 1859), maybe the first-ever limnological publication in France. Lecoq's last book, on the Massif Central waters (Lecoq 1871) provides some descriptions of Pavin, termed by him the *Auvergne Dead Sea*, due to its initial lack of fish. It is one of the first works on general hydrology in France, although also much less quantitative than Belgrand's book on the Seine River basin (1869), which benefited from the support of highest state authorities, including observations of dozens of imperial civil engineers that were lacking to Lecoq. Considering that the Auvergne scientist had interest in many fields of science, botany, hydrology, geology and even global carbon cycling, his limited study of Pavin can raise questions. He was already 57 years old in 1859 and Pavin access from Clermont was still difficult. Most of all, he was missing the technical instruments and the laboratory facilities that were gradually being developed on Lake Geneva (Léman) by François-Alexis Forel, the founder of limnology. Forel had his own laboratory and boat by the lake shore in Morges (Vaud canton) and his monography on Léman (1892), is also much more quantitative than Lecoq's work in hydrology. Modern scientific exploration of French lakes begins with André Delebecque, who worked besides Forel on Léman and comes to Pavin in 1892, fully equipped.

1.4.3 The First Golden Age of Science at Pavin: Berthoule, Delebecque, Martel, Bruyant (1880–1914)

A few years after Lecoq died in 1871, botanists and zoologists from Clermont Faculty of Sciences initiate the first flora and fauna inventories of Auvergne lakes, which correspond to the start of French limnology, particularly at Pavin. These scientists are now well equipped and are essentially focusing on surface waters. Their research is much facilitated by Amédée Berthoule, at this period mayor of Besse (Berthoule 1896). He is well recognized for his master book (*Les Lacs d'Auvergne. Orographie, faune naturelle, faune introduite*, 1890) on Auvergne lakes and their halieutic potential, which he developed in with many quantitative data and fauna lists (Fig. 1.4), a major improvement of Lecoq's work, 20 years before. It can be considered as the first limnology textbook published in France.

1.4.3.1 Clermont Botanists and Zoologists Establish the Limnological Station at Besse

The first hydrobiologist at Pavin is Girod, a zoology professor at the Faculty of Science in Clermont-Ferrand, who makes the first fauna inventory, specifically on sponges (Girod 1877–1878). His students are extending this inventory to the other lakes of the region. They are using the plankton net developed at Monaco by Prince Albert for the ocean research vessel "Hirondelle". Richard studies the cladocerans (Richard 1887), Eusebio the pelagic fauna as entomostracae and rotifers, Bruyant focusses on phytoplankton, making vertical profiles and horizontal catches and studying seiches. Between 1887 and 1912 this team publishes 25 papers, fully listed in the analysis of Wurtz (1945). In total Berthoule, Girod and their colleagues study 20 lakes. Their work demonstrates for the first time the great variety of Auvergne lakes, as foreseen by Lecoq and Berthoule.

TABLEAU SYNOPTIQUE DE LA FAUNE DES LACS.

NOMS DES LACS.	NOMS DES ESPÈCES.											
—	Brême.	Brochet.	Carpe.	Chevenne.	Corégone.	Écrevisse.	Épinoche.	Gardon.	O. Chevalier	Perche.	Tanche.	Truite.
Anglard	✠✠	✠						✠		✠	✠	
Aydat			✠✠	✠		✠		✠		✠✠	✠	O
Chambédaze		✠✠	O					✠✠		✠✠		
Chambon	✠✠	✠				✠		✠✠		✠✠	✠✠	✠
— Couessat	✠✠	✠				✠		✠✠		✠✠	✠✠	
Chauvet					O				O	✠✠		OO
Esclauzes		✠								✠✠	✠✠	
Godivelle inférieur			✠✠							✠	✠✠	✠
— supérieur										✠		
Guéry							✠					
La Crégut										✠	✠	✠✠
La Faye											✠✠	✠
La Landie			✠✠							✠✠	✠✠	O
Pavin						O			O			OO
Servière										✠✠		O
Tazanat	✠	✠	✠							✠✠	✠	

Les signes ✠✠ désignent les espèces appartenant à la faune naturelle.
— OO — — — introduite.
Leur nombre indique la prédominance de chacune d'elles.

Fig. 1.4 Fish distribution in Cezallier and Mont Dore lakes. It only features introduced species at Pavin and one native species at Chauvet Lake (Berthoule's 1890)(BCU library Clermont-Ferrand)

The most innovative scientists is Charles Bruyant (1894). Together with Eusebio, he publishes in 1904 the *"Introduction à l'aquaculture générale. Matériaux pour l'étude des rivières et des lacs d'Auvergne"*. This book goes further than Lecoq's and Berthoule's, including rivers, their hydrological and thermal regimes and their chemistry in relation to the various basin rock types. It is a major analysis realized at a time when all these aquatic systems were still in a quasi-pristine state. In 1908 Bruyant becomes director of the *Pisciculture départementale*, initiated by Lecoq and Rico. He does not belong to the department of zoology of the Faculty of Sciences, as most of his colleagues, but is professor at the School of Medecine and Pharmacy where he studies phytoplankton. Following Eusebio (1896) he develops the Hydrobiological Station at Besse, also named the Limnological Station, on the model of the Marine Research Stations. Its official opening takes place in August 1908 during the meeting of the French Association for the Advancement of Sciences, at Clermont. From there a group of French and foreign scientists comes to Besse, surprised by the new facility and struck by this new scientific center (Reynouard 1909). The limnological station, i.e. the pisciculture and the hydrobiology station, acquires a high scientific visibility. Its objectives combine the fish introduction in rivers and lakes and the study of the complex relations between animal and plant species and their physico-chemical environment. Bruyant also considers enlarging the scope of the station to mountain environments, with geology, mountain vegetation and also local history (Bruyant 1910). He launches a scientific journal, the *Annales de la Station de Limnologie de Besse*, with a first volume of 400 pages including an important selected bibliography. Bruyant also promotes this new science, in an eight-page introduction to limnology for a remarkable guidebook (Cany et al. 1916), probably his latest work as he will disappear in World War one. The Besse Annales will be stopped and will reappear only in the 1950s. Bruyant and Eusebio are two pionners of limnological concepts as for the vertical structure of lakes (Fig. 1.5b, c).

In summer 1892 Pavin is hosting two other prominent French scientists: André Delebecque, a young limnologist who has collaborated on Lake Geneva with Forel, the "father of limnology", and Edouard-Alfred Martel, the "father of speleology", attracted by the Creux de Soucy.

1.4.3.2 André Delebecque at Pavin (1892)

André Delebecque (1861–1947) is a young state civil engineer from the most prestigious schools of engineers, Ecole Polytechnique and Ecole des Ponts et Chaussées. Unlike his comrades he chooses in 1887 to develop the newly established Thonon-les-Bains pisciculture. There, on Lake Geneva shores, he contributes to the bathymetry of Léman (Lake Geneva) and meets Forel. Few years later he initiates the first in-depth survey of French lakes: bathymetry, thermography, sedimentology, water and sediment chemistry. He uses his own portable skiff equipped with a winch and up-to-date sampling equipments (Touchart 2002) as Secchi disk, Negretti and Zambra thermometer, sediment grabs, comparable in many ways to those used in oceanography. His chemical analyses are performed by two specialized laboratories, at the Geneva University (Pr. Duparc) and in Paris (Laboratoire des Ponts et Chaussées). Although he is not making biological inventory, he is the first limnologist, in France and elsewhere, to compare more than 300 lakes of different origins and types.

In Auvergne, the young limnologist makes the sounding and survey of all lakes including the seven major crater lakes of the Massif Central: Pavin (92 m now), Servières (27 m), Chauvet (63.2 m), Godivelle d'en Haut (43.7 m) and Tazenat (66.6 m), all in Auvergne, and Le Bouchet (27.5 m) and Issarlés (108.6 m) in Velay. He also describes the many peat-bog lakes in Pavin region (Chambedaze, Bourdouze, Godivelle d'en Bas). With Ritter, a young geochemist from Geneva, probably the first foreign scientist at Pavin, he works on Pavin from June 19 to 24, 1892, and makes important discoveries (Delebecque 1898):

1. Pavin bottom is flat and regular, with steep slopes, near cliffs on the eastern side; it is the second most "hollow" lake in France, after Issarlés (Fig. 1.5a),
2. The water transparency in June is one of the highest he had measured in French lakes, 8.5 m; nearby peat bogs have less than 1 m,
3. Sediments are highly siliceous, corresponding to a "*randammite*", now *diatomite*, i.e. only composed of diatoms siliceous remains,
4. Two temperature profiles are made, every meter in surface, then every 10 m; below 25 m waters are very cold, less than 5 °C, reaching 4.6 °C at the bottom. Another measurement provides 4.8 °C: Delebecque attributes this slight increase to organic matter decomposition. Pavin is a typical dimictic lake (two mixing periods and two stratification periods),
5. Pavin waters are less mineralized compared to Jura lakes or others, but more mineralized than other Auvergne lakes, reaching record dissolved silica levels (22.1 mg/L at outlet).

Delebecque is a pioneer of comparative limnology. While Forel (1892) has spent his lifetime to study Léman (Geneva Lake), Delebecque completes, in few years time, the first standardized survey of all French lakes for all their physical and chemical attributes: lake morphology with detailed bathymetric map, lake origin, thermal stratification, water chemistry, lake sediments origin and chemistry etc. He sets typologies, which are still used today (Dussard 1966;

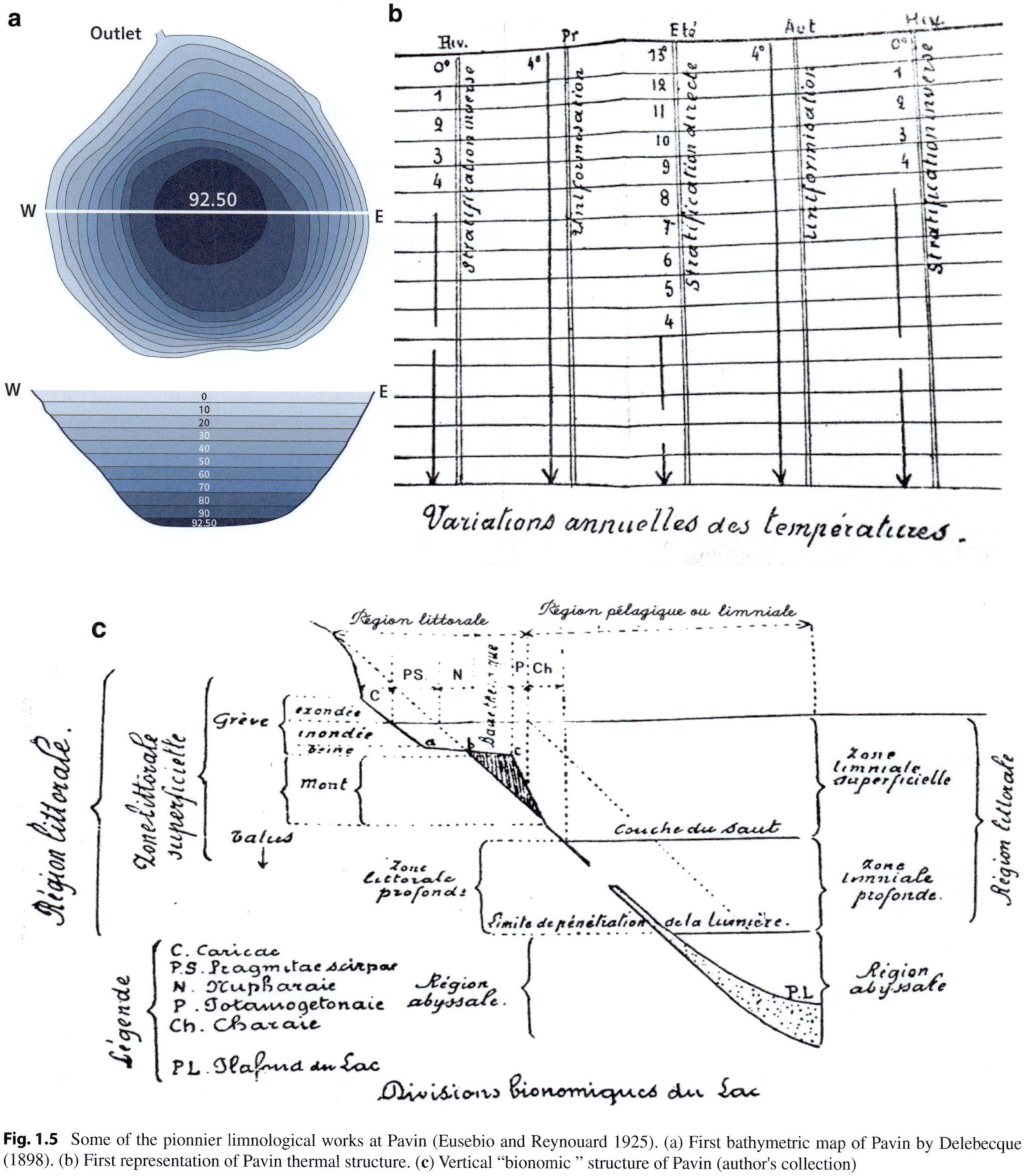

Fig. 1.5 Some of the pionnier limnological works at Pavin (Eusebio and Reynouard 1925). (a) First bathymetric map of Pavin by Delebecque (1898). (b) First representation of Pavin thermal structure. (**c**) Vertical "bionomic " structure of Pavin (author's collection)

Touchart 2002). He does not measure dissolved oxygen, a part of Léman: this new approach initiated by Gérardin in the Seine basin (1875) is used in this period for rivers suspected to be polluted. Delebecque does not focus on bottom waters – limnologists have not yet found that few lakes can be anoxic. At Pavin, he does not record any abnormality, except for the peculiar smell from bottom sediments, and completely misses the absence of 0_2 in the bottom waters, the *meromixis* which will be "discovered" 60 years later.

Delebecque's work on Pavin appears to be complementary to the hydrobiological studies performed by Berthoule, then by the Clermont team. His master work on French lakes did not get the international recognition he deserved, particularly in the history of limnology (Touchart and Dussard 1998) but his work contributed to get Pavin recognized by prominent limnologists such as G.E. Hutchinson (1957). Many of his observations are still used today by French limnologists (Pourriot and Meybeck 1995).

1.4.3.3 Edouard-Alfred Martel at Creux de Soucy (1892)

Edouard-Alfred Martel (1859–1938) comes to Besse in June 1892 with his team and his special equipment including a foldable skiff. He is a friend of Delebecque, with whom he worked in Savoy (Touchart and Dussard 1998). Both have planned to explore the Creux de Soucy cavity together. Delebecque and his young colleague, Ritter, would be in charge of the study of the bizarre *tiny lake, with near freezing waters in summer*, according to Chevalier (1770). They benefit from the full support of Amédée Berthoule, the mayor of Besse, whose father has tried to explore the cave some 30 years before without success, being stopped by a CO_2 layer which caused him headaches and dizziness. The expedition mobilizes eight people.

Martel is attracted by the famous cave, the only one in a volcanic setting and previously described by Belleforest in 1575 (See Sect. 2.3.3). The cave is some 1.3 km south of Pavin, together with few other depressions already mapped by Cassini in the late 1700s. It is 35 m deep and the small lake in the bottom of the cave is believed by all to be related with Pavin (Figs. 1.2c, and 1.3b), a scientific possibility established by Chevalier (1770). The cave is hidden in the Montchal scoriaceous lava flow so that Martel (1894) has difficulties finding the very small opening, only few meters wide and he needs to be guided. Unfortunately he is also stopped at 4 m above the small lake by the absence of oxygen and must go back. The cavity bottom will only be reached that year in August 1892 by Berthoule, and then it will fully be explored by the Clermont naturalists, Gautier and Bruyant, a few months later, in November 1892, when CO_2 is totally absent from the cave. These attempts to unveil the Creux de Soucy mysteries, highly anticipated in Auvergne for more than 300 years, correspond to major scientific expeditions and their preliminary results are published at once.

The small Soucy Lake is very peculiar (Gautier 1892; Martel 1892, 1894; Gautier and Bruyant 1896): its temperature is around 2 °C all year round, *i.e.* it is a "*polar lake*" always below 4 °C (Delebecque 1898; Touchart 2002) although it very rarely freezes, unlike all other polar lakes. In such extreme conditions, without light, with minimal temperature and with very limited nutrients inputs – from the snow melt – aquatic biomass and biodiversity are extremely limited: the aquatic plant community is very simple, essentially the diatom *Asterionella Formosa*, also dominating in Pavin. The lake depth is around 9.5 m and seasonaly variable, with highest levels at the spring melt.

In the late 1960s new underwater exploration revealed a near vertical corridor some 40 m deep, still being investigated. The Soucy Lake level is 45 m above the one of Pavin, making the connexion between the two possible, but the groundwater circulation remains debated today. As for Pavin, the scientific and patrimonial value of the Creux de Soucy is worthy of recognition and full protection.

1.4.4 Pavin Meromixis Discovery by Olivier and Pelletier (1950–1960s)

After 1918 the limnological survey at Pavin is slowed down but does not stop. Eusebio publishes with Reynouard, a local historian and former Besse mayor, the first scientific guidebook devoted to Pavin (Eusebio and Reynouard 1925). In this remarkable brochure they combine the presentation of Pavin long history and pioneer scientific observations as the many vertical divisions of Pavin surface waters, combining temperature, morphology, sedimentology and plant communities (Fig. 1.5b, c). In 1936, Luc Olivier, a Clermont hydrobiologist, makes repeated vertical plankton profiles at Pavin between 0 and 20 m. He is using Richard water sampling bottles with a reversing thermometer. He also innovates, making profiles of pH and of dissolved oxygen, by the Winkler method, between 0 and 50 m. They are among the first profiles at such depth in France, after those made on Léman. But he does not find any peculiarity in the oxygen profile (Olivier 1939). In 1943 another survey by Wurtz (1945) confirms the occurrence of O_2 at 70 m (8.34 cm^3 O_2/L) for 4.4 °C. On May 24, 1951 Olivier makes new profiles, every 10 m, down to 92 m, and finds for the first time a marked O_2 depletion at 65 m, confirmed again on August 16: 2.3 cm^3 O_2/L at 60 m, 0 at 70 m, 9.1 cm^3 at 80 m and at 90 m. He reports at once this extra-ordinary profile to the French Academy of Sciences (Olivier 1952). For him an anoxic layer has just been established around 70 m, "in contrast to Wurtz's opinion".

In summer 1962 Jean Pelletier, a young limnologist from Thonon-les-Bains hydrobiological station, makes another vertical profile for temperature, oxygen, dissolved iron and H_2S. It is followed by five continuous temperature profiles from 1963 to 1967 measured in situ with a Metrix thermal probe (Fig. 1.6): another French premiere at Pavin (Pelletier 1963, 1968).He confirms Olivier findings and adds new extraordinary Pavin features (Fig. 1.6):

1. There is no oxygen below 70 m,
2. There is a marked yet moderate thermal increase in bottom waters, from 4.2 °C at 60 m to 5.2 °C at 90 m (a thermal inversion),
3. H_2S is present below 70 m,
4. Dissolved iron is 100 times more concentrated in deep waters,

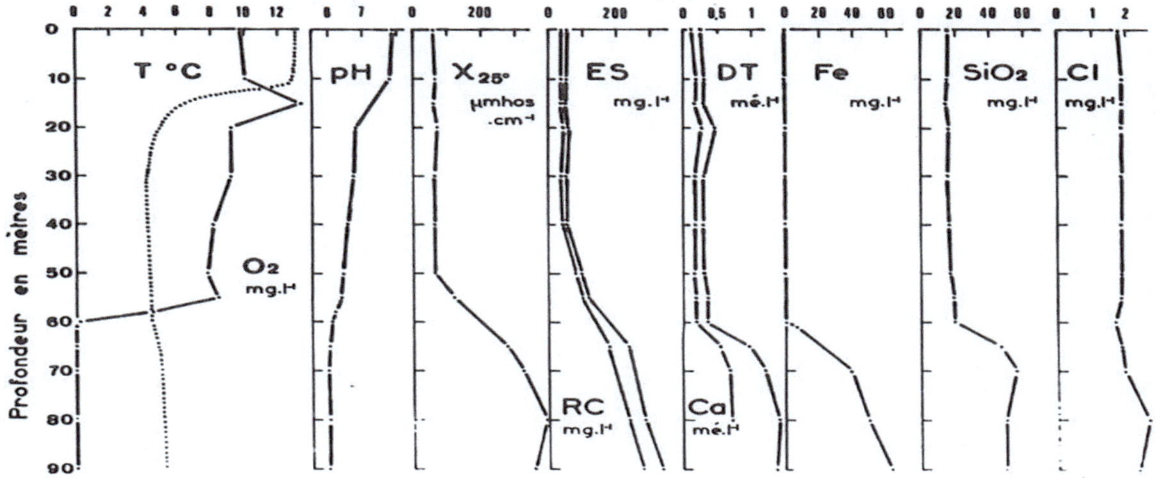

Fig. 1.6 Pelletier's vertical profiles showing Pavin meromixis in October 1967: temperature, O_2, pH, conductivity, total dissolved solids (ES), total hardness (DT), dissolved iron, silica and chloride (1968). (Ann. Stat. Limnologie Besse-en-Chandesse)

5. Bottom waters are very mineralized, with silica exceeding 50 mg/L
6. Pavin is termed "*meromictic*" by Pelletier, with an upper layer, the *mixolimnion* mixed twice a year, and a bottom layer, the *monomolimion*, that is never mixed, according to Findenegg (1935).

The discovery of Pavin meromixis is surprising: according to the current geochemical models the quantity of dissolved salts present in deep waters, which originate from more mineralized hydrothermal inputs, corresponds to multisecular hydrothermal inputs (Jezequel et al. 2010b). This suggests that the meromixis should have occurred long ago but was not correctly identified for lack of focus on deep waters and/or for difficulties in dissolved oxygen measurements. It is also possible that the anoxic bottom layer has been much reduced from 1937 to early 1950s due to a soft degassing and partial mixing event in 1936, still debated (see further). After Pelletier's first profiles, all oxygen profiles will show anoxia and a temperature increase below a certain depth, which may fluctuate.

Meromixis, with its associated mixolimnion and monimolimnion, is a rare limnological feature that can only occur when lakes are very "hollow" and/or when deep waters are denser than surface waters, thus limiting vertical water mixing. The primary cause of meromixis combines lake hydrodynamics and oxygen balance: (i) inputs of waters with higher dissolved salts contents (sea water intrusion, saline groundwater) and/or (ii) inputs to deep waters of organic algal detritus from surface waters, in excess of the mineralization capacity of deep water, thus consuming dissolved oxygen. Many European maar-lakes in Italy (Albano, Nemi, Monticchio, Averno) and in Eifel (Scharf and Bjôrk 1992) are meromictic (Table 1.1), most probably due to hydrothermal inputs, although this is still debated for the Eifel lakes. The Girotte Lake in the French Alps was another meromictic lake (Dussart 1952, 1966), due to gypsum-draining groundwater inputs, but it has been transformed into a hydropower reservoir in the 1950s: Pavin is the only meromictic lake still found today in France.

1.5 Pavin Acquires a Status of International Field Laboratory (1965–2000)

1.5.1 International Projects Select Pavin as a Pristine Lake (1965–1975)

In the 1960s Pavin is still in a quasi-pristine state, with very limited nutrients inputs. Nicole Omaly-Lair who now heads the Clermont limnology group continues the tradition of technical innovations at Pavin. She uses a Cambridge oxygen probe for a continuous in situ 0_2 profile, for the first time in France (Omaly 1968). She defines the transition layer, or *mesolimnion*, later decomposed by geochemists into an *oxicline* and multiple *chemoclines* for each redox couples: sulfur, iron, manganese, arsenic, nitrate etc. She is the only post WWII limnologist to mention the possibility of strange legendary Pavin behaviour, which will be mocked by local historians (Fournier 1971): Pavin stories, as reported by Lecoq, will never be considered or mentioned by scientists until 2009 (See Sect. 1.8). Pavin also becomes part of several national and international programs on lakes.

As field and laboratory facilities are found at Besse and previous limnological surveys are numerous, Pavin becomes a land mark for foreign limnologists, looking for a mountain oligotrophic lake with high transparency and low nutrient levels, particularly nitrate, and a very high natural silica content, as already noted by Delebecque (Persoone et al. 1968; Flik et al. 1973, Paanaker and Hallegraef 1987). Pavin is also selected within the global scale Aqua project of SIL (Societas Internationalis Limnologiae) and Unesco which aims to "*get international recognition for a list of freshwater and brackish water area which are of agreed international importance for research, education, training and conservation*" (Luther and Rzoska 1971). Pavin is put on this world list, along with a handful of French lakes, by Bernard Dussart (1922–2008) the head of French limnology at this time. Unfortunately this project, which was conceived to be managed by UNESCO, will not be effective.

The second program is the famous OECD network of lakes trophic status, established by Richard Vollenweider (1922–2007), the world specialist of lake eutrophication for which he received the Tyler price in 1986. His contact for Pavin is Nicole Omaly-Lair and the Clermont team greatly benefits from this international collaboration. Their work focuses on seasonal variations in Pavin mixolimnion (Fig. 1.7a), the phyto-zooplankton relationships, nutrients cycle, lipid markers, occurring within the *mixolimnion*, between 0 and 50 m (Lair 1975–1976; Restituito 1984; Restituito and Lair 1975–1976; Devaux and Lair 1976; Devaux 1980a, b; Devaux et al 1983; Amblard 1986; Amblard and Bourdier 1990).

In the 1980s Pavin is one of the six lakes included in the French national program on lake research – named the GRECO Lacs – headed by Roger Pourriot (ENS Paris). The results of this program are finalized by a collective textbook on limnology in which Pavin is often highlighted, as the complexity of its vertical structure (Pourriot and Meybeck 1995; Meybeck 1995b).

Many old Pavin descriptions highlight its rare transparency, even in summer during the algal growth period. Delebecque measured an average Secchi depth of 8.5 m in June 1892. The 20 measurements made between 1892 and 1970 (Olivier 1939; Würtz 1945; Omaly 1968; Devaux 1980) show a maximum of 15.5 m and an average of 8.5 m. For Wurtz (1945) Pavin is meso-eutrophic due to the *Anabaena* blooms occurrence he has observed:"Pavin is beginning to get eutrophied". This occurs at a period of absent human impacts and could be linked to internal sources of nutrients, if the 1936 partial mixing of deep waters, unknown until 1986-see further- is accepted.

In the mid-1980s Clermont limnologists begin to attract the attention of authorities on trophic state change at Pavin due to the artificial fertilization of grassland drained to the lake by groundwaters inputs. Nitrogen and phosphorus levels increase subsequently, enhancing the algal development. Pavin was joining other Auvergne eutrophied lakes as Aydat and Chambon (PNRVA 2004). The societal response to Pavin degradation will be slow: the origin and pathway of excess nutrients had to be demonstrated first to farmers. It took more than 20 years for Jean Devaux, from Clermont University, who is studying Pavin since 1970, to convince all authorities to regulate nutrient use. In 2012 a first agreement is finally reached between all parties, land owners, farmers, Besse, Puy-de-Dôme department and Auvergne region with some European funds. When considering the water renewal time in the monimolimnion, the Pavin recovery might take decades.

1.5.2 Pavin, a Laboratory for Innovative Lake Research (1965–1986)

In the 1960s and 1970s Pavin attracts isotope geochemists looking for lake systems suitable to test their stable or radioactive isotopes toolbox: oxygen-18, tritium, cesium-137, lead- 210, silicium –32, plutonium, some of them natural, others being artificial radionuclides. These teams come from Bordeaux University, Ecole Normale Supérieure (ENS-Paris), Centre de Recherches Géodynamiques of Thonon-les-Bains (CRG-University Paris 6) and Tata Institute for Fundamental Research (TIFR- Ahmedabad, India). At Pavin they benefit from the scientific and logistical expertise of their Clermont colleagues, Nicole Omaly-Lair and Jean Devaux. The international reputation of Pavin is once more boosted and new peculiarities are found.

Tritium study in Pavin Lake by Alevinerie et al. (1966), a team from Bordeaux, is one of the first in Europe. This new approach in hydrology is developing fast in the 1960s, following the injection of this radioisotope in the atmosphere by nuclear tests. It also provides relative dating: waters without tritium predate 1950. On December 1965 the first depth profile of tritium is realized at meromictic Pavin. Surface waters from 0 to 50 m are rich in tritium (180–220 Tritium units, TU) while bottom waters have much less (<20 TU), highlighting the great contrast of their residence time: bottom waters are at least aged 12 years, according to authors. This profile is confirmed by the Thonon team (Crouzet et al. 1969; Meybeck et al. 1975) but the very last 20 cm layer of water, sampled with a special Jenkins-Mortimer corer, shows an increase of tritium, suggesting the possible and very limited input of more recent waters at the bottom. The renewal of the monimolimnion is complex and may include hydrothermal inputs (Krishnaswami et al. 1971; Martin 1985; Viollier et al. 1997; Aeschbach-Hertig et al. 2002). The last hydrogeochemical model is considering the input of 1.6 L/s of highly mineralized water (Assayag et al. 2008; Jezequel et al. 2010b) responsible for the peculiar ionic chemistry of deep waters found in the late 1960s (Pelletier 1968; Meybeck et al. 1975), unique in French lakes (Meybeck 1995a).

In 1970, a 20 cm Pavin sediment core and a Lake Geneva core, taken by the ENS team, is shipped to India at TIFR to

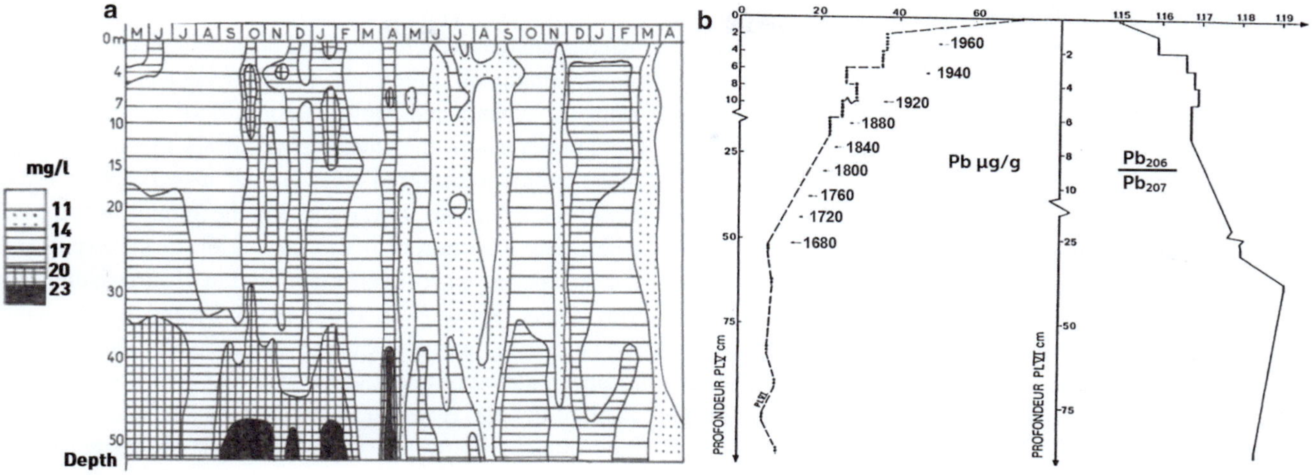

Fig. 1.7 Spatio-temporal complexity at Pavin. (**a**) Time- depth profile of dissolved silica in Pavin mixolimnion (0–50 m) from May 1973 to April 1975 (Restituito and Lair 1975–1976, Ann. Stat. biol. Besse-en-Chandesse.) (**b**) Profiles of total lead content (µg/g) and lead 206/lead 207 isotopic ratio in Pavin deep sediments (Elbaz-Poulichet, 1982)

measure and compare lead-210 and cesium-137. This results in a new method to date recent sediments widely used today (Krishnaswami et al. 1971). Pavin is essentially fed by rain and groundwater inputs and biogenic diatomite sediment is presently forming (Gasse 1969) as in Lake Myvatn in Iceland (Kristmanndottir and Armansson 2004). The silica content in sediments is close to world records with maximum SiO_2 up to 90 % of organic-free material – while Al and Ti, mostly brought to the lake as atmospheric dust, are very low, (Meybeck et al. 1975; Meybeck 1995a), confirming the original findings of Delbecque (1898). As such it is an excellent site for the study of the formation of diatomite, an economic material. This lake system is also very well suited to study the silica cycle and to test the silicium −32 dating, a natural cosmogenic isotope, now used to date Himalayan glaciers (Nijampurkar et al. 1983; Martin et al. 1992). The ENS team also complements the geochemical profiles (e.g. 55, 60, 65, 70, 90 m) with new redox couples as manganese and arsenic (Figuères 1978; Seyler and Martin 1989) and the mercury profile is studied by ocean geochemists (Cossa et al. 1994). These profiles reveal the complex structure of the mesolimnion, the transition layer where each couple is reduced at a different depth (Martin 1985; Viollier et al. 1995; Olive and Boulègue 2004; Chaps. 10, 11, and 12).

Such studies also show that Pavin, as many other maar-lakes, is an exceptionnal giant rain and dust collector with very limited runoff and river inputs, in comparison to all other lake types (Martin 1985; Meybeck 1995a).

The fine laminated sediments archive the evolution of Pavin area climate (Stebich et al. 2005; Schettler et al. 2007; Chapron et al. 2010, 2012; Chassiot et al. 2016; Chaps. 22 and 23). They also register the past atmospheric pollution, as for plutonium and lead: Catherine Jeandel (1981) finds ultra-trace of plutonium contamination, originating from far-away nuclear industry sources and Françoise Elbaz-Poulichet (1982 and in Martin 1985) realizes the first lake profile of lead isotopes in France, showing a two-steps increase of atmospheric Pb from different pollution sources: the first one in the nineteenth century, due to far away industrial sources emissions, and the second in the 1960s due to leaded gasoline, as cars are now frequent at Pavin (Fig. 1.7b).

After 1986 another page of research is opened at Pavin when water chemists from the IPG-Paris, lead by Gil Michard and Gérard Sarazin, begin a long-term study of Pavin, which is still in progress (Michard et al. 1994; Viollier et al. 1995, 1997; Assayag et al. 2008). They are specialized in redox couples analysis and are amazed to find such an exceptional natural laboratory. Their multiple results over the last 25 years are synthetized in this volume (Chaps. 10, 11 and 12). Meanwhile Clermont scientists continue to innovate at Pavin, 100 years after Berthoule, Delebecque and Bruyant: they are now developing new tools in microbiological analysis of the anoxic monimolimnion (Carrias et al. 1996, 2001; Colombet et al. 2006). Their work is revealing a new world, the Archaea, largely unknown so far (Lehours et al. 2005, 2007). Pavin monimolimnion waters, iron-rich and devoided of oxygen and rich in mantle gases (Aeschbach-Hertig et al. 1999), possibly for a long period, are holding communities that are similar to those of past geological time, billions years ago (Lehours et al. 2010; Fonty et al. 2010; Chaps. 14–20).

While this second golden age of Pavin research, started 30 years ago, is developing, another issue suddenly emerged in August 1986, triggered by the Lake Nyos catastrophy in

Cameroon, where a limnic explosion caused extreme Human damages in few hours. Reactions in the scientific community are rapid at Pavin. Let us first consider the evidence of maar-lakes degassing before exposing the degassing controversy at Pavin.

1.6 Maar Lakes Degassing Evidence in Cameroun and Italy

The explosion of Lake Nyos triggers specific research, first at Nyos then on other meromictic maar-lakes in Italy, Germany, South America and in Pavin, leading progressively to a new vision of the long-term functioning of these rare systems. Several Italian teams of volcanologists and limnologists find that in some maar-lakes past degassing activity, 100–1000 years ago, have been previously observed and reported in historical archives, particularly in the Monticchio Lakes and in the Albano Lake- all meromictic maar-lakes.

1.6.1 Nyos (21 August 1986) and Monoun (15 August 1984) Degassing Events and Their Effects on Populations

The Nyos event occurs on August 21, 1986 in Cameroon. In one night at least 1700 people living in valley villages downstream of the maar-lake (1.58 km^2, 208 m deep) were suddenly killed along with 3000–6000 cows (Le Guern and Sigvaldasson 1989; Sigvaldasson 1989). Lake waters degassed violently and emitted CO_2 escaped the crater rim and went down the thalweg over more than 20 km in distance, with an estimated velocity of 70 km/h. Eyewitnesses who were fortunate enough to survive the disaster gave a detailed account (Kling et al. 1987; Le Guern et al. 1992).

The Nyos event began the day before with minor upwelling of hot water. On August 21 a small explosion occurred in Lake Nyos followed in the evening by an intermittent jet of water topped by a white plume. At 10 p.m. a major detonation – a series of rumbling or bubbling sounds lasting 15–20 s for other witnesses – occurred in the lake. One observer walked to the crater rim and saw a white cloud or mist rise from the lake then a large water surge and soon after carbon dioxide invaded the low lying valleys. Although people smelled the odour of rotten eggs or gunpowder, they lost consciousness owing to the lack of oxygen in the dense, ground-hugging cloud of CO_2 mixed with water vapour and droplets, which smelled bitter and acidic. Other animals, birds and insects also died in the Nyos eruption. Vegetation was mechanically damaged but no thermal or chemical damage could be identified with certainty. In addition, the surface of the lake turned reddish brown just after the event owing to the oxidation of dissolved iron mixed to the surface from depth. International teams from USA, France, Italy, Japan and Cameroon converged within few days to Nyos. The mechanism of the event remained disputed for some time. At the international conference on Nyos held in Yaoundé in 1989 (Le Guern and Sigvaldasson 1989) two hypotheses were presented: (i) the volcanic hypothesis, in which a magmatic eruption occurred through the lake (Tazieff 1987, 1989; Tazieff et al. 1987) and (ii) the limnic hypothesis, where gas was stored in the lake prior to the event and released during a lake overturn or mixing of surface and bottom waters (Kling et al. 1987, 1989). Scientists advancing the volcanic hypothesis cited as evidence, the violence and localized nature of the process, the color of the lake and the smell of the gas. However, surveys of the lake and surroundings failed to discover fresh volcanic rock, heat inputs to the lake, volcanic vents. Whether these gas eruptions were volcanic in nature, and therefore impossible to stop, or whether they were limnological in nature, and therefore possible to identify and mitigate through gas removal, had direct political and socioeconomic consequences (Le Guern et al. 1992; Zhang and Kling 2006).

The medical symptoms on the Nyos victims and survivors have been analyzed by a specialized team (Baxter et al. 1989). Their study, focused on 845 hospitalized survivors, out of the 5000 total survivors, recorded: coughing (31 %), headache (26 %), skin lesions resembling burns (19 %), fever (12 %), weakness/malaise (11 %), limb weakness (6 %), severe eye symptoms, vomiting, diarrhea, dyspnoea (5 % each). Many of them had a sudden loss of consciousness with coma lasting few hours, followed by rapid recovery that could be related to anesthetic concentration of CO_2 mixed with air. No persistent respiratory disability was noted in survivors, implying low concentrations of SO_2 or H_2S. The most common abnormalities were erythema, big skin bullae, more than 5 cm, and other skin changes resembling burns, often developed 2 or 3 days after the degassing. The skin lesions were most often present on the face and many over the zygomatic region, in other cases on the leg, the abdomen or back. Only five patients – out of 161 – attributed their skin lesions to burns. Neurological damages could not be estimated.

The Monoun event, also in Cameroon, which occurred 2 years before, in 1984, at the nearby Monoun Lake (0.53 km2, 96 m) and claimed 37 victims, was only described by scientists after the Nyos explosion. Haraldur Sigurdsson, a volcanologist from the University of Rhode Island, made a post-event investigation few months only after the event, performed an array of analyses and found no signs of a volcanic eruption. He detected "*a hitherto unknown natural hazard*" that could wipe out entire towns, and in 1986, a few months before the Nyos disaster, he submitted his study to *Science*, the prominent scientific journal, which rejected the paper as far-fetched (Krajick 2003). Sigurdsson paper was eventually accepted by the Journal of Volcanic and Geothermal Research and therefore published a year after the Nyos event (Sigurdsson et al. 1987) so that no early warning was given to the scientific community about such type of lakes or to local Cameroon authorities. This rejection by the community is exemplary of the difficulties scientists have to conceive such events.

Lake Monoun was then surveyed by scientists from several countries and one of them, Kusakabe, said: "*if the [surface] waters were stirred up by only three feet, the water could start bubbling and trigger an explosion*"; his colleague, Bill Evans, advised caution: "*Let's not go splashing around too much out there*" (Krajick 2003). This corresponds to a near-saturation state (97 %), very sensitive to any perturbation, as throwing a stone. Monoun rescuers, who came few hours after the catastrophic event, observed "*a misty cloud which did not disperse until three to four hours later*" (Baxter et al. 1989). The Monoun eruption also occurred at night and the gas cloud filled a low-lying area near the lake (Sigurdsson et al. 1987). Travelers on a nearby road were killed as they entered the area, as were people who left higher ground to help those who had fallen in the valley. In the lake the water fountains reached 5 m high.

It is now aknowledged that both Monoun and Nyos violently degassed CO_2 from deep stagnant waters, a process termed "*limnic eruption*" (Sabroux 2007). Pre and post eruptions investigations reported that both Cameroun lakes were meromictic (Kling 1987), anoxic and rich in CO_2, originating from mantle degassing (Sigurdsson et al. 1987; Zhang and Kling 2006). Scientists now agree that the event was limnological in nature and represents CO_2-driven water eruptions owing to the rapid exsolution of CO_2 gas bubbles from dissolved CO_2 stored in the lakes. The degassing process may be triggered by an earthquake, a rockfall, a slump of sediments deposited on steep slopes, a spillover of lake water releasing the gas pressure in sediments, and/or their combination.

1.6.2 Ancient Degassing Events in Italian Maar-Lakes, Albano and Monticchio

In the decades following 1986, other maar-lakes in Italy are also checked for risk analysis at Monticchio Lakes and at Albano Lake (Chiodini et al. 1997; Caracausi et al. 2009; Funiciello et al. 2002, 2003, 2010; De Benedetti et al. 2008). During their investigations the Italian scientists are discovering evidence of ancient degassing events from Antiquity period to the early XIXth, which complement the descriptions of Monoun and Nyos events. Meanwhile German scientists do not take reference to past degassing events or legends (Scharf and Björk 1992). This attitude towards lake history may explain the different opinions on degassing risks: Italian volcanologists and geochemists are very cautious and highlight present risks while the risk issue is so far not raised at Eifel.

1.6.2.1 The Albano Catastrophic Degassing and Spillover Event in Latium (398 BC)

Many teams have extensively studied Albano maar-lake in recent years within a major Risk assessment programme. These investigations (Funiciello et al. 2002, 2003, 2010; De Benedetti et al. 2008; Cabassi et al. 2013) shed a light on another type of catastrophic degassing that occurred some 2400 years before. Italian scientists compared their current findings with historical sources from Latin authors who have reported a very famous story of a sudden and catastrophic increase of Albano Lake level, that spilled over its banks and forced Romans to dig out a tunnel to prevent future inundations. This catastrophic event, viewed by Archaic Rome contemporary historians as a legendary event, is now regarded as a major spillover due to degassing by Italian volcanologists, as Franco Barberi, a prominent member of this community. The Albano spillover is probably the earliest and one of the most spectacular lake degassing report.

Funiciello's prime source is Dionysius from Halicarnassus (first century BC), who wrote the early Rome history in 20 books (*Antiquitates Romanae*); he is in turn quoted by Livy, Titus Livius (59 BC-17 AD) (*Historiae Romanae*), then by Plutarch (46–125 AD (*Parallel Lives*, Themistocles-Camillus). The event can be summed up as such:

> *The event occurred between July 23 and August 24, 398 BC. During that period the water level rose at eye sight. But no precursor of the event such as neither water boiling nor climatic cause could be involved. The water reached the lower part of the crater rim which suddenly collapsed under the weight then an enormous mass of water flooded and destroyed the surrounding landscape. The Romans, who were at that time besieging the Etruscans at Veius, were scared by the prodigy, which occured very close to Rome, and sent delegates to the Pythia of Delphi [in Greece] who told them that "Poseidon [the Etruscans protector] was angered against Romans" and required to let the lake waters join freely the sea.*

This story is not fabricated; it has no fantastical tone and is well dated, detailed and considered as a *prodigium* of nature, with very limited intervention of supernatural forces. The surge is gradual yet however rapid, about 70 m in 1 month, i.e. it could have reached a velocity of several meters per day, a rise visible to the naked eye! There is no mention here of an explosive event, like in Nyos. The surge could not be stopped and probably destroyed any existing human settlements within the crater, one of the possible origins for the *Sunken city* legends for lakes (See 3.3.1). The consequences of the spillover are catastrophic and unexpected.

Italian scientists base their belief of the historical sources on their study of stratigraphic sections on the Ciampino plain, on the north side of the lake. Together with archeological excavations, they revealed repeated *lahars*, *i.e.* volcanic mudflows, occurring along the north-western slope of the maar, between 3000 BP and 2400 BP, filling in the paleodrainage network, blanketing the previous ground surface and forming a vast plain. Paleohydraulic analyses on fluvial and lahar deposits indicate that water flows that carried these lahars were in excess of hundreds of cubic meters per second, i.e. more than three orders of magnitude higher than normal streamflows on the lake rims. Lahars, now up to 20 m thick in thalwegs, were known by geologists as the Tavolato Conglomerate but until recently, they had no satisfactory explanation for their formation. The catastrophic inundations would be linked to overspills of the lake from its lowest rim, as in the historical sources. A lake rollover phenomenon, caused by the injection through the lake bottom of hot fluids, from mantellic origin, in the lake bottom rich in volcanic gases, is invoked as the prime cause of this event. According to the paleomorphology of Albano shoreline and to the sublacustrine archeological findings, such events have occurred previously several times during the Holocene. Paleo shores at -20 m and -64 m of present shoreline have been dated 7.1 and 4.1 ka, an indication of very different lake levels in those periods (Anzidei et al. 2008). Over the last 250 years multiple evidences of degassing have been described around lake Albano and the Albian Hills volcanic district (Funiciello et al 2002).

1.6.2.2 Monticchio Lakes (Southern Italy) and Their Pioneer Degassing Studies, 1777–1838

The first scientific study on Monticchio Lakes is realized by Domenico Tata, a Naples ecclesiastic, a noted mineralogist and volcanologist of the late eighteenth century. He reports his work in a remarkable *Lettera sul Monte Vulture* (Tata 1778), dedicated to his friend William Hamilton, representing the British king at the Napoli court, and one of the earliest UK volcanologist, famous for his observations of Etna and Vesuvius eruptions. When Tata visits Vulture in 1777, he has already heard about the marvelous lakes of Monticchio since he comes fully equipped with a sounding line to measure their depths, a water sampling bottle and with instruments to precisely position the latitude and longitude of the lakes. Domenico Tata has a more scientific objective than the one of Chevalier at Pavin, few years before (See 1.4.1) and he has already faced active volcanism and its magmatic degassing processes around Naples.

Tata focus is on Lago Piccolo. There, he does not observe any water movements in the lake and makes some deep water sampling in the middle of the lake but, as he tries to remove its cork with great difficulty, a sudden water and gas fountain rises one meter high above the bottle. It is probably one of the first sampling of deep anoxic waters with dissolved gases. At Monoun the same phenomenon will be reported by Sigurdsson et al. (1987).

After Tata, Monticchio Lakes were studied by two geologists, Giambattista Brocchi (1772–1826) and Riccioli, who analyze the gases emitted by the lakes and identify carbonic gas and hydrogen sulfide (Ciarallo and Capaldo 1995), possibly the first measurement of this kind, 150 years before the "discovery" of CO_2 degassing at Nyos. Some years later, Michele Tenore, a prominent botanist and the founder and director of Napoli Royal Botanical Garden, arrives to Monticchio with his assistant, Giovanni Gussone. They are mandated by the Naples Academia of Science to explore Mount Vulture area, a *terra incognita* so far. Their reports (Gussone and Tenore 1834; Tenore and Gussone 1838) have enormous scientific recognition. At the Lago Piccolo Tenore finds what he was expecting: a "permanent water fountain" coming out at the lake surface that agitates and mixes with the whole lake. According to Tenore and Gussone, this phenomemon has been going on since the early 1800s. During peak degassing the maximum height of the fountain was estimated to be around 5 m in November 1820 according to Father Paolino Tortonella, probably from the nearby San Michele Abbey (see Chap. 3), and other local witnesses.

In 1851, two Naples geologists make an in-depth study of Mount Vulture region (Palmieri and Scacchi, 1852). They, too, are expecting to see the marvelous water fountain; however they only find some bubbles at a given spot in the lake, where water is 11.4 m deep. Palmieri doubts that the fountain could have been so high in the past but accepts the phenomenon: he imagines that deposited rocks and sediments on the lake bottom probably blocked gases where the spring is located, until the high pressure released them.

A preliminary chronology of Monticchio *misbehaviour* events, as abnormal lake events have been termed at Nyos by Shanklin (1989, 2007), can be proposed as such, taking into account the contemporary analyses of Ciarallo and Capaldo (1995), Chiodini et al. (1997), Caracausi et al. (2009):

1770s: Lago Piccolo is meromictic; its bottom waters are rich in gases, including H_2S
1800, June 1: *Water column* of 6 m high in Lago Grande
1810, May 31: Lago Grande *overspills* with 5.5 m *water column*, then general lake agitation (June 1–23); *strong roar* on July 31, perceived by local people
1810, July 31: an *intense roar* was also heard; the column of water reached a height of about 3 m
1820: *water column* 5 m high in Lago Piccolo, lake agitation with *fish kill*
1830s (?): analysis of CO_2 and H_2S by Brocchi
1838: a *permanent boiling and water agitation* is still seen at a given spot in Lago Piccolo
1851: only few bubbles remain emitted in Lago Piccolo

Tata, Brocchi, Tenore and Palmieri are prominent scientists of their time with experience of volcanic degassing in Latium, Campania and Sicilia. All are also aware about Averno, the famous lake without birds in the Campi Flegrei volcanic district, near Naples, renown since the Antiquity

Table 1.2 Sensory descriptors of maar-lake degassing events and their impacts on populations

A. Degassing evidence in the lake and its close surroundings (lake examples)
(i) *Lake surface boiling (Monoun, 1984),*
(ii) *Lake storm* (at Nyos, 1986; Monticchio, 1810; Monoun 1984): the lake surface is very agitated with spurts, occasional jets. At Nyos, the erosion marks on South lake shore suggest a water jet up to 50 m–80 m,
(iii) *Reddish water colour and fluffy aspect* (Nyos, 1986): due to the precipitation of ferrous iron into ferric hydroxide or ferric carbonate,
(iv) *Gun powder and rotten egg smell* (Nyos, 1986): the ambiant air has a characteristic sulfur smell,
(v) *Detonation, thunder, roars, hissing, rumbling sounds* (Nyos, 1986; Monticchio, 1810), reported before, during or after the main event. According to Le Guern and Chevrier, 48 h after the main event, three violent detonations are heard within 5mn (Tazieff et al. 1987); other Cameroon witnesses recall gun shots. Big roars are heard at Monticchio,
(vi) *Lightning* and */or luminescence* (Nyos, 1986; Monticchio, 1810): one detonation at Nyos is accompanied with brief lightning with a powerfull light jet, as a flash light without centre (Tazieff et al. 1987),
(vii) *Degraded lake water taste* (Monticchio, 1770s): deep water tasted by Tata is "awful and one cannot bear it".
(viii) *Corrosive lake water* (Monticchio, 1770s): after drinking the tongue turns black and the water causes skin damages,"as for vitriol" (sulfuric acid)
(ix) *Reduced lake buoyancy* (Nyos, 1986), since the water density can be as low as 0.1 kg/L (Sabroux 2007),
(x) *Misty cloud* (Monoun, 1984), or *white plume* (Nyos, 1986) observed above the lake surface, probably CO_2 clouds,
(xi) *Permanent water fountains* generated from deep waters (Monticchio, 1800s; Monoun, 1984) [Deep water sampled in Monticchio Lago Grande by Tata are degassing when brought to the surface producing a kind of permanent, 1 m high jet. A similar effect is noted at Monoun by Sigurdsson],
(xii) *Lake spillovers* (Albano, 398 BC): during such type of paroxistic degassing events the lake. Level can raise at a rate exceeding 1 m/day and overflow the lower part of maar rims.
(xiii) *Lake fish kills and absence of fishes* (Nyos, 1986; Monticchio, 1810): due to the mixing with anoxic waters, also containing toxic gas or substances, fish kills can be observed. Lake Nyos was devoid of fishes before the catastrophic event (Sabroux, 2007)
B. Spatial extend and temporality of degassing events.
(xiii) *Spillover lahars* (Albano, 398 BC): these mud-flows deposited in downstream thalwegs may exceed 10 km long, they are eventually transformed into conglomerates,
(xiv) *Degassing temporal variability*. Degassing may take various forms intensity and temporality: permanent light bubbling, permanent boiling at specific position in the lake; degassing and spillover episodes (day to months, Monticchio, Albano), sudden violent rollover, explosive events (Nyos, Monoun),
(xv) *Degassing period duration* (Monticchio, Albano): temporal variability of degassing processes can be studied at different scales: subdecadal (Albano, 1997–2008), multi-decadal (Monticchio, 1770–1850), multi-millenary (Albano, Holocene and 3000 BP-2400 BP).
C. Damages to human populations and to the environment
(xvi) *Asphyxia and lethal CO2 intoxication* (Nyos, 1986, Monoun, 1984): CO_2 degassing caused massive animal and human killing, including birds and insects in downstream valleys,
(xvii) *Health damage* to survivors in downstream valleys (Nyos, 1986): skin damages with bullae developed few days after the event, mental shock and temporarily loss of consciouness, sight and speaking capacity,
(xviii) *Mechanical damages to thalweg vegetation* (Nyos, 1986), due to the power of the CO_2 flow.

period for its release of deleterious gases and its absence of birds (See also Chap. 3) and they do not seem so surprised about this very unusual lake behavior at Monticchio. Their descriptions are in many ways similar to those made 160 years later at Nyos, yet the Monticchio degassing is much less violent, it lasts for several decades and no Human victims are reported during that period. Today there is no sign of degassing in neither of both lakes. The degassing evidence at Monticchio has been gradually forgotten by the scientific community until the late 1990s, possibly because it was not catastrophic enough. Any records of Monticchio "misbehaviors" are still not found today in the local touristic information. However older dramatic events may persist in local memories through the exceptional millenary religious history of this remote and isolated place (See Sects. 3.5.3 and 3.7).

1.7 Sensory Grid of Degassing in Maar-Lakes

The Nyos event characteristics from *witness accounts* have been published by many scientists (Tazieff et al. 1987; Sigvaldason 1989; Le Guern et al. 1992; Zhang and Kling 2006), some of them reported at the Besse meeting. Other types of degassing symptoms and intensity have been described at Monoun, Monticchio and Albano. These qualitative data, generated or collected by scientists, have mobilized all senses: sight, vision, smell, taste, touch and hearing. They refer to the lake itself, its waters, its ambient air, its fauna, its surroundings (spillovers), and its impacts on populations, animals and crops, particularly in valleys downstream of the lake. These degassing descriptors, synthetized on Table 1.2, do not require an advanced scientific knowledge or training and can just be observed by riparian populations and lay people. They will be the basis for the re-interpretation of Pavin historical sources (Chap. 2) and of its legends and beliefs (Chap. 3).

1.8 Pavin Degassing Controversy (1986–2016)

The Nyos explosion in 1986 takes all volcanologists and limnologists by surprise, excepted Sigurdsson, and shed a sudden light on meromictic maar- lakes accumulating dissolved gases in deep waters. Lake degassing events, violent or not, are at that time undescribed by the modern scientific community and are not covered in major limnology textbooks (Hutchinson 1957; Wetzel 1975), or in volcanology and volcanic hazards textbooks and guidebooks (Kraft 1974; Hewitt and Burton 1971; Blong 1984). It must be recognized that maar-lakes are quite uncommon: their proportion is of the order of one per 1000 to one per 10,000 of all lakes (Meybeck 1995a).

A few days after the Nyos event (late August 1986), Haroun Tazieff (1914–1998), the famous French volcanologist and former head of the new French Natural Hazard Commissariat, questions the possibility of a degassing risk at Pavin (See quotation at the beginning of this chapter). He sends his team at Nyos at once and rings the alarm bell in the national and Auvergne media, demanding immediate research on this new type of hazard. Few days after Tazieff's call for immediate action, Renaud Vié-Lesage, the Natural Hazard commissioner, also an Earth scientist, asks Clermont volcanologists, Kornprobst and Camus, to perform a preliminary inventory of a potential degassing risk in Auvergne: the answer arrives within two months and the focus is set on Pavin. Meanwhile, from September 1986 to January 1987, local and national French media follow the ongoing observations made at Nyos by the French team and question these local volcanologists about explosion risks at Pavin (La Montagne 1986a, b; 1987a, b, c)

Vié Le Sage asks for an immediate study of Pavin in November 1986, Guy Camus assembles in summer 1987 a multidisciplinary team – volcanologists, water chemists, isotope geochemists – whichmakes in-depth geochemical inventories and innovative dissolved CO_2 measurements, with special samplers to preserve Pavin deep waters from degassing. Preliminary results are as soon communicated to the local press which reports in June 1987: "*The CO2 contents between 80 and 90 m deep are minimal and of the same order than those found at the surface, tells us Renaud Vié Le Sage, professor at Paris 7 University, in charge of Natural Hazards*" (La Montagne 1987c). The full investigation will take much longer to be published and will be much more balanced: " *Pavin does not present any limnic hazard as it receives a little amount of endogenous CO2 [from mantle degassing] and that most CO2 originates from the dismutation of the organic matter produced in situ*" (Camus et al. 1993). Pavin neighbours are reassured, but not for too long.

In 2005–2006, the question of Pavin degassing and risk management is revisited by Pierre Lavina, a volcanologist, and Thierry Del Rosso, an hydrogeologist, while they realize new field observations for an updating of the local geological map of Pavin area and estimate nutrient inputs to Pavin by groundwaters. They found evidence of CO_2 degassing spots in many places in the area (Lavina and Del Rosso 2009). This confirms both the observations on CO_2 fluctuations made before at Creux de Soucy (Martel 1892, 1894, see Sect. 1.4.3.3) and the common knowledge of degassing spots in the area by local Besse people (Boyer-Vidal 1888), an evidence forgotten in the 1990s and recently confirmed (Gal and Gadalia 2011). They also mapped for the first time a 9 km long mudflow deposit in Couze Pavin thalweg, downstream of the lake, which they attribute to a violent lake spillover around 1300 AD (Del Rosso-d'Hers, 2010) (Schematically presented on Fig. 1.2c). Their findings reactivate the risk assessment at Pavin: local authorities suppress the release of this information while installing a set of security measures, landslides surveys within the crater, stabilization of crater slopes to prevent rockfalls, laser-telemetry survey of crater etc. But the findings are eventually leaked to the local press, reactivating the risk concern at Pavin.

In June 2009 Pavin geochemists, limnologists and microbiologists organize a meeting in the town of Besse to discuss Pavin and other meromictic lakes (Jezequiel, Amblard and Fonty, 2010). It is attended by 120 scientists during 5 days, including by members of the Nyos French team, François Le Guern (1942–2011), Tazieff's closest collaborator, and Jean-Christophe Sabroux (Sabroux 2009), and by Michel Halbwachs, the initiator of current Nyos degassing by pipe. Leguern and Sabroux had come at a few days only after the main event, at a time when Lake Nyos was still exhibiting some degassing features and report at the Besse meeting its unusual features with witness accounts from survivors. Lavina presents the mudfow deposits he found in the Couze Pavin talweg and Del Rosso also shares the past "Pavin stories" with this scientific audience and links some legends to past catastrophic events in Pavin area, either of volcanic origin or of limnic origin, a thesis he had just exposed in the local press (Del Rosso 2009a, b) but which is strongly challenged by the rest of scientific community. Pavin legends are actually quite complex (See Chap. 3).

The Besse workshop, published in 2010, includes papers on lake sedimentary archives (Chapron et al. 2010), hydrological balance (Jezequel et al. 2010a), carbon cycle and current risk of gaseous outburst (Jezequel et al. 2010b) microbial communities (Fonty et al. 2010), forest history around Pavin (Lathuillière 2010) and on Pavin patrimonial value and preliminary history

(Meybeck 2010). Few other meromictic lakes were discussed at Besse; however none were maar-lakes with the exception of Lake Nyos. In their analysis of Pavin carbon cycle, Jezequel et al. state that *"no spontaneous gaseous eruption can take place… the present situation is stable and only important landslides, disturbing bottom sediments, could generate a partial degassing"*. After this meeting our preliminary analysis of Pavin history (Meybeck 2010) reveals a long history of possible "misbehaviour" events at Pavin.

Anomalies found in Pavin sediment records led to further investigations and to the evidence of two major catastrophic events (Chapron et al. 2010, 2012; Chap. 23): (i) around 600 AD, corresponding to a large subaquatic slide, a spillover of the lake with a drop of lake level by 13 m, due to the sill erosion, (ii) a second subaquatic slide occurred around AD 1300 in Lake Pavin and several other lakes in the area (Guéry, Chauvet, Moncineyre, see their position in Fig. 1.3a) supporting a regional earthquake triggering (Chassiot et al. 2016). According to Chapron et al. (2012) subaquatic slides in Pavin were *"large enough to be associated with violent waves in the lake. Such exceptional waves can favour erosion at the outlet and a rupture of the Pavin crater ring resulting in: (i) abrupt lake level drop and (ii) the spillover of a debris flow downstream in the Couze Pavin valley. Such violent waves may also result from a limnic eruption. This abrupt lake level drop can be related to a rupture of the crater ring (outburst event) and imply a sudden discharge of ca. 2.8 million m3. Such outburst event would trigger a large debris flow in the Couze Pavin River and could also favour a limnic eruption (abrupt release of gas due to pressure drop)*. The 600 AD and 1300 AD events can be considered as the first paleolimnological evidence of past Pavin degassing and/or catastrophic events, recorded in sediment archives. The long mudflow deposit found in the Couze Pavin thalweg could originate from this Pavin spillover (Lavina and Del Rosso 2009).

1.9 Conclusions

Pavin has a long and rich common history with lake scientists, placing it on a top level of the development of limnology. Scientists have been attracted there since 1770, when an expedition organized by one king engineer succeeded to measure its depth, a major challenge at this time. However during the next 90 years, naturalists were lacking sampling apparatus and other means of observation, faced lake access difficulties, had no boat for their investigation and they mainly observed Pavin from its rim. They probably received limited support by local population who feared the lake to be bottomless and fishless (See the next two chapters). This ended with the construction of a pathway to the lake outlet, the construction of the first boat (1847) then the successful fish introduction in 1859, all by Lecoq, the prominent naturalist of Auvergne.

From 1880 to 1914 scientists from Clermont University, backed by Berthoule, the Besse mayor, are at the forefront of French limnology. They establish the limnological station of Besse, the first of its kind in France. Charles Bruyant models the Station after the oceanographic stations and presents it to an international audience in Besse in 1908, but its development is limited after the death of Bruyant in World War I.

In June1892 Pavin is studied by André Delebecque, the most inventive French limnologist, who develops comparative limnology on more than 300 French lakes for all geographical, physical and chemical aspects. He highlights many of the exceptional Pavin characteristics but does not notice any peculiarity in bottom waters. Meanwhile, his friend Edouard-Alfred Martel, the founder of cave science, explores the tiny Creux de Soucy cave and its unique lake.

In the 1950s, Luc Olivier, a Clermont hydrobiologist, measures anomalous O_2 profiles for the first time at Pavin and, in 1963, Jean Pelletier confirms the permanent absence of oxygen in deep waters, the meromixis, and the inverted thermal gradient in bottom waters. UNESCO selects Pavin among the world's remarkable lakes and OECD, includes it in its network of alpine oligotrophic lakes set up by Richard Vollenweider.

Since the 1970s Pavin attracts scientists, for its unique morphological, hydrological and geochemical characteristics. They come with up-to-date material, test new theories and develop innovative approaches, benefit from the cumulated scientific results and have excellent local working conditions. Today several dozens of scientists from half a dozen countries have chosen Pavin as a natural laboratory, particularly for its chemocline and its microbial diversity, all of them fully convinced about its uniqueness in their own field (Jezequel et al. 2010a; Meybeck 2010).

The Nyos event in 1986, which took the whole volcanologists and limnologists' communities by surprise, has not been the first case of lake degassing described by scientists. Actually, prior degassing observations had been already recorded at the Monticchio Lakes in Italy, 150 years before, by prominent scientists of their time. Italian scientists are now acknowledging another catastrophic event that occurred at Lake Albano near Rome in 398 BC, reported by Greek and Latin historians, as a major spillover due to a degassing process. Both degassing evidences were then forgotten until Italian volcanologists and geochemists from INVG and Roma Tre university reconsidered them as a potential degassing risk (Chiodini et al. 1997; Funiciello et al. 2002, 2003; Caracausi et al. 2009) and took these historical descriptions into consideration (Caracausi et al. 2009; De Benedetti et al 2008; Funiciello et al. 2010).

A first outlook on Nyos, Monticchio and Albano lakes shows that lake degassing may take multiple forms and intensities. Gentle and harmless limited degassing may only be reported by some curious and/or educated persons, particularly when the lake is not found in its normal state. More violent and/or dangerous degassing events, the *misbehavior* (Shanklin 2007), are reported provided they are witnessed by someone, or evidenced by their damages on populations, animals and crops, near the lake or far away in the valley. Degassing events and their effects on the environment and populations are described by witnesses by a set of sensory criteria (Table 1.2). This set shows a wide variety of indicators, depending on the degassing intensity and types, on mobilized senses, on the position of witnesses.

The most catastrophic degassing events, as the Albano overspill (398BC) may be recorded, like other types of natural hazards, by local populations for hundred of years or more. They can also be transformed into myths, legends, fantastic stories and local fears that may leave traces in local iconography and toponymy, as discussed in Chaps. 2 and 3. In their conclusions on risk assessment in the Colli Albani volcano where Lake Albano is located, the Italian volcanologists from Roma Tre state:"*The acquisition and verification of archaeological and historical data are essential to support hazard analysis*" (Funiciello et al. 2010). Meanwhile German scientists working on Eifel maars do not take reference to past degassing events or legends (Scharf and Björk 1992). Today maar-lake degassing is still poorly covered in specialized textbooks and Monticchio and Albano events are not yet reported in them (Smith 2003; Rouwet et al 2015).

At Pavin the degassing issue has been quickly put on the scientific agenda in 1986 and remained a matter of discussion since then. The survey made for the authorities reported that degassing risk was non-existent in normal conditions (Camus et al. 1993). This opinion was confirmed during the 2009 meeting at Besse on Pavin. However, the preliminary historical analysis of Pavin (Meybeck 2010) and the new palaelolimnological findings at Pavin (Chapron et al. 2012; Chassiot et al. 2016) suggest that major degassing events have actually taken place in the past, before the start of regular limnological surveys at Pavin in the 1880s (a possible misbehavior event reported for summer 1936 has not been considered by the scientific community, see the next chapter).

The historical, iconographic and folkloric analyses, realized by us between 2010 and 2016, while paleolimnologists were finding evidence of degassing at Pavin, reveal a very rich and largely ignored corpus of sources, since Antiquity. This corpus, re-interpreted through the sensory indicators grid of degassing unveils many ignored misbehaviour events of Pavin, some confirming the paleolimnological analysis, other apparently not recorded in sediments, a latent fear of the lake over two millennia and a complex interaction between local populations and *Lacus Pavens*, *i.e.* the terrifying lake, as it was actually called since at least 1605. Aside from the complex history of Pavin (Chap. 2), many forgotten or mis-attributed legends and fantastic stories provide new clues of past Pavin misbehavior (Chap. 3).

Aknowledgements of chapters 1, 2, 3 are clustered and presented at the end of chapter 3.

References

Aeschbach-Hertig W, Kipfer R, Hofer M, Imboden DM, Wieler R, Signer P (1996) Quantification of gas fluxes from the subcontinental mantle. The example of Laacher See, a maar-lake in Germany. Geochim Cosmochim Acta 60:31–41

Aeschbach-Hertig W, Hofer M, Kipfer R, Imboden DM, Wieler R (1999) Accumulation of mantle gases in a permanently stratified volcanic lake (Lac Pavin, France). Geochim Cosmochim Acta 63:3357–3372

Alvinerie J, Begot B, Levêque P, Vigneaux M (1966) Activité en tritium et caractéristiques chimiques des eaux du lac Pavin. C R Acad Sci Paris 262:846–849

Amblard C (1986) A study of spatial and temporal variability in the adenosine nucleotides of lake phytoplankton. Hydrobiologica 137:159–173

Amblard C, Bourdier G (1990) The Spring bloom of the diatom Melosira italic, subsp. subarctica in Lake Pavin: biochemical, energetic and metabolic aspects during sedimentation. J Plankt Res 12:645–660

Anzidei M, Carapezza ML, Esposito A, Giordano G, Lelli M, Tarchini L (2008) The Albano Maar Lake high resolution bathymetry and dissolved CO_2 budget (Colli Albani volcano, Italy): constrains to hazard evaluation. J Volc Geoth Res 171:258–268

Assayag N, Jézéquel D, Ader M, Viollier E, Michard G, Prévot F, Agrinier P (2008) Hydrological budget, carbon sources and biogeochemical processes in lac Pavin: constrains from delta 18 O of water and delta 13C of dissolved organic matter. Appl Geochem 23:2800–2816

Bakalowicz M (1971) Le Creux de Soucy. Annales Spéléologie 26(2):387–409

Baxter PJ, Kapita M, Mfondu D (1989) Lake Nyos disaster, Cameroon, 1986: the medical effects of large scale emission of carbon dioxide. British Med J 298(6685):1437–1441

Belgrand E (1869) Histoire générale de Paris. La Seine et le Bassin Parisien aux âges antéhistoriques. Paris

Berthoule A (1890) Les lacs d'Auvergne. Orographie, faune naturelle, faune introduite. Revue Société Nationale d'Acclimatation. Paris, 136 pp

Berthoule A (1896) Les lacs de Besse-en-Chandesse. In: L'Auvergne. IV Congrès Int. Hydroclimatologie Géologie, Clermont-Ferrand

Blong RJ (1984) Volcanic hazards. A sourcebook of the effects of eruptions. Academic Press, Sydney

Boyer-Vidal A (1888) Besse et Vassivière. Notice historique. Imprimerie clermontoise, Clermont-Ferrand, 50 pp, appendices. Réédition ACVAM 2005, 72 pp

van den Bogaard P, Schmincke HU (1995) Laacher See tephra: a widespread isochrones late Quaternary tephra layer in Central and Northern Europe. Geol Soc Am Bull 96(2):1554–1571

Bruyant C (1894) Contribution à l'étude des lacs d'Auvergne. Revue d'Auvergne: 53–57, 129–160, 217–252

Bruyant C (1910) La Station de Besse. La Montagne d'Auvergne 2:3–4

Bruyant C, Eusebio A (1904) Introduction à l'aquaculture générale. Matériaux pour l'étude des rivières et des lacs d'Auvergne. Bellet, Clermont Ferrand, 162 pp

Buchel G (1993) Maars of the Eifel. In: Negendank J F W and Zolitschka B (eds) Paleolimnology of European Maar Lakes. Springer, Lectures Notes in Earth Sciences #49:1–13

Cabassi J, Tassi F, Vaselli O, Fiebig J, Nocentini M, Capeciacchi F, Rouwet D, Bicocchi G (2013) Biogeochemical processes involving dissolved CO_2 and CH_4 at Albano, Averno, and Monticchio meromictic lakes (Central-Southern Italy). Bull Volcano 75:683. doi:10.1007/s00445-012-0683-0

Caliro S, Chiodini G, Minopoli C, Signorini A, Avino R, Granieri D (2008) Geochemical and biochemical evidence of lake overturn and fish kills at Lake Averno. J Volc Geoth Res 178:305–316

Camus G, Michard G, Olive P, Boivin P, Desgranges P, Jézéquel D, Meybeck M, Peyrus J-C (1993) Risques d'éruption gazeuse en Auvergne. Bull Soc Geol Fr 164:767–778

Cany G, Percepied E, Perpère E (1916) Séjour et promenades au Mont Dore et à la Bourboule. Guide des baigneurs, touristes, naturalistes (Collab. Ph Gangeaud, C Bruyant, préface A Lacroix), 394 pp

Caracausi A, Nuccio PM, Favara R, Nicolosi M, Paternoster M (2009) Gas hazard assessment at the Monticchio crater lakes of Mount Vulture, a volcano in Southern Italy. Terra Nova 21(2):83–87

Carapezza ML, Lelli M, Tarchini L (2008) Geochemistry of the Albano and Nemi crater lakes in the volcanic district of the Albian Hills (Roma, Italy). J Volc Geotherm Res 178:297–304

Carrias J-F, Amblard C, Bourdier G (1996) Protistan bacterivory in an ologotrophic lake: importance of attached ciliates and flagellates. Microbiol Ecol 31:249–268

Carrias JF, Thouvenot A, Amblard C, Sime-Ngando T (2001) Dynamics and growth estimates of planktonic protists during early spring at Lake Pavin. France. Aquat Microbiol Ecol 24:163–174

Chapron E, Alberic P, Jezequel D, Ledoux G, Massault M (2010) Les Archives sédimentaires de l'histoire du lac Pavin. Rev Sci Nat Auvergne 74–75:57–66

Chapron E, Ledoux G, Simmoneau A, Alberic P, Saint-Onge G, Lajeunesse P, Boivin P, Desmet M (2012) New evidence of Holocene mass wasting events in recent volcanic lakes from the French Massif Central (lakes Pavin, Montcineyre and Chauvet) and implications for natural hazards. In: 5th International Symposium Submarine Mass Movements and their Consequences, Kyoto. Chap.23, pp 256–264

Chassiot L, Chapron E, Di-Giovanni C, Alberic P, Lajeunesse P, Lehours A-C, Meybeck M. (2016) Extreme events in maar lake Pavin sedimentary record: implication for natural hazards assessment in the French Massif Central, Quat Sci Res, 141, 9–25

Chevalier (1770) Rapport du sondage au Pavin du 28 septembre. In: Delarbre A (abbé), 1805. Notice sur l'Auvergne et en particulier sur la ville de Clermont. Clermont-Ferrand, 256 pp

Chiodini G, Cioni R, Marini L, Principe C, Braco B (1997) Water and gas chemistry of the Lake Piccolo of Monticchio (Mt Vulture, Italy). Curr Res Volcanic Lakes, IAVEI News lett 10:3–9

Chondrogianni C, Ariztegui D, Guilizzoni P (1996) Lakes Albano and Nemi (Central Italy), an overview. In: Guilizzoni P and Oldfield F (eds) Palaeo-environmental analysis of Italian crater lakes and Adriatic sediments. Mem Ist Ital Idrobiol 55:17–22

Ciarallo A, Capaldo L (1995) Viaggo al Vulture (transcription of Tenore et Gussone, 1838: Peregrinazioni effetuate nelle state del 1838 dai Signore Tenore e Gussone… in Principitato Citeriore e di Basilica). Edition Osanna Venosa, 157 pp

Colombet J, Sime-Ngando T, Cauchie HM, Fonty G, Hoffman L, Demeure G (2006) Depth-related gradients of viral activity in Lake Pavin. Appl Envir Microbiol 72:4440–4445

Cossa D, Mason RP, Fitzgerald WF (1994) Chemical speciation of mercury in a meromictic lake. In: Watras CJ and Huckabee JW (eds). Mercury pollution, Lewis Publisher, pp 58–67

Crouzet E, Hubert P, Meybeck M, Olive P (1969) Etude de la dynamique des lacs par des mesures isotopiques. Ann Stat Biol Besse-en-Chandesse, tome 4:367–376

De Benedetti AA, Funiciello R, Giordano G, Didio G, Caprili E, Paterne E (2008) Volcanological history and myths of the Lake Albano maar (Colli Albani volcano, Italy). J Volcan Geoth Res 176:387–406

Del Rosso T (2009a) Les Légendes du Pavin. Oral Communication. Int. Workshop Pavin and other meromictic lakes, May 14–17, 2009, Besse

Del Rosso T (2009b) Interview with Marc Gachon, La Galipote, 37–38

Del Rosso-d'Hers T (2010) Montchal-Pavin, mise au point sur les travaux récents (2004–2010). Eruption 23:41–43

Delarbre A (1795) Notes et observations sur les volcans, puits, mines et fontaines d'Auvergne. Manuscrit Ms 716, University Library (BCIU), Clermont-Ferrand

Delarbre A (abbé), (1805) Notice sur l'Auvergne et en particulier sur la ville de Clermont. Clermont-Ferrand, 256pp

Delbecque A (1898) Les Lacs français. 1 vol., 153 fig., 22 pl., Chamerot et Renouard, Paris, 436 pp

Depping GB (1811) Les merveilles et beautés de la nature en France. P Blanchard, Paris, 634 pp

Devaux J (1980) Contribution à l'étude limnologique du lac Pavin (France) II. Relation entre paramètres abiotiques, phyto et zooplancton dans la zone 0–20 mètres. Hydrobiologia 68:17–34

Devaux J, Lair N (1976) Production primaire et biomasse des populations algales dans un petit lac oligotrophe du Massif Central français. Hydrobiologia 50(3):209–220

Devaux J, Lair N, Amblard C (1983) Un écosystème lacustre profond: le lac Pavin. In: Lamotte et M Bourlière F. (éd) Problèmes d'écologie, structure et fonctionnement des écosystèmes limniques, Masson, 1–49

Dussart B (1952) Contribution à l'étude des lacs de Savoie: le lac de la Girotte. Ann Stat Centr Hydrobiol Appliquée 4:379–391

Dussart B (1966, 1992) Limnologie, l'étude des eaux continentales. Gauthiers-Villars, Paris, 736 pp

Elbaz-Poulichet F (1982) Apports fluviatiles et estuariens de Pb, Cd et Cu aux océans. Thèse d'état, université de Paris VI

Ellwood NTW, Albertano P, Galvez R, Funiciello R, Mosello R (2009) Water chemistry and trophic evaluation of Lake Albano (Central Italy): a four-year water monitoring study. J. Limnology, http://dx.doi.org/10.4081/jlimnol.2009.288

Eusebio A (1896) La Station biologique de Besse. Rev d'Auvergne 13:457–460

Eusebio A, Reynouard J (1925) Le lac Pavin et le Creux de Soucy. Station limnologique de Besse, Besse, 55 pp

Figuères G (1978) Contribution à l'étude géochimique du fer et du manganèse dans les eaux de surface. Thèse 3e cycle, université Paris 6

Findenegg I (1935) Holomiktische und meromiktische Seen. Int Rev Hydrobiol 35:586–610

Flik BJC, Hallegraeff GM, Lingeman R (1973) Limnological notes on Lake Pavin, Auvergne. Ann Stat Biol Besse-en-Chandesse 8:119–144

Fonty G, Lehours A-C, Borrel G, Petit-Biderre C, Peyret P, Morel-Desrosiers N, Morel J-P, Sime-Ngando T, Amblard C (2010) Le Lac Pavin: originalités et intérêts microbiologiques de son compartiment anoxique. Rev Sci Nat Auvergne 74–75:111–130

Forel F-A (1892) Le Léman, monographie limnologique. F. Rouge, Genève, 3 vol

Fournier PF (1971) Le Nom du lac Pavin, Essai d'étymologie. Bull hist Sci Auvergne 85(631):159–164

Funiciello R, Giordano G, De Rita D, Carapezza M-L, Barberi F (2002) L'Attivita recente del cratere del Lago Albano di Castelgandolfo. Rend Fis Acc Linei 9:113–143

Funiciello R, Giordano G, De Rita D (2003) The Albano maar-lake (Colli Albani Volcano, Italy): recent volcanic activity and evidence of pre-Roman Age catastrophic lahars events. J Volc Geoth Res 123:43–61

Funiciello R, Heiken G, De Benedetti A-A, Giordano G (2010) Volcanic activity of the Lake Albano maar in Roman history and mythology. In: Funiciello R and Giordano G (eds) The Colli Albano Volcano, Special Publications, Geological Society, London

Gasse F (1969) Les Sédiments à diatomées du lac Pavin (Auvergne). Ann Stat Biol Besse-en-Chandesse 4:221–238

Gautier P (1892) Observations sur le Creux de Soucy. Comptes Rendus Acad Sci, 22 nov 1892

Gautier P, Bruyant C (1896) Le Creux de Soucy. In: L'Auvergne. IV Congrès Int. Hydroclimatologie Géologie, Clermont-Ferrand, 313–320

Gérardin A (1875) Altération de la Seine aux alentours de Paris depuis novembre 1874 jusqu'en mai 1875. Comptes Rendus Acad Sci Paris 80:1326–1328

Gobin L (1896) Essai sur la géographie de l'Auvergne. Thèse, université de Clermont-Ferrand

Gussone G, Tenore M (1834) Osservazioni fisico-geognostiche fatte e un viaggo per diversi luoghi delle provincie di terra di Lavoro e di Abruzzo, nella state de 1834 dai Signori Gussone e Tenore per disposizione della Reale Academia di Scienze, Napoli, 75 pp (on line)

Hewitt K, Burton (1971) The Hazardousness of a place, a regional ecology of damage events, Toronto, Dept Geography, University Toronto

Hutchinson GE (1957) A treatise on limnology, vol 1: geography, physics, and chemistry. Wiley, p 1010

Jacquet J, Mandelbrot L (1966) Etudes nouvelles entreprises par Electricité de France sur l'évaporation des lacs et des réservoirs. Hydrol Sci J 11(4):44–60

Jeandel C (1981) Comportement géochimique des isotopes du plutonium dans les milieux naturels (lacustre, fluvial, estuarien); Thèse, Laboratoire Géochimie Marine, ENS-Paris, Université Paris 6

Jezequel D, Sarazin G, Prévot F, Viollier E, Groleau A, Michard G, Agrinier P, Binet S, Bergonzini L (2010a) Bilan hydrique du lac Pavin. Rev Sci Nat Auvergne 74–75:67–90

Jezequel D, Michard G, Viollier E, Prévot F, Grleau A, Sarazin G, Lopes F, Agrinier P, Albéric P, Bergonzini L (2010b) Le Cycle du carbone et les risques d'éruption gazeuse au Pavin. Rev Sci Nat Auvergne 74–75:91–110

Juvigné E, Camus G, de Goër de Hervé A (1993) Maars of the Northern Auvergne, state of knowledge. In: Negendank JFW and Zolitschka B (eds) Paleolimnology of European Maar Lakes. Lecture Notes Earth Science 49, Springer, pp 81–94

Kling GW (1987) The comparative limnology of lakes in Cameroon, West Africa. Ph. D. thesis, Duke University

Kling GW, Clark MA, Wagner GN, Compton HR, Humphrey AM, Devine JD, Ewans WC, Lockwood JP, Tuttle ME, Koenigsberg EJ (1987) The 1986 Lake Nyos gas disaster in Cameroon. West Afr Sci 236(4798):169–175

Kling GW, Tuttle ML and Evans WC (1989) The Evolution of thermal structure and water in Lake Nyos. In: Le Guern F and Sigvaldason G (eds) The Lake Nyos Event and Natural C02 Degassing. J. Volcanol Geotherm Res 39:151–165

Kraft M (1974) Guide des volcans d'Europe. Delachaux & Niestlé, 412 pp

Krajick K (2003) Efforts to tame second African "Killer Lake" begin. Science 299:805

Krishnaswami S, Lal D, Martin J-M, Meybeck M (1971) Geochronology of lake sediments. Earth Planet. Sci Lett 11:407–414

Kristmanndottir K, Armansson H (2004) Groundwater in the Lake Myvatn area, Northern Iceland. Chemistry, origins and interactions. Aquatic Ecol 38:115–128

La Montagne (1986a) Risques de dégazage des lacs de cratère. M Kornprobst remettra son pré-rapport fin octobre. La Montagne, 30 September 1986, Clermont-Ferrand

La Montagne (1986b) Deux lacs auvergnats sous la plus haute surveillance. La Montagne, 7 November 1986, Clermont-Ferrand

La Montagne (1987a) Le Pavin: le point de vue des hydrobiologistes. La Montagne, 7 January 1987, Clermont-Ferrand

La Montagne (1987b) Nouvelles explosions au lac Nyos (Cameroun). La Montagne, 8 January 1987, Clermont-Ferrand

La Montagne (1987c) Dix-sept scientifiques au chevet du Pavin pour enterrer le fantôme du lac Nyos. La Montagne, 24 June 1987, Clermont-Ferrand

La Montagne d'Auvergne (1910). Revue illustrée, littéraire, artistique, touristique, sportive et liste des étrangers, Issoire

Lair N (1975–1976) Projet Alpin OCDE pour la lutte contre l'eutrophisation des lacs du Massif Central français. I protocoles de mesure. Ann Stat Biol Besse-en-Chandesse 10:92–145

Lathuillière L (2010) Les forêts autour du lac Pavin. Présentation historique. Rev Sci Nat Auvergne 74–75:135–149

Lavina P, Del Rosso T (2009) New geological findings in Pavin area, Oral communication, Symposium Pavin and other meromictic lakes, Besse-en-Chandesse, May 2009

Le Guern F, Sigvaldason G (eds) (1989) The Lake Nyos event and natural CO_2 degassing. J Volcanol Geotherm Res 39:95–275

Le Guern F, Shanklin E, Tebor S (1992) Witness and accounts of the catastrophic event of August 1986 at Lake Nyos (Cameroon). J Volc Geotherm Res 51:171–184

Lecoq H (1835a) Description de l'Auvergne. Le Mont d'Or et ses environs. J-B Baillère, Paris, 437 pp

Lecoq H (1835b) L'Auvergne pittoresque. Atlas colorié. Le Mont-Dore et ses environs. Seize vues coloriées. Chaillon, Paris

Lecoq H (1859) Observations sur une grande espèce de spongille du lac Pavin. Mem Acad Sci Belles Lettres et Arts de Clermont-Ferrand, tI: 476

Lecoq H (1871) L'Eau sur le Plateau Central de la France. J-B Baillère, 394 pp

Legrand d'Aussy PJB (1788) Voyage en Auvergne. E. Onfroy, Paris, 555 pp

Lehours AC, Bardot C, Thenot A, Debroas D, Fonty G (2005) Anaerobic microbial communities in Lake Pavin, a unique meromictic lake in France. Appl Envir Microbiol 71:7389–7400

Lehours AC, Evans P, Bardot C, Joblin K, Fonty G (2007) Phylogenetic diversity of Archaea and Bacteria in the anoxic layer of a meromictic lale (Lake Pavin, France). Appl Envir Microbiol 73: 2016–2019

Lehours AC, Carrias JF, Amblard C, Sime-Ngando T, Fonty G (2010) Les Bactéries du lac Pavin. Pour la Science, 387:30–35

Luther H, Rzoska J (1971) Project Aqua: a source book of inland waters proposed for conservation. Oxford-Blackwell, 239 pp

Maichado A (2004) An index of naturalness. J Nat Conserv 12:95–110

Mancino G, Nolé A, Urbano V, Amato M, Ferrara A (2009) Assessing water quality by remote sensing in small lakes: the case study of Monticchio Lakesin Southern Italy. Forest Biogeosci Forestry 2:154–161

Margaritora FG (1992) Limnology in Latium: the volcanic lakes. In: GuilizzoniP. Tartari G. and Giussani G. (eds) Limnology in Italy, Mem. Ist. It. Idrobiologia, 50, pp 319–336.

Martel EA (1892) Le Creux de Souci. La Nature, pp 165–167

Martel E-A (1894) Les Abymes. Chap. XXII Le Creux-de Souci. (Puy-de-Dôme). Delagrave, pp 387–393

Martin J-M (1985) The Pavin crater lake. In: Stumm W. (ed) Chemical Processes in Lakes, John Wiley, pp 169–188

Martin J-M, Meybeck M, Nijampurkar VN, Somayajulu BLK (1992) 210Pb, 226Ra, 32Si in Pavin Lake (Massif Central, France). Chemical Geol Isotope Geosci 94:173–181

Meybeck M (1995a) Les lacs et leurs bassins versants. In: Pourriot R and Meybeck M (eds) Limnologie Générale, Masson, Paris

Meybeck M (1995b) Global distribution of lakes. In: Lerman A Imboden DM and Gat JR (eds) Physics and chemistry of lakes. Springer-Verlag, pp 1–35

Meybeck M (2010) Le Pavin et ses voisins, histoire d'un patrimoine européen unique. Rev Sci Nat Auvergne 74–75:7–44
Meybeck M, Martin J-M, Olive P (1975) Géochimie des eaux et des sédiments de quelques lacs volcaniques du Massif Central français. Verh Intern Verein Limnol 19:1150–1164
Michard G, Viollier E, Jézéquel D, Sarazin G (1994) Geochemical study of a crater lake: Lake Pavin, France. Identification, location and quantification of the chemical reactions in the lake. Chem Geol 115:103–115
Michel F (2008) Le Tour de France d'un géologue. BRGM et Delachaux et Niestlé, 383 pp
Niessen F, Lami A, Guilizzoni P (1993) Climatic and tectonic effects of sedimentation in Central Italian volcano lakes (Latium). In: Negendank JFW and Zolitschka B (eds) Paleolimnology of European Maar Lakes. Lecture Notes Earth Sci. 49, Springer, pp 129–148
Nijampurkar V, Martin JM, Meybeck M (1983) Si-32 as a tool for studying silicon behavior in estuaries. Limnol Oceanogr 28:1237–1242
Olive P, Boulègue J (2004) Biogeochemical study of meromictic lake: Pavin Lake, France. Géomorphologie 4:305–316
Olivier L (1939) Matériaux pour la connaissance limnologique des lacs Mont Doriens. Imprimerie de Bussac, Clermont-Ferrand, 79 pp
Olivier L (1952) Sur la présence en été, dans le lac Pavin, d'une couche dépourvue d'oxygène vers 70 m de profondeur. Comptes Rendus Acad Sci Paris (D) 234:743–745
Omaly N (1968) Le lac Pavin, historique et hydrobiologie. Rev Sci Nat Auvergne 34:7–31
Pelletier J-P (1963) Température et oxygène dissous dans la couche profonde du lac Pavin. Comptes Rendus 88ème Congrès Sociétés Savantes 2:653–658
Pelletier J-P (1968) Un Lac méromictique, le Pavin (Auvergne). Ann Stat Biologique, Besse-en-Chandesse 3:147–170
Persoone G, Sorgeloos P, Groslière C, Omaly N (1968) Sur la limnologie de quelques lacs du Puy de Dôme: physicochimie et plankton de surface. Ann Station Biologique Besse en Chandesse 41–114
Pourriot R, Meybeck M (1995) Limnologie générale. Masson, Barcelone
Restituito F (1984) Study of a layer containing hydrogen sulfide in a meromictic lake, Lake Pavin, France. Hydrobiologia 109:235–242
Restituito F, Lair N (1975–1976) Projet alpin OCDE pour la lutte contre l'eutrophisation des lacs du Massif Central français. III Le lac Pavin. Ann Stat Biol Besse-en-Chandesse 10:146–191
Reynouard J (1909) Besse, autrefois, aujourd'hui ? Ann. Station de Limnologie de Besse-en-Chandesse, tome I: –14
Richard J (1887) Sur la faune pélagique de quelques lacs d'Auvergne. Comptes Rendus Acad Sci 14–12 Nov 1887
Rico B (1876) L'Aquiculture en Auvergne. Extrait Bull. Soc. Acclimation:27
Rouwet D et al (eds) (2015) Volcanic lakes. Advances in volcanology. doi:10.1007/978-3-642-36833-2_4, Springer
Sabouraud C (coord.) (2004) Guide de la Géologie de la France. Belin, 814 pp
Sabroux J-C (2007) Lac Nyos, le monstre domestiqué. Eruption 14:31–39
Sabroux J-C (2009) Oral discussion. Int. workshop Pavin and other meromictic lakes, May 14–17, 2009, Besse
Scharf BW, Björk S (eds) (1992) Limnology of Eifel Maar Lakes. Adv Limnology (Archiv. für Hydrobiol.) 38:348
Schettler G, Alberic P (2008) Laghi di Monticchio (Southern Italy, Region Basilicata): genesis of sediments—a geochemical study. J Paleolimnol 40:529–556
Schettler G, Schwab MJ, Stebich M (2007) A 700 years record of climate change based on geochemical and palynological data from varved sediments (Lac Pavin, France). Chem Geol 240:11–35
Seyler P, Martin JM (1989) Biogeochemical processes affecting arsenic species in a permanently stratified lake. Envir Sci Technol 23:1258–1263
Shanklin E (1989) Exploding lakes and maleficent water in Grassfield legends and myths. J Volc Geoth Res 39:233–246
Shanklin E (2007) Exploding lakes, myths and reality, an African case study. In: Piccardi L and Masse WB (eds) Myth and Geology. Geol Soc London Spec Pub 273:165–176
Sigurdsson H, Devine JD, Tcha FM, Presser FM, Pringle MK, Evans WC (1987) Origins of the lethal gas burst from lake Monoun, Cameroon. J Volc Geoth Res 31:1–16
Sigvaldason GE (1989) International Conference on Lake Nyos Disaster, Yaounde, Cameroon 16–20 March, 1987: conclusions and recommendations. J Volcanol Geotherm Res 39:97–107
Smith K (2003) Environmental Hazards (3rd edn). Routledge, 392 pp
Stebich M, Brüchman C, Kulbe T, Negendank JFW (2005) Vegetation history, human impacts and climate change during the last 700y recorded in annually laminated sediments of Lake Pavin. France. Rev Palaeobot Palynol 133(1–2):115–133
Tata D (1778) Lettera sul Monte Volture a sua eccellenza il signor D. Gugliemo Hamilton. Stamperia Simoniania, Napoli, 62 pp
Tazieff MH (1986) La Montagne, 2 décembre 1986, Clermont -Ferrant
Tazieff MH (1987) Lake Nyos new phreatic eruptions. A note distributed at the Hawaii Symposium on How Volcanoes Work, Hilo HI, January 1987
Tazieff H (1989) Mechanisms of the Nyos carbon dioxide disaster and of so-called phreatic steam eruptions. J Volcanol Geotherm Res 39:109–116
Tazieff MH, Faivre-Pierret RX, Le Guern F, Chevrier RM (1987) La Catastrophe de Nyos, République du Cameroun. Centre des Faibles Radioactivités, CNRS Contrib 811, 76 pp
Tenore et Gussone (1838) Peregrinazioni effetuate nelle state del 1838 dai Signore Tenore e Gussone. In: Principitato Citeriore e di Basilica). Edité en 1995 par A Ciarallo et L Capaldo Viaggo al Vulture, Edition Osanna Venosa, 157 pp
Thienemann A (1914–1915) Physikalische und chemische Untersuchungen in der Maaren der Eifel. Verh. Naturhist. Ver. Preuss.Rheinl. Westf., 70: 249–302 and 71: 273–389
Touchart L (2002) Limnologie physique et dynamique. L'Harmattan, 396 pp
Touchart L (2003) Hydrologie. Mers, fleuves et lacs. Armand Collin, Paris, 192 pp
Touchart L, Dussard B (1998) André Delebecque et le centenaire de la limnologie française. Annales Géographie 107:446–453
Touchart L, Ishiguro N (1999) Les lacs japonais. Annales de Géographie 108:115–133
Viollier E, Jézéquel D, Michard G, Pèpe M, Sarazin G, Albéric P (1995) Geochemical study of a crater lake (Lake Pavin, France): trace elements behaviour in the monimolimnion. Chem Geol 125:61–72
Viollier E, Michard G, Jézéquel D, Pèpe M, Sarazin G (1997) Geochemical study of a crater lake: Lake Pavin, Puy-de-Dôme, France. Constrains afforded by the particulate matter distribution in the element cycling within the lake. Chem Geol 142:225–241
Wetzel RG (1975) Limnology, Saunders, 743pp
Wever de P, Le Nechet Y, Cornée A (2006) Vade-mecum pour l'inventaire du patrimoine géologique national. Société Géologique de France, Mémoire 92, 162 pp
Wurtz A (1945) Recherches écologiques sur les lacs, tourbières et étangs d'Auvergne. Ann Stat Centr Hydrobiol Appliquée 1:89–166
Yoshimura S (1937) Abnormal stratification on inland lakes. Proc Imp Acad Jpn 13:316–319
Yoshimura S (1938) Dissolved oxygen of the lakes waters of Japan. Sci Rep Tokyo Bunrika Daig sect C 8:63–277
Zhang Y, Kling GW (2006) Dynamics of lake eruptions and possible oceans eruptions. Annu Rev Earth Planet Sci 34:293–324

Pavin, A Rich but Fragmented History (200 AD–2016)

2

Michel Meybeck

Abstract

"Pavin Stories", told to the lake's visitors seeking its wild beauty, were reported throughout the nineteenth century by famous scholars such as Pierre Larousse, an encyclopedist, and Elysée Reclus, a geographer: Pavin had no depth, no fish, could trigger furious storms and no boat could sail on it. Pavin history before 1770 was not established. The Pavin history puzzle, although still incomplete, is re-interpreted here through the grid of sensorial degassing indicators (Table 1.2). It starts with a famous "treasure stone" retrieved from its waters in 1909, actually a pompeian-type millstone probably carved at Pavin during tranquil periods of the lake, when Roman baths were constructed at the Mont-Dore, 6 km away. Literary sources are lacking afterwards but a small early medieval Christian settlement existed at the nearby mountain of Vassivière, some 1.5 km away from Pavin, possibly succeeding an earlier cult (Chap. 3). In 1547 a miracle occurred there and an extra-ordinary explosion with thunder and lightning is witnessed in 1551. Latent Pavin misbehavior is reported from the mid-sixteenth to the mid-seventeenth century: a storm suddenly triggered by a stone thrown into the lake (François de Belleforest, *Cosmographie Universelle*, 1575). These storms, which caused damages to the surrounding valleys, were reported to the King of France during its visit in Auvergne (1566). At the time, the lake, already termed since long ago *lacus pavens* (The Terrible Lake), was avoided by the local people, who feared its permanent summer fog and occasional storms (Jean Banc *La Mémoire retrouvée des merveilles des eaux naturelles* 1605; Godivel II, late 1600s). Belleforest's description will be copied by other geographers for three hundred years, making Pavin very famous beyond the borders of Auvergne. Another misbehavior event was also witnessed in 1783. However, the highest Auvergne authorities of the time ignored this very clear sudden lake overturn, thereby undermining the validity of the Pavin Stories. A similar event probably occurred in 1936. The "normalization" of Pavin was achieved by the introduction by Lecoq of the first boat (1847) and the first fish (1859), which facilitated its scientific exploration (Chap. 1). Pavin's fame has remained intact since the early 1900s, but its rich history and wealth of legends (Chap. 3), as well as its unique scientific character, are largely ignored by guidebooks. When Pavin historical descriptions are compared to other lake degassing events (Chap. 1), there is no doubt that several degassing periods and/or events of various intensities were witnessed there in the past.

M. Meybeck (✉)
METIS, CNRS-Unversité Paris 6, UMR 7619,
Place Jussieu, 75252 Paris Cedex 05, France
e-mail: michel.meybeck@upmc.fr

T. Sime-Ngando et al. (eds.), *Lake Pavin*, DOI 10.1007/978-3-319-39961-4_2

Keywords
Pavin Lake • Lake history • Lake degassing • Limnic eruption • Lake fear

"A mountain called Puy de Dome on which there is a big abyss from which escapes ordinarily a big storm of thunder and hail, which damages grains in the valleys" (report made by Auvergne officials to King Charles IX during his visit to Clermont, 30 March 1566, Abel Jouan 1566)

"A short distance from Besse, in Auvergne, there is a mountain on which there is a great lake, so deep that one believes it to be bottomless. As far as one can tell, no spring is feeding it. If one throws a stone into it, soon the thunder rumbles, lightning shine, rain and hail hit your face" (Paul Merula, *Cosmo*graphia, 1605)

"Pavin is another such place, vaguely famous, on which hundred stupidities have been told. As so few people have visited Auvergne, no wonder that such fables are accredited, particularly when no one rises to fight against them"(Legrand d'Aussy, *Le Voyage en Auvergne* 1788)

2.1 Introduction

Pavin, as called by locals and visitors, is a small round lake on top of an Auvergne mountain. It is unique in France for its long and complex history, being celebrated at least since the sixteenth century by many geographers (see above, Merula's statement, a Netherland geographer, in 1605). Fears surrounding the lake were already reported by Auvergne Province officials to the king of France, Charles IX in 1566 (see above, report by Abel Jouan, the royal chronicler). These descriptions contrast with the ones naturalists and scientists have made after royal engineer Chevalier's expedition in 1770 to Pavin (see Sect. 1.4.1). In the late eighteenth century, Pavin was a must-see for a Parisian voyager and writer like Legrand d'Aussy in 1788 (see above) and throughout the XIXth Pavin looked absolutely normal to its visitors. But *Pavin stories* were still told by local people to the growing number of visitors from the nearby Mont D'Or (now Mont Dore) spa. Louis Piesse is one of them and states in 1863: *"There are so many tales about Pavin, or Pavens at it was named by Romans, concerning the thick clouds that were blanketing it, the exhalations that one inhaled on its shore, the stones thrown into the surface waters, that will develop from its depth clouds from which the storm would come, and a perpetual bubbling"*. [As all other original citations in French, this text has been translated by us].

How should such a statement be understood? Was *pavens* the original name of the lake? What are the origins of these phenomenons: thick clouds, exhalations, storm, bubbling that were still referred to by local people in the mid-XIXth, while Lecoq was introducing the first trouts to the lake (Chap. 1). Why has this famous Auvergne naturalist ignored the descriptions of Jouan, Merula and all other European geographers and why did he consider they were not grounded? Throughout the twentieth century, why have the literary and scientific communities ignored the old Pavin descriptions, once reported to authorities by educated witnesses then transformed into *Pavin stories*, transmitted orally over centuries?

The scientific community is now more conscious of, and more interested in past historical and mythological events associated with tsunamis, landslides, volcanic eruptions and their related hazardous places and events, centuries to millenia ago, as well as in how they have been transmitted to other generations through time (*Myths and Geology*, Piccardi and Masse 2007). This approach has been initiated 15 years ago by Italian volcanologists, geochemists, and geologists working on Italian maar-lakes, such as Albano and Monticchio, a very rare lake type only found in volcanic regions which Pavin also belongs to (Chap. 1). They found and endorsed for these volcanic crater lakes a very rich history of past degassing events, reported by the first naturalists working on Monticchio, between 1770 and 1850, and by Latin historians working on Archaic Rome history (see Sects.1.6.2.1 and 1.6.2.2). This interest for maar-lakes was suddenly triggered by the catastrophic degassing of Lake Nyos (Cameroon) which cost 1700 lives within one night in August 1986 (Sect. 1.6.1).

In Auvergne, the historical research on maar-lakes has not yet been made. In 1986 the study of degassing risk at Pavin (see Chap. 1) did not involve any Human and Social scientists such as archeologists, geographers, historians, folklorists or anthropologists. In contrast, such historical approach had been used in Auvergne for the earthquake risk analysis by Mallot, a historian. Her research in ecclesiastic archives found many unknown important earthquakes (Mallot 1995), that changed the regional risk assessment and lead to the cancellation of a planned nuclear plant in Limagne.

In recent years several catastrophic events related to Pavin have been described by scientists (Chapron et al. 2010, 2012; Chassiot et al. 2016; Chap. 23). These observations suggest that Pavin had not always been as quiet as today: were these events also perceived or recorded by people near the lake, particularly at Vassivière Mountain, a small religious settlement about 1.5 km away, and at Besse, the nearby town, 5 km away?

Our search for the history of Pavin and its surroundings generated a vast and heterogeneous material, largely unknown, misused or unused by previous Pavin historians. Based on this material, the following questions are raised here: (i) How the extra-ordinary Pavin has been described through centuries, while society's understanding of natural forces and beliefs was evolving, and while the lake state was also modified? (ii) Are these texts sufficiently contextualized, reliable and coherent? (iii) Can these sources be re-interpreted with the grid of sensorial lake degassing descriptors (Table 1.2), (iv) Can we reconstruct the past state of the lake, thus providing contemporary scientists with another set of information?

2.2 Finding Pavin Puzzle Pieces

Pavin's history is a fragmented and incomplete puzzle: some well-known pieces, although linked to Pavin, could not be understood before the description of the degassing process at Nyos in 1986, many pieces have faded and have been gradually forgotten, some pieces were known but not connected to Pavin and finally, some descriptions may be discovered a long time after their initial reporting. Pavin's history analysis was hindered by the fact that Auvergne historians and scientists could not conceive of past degassing events of any kind taking place at Pavin, even after the Nyos degassing evidence. This skeptical attitude was also present in Italy until very recently (Caracausi et al. 2009; Funiciello et al. 2010). Assembling Pavin historical puzzle cumulates these difficulties.

The first analysis of Pavin history is published in 1805 by Abbé Delarbre (1795, 1805) in charge of the Clermont Botanical Gardens: as he cannot provide a scientific explanation of any of the *Pavin stories*, he makes his best to disproof all of them, without citing any of his primary sources. This position is followed by Lecoq (1835a), the main Auvergne naturalist of his time (see Sect. 1.4.2). In 1874 a mid-seventeenth century manuscript describing Pavin in details, the Godivel manuscript, is discovered and published (Jaloustre 1884). This description remains undiscussed until today, except by Eusebio and Reynouard (1925), a limnologist and a historian. They try their best to assemble pieces of the Pavin puzzle and make, for the first time, a groundbreaking hypothesis of a major catastrophic event at Pavin during Antiquity. But well respected Auvergne writers such as Pourrat (1935) ignore their analysis and present the *Pavin stories*, generally those reported by Lecoq, mixed with a recent legend, the *Sunken City* of Besse (see Sect. 3.3.1) without proposing any clue. Local historians and geographers (Crégut 1921; Fournier 1971; Charbonnier et al. 2011) either consider all *Pavin stories* as unfounded or simply ignore them.

Some pieces of the Pavin puzzle have been previously described and analysed separately by specialists of each field, *e.g.* Pelletier (2008) for cartography, Block (2003) for church misericords, Auserve (2004) for religious history, Fournier (1971) for etymology, Vernière (1899–1900) for the history of past voyagers in Auvergne, Pourrat (1935) for Auvergne culture analysis, Sebillot (1904–1906) and Reyt (2000, 2002), for the analysis of legends and tales. However most of these works have been conducted before the Nyos event and none of these specialists integrated the possibility of catastrophic events due to degassing hypothesis.

Our interpretation of Pavin sources has been recursive, through our multiple iterations (Fig. 1.1), as we were progressing on the characteristics of degassing lakes and discovering the rich body of history and legends surrounding other maar-lakes. A coherent image gradually appeared out of this heterogeneous and incomplete puzzle. This chapter first presents our prime puzzle pieces, *i.e.* those well identified, reliable, written by educated authors and with complete context (precise date and hour, location, detailed description), in a chronological order from Antiquity to the nineteenth century. Then the history of Pavin, as reported and used by guidebooks, scholars and authorities in the twentieth and twenty-first century, is summarized.

The other Pavin puzzle pieces, *i.e.* those with less context, unknown authors or unprecised location, legends collected in the nineteenth century, religious sources, featuring dragon and tales of fairies and other fantastic animals and miracles are analysed, in comparison to similar sources found for other European maar-lakes, in the next Chap. 3.

2.3 Pavin's History Highlights

2.3.1 Antiquity: A Pompeian Millstone Retrieved from Pavin Waters in 1909

The Roman presence near Pavin is attested. After the conquest of Gaul by Julius Cesar, Romans have developed intensive agriculture in the fertile Limagne plain, a major granary for Imperial Rome some 40 km north of Pavin. They also established the famous *Mons Aureus* Roman baths – now Mont Dore – on the other side of the eponymic volcanic complex, some 6 h walk north of Pavin across the mountain. Their ruins, containing coins dating from emperors Neron, Augustus and Agrippa, *i.e.* during the first and second centuries AD (Bertrand 1819; Durand-Lefebvre 1926; Chabrol 1931), were found during the construction of the present-day thermal baths in the early 1800s (Fig. 2.1b). Gallo-Roman poteries have also been found in the 1950s in the Landie Lake, some 15 km, SW of Pavin, at an altitude of 1100 m (J. Bernard, pers. com.). At Pavin a Pompeian millstone has been found a century ago in the lake waters.

This so-called *Pierre au Trésor* (Treasure Stone) is a 250 kg volcanic stone, with cylinder and conical shape,

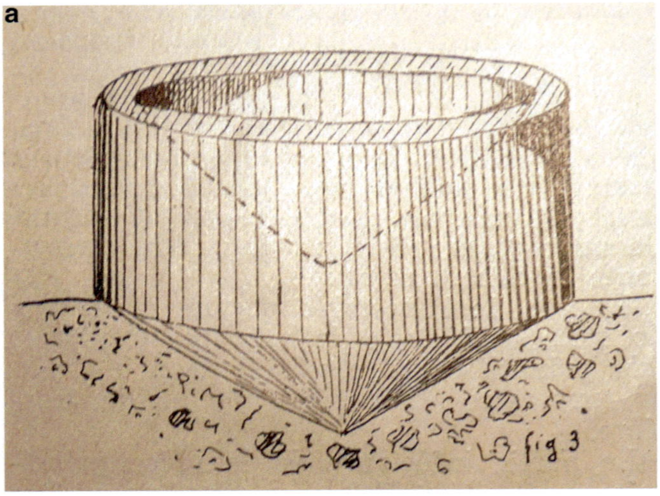

Fig. 2.1 (**a**) The Treasure Stone retrieved from Pavin lake waters, an unfinished Pompeian millstone, (Reynouard 1909). (**b**) Archeological Imperial Rome artifacts dated 100 to 300 AD found in the early 1800s at Mons Aureus Roman baths, now Mont Dore, 10 km from Pavin (Lacour 1850) (BCU library, Clermont-Ferrand)

carved on internal and external side (Fig. 2.1a). It was spotted during the nineteenth century, retrieved from the lake shore in 1909 (Reynouard 1910), brought to Besse that year, then forgotten for 100 years – it was not registered in the archeological inventory – and re-discovered in 2010 (Meybeck 2010). It is now exhibited on the lake shore. Reynouard also mentions few other broken carved stones and concludes they were antique millstones carved from inside Pavin rims. The origin of the stone was confirmed in 2012 by an archaeologist as an unfinished "Pompeian-type meta millstone"(Luc Jaccotey, INRAP pers. com., 2012). Petrographical analyses are currently made to confirm the *in situ* extraction. Until now the dating of the millstone has not been made. Considering its style, a likely hypothesis is that is contemporary of the Mons Aureus settlement.

For Reynouard, a local historian and former Besse mayor, *"a small tribe was disseminated near the springs* [within the crater]*…they were there to extract and carve their millstones…then a cataclysm came: (i) a last eruption from Montchal* [the adjacent nearby volcano], *(ii) an earthquake or (iii) a dreadfull cyclone that would have lifted waters to flood the shore settlements"* (Eusebio and Reynouard 1925). He is the first to make the hypothesis of an early human activity at Pavin and to postulate a catastrophic event to explain its interruption. His third proposition, barely conceivable in 1925, has been completely ignored until recently when paleolimnologists found evidence of important sediment slides in Pavin, dated around 600 AD, that could have triggered a sudden catastrophic lake spillover (Chapron et al. 2012; Chassiot et al. 2016). A text from Gregorius of Tours (539–596), the first historian of Christian Gaul, so far not attributed to Pavin, relates a famous pagan lake cult associated to catastrophic storm events (see Chap. 3).

2.3.2 Early Antique and Medieval Worship Near Pavin, on the Vassivière Mountain

Before the mid-sixteenth century, our Pavin's history puzzle pieces are limited to religious history, most likely linked to Pavin, as further discussed in Chap. 3. The nearby Vassivière Mountain, at an altitude of 1350 m, 1.5 km north-west of the lake (Fig. 1.1c) is an early and unexpected human settlement, that can be described as one of the "defortunate" locations, *i.e.* those abandoned by people and left to the forces of nature, where religious settlements were built to help the astray passer-by, as they have been described in the Alps mountains (Reyt 2000, 2002).

According to late nineteenth century historians, an early medieval church was standing on this mountain until 1321, itself possibly located on a pre-Christian wall (Jaloustre 1910). However the wall has not confirmed by contemporary historians nor subject to archeological research, but the ruins of the primitive church were seen when the present pilgrimage chapel was built around 1550 (Jaloustre 1910; Ph. Auserve 2004 and pers.com.). For Jaloustre, Vassivière toponym could originate from celtic words *"vas"*, temple and *"iver"*, water. In this location, *"a temple built by Celts was dedicated to the Clamouze headwaters deities. People would come here to celebrate feasts. When Christianity took over, the Virgin celebration replaced the one of the Spring water deity. A village became the center of a Christian parish, succeeding to the Gallic dwellings"*. This hypothesis could explain the location of this early church on such defortunate mountain, away from all roads and snowbound for 5 months of the year. Gregorius of Tours also mentions an early Christian settlement established after the pagan lake cult, near a mountain lake, however not traditionally attributed to

Pavin but to Lake St. Andéol in the Gévaudan mountains (see further discussion in Chap. 3).

In 1321 the local lord, Bernard de la Tour d'Auvergne, ordered the destruction of the initial Vassivière medieval church, a very rare decision, and authorized the re-use of its stones, on the grounds that profane events were taking place at this location *["Quam plurina profana in dicto loco comiti et fuisse commissa"*] and granted a very peculiar status to the nearby city of Besse and to its people, an event noted by all historians (Boyer-Vidal 1888; Jaloustre 1910; Auserve 2004). The causes of Vassivière first church destruction and the nature of the profane celebration remain unknown.

Paleolimnology work at Pavin records another major catastrophic event around 1300 AD corresponding to a spillover of the lake with a drop of lake level by 6.5 m, due to the sill erosion (Chapron et al. 2012; Chassiot et al. 2016; Chap. 23). Also, studies on the Couze Pavin talweg sediments show a 9 km-long mudflow originating from the Pavin outlet attributed to this period that could correspond to this catastrophic spillover (Lavina and Del Rosso 2009). The regeneration of some pagan ceremonies at Vassivière, following the 1300 catastrophic event, can be hypothesized. From 1321 to 1550s a statue of the Virgin was left there at a place named locally *Seignadou, i.e.* where people should sign themselves when passing by.

2.3.3 *Lacus pavens* Terrifies the Whole Region Throughout the Sixteenth Century

During the sixteenth century written sources of all kinds on Pavin, actually *lacus pavens,* are numerous. In 1547 the Vassivière Mountain is recognized for a miracle, discussed in the next chapter, soon attributed to the intercession of the Vassivière Virgin, and since that date religious authorities register any evidence of miracles and any odd events in the vicinity of Vassivière, *i.e.* of Pavin. In 1566, Auvergne authorities complain to King Charles IX of France during his stay in Clermont, about an abyss in mountains, presented as a major hazardous area. In 1575, Pavin is featured in one of the first-ever landscape paintings in France and cosmographer Belleforest describes the marvellous misbehaviour of a lake near Besse (1575). At this period the lake is named *pavens*, the terrifying (Banc 1605). Unusual events at/or near Pavin are highlighted here.

2.3.3.1 The Terrible Explosion Witnessed at Vassivière by Besse People (28 August 1551 Pavin Event)

After 1547, all Vassivière miracles are carefully recorded by the church authorities and kept in the church archives. They are presented in chronological order by Father Michel Coyssard (1547–1643), a famous Jesuit lexicographer born in Besse, then detailed by Dom Cladière, a Benedictan also born in Besse (1688, re-edited in 1837), as well as by many Vassivière historians such as Chaix (1869); Jaloustre (1910) and Pourreyron (1935). Coyssard reports a sudden storm that occurred there on 28–29 August 1551, when a mentally sick child was brought to the miraculous statue, the new chapel being under construction:

> *The air, which was at this moment very calm and tranquil, changed so suddenly that all attendants were astonished: an impetuous wind rose all of a sudden with such violence that they were almost all knocked on the ground, scattering their hats and coats. What put them most in fright and fear is that at the same time the whole air was like ignited by the length of terrible lightning, that occurred without stopping, and by terrible thunder that could be heard around them…this was just the effects of the furiousness of the demon who was being exorcized, as one could smell for a long time after an awful sulfur smell. The weather went back to clear state as it was before, lightning and thunder stopped and the atmosphere went tranquil again.*

This violent and very sudden atmospheric event was witnessed by one hundred Besse people. Vicar Boyer, Etienne Desserres, Jean Libinoux, Etienne Ribeyre, Guillaume Lafont, Pierre Blanchier, all witnesses, made a written juridical statement to the religious authorities. These people were standing on the mountain about 1.5 km away from the Pavin, at the same altitude as the crater rim, but not in the direction of the Pavin outlet. This event is very sudden, violent and, probably, unexpected for those who witness it, which explains why it is officially registered.

There is little doubt that this event was actually a violent degassing event of Pavin. Although there is no description of the lake proper, as it cannot be seen from Vassivière, one notes several descriptors of a violent degassing event similar to the one witnessed by the French team at Nyos, few days after the main degassing event or at Monticchio (Table 1.2): the Vassivière event starts during fair weather, then a very violent and sudden wind occurs with *"thunder noise"*, *"ignited atmosphere*"; it terminates abruptly while a *"sulfur stench"* remains. These extra-ordinary atmospheric characteristics were interpreted in 1551 as the presence of the evil being exorcised from the sick child (Don Cladière 1688). Pavin historians, such as Delarbre (1805); Lecoq (1835a); Eusebio and Reynouard (1925), have missed or ignored this event which clearly links Vassivière and Pavin. We also missed it in our preliminary analysis (Meybeck 2010). Other events featured in Vassivière register from 1550 to1650 present well contextualized eventswhich can be interpreted as loss of consciousness of individuals due to CO_2 intoxications.

2.3.3.2 A Hazardous Abyss, Generating Storm, Thunder and Hail, Presented to Charles IX (1566)

On March 31 1566, Charles IX, touring his kingdom of France with his mother Catherine de Medicis and his brothers and sister, is hosted at Clermont for 3 days, located at 3-day trip from Pavin. The royal chronicler Abel Jouan (1566) makes a short account of this brief stay in the Auvergne main city in which the king and his followers were informed about ***"a mountain, called Puy de Dome, on which there is a big abyss from which escapes ordinarily a big storm of thunder and hail, which damages grains in the valleys".***

Although Jouan's description wrongly places the terrible abyss – a lake is not mentioned – on top of the Puy de Dôme, a famous mountain near Clermont, the attribution is most likely at the Mont Dore mountain and concerns Pavin Lake [the Puy de Dôme mountain has no crater], as understood by the very few scholars who reported this text as a legend (Legrand d'Aussy 1788; A. Pourrat 1973). It remained undiscussed by Delarbre and Lecoq who were aware of this famous Royal visit (Jouan's chronicle is listed as a source on Auvergne history in one of Lecoq's work). The lake is causing so many damages and fear that the case is brought to the attention of the King and his mother, who is also the ruler of La Tour d'Auvergne and Besse at that time. According to this official complain, Pavin misbehaviour is permanent, damaging grains in the downstream valleys, which could correspond to at least 20 to 30 km downstream. It suggests that lake degassing events associated with storm, thunder and hail as encountered in the 1551 event were common at this period. This description remained largely ignored just like the August 1551 event.

2.3.3.3 Pavin Painted on the First Realistic Landscape Picture in France (1571–1579)

The Queen Mother Catherine de Medicis, and lord of Besse, orders from her royal notary, Antoine Godivel from Besse, a magnificent illuminated manuscript, the royal land register of Besse (Godivel I 1579a) today kept at the Chantilly Castle library. As many members of the Godivel family of Besse have played a role in Pavin's history, we shall name him Godivel I. One key feature of this unique register is the hand-painted panorama of the Besse countryside (about 40 cm × 80 cm), in which Pavin is represented on the upper right corner, overlooking the town. It can be spotted by its position, its crater-like shape and its cascading outlet reaching the Couze Pavin River. The surface of the lake seems still. The lake is enlarged in this picture considering its actual size (about 400 m in diameter), an indication of its importance. The Vassivière Mountain is marginally represented in the very upper part, identifiable by a small cross. There is no farm featured in the upper Couze Pavin River valley upstream of Besse, which looks very isolated, in contrast to other Besse valleys. It seems that the diminutive pathway from Besse to Vassivière is actually avoiding the lake outlet. The Montchal Mountain, another recent volcano adjacent to Pavin (Fig. 1.2c), is also featured, slightly above the crater lake. The gothic-written illuminated Chantilly manuscript register and its copy, not illuminated and now kept in the Clermont regional archives (Godivel I 1579b), remained to be checked for the mention of Pavin.

It is without doubt one of the first representation ever made in France of a landscape from an aerial viewpoint (Pelletier-Decitre 2008). It is extremely accurate: the street map of the medieval city of Besse is still the same as today. At the same period, the first accurate map of the kingdom, *La Charte de France*, by François La Guillotière (1575–1590) features the *Vaucivière Mountain* (Vassivière), an indicator of the fame of the new pilgrimage, but not Pavin.

2.3.3.4 Pavin Marvelous Response to a Thrown Stone in Belleforest's Cosmographia Universalis (1575)

In the description of Auvergne made by the cosmographer François de Belleforest (1575), there is still no direct mention of Pavin, nor of Paven, the common spelling until 1800, neither in the index nor in the text; but one finds a ***"Lac merveilleux, en haut d'une montagne en Auvergne"*** [Marvellous lake located on top of an Auvergne mountain] as well as a "Marvellous abyss named des Soulcis" *[Abyme merveilleux nommé des Soulcis]*, a small cavity in the nearby lava flow. These sites are less than 1 km away from each other (Fig. 1.2c). The Creux de Soucy, as this cavity is now called, allows to precisely locate the un-named Belleforest's lake and their description follows the one of Mont d'or famous thermal baths, formerly Mons Aureus (Fig. 2.2).

> *"…Near this Mont D'or stands the city of Besse, at a half quarter league from which one can see a great lake located nearly on top of a mountain, without a measurable depth and without any water inlet. Actually it is both admirable and terrible to see as, if a stone is thrown into it, it is ascertained that soon thunder, lightning, rainstorm and hail would occur. Not far from this lake there is an abyss named Creux des Soulcis…"*

François de Belleforest (1530–1583) is a prolific translator of historical and literature works written in Latin. After having been a royal historiographer for King Henri III, he is commissioned by Parisian publishers Nicolas Chesneau and Michel Sonnius to translate the *Cosmographia Universalis – i.e.* geography – of Sebastian Münster (1488–1552). Münster Cosmographia (1565–1566), originally in German, is a very famous work of its time, translated into French, English, Italian and other languages. In his version, Belleforest adds a new and very detailed description of the Kingdom of France and its provinces, in specific chapters that are missing from Münster's. The Auvergne description by Belleforest is one such addition: it is the most complete description of this province at the time. It has often been misquoted by the very few Auvergne historians referring back to it as "Münster-Belleforest 1575" *(e.g.* Fournier 1971), and we have reproduced this error (Meybeck 2010). For each new chapter, Belleforest mobilized a set of local informers to whom questionnaires were sent (Simonin

a b

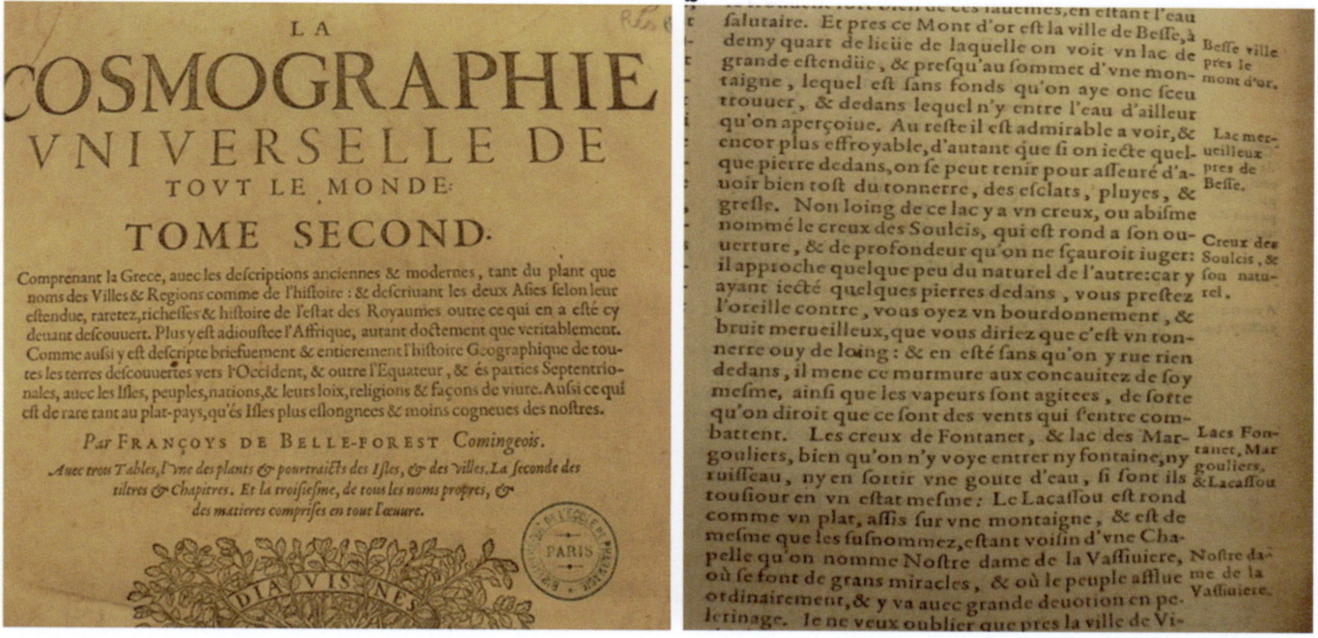

LA
COSMOGRAPHIE
VNIVERSELLE DE
TOVT LE MONDE:
TOME SECOND.

Comprenant la Grece, auec les deſcriptions anciennes & modernes, tant du plant que noms des Villes & Regions comme de l'hiſtoire : & deſcriuant les deux Aſies ſelon leur eſtendue, raretez, richeſſes & hiſtoire de l'eſtat des Royaumes outre ce qui en a eſté cy deuant deſcouuert. Plus y eſt adiouſtee l'Affrique, autant doctement que veritablement. Comme auſsi y eſt deſcripte briefuement & entierement l'hiſtoire Geographique de toutes les terres deſcouuertes vers l'Occident, & outre l'Equateur, & és parties Septentrionales, auec les Iſles, peuples, nations, & leurs loix, religions & façons de viure. Auſsi ce qui eſt de rare tant au plat-pays, qu'és Iſles plus eſlongnees & moins cogneues des noſtres.

Par FRANÇOYS DE BELLE-FOREST *Comingeois.*

Auec trois Tables, l'vne des plants & pourtraicts des Iſles, & des villes. La ſeconde des tiltres & Chapitres. Et la troiſieſme, de tous les noms propres, & des matieres compriſes en tout l'œuure.

DIA VIS NES

PARIS

ſalutaire. Et pres ce Mont d'or eſt la ville de Beſſe, à demy quart de lieüe de laquelle on voit vn lac de grande eſtendüe, & preſqu'au ſommet d'vne montaigne, lequel eſt ſans fonds qu'on aye onc ſceu trouuer, & dedans lequel n'y entre l'eau d'ailleur qu'on aperçoiue. Au reſte il eſt admirable a voir, & encor plus effroyable, d'autant que ſi on iecte quelque pierre dedans, on ſe peut tenir pour aſſeuré d'auoir bien toſt du tonnerre, des eſclats, pluyes, & greſle. Non loing de ce lac y a vn creux, ou abiſme nommé le creux des Soulcis, qui eſt rond a ſon ouuerture, & de profondeur qu'on ne ſçauroit iuger: il approche quelque peu du naturel de l'autre: car y ayant iecté quelques pierres dedans, vous preſtez l'oreille contre, vous oyez vn bourdonnement, & bruit merueilleux, que vous diriez que c'eſt vn tonnerre ouy de loing : & en eſté ſans qu'on y rue rien dedans, il mene ce murmure aux concauitez de ſoy meſme, ainſi que les vapeurs ſont agitees, de ſorte qu'on diroit que ce ſont des vents qui ſ'entre combattent. Les creux de Fontanet, & lac des Margouliers, bien qu'on n'y voye entrer ny fontaine, ny ruiſſeau, ny en ſortir vne goute d'eau, ſi ſont ils touſiour en vn eſtat meſme: Le Lacaſſou eſt rond comme vn plat, aſſis ſur vne montaigne, & eſt de meſme que les ſuſnommez, eſtant voiſin d'vne Chapelle qu'on nomme Noſtre dame de la Vaſſiuiere, où ſe font de grans miracles, & où le peuple afflue ordinairement, & y va auec grande deuotion en pelerinage. Ie ne veux oublier que pres la ville de Vi-

Beſſe ville pres le mont d'or.

Lac merueilleux pres de Beſſe.

Creux des Soulcis, & ſon naturel.

Lacs Fontanet, Margouliers, & Lacaſſou

Noſtre dame de la Vaſſiuiere.

Fig. 2.2 Belleforest's Cosmographia (1575) with mention of a "marvellous lake" near Besse where a thrown stone triggers thunder, lightning and storms, and of the nearby Creux de Soucy. It is the first written account of Pavin and Creux de Soucy (Library of the Faculté de pharmacie ,Paris)

1992; Deboffe 1995). His cosmography in two volumes is very well organized with a final index of topics and locations featuring about 10 000 entries.

Pavin (44 ha) and "Creux des Soulcis" (10 m^2, cavity opening) as well as two other nearby ponds less than an acre in size, are featured in this Auvergne description: they are indeed very modest water bodies to be featured in a Universal Geography! Until the nineteenth century, these places were difficult to access and/ or find (Chap. 1). Their description could only have been written first-hand by a local and educated informer who remained unknown. He could be Antoine Godivel the Queen notary who knew this territory so well, as supervisor of the Besse land register, and who could have been an acquaintance of Belleforest at the king's court.

In his *Cosmographia*, Belleforest describes only a few lakes, some major European ones such as Léman and Bodan (Geneva and Constance lakes), as well as a few others with remarkable features like the Asphalt Lake (Dead Sea, famous since the Antiquity), a cenote-type pit lake (a dissolution pit in limestone) in Haiti or the seasonal – *i.e.* karstic – lakes in Illyria (today Slovenia and Croatia). All these uncommon lake types indeed exist, and will only be understood centuries after his description. Their mention in the *Cosmographia* demonstrates that Belleforest was well aware of the natural wonders of his time. Furthermore, he did not try to explain them with fantastic stories, legends or miracles: although he is a near-fanatic catholic, he remains respectuous of text material (Deboffe 1995). Belleforest's work is attacked, as soon as it is published, by André Thevet (1516–1590), another cosmographer and competitor (Lestringant 1991), who claims that a true cosmographer should not rely on second-hand material but should instead, be on site (he himself went to South America and Italy). In Thevet's original description of Auvergne, published the same year (1575), Mont d'Or area and Pavin, are not featured although, as a royal historiographer, he could not be ignorant of the complains made by Auvergne people to Charles IX, as mentioned by Jouan a few years before. Other early cosmographers like Charles Etienne (*Dictionarium Historicum et Geographicum* 1553); Petrus Apianus (1581); and Simeoni (1559) who describes the Limagne Plain, do not feature Pavin: Belleforest's mention of Pavin could be the first one made in a cosmographia.

Considering the previous description of Jouan (1566) and the 1551 event, Belleforest's concise yet detailed location and description of Pavin seems to be quite reliable: the lake is on top of a mountain, it is bottomless and without visible water input. In addition, the description confirms the three Nyos-like violent attributes (Table 1.2): thunder, lightning and storms, already described in 1551, and adds that they can be triggered by a stone thrown into the lake and that they results in hail formation. Whether the stone is a metaphor for a rockslide or should be understood literally remains uncertain. In the second hypothesis, the lake would be close to saturation, until the stone is thrown, as observed by contemporary geochemists at Monoun Lake, in Cameroon (see Sect. 1.6.1). Belleforest also mentions few other *Thrown*

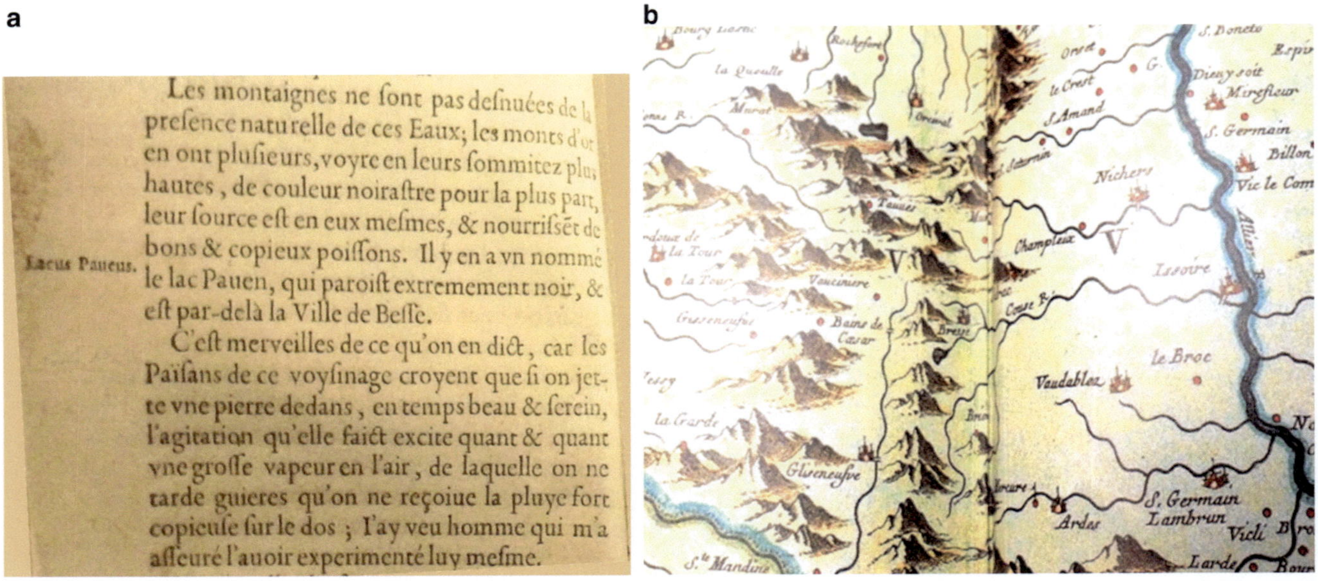
a

Lacus Paueus.

Les montaignes ne sont pas desnuées de la
presence naturelle de ces Eaux; les monts d'or
en ont plusieurs, voyre en leurs sommitez plus
hautes, de couleur noirastre pour la plus part,
leur source est en eux mesmes, & nourrissẽt de
bons & copieux poissons. Il y en a vn nommé
le lac Pauen, qui paroist extremement noir, &
est par-delà la Ville de Besse.
C'est merveilles de ce qu'on en dict, car les
Paisans de ce voysinage croyent que si on iet-
te vne pierre dedans, en temps beau & serein,
l'agitation qu'elle faict excite quant & quant
vne grosse vapeur en l'air, de laquelle on ne
tarde guieres qu'on ne reçoiue la pluye fort
copieuse sur le dos ; I'ay veu homme qui m'a
asseuré l'auoir experimenté luy mesme.

b

Fig. 2.3 (**a**) Lacus pavens, the terrifying lake, mentioned for the first time by Banc (1605) (BCU library, Clermont-Ferrand). (**b**) "Paven" Lake is mapped for the first time, although not named, in Amsterdam by Joan Blaeu (1663) in his world atlas featuring also Vassivière mountain (Vaucinière), Mons Aurus baths (Bains de César, wrongly located), i.e. Mont Dore mountain, and Besse city (Bresse). Other lakes of the region are not featured. The lake is on the N-S watershed divide between Dordogne and Allier

Stone stories in some mountain lakes, in particular in the Canigou Mountain (Pyrenees) during the Middle Age (See a discussion in Chap. 3). However, when considering the 1551 Pavin event, the Jouan's 1566 record and the accurate description of Pavin's area in Belleforest, his account of Pavin seems reliable, even if the *Thrown Stone* part seems to be a stereotypical presentation:

2.3.3.5 *Lacus pavens*, the Terrifying Lake, Is the Original Pavin Name (Banc 1605)

Pavin naming is found for the first time in 1605, in the first census of Auvergne and Bourbonnais thermal waters, by Jean Banc, medical doctor in Moulins *(La Mémoire renouvelée des merveilles des eaux naturelles* 1605). In his general chapter on thermal waters, he starts with two examples of water-related marvels in Auvergne: the first is a *Lacus pavens*, as printed in the column margin (Fig. 2.3a), the second are the famous St. Alyre petrifying springs near Clermont. Banc's description, used here for the first time about Pavin, is presented as such:

> *There is one lake named the Pauen [Paven] Lake which looks extremely black and is above the town of Besse. It is marvellous, as they say, as the local peasants think that if a stone is thrown into it, when the weather is fair and serene, the resulting agitation excites eventually a big vapour in the air, from which soon one receives an abundant rain on the back. I have seen a man who ensured me that he had experienced this himself.*

According to Banc the *Lacus pavens* naming is clear: Paven's means "terrifying" in Latin and its origin is ancient, possibly a legacy from Roman times, which has to be confirmed. Roman toponyms remained in use locally until the Middle Age: the center of the Mont d'Or village was still named *Pantheon* in 1420, in relation to the big temple established during the first–second century AD (Durand-Lefebvre 1926).

Banc is considered by hydrothermalism historians as a prime historical source on Auvergne thermal waters and, unlike many seventeenth century geographers, he has visited this province and went to Besse. It is likely, although not expressedly stated, that he visited the lake, and his description includes the statement of a trusted local witnes. It is therefore probable that this particular characteristic was not just a stereotype added by Belleforest a few years before and borrowed from other lakes. Banc's description is original and clearly independent from the previous ones made in 1551, 1566 and 1575: he does not explicitly mention thunder and lightning, as in Belleforest's text. He describes the frequent degassing, the *"big vapour emitted in the air"*, which results in *"rain"*, as the hail triggered by the thrown stone. This process is remarkable as it also happens here when the weather is fair and serene, as during the 1551 event and as suggested by the report to King Charles IX in 1566. It also appears less violent than in previous descriptions.

Following Banc's account, the second time the lake is named, in our Pavin corpus, is in another general work, *Les Rivières de France* by Louis Coulon (1644) as *Pauen* and *Pauin*. We then find *Pauain* (Du Bouchet (1645), *Pavent* (Godivel II, mid-XVIIth) and *Paven* (Jaillot 1715; Cassini 1815; Donnet et al 1820). During the same period, the lake is mapped for the first time in Joan Blaeu's (1598–1673) world atlas published in Amsterdam (1644), in which the tiny lake is the only one in Auvergne to be featured on the map, another indication of its extra-ordinary fame (Fig. 2.3b). After 1800, the *Pavin* spelling gradually takes over.

The origin of Pavin's name has been, and still is, a subject of much debate among Auvergne specialists which illustrates the entangled analyses of Pavin history. Fournier (1971), a well respected Auvergne historian and etymologist, states: *"I do not know any ancient naming of the lake as Pavin. The first mention of a lake is found in Münster and Belleforest in 1575"* (as we've seen, Münster is not the author of this mention). Fournier also considers that Pavin was first named as such by Chevalier (1770), and then by Dulaure (1788); Legrand d'Aussy (1788); and Delarbre (1805), therefore ignoring Banc and all other previous spellings used. For him, the possible origins of Pavin are (i) *paveo*, meaning""I terrify" in Latin, an hypothesis rapidly ruled out by him, (ii) *pavage* – paving – as there are rocks on the shoreline that mimic paved ground, (iii) *palenc*, a palisade in Occitan language, as some cliffs within the inner crater look like it. Fournier favors this last etymology as Herillier (2011). For Fournier, the *pavens lacus* expression is expressed for the first time in 1921 (Crégut 1921). Crégut found the Pavin description of Briet (1648) with the Trown stone story and mocks those "pleasant etymologists who derive Pavin from lacus pavens", although Canon Crégut, as Vicar Delarbre in 1805, does not mention who were they. A further research shows that Briet was using André Du Chesne (1614), himself reproducing Belleforest text without quoting him.

Actually, all these possible etymologies had already been mentioned in the mid-1600s manuscript of Godivel II (1650s, see further), but for this scholar from Besse, there is no doubt about the name, *"The lake has always been called Pavent* (sic)*...It is likely that such denomination is due to the fear and fright generated to all when one looks at it, even everyday of one's life"*.

2.3.4 The Admirable and Terrifying Pavens Gets Famous During the Seventeenth Century

Belleforest's stereotypical *Thrown Stone* story, featuring the stone that trigger thunder, lightning and subsequent storm with hail, will be reproduced for 200 years and more without much modification by many cosmographers, clerks and scholars. Most of them do not mention its origin nor develop an analysis; none of them have visited the lake, as they do not add their personal touch concerning this exceptional phenomenon, but they increase Pavin's fame among scholars, well outside the Auvergne province and the kingdom.

Bessin (1604) is one of the first to reproduce the *Thrown Stone* story, followed by the Dutch geographer Paul Merula (1558–1607) in his 1605 *Cosmographia*, by André du Chesne (1584–1640) in his *Description of France* (1614), Pierre d'Avity (1573–1635) in his *Cosmographia* (1643), Louis Coulon (1605–1664) in his *Hydrographical Description of France* (1644 and 1654), de La Planche in a manuscript description of French cities (1646), French Jesuit Philippe Briet (1601–1668) in his *Parallela geographiae et novis* (1648) and Louis Moreri (1643–1680) in his *Great dictionary* (1712).

Father Fodéré (1619), an Auvergne clerk, qualifies Pavin as " ***admirable and terrible*** "; his visit to the lake is unconfirmed. According to Audigier (1720), Athanasius Kircher, the famous German scholar (1601–1680), also mentioned the same description of Pavin. However we failed to find this quotation in Kircher's 40 books, likely to be in his *Mundus Subterraneus* (1664–1678). Most of these mentions of Pavin remain unknown – or voluntary undisclosed – by Delarbre (1780, 1805); Lecoq and Bouillet 1831; Lecoq (1835a); Eusebio and Reynouard (1925) and Fournier (1971).

2.3.4.1 The Godivel Manuscript, Late Seventeenth Century, Details Permanent and Occasional "Paven" Degassing

The manuscript is uncovered at Besse in 1874 in the Godivel family archives, when the old and famous family lineage terminated in Besse, and published in 1884 by Jaloustre, the Besse historian. The author of the manuscript, a master of Latin and Greek, could be Nicolas Godivel, Besse chatelain and lawyer at the Auvergne Parliament in 1686, according to Cladière (1687); we will name him Godivel II. The text has been written, according to Jaloustre, in the middle of the seventeenth century, maybe at the turn of this century, as *Remarques touchant la Ville de Besse.* Pavin's three-page-long detailed description is an important part of this manuscript. As seen previously Godivel also discuss *Pavent* name. Pavin can be summarized as such:

> *The lake is one of the most beautiful to be found in Europe. Its depth is thought to be unlimited: some courageous and curious [men] went with a craft in the middle of the lake and let run 313 fathoms of ropes* [without reaching the bottom]. *The outlet discharge is important even during summer drought. The water is clear and smells like ferrugineous springs in the Pyrenees. Fish has never been seen* [in the lake] *nor any plant. Birds are rare in the surrounding woods: the reason is that the fog which is seen escaping at any time from the lake keeps them out and makes this place unhabitable...This place is feared because in summer several exhalations and vapours are emitted and eventually result in clouds and storms that threaten the whole area.*

This text is fully coherent with the previous sixteenth century descriptions and it corresponds to different intensities or types of degassing (Table 1.2). The first one is a permanent soft degassing – the fog – when the lake is not frozen, with a sulfur odour and a ferrugineous taste. The absence of fish, birds and even of aquatic vegetation is a noted feature of

Pavin. The second one, which occurs a few times each summer, is characterized by violent exhalations and vapours, with clouds that are threatening the whole area. It corresponds to a sudden, spontaneous and intense degassing, although not as violent as the ones either observed by the Vassivière pilgrims in 1551 or reported by Belleforest, as thunder and lightning are not mentioned here, but is quite similar to Jouan's description (1566) and to Banc's description (1605). It also says that attemps by local courageous people failed to determine the depth of the lake, a finding similar to Belleforest's text. According to Jaloustre (1884) this manuscript is from the second half of the XVIIth century.

The Godivel manuscript is a major source on Besse history, used by few local historians working on the town's past (Boyer-Vidal 1888; Eusebio and Reynouard 1925; Gomis 2006), ignored by Crégut (1921), unused by Fournier (1971), briefly quoted by a limnologist (Omaly 1968). Before our preliminary work (Meybeck 2010), this manuscript was considered by Eusebio and Reynouard (1925) to be more of a record of legends than a reliable description.

2.3.5 Pavin's State Is Gradually Presented as Normal (Eighteenth Century)

Belleforest's followers are still found during the eighteenth century like Patrick Gordon in his *Geographic Grammar* (1743). Louis de Mailly (1689–1767), in his work on *Nature Principal Wonders* (1723), reports other *Thrown Stone* stories generating dangerous storms for one lake in Germany and one in Switzerland, and states that another lake in Auvergne produces similar effects (see discussion in Chap. 3). It is unlikely that these scholars have visited Pavin. Other scholars are now ignoring Belleforest, as Bruzen de la Martinière (1683–1746) in his *Dictionnaire géographique, historique et critique* (1739–1741), and Piganiol de la Force in his *Nouvelle description de la France* (1754).

Canon Pierre Audigier (1659–1744), often considered as one of the first Auvergne historians, cites both Pavin and Creux de Soucis in his manuscript on Auvergne History (Audigier 1720, only published in 1894, p. 139); it is an early critics of the *Thrown Stone* story:

> *…Its water is awful to see as its depth is difficult to find…If one makes an effort to shoot a gun, one can hear echos as those made by thunder when it bursts out several times; this lead people to believe that, when a stone is thrown into the lake, in a calm and blue day, the agitation of the water caused by the stone, generates a vapour followed by rain, storms with hail and thunder. Despite the publication of this wonder by our geographers, as well by Father Kircher, based on some authors, it is certain that this does not happen.*

It is difficult to check if Audigier himself has visited the lake. His statement confirms a previous opinion by Jesuit de Frétat (1672) on the prime importance of Pavin in Auvergne and on the failure of the *Thrown Stone* experiment. The thunder is now explained by gunshots echos. The stone and the gunshot experiments will be performed by each visitor to Pavin, particularly when guided by local people, until the end of nineteenth century.

After the mid-1700s, Auvergne naturalists fiercely deny the Belleforest's description and its repetition by seventeenth century geographers, which they now consider to be *Pavin stories* since they are unable to observe any of the former Pavin attributes when they visit the lake. The first disproof of these "stories" relates to Pavin's bottomless reputation.

2.3.5.1 Finding Paven Depth, a Major Challenge (1672–1770)

Pavin is thought to be bottomless. This is not the least of its many mysteries. In 1672, Amable de Frêtat, a Jesuit, draws the first accurate map of the Pavin area. He considers *Paven* as the VIIIth wonder of Auvergne in a list of XIII, which he briefly develops in the margin of his map. There he notes that the lake water is very beautiful and clear and *"it is not true that when throwing a stone one triggers a great vapour and thunder as reported by some authors. The lake is 60 fathom deep"* [110 m]. This is the first disproof of Belleforest's statement by a scholar and a visitor to the lake who is the first to mention the *Thrown Stone* experiment. Frétat does not provide any other detail on Pavin's state, nor does he mentions how he has measured the depth. A permanent soft degassing in 1672, as stated in the Godivel II manuscript, cannot be totally ruled out as Frêtat only refutes the *Thrown Stone* effect.

On March 9 1726, another Godivel from Besse (Godivel III) failed to find Pavin depth while the lake is frozen, but measures for the first time its length and width: 1000×927 geometric steps (five feet), which probably required a sustained effort by several people (Legrand d'Aussy 1788; Delarbre 1805). In 1760 Domeiron, a voyager, visits Pavin but also fails to generate a storm while throwing a stone (La Porte et al. 1790). However Auvergne historians do not mention de Frêtat's work: the first successful sounding of Pavin [96 m] is attributed instead to Chevalier in 1770 (Delarbre 1805). It can be considered as the first scientific approach on Pavin (see Sect. 1.4.1).and will be a major disproof of Pavin stories. Few years after Chevalier's expedition another misbehavior event is witnessed at Pavin but will remain ignored for 200 years.

2.3.6 A Sudden, Unexpected and Moderate Degassing Event (21 August 1783 Pavin Event)

This remarkable and non-violent event is witnessed and published by Besse chatelain Godivel (Godivel IV, grand son of Godivel III). It took place after Chevalier's expedition.

Godivel IV, who knew Pavin very well, found his observation worth publishing, and only a week later sent a two-page "letter to the editor"to the *Feuille Hebdomadaire de la Haute et Basse Auvergne*, a new pre-revolutionary weekly magazine. The abridged version follows:

> *Paven Lake, at half league from Besse, is one of Auvergne marvels. It represents a terrible gorge in a mountain…Its water has always been very clear, never clouded by rains nor by storms or inlets, which are invisible. The outlet discharge is very stable and could work a watermill. On August 21* [1783], *on a very clear day I have crossed this stream at eight in the morning to go to Vassivière. When going back at four in the evening with three priests, a deacon and my son, scholar in philosophy, I found the outlet water very turbid and yellowish…My son climbed to the lake, some 150 toises* [180 m] *from there, and found the lake surface completely troubled, carrying a sort of yellowish silt* [limon glutineux]. *He took the water in his hand and when this substance dried it produced on his hand some pimples with a slight pain that lasted some time. The same day the land keeper came to tell me: "Sir, there is something extraordinary in Paven which I have never seen, the water is boiling everywhere but there is no wind, it is yellowish and even rusty in some places". The day after, I called my farmer Gelac who stays close to the lake. He told me that the water had been dirty two or three times this summer, in the evening…I suppose that there has been inside the lake some slide, so important that such volume of water was clouded and infected by some venomous material…*

This important text on Pavin is much contextualized, with the exact date as well as named witnesses. One notes once more the initial spelling of the lake, the celebration of Pavin as a wonder of Auvergne, as in Frétat's list (1672), and the latent fear *(a terrible gorge in the mountain)*. At the time, the path from Besse to Vassivière avoids the lake outlet and is located on the opposite side of the valley. The witnesses immediately notice Pavin's abnormal behaviour, a sudden change of colour and nature in lake's water. It is very different and much less violent than previous descriptions of degassing at Pavin (the atmospheric event of 1551 or the descriptions made by Belleforest, Banc and Godivel II). Godivel details: (i) a colour change, from very clear water to yellowish, sometimes rusty, (ii) a modification of the nature of the waters which look silty, (iii) the phenomenon occurred suddenly, within a few hours, between morning and afternoon, (iv) the whole lake surface is affected, (v) this abnormal behaviour had been observed by a third witness several times during the summer of 1783, (vi) it was associated with the whole lake boiling in the land keeper testimony, (vii) the water is slightly corrosive and can cause skin desease. All these observations have been made by scientists in other degassing maar-lakes (Table 1.2).

Godivel IV's observation of this phenomenon is very factual an detailed, it includes a rational explanation: he attributes this event to an internal slide, a processus that had actually occurred several times at Pavin, as made evident today by sedimentologists (Chapron et al. 2010, 2012; Chassiot et al. 2016). Considering a probable meromictic nature of Pavin maar-lake at that time, *i.e.* the permanent occurrence of an anoxic bottom water layer (see Chap. 1), these changes of water colour and aspect are likely due to rollovers of lake deep waters with precipitation of iron hydroxide when mixed with oxygenated surface waters. Such hydro-dynamic process is also invoked for the Albano maar-lake in Italy (Ellwood et al. 2009; Funiciello et al. 2010).

The journal in which Godivel IV is publishing is the first news magazine in Auvergne. It aims to be read by nobles, bourgeois and ecclesiastic and has a section devoted to "Arts and science". In her analysis of this journal, Labarre (1977) is indexing this sudden modification of the river as a "pollution in Vassivière stream". Despite its definitive scientific value, this description, made outside of the academic world, will remain ignored until…1987! It is unquoted by contemporary Auvergne authorities (Chabrol 1786) and naturalists (Delarbre 1795, 1805; Lecoq 1835a), by limnologists and historians (Eusebio and Reynouard 1925; Fournier 1971), only to be mentioned for the first time by journalists working on Pavin's history after the Nyos event (see further). It remained to this day un-analysed by historians, volcanologists and limnologists. In a 1877 manuscript one finds a possible veiled reference to this event by Jean-Baptiste Bouillet, a noted archaeologist and former co-author of Lecoq, who mentions that during the year of the Calabria earthquake in 1783 (now scaled 7.0 on the Richter scale), *the waters of Pavin turned red during 3 weeks.* However this earthquake occurred in February and March: could the destabilization of Pavin sediments have occurred some months later and triggered a rollover?

This soft degassing is by many aspects similar to that of Lake Monticchio, observed between 1800 and 1840 including the corrosive nature of Pavin's waters, similar to the "vitriol-like" water taken from Monticchio's deep waters by Tata in 1777 (see Sect. 1.6.2.2). This event is also omited in the following statement by Chabrol.

2.3.6.1 An Official Statement Denies Pavin Marvels (Chabrol 1786)

Guillaume-Michel Chabrol (1714–1792), a King's state councellor, establishes in 1786 the Auvergne register of local rights, a remarkable work – 858p with multiple indexes – and a major historical ressource for this province. All villages are listed one by one with the history of their owners' rights, sometimes as far back as the thirteenth century. It is likely that Chabrol has been to Besse and he curiously starts this legal register for this town by a detailed description of Pavin, warning local people and visitors about *Pavin's stories*, which he then disprooves one by one. It is actually one of the most complete lists of Pavin's marvels:

> *The town of Besse is near the Monts-Dors: there is a very large community of priests and this town is famous for its devotion to the Virgin, the image of which stays in summer at Vassivière and the winter at Besse…There is in Besse territory a lake named the*

Pavin and the Creux de Soucy about which some marvellous events are reported [outside Auvergne], *which have nothing remarkable, except that this lake has no fish; the depth is fifty-seven fathoms but the stones thrown into it do not raise any storm, nor hail nor thunder as it has been written by many geographers; guns shot ashore have no other effects than the repetition of various echoes, and there is no smoke nor hot exhalations; the lake water does not generate any bad effect when in contact with it; it has been wrongly said that it was bottomless; there is no external input of water but there is a small outlet because of internal springs. It is told that waters from the Creux de Soucy are communicating with it. It may or may not be so but there is no interest in this.*

This looks like an official statement from Auvergne authorities to end the reporting of Pavin's stories. Chabrol does not mention chatelain Godivel's publication of the odd event at Pavin, 3 years prior, even though his veiled reference to the corrosive nature of Pavin's water suggests that Chabrol knows this description, or that it has been reported to him.

2.3.6.2 Pavin Is at Last Visited by Naturalists (1770–1820)

In the second half of the eighteenth century, interest in natural sciences is developing, particularly in the Auvergne Mountains where so many geological, botanical and hydrological curiosities remain to be discovered and studied for their picturesque and scientific values (Babeau 1928). The successful expedition of Chevalier in 1770 has piqued the attention of many naturalists: Monnet, an hydrologist and a chemist, Delarbre, the Clermont botanist, Dulaure and Legrand d'Aussy, historians and writers, Montlosier and Lacoste de Plaisance, volcanologists, Desmarets, geographer and geologist or La Porte, a traveler. These visitors to Pavin do not observe any of the lake's characteristics described by so many authors during the previous centuries, and they ignore Godivel IV's recent publication. These naturalists are not equipped to extend Chevalier's scientific investigations in 1770. None of them attempts to take a sample of surface and deep water for a chemical analysis, as done in 1777 by Tata at the Monticchio Lakes in Naples kingdom (Sect. 1.6.2.2). Likewise, they do not try to make a thermal profile as done by de Saussure in Lake Geneva (see Chap. 1). They are usually coming from the Bains du Mont d'Or village, to day Mont Dore, across the mountain and are guided by local people who continue to tell their stories about the lake, despite Chabrol's recommendation. Guides carry guns to perform the "Pavin experiment" in front of visitors, reproducing the thunder effect with the shooting echo and throwing stones into the lake, without success. In contrast with the stereotypical seventeenth century repetitions of the *Thrown Stone* story, these new descriptions are nows detailed, personalized and generally combined with those of Creux de Soucy. Only some of them are reported here.

A.G. Monnet, a royal mines inspector, comes twice to Auvergne to evaluate its hydro- thermal capacities. He makes a geological and hydrological description of Pavin (1788). He celebrates *Paven*, the Auvergne wonder and the nearby Vassivière pilgrimage, so respected by local people, recognizes Pavin as a volcanic crater and takes note of its paved-liked banks, which, according to him, could be the origin of its name. He attributes the absence of fish to the coldness of the waters, a hypothesis also made by many visitors after him. He does not mention other *Pavin stories.*

Legrand d'Aussy (1737–1800), an historian and a naturalist, member of the Academy, describes Pavin in ten pages in his *Voyage en Auvergne* (1788). He is quite smitten: *"maybe the most beautiful lake in Europe, for sure the most singular"*. He then disproves the old tales one by one: (i) he sees some fish [there are no big fish, only few roaches and bleaks are present in Pavin before 1859], (ii) he carries pistols and experiments with the gunshot echos, (iii) he throws a stone… nothing happens. However he is fascinated by Pavin's mysteries and would like to dive into the deep waters to unveil its multiple secrets. As he is not born in Auvergne, his work is much criticized by local scholars such as Delarbre, but for others, his work is considered as the most relevant description of this province.

Antoine Delarbre (1724–1807), the vicar of Clermont's cathedral and founder of the Clermont Botanical Gardens, visits Pavin for the first time in 1777 with his friends Ozy and Monnier, a naturalist and a pharmacist (Vernière 1899). His first report on Pavin is in an unpublished manuscript (1795); it is followed by his presentation of Auvergne (1805). He holds the most skeptical attitude about *Pavin's stories*, which he vaguely presents and then disprooves one by one. He does not mention Jouan, Belleforest, Fodéré or chatelain Godivel and labels all geographers of the seventeenth century as *"compilators who made together an arrangement to tell old stories that lack nothing but the truth"*. Therefore he does not mention any of them in his 1805 version, but quotes Moeri 1712 and Gordon 1748 in his 1795 manuscript! For him the sound produced by cracking ice can be similar to thunder, thus explaining this legend (even though all thunder noise accounts are reported for the summer season). Delarbre also observed *"some fish"* in the lake and as far as he is concerned, this is the ultimate proof of the lack of grounds of all other *Pavin stories.*

Volcanology pioneers' only concerns are the origins of Pavin and its neighbour, the Montchal volcano. For the most prominent among them, Montlosier (1755–1838), Pavin does not originate from a *roasting eruption* [with lava flow emission] but from a *pulverulent eruption* due to the action of water or air (1789).

Lacoste de Plaisance (1755–1828) reports in 1803: *"very close to Puy de Montchal, [in French mont-chaud, brûlant* – scorching mountain – *a denomination so expressive which is a proof that those who named it must have seen this volcano burning. Such a name would never have been given to a mountain that would not burn], there is a lake named Pavin, which should owe its existence to a volcanic explosion".* Lacoste is the only one to place such importance on Montchal's toponymy, the crater adjacent to Pavin. He follows Montlosier regarding Pavin's origin, a position also endorsed by Poulett-Scrope (1866). However, the origin of the lake remains in debate until 1900: Julien (1869, in Vimont 1874) favours a glacial origin, which explains the great depth of Pavin, followed by a damming of the glacial valley by the Montchal volcano; Glangeaud proposes a collapse of the Pavin volcano after its eruption (in Boule et al. 1901). The history of volcanology of the Pavin area is fully developed in Chap. 5.

Dulaure (1755–1835) publishes his *Description of France* in six volumes in 1788. Born in Auvergne, he visited Pavin and has been hearing about its stories for a long time. In his section on Auvergne, he makes a balanced analysis of past Pavin descriptions. He is one of the few to reproduce the statements of Jouan (1566) and Fodéré (1619). Then, based on his visit to Pavin and on the findings of Chevalier, he strongly challenges the *"marvellous phenomenon"* these authors have described. For him the *Thrown Stone* story is probably due to the remembrance of a former volcanic explosion, which could explain that the lake's name derives from *pavens*, the terrifying.

Until now, Delarbre has been considered as the main historical source on Pavin, although he seems more biased than Dulaure or Legrand d'Aussy. After Delarbre's publication in 1805, all of Pavin visitors, like Lecoq and Bouillet 1831; Lecoq (1835a) will share his opinion.

2.3.7 The Attractive and Repulsive Dual Nature of Pavin Throughout the Nineteenth Century

During the nineteenth century, Pavin is fully normalized, mostly by Henri Lecoq who introduces the first large fish (char and salmon) in the lake (1859), which serves as the ultimate proof to dispel all of Pavin's marvellous effects. On the other side, he reports in details what he calls the *Pavin stories*, which he heard from his guides, thus ironically promoting these tales throughout the nineteenth century. Excursions from Mont Dore to Pavin are becoming a must-do and dozens of elaborate Pavin descriptions, with new romantic overtones, are recorded in the travel memoires of the wealthy people coming to the Mont Dore spas. Some of them are famous writers and they choose Pavin as a central component in their novels and short stories, as Scribe and Assolant. After Lecoq's death in 1871, Clermont scientists start to study Pavin with their specialized equipment and will not pay any attention to the old stories.

2.3.7.1 Romantic Visitors Celebrate the Latent Fears of Pavin and Its Unique Atmosphere

At the beginning of the XIXth century geographers still report Belleforest's description (Prudhomme, 1807). Pre-romantic writers celebrating France's natural wonders are very impressed by the scenery and by the volcanic origin of Pavin. Depping (1811) spends three pages on Pavin:

> *"…It is one of the most beautiful and most singular lakes of our country and should be added to the beautiful monuments enriched by nature on the land of Auvergne …At times when the volcano was in action, there was in its crown an indentation through which the liquid and fluid substances were vomited… The shore of the lake forms a kind of horizontal bench which is covered by lavas rocks placed jointly, as would be a natural pavement…"* Depping also suggests that *"the study of deep waters would reveal many singularities"*, a statement probably borrowed from Legrand d'Aussy.

Jean-Charles Nodier (1780–1844), Isidore Taylor (1789–1879) and Alphonse De Cailleux (1788–1876) describe Auvergne in 1829 in their *Voyages pittoresques et romantiques dans l'ancienne France* in 23 volumes, published from 1820 to 1878. This monumental work, the first of its kind, celebrates the archeological and picturesque patrimony of France. Their volume on Auvergne is fully illustrated by splendid pre-romantic engraving, including the earliest Pavin engraving (Fig. 2.4). They dedicate many elogious pages to Pavin that can be abridged as such:

> *Pavin is the ruin of a volcano and the location of one of the most famous wonder as told by the people…Here a stone thrown by an imprudent person…would generate, from the deep, thunder clouds that vomit a storm…Often thick clouds and hot exhalations would spread on its shore…The hands of the voyager refreshed* [in the lake] *would be covered by swellings and pimples as if they were burned…Its name, Pavens, imposed by the Romans, tells well that it was a threat in this area.*

Nodier has a background in Natural Sciences and baron Taylor has personally experienced active volcanism during previous visits to Vesuvius and Etna, qualities so far rarely found in visitors to the Pavin. Most of this description is related to what they heard from local guides, either from Mont Dore or from Besse: thick clouds and hot exhalations are explicitely mentioned and the *Thrown Stone* story triggering thunder and storms is reported once more. The mention of the skin disease, swelling and pimples, is identical to the one reported by Godivel IV in 1783. Nodier also links the Vassivière pilgrimage to a latent fear of atmospheric phenomena at Besse (see Chap. 3) and concludes: *"there is always a positive and undisputable truth behind all popular superstitions, for the one who knows how to find it".*

In the 1820s, the new road between La Tour d'Auvergne and Besse is established in the Couze Pavin valley, according to maps by Desmarest (1798–1828) and Maury (1844). Before that period, the road avoids the Pavin valley and is located on the plateau, some kilometers further south, near

Fig. 2.4 First Pavin drawing, with a marked romantic vision (Nodier et al, *Voyages pittoresques et romantiques dans l'ancienne France*, 1829) featuring a marked outlet incision (Author's collection)

the Creux de Soucy (Fig. 1.2c). At this period only few geographers only mention Pavin as a "natural curiosity" (Kilian and Piquet 1823)

Throughout the nineteenth century, Pavin will be high on the list of excursions to be taken from the Mont Dore area, first by foot or on mule, and much later by car. Famous writers are coming to Pavin, spreading its fame, and sometimes they include it as a major component in their novels, as Gibory (1840) in *La Vierge aux œillets*, Scribe (1791–1861) in his *Roi de Carreau* (1843), George Sand (1859). The most famous author is Guy de Maupassant (1850–1893) who places *Humble Drame*, a short-story, at the Sancy Mountain and creates a very attractive vision of Pavin *"so clear and so blue…so charming that one would like to live in a hut in the wood dominating this crater where cold and tranquil water is resting"* (Maupassant 1883). The last novelist to feature a major scene at Pavin is Paul Bourget (1852–1935) in the *Démon de midi* (Bourget 1914).

2.3.7.2 The Sunken Castle of Pavin Featured in a Novel (Le Puy de Montchalm, Assolant, 1875)

One of the novels featuring Pavin, *Le Puy de Montchalm* by Assolant (1875), counts among the many origins of the legend of the *Sunken City*. The author places the action at Pavin in 1665, a dark period of Auvergne history when nobles were behaving like ruffians. A big castle is featured on the lake shore: *"the lake is subject to terrible storms and the former volcano crater seems to still breathe and moves like a wild animal"* (p.24), *"local people say the lake is the inferno gate and that if a stone was thrown a storm would occur, lifting the lake and the devil would appear at the lake surface"* (p.184). The owners of the castle, feasting and behaving badly, terrify the entire region, but there is a happy ending when the whole castle is sunk into the lake, troublemakers and all. The city sunk into a lake for its trespassing is a recurrent theme in lake legends and many interpretations can be given (see Chap. 3). As regards to Pavin, the sunken city story does not appear, in the analysed texts, before the end of the nineteenth century.

2.3.7.3 Pavin Stories Normalized by Lecoq (1831–1870)

Henri Lecoq (1802–1871) is the most famous Auvergne scientist of his time, being a botanist, a geologist and a hydrologist (see Sect. 1.4.2). When he visits Pavin for the first time in 1831 with his companion Jean-Baptiste Bouillet, a mineralogist, he reports a *"charming picture, made even more magestic by an absolute silence"* (Lecoq and Bouillet 1831). His illustrated account of his visit shows the near absence of vegetation within the crater, the pristine and impressive state of the rims and of the outlet (Fig. 2.5) (Lecoq 1835a, b). He

Fig. 2.5 Lecoq's drawing of Pavin in pristine state (1835b) showing the "terrible gorge" described by chatelain Godivel in 1783 and the explosive nature of the crater. The access to the lake is from the left rim of the cascading outlet. The Montchal volcano, not vegetated, is topping Pavin explosion crater. The barns are not in the Couze Pavin talweg but on its North side. (Musée Lecoq, Clermont-Ferrand)

mentions all the dreadful stories, still told by his local guides despite Chabrol's official warning: the lake has no depth, no fish, a stone thrown *"makes the water boil and generates a storm"…;* there is *"no boat, no wherry comes across these deep waters, no one would dare to linger on the liquid plain for there is in the middle a whirl that would swallow the careless"* (the Whirl and Storm legend, Eusebio an Reynouard 1925). He labels these as *"absurd tales and legends invented by fear and transmitted by credulity".* Some key characteristics of lake degassing (Table 1.2) are once more present in this list, whith the addition of a lack of floatabilty that fully explains the absence of boats. Lecoq's mention of boiling waters could be related to the summer 1783 event described by chatelain Godivel, not mentioned by him.

The lack of boats on the lake will at last be resolved in 1844: Lecoq himself manages to have one constructed near the lake and he uses it several times for his observations. The lack of fish – excepted two small species – comes to an end in 1859 when he successfully introduces into Pavin thousands of chars and salmons juveniles – obtained from a new fish-breeding technique of which he is a pioneer (see Sect. 1.4.2). The fish introduction convinces the local people of the stupidity of their tales, chars are sent to all restaurants of the region and fishing soon becomes an attraction for the Mont Dore visitors who are now securely sailing the lake on few small boats. Lecoq confirms Chabrol and Delarbre's rebuttals of *Pavin stories* such as: the origins of its waters are not mysterious any more as springs have been found by Chevalier in 1770; thunder is caused by echos and/or by ice cracking in winter etc (Lecoq 1835a).

Lecoq has not deciphered the complex history of Pavin. Although he is familiar with the works of Belleforest and Banc, and of Jouan's chronicle, which are listed among his references on Auvergne, he does not refer to them, nor to Godivel IV's letter to the editor of the *Feuille d'Auvergne* in 1783, which was also ignored by Chabrol and Delarbre. Finally it must be stated that he could not be aware of the Godivel II manuscript, only discovered in 1874, nor compare Pavin to Nyos descriptions. Every time Lecoq comes to Pavin, the lake seems normal. His attitude is in contrast with the one of his contemporary colleagues from Naples, Tenore and Gussone (1838) and Palmeri and Sachi (1852), who could conceive of extra-ordinary events in active volcanic areas and described the degassing of the Monticchio Lakes, which occurred two decades before their visit, on the basis of testimonies and previous observations (see Sect. 1.6.2.2). We ignore if Lecoq was aware of the descriptions of Monticchio degassing made by Naples scientists between 1777 and 1850s.

Lecoq's attitude about Pavin stories is shared by his companion, Jean Baptiste Bouillet, a mineralogist and archaeologist: *"A thousand tales have been made about this lake; many prodigies are attributed to it that we will refrain to report here"* (Guide du voyageur à Clermont-Ferrand 1836). This statement illustrates the position of all nineteenth and twentieth century naturalists.

2.3.7.4 Pavin Stories Diffusion in Guidebooks, Larousse Dictionary and Reclus Textbooks

Very early in the nineteenth century the *Thrown Stone* story by Belleforest is still mentioned in most Auvergne and Mont Dore guidebooks and travel acounts (Taillandier 1826; Stuart-Costello 1841). Then Lecoq's reporting gets a prominent place among scholars and guidebooks authors, thus promoting the centuries-old Pavin's stories and drawing again the attention to the lake throughout the second half of the XIXth. After Lecoq and Rico have opened Pavin to visitors and introduced fish (see Sect. 1.4.2), Pavin becomes very popular and no one truly believes in those marvels any longer.

Adolphe Joanne (1813–1881), a geographer and initiator of a famous guidebook series, is one of the first to quote of Lecoq, upon his visit Pavin: *"none of the local people, says M. Lecoq (Description de l'Auvergne) would dare to venture on this liquid plain. A whirl exists in the middle that swallows the unfortunate who would lead his boat there. A stone, even thrown from far so as to reach the abyss, would make the water boil and produce a storm. The sounding lead melts in the middle of the lake and its depth has no limits"*… and remarks *"an elegant boat sails on Pavin waters, despite the terrible legends told in this country"*. (Joanne 1867). In the 1894 version of the guide Pavin terrible stories will just be evoked.

Pierre Larousse (1817–1875), the famous encyclopaedist, in turn includes Pavin in the first edition of his *Grand Dictionnaire Universel* (1866–1877) using Joanne's citation: *"Pavin Lake, says M. Joanne, played a great role in Auvergne legends: it was enough, he was told, to throw a stone into it to trigger terrible storms, therefore its name Pavin (pavens, frightening)"*.

The *Thrown Stone* story is also found in Cosse (1857) who links the ancestral fear – *pavens* – with the remembrance of past volcanic eruption, Piette (1863); Nadeau (1863); Labesse and Pierret (1892); Tardieu (1888). Many visitors are still performing the "Pavin experiments" with guns and stones (Chaumette 1857; Cosse 1857; Clugnet and Alvin 1877; Bouillet 1878). In other guidebooks the Pavin's stories become less detailed, the existence of "terrifying tales" is only just mentioned and then finally dropped (Neulat 1846; Chabory 1894; Bouillet 1878; Joanne 1881; Lacroix de Marlès 1884; Larive-Fleury 1889).

Elysée Reclus (1830–1905), the prominent French geographer, makes this statement in volume II of his Géographie Universelle (1885):

"Pavin was much feared in the old times and, according to the tradition also found in the Aubrac plateau and in mountainous regions, it was believed that marvelous events would generate storms: a simple stone into the water would be enough to call for a storm. However, nowadays it is transformed into trout fisheries and it has lost its terrifying character. Once judged bottomless, it is in fact more concave than any other Auvergne lakes, as the sounding line touches its center at only 94 m". Recluz is also referring to the description made by Gregorius of Tours of an Aubrac lake, that could be re-attributed to Pavin as discussed in Chap. 3. His brother Onésime Reclus (1837–1916), another noted geographer, also mentions the *Thrown Stone* story in his works on France's geography (1888, 1899).

Jean Ajalbert (1863–1947), a famous writer and critic, gives one of the last romantic descriptions of Pavin: *"terror lake, tragic horror, fantastic vision, gloomy lake, inferno lake"*. His illustration does not promote this vision (Fig. 2.6) but this photo-engraving shows landslide scar within the crater rim, opposite the outlet. In his book on Auvergne, a whole chapter is dedicated to Pavin where he mixes for the first time the *Thrown Stone* story, the *Whirl* story coined by Lecoq and a *Sunken City* story (Ajalbert 1896). Meanwhile, the town council of Besse modifies the lake outlet, which is shifted by few meters to the NE, and the lake level is lowered by 4 m to its current level. The modern technology almost arrives at Pavin: a project, promoted by the Besse town council and supported by the Onésime Recluz (1886), of using the lake for water resource and hydropower – 3 to 4 m3/s diversion particularly during the summer drought – is well underway (Reynouard 1909). Fortunately, the project, which would have generated marked water levels fluctuations, is abandoned. In 1909 the circular pathways around the terrible lake is achieved with the financial help of the Touring Club de France and tourists now have unrestricted access to Pavin.

A few years after Lecoq died in 1871, botanists and zoologists from the Clermont Faculty of Sciences initiate the first flora and fauna inventories of the Auvergne lakes (see Sect. 1.4.3). They now have a boat and use Rico's cabin by the lakeshore (Fig. 2.7). They do not mention the old Pavin's tales, as recommended by Bouillet. Amédée Berthoule, the mayor of Besse and a recognized specialist of fish introduction, does not report Lecoq's Pavin stories but is the first to evoke a sunken city (Berthoule 1890, 1896), a new legend at Pavin which will have a great success in the next century.

2.4 Pavin's Representation in the Twentieth and Twenty-First Centuries

After the turn of the nineteenth century, Pavin begins to be considered a major asset in the development of "climatic" and automobile tourism for the Besse area (Fig. 2.7). The first Auvergne scientific guidebook, published with major local scientists such as the geologist Philippe Glangeaud (Boule et al. 1901), does not mention the *Pavin stories*. They are however still present in many others, like one by Cany

Fig. 2.6 Pavin and its new regulated outlet in 1896 (in Ajalbert 1896). Scars due to possible landslide within the crater rim are visible, opposite the outlet (Author's collection)

Fig. 2.7 Pavin Lake in the 1930s after the excavation of the outlet. The Rico's fisherman cabin is now used by limnologists. (G.D'O Post card, author's collection)

et al. (1916) in which Bruyant, the prominent Clermont limnologist, also highlights the major limnological features of Pavin. This is a rare example of the involvement of scientists in the Pavin mediation, until the 2011 exhibition established by city of Besse at the lake outlet. Several more targeted guidebooks on Pavin and /or Besse are published. The first one written by Abbé Blot (1924) explores the rich architectural heritage of Besse and briefly describes the first excursion around Pavin on the new pedestrian pathway set up in 1909. He too, does not include the *Pavin stories.* The second one, produced by Eusebio, the director of the Besse limnology station, and Reynouard, a local historian and former Besse mayor (1925), is the first to be entirely devoted to Pavin. It synthetizes the current research on the lake and presents the first thermal regime figure and a bathymetric map (see Sect. 1.4.3 and Fig. 1.5).

Reynouard also makes an in-depth analysis of Pavin's history and legends and is the first to clearly recognize that the *Sunken City* legend, probably promoted by local authorities, is recent and rooted in both Assolant's novel, Lecoq's Whirl and Storm legend and Berthoule's new naming of Pavin as the "Dead Sea of Auvergne" (see Sect. 3.3.1). He reports the Whirl and Storm legend, but he does not have any clues to base an interpretation on. His historical references do not go beyond Chevalier's sounding (1770) as reported by Delarbre, his major historical source on Pavin. Despite the limitations of historical sources, due partly to Delarbre and Lecoq who withhold information, Reynouard makes an in-depth analysis : he is the first to describe the *Treasure Stone* (Fig. 2.1), placing its origin some time during the Roman era, and to formulate the hypothesis that catastrophic events at Pavin could have stopped the production of these Antique millstones. His intuition remained unnoticed until recently (Meybeck 2010).

2.4.1 A Sudden and Moderate Degassing Event? (June 1936 Pavin Event)

Another eye-witnessed Pavin misbehavior event occurs in 1936, but is only reported in 1986 when local newspapers are asking scientists about the possibility of Nyos-like degassing in Pavin (see 1.8). The witness, Paul Joanny, anoted pharmacist in Clermont, is questioned by the media about what he saw at Pavin (Joanny 1986 and 1987):

> *"**In June 1936, I was fifteen…I went fishing at the lake for crayfish…and I left my heavy catch in a fish trap. When I came back the morning after, Pavin was fully coloured, from yellow to orange with streaks of blue…One fisherman named Boccard came by and told me he already saw this one or twice. We lifted the grids*** [that have been put at the oulet before 1890, see Figs. 2.6 and 2.7] ***but the escaping water looked like spongy mud …"*** He also identifies the special smell of the mud as *"colloidal sulfur, used in arsenic analysis"*. During our interview in August 2011, Joanny spoke of ***"a yellowish thick foam, a gelatinous mass with a strong sulfur smell, floating on half of the lake, the water itself did not change colour: it was the yellow-orange deposit that did so "***.

During the scientific controversy in 1986–87 surrounding a potential Nyos-like event at Pavin the degassing interpretation of this event was put aside by Clermont hydrobiologists in the local press (La Montagne, 1987a), suggesting a planktonic bloom (*Anabaena* blooms have been reported by Wurtz 1945) or a pollen rain but in no case a foam of sulfur at the surface. The statement of the scientists is highlighted in one of the rare guidebooks mentioning degassing risks at Pavin (Cassan 1998). Joanny is also the first one to mention the Pavin event reported by chatelain Godivel in 1783, passed on to him in 1986 by a well-informed reader of *La Montagne*, the local daily newspaper.

Although there is no explicit mention by Joanny of degassing, the likelihood is high: the suddenness, the sulfur smell, the whole-lake change, the multiple occurrences during summer, the nature of the materials, the night-time sudden occurrence which does not fit with a plankton bloom, all allow to attribute this event to another rollover of the deep waters, a process associated with precipitation of ferric hydroxide, manganese hydroxide and colloidal sulfur, hence some degassing (see Table 1.2). Also, the 1783 and 1936 events are indeed very similar: Pavin may completely change its aspect within a few hours, waters turn yellow, orange or rusty, with traces of blue, the same kind of fluffy or spongy material is found. In both cases these are well-contextualized events and witnesses mention repeated occurences over the summer. Evidence of the 1783 and 1936 events are now found in sedimentary archives of Pavin (Chassiot et al. 2016)

2.4.2 Guidebooks and Textbooks Ignore Scientific Findings at Pavin (1925–2011)

After 1925 different attitudes about Pavin can be found in guidebooks. All of them report the lake's altitude, depth and size, and highlight its wonderful setting and its fish yield, while its limnological characteristics, which make Pavin unique in France and even in Europe (see Sect. 1.3), are very seldom put forward. Boule et al. (1901); Cany et al. (1916); Eusebio and Reynouard (1925); Olivier (1954); and Cassan (1998) are among the rare authors who provide some details on the exceptional nature of the lake, progressively unveiled by scientists since 1880.

In contrast, the terrible *Pavin stories* or legends are very often mentioned, generally supported by the lake etymology, largely chosen as *pavens*=frightening. Many guidebooks and authors mix the *Whirl and Storm* legend, the *Thrown Stone* legend and the *Sunken City* legend (Cany 1916; Pourrat

1935; Guide Michelin 1954; CNSGR 1974; Chamina 1989; 1996; Burel and Debaisieux 1995, Guide du Routard 2010). The tourism industry favours the new *Sunken City* legend, the only one presented by Vazeille (1957), by Graveline and Debaisieux (1984), and by the Office de Tourisme de Besse (2012), probably a more appropriate story for the development of mass tourism at Pavin. Other guidebooks completely ignore all *Pavin stories*, such as the Guides Bleus (Montmarché 1924) and Guide Michelin (1937), Olivier (1954) or Balme (1954), a stark contrast with the position of Joanny and Larousse a 100 years ago.

The absence of any mention of Pavin's rich history and popular fear accounts is striking in most geography textbooks about Auvergne (Semonsous 1949; Balme 1954; De La Torre 1979; Hureau 1978; Tournilhac and Aleil 1988; Bonnaud 1995; Charbonnier et al. 2011). These scholars no longer mention the Pavin's stories, except for Jamot (1983) who has a very critical tone: *"with these legends, the irrational world is opened into the earth's depth"*. The Auvergne Volcanoes Natural Park issues a brochure on all Auvergne lakes (PNRVA 2004), in which Pavin scientific peculiarities are vaguely mentioned, and legends labeled as being *"extracted from the Bible and Auvergne's transposition of the Sodom and Gomorrah myth"*, a statement based on the analysis of the recent Sunken City legend by Reyt (2000, 2002).

Until very recently, limnologists and other scientists working on Pavin were not aware of its rich and complex history, which only became a matter of interest in the last 10 years. Attention to the *Thrown Stone* story was drawn for the first time by the hydrogeologist Del Rosso (2009a, b) in the local press and during a presentation on Pavin's legends at the very end of the Besse Meeting on Pavin and Meromictic Lakes, held in 2009 (see 1.8). He also placed emphasis on the possibility of active volcanism during the Middle Ages, not observed so far by volcanologists, on the basis of other legends. His plea for the scientific value of the Pavin's legends was strongly denied at that meeting by other scientists, based on the sole *Sunken City* tale. Very few recent guidebooks mention the degassing risk at Pavin and their ironic tone suggests that this question should not be seriously addressed: Cassan's guidebook (1998) minimizes the degassing risks and gives the floor to Pavin itself: *"these persistent legends irritates me very much"*, and *Eruptions*, a volcanological magazine, titles a scientific paper as such: *"Is Nyos monster hiding in Pavin's bottom?"* (Michard 2010).

However this attitude is changing: the city of Besse has now recognized the scientific, historical and cultural values of Pavin which are presented since 2012 in a permanent exhibition at the lake outlet with the full cooperation of Pavin scientists. Some very recent guidebooks are now integrating a new vision of Pavin, with the recent scientific findings (Burel and Debaisieux 1995; Plane 2011). Also, the new findings made on Pavin paleolimnology (Chapron et al. 2010, 2012; Chassiot et al. 2016), confirm our preliminary analysis (Meybeck 2010), and are broking through the reluctance of the scientific community regarding past degassing events at Pavin for the first time,

2.5 Conclusions

2.5.1 Why Is Pavin's History So Difficult?

The first difficulty was to go beyond Delarbre's 1805 analysis, which was biased, deliberately (?), ignoring Jouan (1566), Belleforest (1575), Banc (1605), and chatelain Godivel (1783) observations and truncating all references to "compilators". This made it very difficult to go beyond Chevalier's sounding of 1770.

The second difficulty lies in the late discovery of some key source materials: 1884 for the mid-seventeenth century Godivel II manuscript and 1987 for the descriptions of the 1783 and 1936 events.

The third difficulty is the lack of scientific awareness of degassing until 1986: this very rare behaviour popped out suddenly that year among the contemporary scientific community. Therefore, the bizarre descriptions made by the sixteenth century scholars could not be understood by any naturalist of their time nor for centuries thereafter. Although detailed degassing observations had been made from 1778 to 1840 at Monticchio by Naples scientists, they remained unknown to the volcanologists and limnologists of the twentieth century, even in Italy.

The fourth difficulty is the lack of interdisciplinary approach. For instance, in Auvergne, the scientific community was barely aware of Lecoq's *Pavin stories* or, if those were known (Omaly 1968), their historical roots were not given, resulting in Lecoq's opinion being unquestioned: everybody believed they were groundless tales and legends. Some historians such as Reynouard (1925) for Pavin, and Jaloustre (1910) for the adjacent Vassivière pilgrimage site, sensed very old relations between ancient human settlements, ancient cults and possible catastrophic lake events but not Blot (1910). However when questions of past degassing risks at Pavin were raised in 1986, their hypotheses were not revised by historians, geographers or archeologists (e.g. Auserve 2004, 2013). These disciplines were not mobilized by the Risk commissariat at the time and did not express concern about this topic..

The fifth difficulty is re-assessing known texts: in this perspective we had to create a set of sensory degassing descriptors using other lakes in Africa and Europe where degassing had been observed or recently disclosed (Table 1.2). This approach is once more used in Chap. 3 for the re-analysis of several other enigmatic descriptions, fantastical stories and legends, adding a new set of past Pavin information.

Finally, we found that, throughout the nineteenth and twentieth centuries, the past historical descriptions of Pavin, such as the very official report to King Charles IX mentioned by Jouan (1566), have been ignored or completely entangled with local stories told by guides and with the recent forged legend, the Sunken City tale of the early 1900s. This forged legend, well analysed by Reyt (2000, 2002), has been favoured by regional authorities, well ahead of Lecoq's *Pavin stories* (PNRVA 2004), so that grounded and un-grounded Pavin stories were totally mixed.

2.5.2 The Pavin Name

The debate on Pavin's name, started by Godivel II (mid-seventeenth century) and which has been recently re-activated (Fournier 1971; Herillier 2011) seems irrelevant. We found that, in contradiction to Fournier's statement, the original name of the lake is Paven, itself derived from *Lacus pavens*, terrifying lake, as mentioned as early as 1605 by Jean Banc. Paven was officially used until Cassini's map in 1815. The past degassing character of Pavin and its related misbehaviours can fully explain this traditional naming. One question arises: why give a Latin toponym to a local place in Auvergne? We know that millstones workers have already settled by the lake in Antiquity, probably when the great thermal baths of Mont Dore (Mons Aureus) were established at the Isrt and IInd centuries and that other Latin toponyms such as *Pantheon* have lasted until the fifteenth century in Mont Dore. If the lake had already been identified by local population as a possible threat during that period, a latin denomination would then be possible, particularly when considering the very ancient fear of Romans for another similar maar-lake near Rome, Lake Albano (Chap. 3). Another hypothesis would be that *Lacus pavens* is much more recent, a name given by Middle Age or Renaissance scholars who were writing in Latin although, according to Godivel II manuscript – mid-1600s, the lake has "always" been called in such a way.

2.5.3 Historical Evolution of Pavin's Fame and Recognition

Considering their limited size and poor accessibility (located at 1300 m altitude, no road until the 1820s) Pavin Lake – 44 ha – and adjacent Creux de Soucy cavity- 10 m^2-, had an outstanding recognition since the sixteenth century in official reports, local, provincial or national. Both Pavin and Soucy are found among the lists of wonders or curiosities worthy of mention in many universal geography works, from Belleforest's cosmographia in 1575, to Larousse (1866–1877) and Recluz (1885), a recognition today largely ignored. In 1566, provincial authorities are officially complaining to the King of France during his visit in Auvergne about Pavin misbehaviour that damages crops and the terrifying lake is figured in the Queen Mother land register in 1575, which makes it one of the first lakes to be represented in a landscape figure.

Belleforest's 1575 Pavin description – the thrown stone that triggers thunder, lightning and storms – was then reproduced by many French, Netherland, German, English geographers in the next 300 years. The cosmographer probably obtained it from one of his many informers and, considering the level of details given on the Creux de Soucy area, there is little doubt that his informer is both local and well educated. Such stereotypical lake description has been used before (see Chap. 3). However this marvellous process at Pavin is most likely true, as other reliable, non-stereotypical and independent descriptions made at the same period, depict similar Pavin misbehaviour (Vassivière pilgrims in 1551, Jouan in 1566; Banc 1605). In the mid-seventeenth century the Godivel II manuscript, unpublished until 1884, gives another very detailed account of Pavin from a local and reliable witness.

The fame of Pavin seems to then slow down and become more regional. Jesuit cartographer de Frétat (1672) refutes the *Thrown Stone* effect for the first time. Royal civil engineer Chevalier makes a successful sounding in 1770, after many previous failed attempts, an achievement considered by all as the first for Pavin despite de Frétat's earlier depth estimate. This sounding largely contributes to the normalization of the lake by the naturalists and voyagers of the Enlightments' era who now flock to the lake. After 1770, as they visit Pavin, they throw a stone, nothing happens: Belleforest's description is refuted by naturalists who are now accessing to Pavin.

But stories continue to be told by local guides, despite the efforts made by the highest political, ecclesiastic and scientific authorities (Chabrol 1786; Delarbre 1795; Lecoq and Bouillet 1831; Lecoq 1835a): for local people Pavin is still bottomless, devoid of a water inlet, without fish, connected with the Creux de Soucy; its waters can cause pimples on the skin after contact; there is a whirl in the middle so that no boat can sail on it, it had exhalations and, from time to time, violent events.

Throughout the nineteenth century, these *Pavin stories* are widely quoted in Auvergne or Mont Dore guidebooks and used by novelists coming to the Mont Dore spa resulting in a second period of fame for Pavin beyond Auvergne's borders. After Lecoq's normalization of Pavin through the introduction of boats and fish in 1859, the stories are gradually presented as legends and the *Sunken City*, a new legend, appears and is finalized in the 1900s (Chap. 3).

Today the lake receives 200 000 annual visitors and is still lauded, through guidebooks and other touristic material, for its excellent fish and its nice circular pathway around the shore where one can see the devil's chair and a sketch of the city of Besse, sunk into the lake for its trespassing. Pavin's rich history and recent degassing controversy (see Sect. 1.8) remain absent from geography textbooks (Charbonnier et al. 2011). However the general attitude about Pavin is now changing. Plane's guidebook (2011) is the first one to mention the uniqueness of Pavin's deep waters as well as the seriousness with which its degassing risks are now taken. The town of Besse is now offering visitors to Pavin a remarkable permanent exhibit on the lake, its surroundings and Creux de Soucy, and on the most recent scientific results. Belleforest and Lecoq's descriptions are presented in panel on the legends. The recent findings on past catastrophic spillover events near 600 AD and 1300 AD by paleolimnologists and geologists (Chapron et al. 2012; Chassiot et al. 2016; chapter 23) confirm that past Pavin spillover events have occurred. They probably will change the attitude of the many Pavin actors (historians, scientists, tourism industry representatives, risk-management authorities and the media).

2.5.4 Past Pavin Degassing Events and Past Pavin State

There are at least three accounts of sudden Pavin's misbehaviour events of various intensities at Pavin, reported by Vassivière pilgrims in 1551, Godivel IV in 1783 and Joanny in 1936. They are fully contextualized with an exact date, identification and position of witnesses as well as detailed descriptions. They come from local and educated people from Besse who voluntarily record what they have experienced. None of them are apparently aware of the previous events. All witnesses are familiar with Pavin and are common visitors to this area. As such they are more prone to notice disorders that can either be missed by occasional visitors or left un-reported by uneducated local people. For instance, the 1783 events that have occurred several times during this summer remained un-reported until they were mentioned on August 21 to chatelain Godivel, a jurist. In addition to these events, other descriptions of Pavin's state at the sixteenth–seventeeth centuries feature extended degassing periods of various types and intensity (Jouan 1566; Belleforest 1575; Banc 1605; Godivel II manuscript, 1650s). The occasional visits to Pavin by outside visitors such as de Frêtat (1672), Chevalier (1770) and their many followers provide information on the state of the lake the day of their visit: with the exception of the 1783 and 1936 event, none report unusual Pavin behavior after mid-seventeenth century. But witnesses of possible occasional Pavin disorders – so marvelous- may have withheld their observations as for the 1936 event, only disclosed in 1986.

There are many other sources about Pavin, although so far un-attributed to the terrifying lake, such as local religious history and iconography, a fantastical dragon story published in 1632 in Paris and pagan times' dragons and fairies legends collected during the nineteenth century. They are fully detailed and discussed in the next chapter in comparison with similar myths, legends and beliefs found in other European maar-lakes. Even if they are less contextualized and can be hypothetical, they still enrich our analysis of Pavin's history and confirm the latent fear surrounding the lake since pagan times, *i.e.* before the fourth-fifth centuries.

Acknowledgments for chapters 1 and 2 are combined with those of chapter 3.

References

Ajalbert J (1896) L'Auvergne. Ancienne maison Quentin, Paris, 362 pp
Apianus Petrus (1581) Cosmographie, description des quatre parties du monde
Assollant A (1875) Le Puy de Montchalm. Librairie de la société des gens de lettres, Paris, 370 pp
Audigier P (chanoine) (1720) Histoire d'Auvergne, tome I. Projet de l'Histoire d'Auvergne. Manuscript published in 1894, Louis Bellet, Clermont Ferrand, 139 pp
Auserve Ph (2004) Histoire du pèlerinage de N-D de Vassivière, Impr. de A. Dumont, Clermont-Ferrand, 46 pp
Auserve Ph (2013) Pers Com.
Babeau A (1928) Les Voyageurs en France depuis la Renaissance jusqu'à la Révolution, Alfred Mane Tours, 294 pp [Re-édition Slatkine, Geneva, 1970]
Balme P (1954) A travers le département du Puy-de-Dôme. Coll. Le Tourisme en Auvergne, De Bussac, Clermont-Ferrand, 184 pp
Banc J (1605) La Mémoire retrouvée des merveilles des eaux naturelles en faveur de nos Nymphes Françaises et des maladies qui ont recours à leurs emplois salutaires. L Pierre Sevestre, Paris (BCIU library, Clermont-Ferrand)
Bar E (1907) Guide pratique Le Mont-Dore, La Bourboule, Saint-Nectaire. Guides Pol, 112 pp
Belleforest (de) F (1575) Cosmographie Universelle, 3 volumes, Nicolas Chesneau, imprimeur-libraire, Paris. Vol 2, 1127 pp (Bibliothèque Sainte Geneviève-Sorbonne, Paris, réserve, cote RXVIB 3=321-1)
Berthoule A (1890) Les Lacs d'Auvergne. Orographie, faune naturelle, faune introduite. Revue Société Nationale d'Acclimatation, Paris, 136 pp
Berthoule A (1896) Les Lacs de Besse-en-Chandesse. In: L'Auvergne. IV Congrès Int. Hydroclimatologie Géologie, Clermont-Ferrand
Bertrand M (1819) Mémoire sur l'Etablissement Thermal du Mont-d'Or et les Antiquités qu'on vient d'y trouver. Librairie Landriot, Clermond-Ferrand, 47 pp
Blaeu Johan (1663) Atlas Major. Province d'Auvergne, Amsterdam
Block E (2003) Corpus of Medieval Misericords, France, XIII-XIVth, Brepols, Turnhout, 444 pp
Blot A (1910) Histoire du pèlerinage de Vassivière. A. Dumont Clermont-Ferrand, 157 pp
Blot J, abbé (1924) L'Eglise de Besse. Monument historique. Guide Illustré, 63 pp

Bonnaud P (1995) Vivre sur nos volcans. Des appellations plus ou moins contrôlées. In: Volcanisme et Volcans d'Auvergne, Parc Naturel Régional des Volcans d'Auvergne (PNRVA), La Dépêche scientifique du parc n°8-9: 36–37

Bouchet (du) (1645) Carte d'Auvergne (map) (BCIU library, Clermont-Ferrand)

Bouillet JB (1836) Guide du voyageur à Clermont-Ferrand et des localités les plus remarquables au départ du Puy-de-Dôme. Auguste Veysset, Clermont-Ferrand, 355 pp

Bouillet J (1877) Description historique [de Besse], Clermont-Ferrand, University Library (BCIU), Manuscrit, 40 f, Ms 977-fo: 39–40

Bouillet M.N (1878) Dictionnaire Universel d'histoire et de géographie, 26e édition, Hachette, 2054pp

Boule M, Ronchon G, Glangeaud Ph, Vernière A (1901) Le Puy-de-Dôme et Vichy. Guide du Touriste, du Naturaliste et de l'Archéologue

Bourget P (1914) Le démon de midi. Published in: La Petite Illustration, 1921, 2 volumes, Plon-Nourrit. vol 1, 251pp

Boyer-Vidal A (1888) Besse et Vassivière. Notice historique. Imprimerie clermontoise, Clermont-Ferrand, 50 pp, appendices. Réédition ACVAM 2005, 72 pp

Briet (1648) Parallela Geographia Veteris et Novae, tome 1, chap 7, [in Cregut, 1921]

Bruzen de la Martinière A-A (1739–1741) Le Grand dictionnaire géographique, historique et critique. Paris, 6 volumes

Burel J, Debaisieux F (1995) Les Volcans du Mont-Dore, du Cézalier et de l'Artense. Ed. Debaisieux, 64pp

Cany G, Percepied E, Perpère E (1916) Séjour et promenades au Mont Dore et à la Bourboule .Guide des baigneurs, touristes, naturalistes (Collab. Ph Gangeaud, C Bruyant, préface A Lacroix), 394 pp

Caracausi A, Nuccio PM, Favara R, Nicolosi M, Paternoster M (2009) Gas hazard assessment at the Monticchio crater lakes of Mount Vulture, a volcano in Southern Italy. Terra Nova 21(2):83–87

Cassan P (1998) Le lac Pavin. Office National des Forêts, Aulnat, 48 pp

Cassini (César François Cassini de Thury) (1815) Feuille Brioude-Issoire, Carte de France, n°53

Chabory F (1894) Guide complet du promeneur au Mont-Dore et à la Bourboule suivi du Guide du Baigneur aux eaux du Mont-Dore

Chabrol (1786) Coutumes locales de la Haute et Basse Auvergne. Index des paroisses, index des seigneurs et propriétaires (tome IV), Martin Dégoutte, Paris, 858 pp

Chabrol E (1931) L'Auvergne thermale à l'époque gallo-romaine. Ed. L'Auvergne littéraire, 31pp

Chaix LA (abbé) 1869 Histoire de Notre-Dame de Vassivière depuis ses origines jusqu'à nos jours, Clermont-Ferrand, 359p

Chamina (1989) Balades à pied en Auvergne, Massif du Sancy et Artense, Chamina, Clermont-Ferrand, 127pp

Chamina (1996) Lacs et volcans d'Auvergne, 226 pp

Chapron E, Alberic P, Jezequel D, Ledoux G, Massault M (2010) Les Archives sédimentaires de l'histoire du lac Pavin. Rev Sci Nat Auvergne 74–75:57–66

Chapron E, Ledoux G, Simmoneau A, Alberic P, Saint-Onge G, Lajeunesse P, Boivin P, Desmet M (2012) New evidence of Holocene mass wasting events in recent volcanic lakes from the French Massif Central (lakes Pavin, Montcineyre and Chauvet) and implications for natural hazards. 5th international. symposium on submarine mass movements and their consequences, Kyoto. Chap.23: 256–264

Charbonnier P et al (2011) Le Puy-de-Dôme. Série de géographie régionale, Christine Bonneton éditeur, 320pp

Chassiot L, Chapron E, Di Giovanni C, Alberic P, Lajeunesse P, Lehours A-C, Meybeck M (2016) Extreme events in maar lake Pavin sedimentary record: implication for natural hazards assessment in the French Massif Central. Quat Sci Rev 141:9–25

Chaumette EJ, abbé (1857) Souvenirs de voyage ou les vacances en Auvergne, Itinéraire du Puy-de-Dôme, 375pp

Chevalier (1770) Rapport du sondage au Pavin du 28 septembre. In: Delarbre A (abbé), 1805. Notice sur l'Auvergne et en particulier sur la ville de Clermont. Clermont-Ferrand, 256pp

Cladière Jean-Joseph (1688 ou 1687) L'histoire de la Sainte Chapelle de Vassivière près du Mont-d'Or en Auvergne Célèbre par les miracles que Dieu y fait continuellement par l'intercession de sa très-pure et glorieuse Mère Par un religieux Bénédictin de la congrégation de St Maur. Damien Boujon, Clermont. Re-imprimé 1837, Clermont Ferrand

Clugnet L, Dr Alvin (1877) Nouveau guide du baigneur et du touriste au Mont-Dore, 203 pp

CNSGR (1974) Topoguide du GR 30. Comité national des sentiers de grande randonnée

Cosse Etienne-Joseph (1857) Souvenirs de voyage ou Les vacances en Auvergne: Itinéraire du Puy-de-Dôme

Coulon L (1644) Les Rivières de France ou description géographique er historique du cours et du débordement des fleuves, rivières, fontaines, lacs et étangs qui arrosent les provinces du Royaume de France, François Clousier, Paris, 442pp

Coulon L (1654) Le Fidèle conducteur pour les voyages de France, d'Angleterre, d'Allemagne et d'Espagne. Gervais Clouzier, Paris, 239 pp

Crégut R (chanoine) (1921) En feuilletant les vieux livres. Bull. Hist. Sci. Auvergne, 120pp

D'Avity Pierre (1643) Le monde ou la description de ses quatre parties. Tome 2: L'Europe avec tous ses empires, royaumes, états et républiques, 521 pp

Deboffe P (1995) François de Belleforest. Assoc. Autour de F. de Belleforest, Samatan, 224pp

Delarbre A (1795) Notes et observations sur les volcans, puits, mines et fontaines d'Auvergne. Manuscrit Ms 716, University Library (BCIU), Clermont-Ferrand

Delarbre A (abbé) (1805) Notice sur l'Auvergne et en particulier sur la ville de Clermont. Clermont-Ferrand, 256pp

Del Rosso T (2009) Les Légendes du Pavin. Oral communication. Int. workshop Pavin and other meromictic lakes, May 14–17, 2009, Besse

Del Rosso T (2009) (printemps-été). Interview with Marc Gachon, La Galipote: 37–38

Depping GB (1811) Les Merveilles et beautés de la nature en France. P Blanchard, Paris, 634 pp

Desmarest N (1798–1828) Lacs volcaniques. In: Dictionnaire de géographie physique, vol 5: 687–688: lacs volcaniques)

Donnet FL (1820s) Carte du Département du Puy de Dôme. Bibliothèque Universitaire Clermont-Ferrand (CA 10 070)

Du Chesne André (1614) Les Antiquités et recherches des villes, chasteaux et places les plus remarquables de France, 2ème édition, Jean Petit-Pas, Paris, 1025pp

Dulaure Jacques Antoine (1784) Description des principaux lieux de France. Part V. Auvergne

Durand-Lefebvre M (1926) Les Vestiges antiques et le culte des sources au Mont-Dore. Le Puy, 64 pp, ill

Ellwood NTW, Albertano P, Galvez R, Funiciello R, Mosello R (2009) Water chemistry and trophic evaluation of Lake Albano (Central Italy): a four-year water monitoring study. J. Limnology, DOI: http://dx.doi.org/10.4081/jlimnol.2009.288

Etienne Charles (1553) *Dictionarium Historicum et Geographicum*

Eusebio A, Reynouard J (1925) Le lac Pavin et le Creux de Soucy. Station limnologique de Besse, Besse, 55pp

Fodéré Jacques (1619) Narration historique et topographique des convens de l'Ordre de Saint-François et monastère Sainte-Claire, Lyon. Edited by Abbé Randamme, 1861, Mémoires Académie de Clermont, Ferdinand Thibaut, Clermont-Ferrand, 160pp

Fournier PF (1971) Le Nom du lac Pavin. Essai d'étymologie. Bull hist Sci Auvergne 85(631):159–164

Frêtat (de) Aimable (1672) Carte de la Haute et Basse Auvergne (map)

Funiciello R, Giordano G, De Rita D, Carapezza M-L, Barberi F (2002) L'Attivita recente del cratere del Lago Albano di Castelgandolfo. Rend Fis Acc Linei 9:113–143
Funiciello R, Giordano G, De Rita D (2003) The Albano maar-lake (Colli Albani Volcano, Italy): recent volcanic activity and evidence of pre-Roman Age catastrophic lahars events. J Volcanol Geotherm Res 123:43–61
Funiciello R, Heiken G, De Benedetti A-A, Giordano G (2010) Volcanic activity of the Lake Albano maar in Roman history and mythology. In: Funiciello R, Giordano G (eds) the Colli Albano Volcano, Geol. Soc; London, Special. Publ. IAVCEI
Gibory Ferdinand (pseudonym Saint-Sard) (1840) La Vierge aux œillets. Hyppolite Souverain, Paris
Godivel Antoine (Godivel I) (1579a) "Terrier pour la Royne mère du Roy, de sa terre, ville et mandement de Besse, faict par Antoine Godivel, notaire royal en la dite ville". In –f°, 30 ff.lim.et 529 ff chiffrés, Manuscript, Bibliothèque du Musée Condé, Chantilly
Godivel Antoine (Godivel I) (1579b) Terrier de Besse-en-Chandesse. Manuscrit, Non-illustrated copy of the Terrier de Catherine de Médicis. Archives Régionales, Clermont-Ferrrand, 1 E 0154
Godivel II (1650) "Remarques touchant la Ville de Besse". In: Le Beffroi de Besse, Jaloustre, 1884. Mém. Acad. Sci., Belles-Lettres et Arts, Clermont-Ferrand
Godivel III (subdélégué) (1726) Sondage infructueux du 7 mars 1726. In: Legrand d'Aussy (1788)
Godivel IV (Chatelain) (1783) Lettre ouverte au Rédacteur de la feuille. Feuille Hebdomadaire de la Haute et Basse Auvergne, contenant ce qui peut intéresser cette province (Chabrit M ed.), Antoine Delcros imprimeur, Clermont-Ferrand, 27 septembre 1783, p 152
Gomis S (2006) Les « enfants prêtres » des paroisses d'Auvergne. XVIe-XVIIIe siècles, Presses Universitaires Blaise Pascal, Clermont-Ferrand
Gordon P (1748) Grammaire historique, 16ème edition traduite de l'anglais, Durand, Paris (quoted by Delarbre, 1795)
Graveline F, Debaisieux F (1984) Les Volcans d'Auvergne et du Velay, Solar, 62pp
Gregorius from Tours (Florentius Georgius Gregorius) (c 600) Histoire de France. Les Sept livres des miracles. La Gloire des confesseurs, VII, II, traduction Gérard Pradalié
Guide du Routard (2010) Guide du Routard, Auvergne
Guide Michelin (1937) Guide Michelin d'Auvergne
Guide Michelin (1954) Guide vert du pneu Michelin d'Auvergne, 118 pp
Herillier C (2011) Détectives au pays des mots. Christian et Marie Jeanne Hérilier (Interview). In: L'eau d'Auvergne: 164–169
Hureau J (1978) L'Auvergne, le massif central d'aujourd'hui. Edition Jeune Afrique, 237pp
Jaillot (1715) La Province d'Auvergne (map) (BCU library, Clermont-Ferrand)
Jaloustre E (1884) Le Beffroi de Besse. Mém. Acad. Sci., Belles-Lettres et Arts, Clermont-Ferrand (BMIU, Clermont A34 425)
Jaloustre E (1910) Histoire du pèlerinage de Notre-Dame de Vassivière. Impr.Dumont, Clermont-Ferrand, 150pp
Jamot C (1983) Le Tourisme. In: Brossolette P (dir.), Les Monts d'Auvergne. Privat, 471 pp
Joanne A (1865) Itinéraire général de la France. Deuxième partie, Auvergne, Dauphiné, Provence. Hachette, 863 pp
Joanne A (1881) Le Mont D'Or et les eaux minérales d'Auvergne, 200 pp
Joanne A (1894) Le Mont-Dore, La Bourboule, etc. Guides Joanne
Joanny P (1986) De l'eau qui sent (parfois) le soufre, La Montagne, 2 septembre 1986
Joanny P (1987) Un Pharmacien clermontois raconte un souvenir de jeunesse. Interview de J Gandebeuf. Le Républicain Lorrain, Janvier 1987
Jouan Abel (1566) Recueil et discours du voyage du Roy Charles IX, Charles Etienne, imprimeur du Roy, Paris
Kircher A (1678) *Mundus subterraneus*
Kilian A-J, Piquet Ch (1823–1833) Dictionnaire géographique universel, Paris
Labarre S (1977) La Feuille hebdomadaire pour la Province d'Auvergne, Mémoires Institut d'Histoire de Clermont-Ferrand, 137 pp, annexes
Labesse E-D, Pierret H (1892) Notre pays de France. En cheminant (Auvergne). Ducrocq, Paris, 246 pp
Lacoste de Plaisance (Abbé) (1803) Observations sur les volcans de l'Auvergne. Clermont, 1 vol
Lacroix de Marlès, Jules (1884) Merveilles et phénomènes de la nature
La Guillotière François (1575–1590) La Charte de France
La Planche (de) Pierre (1640–1646) Description des villes de France. Manuscrit appartenant à M Meurtey de Tupigny. Quoted in: P Duval (1667) La France sous le Roy Louis XIV: 176
La Porte (de) J (1790) Le Voyageur françois ou la connaissance de l'ancien et du nouveau monde. Tome XXXI, Lettre CDXII (Laporte, Fontenay, Domeiron, 1760. Le Voyageur Français), Vincent, Paris, vol 32: 19–23
Larive, Fleury (1889) Dictionnaire français illustré des mots et des choses, Tome 2, G Chamerot
Larousse P (1866–1877) Grand dictionnaire universel du XIXème siècle, français, historique, géographique, mythologique, bibliographique, littéraire, artistique, scientifique, etc., Paris, Tome 12, 1560 pp
La Torre (de) M (1979) L'Auvergne. Le Guide de la France, Berger Levrault, 193 pp
Lauras-Pourrat A (1973) Guide de l'Auvergne mystérieuse. Tchou, Paris
Lavina P, Del Rosso T (2009) New geological findings in Pavin area, Oral communication, Symposium Pavin and other meromictic lakes, Besse-en-Chandesse, May 2009
Lecoq H, Bouillet J-B (1831) Itinéraire du Département du Puy-de-Dôme, 178 pp
Lecoq H (1835a) Description de l'Auvergne. Le Mont d'Or et ses environs. J-B Baillère, Paris, 437pp
Lecoq H (1835b) L'Auvergne pittoresque. Atlas colorié. Le Mont-Dore et ses environs. Seize vues coloriées b. Chaillon, Paris
Legrand d'Aussy P JB (1788) Voyage en Auvergne. E. Onfroy, Paris, 555pp
Lestringant F (1991) André Thevet, cosmographe des derniers Valois. Genève, Droz, Travaux humanisme et Renaissance n° 251, 430
Mailly (de) L (1723) Principales Merveilles de la Nature où l'on traite de la substance de la terre, de la mer, des fleuves, lacs, rivières, montagnes, rochers, etc. avec un précis des choses les plus rares et les plus curieuses qui s'y voient. Rouen, 330 pp
Mallot F (1995) Les tremblements de terre en Basse-Auvergne aux XV et XVIème siècles. Histoire de l'Auvergne, CREA, Université Blaise Pascal, Clermont-ferrand, pp 86–90
Maupassant G (1883) Humble drame, Gil Blas. Re-edited by La Pléiade édition, Gallimard, 1974
Maury (1844) Carte du Puy de Dôme (map) (BCU library, Clermont-Ferrand)
Merula P (1605) Cosmographia, caput X, In: B. Thalès , Voyage dans la Vieille France par Jodocus Sincerus(1614), 1859, Paris, 362 pp
Meybeck M (2010) Le Pavin et ses voisins, histoire d'un patrimoine européen unique. Rev Sci Nat Auvergne 74–75:7–44
Michard G (2010) Le Monstre du lac Nyos se cache t-il dans le Pavin ?, Eruptions
Monmarché M, Doré R (1924, 1931, 1936) Auvergne et Centre. Guide Bleu. Hachette, 596pp
Monnet AG (1788) Voyages minéralogiques faits en Auvergne dans les années 1772, 1784 et Observations sur la Physique, sur l'Histoire Naturelle et sur les Arts. Tome XXXII: 179–199

Montlosier (de Reynaud de) (1789) Essai sur la théorie des volcans d'Auvergne. Re-published in 1802,184 pp
Moreri (1712) Grand dictionnaire géographique (Quoted by Delarbre 1795)
Münster S (1565–1566) Cosmographie Universelle, 1430 pp (Paris faculté de pharmacie library)
Nadeau L (1863) Voyage en Auvergne (département du Puy-de-Dôme), Dentu, Paris. (re-edited: La Tour Gile)
Nodier Ch, Taylor JD, De Cailleux J (1829) Voyages pittoresques et romantiques dans l'ancienne France (ré-édition 1987, Presse du Languedoc, Montpellier Tome 1, 110 pp)
Olivier L (1954) Saint Nectaire, Besse, Murols, les lacs Montdoriens. De Bussac, Clermont-Ferrand, 64pp
Omaly N (1968) Le lac Pavin, historique et hydrobiologie. Rev Sci Nat Auvergne 34:7–31
Palmeri L, Scacchi A (1852) Della regione vulcanica del Monte Vulture. Gaetano Mobile, Naples, 160pp
Pelletier-Decitre M (2008) De Ptolémée à la Guillotière (XV-XVIème siècles). Des cartes pour la France, pourquoi, comment. Comité des Travaux Historiques et Géographiques, Paris
Piccardi L, Masse WB (eds) (2007) Myths and geology. Geol. Soc. London Special Public 273: 51–59
Piesse L (1863) Guide des eaux thermales du Mont Dore. Guides Joanne, L Hachette, Paris, 220pp
Piganiol de la Force (1754) Nouvelle description de la France (Tome XI, chap XXII, Descrition de l'Auvergne), 571 pp
Plane J-Th (2011) Massif du Sancy. Les 15 plus belles balades et randonnées. Rando Editions, 56 pp
PNRVA (éd.) (2004) Les Lacs des volcans d'Auvergne. Parc Naturel Régional des Volcans d'Auvergne
Poulet-Scroppe G (1866) Géologie et volcans éteints du centre de la France. Thibaud, Clermont, 261 pp
Pourrat H (1935) Au fort de l'Auvergne. Arthaud, 170pp
Pourreyron C (Chanoine) (1935) Le Culte de Notre-Dame au diocèse de Clermont en Auvergne. Bost, Nancy, 303 pp
Prudhomme LM (1804) Dictionnaire universel de la France. Laporte, Paris 5 vol
Sand (George) (1859) Voyages en Auvergne.1827-1859-1873. Editions Paleo, 168 pp
Reclus E (1885) Géographie universelle, Vol II, France. Hachette, Paris, 1017 pp
Reclus O (1886) France, Algérie, colonies. Hachette.Paris, 802 pp
Reclus O (1899) Le plus beau royaume sous le soleil, Brodart, Paris, 861 pp
Reynouard J (1909) Besse, autrefois, aujourd'hui ? Ann. Stat. Limnologie Besse-en-Chandesse, tome I:1 –14
Reynouard J (1910) La Légende du Pavin. La Montagne d'Auvergne, 28 Aug 1910: 1–4
Reyt P (2000) Les Dragons de la crue. Cahiers de Géographie du Québec 44(122):127–145
Reyt P (2002) Nous et les autres : l'égalité face au risque. Cahiers de Géographie du Québec 46(127):7–24
Scribe Eugène (1848) Le Roi de carreau, de Vigny (quoted by Eusebio and Reynouard, 1925)
Sébillot Paul (1904–1906) Les Eaux douces. In: Folklore de France, Ré-Ed. 1983, Imago, Paris, 304 pp
Semonsous J (1949) L'Auvergne, sa géographie, son histoire. Charles-Lavauzelle, Paris, 256 pp
Shanklin E (1989) Exploding lakes and maleficent water in Grassfield legends and myths. J Volcanol Geotherm Res 39:233–246
Shanklin E (2007) Exploding lakes, myths and reality, an African case study. In: Piccardi L, Masse WB (eds) Myth and geology. Geol Soc London Spec Pub vol 273: 165–176
Simonin M (1983) Les Elites chorographiques ou la Description de la France dans la chorographie universelle de Belleforest. In: Voyager à la Renaissance, Edit: Jean Gérard et J CH Margolin, Colloque de Tours: 433–441
Simonin M (1992) Vivre de sa plume au XVIème siècle ou La Carrière de François de Belleforest, Droz, Genève, 328 pp
Stuart-Costello L (1841) Voyage en Auvergne, dans le Velay et en Bourbonnais fait en 1841. Ré-édition 1945, De Bussac, Clermont-Ferrand, 237pp
Taillandier (1826) Résumé de l'Histoire d'Auvergne par un Auvergnat. Lecointe et Durey, Paris, 554pp
Tardieu A (1888) Puy-de-Dôme, Guide complet illustré, Monographies des villes et villages de France, Ré-édition Res. Universis 1993, 329pp
Tata D (1778) Lettera sul Monte Vulture a sua eccellenza il signor D. Gugliemo Hamilton. Stamperia Simoniania, Napoli, 62pp
Taylor J, Nodier C, de Cailleux A (1829) Voyages pittoresques et romantiques dans l'ancienne France. 3, Auvergne. 1829, Didot, Paris
Tenore et Gussone (1838) Peregrinazioni effetuate nelle state del 1838 dai Signore Tenore e Gussone. In: Principitato Citeriore e di Basilica). Edité en 1995 par A Ciarallo et L Capaldo Viaggo al Vulture, Edition Osanna Venosa, 157 pp
Thevet A (1575) Cosmographie universelle. Tome II Europe (Paris Faculté de pharmacie library)
Tournilhac B, Aleil PF (1988) Besse. In: Histoire des communes du Puy-de-Dôme, Arrondissement d'Issoire: 97–106
Vazeille A (1957) La Tour d'Auvergne et sa région. Le Tourisme en Auvergne n° 28, De Bussac, Clermont-Ferrand, 64 pp
Vernière A (1900) Les voyageurs et les naturalistes dans l'Auvergne et dans le Velay. Typ. et lit. G. Mont-Louis
Vimont E (1874) Les Lacs Pavin, de la Montcineyre et de la Godivelle. Annuaire du Club Alpin Français. Paris
Wurtz A (1945) Recherches écologiques sur les lacs, tourbières et étangs d'Auvergne. Ann Stat Centr Hydrobiol Appliquée 1:89–166

Dragons, Fairies, Miracles and Worship at Pavin and Other European Maar-Lakes

3

Michel Meybeck

Abstract

The legends, miracles and fantastic stories of Pavin and around are numerous, although the key ones are so far not attributed to it. The current legend, that of the *Sunken City*, common to many lakes, was created in the late XIXth century. The *Thrown Stone* and the *Whirl and storm* stories reported in the XIXth century are the transposition of the descriptions made since the XVIth century (Chap. 2), although the Thrown Stone motif also existed for other mountain lakes, before being used for Pavin. The *1547 miracle* that occurred near Pavin can be understood as a passer-by loosing and recovering his sight during a Pavin CO_2 degassing event. It is followed by many abnormal events during 100 years, recorded in the Vassière pilgrimage registers, that can be attributed to Pavin misbehavior. The *Gloomy Lake Dragon* story, reported in 1632, can be related to a mudflow produced by Pavin overspill.

The *Fairies Garden* legend, found in church archives in the XIXth century and here attributed to Pavin, takes place during the pagan era when another type of lake-bound dragon was regularly worshipped by local tribes, including human sacrifices, until eventually defeated by the first evangelizator of the area (Vth century). Gregorius of Tours (before 600 AD) mentions another *dragon attack of hermit Caluppa*, around 590 AD, and details a regular pagan lake cult, followed by an early Christian settlement, both could be re-attributed to Pavin. The fear of *lacus pavens* is possibly featured in the iconography of the medieval Besse church, some 4 km away (XIIth century capitals, XVIth century misericords, 1670 Santa Martha chapel).

The Longue Durée comparison of Pavin legends with those of other maar lakes in Italy and Germany highlights multiple signs of misbehaviors in most of these meromictic lakes, which can be attributed to degassing features. Since the early Antiquity regular and important pagan cults were established on their rims (Averno, Albano and Nemi Lakes), followed by Christian cults (Defeat of *pope Silvester dragon* -330s AD- and conversion of pagans, possibly at Albano Lake; Monticchio, Laachersee Lakes). Tunnels or canals were excavated to avoid lake overspills (Albano, Nemi, Averno, Monticchio, and Laachersee). Descriptions of dragons (Albano or Nemi), fantastic animals (Ulmen Maar), access to Inferno (Averno), legendary records of village destructions and sudden death (Toten Maar, Pulver Maar) are some of the means of risk transmission across generations that started at least 2700 years ago. Degassing maar-lakes may be a major point of origin for some European myths, lake legends and fantastic creatures, a point worth further investigation.

M. Meybeck (✉)
METIS, CNRS-Université Paris 6, UMR 7619,
Place Jussieu, 75252 Paris Cedex 05, France
e-mail: michel.meybeck@upmc.fr

T. Sime-Ngando et al. (eds.), *Lake Pavin*, DOI 10.1007/978-3-319-39961-4_3

Keywords
Pavin lake • Maar-lake • Lake degassing • Lake legends • Dragon • Geo-mythology

There are so many tales about Pavin, or Pavens at it was named by Romans, concerning the thick clouds that were blanketing it, the exhalations that one inhaled on its shore, the stones thrown into the surface waters, that will develop from its depth clouds from which the storm would come, and a perpetual bubbling. (Piesse 1863)

In those times it was told that its waters were covering a damned city which should have sunk into the infernos, as no one ever found its depth. No boat could sail on it without sinking and no fish, carp, perch or trout, could live there... They also said that, if one was throwing a stone into the lake, it triggered winds and storm... This lake without boats, as a perpetual whirl was swallowing them, this bottomless lake, which should communicate with the infernos, remained for a very long time surrounded by fears. (Henri Pourrat 1935)

"Still in the 16th century many natural events were prodigia"... "a myth also trades definite and conclusive statements concerning a true state of the case, which are founded on real events. These sentences are transmitted through time to confirm life continuance becoming a permanent structure which simultaneously refer to the past, present and future... ancient historiographers, not having all the answers to every possible questions, solved the problem by introducing narration statements belonging to divine actions, fate and myth". (Vittori et al. 2007)

3.1 Introduction

Marvellous stories without anything worth mentioning (Chabrol 1786), *one hundred absurd nonsenses* (Legrand d'Aussy 1788), *tradition* (La Porte et al. 1790), *fables* (Lavallée 1796), *old time stories and women's tales* (Delarbre 1805): these terms have qualified the old descriptions of Pavin made during the XVIth and XVIIth centuries (see Chap. 2). Their collection and publication by Lecoq in 1835 will unexpectedly promote these *Pavin stories* throughout the XIXth century (see Sect. 2.3.6 and above quote of Piesse 1863). The fear of Pavin by the local population, latent in the earliest descriptions, may be much older. According to Taylor, Nodier and Cailleux, the precursors of the Romanticism, the Vassivière Virgin, on her mountain sanctuary near Pavin, protects Besse people from threats generated by telluric forces (Nodier et al. 1829). At the turn of the XIXth century, Jaloustre (1910), an Auvergne historian, suggested that a pre-Christian cult at or near Pavin could have been present at Vassivère. For Reynouard, a local historian, Pavin itself was home to a tribe, eventually expelled by a catastrophe (Eusebio and Reynouard 1925). Camille Julian (1901), a well-respected Antiquity historian, stated about people living near Pavin during Antiquity: "*They were placing in these silent and hypocritical waters the impregnable asylum of a divinity of the deep, which should not be troubled, only with presents.*"

Our analysis of Pavin legends here is inscribed within a geo-mythology approach. The pioneer work of Back (1981) on hydromythology and ethnohydrology opened interdisciplinary approaches in Earth Sciences and the perception of geological hazards by local people, as transmitted across generations through myths and legends, has allowed for the reconstruction of past events (*Myths and Geology*, Piccardi and Masse 2007). This community works mostly on extreme and catastrophic events – *e.g.* tsunamis, landslides, volcanic eruptions, earthquakes. Convincing examples of combined interpretations of legends, myths or beliefs with earth system dynamics, have already been given: for earthquakes in Gargano, Italy (Piccardi 2005), for submarine inflow of Greek karstic freshwaters, a marvel described by Antiquity historians (Clendenon 2009a, b, 2010). The Italian geochemists and volcanologists were the first to integrate the so-called mythological stories of lake Albano in which they found actual descriptions of degassing (Funiciello et al. 2002, 2003, 2010).

So far the community of anthropologists, folklorists and mythology specialists has only worked in concert on well identified natural wonders, such as volcanoes, earthquakes or meteorites, a specific example being the historical Ensisheim meteorite fallout in Alsace, in 1492 (Kammerer 1994), but has not yet included lake degassing as a grounded basis for lake legends, excepted for Eugenia Shanklin, an anthropologist from the College of New Jersey, USA. She studied Nyos and Monoun lakes legends in Cameroun soon after the Nyos event of 1986 (Shanklin 1989, 2007) and found that hazard perception of these degassing lakes was quite developed among the first settled populations. In her euhemeristic interpretation of the legends, she postulated that old lake stories can be explained very precisely by referring to actual events and that lake stories can provide contemporary scientists with insight on past lake behavior and concluded that lake stories were worth exploring at length. The recent findings of Italian earth scientists (Funiciello et al. 2002), about a past catastrophic degassing event at Lake Albano in 398 BC, so far denied by Archaic Rome historians, are a first confirmation of Shanklin's approach (See Sect. 1.6.2.1).

The following questions are addressed here: (i) What are the existing tales or legends associated with Pavin, with their authors and/or contexts? (ii) Are there other representations, religious, legendary and/or fantastic, that could be related to Pavin? (iii) How has the fear of Pavin been expressed in pagan and Christian beliefs? (iv) Are these legends, beliefs and stories different from the ones found in other European maar-lakes, *i.e.* is there a common frame in Auvergne, Italy and Eifel?

3.2 Approach and Sources

The methodology used here is definitively interdisciplinary and iterative (see Fig. 1.1), combining historical sources on Pavin (VIth and XVIth to XVIIth centuries), XIXth century folklore registers, religious history and iconography (XIIth to XVIIth), with limnology, history, mythology and folklore of other meromictic maar-lakes in Italy and Germany: the Colli Albani near Rome holds Albano and Nemi Lakes, the Campi Phlegrei, west of Naples, holds Averno Lake and the Monte Vulture, between Campania and Puglia, holds the two Monticchio Lakes; the Eifel region, west of the Rhine south of Aachen, holds a dozen lakes (See their characteristics on Table 1.1). In addition to these maar-lakes some other lakes with specific legends are briefly considered (Léman, Pilatus Lake near Lucerne, Switzerland, and Nohedes Lake in the Canigou Mountain massif of the Pyrenees). The studied period goes as far back as possible, *i.e.* to pre-Roman period, for the Latium lakes and Averno.

The use of legends as part of the risk assessment process at Pavin has been orally advocated by Thierry Del Rosso, a hydrogeologist, in a long interview published by the Galipote, a local satiric magazine. He argued on the possibility of violent events at Pavin, based on new scientific findings and on some Pavin tales: *"the fairy of the waters and its associated phenomena are without doubts geysers…other tales mention lava fountains and fissures, cinder ash mud that fill the valleys, very certainly in the mid-1200s, the city [of Besse] has been considerably destroyed by a water mass originating from Pavin"* (Del Rosso 2009a). He also presented his argument in May 2009 in the final and public session of the international Conference on *Pavin and other meromictic lakes* held in Besse (Jezequel et al. 2010) mentioning fairies and dragons that could be related to past limnic and even volcanic events, although he did not present his sources (Del Rosso 2009b). The occurrence of recent volcanic events was much questioned by other scientists who also replied that (i) there was only one tale at Pavin, the *Sunken City*, (ii) this tale was not specific to Pavin but a common tale reported in many lakes, (iii) it had been already investigated and interpreted by Reyt (2000, 2002) as a biblical legacy. So far Del Rosso has not published his thesis, except as a brief summary of *Pavin stories* (Del Rosso-d'Hers 2010). We argued, in a preliminary analysis (Meybeck 2010), that Pavin had been greatly feared in the past, that the so-called "Pavin stories" could be related to past degassing events and that local religious history- the famous Vassivière pilgrimage- could be linked to Pavin.

Our interpretation of the legends, stories, religious history and iconography at Pavin has then been checked in folkloric registers in which lake tales are common. Many of them have been collected, reported and partially analyzed by Paul Sébillot (1843–1918), the French ethnographer and folklorist, in his *Tales on the Earth and Underground World* (1904–1906) which analyses 15,000–16,000 items and has a chapter dedicated to *Still Waters*, by the American Stith Thompson (1855–1975), the founder of the international Motiv-Index of Folk Literature, the *Thompson register of folk tales* (1955–1958), a master register collecting a thousand of items (tales, legends, fabliaux, etc.) and by Van Gennep in his "*Manuel de folklore français*" (1949). Lakes have also long been mentioned for their association with religious practices: according to Lavroff (1872), *"Cicero and Strabon reported that there was a sacred lake near Toulouse where local tribes were offering gold and silver offerings and Gregorius also mentioned a sacred lake"*. [As all other original citations in French, this text has been translated by us].

Pavin legends are first compared and discussed here one by one with similar legends from other locations, adopting a reversed chronological order from the most recent sources to the older ones, which are often more difficult to interpret. The first analyzed legend is the *Sunken City*, a contemporary legend, then the *Thrown Stone* story and the *Whirl and storm*, reported many times during the XIXth century. These legends are now often jumbled in Auvergne presentations, as those written by Henri Pourrat (1935) (see quote above) as well as in guidebooks (see Chap. 2).

The core of our research concerns the re-attribution to Pavin of texts that have been known to historians, scholars and folklorists for a long time and that have been re-analyzed on the basis of the typology of sensory perception of degassing (See Table 1.2) as well as their historical context (See Chap. 2). The analysis of beliefs and religious iconography is conducted on the Besse church, built from the XIth to the XVIIth century, and on the nearby Vassivière pilgrimage, initiated by a miracle in 1547. Finally Pavin legends and religious history are discussed in parallel with the rich Antiquity and Christian era history found for several other maar-lakes in Latium (Italy) and in Eifel (Germany).

3.3 The Principal Pavin Legends, as Recognized Today

3.3.1 The *Sunken City* and Other Similar Tales at Pavin (Late XIXth Century) and Other Lakes

3.3.1.1 The Progressive Elaboration of the Tale at Pavin (1871–1935)

Until 1875 there was no indication of a sunken city tale at Pavin: out of the dozens of mentions of Pavin stories found since 1566 there is only one featuring this theme (Monnet 1788). However it was present at the Tazenat Lake, another Auvergne crater lake north of Clermont where such tales were commonly reported to visitors (Nadeau 1862). For instance, neither Nodier et al. (1829) nor Sébillot (1904–06) mention this particular tale at Pavin.

The *Sunken City* tale was gradually forged for Pavin at the end of the XIXth century. When Pavin access was opened to tourism by Lecoq, the major Auvergne naturalist (see Sects. 1.4.2 and 2.3.6), visitors sailing on the lake soon discovered in its waters a massive sunken carved stone of about 250 kg, the *Treasure Stone*, taken for a proof of a sunken settlement (Eusebio and Reynouard 1925). The stone is now recognized as an antique millstone (Fig. 2.1). Meanwhile, in his last book on *Waters of Massif Central* (1871), Henri Lecoq qualifies Pavin as the *Auvergne Dead Sea*, due to its original absence of fish before 1859. In 1875 Alfred Assolant in his novel located at Pavin (*The Puy de Montchalm*, 1875) features a big castle with immodest lords who terrify the country (see Sect. 2.3.6). At the end of the novel the castle and its residents are sunk into the lake, the gate of inferno. Soon after Pavin is termed the *Sodom and Gomorrah of Auvergne* by Berthoule, a Besse mayor (1890, 1896), and Vimont (1874), the head of Clermont public library. In turn, Ajalbert, a writer, makes a detailed description of Pavin (1896), featuring the Besse city sunk into the lake for misconduct, according to local belief. Thus does a Sunken *City* tale start to spread among the local people.

A mixed Pavin tale is finally formalized by Emile Roux-Parassac (1874–1940), poet, writer, journalist and local tourism promoter, in his guidebook for automobile drivers on the Couze Pavin and Couze Chambon valleys (Roux-Parassac 1910). Pavin legend appears under the title: *"the Lake of Terror"* (Le Lac de l'épouvante), a direct transcription of its latin name, *lacus pavens* (see Sect. 2.3.3). The volcanic crater is due to the anger of Lucifer defeated by God, glaciers recovered the land then melted. Lucifer raging tears filled the crater. There, terrible storms and whirls may occur. In an article published in *La Montagne d'Auvergne*, a local tourism magazine, Roux-Parassac adds (1910): *"Its name means the terror…In this place demons are coming by the inferno abyss, they bathe and their black souls trouble its waters. Their cursed breath lift storms while leaves are not moved on branches, and their breath attracts the careless who would dare to venture on the abyss"*. Here Roux-Parassac mixes here several elements of Lecoq's Pavin stories – the whirl – with other elements such as the specific odour emitted by Pavin or the leaves that do not move during those storms, an indication that the storm originates from the deep waters, as mentioned in the Godivel II manuscript (mid-XVIIth), just published by Jaloustre (1884) and probably known by Roux-Parassac. Another similar version of the tale will be developed by Murat-Vaché (1925), then by Henri Pourrat (1887–1959), the Auvergne writer, in 1935 (see above quote). One notes the multiple references to deep underground influences on Pavin, including the origins of its waters, an influence largely recognized today by scientists (see this volume) but unknown in 1910.

Today, this complex tale is largely mixed with the sunken city and is presented in most guidebooks as the original and only legend (see Sect. 2.4) so that, when Reyt made his analysis of Pavin legends, among others lake legends in France (Reyt 2000, 2002), he only found and analyzed the Sunken City in which the lake misconduct has been gradually replaced by the people misconduct.

3.3.1.2 Other Sunken City Tales in Auvergne

Sébillot, the initiator of folkloric studies in France, quotes a dozen of sunken cities tales in his 88-page chapter on still waters (1904–1906). In Auvergne three other crater lakes are mentioned for their sunken cities tales by Sébillot, Annette Lauras-Pourrat (1973) and Graveline (2011): (i) Issarlés, (ii) du Bouchet with its central whirl and its early evangelisators: a bucket dropped into its deep waters would be retrieved full of blood and (iii) Tazenat where a petrified woman rock is shown to visitors, "a proof of this Auvergne Sodom" (Nadeau 1863). The Pavin absence in Sébillot confirms that the *Sunken City* tale was not the tradition at Pavin before the end of the XIXth century.

For Sébillot these sunken city legends can be classified under the *"impiety penalty, in relation to the tradition of corrupted Dead Sea Cities"*. This interpretation shared later by Reyt (2002), a geographer of religious history: the Sodom and Gomorrah myth spreads during the Middle Age across Europe as *exemplum* containing four themes: transgression, punishment, purification and renaissance of a better society through the survival of the Righteous. They were then transposed into the *Sunken City* tale, found in many French lakes, *e.g.* St. Point Lake in Jura (Defrasne 1951). Reyt notes that wind and/or storm occurrence is always found together with supernatural phenomena related to evil forces. This myth provides a sense and a dimension to lake landscapes.

3.3.1.3 Destroyed Cities on Lake Geneva, Albano Lake and Totenmaar: Precursors of Sunken Cities Tales?

Tauredunum Event on Léman lake (563 AD)

The fear of lakes sinking cities may be grounded on several natural phenomena. A text from Gregorius of Tours (Florentius Georgius Gregorius, around 539–596), well known for his history of the first Christian predicators in Gaul, reports in much detail a catastrophic event, the Tauredunum event, followed by a giant wave, which suddenly destroyed all riparian cities (Lausanne, Geneva) around *Lacus Lemanus* (Geneva Lake) in 563 AD. This event, still told in the XIXth century by Léman people, has been thoroughly analyzed by F.A. Forel (1892, vol 3, chap IV, pp 496–507). But the founder of limnology and lake physics specialist could not find any phenomenon that would generate waves higher than 0.5-1 m. He concluded that the catastrophic event was very overstated by Gregorius. Recent investigations by Geneva limnologists (Kremer et al. 2012), seem to end this long-lasting mystery. They propose an exceptional giant lake tsunami, 13 m high in Lausanne and 8 m in Geneva, caused by a massive delta landslide generated by the earthquake associated with the Tauredunum landslide, which occurred in the Upper Rhône Valley, making Gregorius description perfectly coherent with this event. Geneva authorities are now assessing this new type of risk, occurring very rarely but with devastating consequences. Forel also described sudden destructions of villages associated with the collapse of lake dams in alpine valleys. It must be noted that Forel's concern for lake history is very rare, only shared by a handful of limnologists such as G.E. Hutchinson who studied the paleolimnology of an Italian lake during Roman times (Hutchinson and Cowgill 1970).

Albano Lake Overspill (398 AD)

The Lake Albano rise by 70 m in 2 months which occurred in 398 BC (Funiciello et al. 2002, 2003, 2010) certainly flooded all Human settlements within this large crater and threatened Rome (see Sect. 1.6.2.1). This catastrophic event, which had an enormous impact- the Delphi Sybil was consulted- could be another origin of the widespread Sunken City legend. This type of event could have occurred again 2000 years later in Eifel, destroying riparian villages at Toten Maar (see further).

The Sunken Castle and the Lake of the Deaths (Toten Maar) (1562 Degassing Event?)

According to Alois Mayer, the specialist of Eifel legends (2008), where one finds today the gentle surface of Lake Weinfelder there was in the old times a landscape and a beautifull castle, ruled by a rich but good hearted count but his wife's heart was as hard as stone …

One day the countess refuses charity on a poor starving old man and she unleaches her ferocious dogs on him. Meanwhile the count was hunting in the forest with his horse, Falchert, away from his castle. ***Suddenly the sky over the town turned black. From these dark clouds lightnings were bursting out and thunder was rolling. With a deafening noise the ground splitted. An incredible mass of water was spurting and entered the castle*** *. The castle people, full of fear, took refuge under the roof and the countess climbed rapidly the steps to the dungeon. But any escape was useless.* ***The flow continued to rise and completely sunked the castle*** *. The count sents his squire to get his forgotten hunting glove at the castle. The squire does not find any castle but a water body. He comes back and reports "Master, your castle has vanished, now there is a deep lake". The count was striked, he could not believe it: "You are talking non sense. What you say is impossible. Where could the water come from in such sandy and rocky place? It is like if my horse Falchert, on which I am standing now, was able to generate a spring while stamping on the ground. As soon he pronounced these words, the horse starts to stamp the ground and soon a new spring came by. The count rushes to his castle and only finds one survivor, his baby floating in his cradle. The count goes to a monastery for expiation. The spring at the corner of the forest still cointinues to fizz and is now called Falchertborn by the local people.* (abridged version of the tale)

This is a typical legend of a city sunk for misconduct, which can be re-interpreted as a sudden and violent spillover of the Weinfelder Lake possibly due to the degassing of this meromictic maar-lake. The lake is known today as Totensee, the lake of the Dead. According to historians the Weinfeld village and its population have been totally destroyed in 1562, alledgely by the Plague. This should be checked as "the plague" could be another representation of massive damages caused at maar-lakes by CO_2 degassing to riparian populations, as observed at Lake Nyos (See Sect. 1.6.1). There is an explicit mention of lake overspill and lightnings and storms are precursors of the event, as quoted many times (See Table 1.2). The earth is split, with deafening noise, suggesting that the overspill was associated with an earthquake. Bubbles are still mentioned today at the Falchertborn spring.

Finally, Sunken Cities tales can also originate from: (i) the bursting of natural lake dams, in alpine valleys (Forel 1892) and in the Middle Rhine (Park and Schmincke 1997), (ii) the presence of Neolithic and Bronze ages materials found underwater in many Jura lakes at very low water stands.

3.3.2 The *Thrown Stone* Story at Pavin (XIXth) and in Other Mountain Lakes

3.3.2.1 *Pavin* Thrown Stone *and* Whirl and Storm *Stories as Reported to Lecoq (1835)*

When Lecoq makes his first visit to Pavin in July 1831, he reports what he hears from local people (Lecoq and Bouillet 1831; Lecoq 1835a; see Sect. 2.3.6). This legend will be qualified as the *Whirl and Storm* story by Eusebio and Reynouard (1925) (Fig. 3.1): *"No boat or wherry is crossing these deep waters. No one would dare to linger on this liquid plain. In the middle there is a whirl that would swallow the daredevil who would lead his boat there. A stone thrown far enough into the lake would make the water boil and produce a storm. The sounding lead would melt in the middle of the lake and its depth has no limit.* These are the absurd stories we were told while asking for a boat to cross the lake" (Lecoq 1835a).

The first occurrence of the *Thrown Stone* story, actually much older, is found in Belleforest's Cosmographia (1575):

"…Near this Mont D'or stands the city of Besse, at a half quarter league from which one can see a great lake located nearly on top of a mountain, without a measurable depth and without any water inlet. Actually it is both admirable and terrible to see as, if a stone is thrown into it, it is ascertained that soon thunder, lightning, rainstorm and hail would occur". This is most probably an actual description of Pavin at that time, which will be copied for the next 200 years by many geographers and scholars (see Sect. 2.3.2).

Some of the many Pavin mysteries (lake depth, origin of waters) were resolved in 1770 by Chevalier's exploration of the lake (See Sect. 1.4.1). Others, as the whirl, the boiling waters, the absence of fish, the sudden thunder and lightning, the sulfur odour and the aggressive waters can be compared to the grid of sensory degassing indicators (see Table 1.2). They actually correspond to various forms and intensities of Pavin Lake degassing, as described in the XVIth and early XVIIth centuries, which so far had not been identified as such (Chap. 2).

The descriptions of the lake by Ajalbert (1896) and by Henri Pourrat (1935), one of the prominent Auvergne writers, are excellent examples of the literary version of *Pavin stories* (see Pourrat's quote at the head of this chapter).

Fig. 3.1 The representation of Pavin Whirl and Storm legend, featuring Satan, in Eusebio and Reynouard Pavin guide (1925) (Author's collection)

3.3.2.2 Other Thrown Stones Lake Stories (Lucerne, Canigou and a German Lake)

This story of the throwing of a stone into a lake triggering thunder and storms is not specific to Pavin. Similar stories are found in Catalunia, Switzerland and in Germany. Sébillot himself, quoting Felieu de la Pena (1609), makes the comparison between "the lake near Besse" (Pavin) and a similar story told about the Nohedes Lake in the Canigou Mountain (Pyrenees) (Sébillot 1904–1906). Actually the original description of this other marvelous lake is much older and already found in Gervais of Tilbury (Gervasius Tilberiensis (115?–122?) in his *Otia Imperialia, Liber de Mirabilis Mundi*, the Encyclopedic Miscellany of Wonders (1215). In chapter LXVI, a wonderful lake is described in the Canigou Mountain, which is so high that it is inhabitable. There, in the bottomless Blackwater Lake, *"one can find a fallen castle inhabited by devils. When a stone or any other solid material is thrown into the lake a big storm gets out of it as if the devils were angry".* The Nohedes Lake is part of several mountain lakes in the Canigou Mountain region, probably of glacial origins, certainly not maar-lakes. Their mixing status is undetermined. Lakes boiling, emitted vapors, typical of degassing lakes, are very difficult to conceive at Nohedes, unless they are a re-use of anterior legends, possibly in maar-lakes, such as Lake Albano or Lake Averno. Some of the details of the Nohedes legend are similar to the description of Enaeus descend in Inferno, described by Virgil. A sunken castle is featured at Nohedes, as in Assolant's novel.

Another Thrown Stone story, told before Belleforest Pavin description, is found for a diminutive mountain lake, Pilatus, which gave its name to a famous alpine summit near Lucerne (Switzerland). According to a long tradition from the XVth century, the lake could suddenly change its color to red, *i.e.* the blood of Christ on Pilatus' hands that he finally managed to wash there (Cuvelier 1934; Seewer 2013). Other Pilatus stories reported by Gaspar Schott, assistant to the great German scholar Athanasius Kircher (1601–1680), feature winged dragons (Schott, *Physica Curiosa*, 1662) and a "dragonstone", still kept in a Swiss museum. Lucerne authorities officially forbid access to the lake to visitors for centuries (Seewer 2013). Pilatus Lake legends show some similarities with Nohedes and Pavin: this small lake is near the top of a mountain, has no depth and may change its colour. It is also feared for centuries by local people even far away down in the valley and local religious and political authorities, backed by Jesuits, are well aware of its recurrent misconduct events, featured as dragons either winged or of the river type. However Pilatus Lake is not in a volcanic region. It was drained by the Lucerne people, so that its possible meromictic character is difficult to assess.

Another unknown lake in Germany, associated with a similar Thrown Stone story, is described by Louis de Mailly (1657–1724) in his *Principal Wonders of Nature* (1723):

> *In Baden principality, at four leagues from the Prince's residence,* ***there is a lake and when someone dips something heavy into it, at once the sky becomes overcast and a storm with thunder and lightning occurs causing great damages to the surroundings.*** *Some foreign Jesuits having reached the Baden Collegium heard about this wonder and got curious to see it. They came with some bourgeois of the city with a water hunting dog. As he would not enter the lake they threw him by force but he escaped with great scream. After having thrown blessed wax, they threw stones: they did not see any change, even after hours of waiting.* ***In the evening came a terrible storm with thunder, rain and lightning, which lasted fifteen days*** *(Gasp.Scott, Physica Curiosa).* ***There is in Auvergne a lake on a mountain that does the same.***

The original quote is also from Schott, the Kircher's assistant (1662). De Mailly adds the comparison between the German lake and an Auvergne lake on a mountain. According to Canon Audigier (1720) Kircher attributed this Thrown Stone story to Pavin. The period is around 1600s, as the Society of Jesus was already funded. The misconduct signs of both German and Auvergne lakes are very similar: thrown stone resulting in thunder, lightning and storm that devastate the surroundings. We learn here that the German lake was feared in this time by local authorities who enlisted the help of foreign Jesuits to try to understand this wonder. They used the blessed wax as a possible form of exorcism and the poor dog probably encountered corrosive water or sulfurous gaz traces (Table 1.2; Sect. 2.3.3). The Jesuits also experimented with throwing stones into the lake, as was done at Pavin. Although the exact location of the German lake is not given and neither is the exact date of this visit, it could be located in Eifel (Weinfelder Maar, Ulmen Maar, Laacher See?), the volcanic region where meromictic maar-lakes and legends of lake catastrophs are common but which does not belong to the Prince of Baden. This lake description is nearly identical to the one made by Gregorius of Tours about a lake in Auvergne mountains, worshipped with presents and producing thunder, rain and lightning (see further).

Sébillot in his book on *Folklore of France* (1904–1906) which analyses 15,000–16,000 items observes in chapter V (*Still Waters*, p. 205–293) that: *"people living near lakes and ponds think that one should not throw anything into these water bodies so as not to irritate the spirits living there or to trigger a storm. This belief was reported by Gervaise of Tilbury in the XIIIth century and is stated by Belleforest at the XVIth century"*. The references to Nohedes Lake and Pavin are there however Sébillot does not give any clue to the Thrown Stone story.

The CO_2 degassing of near-saturated maar-lakes can be triggered by a minor disruption of the lake surface, as a thrown stone, as experienced by volcanologists at the Monoun crater lake in Cameroun (See Sect. 1.6.1). Such process could have been witnessed very occasionnally in few European maar-lakes, in Latium, Eifel and Pavin, kept in local memories for very long and these marvelous and frightening lake response, gradually applied to lakes of similar aspects, as at Nohedes and Pilatus, spreading the Thrown stone legend.

3.4 Encounters with Fantastic Cratures at Pavin and Other Maar Lakes

These encounters generally found in older sources, are not explicit about lake disorders, poorly contextualized, use metaphors and have been missed by modern historians. Many of these texts, known to historians, scholars and folklorists, placed within their historical context (See Chap. 2), can be re-analyzed on the basis of possible degassing events (Table 1.2), Many of them are presented along with fantastic and threatening creatures. Their attribution to the *Lacus pavens*, the terrifying lake, is first discussed then each type is compared to similar stories, beliefs or perception, regarding other European maar-lakes. Four texts are presented in an inverse chronological order: a fantastic dragon story of 1632, the *Gloomy Lake Dragon*, two legends collected during the XIXth century featuring pre-Christian fairies and dragon stories, the *Fairy Garden*, and the *Fairy Dance*, and two texts by Gregorius of Tours (late VIth century). They provide important potential information on past state of Pavin, particularly during Antiquity.

3.4.1 Dragon Encounter at Pavin (Pavin event, May 1632?)

This fantastic story entitled "Le Dragon du lac Ténébreux" (*Dragon of the Gloomy Lake")*, authored by de Beaufort, was published in Paris in 1632 by Matthieu Colombei (Fig. 3.2). It is a 16-page story presented by its author as *"to expose actions that belong to marvels and going beyond the ordinary displays of nature", i.e.* it should be regarded as another detailed reporting of a kind of *prodigium*, as defined in Antique Rome (Lycosthenes 1557; Rasmussen 2003) which brings precious new elements on Pavin behavior.

The first three pages present a complex personal background of the hero, Jean de La Brière. Then the action is presented: it takes place near the town of Besse in Auvergne, at a village named Chaufour ("hot oven" which could be a transposition of the volcanic crater that is topping Pavin, the Montchal, "hot mountain") on the way to the famous mountain pilgrimage of *"Notre Dame d'Orsival where numerous miracles are produced"*, a clear transposition of the nearby Vassivière pilgrimage (see Chap. 2). The action can be dated

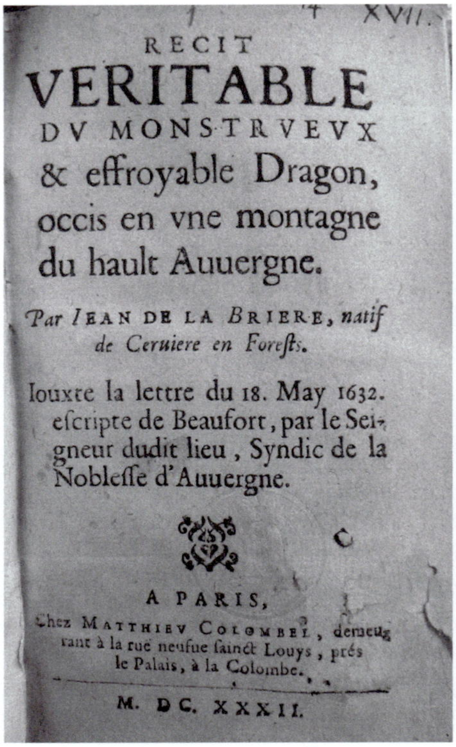
RECIT
VERITABLE
DV MONSTRVEVX
& effroyable Dragon,
occis en vne montagne
du hault Auuergne.

Par IEAN DE LA BRIERE, *natif de Ceruiere en Foreſts.*

Iouxte la lettre du 18. May 1632.
eſcripte de Beaufort, par le Seigneur dudit lieu, Syndic de la
Nobleſſe d'Auuergne.

A PARIS,
Chez Matthiev Colombel, demeurant à la ruë neufue ſainct Louys, prés le Palais, à la Colombe.
M. DC. XXXII.

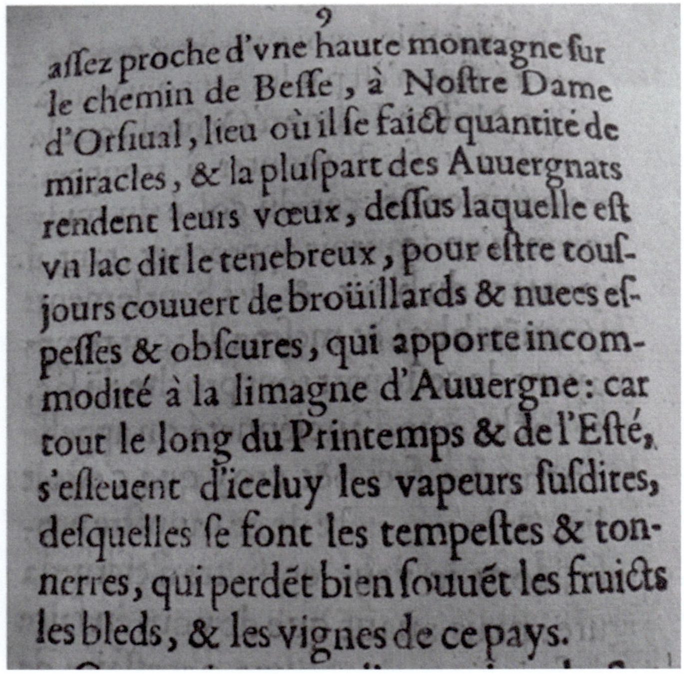
9
aſſez proche d'vne haute montagne ſur
le chemin de Beſſe, à Noſtre Dame
d'Orſiual, lieu où il ſe faict quantité de
miracles, & la pluſpart des Auuergnats
rendent leurs vœux, deſſus laquelle eſt
vn lac dit le tenebreux, pour eſtre touſ-
jours couuert de broüillards & nuees eſ-
peſſes & obſcures, qui apporte incom-
modité à la limagne d'Auuergne: car
tout le long du Printemps & de l'Eſté,
s'eſleuent d'iceluy les vapeurs ſuſdites,
deſquelles ſe font les tempeſtes & ton-
nerres, qui perdēt bien ſouuēt les fruicts
les bleds, & les vignes de ce pays.

Fig. 3.2 The Gloomy Lake dragon story published by de Beaufort in 1632 at Paris, so far not attributed to Pavin, describes with precision a sudden mudflow, one of the multiple misbehaviour of *lacus pavens*. (Bibliothèque Ste Geneviève, Paris). It has been re-published again in 1832. It was not attributed so far to Pavin by folklorists

between the mid-XVIth, since it refers to the pilgrimage, and 1632 and it is located by the narrator exactly at the outlet of the Lac Ténébreux (Gloomy Lake), an obvious transposition of *Lacus pavens*. The action can be abridged into several parts:

> (i) ***"There is a lake, 2½ leagues from Besse, named "Ténébreux" because it is always covered with fog and thick and dark clouds that bring damages to the limagne*** **[valleys]** ***of Auvergne: because during the whole spring and summer, these vapours escape from the lake, causing storms and thunder that very often damage fruits, grains and vineyards in this country…"***

This description features thunder, storms and damages on crops, including lowland vineyards that are far downstream of Pavin, in the valleys (limagne) (Fig. 1.2b, c). All these Pavin attributes are already present in several other witnessed accounts of Pavin misbehaviour between 1551 and the mid-XVIIth century (See Sects. 2.3.3 and 2.3.4). The storm is not triggered by a thrown stone in this story.

> (ii) ***"The Ténébreux Lake is wide and its water is green as spring grass and its sand is as verdigris. There are monster fishes that have never been fished because this place is inaccessible and uninhabitable, and only visited by the pilgrims attracted by their curiosity; however*** **[they come]** ***on one side only, since on the south side*** **[of the valley, where the lake outlet is]** ***no one would dare to come, as one can hear such terrible sound and screams, from immemorial time".***

The grass-green water colour is an unusual colour for common lakes but it is observed today in some acidic volcanic lakes. The inaccessibility and inhabitability of this place is outlined. Pilgrims fearing the terrible lake sounds and screams, avoid the direct route from Besse to the Vassivière mountain (see schematic pathway on Fig. 1.2c). The cascading lake outlet and the pathway between Besse and Vassivière are represented in the 1575 picture of the Land Register of Besse (Godivel I 1579a) (Chap. 2).

> (iii) ***"From the beginning of time there is, adjacent to the lake, a rock named the Fairies Rock*** **(Roche des fées)** ***which is believed to be the hiding place of the monster which I will describe".***

The Fairies Rock refers to the Montchal volcano, *i.e.* the burning mountain, adjacent to Pavin. At this stage the story becomes fantastical with the dragon encounter, dated on the "15th of last May", possibly in 1632, at midnight when La Brière comes back from Chaufour to Besse.

> (iv) ***"The monster has an excessive length*** *…a head like an ox…but longer. His eyes are sky-blue with red like a turkey cock crest, with peacock feather between the horns and white feather wings like a swan; his neck is 3 to 4 feet long…his back is covered with strong scales…the tail is 15 to 16 feet long; he has big claws"*, etc.

This dragon portrait is an archetype combining multiple attributes: as long like a snake, powerful as an ox, rapid and

Fig. 3.3 Fantastic creatures which may be associated with degassing maar-lakes misbehaviours. (*left*): The fight against the dragon (Schott 1662)(Bibliothèque Ste Geneviève, Paris): several types of dragons are described at Pavin, the major one with ferocious eyes shining and terrible teeth grinding, is located in the *lacus pavens* (Fairies Garden legend), another is assaulting the passer-by at the lake outlet (Fairies Garden legend, Dragon of the Gloomy Lake story). (*right*): Giant fishes described at Ulmen Lake, Eifel (*Ulmu lacus in quo piscis magnus spectatus aliquando*, Münster 1566) (library of the Faculté de pharmacie, Paris). Such "monster fishes", also described at Pavin (de Beaufort, 1632), are interpreted here as water fountain

winged like a bird, ferocious with mighty claws, and has a rugged surface. There is no image of the dragon in this 1632 version, but the dragon picture presented by Schott (1662) fits well with this description (see Fig. 3.3a). Attacked, La Brière fights back and breaks his sword on the ferocious monster; he is nearly engulfed by the animal. At this moment, he remembers the miracles that Notre Dame [of Vassivière] makes "almost every day": he seizes the animal by its neck and squeezes it strongly, shouting for help, but he is finally overhelmed by the dragon.

(v) On the morning after, La Brière is found by four herdsmen " ***under the dreadful snake; they unload on him the wood they brought and take him out from beneath this weighty mass*** ".

He can only be rescued with wood pieces unloaded by the four shepherds.

(vi) La Brière story is not yet over:

He is brought breathless to the Chapelle des Transis... ***he is showing buboes and pimples, similar to those caused by the Plague.***

This dragon encounter most probably refers to a sudden mudflow generated by a lake surge of the lake. One must note, as in other stories of river dragons, the scaly body, possibly referring to the much debris found at the surface of mudflows. Wood brought by peasants seems appropriate to approach and rescue the victim from the fresh mud. In some alpine passes, *Transis* chapels refers to repositories of dead corpses found after winter in the mountain; here it refers to the one of Vassivière and could be related to the natural violence of this area (Ph. Reyt, pers.com, 2013). The strange skin diseases found on the rescued La Brière, refered here as the Plague [*l' Epidemie*], are similar to the pimples mentioned at Pavin in 1783 by Godivel IV, after his son had soaked his hand in Pavin waters, during another misconduct event and by Nodier et al. (1829) (see Sect. 2.3.5). Buboes and loss of consciousness have been reported about victims and/or survivors of the CO_2 degassing at Nyos (Baxter et al 1989) (see Sect. 1.6.1).

This description of *Lac Ténébreux* is very precise-e.g.describing pilgrims pathway in the Couze valley-suggesting that the author has either witnessed the event on May 15th 1632, or is someone with perfect knowledge of the area. The style is similar to those of "canards", small brochures of few pages distributed door-to-door, reporting extraordinary events during the XVIth and XVIIth. They were occasional publications that prefigured future newspapers. It is published when fear of Pavin is at its maximum. This story, re-printed in Lyon in 1832, for unknown reasons, has not been attributed to Pavin, excepted by Reynouard (1910a, b). It is found today in a few registers of Auvergne tales in the dragon sections (Laconche 1994) and has been presented by the writer Samivel as a typical mountain dragon story in his chapter on mountain myths in the reference book *La Montagne* (Samivel 1956). Pavin's spillovers and mudflows are now documented by scientists, but not yet for this period (see Chap. 1).

3.4.1.1 River Dragons

The dragon metaphor is widely used during the Middle Age to represent dangerous river events (Reyt 2002) such as floods and/or mudflows. These dragons are elongated, powerful, they have scales, they move in valleys and they should not be confused with winged dragons, which move through the air (although river dragons may occasionally be represented with wings). The archetype of river dragons is the *Tarasque*, found in the Rhone River valley, at Tarascon (France) where the torrential rivers coming from the Cevennes may generate monstrous floods. The Tarasque is said to have been defeated by Santa Martha, the patron saint of Tarascon, according to the Golden Legend tradition reported by Jacobus Voragine (1265). Another river dragon attack is well documented; it occured in 1304 at the famous Murbach Abbey, located in the steep Belchental valley (Alsace), during a terrible storm: *"on the waters was floating a dreadful dragon...when the flood retired the reptile was stranded on the ground,* [few villages downstream], *where again it caused damages among people and livestock... finally courageous men attacked it and killed it after many efforts"* (Legin 2003). It refers most probably to a river mud flow, it shows how populations could describe rare and terrifying events they could not understand, as for the 1632 Pavin dragon.

Dragons were already common in Antiquity: Conrad Lycosthenus, the Alsatian encyclopaedist (1518–1561) who transcribed Julius Obsequens *prodigii* observed during the Republican period of Roma, mentions several of them in his *Prodigiorum ac ostentorum chronicon* (1557). Dragon representation in art and literature is particularly developed from the XIVth to the XVIIth as illustrated by Schott in 1662 (Fig. 3.3a).

River dragons stories and representations lasted until the XIXth century: one Epinal image publisher (Pellerin 1829) is still featuring a dramatic flood event, that occurred on October 6th 1829 at Fagna in Chile, as a dragon: *"the amphibian monster came...He was covered with scales, bullets were bouncing off him...He devastated the town claiming 80 human victims and many bulls, cow, pigs"*. Then the "Fagna harpy" jumps in a lake and is finally caught: it is 23 ft long, has a bull's head, a scaly body, wings and two tails. The coloured representation on this famous Pellerin image, circulating door-to-door in French villages in the XIXth century, is close to the classic representation of the Tarasque defeated by Santa Martha.

3.4.2 Monster Fishes Observed at Pavin (1632) and in Ulmen Maar, Eifel (Münster 1566)

The Dragon of the Gloomy Lake story mentions that the lake holds *"monster fishes actually never fished"*. This should be interpreted as fountains of degassing waters, *i.e.* little spurting geysers as those produced by blowing whales. Such degassing figures have been observed by scientists at Monticchio Lakes and at Nyos (Table 1.2). Another "encounter with whales" in crater lakes is reported and illustrated (Fig. 3.3b) in 1556 by Sebastian Münster (1458–1552) at Ulmen Maar, one of the many Eifel maar-lakes. In his *Cosmographia Universalis*, 1430 pages, the first ever to describe and map the New World, Münster develops the description of *Eyfalia* lakes:

> *"In Eifel there are two big lakes, one near Ulmen castle [Ulmen Maar], and the other near the Laach monastery [Laacher See]: both are very deep although they have no water inlets but outlets. They are called locally maars and are rich in fishes...* ***It is said that at Ulmen one finds a fish, or if you want, a whale, which has been seen by several people, one 30 feet long, the other one 12 feet. And when they appear it is a certain sign, as they said, that one of the children of the place named Ganerben will die*** *, and they maintain that they have experienced this several times. They also write that Ulmen is so deep that one cannot find its depth, even after having sent the sounding line for three hundred fathoms."*

Münster illustrates his Eifel description by a wood engraving of a blowing whale (Fig. 3.3b) with open mouth, with the legend: *"Ulmu lacus in quo piscis magnus spectatus aliquando"* [Ulmen Lake where a great fish can be seen].

Ulmen Maar today is meromictic, at least since 1912 when it was first studied by Thienemann, the founder of German limnology (Thienemann 1914–1915). It has neither aerial inlet nor outlet. Close to the lake there is a spring in which Celtic and Roman votive artifacts have been found. For centuries this place was frequented by woman praying for healthy children. It is important to note that the occurrence of monster fishes is connected, according to Münster, to the death of children in the vicinity, which could be interpreted as damages caused to those sleeping on the ground during such degassing events. There are at least three Eifel maar-lakes where damage to Humans have been reported: Weinfelder Maar (Totensee), Ulmen Maar and Pulver Maar (see further).

3.4.3 The *Fairies Dance* Legend: Past Fumaroles at Pavin-Montchal?

The *Fairies Dance* tale has been collected and published in 1856 by de Ribier du Chatelet (1779–1844), the historian of Haute-Auvergne, now Cantal. It features a young boy in the mountain in early or pre-Christian times:

> *Irald is passing below the Fairies Lake at 11 at night...He sees on the summit aerial beings. He comes close to the dance without being noticed but the dance is stopped and he is invited by two fairies. The boy rushes but when his hands join those of the beautiful creatures,* ***something dry and icy catches him in a vice. Then starts an infernal round dance. Irald is pulled away***

***at great speed…he falls down** . The dance is going on and on. At midnight the moon is unveiled and the beautiful girls are transformed into ugly skeletons whose empty skulls are throwing flames. The rotten body of a child, dead before having been baptized, is brought in and the hideous group makes a terrible feast. Irald recommends himself to St. Geraud and soon the infernal band is disrupted. The fairy that was once the most seducing exhales on him an inflamed breath, the fire burns his hair and prints on his cheeks a stigma in shades of blood. Irald lost conscience…when he woke up the mountain got its normal aspect back but he kept his wounds.* [abridged version].

Taking into account that the Fairies Lake and the Fairies Mountain correspond to the Pavin-Montchal ensemble (see Sect. 3.4.1) and that Pavin area has been subject to past degassing activity (Chap. 1 and 2) the following re-interpretation can be proposed: (i) The *aerial beings*, seen when looking up towards the Fairies lake, correspond to gas exhalations from Pavin area,(ii) the fairy, suddenly catching the hero in a *dry and icy vice which makes him loose conscience* can be interpreted as a loss of consciousness, as observed on Nyos survivors (see Sect. 1.6.1), (iii) dead babies brought to the place can represent young victims of this deadly place – as in Eifel – and/or human sacrifices during pagan times,(iv) *the ugly skeletons whose empty skulls are throwing flames and the flame-exhaling fairy* could be fumaroles on the adjacent Montchal, *i.e.* the *scorching mountain* for local people, according to volcanologist Lacoste de Plaisance (1803). After the victory of the Christian predicator the place gradually went back to normal but remained bareland and without snow.

This legend is localized by de Ribier (1856), a specialist of the Haute Auvergne history, at the Chamaroux Mountain, a baren volcanic cone without any lake (1476 m), some 30 km South of Pavin, at the limit between Puy-de-Dôme and Cantal (*i.e.* between Haute and Basse Auvergne). In this legend fumaroles fairies lasted at least until St. Géraud's time (855–909), the founder of Aurillac Abbey in Cantal.

3.4.4 The *Fairies Garden* Legend: The Original Description of Maleficent *Pavens* at Pagan Times

This legend is published 50 years after the *Fairy Dance* by Ludovic Soubrier (1846–1880), a Cantal priest and local folklorist publishing under the pseudonym of Louis Boissière (Moulier 2010), in his 27 tales and legends from the Haute Auvergne (1893) (Fig. 3.4). Soubrier possibly found this one in church archives as he notes that *"this tale has been transmitted thanks to the religious care with which our fathers were keeping memories from the past"*. His legend features with much details the Fairies Garden as a marvellous and terrifying place:

(i) *"a wonderful palace with ramparts that seemed to be forged into incandescent cast iron…a flow of molten gold was running in ditches…everywhere **marvelous waters were spurting from huge silver ponds. Morning and night an extraordinary breeze was softly affecting the trees, shaking the spurting water plumes** and its voice was covering the sound of nature. This was* the Fairies Garden, *left by them in winter…"*
(ii) *"Nearby people were bowing down in front of this wonder as if they were bowing in front of their gods. Fairies were accepting with pride this sacrilegious worship. Those invoking the fairies could see, when raising their heads, a **white phantom smiling at them while continuing his flight** : it was the fairy. The surrounding wall was indeed stopping any daredevil: it was an **immense dragon curled three times around the mountain. From far away one could see his monstrous rump, his ferocious eyes shining and terrible teeth grinding. The monster was voracious and, when someone was passing under in the plain, a fairy rushed forward in the air, ran to meet the misfortunate one, enthralled him and misled him.** Then the monster was loosening its many coils, rushed suddenly on his victim in a cast-iron noise and, after a rough roar, went back to its observer position…"*
(iii) *"To calm him down and to please the fairies, in charge of his subsistence, **people used to offer him human victims.** Each month, during the short sleep of the monster, a young boy was left at the bottom of the mountain where the dragon came to eat him when waking up…"*
(iv) ***"For hundreds of years the hydra was fed with children until the coming of one predicator of a new god** . He reproached them for their idolatry and promised, in the name of God, to get them rid of the dragon. He takes a spear and walks to the mountain. The **dragon opens a terrible mouth, rolls his blood-injected eyes, slides down on the grass slopes, smashing everything on his way** and making sinister roars. At few feet from him, the apostle raises his hand in the name of Christ and makes the Cross sign, the monster suddenly stops, as if he was hitting a cast-iron wall; the Unknown pushes his spear in the monster head, despite his thick scales that offered him impenetrable body armor. **The dragon gave out a terrible cry…he rolls lifeless at the bottom of the mountain.** The apostle is winning the heart of the grateful barbarians. Soon the Fairy palace fades out, stops shining, **there is only a sterile rock rearing its bald head**. Until now the green grass did not dare to overrun this damned place which dries up snow and rain so quickly".* [abridged version].

The Fairy Garden is left in winter. This toponym has been encountered in the previous legend and the Fairy Rock is also the location of the 1632 dragon story (see Sect. 3.3.2), which certainly corresponds to Pavin. In addition there is a large body of scientific and historical evidences of past Pavin "misbehaviour" (see Chaps. 1 and 2). A re-attribution of the *Fairy garden* legend to Pavin is therefore proposed, adding precious details on Pavin behaviour and its perception by local people at pagan times:

1. The *dragon* guards the palace, *i.e.* the lake and its rims. He can be perceived from far away by his *ferocious eyes* (lightning), *grinding teeth* (terrible noise and/or thunder sound, reported many times at Pavin) and *monstrous rump* (possibly the crater wall or the horizontal lava outcrops, as drawn by Lecoq 1835b, see Fig. 2.5). Lightning and thunder, also described at Vassivière in 1551 and at

Fig. 3.4 The original and oldest Pavin legend, so far ignored: the Fairies Garden legend (Soubrier 1893) (BCU library, Clermont-Ferrand). It features a maleficent and powerful lake dragon, worshipped by pagan cult for centuries then defeated by the first local evangelisator, i.e. around the Vth century. The legend details Pavin misbehavior, i.e. various forms of degassing, during Antiquity

Nyos in 1986, are the story-motiv of Belleforest's description of Pavin (see Sect. 3.3.2).

2. The *marvelous spurting waters* associated with soft breeze are the water fountains, taken for the water blown by giant fishes also present in the 1632 dragon story (see Sect. 3.3.2).
3. The walls of the castle, with their *incandescent cast-iron* and *molten gold* colours could be hydrothermal springs rich in oxides and sulfur-rich vents, as found today in some active acidic maar-lakes. [These could have been interpreted by Del Rosso (2009b) as lava flows, a hypothesis strongly debated among volcanologists at the 2009 Besse meeting.]
4. A regular breeze or fog occurs in the morning and in the evening. Such permanent fog, characteristic of active degassing, is also described at Pavin by Godivel II in the mid-XVIIth (see Sect. 2.3.4) and in the 1632 dragon story (Sect. 3.3.2).
5. The *fairy who rushes forward in the air, runs to meet, under the lake, the misfortunate one, fascinates him*, and misleads him, also present in the previous legend and in the 1632 dragon story, is interpreted as CO_2 overflow and related loss of conscience of victims.
6. The *powerful dragon* slides from the mountain with a cast-iron noise, destructing everything, and stops after a rough roar, also featured in the 1632 dragon story, could correspond to major water spillovers and/or mudflows event observed by geologists and paleolimnologists (Lavina and Del Rosso 2009; Chapron et al. 2012; Chap. 23).

7. Regular (monthly?) pagan cults and human sacrifices of young boy were performed at the lake. Such lake cult has also been performed in Antiquity at the Albano Lake (See further).
8. The terrifying monster and its associated cult have been active for "hundred of years". Then they are stopped by a predicator – as in the previous legend. In Pavin area the first Christian predicators are *Stremonius* – now Saint Austremoine – in Issoire, *Baudimius and Sennectarius* in Saint-Nectaire, all in the IVth century. Auvergne ermit Caluppa, who encountered two dragons before his death in 594, according to Gregorius of Tours, could also be the predicator who defeated this dragon (See further).
9. After the defeat of the dragon the rock is left sterile, as in the previous legend. Until 1830 the Montchal volcano was a baren ground (see Lecoq's drawing, Fig. 2.5).

The *Fairies Garden appears to be the original and oldest legend on Pavin*. It is very detailed and fully coherent with what we have learnt so far about the lake. Ironically it is still totally ignored from folklorists, historians and scientists. In *Tales on the Earth and Underground World*, section 3 on Lakes and fairies, 1904–1906, Paul Sébillot only mentions the *Fairies Dance* (A 920 1.7), actually not attributed to Pavin, but not the *Fairies Garden*, published in 1893, nor *"the Lake of Terror"* published in 1910 by Roux-Parassac. These three tales are now present in the remarquable register of Auvergne tales collected by Gilbert Laconche (1994) where only *the Lake of Terror* is attributed to Pavin while the Dragon of the Gloomy Lake is in the section on "Stories of animals, wolves and dragons". The Thompson world register of folk tales (1955–1958), which does not mention any of these Pavin tales, quotes only one legend concerning a village under a lake (F. 725.5.1) out of thirty lake legends. Lake dragon is also very rare in Thompson with only one occurrence (B.117.2) out of 60 dragon occurrences, for an Irish lake that presents many similarities with the Pavin Fairies Garden tale. In conclusion, neither Sébillot nor Stith Thompson or Van Gennep (1949) featured and/or interpreted the three Pavin legends and stories. The comparison of Pavin legends with myths and legends collected from other maar-lakes has greatly facilitated their interpretation, an approach confirmed by the consideration of religious cults at these lakes during pagan then Christian times.

3.5 Pre/ Early Christian Lake Cults in Maar-Lakes, Pavin, Averno, Albano and Nemi

The Fairies Garden legend presents multisecular pagan lake cult at Pavin. A famous text by Gregorius, the first historian of Merovingians (538–594) features a recurrent lake cult during Pagan times, which could be re-attributed to Pavin. Very important cults were also taking place during Greek colony and Roman times in Averno, Albano and Nemi maar-lakes

3.5.1 Gregorius' Famous Lake Cult in Gevaudan: A Re-attribution to Pavin?

Gregory of Tours, already mentioned for his account of the Tauredunum event on Léman, is born in *Urs Averna* – now Clermont-Ferrand – at 50 km from Pavin. He describes in his famous account of the life of Saint Hilarius from Poitiers – written inthe late VIth century- the very unusual behaviour of one lake in Auvergne (Gregorius 600s):

> *"At this time there was in Gevaudan, on a mountain named Helanus, a big lake. There, sometimes ago,* ***a crowd coming from the countryside was celebrating this place; they threw man's pieces of cloth and fabrics, others wool fleeces; most of them were throwing cheeses, candles, and various artifacts, according to their wealth, that would be too long to account. They were coming with chariots, bringing food and drink, slaughtering cattle and were feasting during three days. The fourth day, when they were about to leave, they were assaulted by a big storm with thunder and immense lightning and from heaven there came an intense storm of rain and hail, so violent that everybody doubted they would escape*** *. This happened every year and the superstition was deeply present in these mindless people. After a long period came a priest, made bishop at the city. He preached the crowd not to celebrate this anymore or they will be hit by heaven's anger but his preaching was not affecting these simple-minded people. Then* ***the man of God, inspired by Him, built, away from the lake shore, a chapel*** *dedicated to the blessed Hilarius of Poitiers and placed the relics of the holy person into it, saying to the people: "Have fear, my sons, fear to sin in front of our Lord for there is nothing to be venerated in this lake". These people were deeply touched, converted themselves and quit the lake. What they used to offer they brought it to the holy place and were freed from their errors".* (our translation from a French version of *De La Gloire des Confesseurs, Livre VII, De St Hilaire*, http://remacle.org/bloodwolf/historiens/gregoire/miracles3.htm

This text features pagan times when a lake, localized in the Gabales people territory, here attributed to Gevaudan, was worshipped each year by local people, partaking in three-day feast, presenting offerings and sacrificing animals. Their worship is interrupted by sudden storm with thunder and lightning of unusual size and type, and an intense rainstorm with hail is mentioned. The continuous cult and the lake misbehavior are finally interrupted by a Christian predicator who builds a chapel dedicated to Saint Hilarius. The storm description is very close to a violent lake degassing event (Table 1.2).

Traditionally (*e.g.* Sébillot 1904–1906, chapter V, p. 286) this text is not attributed to Pavin but to another lake, Saint Andéol, in Gevaudan, some 150 km away. Saint Andeol Lake is another isolated high plateau lake, at 1223 m; it is shallow, probably originates from glacial scour on crystalline rocks and is not meromictic. As for the Vassivière pilgrimage at 1.5 km from Pavin, another famous mountain pilgrimage has been active at this lake until 1867. Its origin is uncertain but ancient: on July 8, 1640 there were 4 000 pilgrims on the mountain; the cult implied to go around the lake up to three times for sins expiation and it lasted until 1867 (Ténèze 1978). The 1950s excavations near the lake found the ruins of a Gallo-roman temple.Recent research confirms the existence of a proto historical lake cult, a confirmation for archeologists that Gregorius lake is indeed localized at St. Andéol, an interpretation which originates at least from the XVIIIth, as discussed in Fau et al. (2010). Gregorius text is also now reported as a Gevaudan legend (Ténèze 1978).

A re-attribution of Gregorius text to Pavin is however proposed here for the following reasons: (i) there are striking similarities concerning the description of lake cults in this text and in the *Fairies Garden* legend, attributed to Pavin, (ii) the exceptional storm with thunder, lightning and hail described by Gregorius is much similar to the many degassing descriptions made at Pavin between 1551 and 1650) (see Sects. 2.3.3 and 2.3.4), (iii) St. Andeol Lake does not present any of the characteristics of other degassing lakes, *i.e.* deep meromictic maars or crater lakes (*e.g.* Albano, Averno, Monticchio, Pulver, Nyos), (iv) There is no other historical sources of violent events at St Andéol, in contrast to Pavin, although it must be recognized that "normal" thunder and storms occur each summer on Gevaudan plateau.

Such re-attribution is coherent with what we now know on Pavin: (i) Pavin has been settled during Antiquity, as evidenced by the Pompeian millstone found in its waters (Fig. 2.1), (ii) a pagan celebration at Pavin is evoked in 1615 by Jesuit Father Coyssard in his first record of the new Vassivière pilgrimage (in Cladière 1688): the XVIth crowd throwing coins into the sacred spring of Vassivière *"were continuing the rites of Gaul people, stopping by the Pavin shore to offer to its water silver coins and precious vases"*. According to Jean Baptiste Bouillet, the XIXth century archeologist, several families from Besse possessed antique artifacts retrieved from the lake, although these have not been described (Bouillet, manuscript, 1875),(iii) Such pagan lake cult, *i.e.* before the V-VIth century, could explain the original Latin name of Pavin, *Lacus pavens* (Banc 1605; Godivel II, 1650s), (iv) Hilarius (c300–c368) is the first bishop of Poitiers who Christianized the Aquitania province to which the Pavin area was a part of in the IVth, (v) the construction of a chapel by the Christian preacher, away from the lake, could correspond to the early Christian settlement at Vassivière (see location on Fig. 1.3c), mentioned by Jaloustre (1910): *"authors thought that this place could be the Celtic sanctuary offered to the Clamouze River deity. People should go there for celebrations. When Christianity was established the worship of the Virgin replaced the one of springs…It is likely that a small Christian parish succeeded to the Gallic settlement established close to the sacred springs"*. Jaloustre adds that, when the present chapel was built, huge logs – 10 m long and 80 cm in diameter – were found in peat, in addition to the ruins of the medieval church, destructed in 1321. So far no archeological excavation has been realized at Vassivière.

Archeological research at Pavin and in the Vassivière Mountain, as the one that is currently made with success at St Andéol, could confirm our hypothesis and establish possible connexions between St Andéol and Pavin pagan lake cults. This association has been already stated by a reknown historian of Gaul, Camille Julian (1859–1933), who made a direct parallel between pre-Christian lake cults at Averno-Cumeo in Italy, at Pavin and at "a lake in Gevaudan" (Julian 1901):

> *"The worshipping of Pavin and a lake in Gevaudan, by Gaul Arverns people, in the same way that Averno Lake, where the famous Cuma Sybilla was standing in lake vapours, was worshipped by pre-Roman people, near Naples"* and: *"In Auvergne, as in Campania, lakes have attracted the fearful imaginations since a long time ago. I do not know if the Gauls were seing shadows going out from the bottomless Pavin Lake, as the Greeks from Cuma were invoking shadows on the Lake Averno shore, but these people were placing in these silent and hypocrite waters the impregnable asylum of a chthonic deity that should not be disturbed, only with presents. For three consecutive days, on the shores of a Gevaudan lake, the peasant people were gathered to make libations and sacrifices: they were throwing into the waters fabrics, sheep fleeces, wax cakes, bread, not to mention more costly presents. And during three days they were feasting and having orgy, finally interrupted by the angry god. Gregorius of Tours stated that a holy predicator ended this superstition"*.

Julian statement pushes for a re-analysis of all the pre-Christian lake cults, particularly for maar-lakes, in consideration of the new hypothesis of possible violent lake degassing. A preliminary analysis in Italy provides promising results.

3.5.2 Hermit Caluppa Defeats Two Dragons: A Record of the Catastrophic Pavin Event (590s)?

The story of hermit Caluppa dragon is also reported by Gregorius (*Vitae Patrum*, 11), as a remarquable encounter with the evil. Caluppa was a hermit living in a grotto at the "Monasterium Melitinsin, in a 500 foot cliff in Auvergne". The story has been unearthed in 1975 by Fournier, an Auvergne historian. At the end of his life Caluppa, who died in 594 according to Fournier (1975), is attacked by two powerful dragons, which can be summed up as such:

> **Two dragons of enormous size progressed in his direction and stopped very close. One of them was stronger than the other. His chest was projected forward, his face close from the Saint's face. He rose to speak to him softer. The Saint,**

terrified, icy and stiffened as if he was in bronze, could not move any limb, even raising his hand to sign himself. As both remain immobile for a long time, the Saint prays in his heart then manage to move one hand, signs in the name of the cross and addresses pious words to the main dragon. At each word the dragon went deeper into the ground. Meanwhile the other dragon gradually coiled around Caluppa's legs by surprise. The Saint says to him: "go away, Satan", and the dragon withdraws at the sill of the hermit's cell, emit by its lower part an astounding noise and fills the cell with an unbearable stench that could not be attributed to anyone but to the devil himself. (our summary of Fournier's version of Gregorius text)

Fournier does not propose any explanation to the dragons encounter. He locates the monasterium at Meallet in Cantal, the Haute Auvergne, South of Pavin (Fig. 1.2b). Another location and interpretation are proposed here. The cliff with the hermit cell, is the Jonas caves, a famous and unique ensemble of caves, dug-out by man in a 150 m high cliff, allegedly occupied for about 2000 years and located a dozen km downstream Pavin in the Couze Pavin thalweg (Fig. 1.2c).

This account, very detailed indeed and probably transcribed by witnesses, may be Caluppa himself, can be re-interpreted as the enormous Pavin Lake overspill, dated around 600 AD, which is evidenced in the paleosediment records of the lake (Chassiot et al. 2016; Chap. 2.3). The volume of water emitted by the lake is estimated to 4 millions m^3, and the maximum outlet discharge to 1500 m^3/s. It generated an enormous mud flow which could have reached the foot of the Jonas cave. The first dragon could be attributed to the CO_2 flow along the valley floor, causing the sudden and long-lasting paralisis, as in other degassing events (Table 1.2). The second one could be the mud flow with a sulfurous smell, also common at other degassing events and perceived near Pavin at other occasions. The sound of the explosion could be heard very far away. The degassing and overspill event could have been triggered by a major earthquake that occurred in 582 AD in the Pyrénées and felt in Auvergne, according to Lecoq (1835), or to a less known earthquake in 580 AD near Angers.

3.5.3 Lake Cults at Italian Maar-Lakes in Antiquity: Entrance to *Hades*, *Feriae Latinae*, *Diana Nemorensis*

Lake cults in Antiquity are described by several Antiquity historians as Strabon, according to Lavroff (1872) who connected the Sunken City legends to these customs. At least three Italian maar-lakes of Latium (Albano and Nemi) and Campania (Averno) had specific deities and cults that have lasted throughout the Antiquity. The new outlook on degassing lakes, started by Italian earth scientists, fosters a new vision on these celebrations, some of which were very similar to the one described in Auvergne by Gregorius.

3.5.3.1 Averno Lake, the Greek Entrance to Hades

The *lake without birds* for its Greek settlers is a small meromictic maar- lake located in the Phlegrean Fields, near Naples (see Chap. 1). Averno was a very famous site during the Greek colonization of Italy. Latin authors as Virgil describe Averno as a lake with constant fog and vapours, hence its permanent absence of birds. Averno Lake is one of the smallest geographic features represented on Peutinger map, attributed to 350 AD. Ruins of a temple, possibly dedicated to Mephitis, the goddess of noxious vapors, are still visible today on the lake shore. It has been visited by many travelers such as the French cosmographer Thevet (1575) who reports: "*It was the entrance to Inferno, its waters were sulfur-rich and the Ancient thought they came from where the souls of evil persons were tormented..., it was named Paluz-Acherusia where in the past humans were immolated to Inferno gods*". Ephraim Chambers (1680–1740) describes the "Averni sites "in his encyclopedia (1741) as such:

> *"Certain lakes, grottos, places which infect the Air with poisonous Steams or Vapours also called Mephites...Birds could not fly over them but dropped down dead". The most celebrated, Avernus, was a lake near Baya in Campania, by Strabo called the Lucrina Lake and by Italian geographers, Lago di Tripergola...The Circumstance combined with the great depth of the lake occasioned them to take it for the Gateway of to Hell and, accordingly, Virgil makes Aeneas descend this way to the Inferi. Vibius Sequester says there was no bottom to be found of it: "Immensa altitudinis cujus ima part apprehendi non potest".*

3.5.3.2 Albano Lake and the Long Lasting Feriae Latinae Celebrations (700s BC- 400s AD)

The biggest European maar-lake at 30 km south of Rome (Table 1.1) has been the location of a major and long-lasting cult, the *Feriae Latinae* which celebrated the major god of Latium, *Jupiter Latiaris*, "the one that triggers storms". His temple was located on Monte Cavo (950 m a.s.l.), the highest portion of the rim between Albano and Nemi Lakes. According to the Dictionary of Greek and Roman Antiquities of Charles Daremberg and Edmond Saglio (1877–1919, p. 1066) this annual celebration in Monte Cavo lasted 3 days, bringing in people from all Latium League cities. The first 2 days were festive, every one forgetting who they were, master or slave. Each city was represented and praying, people were bringing milk, cheese, any kind of food, including lambs. The last day a white bull was sacrificed and its meat was shared among people from each city. The *Feriae Latinae*, established by the Latium League, was one of the earliest celebrations and remained the most important one during at least twelve centuries, from the destruction of Alba Longa, in 665 BC, to 400s AD when it was stopped by the Christian Emperor Theodose. This cult may already have existed when Alba Longa was first established, around 1200 BC according to the tradition. The *thunder and lightning* were the major attributes of *Jupiter Latiarius, tempestatium divinarium potens* [the one who could predict storms]. According to some authors human blood was also used in the ceremonies, although this point is now being contested (Mahieu 2010). Jupiter Latiaris has been the most and longest worshipped god in Italy, requiring the greatest number of human sacrifices.

Lake Albano anger was known and feared since at least its catastrophic overspill in 398 BC (see Sect. 1.6.2.1): few years later the Romans excavated a tunnel through the lake rim to prevent such surges. This technical response, an engineering marvel (Castellani and Dragoni 1997), was 1450 m long, 1.2 m wide and 2 m high. It remained in operation for more than 2300 years, maintaining the lake level at 70 m below the ridge. It can be considered as the "first prevention device" in overspill risk management (De Benedetti et al. 2008).

The many similarities between the pagan lake cults described by Gregorius of Tours and in the Fairies Garden legend – and the *Feriae Latinae* are striking: yearly gathering of all local populations, 3-day feast, type of offerings, thunder and lightning. The Roman builders of the Mons Aureus thermal baths (Mont Dore) were certainly aware of the famous Latium celebration, held each year on the rim of a round, deep and frightening lake, very similar to Pavin. Lake cult celebrations at Pavin, evoked by several authors, which are to be confirmed, may have been influenced by the *Feriae Latinae*.

3.5.3.3 The Terrible Pit-Dragon Defeated by Pope Silvester: Another Degassing Event at Albano, 330s AD?

Many dragons are present in Jacobus Voragine's *Golden Legend* (1260–1275) which details the life of early Christian saints and has been widely used throughout the Catholic Church. Many of these early predicators are fighting against pagan cults and some of them are also presented as dragon killers or *sauroctonous* saints who defeat the supernatural forces in the name of the Cross. About forty of such saints have been described (Sergent 1997), the most famous ones being today Santa Martha and Saint Georges. The story of Saint Silvester, pope of Rome (314–335) under Constantine emperor, features his conversion of pagan priests after deafeating a pit-dragon:

> *"In this time it happened that* ***there was in Rome a dragon in a pit which every day slew with his breath more than three hundred men.*** *Then came the bishops of the idols unto the emperor [Constantine] and said unto him: "O thou most holy emperor, sith the time that thou hast received Christian faith the dragon which is in yonder fosse or pit ..."* Then S. Silvester seeks advice from S. Peter how to defeat the dragon.

> *"When* [S. Silvester] *came to the pit,* ***he descended down one hundred and fifty steps, bearing with him two lanterns****, and found the dragon, and said the words that S. Peter had said to him, and bound his mouth with the thread, and sealed it, and after returned, and as he came upward again* ***he met with two enchanters...which were almost dead of the stench of the dragon****, whom he brought with him whole and sound, which anon were baptized, with a great multitude of people with them. Thus was the city of Rome delivered from double death that was from the culture and worshipping of false idols, and from the venom of the dragon".*

As for the Caluppa dragon story a re-interpretation can be proposed. The dragon is living in a pit, *i.e.* a crater. It is worshipped since long by local people in a dedicated sanctuary. It is venomous, expelling a noxious and stinking breath; the emitted gases are probably sulfurous, which produces at this moment, massive casualties, killing 300 people each day. The catastrophic event forces the pagan priests to seek help from Emperor Constantin, recently converted to Christian faith, who mandates Silvester, the leader of Rome Christian community, to solve this pressing issue. The latter enter the crater with lanterns [so that he could determine if the CO_2 was present or not], and protects himself by a thread. When coming back from the deep crater, Silvester rescues two pagan priests – the enchanters – who are in a state of shock [same symptom as Caluppa]. They recover and are baptized at once, soon followed by many other worshippers of the false idols.

This story is similar to a degassing event. The exact location is not given but it took place near Rome. Lake Albano, already mentioned for its misbehavior, is a likely candidate: there a catastrophic lake surge already occurred seven centuries before, in 398 BC (see Sect. 1.6.2.1). If the pit-dragon event is correctly attributed by Voragine to Silvester, according to the tradition, it would have happened around 330 AD. At this epoch the *Feriae Latinae* was still celebrated on the lake rim at Monte Cavo. If this re-interpretation is confirmed, Lake Albano and its degassing would have played a certain role in the conversion of local people to Christian faith and this major degassing event should be added to the 398 BC event. In 330s AD the lake level could have been some 20 m higher than today, judging by the 150 ft descends of Silvester.

This story presents great similarities with the Gregorius lake cult text, with the *Fairy Garden* legend, and with the

Caluppa dragon encounter: all highlight the conversion of local people to Christian faith and present the lake disorders dragons as the evil, defeated by the Christian faith.

3.5.3.4 Nemi Lake, Home of Diana and Its Giant Roman Vessels

Few kilometers south of Albano, was the location of another important early Roman sanctuary – fifth century BC – devoted to *Diana Nemorensis*, living by maar- lake Nemi in her sacred wood (Bersani and Castellani 2005). The presence at Nemi of another tunnel, built around the IVth-Vth BC through the crater rim (Castellani and Dragoni 1997), suggests the prevention of a lake surge, as for Lake Albano (see Sect. 1.6.2.1). Nemi is also famous for its giant Roman sunken ships containing bronze artifacts: their retrieval was attempted as early as 1538 as well as in the late 1800s, both in vain. In 1928–1932, the lake level was artificially lowered by 23 m using pumps to allow for the archeological exploration during which a second giant wooden ship was found. In 1943 the lake level was re-established and a museum was built on the lake shore but was burnt during the German retreat in 1945. The Nemi Roman ships, built under Emperor Caligula (37–41 AD), are exceptional by their size, 70 m×20 m for the *Prima Nave* and 73 m×24 m for the *Secunda Nave* as well in their decoration which includes precious marbles, many bronze artifacts as lions and wolves heads and one *Medusa* head, which survived the fire and are now kept at the National Roman Museum in Rome. The ships, among the greatest wooden ships ever produced, were probably sunk after Caligula assassination. The *Prima Nave*, which had no visible means of propulsion, was likely towed to the center of the lake when needed. The distribution of its super-structures gives the ship a discontinuous look and is radically different from any other ancient construction. The exact function of the Nemi ships is still debated. Given that Latium people have witnessed lake disorders and catastrophic events at Albano and possibly at Nemi itself, five centuries before, the Nemi ships could well be connected to celebrations for the prevention of such events.

3.6 Christian Celebrations Protect Populations from Pavin Misbehavior

Pavin is located between Besse medieval city and the Vassivière Mountain, a religious site of regional importance, the some 1.5 km from the lake, at the same altitude – about 1350 m – but away from its possible disorders. Besse, about one thousand inhabitants at medieval times, is 3 km from Pavin (See Fig. 1.2c). Its St. Andrё Church dates from the XIth–XIIth century. During the Middle Age there was a small church at Vassivière, which was dismantled in 1321 (see Sect. 2.3.2). Then a famous pilgrimage was started in the mid-XVIth century after a remarquable miracle which took place in 1547 near Pavin. Already in the XIIth there was at Besse a college of 20–30 priests at Besse, which was increased to 60 priests in 1498 by a Papal bull from Alexander VI (Boyer-Vidal 1888; Blot 1910; Gomis 2006). Such density of clergymen in this remote Auvergne village, of difficult access during the 4 to 5-months-long winter, is striking and it is likely that a great number of masses were celebrated each day. A re-interpretation of rich religious iconography and history of Besse-Vassivière, taking into account the latent fear in Pavin area and the historical sources of Pavin degassing misbehavior (Chap. 2) is proposed here in a chronological order.

3.6.1 Besse Church Medieval Capitals (XIIth Century) Present Violent Scenes

The rich Romanesque iconography of Besse Church is still intact. The capitals considered as one of the most various and vigorous suite in Auvergne are well described by Blot (1924). The capitals "denominations" by de Bussac (1961), are complemented here (*in parenthesis*) by other possible interpretations: (i) "Eagles grabbing sheeps in their claws" *(loss of sheeps during degassing events)*, (ii) "Drunkar damnation": a man with a great beard lying in a bed with closed eyes. His right hand is taken by a horned devil which is holding a young man upside down. This young man is also seized by a second evil holding his arm and leg. In the middle a third evil is compressing his chest. Under the bed, a snake and a small wine keg are featured (Blot 1924), *(a figuration of someone taken by Pavin death in his sleep?)*, (iii) "Ox sacrifice scene" *(figuration of former pagan rites at Pavin)*, (iv) "Horned centaurs with *Minorta Urs* inscription (minotaurus)" *(another metaphoric representation of Pavin threats)*. This new interpretation is still very hypothetical as other Auvergne Romanesque capitals often feature similar creatures, such as centaurus (Guyon 2010), but the representation of many fantastic beings, attacks by animals and humans or animal sacrifices is striking.

3.6.2 The First Vassière Miracle and the Ensuing Pilgrimage (June 1547 Pavin Degassing Event?)

In 1547 a miracle is observed between Pavin and the Vassivière Mountain, some 1.5 km away from the lake. At this time the Catholic faith is threatened by the Reform, very present at Issoire, some 40 km from Besse, so that the event is reported at once to the Catholic religious authorities, archived with the witnesses' names and declared as a miracle illustrating the power of the Virgin. Father Coyssard, a Jesuit born in Besse that year (1547–1623) and famous musicographer of his time, has gathered in 1615 the information about this miracle and the many others that followed. His detailed account is very precise and has been reproduced many times by the Vassivière historians (Cladière 1688; Chaix 1869; Jaloustre 1910; Pourreyron 1935; Auserve 2004, 2013):

"In June 1547 a few Besse merchants going to the LaTour market were following the path which, across the mountain, lies very close to Vassivière [i.e. near Pavin, see Fig. 1.2, Chap. 1]. *They stop in front of the venerated image* [a small statue of the Virgin, left after the destruction of a previous church, see Chap. 2] *to pay their tribute. One of them stays aside…Pierre Guelf goes on his way and reaches the streamlet which, very close from there, flows in a ravine, but* ***as he arrived there he cannot go further. He has suddenly lost his sight. His comrades gather around him, trying to rescue him, in vain*** . *Pierre understands, retraces his steps to the image of Notre Dame, and there faith fully and with repentance, he cries his misery, confesses his fault and promises to fix his mistake: immediately he is able to see again".*

The first part of this event is reported in details (date, name of person, context, location): although it is not stated, it most problably occurred in the Couze River thalweg at the Pavin outlet ravine. It is a sudden phenomenon and the victim is somewhat stunned and losses his sight. The rapidity of the phenomenon, the loss of consciousness of the victim, already mentioned here many times, his temporary loss of sight with recovery, have also been documented for the Nyos survivors (Baxter et al. 1989; see Sect. 1.6.1). In the context of a multisecular Pavin fear, it is not surprising that this recovery was immediately interpreted as a miracle operated by the Virgin, particularly within the exceptional priests' community of Besse, when considering the major and violent religious conflict between Catholics and Protestants.

Soon after the 1547 miracle a chapel is re-built on the Vassivière Mountain between 1551 and 1555, with the authorization of the Queen Mother Catherine de Medicis, and with funds collected from the whole Province of Auvergne, as other miracles have followed. This pilgrimage, dedicated to Notre-Dame and one of the major Marial celebrations in Auvergne, links Besse to the Pavin area. Twice a year, processions carrying the statue of the Virgin across the mountain, first from the town church to the mountain chapel in early July and then back at the end of September where the Virgin rests in winter. In the XVIIth century the pilgrimage reached its highest attendance, 6000 people including delegates from all nearby villages in a 20 km radius (Cladière 1688; Ajalbert 1896). The success of this ceremony is a measure of the gratitude felt by the people for the defeat of the local evil. The remaining funds permitted to build a new church choir in St André and equip the church with the magnificent stalls, soon followed by the Santa Martha chapel. The modern Christian pilgrimage, particularly celebrated in the second half of the XIXth century (Ajalbert 1896; Jaloustre 1910, Fig. 3.6b) has been in place for the last 460 years, possibly succeeding to 1500 years of local cult.

Until now, neither the Vassivière pilgrimage nor the local religious history and iconography have not been connected to Pavin, except in our preliminary paper (Meybeck 2010) and in the *"Voyages pittoresques et romantiques dans l'ancienne France"* (1829) where Nodier, Taylor and Cailleux report, although not explicitly (p. 130), how "tradition" clearly connects Pavin to the Vassivière pilgrimage:

"Some would also tell you that if she **[the Vassivière Virgin]** ***was not brought back*** **[from Besse to Vassivière],** ***one could see at once a luminous vapour surging from the earth and wrap the church, with gliding cherubs, musicians archangels and all Christian mythology spirits, who bring her back with pomp in her uninhabited land".*** This "luminous vapour", reported in 1551 in the Vassivière register, is described in other lake degassing events (Table 1.2). Today the Vassivière pilgrimmage, still very active, attracts pilgrims with multiple motivations, some of them still quoting celtic roots for this peculiar celebration (Bernard, 2013).

3.6.3 Besse Church Misericords (1570s): A Unique Set of Fairies, Dragon and Amazed Human Heads

The St. André Church in Besse holds 29 remarkable carved misericords (Fig. 3.5), *i.e.* the lower part of church stalls, dating from the late XVIth century (Blot 1924). Misericords usually *"contain a wealth of information on the lives and the beliefs of people who created them and those for whom they were designed"*, according to Elaine Block (2003) in her comprehensive analysis of French stalls. Misericords very seldom represent religious iconography, since priests sit on them, and popular and local scenes are commonly depicted.

Taking into account the Pavin legends, previously presented, and what we know of the damages suffered by victims of CO_2 intoxication (Baxter et al. 1989) and of the degassing misbehavior processes at Pavin, as analysed here and in Chap. 2, we can add to Block's "denominations" the following interpretations: impaired vision (as for the 1547 Vassivière miracle), child with buboes and skin diseases, amazement of victims, sleep of death, fairies who lure passers-by (Fig. 3.5a), amazed Medusa head, face affected by CO_2 intoxication (Fig. 3.5c, d), loss of consciousness, dragon (Fig. 3.5b) etc.

As for the capitals, these tentative re-interpretations have now to be discussed by specialists, taking into account the latent fear present in Pavin-Besse area and the Pavin misconducts during that period. Block (2003) lists 1437 French credences in which she found only 16 devils figures. None of the 1200 figures she has reproduced is similar to those of Besse which are definitely very peculiar, except for the village craftsmen

Fig. 3.5 Misericords of Saint André Church (Besse, late XVIth) with fantastic creatures, bizarre and suffering heads. This exceptional set is re-attributed to representations of Pavin misbehaviors (**a**: dragon; **b**: fairy) and of damages to Humans (**c** and **d**) (Author's photographs)

(Block, SB03 to SB06 and SH01 to SH03). She mentions another "monster figure swallowing a man" in Lautenbach credences in Alsace, where a flood-dragon has also been reported (Leggin, 2003) (see previously). The Besse Abbé Blot, in his description of the church (1924), had an inking that these stalls were uncommon, describing *"bizarre heads with closed eyelids, bulging eyes, mouth dilated by pain, scaly tongue and also a flying woman without arms, an allured flying woman"*, [the NB07, Block).

3.6.4 The Santa Martha Chapel, in Besse: Another Protection Against Pavin Disorders (1670)

The Santa Martha chapel was built in 1670 in St. André Church. There, her statue, about 1 m high, was originally standing in the middle of a dedicated wooden altar, reproduced by Blot (1924) (Fig. 3.6a). Santa Martha can be easily identified by the defeated Tarasque lying at her feet.

Santa Martha is often found in churches of cities along major rivers: she is one of the most prominent sauroctonous saints and is commonly invoked against river floods and other water disorders (Reyt 2000, 2002). As the city of Besse was built high above the thalweg of the Couze Pavin, in the headwaters area, the flood risk is limited and her presence most likely guarantees protection against the catastrophic events produced by the lake as the dragon featured in the Gloomy Lake Dragon story (Section 3.4.1) dated 1632.

3.7 Social Responses to Maar-Lakes Disorders

The social responses to lake disorders generate a second set of indicators of past maar-lakes degassing and their related latent fear. They are also very diverse and include specific constructions, still visible and/or functioning today, lake cults with specific deities and Christian communities, lake-related legends and fantastic stories featuring fantastic ani-

Fig. 3.6 Christian protection against Pavin misbehavior. (**a**) Santa Martha from Tarascon present at Besse Saint André church (1670) (Author's photograph). The Provence saint is a major sauroctonous – dragon-killer saint. (**b**) The chapel of the Vassivière pilgrimage, built in 1550–1555 after the 1547 miracle event which took place below Lake Pavin outlet, 1.5 km from there. This mountain site (1300 m elevation) may have been an antique pagan cult, possibly evoqued by Gregorius of Tours (600 AD); the fallen rocks could be the ruins of a celtic wall mentioned by Jaloustre (1910) (XIXth century engraving, Clermont BCU library)

mals, as dragons, and/or marvelous events. They complement the paleolimnological approach (e.g. Chap. 23) and the historical approach (See Chap. 2).

3.7.1 The Technical Response: Tunnels and Canals

The technical response, through tunnels or canals, generated some engineering marvels (Castellani and Dragoni 1997). In Latium, as early as 400–500 BC, repeated catastrophic lake surges and overspills during degassing were avoided by tunnels constructed across crater rims at Albano and Nemi. At Albano the tunnel excavated in 394 BC is a technical master of its time and remained in operation for more than 2300 years, maintaining the lake level at 70 m below the ridge. It can be considered as the "first prevention device" in risk management (de Benedetti et al. 2008). At Nemi a similar antique tunnel was re-used in the 1930s to lower the lake level, suggesting a similar lake overspill threat around Vth BC. The Laachersee tunnel in Eifel, dated around 1 100 AD, has a construction similar with the previous ones and serves the same function (Leidinger 2007). Other tunnels and a canal have been constructed at Averno during Antiquity, but their exact function is debated. Lake Monticchio outlet has also been connected to the nearby river by a canal excavated through the crater rim and several Eifel maars have canals (Scharf et Björk 1992). The degassing pipe installed successfully at Nyos after the 1986 event follows this remediation approach (Halbwachs et al. 2004).

3.7.2 Lake Cults and Other Religious Responses

The pre-Christian worshipping response has been important in Italy during Antiquity. Long before the birth of Rome, in the VIIth century BC, there were major sanctuaries in the Lake Nemi lake crater, dedicated to *Diana Nemorensis*, and on the rim between Nemi and Albano craters another temple was dedicated to *Jupiter Latiaris* (Carandini 1997). There, the 3 day *Feriae Latinae* was celebrated each year by all Latium cities and sacrifices were made to the master of thunders and lightning. Considering the possible occurrence of lake disorders at Nemi, the famous gigantic Nemi ships built by Emperor Caligula could also have had a cultual function. The Averno-Cumeo site, presented as an entrance to the Hades, was also an early sanctuary and place of the Sibyl divination, which lasted from the Greek colonization of Italy to Imperial Rome. On the shores of Monticchio Piccolo Lake antique ex-votos have been found in San Michele grotto. Celtic archeological artifacts have been found on the shores of some Eifel lakes and their relation to lake cult could be checked. At Pavin pre-Christian cults and offerings are mentioned by some XIXth century authors, possibly described by Gregorius and present in the Fairies Garden legend; they remain to be found by archeological exploration.

The Christian reaction is in continuity with the previous one. Sanctuaries are also established, as early as 700s AD within the Monticchio crater. At Vassivière – Pavin an early Middle –Ages sanctuary existed. The Maria- Laach abbey was founded at Laacher See in 1093, also within the crater (Leidinger 2007). Other maar-lakes with reported disorders should be checked for such early Christian sanctuaries. They are dedicated to new powerful protectors, Santa Maria and Saint Michael. The Supreme Seraphin's miraculous attributes, lightning and thunder, were inscribed in heaven at Monticchio in the VIIth century, according to the tradition, it could be associated with degassing events. The protection of populations is also ensured by Round-the-lake yearly celebrations at Monticchio until today and at Pulver Maar in the past (Middle Ages?), according to the Shepherd Legend, Mayer 2008, 2013; Projekt Gutenberg 2016) which can be compared to the bi- annual procession of the Virgin between Besse and Pavin. When celebrations are stopped the disorders start again (Pulver). The memory of the lake damages is also preserved for the next generations through the religious iconography as with the Besse church.

3.7.3 Maar-Lakes, one Possible Origin of Dragon Myths and Legends in Western World

Past maar- lake disorders are often personalized through fantastic creatures in legends and myths, as for other natural disorders, e.g. volcanic eruptions, tsunamis (Piccardi and Masse 2007; Vittoni et al. 2007). Lake disorders can combine atmospheric, aquatic and telluric processes; therefore they are represented by various dreadful and powerful creatures, each of them having a peculiar character and/or power.

Dragons, which combine these attributes, are commonly featured. They are powerful, they hit their victims hard, often silently and without any notice. In some cases lake degassing can produce atmospheric lightnings or methane ignition that could be taken for the famous dragon flammes, a point which should be further investigated. We found at least two dragons living in maar-lake (Pavin, and possibly Albano). They may present aerial or telluric features (Caluppa's dragons). Lake dragons are still featured in 1632 at Pavin (aerial, aquatic and telluric dragon). In the early Christian tradition dragons representing Satan are defeated by the first Christian evangelisator, the saurochtonous saint. The Voragine's Silvester dragon is defeated by Silvester – around 330 AD – which generates a mass conversion of the local pagan people, still worshipping their idols (*Jupiter Latiaris* cult?), to the new Christian faith. Conversion events by a sauroctonous predicator are reported at Pavin, in the Fairies Garden legend and possibly in the lake-cult story reported by Gregorius. Near 590 AD the hermit Caluppa defeats two dragons in Auvergne, according to Gregorius, a story most likely related to a major Pavin overspill.

Saint George's dragon is probably also a maar-lake dragon: according to the *Legenda Aurea* of Voragine (1265), it was living in a lake, had a poisonous breath and had many times killed every one close to him, threatening the nearby local town. To calm the furor of the beast, the people made everyday animal and human sacrifices… The daughter of the king is chosen to be sacrificed… Suddenly the monster raised its head above the lake … but Georges went up on his horse and, with the sign of the Cross, wound the monster with his lance. The young girl adopts the Christian faith, tames the dragon and carries him by her belt as a lead. The similarities with the known maar-lake dragons are striking; however the action is not fully dated or located by Voragine. This opens perspectives on the re-interpretation of some dragon-killer stories at the transition between pagan and Christian times. A similar dragon story is found in the Rhine Valley tales, featuring the conversion of pagan people to Christian faith, although its connexion with Eifel maars is not made.

During our analysis other fantastic creatures were found in descriptions and/or iconography: *Medusa* (Nemi, Pavin-Besse), *Hydra* (Pavin-Besse), *Mephytes* (Averno), fairies, griffons, centaurs and minotaurus (Pavin-Besse), giant fishes (Pavin, Ulmen). They could also be various representations of lake misbehaviors

3.7.4 Legends

Legends about lake disorders have a memorial function; they are the long term transcription of past disorders and another way for succeeding generations to remember lake misbehaviours and their catastrophic events. They are sometimes converted to Christian legends involving saurochtonous saints (Saint Silvester, Saint Georges). The oldest ones (e.g. Pavin Fairy garden), which refer to fantastic creatures as fairies and dragons, are more than fifteen centuries old. When lake misbehaviors occurred later during the Christian times, legends do not any more feature fantastic creatures but simply marvelous descriptions, as in Eifel maars (Pulver, Weinfelden-Toten).

The original *Pavin legends*, the Fairies Garden and Fairy Dance, are no loger fulfilling a memorial function, despite their terrifying descriptions, including Human victims: they are not localized at Pavin any more, are forgotten and even absent from all Auvergne guidebooks (see Chap. 2). In contrast, the *Pavin stories* fulfilled this remembrance function until the late XIXth century, when they were replaced by the Sunken City tale. This transfer of the past lake disorders to Besse people disorders has nearly erased Pavin's risk history and its awareness potential. In Eifel, many legends can be interpreted as maar-lakes misbehaviours as "Der Schâfer am pilvermaar" (Projekt Gutenberg #14), "Der versunkene Schloss" (Toten Maar) (#110) and "von Ulmener Maar" (#109). A systematic review of these Eifel and Rhine legends would probably reveal more of them.

3.8 General Conclusions

3.8.1 Perception and Representation of Pavin Misbehavior

3.8.1.1 Present Pavin Legends and Stories Can Be Disentangled

"*Pavin stories*" widely reported throughout the XIXth century actually results from observations made earlier, during periods of permanent degassing, such as the XVIth and early XVIIth centuries and, possibly, during singular degassing events. They included the *Thrown Stone* story and the *Whirl and Storm* story, still related by local population to visitors until the mid-XIXth century, at a time when there was no more permanent degassing evidence. These stories never called any fantastic creatures, divine or evil interventions, but instead presented sets of marvelous Pavin attributes that made this lake unique, at least in the XVIth and XVIIth centuries: a lake on top of a mountain, the absence of fish and birds, the permanent fog, the occasional boiling of water, the sudden storms with thunder and lightning triggered or not by a stone thrown into the lake. Some of them were refuted by Chevalier in 1770, such as the "bottomless" lake and the "lake fed by unknown waters inputs", then by scholars and authorities who started to label these stories as tales and legends, and dismantled them one by one. All ignored the rare misbehaviour event officially reported in 1783 by chatelain Godivel. This position will be endorsed by all visitors to Pavin, may they be guidebooks authors, famous geographers or encyclopaedists during the XIXth century. Absolute refutation of all Pavin stories was achieved through the introduction of fish by Lecoq in 1859 and the uncessful Thrown stone experiment, performed since the late XVIIth until the mid-XIXth centuries and presented as absolute proofs of the fallacy of all Pavin stories.

The *Sunken City*, the most commonly quoted Pavin legend today is very recent; it is absent from Sébillot's main work on French folklore (1904–1906). Prompted by the absence of fish in this "Auvergne Dead Sea" (Lecoq 1871), by the findings of a carved stone within the lake and by a historical novel featuring a former castle on the lakeshore, it was generated in the early 1900s and rapidly promoted in guidebooks as a touristic asset. In this legend the lake misconduct has been shifted to the people misconduct. Sébillot and others after him (Reyt 2000, 2002) postulated that this tale, common in many lakes, refers to the biblical myth of Sodom and Gomorrah being punished for their transgressions. Our analysis shows Sunken Cities tales may also be grounded on other natural phenomena.

The *Thrown Stone* that triggers thunder, lightning and storms at Pavin, as reported by cosmographer Belleforest in 1575, is the most quoted Pavin story during the XVIIth and XVIIIth century. When comparing with other available historical sources on Pavin, this description fits well with the state of the lake at the time, probably close to CO_2 saturation. However, Belleforest used a stereotyped story-motif that appears to be much older, used for at least two high mountain lakes, very different from maar-lakes: in Canigou (Pyrenees) by Gervaise of Tilbury (c. 1155–1234) and Pilatus in the XVth century (Swiss Alps). The occurrence of the Thrown stone story in these lakes remains to be further addressed.

The *Dragon of the Gloomy Lake* (1632), which describes, in a remarkable fantastic tone, the encounter of a passer-by with a dragon can be precisely located near the Pavin outlet ravine, as probably for the 1547 founding miracle of Vassivière. The details of this encounter, the rescue of the hero and the buboes legacy are very coherent with a sudden lake overspill and a mudflow emitted by the lake, here represented as a furious dragon, a representation of river disorders and other natural risks still wide-spread at that time. It could be a "canard", the short-stories that pre-figured newspapers.

3.8.1.2 Pavin Fairies and Dragons Are Anchored in Pre and Early Christian Times

Two original Pavin legends have been found: the *Fairies Dance* and the *Fairies Garden*. Both present Pavin and its area before the Christian era. They report its marvelous events as fantastic creatures, fairies and dragon. The *Fairies Garden*, kept in church archives (Soubrier 1893), appears to be the genuine Pavin legend. It remained so far un-analysed. It describes Pavin with great details, although with a fantastic tone appropriate to the pre-christian period: the lake crater was an extraordinary castle guarded by different sorts of fairies and a powerful dragons, each of them corresponding to various intensities of degassing processes: permanent vapours, water fountains, gas emission, lake overspills through the lake outlet of toxic gas, water and mud (a fairy followed by a dragon). Other famous Pavin attributes, as thunder and lightning, are also present as the grinding teeth and ferocious eyes of the dragon. The castle and its creatures were much feared and were worshipped each year by local people for hundred of years through offerings and human sacrifices, until the first evangelisator in this area puts an end to this pagan cult. The *Fairy Dance* legend features only fairies, some lure then enthralls the passer-by (possibly the CO_2 emission through the outlet), while others with flaming heads, can cause burns, suggesting fumaroles which could have given its name to *Montchal* volcano adjacent to Pavin (scorching mountain).

The *Caluppa dragons* encounter, dated around 580–590 AD and reported by Gregorius of Tours refers to a major overspill of Pavin, probably triggered by an earthquake and recorded in sedimentary archives (Chap. 23). The encounter location could be the Jonas cave. The first dragon corresponds to the CO_2 cloud, followed by the mud-flow (second dragon).

The *Pagan Lake Cult* text by Gregorius of Tours (600s AD) features a worshipped lake in Helanus Mountain in the Gabbales territory. There, the lake cult, performed each year by all local people over 3 days, with offerings and feasting, ends with great thunder, lightning and storm. Until now this well-known text was attributed to Lake St. Andéol, a high plateau lake in Gevaudan, some 150 km away from Pavin, not likely to have any lake disorders, but where Lake Cult has also been performed. Many arguments can be raised to re-attribute it to Pavin. If so, this text could correspond to Pavin celebrations during Antiquity, as mentioned by Father Coyssard (1615) and by historians at the end of the XIXth century (Julian 1901; Jaloustre 1910) and would be very coherent with the Fairies Garden legend.

This type of lake cult at Pavin is very similar to the *Feriae Latinae* held each year by all Latium cities, from the VIIth century BC to the IVth century AD, at the *Jupiter Latiaris* temple located at Monte Cavo, on the rim between two Albano and Nemi maar- Lakes, near Rome. There, in Latium, Jupiter's attributes are thunder and lightning, also frequently mentioned at Pavin. The connection between the Feriae Latinae, certainly known in Auvergne by the Roman builders of the *Mons Aureus* baths (now Mont Dore), with a lake cult at Pavin during the Roman period remains to be investigated. It is likely that *Lacus pavens*, the terrible lake, the ancient original name of Pavin since a long time, according to Banc (1605) and Godivel II (1650s), has acquired its reputation during the Roman period.

3.8.1.3 Pavin Fear Is Omni-Present in the Besse-Vassivière Christian Tradition

The peculiar history and iconography of the Besse-Vassivière Christian sanctuaries are probably in continuity of an early pagan cult of Pavin. At Vassivière mountain, 1.5 km from Pavin (1300 m asl elevation), there was an early Christian sanctuary, dismantled in 1321 AD by the local lord of La Tour due to its "profane" use. According to some XIXth century historians it could have been built on previous celtic settlement, although this hypothesis, not retained by contemporary historians, should be confirmed by archaeological research. If the Gregorius lake cult text is re-attributed to Pavin, this first chapel could correspond to the one built "away from the lake". During the Middle Ages few dozens of priests were present in the small city of Besse, about one thousand inhabitants, 4 km from Pavin. They were authorized to grow to sixty in 1498, by Pope Alexander VI, a possible sign of the religious importance for this area.

In June 1547 a degassing event occurring at Pavin lake outlet – in our interpretation – shocked one person blind (he recovered his sight later). At a time of regional tension between Protestants and Catholics, this was quickly considered as a miracle which permitted the construction of a chapel on Vassivière Mountain in the 1550s and the start of a famous pilgrimage, recognized by the Queen Catherine de Medicis and still attended today. From 1547 to the Revolution the ecclesiastic authorities kept a register of all "miracles" attributed to the Vassivière Virgin. Some of these miracles, highly contextualized, feature odd events near Pavin and/or damages to populations that can be linked to Pavin disorders.

The iconography of Besse Saint André Church (XIIth–XVIIth century) features many brutal, bizarre or fantastic scenes that can be re-interpreted in light of Pavin rich and violent history. Its set of misericords (1570s) is unique in France, featuring various damages to Humans, a set of suffering heads, while other present fantastic creatures, flying or terrestrial, some described previously by scholars as *Medusa* and *Hydra,Tarasque* (the dragon of aquatic disorders), all possible metaphors of Pavin disorders. The chapel dedicated to Santa Martha (1670s), the patron saint against natural aquatic disorders, completes the protection package. The close relation between Pavin, the Vassivière pilgrimage and Besse was still told by local people to Taylor and Nodier in 1823, but remained undescribed in latter Pavin descriptions.

3.8.1.4 New Insight on Pavin State and Behaviour: Antiquity and Middle Age

Legends and religious iconography add new information on Pavin state during the Antiquity and Middle Ages, and confirm the historical sources (Chap. 2). There is a convergence of independent sources indicating that *Lacus pavens* was already a dreadful place in Antiquity, possibly before the conquest of Gaul by the Romans and most certainly after. Major storm events were taking place at that time, together with some periods of quiescence: when Pompeian-type millstones were extracted within the crater. However it did not last long when considering the major catastrophic event – the Caluppa dragons' encounter- around 590 AD.

The "Profane celebrations" which took place at Vassivière, resulting in the early church being dismantled in 1321 AD by order of the local lord, could be related to some re-activation of Pagan rites following the major lake overspill -dated around 1300- and observed in paleolimnology records and from the Couze Pavin valley mudflow sediments. The analysis of Besse and Couze Pavin valley archives for the Middle Ages period remains to be realized.

The Vassivière register lists multiple events that could be related to Pavin disorders (e.g. 1547, 1551). The 1632 fantastic dragon story complements the list. They are in coherence with the historical sources describing Pavin for that period dated 1566, 1575, 1605, and mid-XVIIth century (Godivel manuscript). For more than 100 years Pavin had a continuous, although not so violent, degassing activity: daily emission of vapors, frequent thunder and lightning and occasionally (1632?) an overspill in the Couze Pavin outlet.

3.8.2 Pavin Multiple Patrimonial Value

We have already pointed out the Pavin naturalness (Meybeck 2010), its prominent position in the development of French limnology (see Chap. 1) and its specificity among all other Auvergne lakes (Chaps. 1 and 2). In comparison to the Eifel and Latium lakes, the Pavin basin is still in a subpristine condition, but it is now impacted by recent nutrient inputs. Monticchio lakes present a limited Human impact but are more eutrophic than Pavin. Other exceptional characteristics of Pavin are unveiled in this volume: maar-lake origin, very "hollow" lake, water budget dominated by rain and groundwater inputs, near-total absence of detrital material input, very simple trophic networks, meromixis, sub-lacustrine hydrothermal inputs of CO_2 and dissolved salts, inversed temperature gradient in deep waters, stable and complex chemoclines, specific Archaea and Bacteria communities, laminated diatomite etc.

The extraordinary position of Pavin-Vassivière within local Auvergne history, culture, religion, science and folklore adds another dimension to the exceptional importance of this area. This high patrimonial value at the European level, as a natural and cultural site, is not recognized so far. Today, Pavin is only mentioned in regional guidebooks for its beautiful landscape, its fish and its forged Sunken City legend. The scientific and/or cultural richness of Pavin, highly celebrated 100 years ago (Ajalbert 1896; Cany et al. 1916; Pourrat 1935), is now omitted. Pavin is not even recognized as a geological marvel: it is absent from most national geological and/or volcanological registers established in France (de Wever et al. 2006) and from international geotourism registers (Newsome and Dowling 2010). The marvelous Fairies Garden legend or the Dragon of the Gloomy Lake story should be recognized and replace the Sunken city tale. Pavin classification as a remarquable European natural and cultural site has been proposed (Meybeck 2010).

3.8.3 Ancient Descriptions of Maar-Lakes Events Should Be Considered in Present Risk Analysis

Our study registered or postulated Lake misconducts in a dozen volcanic maar-lakes, in Italy (five lakes), Germany (Eifel, four lakes) and in Auvergne (Pavin), and in Cameroun (Monoun and Nyos) *i.e.* of the order of one lake out of 10 000 in Europe. At each of these locations past lake "misbehaviors", *i.e.* a threatening deviation from the normal functioning of a still water body (Shanklin 1989), have been perceived, reported, and often memorized across ages. These disorders present a great variability of symptoms, intensity and temporality. They can be frequent or permanent, with vapour emissions such as in Averno's case and/or with small water fountains, as at Ulmen. Sometimes these disorders are sudden, powerful or violent causing great damages in the riparian and/or downstream populations, their settlements, crops or livestocks, as is noted at Albano, Pavin and Pulver. When possible, the Longue durée analysis (1000 years) shows evidence of long secular periods of quiescence between periods of activity and/or sudden burst. These past lake misconducts have been described, sometimes not explicitly, over 2000 years by many major European authors: Plutarch, Virgil, Julius Obsequens, Gregorius, Jacobus de Voragine, Belleforest, Münster etc. Most of the time they simply presented lake disorders, as Nature *prodigii* or "marvels". Sometimes they featured them as powerful fantastic creatures.

Modern scientists and historians have so far failed to collect, compare or explain these extra-ordinary behaviors which occured rarely, at different periods and in different places over 2500 years or more. Even the best volcanologists of their times, the Naples scientists, did not make the relation between lake legends and myths at Averno or Albano and the degassing they were observing at the Monticchio lakes, between 1770 and 1830. When they stopped their studies on Monticchio lakes, their observations were gradually forgotten. When the Nios event occurred in 1986, there was no scientific ground to interprete lake misbehaviors (the 1984 Monoun event analysis by Sigurdson was still in press): limnology and volcanology textbooks were totally ignorant of limnic eruptions and other degassing processes before 2000s and contemporary historians ignored, unanalyzed or untrusted the past lake stories, as for Pavin, until some of them were unearthed in the 2000s by Italian Earth scientists (De Benedetti et al. 2008).

3.8.4 The Necessary Interdisciplinary Interpretation in Maar-Lakes Study

The historical research on maar-lakes is necessary. In 2009, Caracausi, a volcanologists from INVG in Palermo working on Monticchio lakes concluded: "*Our historical investigations have allowed us more completely to assess the gas-hazard scenario of this [Monticchio] area, and we have come to the conclusion that triggering processes totally different from a landslide and/or a seismic shock should also be taken into due consideration. In fact, there have been gas bursts in both the Monticchio Lakes up to no later than 200 years ago, which indicates that they have a recurring character, even if the temporal intervals have not been constant…The possibilities of gas eruption, triggered by crustal accumulation of mantle-derived CO_2 and/or the resuming of explosive volcanic activity* *cannot be ruled out. For those reasons, in our opinion, further investigations are strongly recommended. Therefore, we hope that our results will draw attention towards all those crater lakes located in volcanic areas apparently no longer active with a special attention to those situated in areas of tourist interest, where gas risk is probably underestimated or even ignored*" (Caracausi et al. 2009).

Our research on Pavin Longue Durée evolution and its perception by local people has mixed various disciplinary approaches and we are away of its limitations, being untrained in many of them. Our analysis greatly benefitted from the intercomparison between Pavin and other maar-lakes: all of them present evidence or suspicion of past misbehavior! We also greatly benefited from the findings, in the last 3 years, of paleolimnologists (See Chap. 23), which confirmed our 2010 hypothesis on Pavin misbehavior and vice-versa (Chassiot et al. 2016).

The Pavin case study is also an illustration of the importance of geomythology in maar-lakes. The folklore historians of the XIXth and XXth centuries who collected lake legends (*e.g.* Soubrier, De Ribier, Reynouard, Pourrat in Auvergne, Sébillot in France, Thompson, Van Gennep in Europe), could not interpret them as signs of past lake disorders, as these were discovered only after 1986. Until recently (Piccardi and Masse 2007) lake scientists had no shared interest nor means communication of with folklore specialists or historians and vice-versa. An interdisciplinary approach is therefore needed. It was initiated in the hydrology field by Back (1981) and gave remarkable results for the analysis of descriptions of the marvelous coastal freshwater springs in Antique Greece (karsts inputs) (Clendenon 2009a, b, 2010) and started to be used at Nyos (Le Guern et al. 1992; Shanklin 1989, 2007). Geomythology is now well developed in Italy (*e.g.* Piccardi 2005; Funiciello et al. 2003). Such interdisciplinary approach has already been used with great success in other fields as the ethno-botanical and ethno-pharmaceutical research.

The systematic analysis of maar-lakes legends, myths, folklore, religious history and their associated fears, beliefs and representations should now be continued in Europe and also in Oceania, South America, Asia, and East Africa. It would probably bring evidence of past lake disorders in these regions and therefore of their degassing risks and processes. This needs an interdisciplinary approach and intercomparisons across time and space.

Acknowledgements Chapters 1, 2 and 3 have greatly benefitted from the support of many friends and colleagues: Florence and Jay Bautista, for the editing, Jacqueline Grimmer, for the page set-up, and Aurelien Baro for the graphics. Several colleagues have read earlier versions of this work and made precious and supportive comments that lead to many changes: Laurent Touchart (Geography, Orléans), Laurence Lestel (Environmental history, Paris 6), Marie-Françoise André (Geography, Clermont-Ferrand), Andrew Manns (Cultural History, Warburg Institute, London), Geneviève Fraisse-Mouillaud (Literature, Aix en Provence), Emmanuel Chapron (Geosciences, Toulouse), Marc Prival (Auvergne geographer) and Arnaldo De Benedetti (Geosciences, Roma Tre). Exchanges with Philippe Reyt and Aloys Mayer concerning legends of French and Eifel lakes and with Annie Dumont (Archeology, Dijon) concerning Gregorius have been essential. Luc Jaccotey (INRAP), Nicole Omaly-Lair, Jean Pelletier and Jean Devaux (Clermont-Ferrand), the late François Le Guern (Orsay) and Michel Halbwachs (Chambéry), have shared with me their experience of Pavin and Nyos. I have received precious help for the translation of some Italian and Latin texts from Catherine Siroux (CNRS, Paris) and Pierrick Tigreat (Rouen). I have been preceeded in my research on degassing lakes history and mythology by the late Evguenia Shanklin (Trenton State College, NJ), Arnaldo De Benedetti, Roberto Funiciello, Guido Giordano (Geological Sciences, Roma Tre), Antonio Caracausi (INVG, Palermo), Marco Anzidei (INVG Roma), Luigi Piccardi (CNR Geosciences, Florence) and other Italian colleagues: their works have continuously inspired me and some of them shared with me their historical references on degassing lakes. The discussions with Léo Chassiot and Emmanuel Chapron (Geosciences, Orléans) about past Pavin events in sediment archives have been decisive. I address my sincere thanks to the many librarians who helped me, from the BCIU library and the Musée Lecoq (Clermont-Ferrand) and the Faculté de Pharmacie (Paris) for their help. Some of the illustrations of these chapters also originate from their collections. My wife Agnès and many friends of Pavin, at Egliseneuve and Besse, have shown a continuous interest for our "admirable and terrifying" lake Pavin. To all of them I express my deep gratitude.

References

Ajalbert J (1896) L'Auvergne. Ancienne maison Quentin, Paris, 362pp

Assollant A (1875) Le Puy de Montchalm. Librairie de la société des gens de lettres, Paris, 370pp

Audigier P (chanoine) (1720) Histoire d'Auvergne, tome I. Projet de l'Histoire d'Auvergne. Manuscript published in 1894, Louis Bellet, Clermont Ferrand, 139 pp

Auserve PH (2004) Histoire du pèlerinage de N-D de Vassivière. Besse, Impr. Morillat, 46pp.

Auserve PH (2013) Personnal Communication

Back W (1981) Hydromythology and ethnohydrology of the new world. Water Res Res 17:257–287

Banc J (1605) La Mémoire retrouvée des merveilles des eaux naturelles en faveur de nos Nymphes Françaises et des maladies qui ont recours à leurs emplois salutaires. L Pierre Sevestre, Paris

Baxter PJ, Kapita M, Mfondu D (1989) Lake Nyos disaster, Cameroon, 1986: the medical effects of large scale emission of carbon dioxide. British Med J 298(6685):1437–1441

Beaufort de (1632) Récit véritable du monstrueux et effroyable dragon occis en une Montagne de Haute Auvergne, par Jean de La Brière, natif de Cervière en Forest, jouxte la lettre du 18 mai 1632 escripte par De Beaufort, seigneur dudit lieu, syndic de la noblesse d'Auvergne. Livre chez Colombel, 16 pp [Bibl. Sainte Geneviève, Paris, 8 Q 17 (bis) INV 743 Res (P.1)

Belleforest F de (1575) Cosmographie Universelle, 3 volumes. Nicolas Chesneau, imprimeur-libraire, Paris, vol 2, 1127 pp (Bibliothèque Sainte Geneviève, Paris, réserve, cote RXVIB 3=321-1)

Bernard A (2013) Le pèlerinage aujourd'hui: entre socialité et hagiothérapie. Quest de commu 23:57–78. doi:10.4000/questionsdecommunication.8343, www.cairn.info/revue-questions-de-communication-2013-1-page-57.htm

Bersani P, Castellani V (2005) Considerations on water flow regulation in ancient time in the Alban Hills. Geologia, Tecnica & Ambientale 1:59–90

Berthoule A (1890) Les lacs d'Auvergne. Orographie, faune naturelle, faune introduite. Revue Société Nationale d'Acclimatation. Paris, 136 pp

Berthoule A (1896) Les lacs de Besse-en-Chandesse. In: L'Auvergne. IV Congrès Int. Hydroclimatologie Géologie, Clermont-Ferrand
Block EC (2003) Corpus of Medieval Misericords, France, XIII–XIVth, Brepols, Turnhout, 444 pp
Blot A (1910) Histoire du pèlerinage de Vassivière. A. Dumont Clermont-Ferrand, 157 pp
Blot J, abbé (1924) L'Eglise de Besse. Monument historique. Guide Illustré. 63 pp
Boyer-Vidal A (1888) Besse et Vassivière. Notice historique. Imprimerie clermontoise, Clermont-Ferrand, 50 pp, appendices. Réédition ACVAM 2005, 72 pp
Caracausi A, Nuccio PM, Favara R, Nicolosi M, Paternoster M (2009) Gas hazard assessment at the Monticchio crater lakes of Mount Vulture, a volcano in Southern Italy. Terra Nova 21(2):83–87
Carandini A (1997) La nascita di Roma: dei, Lari, e roi e uomini all'alba di una civilta. Einaudi ed., Torino, 766
Castellani V, Dragoni W (1997) Ancient tunnels: from Roman outlets back to the early Greek civilization. In: Proceedings of the 12th international congress of speology, 1997, La Chaux de Fonds, Switzerland, volume 3. Symposium 3: Mines and Speology: 265–268
Chabrol (1786) Coutumes locales de la Haute et Basse Auvergne. Index des paroisses, index des seigneurs et propriétaires (tome IV), Martin Dégoutte, Paris, 858 pp
Chaix LA (abbé) (1869) Histoire de Notre-Dame de Vassivière depuis ses origines jusqu'à nos jours. Clermont-Ferrand, 359p
Chambers E (1741) Cyclopedia. An universal dictionnary of arts and sciences. Midwinter et al, London
Chapron E, Ledoux G, Simmoneau A, Alberic P, Saint-Onge G, Lajeunesse P, Boivin P, Desmet M (2012) New evidence of Holocene mass wasting events in recent volcanic lakes from the French Massif Central (lakes Pavin, Montcineyre and Chauvet) and implications for natural hazards. In: 5th International symposium on submarine mass movements and their consequences, Kyoto, Chapter 23:256–264
Chassiot L, Chapron E, Di Giovanni C, Alberic P, Lajeunesse P, Lehours A-C, Meybeck M (2016) Extreme events in maar lake Pavin sedimentary record: implication for natural hazards assessment in the French Massif Central. Quat Sci Rev 141:9–25
Cladière Jean-Joseph (1688 ou 1687) L'histoire de la Sainte Chapelle de Vassivière près du Mont-d'Or en Auvergne Célèbre par les miracles que Dieu y fait continuellement par l'intercession de sa três-pure et glorieuse Mère Par un religieux Bénédictin de la congrégation de St Maur. Damien Boujon, Clermont. Re-edited 1837, Clermont Ferrand
Clendenon CJ (2009a) Hydromythology and the ancient Greek world: an earth science perspective emphasizing karst hydrology. Fineline Sci Press, Lansing, MI, xii–xvi:305–314, 318–319, 347–345
Clendenon CJ (2009b) Ancient Greek hydromyths about the submarine transport of terrestrial freshwater through seabed offshore of karstic regions. Acta Karstologica 38(2/3):293–302
Clendenon CJ (2010) Ancient Greek geographer Pausanias as a qualitative karst hydrogeologist. Groundwater 48(3):465–470
Coyssard (1615) Histoire du pèlerinage de Notre-Dame de Vassivière
Cuvelier A (1934) A l'ombre du Mont-Pilate. In: Contes et légendes de Suisse, Fernand Nathan, pp 6–20
Daremberg C, Saglio E (1877–1919) Dictionnaire des antiquités grecques et romaines d'après les textes et les monuments contenant l'explication des termes qui se rapportent aux mœurs, aux institutions, à la religion, aux arts, aux sciences, au costume, au mobilier, à la guerre, à la marine, aux métiers, au monnaies, poids et mesures, etc. et en général à la vie publique et privée des anciens. Hachette, Paris
De Benedetti AA, Funiciello R, GiordanoG DG, Caprili E, Paterne E (2008) Volcanological history and myths of the Lake Albano maar (Colli Albani volcano, Italy). J Volcan Geoth Res 176:387–406
Defrasne J (1951) Le Lac de Saint-Point. In: Contes et Légendes de Franche-Comté, Fernand Nathan, pp 45–51
De La Pena F (1609) Cronica de Catalunya (quoted by Sebillot 1904–1906)
Delarbre A (abbé) (1805) Notice sur l'Auvergne et en particulier sur la ville de Clermont. Clermont-Ferrand. 256 pp
Del Rosso T (2009a) Interview with Marc Gachon, La Galipote (printemps-été):37–38
Del Rosso T (2009b) Les Légendes du Pavin. Oral communication. International workshop Pavin and other meromictic lakes, May 14–17, 2009, Besse
Del Rosso-d'Hers T (2010) Montchal-Pavin, mise au point sur les travaux récents (2004–2010). Eruption 23:41–43
Eusebio A, Reynouard J (1925) Le lac Pavin et le Creux de Soucy. Station limnologique de Besse, Besse, 55 pp
Fau L, Cantournet C, Crescentini D, Dieulafait C et F, Izac-Imbert L, Pradalié G (2010) Le lac de Saint-Andéol en Aubrac (Lozère): essai d'interprétation de l'ensemble cultuel. Archéologie du Midi médiéval, tome 28:3–32
Forel F-A (1892) Le Léman, monographie limnologique, vol 3. F. Rouge, Genève
Fournier PF (1975) La Haute Auvergne du temps de Grégoire de Tours. Revue de la Haute Auvergne 45:233–249
Funiciello R, Giordano G, De Rita D, Carapezza M-L, Barberi F (2002) L'Attivita recente del cratere del Lago Albano di Castelgandolfo. Rend Fis Acc Linei 9:113–143
Funiciello R, Giordano G, De Rita D (2003) The Albano maar-lake (Colli Albani Volcano, Italy): recent volcanic activity and evidence of pre-Roman Age catastrophic lahars events. J Volc Geoth Res 123:43–61
Funiciello R, Heiken G, De Benedetti A-A, Giordano G (2010) Volcanic activity of the Lake Albano maar in Roman history and mythology. In: Funiciello R, Giordano G (eds) The Colli Albano Volcano, Geological Society of London Special Publications, pp 331–308
Gervais of Tilbury (1215) *Otia Imperiala*. French translation by Jean d'Antioche and Jean de Vignay, edited by C. Pignatelli and D. Gessner. Droz, Geneva
Godivel Antoine (Godivel I) (1579a) "Terrier pour la Royne mère du Roy, de sa terre, ville et mandement de Besse, faict par Antoine Godivel, notaire royal en la dite ville". In –f°, 30 ff.lim.et 529 ff chiffrés, Manuscript, Bibliothèque du Musée Condé, Chantilly
Godivel Antoine (Godivel I) (1579b) Terrier de Besse-en-Chandesse. Manuscrit, Copie d'époque du Terrier de Catherine de Médicis. Archives Régionales, Clermont-Ferrrand, 1 E 0154
Gomis S (2006) Les « enfants prêtres des montagnes d'Auvergne, XVI–XVIII ième siècles. Presses universitaires Blaise Pascal, Clermont-Ferrand, 546pp
Graveline N (2011) En Auvergne. Loisirs, découverte, art de vivre, n° 19:16–28
Gregorius from Tours (Florentius Gergius Gregorius) (600) Histoire de France. Les sept livres des miracles. La gloire des confesseurs, VII, II, traduction Gérard Pradalié in Fau et al (2010)
Guyon D (2010) Les chapiteaux des églises du Brivadois. Jeanne d'Arc editions, Le Puy, 155 pp
Halbwachs M, Sabroux J-C, Grangeon J, Kayser G, Touchon-Danguy J-C, Felix A, Béard J-C, Villevieille A, Vitter G, Richon P, et al (2004) Degassing the "Killer Lakes" Nyos and Monoun, Cameroon. EOS 85:281–285
Hutchinson GE, Cowgill U (1970) Ianula, an account of the history and development of the lago di Monterosi, Latium. Trans Am Phil Soc 60(4):163–170
Jacobus V (1265s) The Golden Legend (*Legenda Aurea*) La Légende Dorée, liber4, #64 Santa Martha; Liber 2#88 Saint Silvester
Jaloustre E (1910) Histoire du pèlerinage de Notre-Dame de Vassivière. Impr. Dumont, Clermont-Ferrand, 150 pp
Jezequel D, Amblard C, Fonty G (eds) 2010 Le Lac Pavin. Rev Sci Nat Auvergne 74–75, 150 pp
Jullian C (1901) Vercingetorix. Re-edited, Texto, 398 pp
Kammerer O (1994) Un prodige en Alsace à la fin du XVème siècle: la météorite d'Ensisheim. Actes des congrès de la Société des historiens médiévistes de l'enseignement supérieur public 25:293–315

Kremer K, Simpson G, Girardclos S (2012) Giant Lake Geneva tsunami in AD 563. Nat Geosci 5:756–757
Laconche G (1994) Légendes et diableries du Puy-de-Dôme. Ed. Verso, Guéret, 651 pp
Lacoste de Plaisance (Abbé) (1803) Observations sur les volcans de l'Auvergne, vol 1. Clermont
Lauras-Pourrat A (1973, 1989, 1998) Guide de l'Auvergne mystérieuse. Tchou
Lavina P, Del Rosso T (2009) New findings in Pavin area. Oral com., International Symposium. Pavin and other meromictic lales (Jezequel D., Amblard C., Fonty G., conv.), Besse, May 2009
Lavallée J (1796) Voyage dans les départements de la France. n°53, Paris, 57pp
Lavroff P (1872) Sur l'adoration des lacs et des cours d'eau, et sur les légendes de villes submergées. Bull Soc Anthropologie Paris, II série 7:594–602
Lecoq H, Bouillet J-B (1831) Itinéraire du Département du Puy-de-Dôme, 178 pp
Lecoq H (1835a) Description de l'Auvergne. Le Mont d'Or et ses environs. J-B Baillère, Paris, 437 pp
Lecoq H (1835b) L'Auvergne pittoresque. Atlas colorié. Le Mont-Dore et ses environs. Seize vues coloriées b. Chaillon, Paris
Lecoq H (1871) L'Eau sur le Plateau Central de la France. J-B Baillère, 394 pp
Legin PH (2003) La Chronique des Dominicains à Guebwiller. In: La Galette (ed) L'Abbaye de Murbach en Haute Alsace, 86 pp
Legrand d'Aussy PJB (1788) Voyage en Auvergne. E. Onfroy, Paris, 555 pp
Le Guern F, Shanklin E, Tebor S (1992) Witness and accounts of the catastrophic event of August 1986 at Lake Nyos (Cameroon). J Volc Geotherm Res 51:171–184
Leidinger A (2007) Münster am See. 850 Jahre Abteikirche Maria Laach. Erbe und Auftrag 83(1):87
Lycosthenes C (1557) *Prodigiorum ac ostentorum chronicon, quae praeter naturae ordinem, et in superioribus et his inferioribus mundi regionibus, ab exordio mundi usque ad haec nostra tempora acciderunt.* Basileae per H. Petri, fol, 672 pp
Mailly (de) L (1723) Principales Merveilles de la Nature où l'on traite de la substance de la terre, de la mer, des fleuves, lacs, rivières, montagnes, rochers, etc. avec un précis des choses les plus rares et les plus curieuses qui s'y voient. Rouen, 330 pp
Mahieu V (2010) Sacrifice humains à Jupiter Latiaris. Revue Belge Philologie Histoire 88:93–98
Mayer A (2013) Personnal Communication
Mayer Al (2008) Sagenhafter Eifelsteig. Die schönsten Sagen rund um den Eifelsteig, 164 pp
Meybeck M (2010) Le Pavin et ses voisins, histoire d'un patrimoine européen unique. Rev Sci Nat Auvergne 74–75:7–44
Murat-Vaché G (1925) La Légende du lac Pavin. L'Auvergne Littéraire, Artistique et Félibréenne
Monnet AG (1788) Voyages minéralogiques faits en Auvergne dans les années 1772, 1784 et Observations sur la Physique, sur l'Histoire Naturelle et sur les Arts. Tome XXXII:179–199
Moulier P (2010) Notes de lecture sur « Légendes d'Auvergne » de Ludovic Soubrier. Rev Patrimoine en Haute Auvergne 20:49–51
Münster S (1565–1566) Cosmographie Universelle, 1430 pp (Library of faculté de pharmacie, Paris)
Nadeau L (1862) Voyage en Auvergne (département du Puy-de-Dôme), Dentu, Paris. (ré-édition: La Tour Gile)
Newsome D, Dowling RK (2010) *Geotourism*. Goodfellow pub., 246 pp
Nodier Ch, Taylor JD, De Cailleux J (1829) Voyages pittoresques et romantiques dans l'ancienne France (ré-édition 1987, Presse du Languedoc, Montpellier Tome 1, 110 pp)
Park C, Schmincke HU (1997) Lake formation and catastrophic dam burst during the Late Pleistocene Laacher See eruption (Germany). Naturwissenschaften 84(12):521–525
Pellerin (1829) La Harpie du lac de Fagna. Image imprimée, Imprimerie Pellerin, Epinal
Piccardi L (2005) Paleoseismic evidence of legendary earthquakes: the apparition of Archangel Michael at Monte Sant'Angelo (Italy). Tectonophysics 408:113–128
Piccardi L, Masse WB (eds) (2007) Myths and geology. Geol Soc London Spec Pub 273:51–59
Piesse L (1863) Guide des eaux thermales du Mont Dore. Guides Joanne, L Hachette, Paris, 220 pp
Pourrat H (1935) Au fort de l'Auvergne. Arthaud, 170 pp. Re-edited in 2002
Pourreyron C (Chanoine) (1935) Le Culte de Notre-Dame au diocèse de Clermont en Auvergne
Projekt Gutenberg (2016) Sagen aus dem Rheinland. Der Schâfer am Pulvermaar, Der versunkene Schloss, von Ulmener Maar(on line)
Rasmussen SW (2003) Public portents in Republican Rome. Analecta Romana Instituti Danici, 34, 296 pp
Reynouard J (alias Jean de la Couze) (1910a) Le Dragon de Besse.La Montagne d'Auvergne, 21 Aout 1910, pp 10
Reynouard J (1910b) La Légende du Pavin. La Montagne d'Auvergne, 28 Aug 1910:1–4
Reyt P (2000) Les Dragons de la crue. Cahiers de Géographie du Québec 44(122):127–145
Reyt P (2002) Nous et les autres: l'égalité face au risque. Cahiers de Géographie du Québec 46(127):7–24
Ribier du Chatelet (de) J-B (1856) Dictionnaire statistique ou Histoire, description et statistique du Cantal, vol IV, pp 365 (re-edited in 1964 by Assoc. Cantalienne)
Roux-Parassac E (1910) Dans la vallée de la Couze (vallées du Chambon et du Pavin). A. Vessely, Issoire
Samivel (1956) Les Mythes. In: Herzog M (ed) *La Montagne*. Larousse, 476 pp
Scharf BW et Björk S (eds) (1992) Limnology of Eifel Maar Lakes. Advances in Limnology (Archiv. für Hydrobiol.) 38, 348 pp.
Schott G (1662) Physica curiosa: sive mirabilia naturae et artis libris XII comprehensa, quibus pleraque, quae de angelis, daemonibus, hominibus, spectris, energumensis, monstris, portentis, animalibus, meteoris, etc. rara, arcana, curiosaque: circumferuntur, ad veritatis trutinam expenduntur, variis ex historia ac philosophia petitis disquisitionibus excutiuntur, & innumeris exemplis illustrantur.../Kaspar Schott. – Editio tertia juxta exemplar secundae editionis auctioris. Vol 2
Sébillot P (1904–1906) Les Eaux douces. In: Folklore de France, Ré-Ed. 1983, Imago, Paris, 304 pp
Seewer M (2013) How Pilatus came by its name. Lifa Verlag, Luzerne, 64 pp
Sergent B (1997) Saints sauroctones et fêtes celtiques, Rôles des traditions populaires dans la construction de l'Europe. Saints et dragons, Cahiers internationaux du symbolisme
Shanklin E (1989) Exploding lakes and maleficent water in Grassfield legends and myths. J Volc Geoth Res 39:233–246
Shanklin E (2007) Exploding lakes, myths and reality, an African case study. In: Piccardi L and Masse WB (eds) Myth and Geology. Geol Soc London Spec Pub 273:165–176
Soubrier L (1893 ou 1891 ??). Le Jardin des fées. In: Légendes d'Auvergne. Ill. Marel J:187–193. Réédité 1998, Editions Des Traboules
Tenèze ML (1978) Le Lac de Saint-Andéol. In Récits et Contes populaires d'Auvergne. Gallimard, 189 pp
Thevet A (1575) Cosmographie universelle. Tome II Europe
Van Gennep A (1949) Cycle de mai, Cycle de St Jean et St Pierre. In: Manuel de folklore français, Tome 1, vol 4, 2135 pp
Vittori E, Fulloni S, Piccardi L (2007) Environment and natural hazards in Roman and Medieval texts: presentation of the CLEMENS data base project. In: Piccardi L, Masse WB (eds) Myths and geology. Geol Soc London Special Public, 273:51–59
Voragine J (1265) The Golden Legend (Legenda Aurea)_La Légende Dorée, liber 4, #64 Santa Maria; liber 2,#88 Saint Silvester
Wever (de) P, Le Nechet Y, Cornée A (2006) Vade-mecum pour l'inventaire du patrimoine géologique national. Société Géologique de France, Mémoire 92, 162 pp

Part II

Origin, Volcanology and Geological Environment

Pierre Boivin

Volcanism of the Monts Dore (French Massif Central)

4

Alain Gourgaud

Abstract

Pavin volcano is located both in the southern part of Chaîne des Puys and in the Monts Dore volcanic area. According to the very young age of Pavin volcano (about 7000 y BP), Pavin volcanic area is considered to belong to the Chaîne des Puys, while the activity of the Monts Dore stratovolcano ended about 200 ka ago.

The Monts Dore composite volcano (maximum elevation: 1886 m) has a surface of 500 km^2 and a volume of about 200 km^3. It is composed of two distinct volcanic stratovolcanoes:

- The Monts Dore *s.s.* is 3.09–1.46 Ma old. Its activity began with a caldera collapse resulting from the eruption of a rhyolitic ash-and-pumice flow. In the vicinity of the caldera, domes (and associated *nuée ardente* deposits) and lava flows were emplaced. They are rhyolitic, trachytic, trachyandesitic, basaltic, tephritic and phonolitic in composition.
- The Sancy stratovolcano is younger, 1.1–0.23 Ma old. It is centred on a small caldera related to a trachytic ash-and-pumice flow eruption, formed about 720 ka ago. Predominant trachyandesites (75 % in volume) occur as domes, pyroclastic deposits, lava flows and dykes. Basaltic lava flows are located on the distal part of this volcano.

The petrology of the two volcanoes is almost identical and involves the coexistence of two types of alkaline series, which evolved towards either silica-oversaturated (rhyolitic) or silica-undersaturated (phonolitic) residual liquids. The main magmatic processes controlling this evolution are fractional crystallization, magma mixing and crustal contamination.

Pavin volcano is located within the Sancy area. Therefore, the products of the volcanic activity of Pavin may include xenoliths derived from the Sancy volcano, e.g. trachyandesite, basalt, trachyte, and scarce phonolite. According to its young age, Pavin is best regarded as a satellite of the Chaîne des Puys. But the Pavin volcano was built on the south-east flank of Sancy stratovolcano, which itself was located SE of the previous Monts Dore *s.s.* Thus, the Pavin could also be interpreted as witnessing the first activity of a future stratovolcano.

Keywords

Lake Pavin • Monts Dore • Stratovolcano • Caldera

A. Gourgaud (✉)
LMV, Université Clermont-Auvergne - CNRS - IRD - OPGC, Aubière, France
e-mail: A.Gourgaud@opgc.univ-bpclermont.fr

T. Sime-Ngando et al. (eds.), *Lake Pavin*, DOI 10.1007/978-3-319-39961-4_4

4.1 Introduction

The Monts Dore massif (maximum elevation 1886 m a.s.l. at Puy de Sancy) covers an area about 500 km^2. Its volcanic activity produced about 200 km^3 of eruptive material. This massif consists of two distinct edifices, the Monts Dore *s.s.* and the Sancy, according to Baubron and Cantagrel 1980; Cantagrel and Baubron 1983 (Fig. 4.1).

The Monts Dore *s.s.* stratovolcano is 3.09–1.46 Ma old. Its volcanic activity began by an ignimbritic eruption of rhyolitic ash-and-pumice flows, 3 Ma ago. As a result of this paroxysmic eruption, a caldera collapse event occurred, forming the "Haute Dordogne" caldera. In the vicinity of this caldera, lava flows, dykes, domes (and associated block-and-ash flow deposits) and minor pumice flows were emplaced. They are rhyolitic, trachytic, trachyandesitic, basaltic, tephritic and phonolitic in composition.

The Sancy stratovolcano is younger, 1.1–0.23 Ma old. It is centred on a smaller caldera related to the outpouring of trachytic ash-and-pumice flows, about 720 ka ago. Trachyandesites are predominant (75 % in volume) and were emplaced as domes, pyroclastites, lava flows and dykes. Basaltic lava flows mainly erupted in the distal part of the stratovolcano.

The petrology of the two distinct volcanoes is almost identical and involves the coexistence of two types of alkaline series, which evolved either to silica-oversaturated (rhyolitic) or to silica-undersaturated (phonolitic) residual liquids. The main magmatic processes controlling their evolution are fractional crystallization, magma mixing and crustal contamination.

Pavin volcano is located in the Sancy area. So the products of the volcanic activity of Pavin may include xenolithic material mainly originated from the Sancy volcano: trachyandesite, basalt, trachyte and scarce phonolite.

4.2 Monts Dore Magmas

Magmas of the Monts Dore are of alkaline intraplate type and are potassic in nature. There are two co-existing magmas series: the silica-oversaturated series (from basalt to rhyolite) and the silica-undersaturated series (basanite to phonolite). Three kinds of processes controlled the genesis of the Monts Dore lavas: fractional crystallisation, crustal contamination and magma mixing.

Evidence for fractional crystallisation is provided by the presence of cumulate enclaves in some lavas, the mineral

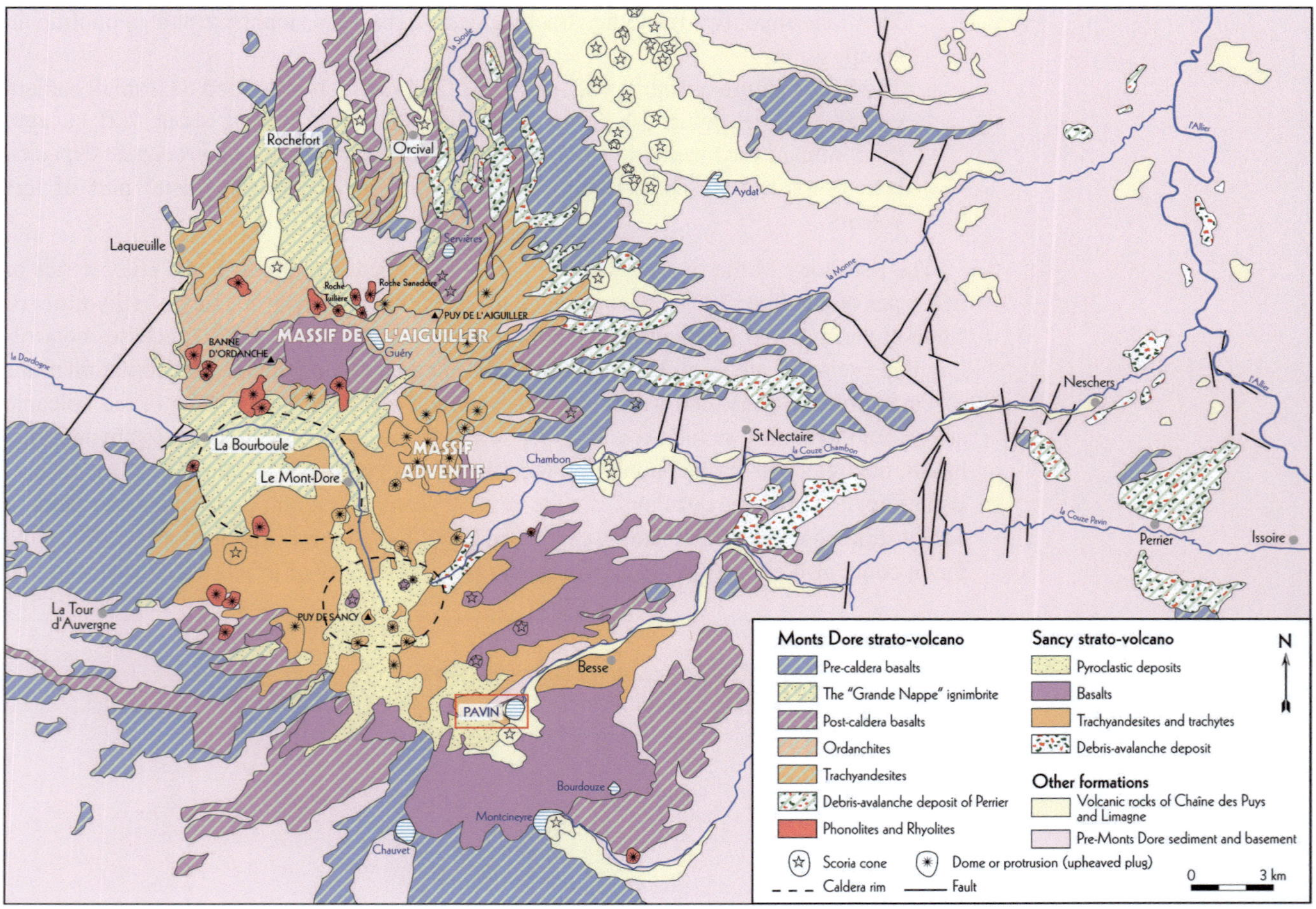

Fig. 4.1 Geological map of the Monts Dore massif (After Cantagrel and Baubron 1983, modified)

compositions (Brousse1961a, b, 1971) and the geochemistry of lavas (Maury 1976; Villemant 1979, 1985).

The role of contamination (Glangeaud 1943; Javoy 1971) is deduced mainly from geochemistry and, in particular, from isotopes (Briot 1988, 1990).

The importance of magma mixing (Glangeaud 1943; Gourgaud 1985; Gourgaud and Maury; 1984) is documented by the presence, in some lavas, of two or more magmas mechanically interstreaked together. Other evidences include the presence, in a single lava, of phenocrysts crystallized from two (or more) different magmas (e.g. basaltic and trachytic) as well as the shape of geochemical trends (e.g. linear correlation between elements with distinct behaviours).

The Monts Dore massif is dominated by lavas of intermediate composition, the so-called trachyandesites of the oversaturated and undersaturated series. These are explained by the mixing of mafic magmas such as basalts from deep levels with more differentiated trachytic or phonolitic melts from shallow levels in the volcanic plumbing system. Two discrete magma reservoirs are believed to have existed beneath the Monts Dore: a deep reservoir where basaltic magma was stored, and a shallow-level one where more differentiated magmas, such as rhyolitic, trachytic or phonolitic ones evolved by fractional crystallization coupled with crustal contamination. In some cases, mafic magmas ascending from depth were injected into the base of the shallow reservoir, causing mixing of magmas of disparate compositions and temperatures. If mixing was incomplete, the resulting lava is highly heterogeneous, with occurrence of cognate enclaves, banded and emulsified textures (Gourgaud 1985) such as observed at the Grande Cascade (Gourgaud and Villemant 1992). More complete mixing generated hybrid lavas that are macroscopically homogeneous. A majority of Monts Dore trachyandesites are believed to originate from such mixing processes.

4.3 Eruptive Styles

4.3.1 Eruption of Effusive (Flows), Extrusive (Domes) and Intrusive (Dykes) Lavas

The products of these styles may or may not exhibit columnar jointing. Sub vertical columns are characteristics of the flat-lying flows such as the Clé du Lac basalt, near lac de Guéry. Horizontal columns are characteristics of dyke margins, such as the trachyte of the Monnéron quarry near la Bourboule. The inner part of eroded phonolitic domes also exhibit large vertical columns, for example Roche Tuilière. The margins of phonolitic domes exhibit columns arranged in a fan shape, as seen at the base of Monnéron quarry.

4.3.2 Explosive Eruptions

The products of explosive eruptions include pyroclastic flows, fall and surges. Debris avalanches could also be related to explosive activity, even though the juvenile magma is not identified in their deposits.

The main types of pyroclastic flows found in the Monts Dore include ignimbrites (ash-and-pumice flows) and *nuées ardentes* (block-and-ash flows). The deposit from the largest eruption is the "Grande nappe" ignimbrite, with a wide extension and an estimated volume of about 8–10 km^3 (Brousse 1961a, b; Mossand 1983). This ignimbrite is composed of ash, fibrous rhyolitic pumice and xenolithic fragments issued from the basement. The smaller pumice flows of Rioubes Haut and Neschers are trachytic in composition and were erupted from the Sancy volcano. The nuée ardente deposits of Puy de la Tache, Gacherie and Durbise were formed by lateral explosions from lava domes.

The ash fall deposits of the so-called Guéry series (Morel 1987; Cantagrel and Briot 1990) are locally reworked by water (cinerites). They contain layers of well stratified ash and are associated to pyroclastic flows. These eruptions were emplaced between 2.19 and 2.08 Ma (Nomade et al. 2014b) and found as far as the Senèze maar sequence 60 km SE of the Monts Dore massif (Nomade et al. 2014a). Pyroclastic surges of the Guéry series were formed by hydromagmatic eruptions (Morel et al. 1992), probably in a subglacial environment.

Debris avalanche deposits were formed by the instability and sliding of the Mont Dore s.s stratovolcano (Aiguiller area). Such events have been recently dated at 2.58 Ma (Nomade et al. 2014a, b). These avalanches reached to the east as far as the Allier valley at Perrier, near Issoire. The last one continued to the north as far as 50 km. Probably, these avalanches transformed progressively downstream into mudflows (lahars). Traces of them are found among the ancient alluvial terraces of the Allier river near Vichy, as far as 60 km from Perrier. In the Sancy volcano, debris avalanche deposits occurred too, and were related to the collapse of the eastern flank of the volcano, more than 900 ka ago.

4.4 Geological History of the Monts Dore Massif

The complex history of the Monts Dore massif is related to a long-lasting activity over about 3.1 Ma, to the variety of erupted lavas (two complete alkaline series) and the diversity of eruptive styles. The main events are illustrated in Figs. 4.1 and 4.2.

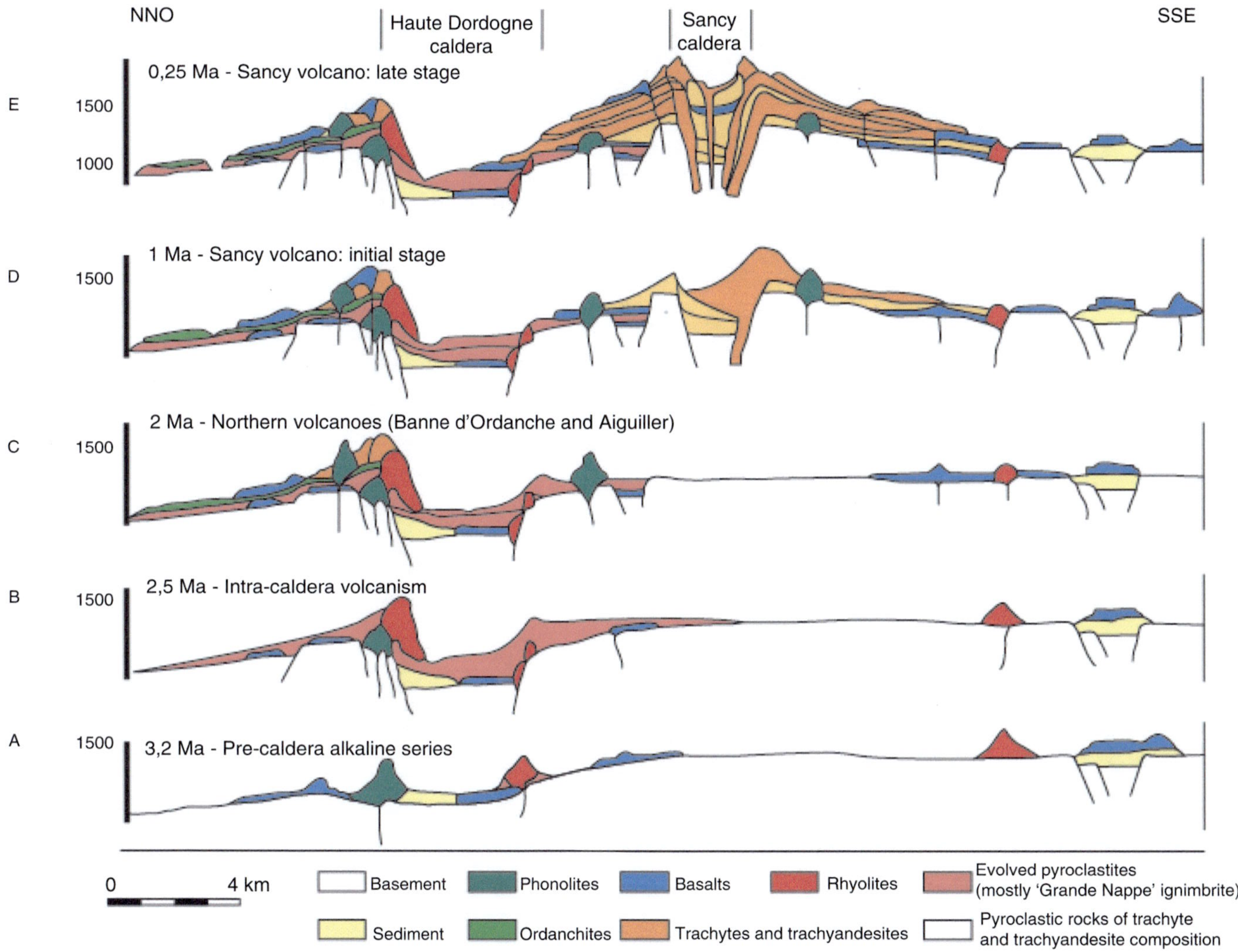

Fig. 4.2 Evolution of the growth and destruction of the Monts Dore massif (After Cantagrel and Baubron 1983, modified). Ordanchite is tephrite related to the undersaturated series. A to E corresponds to the five main cycles of activity

Despite the good geologic knowledge acquired since the 1960s (Brousse 1961a, b, 1963, 1974; Vincent 1985; Pastre and Cantagrel 2001) and developed in numerous theses from Paris-Orsay and Clermont-Ferrand universities, a complete consensus has not emerged yet (Nomade et al. 2012, 2014a, b). The debate still focuses on the tephrochronology of the numerous pyroclastic deposits (mainly pumice fall and flows), on the location and shape of the Haute-Dordogne caldera, the significance of the pumiceous linked to its collapse, and on the debris-avalanche deposits produced by two generations of flank failures in the Monts Dore (Aiguiller area) and Sancy stratovolcanoes.

The pre-volcanic basement of Monts Dore is essentially metamorphic and granitic in nature and Paleozoic (Variscian) in age, but it includes several small Permian, Oligocene and Miocene sedimentary basins. Before the Monts Dore growth, the area was already uplifted as shown by basement outcrops between 900 and 1200 m asl, below the volcanic rocks. Volcanic activity prior to the Monts Dore massif (20–3.2 Ma) emplaced Miocene alkali basaltic and basanitic lava flows similar to those of the Limagne basin as well as a few trachytes, phonolites, and rhyolites, forming a widely scattered alkaline series of Pliocene age (Fig. 4.2a).

4.4.1 The Monts Dore Stratovolcano (3.07–1.46 Ma)

The history of the Monts Dore *s.s.* stratovolcano began with the collapse of the Haute Dordogne caldera, a depression 5 km wide and 250 m deep. Now, it is no longer visible due to infilling by later activity. Distinct caldera boundaries have been delineated (Brousse 1961b, Mossand et al. 1982; Vincent 1981, 1985; Cantagrel and Briot 1990) based on geologic evidence and tectonic and geophysical data (Bayer and Cuer 1981; Nercessian et al. 1984). However, a common boundary has been drawn by all authors around two localities: La Bourboule, to the NW, at the contact between granite and pyroclastic formations (considered as hydrothermalized pumiceous infilling of the caldera), and at the foot of the Charlannes plateau, to the SW, at the contact between the pre-caldera lavas and similar pyroclastic formations.

The Haute Dordogne caldera collapse is attributed to the discharge of voluminous rhyolitic pyroclastic flows that formed the 8–10 km^3 (bulk volume) "Grande Nappe" ignimbrite (Boudon and Mossand 1984). This eruption occurred about 3.07 Ma ago. The ignimbrite, with typical rhyolitic fibrous pumices and quartz phenocrysts, has been found 30 km away from the centre of the caldera, and crops out in the quarry at Rochefort-Montagne, where its thickness exceeds 10 m, more than 10 km away from its postulated source.

The eruption of quartz-rich porphyritic pumices (essentially plinian falls), rich in quartz and termed "Roca Neyra" (3.09 Ma, Nomade et al. 2014a, b), pre-dated that of the Grande Nappe ignimbrite, although age on both formations overlap. Quartz phenocrysts are similar to those of the Grande Nappe, but the "Roca Neyra" extension is limited towards the eastern part of Monts Dore area. Both deposits of Grande Nappe and Roca Neyra may belong to the same eruptive cycle, firstly plinian, then ignimbritic, a feature which suggests an incremental caldera collapse or possibly two successive caldera-forming eruptions.

Following the establishment of a lake, the Haute Dordogne caldera depression was filled with lacustrine sediments. Then followed a period of intra-caldera volcanism involving the eruption of rhyolitic domes (and associated block-and-ash flows) of trachy-phonolitic domes, trachandesitic lava flows, dated between 2.86 ± 0.02 Ma and 2.60 ± 0.02 Ma (Nomade et al. 2014a, b). This period of intense activity ended by four large debris-avalanche (and associated lahars) that formed the actual Perrier Plateau near Issoire (Besson 1978; Ly 1982; Pastre 1987, 2004; Pastre and Cantagrel 2001). These events were followed by the emplacement (dated between 2.40 and 2.10 Ma) of large rhyolitic and phonolitic intrusions in the central part of the volcano (e.g. Puy de Chantauzet, Roche Sanadoire; Mossand et al. 1982; Cantagrel and Baubron 1983).

Major phreatomagmatic eruptions from the Aiguiller vents (Morel et al. 1992) generated the pyroclastic-flow and plinian-fall deposits of the "Guery series" (Cantagrel and Briot 1990), dated at 2.19–2.08 Ma (Nomade et al. 2014a, b).

The activity of this first strato-volcano, the Monts Dore *s.s.*, ended around 1.46 Ma ago (Fig. 4.2c).

Recent tephrochronological data (Nomade et al. 2014a, b) emphasized four cycles of explosive activity (mainly pumices), according to proximal and distal records, between the caldera collapse and rhyolitic ignimbrite (G.I.: 3.09–3.02 Ma) and the end of activity (1.46 Ma). G.II occurred between 2.86 and 2.58 Ma, and G.III, around the Guery area, between 2.36 and 1.91 Ma. G.IV corresponds to a small-volume unit dated at 1.46 Ma.

4.4.2 The Sancy Stratovolcano (1.1–0.23 Ma)

After about 400,000 years of dormancy, a new stratovolcano grew, 10 km south of the previous volcanic centre, on the southeastern flank of the Monts Dore *s.s.* stratovolcano. The Sancy covers about 16 km^2 with a present-day summit at 1886 m (Puy de Sancy). The Sancy was the most active volcano of the French Massif Central during the late Early to Middle Pleistocene periods (Pastre and Cantagrel 2001). Its activity ended around 230 ka ago according to K/Ar ages

(Cantagrel and Baubron 1983), before the beginning of the Chaîne des Puys activity. Pyroclastic deposits, including ignimbrites (pumice-and-ash flows), *nuées ardentes* (block-and-ash flows), pumice falls (Plinian deposits) and surges represent about 20–30 % of the volcanic products (Brousse 1963). Volcanic products belong to both silica-undersaturated (basanites and phonolites, which are poorly represented in Sancy), and –oversaturated (basalts to trachytes) alkaline series. About 75 % of the lavas have an intermediate composition (trachyandesites).

The volcanic activity started about 1.1 Ma ago with a sequence of pyroclastic eruptions dated between, 1.1 ± 0.01 Ma and 1.01 ± 0.01 Ma. This period of intense activity seems to be followed by debris avalanche deposits well exposed in Le Cheix quarries, in the eastern part of the volcano. These deposits are capped by basaltic lava flows dated at 900 ka. Therefore, early as 1.0 Ma, the Sancy stratovolcano was already partly edified (Fig. 4.2d).

The collapse of a small summit caldera (Fig. 4.2e) accompanied the eruption of trachytic pumiceous flows (Lavina 1985) and air falls, about 0.72 Ma ago (Féraud et al. 1990; Nomade et al. 2012). The caldera was later infilled during construction of the present-day summit. The period immediately following caldera collapse was characterized by mafic hydromagmatic activity and deposition of lacustrine sediments. Subsequent activity was predominantly trachyandesitic. It involved the emplacement of several domes with associated pyroclastic flows, the lava succession of the Grande Cascade (0.38 Ma), the basaltic trachyandesite flow of the Plateau de l'Angle (0.36 Ma) and the dykes which today crop out around the summit of Sancy and in the valley of Chaudefour. Basaltic lava flows were also emitted in distal position, such as at La Clé du Lac (0.5 Ma).

The global tephrostratigraphy of the Sancy volcano has been recently revised and updated by Nomade et al. (2012), on the basis of 13 new 40Ar/39Ar age determinations and re-interpretation of the Sancy pyroclastic deposits. Four cycles, termed C.I to C.IV, of pyroclastic activity have been determined, the duration of each being about 100 ka. The volcanic activity started with a major pyroclastic phase 1.1 Ma ago, i.e 100 ka earlier than previously thought. Following Cycle I (1.1–1.0 Ma), the main pyroclastic cycle II (818–685 ka) encompassed at least 8 large eruptions, while the most important occurred c. 719 ka ago. This major event recorded as a 1.4 m-thick layer as far away as 60 km SE of Sancy (Alleret maar) may be correlated to the formation of the small summit caldera ("C2" event of Lavina 1985; "Rioubes haut" pumice-flow deposit c. 718 ka: Féraud et al. 1990). Cycle III (642–537 ka), including the 0.58 Ma-old Neschers pumice-flow deposit (Lo Bello et al. 1987), and Cycle IV (392–280 ka) produced individual tephra deposits mostly towards the north and NE of Sancy.

4.5 Conclusions

This brief summary shows that there never existed a single Monts Dore stratovolcano, but at least two composite, long-lived, and independent edifices separated in space and in time. The Aiguiller massif, as well as the Massif Adventif, may also represent distinct eruptive centres, which were more or less independent and constructed above discrete magma reservoirs.

At the present time, only 230 ka lapsed since the last eruption of Sancy. This is less than half that of the previous period of dormancy in the region. It is therefore quite possible that a new volcano might one day be constructed. Moreover, the volcanic history of the last 100 ka has been marked by the construction of scattered scoria cones and maars in the Monts Dore area. These are probably best regarded as satellites of the Chaîne des Puys.

The Pavin was built on the south-east flank of Sancy stratovolcano, which itself was located SE of the previous Monts Dore *s.s.* Thus, the Pavin young volcanic group (including scoria cones such as Montchal and Montcineyre, and associated lava flows), located at SE of the previous Sancy volcano, could also be interpreted as witnessing the first activity of a future stratovolcano.

Acknowledgments R-C Maury and S. Nomade are thanked for reviewing this chapter.

References

Baubron JC, Cantagrel JM (1980) Les deux volcans des Monts Dore (Massif Central français): arguments chronologiques. C R Acad Sc Paris 290(D):1409–1412

Bayer R, Cuer M (1981) Inversion tri-dimensionnelle des données aéromagnétiques sur le massif volcanique du Mont-Dore; implications structurales et géothermiques. Ann Geophys 37(2):347–365

Besson JC (1978) Les formations volcaniques du versant oriental du massif des Monts Dore (Massif central français); Feuille 1/25 000 Veyre-Monton No 5–6. – Thèse 3ème cycle, Université de Clermont Fd II, 167 p

Boudon G, Mossand P (1984) Les facès d'écoulement et de déferlante: deux composants d'un dépôt cendro-ponceux montdorien (Massif Central français). Société Géologique de France (Editor), 10è RAST, Bordeaux, 78 p

Briot D (1988) Génèse d'une série alcaline intraplaque continentale. Etude géochimique (éléments traces et isotopes Sr-Nd-O) du volcan des Monts-Dore (Massif Central français). Thèse Doctorat Université Clermont-Fd: 195 p

Briot D (1990) Magma mixing versus xenocryst assimilation: the genesis of trachyandesites in Sancy volcano, Massif Central, France. Lithos 25:227–241

Brousse R (1961a) Minéralogie et pétrographie des roches volcaniques du Massif du Mont-Dore (Auvergne) I: Les minéraux; II: Pétrographie systématique. Bull Soc Fr Min Crist LXXXIV:131–186

Brousse R (1961b) Minéralogie et pétrographie des roches volcaniques du massif du Mont-Dore (Auvergne) III. Les deux associations magmatiques. Bull Soc Franç Miner Crist LXXXIV:245–259

Brousse R (1963) Identification de deux coulées de ponces dans le massif volcanique du Mont-Dore. C R Acad Sci 257:2869–2871

Brousse R (1971) Magmatologie du volcanisme néogène et quaternaire du Massif Central. Symposium Jean Jung, Plein-Air Service, Clermont-Fd: 377–478

Brousse R (1974) Le massif du Mont-Dore. Edition G. de Bussac, 167 p

Cantagrel JM et Baubron JC (1983) Chronologie des éruptions dans le massif volcanique des Monts-Dore (méthode potassium-argon). Implications volcanologiques. Géologie de la France (2), I, n 1–2:123–142

Cantagrel JM et Briot D (1990) Avalanches et coulées de débris: le volcan du Guéry; où est la caldéra d'effondrement dans le Massif des Monts Dore ? C R Acad Sci Paris 311:219–225

Feraud G, Bello P, Hall CM, Cantagrel JM, York D, Bernart M (1990) Direct dating of Plio-Quaternary pumices by 40Ar/39Ar step-beating and single-grain laser fusion methods: the example of the Monts-Dore massif (Massif Central, France). J Volcanol Geoth Res 40:39–53

Glangeaud L (1943) Evolution des magmas du massif volcanique du Mont-Dore. Essai d'interprétation géophysique. Bull Soc Geol France 3:419–440

Gourgaud A (1985) Mélanges de magmas dans les séries alcalines et calco-alcalines: leur rôle dans la génèse des laves intermédiaires et leur influence sur les mécanismes éruptifs. Thèse Doctorat d'état, Clermont-Fd, 522 p. et annales Scientifiques de l'Université de Clermont-Fd, n 86

Gourgaud A, Maury RC (1984) Magma mixing in alkaline series: an example from the Sancy Volcano (Monts Dore, French Massif Central). Bull Volcanol 47(4):827–847

Gourgaud A, Villemant B (1992) Evolution of magma mixing in an alkaline suite: the Grande Cascade sequence (Monts-Dore, French Massif Central). Geochemical modelling. J Volcanol Geoth Res 52:255–275

Javoy M (1971) Utilisation des isotopes de l'oxygène en minéralogie. Thèse, Paris, 196p

Lavina P (1985) Le volcan du Sancy et le "Massif Adventif" (Massif des Monts Dore, Massif Central Français). Etudes volcanologiques et structurales. Implications géothermiques. Thèse 3e cycle, Clermont-Fd: 197 p

Lo Bello PL, Féraud G, Hall CM, York D, Lavina P, Bernat M (1987) 40Ar/39Ar step-heating and laser fusion dating of a Quaternary pumice from Neschers, Massif Central, France: The defeat of xenocrystic contamination. Chem Geol 66:61–71

Ly MH (1982) Le plateau de Perrier et la Limagne du Sud: études volcanologiques et chronologiques des produits montdoriens. – Thèse 3ème cycle, Université de Clermont-Fd II, 180 p.

Maury RC (1976) Contamination (par l'encaissant et les enclaves) et cristallisation fractionnée de séries volcaniques alcalines, continentales (Massif Central français) et océaniques (Pacifique Central): l'origine des laves acides. Thèse d'état, Orsay, 455 p

Morel JM (1987) Volcanologie du massif de l'Aiguiller (Monts Dore, Massif central français): Etude pétrographique, dynamique, structurale et rhéologie des coulées de boue associées. – Thèse de l'Université Blaise Pascal, Clermont Fd, 199 p

Morel JM, Gourgaud A, et Vincent PM (1992) Un épisode hydromagmatique majeur dans l'histoire des Monts Dore (Massif Central Français): le massif de l'Aiguiller (Massif Central Français). Bull Soc Géol Fr 163:625–634

Mossand P (1983) Le volcanisme anté et syn-caldera des Monts-Dore (Massif Central français). Implications géothermiques. Thèse Doctorat 3e cycle, Clermont-Fd: 197 p

Mossand P, Cantagrel JM, Vincent PM (1982) La caldéra de haute Dordogne: âge et limites (Massif des Monts Dore, France). Bull Soc Geol Fr Paris XXIV(7):727–738

Nercessian A, Hirn A, Tarantola A (1984) Three-dimensional seismic transmission prospecting of the Mont Dore volcano, France. Geophys J R Astron Soc 76(2):307–315

Nomade S, Scaillet S, Pastre J-F, Nehlig P (2012) Pyroclastic chronology of the Sancy stratovolcano (Mont-Dore, French Massif Central): New high-precision 40Ar/39Ar constraints. J Volcanol Geoth Res 225–226:1–12

Nomade S, Pastre J-F, Nehlig P, Guillou H, Scao V, Scaillet S (2014a) Tephrochronology of the Mont-Dore volcanic massif (Massif Central, France): new 40Ar/39AR constraints of the late Pliocene and early Pleistocene activity. Bull Volcanol 76:798

Nomade S, Pastre JF, Guillou H, Faure M, Guérin G, Delson E, Debard E, Voinchet P, Messager E (2014b) 40Ar/39Ar constraints on some French landmark Late Pliocene to Early Pleistocene large mammalian paleofauna: paleoenvironmental and paleoecological implications. Quat Geochronol 21:2–15

Pastre J-F (1987) Les formations plio-quaternaires du bassin de l'Allier et le volcanisme régional (Massif Central, France). Rapports géodynamiques et stratigraphiques, corrélations téphrostratigraphiques, implications. PhD thesis (unpubl.), Université Paris 6, 733 p

Pastre J-F (2004) The Perrier plateau: a plio-pleistocene long fluvial record in the river Allier Basin, Massif Central, France. Quaternaire 15(1):87–101

Pastre JF, Cantagrel JM (2001) Téphrostratigraphie du Mont Dore (Massif Central, France). Quaternaire 12(4):249–267

Villemant B (1979) Etude géochimique des éléments en traces dans les séries volcaniques du Massif Central. Thèse 3e cycle, Paris: 347 p

Villemant B (1985) La différenciation des séries volcaniques: géochimie des éléments traces dans les séries du Massif Central et d'Italie Centrale. Thèse Doctorat d'état, Paris, 445p

Vincent PM (1981) Volcanisme et chambres magmatiques: l'exemple des Monts-Dore. Livre jubilaire de la Société géologique de France. Mémoire hors-série n°10, 71–85

Vincent PM (1985) Connaissance détaillée du volcan du Sancy, mise en évidence d'une éventuelle caldéra (géométrie, remplissage, âge). Etablissement d'une coupe prévisionnelle en vue d'un forage scientifique de reconnaissance. In: Brousse R, Varet J (eds) Volcanisme récent du Massif Central (Monts Dore), Documents du BRGM, Orléans, 71–84, 95–99

The Origin of Lake Pavin: Two Centuries of Controversy

5

Pierre Boivin and Sylvaine Jouhannel

Abstract

The Lake Pavin, having a circular shape, unusual size and depth, pure water and dark forest draping the slopes down to the shore, has always impressed visitors. This unusual lake garnered a fame that attracted the attention of geologists very early. After the discovery of extinct volcanoes of Auvergne by Guettard (Mém Acad Roy Sci, Paris:27–59, 1752a) in the Chaîne des Puys, naturalists and geologists quickly recognized that most of the Auvergne mountains were volcanic edifices (Desmarest, Mém Acad Roy Sci, Paris (parts 1 & 2 [published in 1774]):705–775, 1771; Dolomieu, J des Mines 7:393–420, 1797). Lake Pavin was also supposed to be of volcanic origin, but the cause of its unusual shape was controversial: according to Montlosier (Essai sur la théorie des volcans d'Auvergne. Landriot, Riom, 1788), the lake was the product of a large explosion but Legrand-d'Aussy (Voyage fait en 1787 et 1788, dans la ci-devant haute et basse Auvergne, aujourd'hui départemens du Puy-de-Dôme, du Cantal et partie de celui de la Haute-Loire. Edn. Chez le Directeur de l'Imprimerie des Sciences et arts, Rue Thérèse, près la rue Helvétius, Paris. 3 volumes 1794–1795) considered it as the result of a collapse. These two assumptions have been defended throughout the nineteenth century: Poulett Scrope (The geology and extinct volcanoes of central France., vol in-8°. Pl. and maps. 2nd edition 1858 edn. Murray, London 1827), Lecoq (Description pittoresque de l'Auvergne. Baillière, Paris, 1835) and Vimont (Annuaire du Club Alpin:337–349, 1874) advocated the explosive origin, while Lacoste (Observations sur les volcans de l'Auvergne suivies de Notes sur divers objets; recueillies dans une course minéralogique faite l'année dernière, an X. Delcros, Granier et Froin, Clermont-Ferrand, An XI, 1802–1803), Dufrénoy and Elie de Beaumont (Ann. des Mines. 3, 3, 1833) and Boule (Bull Soc Géol France XXIV:759, 1896) supported the collapse model. Other explanations have also been proposed such as a local geomorphic anomaly (Huot, Volcans in Desmarest, Nicolas: Encyclopédie méthodique. Géographie-physique, vol V. Chez Mme veuve Agasse, an III, Paris, 1828) or a glacial origin (Julien, Des phénomènes glaciaires dans le Plateau Central de la France et, en particulier, dans le Puy-de-Dôme et le Cantal. Paris, 1869; Giraud, Annales de la Station Limnologique de Besse 1:147–161, 1909). As he studied the deposits associated to the Lake Pavin, Glangeaud (C R Acad Sci Paris 162:428–430, 1916) demonstrated that explosions did produce them, an interpretation accepted by most geologists for the twentieth century and thereafter. It was not before 1973 that Camus et al. (C R Acad Sci Paris, sér D 277:629–632) reconciled the two explanations in the light of modern maar models, in which violent explosions trigger concentric collapses, which in turn enlarged the crater.

P. Boivin (✉) • S. Jouhannel
LMV, Université Clermont-Auvergne - CNRS - IRD - OPGC, Aubière, France
e-mail: P.Boivin@opgc.univ-bpclermont.fr

T. Sime-Ngando et al. (eds.), *Lake Pavin*, DOI 10.1007/978-3-319-39961-4_5

Keywords
France • Massif central • Lake Pavin • History • Volcanology • Maar

The circular shape, unusual size and depth, and pure water of the Lake Pavin as well as the dark forest draping its slopes down to the shore have always impressed visitors. As Michel Meybeck pointed out in chapter 3 (this volume), these characteristics have fuelled many legends. This garnered a great fame to the lake, which attracted the curiosity of many geologists although they defended a more scientific approach.

This article does not aim to describe the side-issue of research carried out in Auvergne. Rivalry and mutual charges of theft of discoveries have long beset relationships between researchers since the beginning of the history of science and still continues nowadays. Here we deal only with officially published works and we focus on debates related to the formation of the Lake Pavin.

The translation of the quotation was carried out by the authors, with the invaluable assistance of J.C. Thouret.

5.1 The Discovery of the Auvergne Volcanoes

Jean-Etienne Guettard, at the time of his trip in 1751, discovered that Auvergne encompassed many extinct volcanoes. Two criteria guided his hypothesis, namely the texture of rocks "*des scories qu'ils sont d'un rouge obscur ou d'un noir sale & matte/scoriae that have a dark red or dirty and mat black color*" and hill shape, such as, for example, the 'puy of La Bannière' upstream of Volvic, "*Cette montagne a la figure qui est assignée aux volcans dans les descriptions que nous en avons, elle est conique/This mountain has the shape that is commonly assigned to volcanoes in descriptions that we obtained, it is conical*" and "*Un peu avant d'arriver au sommet, on entre dans un trou large de quelques toises, d'une forme conique, & qui approche d'un entonnoir; c'est aussi le nom que l'on donne ordinairement à la bouche des volcans actuellement enflammés/Shortly before arriving at the summit, one enters a large hole a few 'toises' (a toise is 6 feet long) wide and with a conical form, it is also the name that one usually gives to the mouth of the currently ignited volcanoes*" (Guettard 1752a, p. 32). This document, published in 1752, was quickly disseminated through the community of naturalists (Ellenberger 1978). For example, Montet (based on his communication in 1754 according to Dulieu 1955) published an inventory of the volcanoes of Herault (Montet 1760). However, these authors applied the shape of conical volcano with crater that they always restricted to a landform. Desmarest in 1763, according to the author himself, introduced a new concept as he identified jointed basaltic lava flows and showed that the "*courants de laves/currents of lava*" always resulted from volcanoes (Desmarest 1771).

According to this principle and with the assistance of Pasumot and Dailley, both engineering geographers of the King at this time, he drew the first map of the volcanoes of Auvergne, which appeared in two editions. The first edition in 1771 covered the 'Chain of Puys' and the 'Monts-Dore-Sancy Mountain'. The second, posthumous edition in 1823 focused on the Puy-de-Dôme department and extended from the Comté d'Auvergne to the East up to the Sioule River Valley to the West (Figs. 5.1 and 5.2).

In all reports associated with these geological maps, Desmarest never quote the Lake Pavin. The latter was mapped as a common lake with the name of Paven, and not as a volcano because no 'courant de lave/*current of lava*' poured out from this hole in the ground. However, reports state that Desmaret spread the idea that Lake Pavin was a 'bouche volcanique/volcanic mouth' (Monnet 1787).

At the end of the eighteenth century, while observations on the eruptions of Vesuvius made the understanding of the growth of volcanic edifices possible (Hamilton 1767, 1776), the most popular and relevant question was the origin of the volcanoes in general (Taylor 2007) and in Auvergne in particular. The majority of authors, including Guettard, proposed the hypothesis involving the combustion of asphalt or coal: "*La matière nécessaire pour servir d'aliment au feu qui a brûlé, & qui peut-être brûle encore dans nos volcans* [d'Auvergne], *se présente presque d'elle-même; tous les environs de ces montagnes sont remplis d'huile de Pétrole, de charbon de terre & de bitume/The requested material to feed the fire which burned, and maybe still burns in our volcanoes* [of Auvergne], *is almost self explanatory; all the*

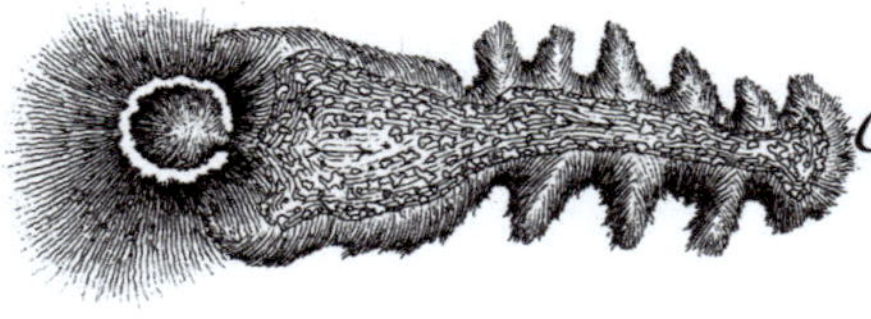

Fig. 5.1 Caption of the posthumous map of Desmarest (1823): Crater and Modern Current in a small Valley

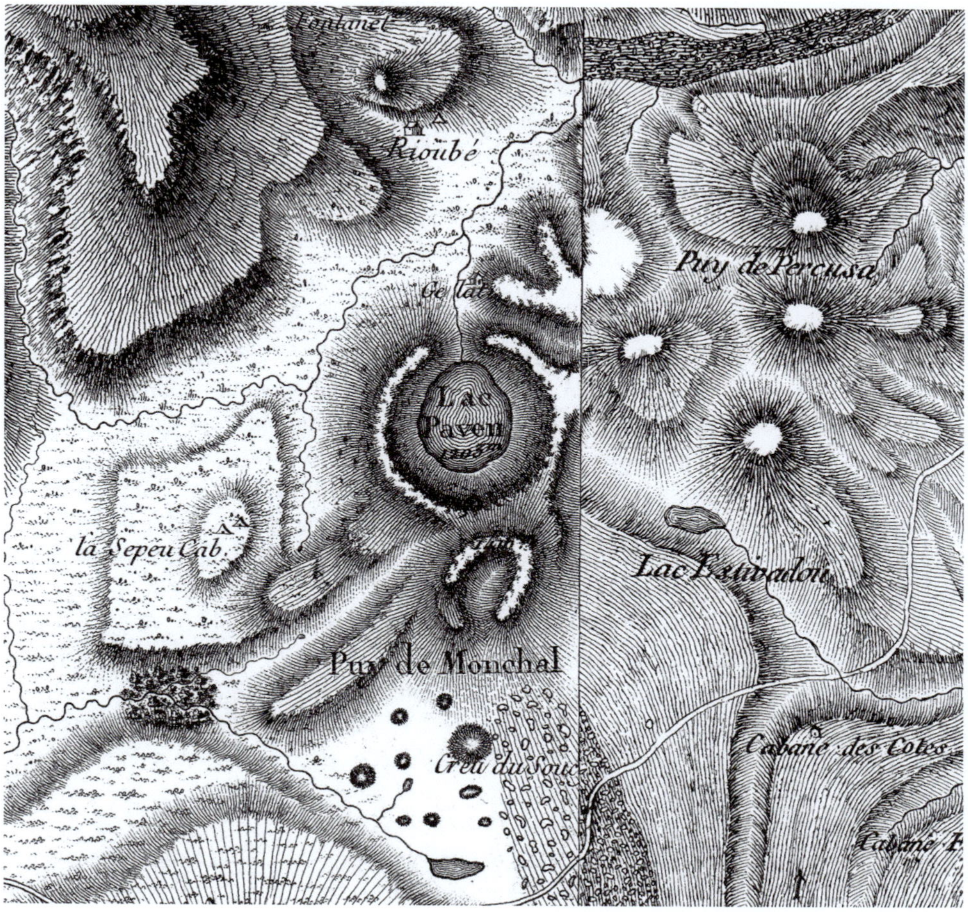

Fig. 5.2 Detail of the posthumous map after Desmarest (1823), showing the lake Paven [Pavin] not associated to any given "*courant de lave/current of lava*". This map was published in 6 sheets. The document presented here is the juxtaposition of sheets IV and V, with different drawn symbols

neighborhood of these mountains are filled with mineral oil, coal and asphalt" (Guettard 1752b, p.8). Instead, other authors, such as the chemist Montet, proposed, after Lemery in the preceding century, the reaction of iron with sulphur and water. Dolomieu, after a visit in Auvergne, noted that the volcanic edifices were aligned side by side but not piled up as at Etna or Vesuvius, and that they all laid on the granite bedrock. He therefore wrote: "*je pense que partout, c'est à de grandes profondeurs dedans ou au-dessous de l'écorce solide du globe que résident les agens volcaniques, ainsi que les bases de toutes les déjections; que là restent cachées les causes qui contribuent à l'inflammation dont sont accompagnées les irruptions, et celles qui produisent la fluidité des laves/I think that everywhere, it is at great depth inside or below the solid earth's crust that the volcanic agents lie as well as the bases of all the excretions; there the causes remain hidden to what contributes to the ignition by which the irruptions are accompanied, and also to what produces the fluidity of lava*" (Dolomieu 1797, p. 398).

At that time the prevailing paradigm of a volcano is that of a conical landform having been built and composed by red, black scoriae or pumice, adorned with a crater, and having emitted 'floods of molten lava' from the depth of the Earth. Hence, it is obvious that this concept could not apply to the Lake Pavin.

5.2 First Research Reports on Lake Pavin at the End of the Eighteenth Century

However, at the same time, the Earl of Montlosier, a famous naturalist from Auvergne (Larouzière 2003), who was facing a number of similar lakes in Auvergne: Pavin, Tazenat et Servière, stressed that their common characters did not fit the model of volcano that was otherwise shared: "*On les a jugés tout de suite, assez vaguement, des cratères : Des cratères, à la bonne heure; mais encore tous nos cratères en général sont à sec, et ceux-ci sont plein d'eau, et cette eau est d'une profondeur immense. Voilà, ce me semble, un caractère bien tranchant, et dont il eût été assez intéressant de chercher à assigner la cause. /They were considered immediately, rather vaguely, as being craters: well, craters, this would be fine, but in contrast with all our craters which are usually dry, those are full of water, and this water shows an immense depth. That, I think, is a sharp character, and he would have been quite interesting to try to assign the cause.* [...]

Ils ont tous les trois des traits de ressemblance bien frappans; mais d'abord il faut y distinguer deux parties importantes. L'abîme qui les constitue lacs, et qui est d'une profondeur qu'on a crue pendant longtemps incommensurable, et la partie évasée au-dessus du niveau de leurs eaux, qui en se repliant autour d'eux, compose ce qu'on appelle leur cratère./All three lakes exhibit striking similar features; but at first it is necessary to distinguish two important issues there. The abyss which render them lakes, and which is a depth that one had long thought was not measurable, and the widened part above the water level, which, being bent around the water, creates the so-called crater. [...]

Il faut ici surtout faire bien attention à trois choses : 1° que ces lacs ne présentent aucun indice de ces matières violemment torréfiées, telles qu'on les trouve dans les autres cratères de nos volcans; 2° qu'ils n'ont fourni aucun courant de lave; 3° (et ce caractère aurait dû être, je crois, placé le premier), que si on doit les regarder comme d'anciennes bouches de volcans, du moins le cratère, et surtout l'abîme ou la voragine, sont infiniment plus évasés, ont infiniment plus de circonférence, que n'en ont tous nos autres volcans; ce qui, en leur supposant plus d'action, contraste singulièrement avec le peu d'effet qu'ils ont produit. Or, ces trois caractères bien extraordinaires, et pourtant bien constatés, me font croire que leur théorie ne doit pas être tout à fait la même que celle des volcans ordinaires./Here we need to to pay particular attention to three issues, namely: 1° these lakes do not present any indication for the violently roasted materials as those found in other craters of our volcanoes; 2°they did not supply any current of lava; 3° (and, I believe, this characteristic should have been placed in the first place), if one must consider them as old mouths of volcanoes, at least the crater, and especially the abyss or the voragine, are infinitely widened, have infinitely more circumference that all our other volcanoes; thus, if we suppose that they had more action, these peculiar lakes singularly contrast with the little effect that they produced. However, these three extraordinary characters, yet well recognized, make me believe that their theory should not be quite the same as that of ordinary volcanoes. [...]

Voici donc ce que je pense sur l'origine de ces lacs : je suis sûr qu'ils n'appartiennent nullement à la classe des volcans ordinaires, que leur volcanisation n'a été qu'incomplète; qu'enfin l'éruption qui a découvert ces abîmes, en composant autour d'eux les vases qui leur servent comme de rempart et d'abri, n'a point été une éruption torréfiante, mais une explosion pulvérulente, causée par l'action de l'air, ou de l'eau condensée en vapeurs dans ces vastes souterrains; et alors je pense qu'un tel effet a pu avoir lieu, lorsque le vrai volcan, une fois éteint, son cratère sera tombé dans la voragine, et en aura tellement bouché l'orifice, que le reste de la force volcanique emprisonnée dans ces cavernes, aura fait un violent et dernier effort, dans le premier endroit où elle aura pu trouver une issue./Hence this is how I envisage the origin of these lakes: I am sure that they do not belong to the category of ordinary volcanoes, that their 'volcanisation' was only incomplete; that finally the eruption which unveiled these abysses, by creating around them the mud and silt which are used as rampart and shelter, was not a torrefying eruption, but a pulverulent explosion, caused by the action of the air or by condensate out of vapors in these vast underground passages; and then I think that such an effect could take place when the true volcano is extinct and its crater falls into the voragine, and will have so much tapped the opening, that the rest of the volcanic strength imprisoned into these caves, will have made a violent and last effort, in the very place where it will have been able to find an exit." (Montlosier 1788, p.159–163).

As depicted here, Montlosier founds his theory only on geomorphological criteria, the absence of volcanic products usually found in the region: "*torrefied*" scoriae and "*currents*" of lava, as well as on chronicles relating the eruptive activity of Etna, which mention the periodic collapse of its crater. On this basis, he proposed a new eruptive mechanism that included three stages:

1. Eruption of a 'common' volcano,
2. Collapse of this volcano in its chimney leading to its obturation,
3. Accumulation of pressure and explosion without molten magma: a "pulverulent explosion" creating one immense crater.

This model presents however a weakness, as the debris of this unique and gigantic explosion are neither visible, nor they have been looked for.

Here steps in Pierre-Jean-Baptiste Legrand d'Aussy, a brilliant spirit, Parisian and academician, whose three volumes of "Voyage en Auvergne" (1788, 1794–1795) were a best-seller back then. At the time of his visit in Auvergne, he naturally met Montlosier who exchanged his ideas with him. While the Earl may have shown a thorough knowledge of the field, Legrand d'Aussy, thanks to his acute sense of observation, has quickly made his own opinion. In fact, Montlosier did not appreciate this character, who nevertheless acted as a stimulus on him to the point that he decided to write his essay "Essai sur la théorie des volcans d'Auvergne" in 1788.

In front of Pavin, which he described accurately, Legrand d'Aussy alo reported, as Montlosier did beforehand, the absence of volcanic products: "*Les cratères ordinaires sont des issues qu'un volcan s'ouvre vers sa calotte pour l'écoulement ou pour l'éjection des matières qu'il pousse hors de son foyer. Celles de ces substances qui sortent liquéfiées ou fondues s'épanchent au dehors, sous la forme de fleuves, et ils produisent d'immenses traînées, que souvent on peut suivre, depuis le foyer d'où elles se sont élancées jusqu'au terme où elles cessent. Pavin n'a rien de semblable. A la vérité, des laves s'étendent sur ses rebords, et même au-delà; mais loin d'avoir comme les autres des coulées de sortie, il reçoit, au contraire, des coulées étrangères./The ordinary craters are exits that a volcano opens towards its cap for the flow or the ejection of the material that it pushes out of its hearth. Those substances that come liquefied or melted pour out in the form of rivers, and they produce huge trails that one can often follow from the hearth where they came out as far as the front where they stop. Pavin does not have anything similar. In fact, lava extends on its edges and even beyond; but far from having produced lava flows like other craters, Lake Pavin, on the contrary, received lava flows from outside.*" (Legrand d'Aussy 1794–1795, T2, p. 342–343)

But the weakness of Montlosier's model did not escape to Legrand d'Aussy's mind, who chose the hypothesis of collapse including the piracy of the lava flows of puy de Montchal, then considered as post-lake Pavin in age:

> *Quelle est donc la nature de ce cratère, si différent des autres ? Si sa vaste et profonde ouverture n'a pu avoir lieu que parce qu'il a lancé au dehors toutes les substances qu'elle contenait, qu'est devenue cette immensurable quantité de matières expulsées ? Pourquoi, au lieu d'en voir sortir des coulées de lave ne voit-on que des coulées qui s'y jettent ? Pavin, au lieu d'être un cratère véritable n'aurait-il donc été qu'un soupirail de volcan ? Ou plutôt, ne serait-il pas, ainsi que Servières qui lui ressemble en petit, un effondrement, un écroulement volcanique./What is the nature of this crater so different from others? If its vast and deep opening could take place only because it expelled outside all the substances which were contained, what was the fate of the immeasurable quantity of expelled material? Why, instead of seeing lava flows leaving the crater, one sees them only entering the lake? Instead of being a true crater, would have Pavin thus been only one window of a volcano? Or rather, would not it be, as Servières which resembles it at a smaller scale, a collapse, a volcanic collapse?* (Legrand d'Aussy 1795, T2, p 343).

Obviously, if nothing was expelled, it was necessary that it had a pre-existing empty space to receive the collapse. Legrand d'Aussy proposed a common sense solution, which prefigured pit-craters and even calderas:

> "*C'est dans l'intérieur des montagnes, et aux dépens de leur masse, que se forment ces longs et larges fleuves de lave, qui vont ensuite s'épancher dans les plaines et les vallées. Ils ne peuvent sortir des entrailles de la terre, sans y laisser, par leur éruption, des vides et des cavités énormes. Or il me semble possible, que pendant la grande durée de sa volcanisation, une montagne ainsi excavée, se soit effondrée sur elle-même. Il me paraît possible qu'en croulant dans ses propres abîmes, elle ait produit un gouffre tel que celui de Pavin; et qu'alors les montagnes voisines, devenues supérieures, y aient fait couler leurs laves et l'aient comblé en partie. / It is in the interior of the mountains, and at the expense of their mass, that these long and broad rivers of lava are formed, which then will spread out in plains and valleys. They cannot go out the bowels of the Earth without leaving, due to their eruption, some space and enormous cavities. Yet, it seems to me possible that during the long duration of its volcanisation, a mountain that was excavated this way, collapsed on itself. It appears possible to me that while collapsing in its own abysses, it produced a pit alike that of Pavin; and then the neighbouring mountains, which became higher, poured their lavas that filled part of it.*" (Legrand d'Aussy 1795, T2, p 343–344).

5.3 Multiple Hypotheses Flourised During the Nineteenth Century

These two assumptions, the first reduced to the third event described by Montlosier, i.e a non-magmatic gigantic explosion, and the second, a collapse, described by Legrand d'Aussy, will be defended throughout the nineteenth century by geologists who visited – or did not visit! – the site of Lake Pavin.

For 30 years after these pionneer publications, no other publication called into question the volcanic origin of the lake Pavin (Delarbre 1795, 1805), nor they brought about new facts. For example, Lacoste (1802–1803), to whom the logic of reasoning often preceded observations, wrote: "*Tous [*les lacs*] n'ont pas la même origine, il en est qui paraîssent avoir été formés par un écroulement du sol dans les gouffres du volcan; écroulement qui peut avoir été produit par le défaut de solidité de la base sur laquelle reposait le sol, ou par une secousse intérieure qui a ébranlé et renversé les fondemens sur lesquels il portait; secousse qui cependant n'a pas été assez forte pour produire aucune déjection au dehors et pour déranger le gisement des substances minérales supérieures. Ces secousses dans l'intérieur des volcans, capables d'entrouvrir la terre, se font sentir principalement lorsque les soupiraux ordinaires fermés, il ne reste plus aucune issue libre aux substances aëriformes, mises en expansion par la chaleur. J'attribue à cette cause la formation du lac Pavin que tout annonce n'être pas un cratère de volcan*[1]*; comme l'a fort bien remarqué M. de Montlosier : il paraît s'être ouvert sous une vaste coulée basaltique. Il est à remarquer que les laves latérale conservent leur position naturelle : ce qui ne serait pas si l'ouverture de ce lac était due à une forte explosion qui eu rejeté au dehors des matières volcaniques, comme le pense l'auteur célèbre que je viens de citer.*

All *[the lakes]* do not have the same origin, some of them appear to be formed by a collapse of the ground in the abyss of the volcano; a collapse which can be produced by the defect of solidity of the basement on which the ground lies,

[1] *Il ne laisse apercevoir aucune trace de coulée, à laquelle il ait donné naissance : le basalte qui le borde n'étant pas sa production. Cependant une immense bouche comme celle-là suppose des déjections immenses : les quantités de celles-ci sont en rapport avec la grandeur du cratère.—On ne voit point de scories sur ses bords.—Il règne tout autour une espèce de banquette dont la formation me paraît impossible à expliquer, en supposant qu'il eut été un cratère, et d'une explication au contraire facile, en admettant qu'il doit son origine à une explosion accidentelle d'un volcan : on voit que non seulement la partie de la coulée qui correspondait au gouffre qui s'ouvrait, a dû se détacher de la masse; mais encore les portions voisines.*

Fig. 5.3 "*Vue du lac Pavin prise des Rampans des Monts d'Or, près de Vassivière dans la direction du Nord Ouest à l'Est Sud. Le Cratère sur lequel sont contenues les Eaux se marie avec la masse d'un ancien volcan qui porte le nom de puy Montchal. A sa cime on retrouve encore les formes émoussées d'un cratère détruit. Le bord du Cratère qui renferme les eaux du lac ont cent vingt pieds d'élévation à partir du niveau de l'Eau qui s'enfonce jusqu'à 288 pieds au dessous. Une source très abondante coule de dessous un courant de Basalte de 47 pieds d'épaisseur. On l'aperçoit à la face intérieure du Rebord qui fait face au spectateur auprès du Puy Montchal/Lake Pavin as seen from Rampans of the Monts d'Or near Vassivière from northwest to east-southeast. The Crater in which Waters are contained does harmonize with the mass of an old volcano that bears the name of puy Montchal. On its summit one still finds the blunted shape of an eroded crater. The rim of the Crater which contains water of the lake is hundred and twenty feet in height above the water level, which sinks down to 288 feet below. A very abundant spring pours from the bottom of a basaltic current 47 feet thick. One sees it at the interior face of the Edge that faces the spectator near the Puy Montchal*". Watercolour board 36 of Delécluze's album and notes (1821). The detailed analysis of the drawing and its comparison with the digital elevation model of the present ground surface shows that the artist deliberately lowered the edge of the crater to make the lake visible

or by an interior shake which shook and reversed the basis on which it was built up; a shake which nevertheless was not strong enough to produce any excretion outside and to disturb the higher layer of the mineral substances. These shakes in the interior of the volcanoes, able to half-open the ground, are felt mainly when the usual air windows are closed, any more free exit remains for the gaseous substances, which are expanded by heat. I assign this cause to the formation of the lake Pavin that everything confirms that it is not a crater of a volcano[2]; as Mr. de Montlosier noted very well: it appears to be open under a vast basaltic lava flow. It is noteworthy that the lava side preserves their natural position, which would not be the case if the opening of this lake was due to a strong explosion, which would have expelled the volcanic material

[2]*It does not allow any evidence of lava flow, which it would have given birth, to be seen: the basalt of the edge is not its product. However such an immense mouth implies widespread excretions, the quantities of which are related to the size of the crater –as no scoriae are visible on its edges. – Some kind of bench prevails around it whose formation, either appears impossible to explain if one assumes that it had been a crater or, on the contrary, an easy explanation consists in admitting that it owes its origin to an accidental explosion of a volcano: one sees that not only the part of the lava flow, which corresponded to the pit which opened, probably detached from the mass, but also the neighbouring lava parts.*"

outside, as suggested by the famous author whom I have just quoted.

Lacoste rallied behind the hypothesis submitted by Montlosier but he did not offer any explanation nor additional argument (Lacoste 1805).

Particular mention must be made to illustrations painted by the artist named Delécluze, who was accompanied to the field by a local guide recommended by the clergyman Lacoste (Fig. 5.3). His watercolours, which aimed to be as faithful as possible to the landscape, were certainly inspired by what the geologists were able to observe on site at that time. The chronicles also reported the constant presence of knowledgeable local guides who used to lead the foreign visitors to the interesting sites (Taylor 2007). These guides were certainly the source of ideas that spread from one visitor to another. In this context, Delécluze's notes that complement his drawings can be read like the transcription of comments obtained from his guide. In this context, his following remark is surprising on behalf of a non geologist: "*Ce lac a du rapport avec plusieurs de ceux qui se trouvent dans le Royaume des Naples. Dolomieu donne la Description de celui qui est sur une des montagnes de l'Ile Pentellaria : l'analogie est frappante (Voyage aux Iles Lipari, page 144) / This lake has relationships with several lakes located in the Kingdom of Naples. Dolomieu provides the description of that lake which is located on one of the mountains of the Pentellaria Island: the analogy is striking (Voyage to the Lipari Islands, page 144)*" (Delécluze 1821, comment on watercolour board 36). In fact, this remark is better explained if it had been prompted to him by an expert.

The first author who published detailed observations included in a work of great fame is George Poulett Scrope (1827, 2nd edition in 1858). Regarding the lake Pavin, he likely accepted the hypothesis provided by Montlosier without any further questioning:

> "*One remarkable and peculiar circumstance attends these cones* [*Mont Sineire and Montchal*]; *viz. the existence of a deep, large, and nearly circular hollow immediately at the foot of each. The bottom is covered with water, and they bear the names of the Lakes Pavin and Mont Sineire; both are bordered by nearly perpendicular rocks of ancient basalt. Their position announces them to be contemporary with the eruption of the neighbouring cones; and it seems probable that, like the Gour de Tazana already described amongst the Monts Dome, they owe their formation to a series of extremely rapid and violent explosions.*" Poulett Scrope (1827, 2nd edt 1858, p 118).

However, and this was written explicitly for the first time, he proposed a chronological relationship with the neighbouring strombolian cone of Montchal: both volcanoes would be contemporary. But Scrope also referred to his description of the Gour de Tazenat that he considered as an analogue:

> " [...] *is a circular lake, called the Gour de Tazana, about half a mile in diameter, and from 30 to 40 feet deep. Its margin for a fourth of the circumference is flat, and elevated above the valley into which the lake discharges itself. Every where else it is environed by steep granitic rocks, thickly sprinkled with small scoriæ and puzzolana, and rising about 200 feet from the level of water. These fragments are all that indicate the volcanic origin of this gulf-like basin, but these are sufficiently decisive. No stream of lava or even fragments of any large size are perceivable.*
>
> *This curious and, in Auvergne, rare variety of crater is identical in characters with some of the largest and most remarkable of the volcanic maars in the Eiffel (particularly that of Meerfeld); wilh this only difference, that the former has been drilled by the volcanic explosions through granite, the latter through a superficies at least, of grauwacke slate and secondary sandstones.*" Poulett Scrope (1827, 2d edt 1858, p 74–75).

The author acknowledged the presence of a deposit of scoriae and pozzolanas scattered on the the surrounding granitic basement. Although he also mentioned the absence of lava flow, he considered that the scattered volcanic materials were enough evidence in favour of the volcanic origin of the lake. He strengthened his diagnosis by using a comparison with the maars of Eifel (in Germany). Subsequently, the link with Lake Pavin remained established only on their morphological similarity. Unfortunately, and not better than Montlosier, he did not recognize the deposits resulting from the eruption of Pavin as he was looking for scorias and pozzolanas only.

The following geologist who became interested in the lake Pavin is Jean-Jacques Nicolas Huot who drafted the article "*Volcanoes*" in Nicolas Desmaret's posthumous encyclopedia in 1828. From an updated extract of the map of N. Desmarest published in 1827, he proposed a very surprising explanation: lava flows, which were produced by the puy de Montchal, would have by-passed an unknown obstacle, which later disappeared, and the lake would then have taken the place of this obstacle:

> "*Le lac Pavin produit une illusion complète au premier abord* (see board 44 [Fig. 5.4]); *sa forme circulaire, ses contours élevés, formés de laves & disposés en entonnoir, donnent tout-à-fait l'idée d'un ancien cratère; mais l'ouverture par laquelle s'échappe au nord la petite rivière de la Couse, les pentes, par lesquelles se terminent les deux extrémités, de sa masse volcanique circulaire, de chaque côté de l'ouverture, n'offrent point les traces d'une rupture opérée par la lave, puisque le cours de la rivière n'est à la sortie de ce lac encombré par aucun reste de coulées. D'ailleurs le diamètre des parois du lac Pavin ne seroit point en rapport avec leur élévation absolue au-dessus du sol. Il faut donc, pour expliquer la formation de ces parois, admettre, comme l'inspection des lieux semble le prouver qu'un courant de lave sorti du puy de Monchal, beaucoup plus élevé que les laves du lac, s'est partagé en deux branches circulaires qui se sont terminées à l'endroit même d'où sort la petite rivière : un effet aussi singulier dans la marche d'une lave en fusion a quelque chose qui doit surprendre, mais sans doute un obstacle aujourd'hui invisible a déterminé la coulée à se partager ainsi. Quant à la profondeur du lac lui même, abstraction faite de la hauteur des parois, on peut l'expliquer en disant que la coulée qui les a formées s'est déposée sur un sol qui forme encore le fond du lac, dont la profondeur est cependant de 120 pieds; les laves se seront répandues tout autour en exhaussant le sol jusqu'à une assez grande distance. La petite rivière coule à la*

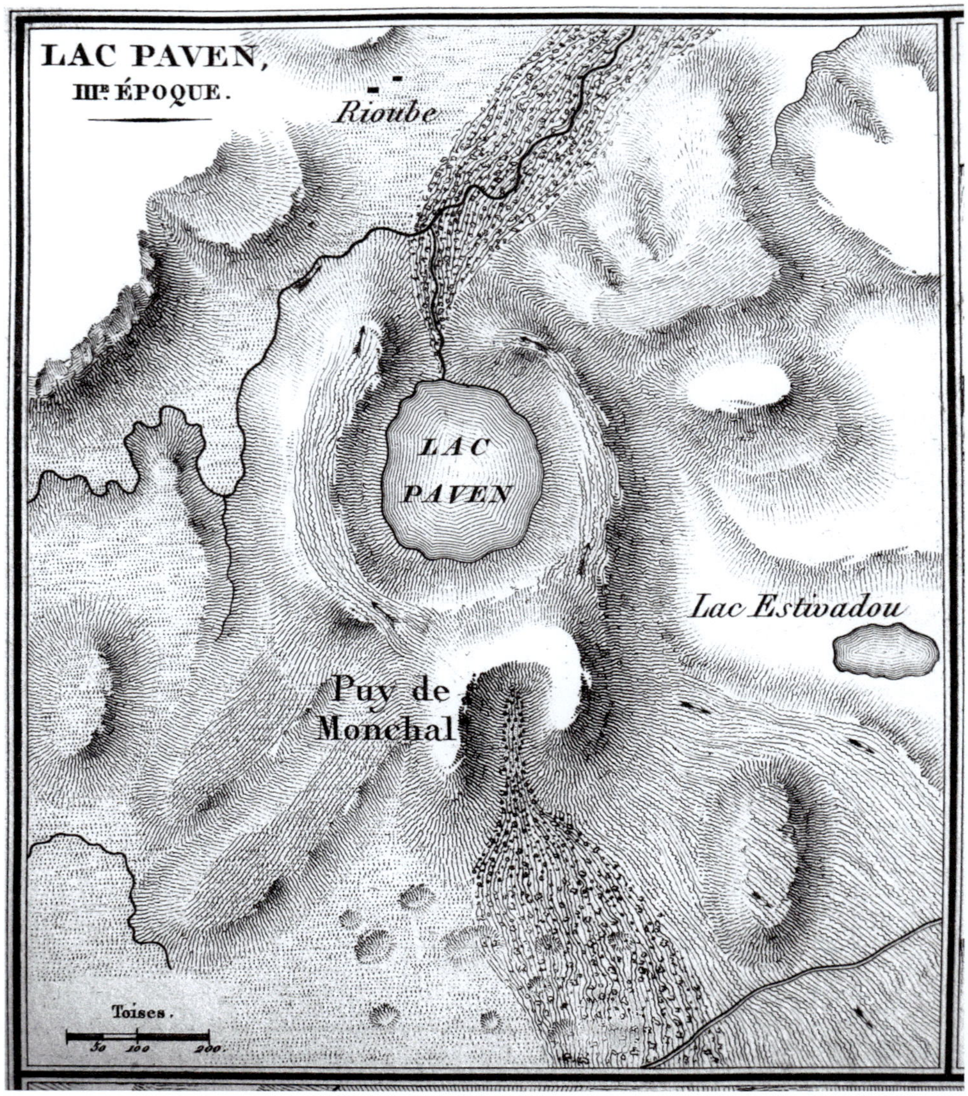

Fig. 5.4 Extract of engraving board 44 of N. Desmarest (1827) used by J.-J.E. Huot (1828) to propose a new explanation. Compared to the version as of 1823, Fig. 5.2, currents of lava that come out the lake Pavin were added as well as arrows that indicate the direction of lava flows emitted by the puy of Montchal

sortie du lac sur les restes du dépôt même de cette lave, dont les derniers jets moins-liquides se seront refroidis en formant les deux murs qui s'élèvent au-dessus des eaux. Le résultat de cette coulée aura été un trou-dans lequel toutes les sources d'alentour ont accumulé leurs eaux, & l'auront rempli jusqu'à la hauteur du point où la lave forme les deux extrémités en pente de la double coulée; car l'eau qui sort du lac n'est & ne peut être que le trop plein de son lit. / The lake Pavin produces a complete illusion at first sight (see board 44 [Fig. 5.4])*; its circular shape, its high edges formed by lavas and shaped as a funnel absolutely provide the idea of an old crater; but the opening by which the small Couze River escapes towards the north, and hillslopes by which both ends of its circular volcanic mass close on each side of the opening do not show the mark of some failure created by the lava, because the course of the river at the exit of this lake is not blocked by any remnant of lava flow. Moreover the diameter of the walls of the lake Pavin would not correspond to their absolute rise above ground level. If one wants to explain the formation of these walls, one needs to admit, as the inspection of the site seems to prove it, that a current of lava came out the volcanic hill of Monchal, much higher that the lavas of the lake, and was divided into two circular branches which ended in the place where the small river escapes: an effect so singular in the behaviour of a molten lava that it comes as a surprise, but without any doubt, an obstacle, nowadays invisible, made the lava flow to be divided this way. As for the depth of the lake itself, and we do not take the height of walls into account, we can explain it by saying that the lava flow which formed it settled on a ground which still makes the bottom of the lake, whose depth is, however, 120 feet; lavas would have spread all around by heightening the ground to a sufficiently great distance. The small river runs at the exit of the lake on the remaining deposit of this lava, whose last, less liquid streams would have cooled by forming the two walls which rise above water. The result of this lava flow would have been a hole in which all the surrounding sources accumulated their water, and would have filled it until the elevation of the point where the lava forms the two steep ends of the double lava flow; because the water which leaves the lake is and can only be the overflow above its bed.* [...]" *(Huot* 1828*, p.688).*

Shortly after, Dufresnoy, the defender of the "craters of elevation", mentioned both lakes Pavin and Laach-See, then said to be "*well known*", as having the same origin, but did not choose between the two explanations, i.e explosion or collapse. What interested him is that Pavin did not belong to his 'uplifted crater' category:

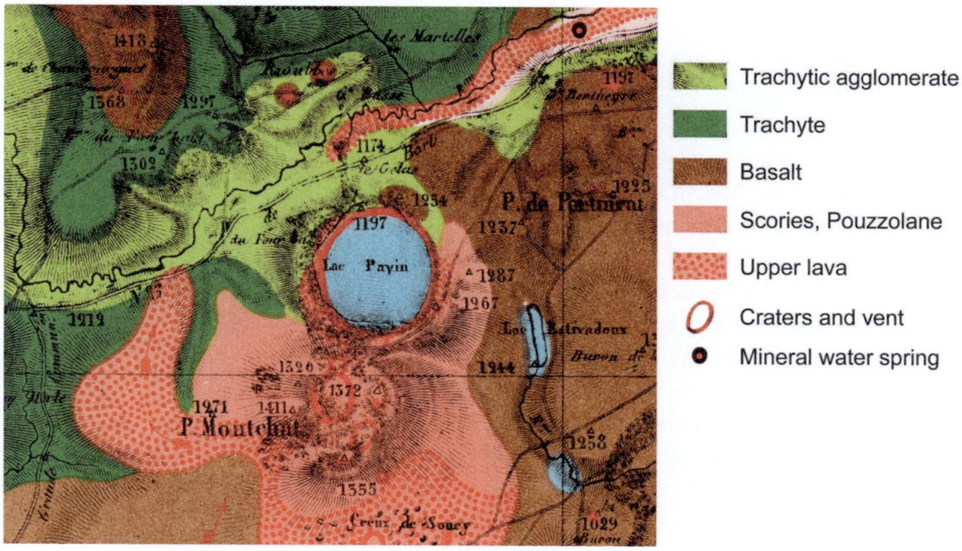

Fig. 5.5 Extract of the geological map of the department of the Puy-de-Dôme (Lecoq 1861)

"On sait que dans les régions volcaniques, indépendamment des cratères d'éruption, il existe des cavités circulaires creusées dans des roches de diverses natures, même dans des schistes et des grauwackes. On leur a donné le nom de cratères lacs; le lac Paven et le lac de Laach en sont des exemples bien connus. Ici la croûte du globe a été seulement percée sans être en même temps soulevée d'une manière sensible, et cette circonstance est ce qui distingue principalement ces cratères lacs des cratères de soulèvement proprement dits. Soit que ces cratères lacs se soient formés par explosion ou par écroulement, aucuns points de la croûte terrestre n'auraient dû être plus sujets à voir se former des cirques de ce genre que ceux où un relèvement circulaire indique que cette croûte a cédé à une action du dedans au dehors. / We know that in the volcanic regions, independently of the eruption craters, there are circular cavities excavated in rocks of diverse composition, including schists and grauwackes. We gave them the name of craters lakes; The lake Paven and the lake of Laach are well known examples. Here the crust of the globe was only broken through without being perceptibly raised at the same time, and this circumstance is what distinguishes mainly these crater lakes from the craters created by uplift. As the craters lakes may have been formed either by explosion or collapse, no spot of the earth's crust should have been more subject to see cirques of this kind to be formed that those where a circular uprise indicates that this crust broke up due to an action from inside towards outside." (Dufresnoy and Elie de Beaumont 1833, p. 555).

Henri Lecoq deserves credit for the acquisition of new data, as he paced the Auvergne region up and down for 40 years to describe all the rocks and draw the first detailed geologic map of the department of Puy de Dôme (Fig. 5.5). His descriptions, stemming from a true naturalist, encompassing rocks as well as plants from one sentence to another, do not provide a feeling for scientific rigor and would often be neglected or sometimes depreciated by his successors. Nevertheless, his observations were always accurate and especially objective as thay were not biased by prejudice.

Lecoq began by describing the stratigraphy of the banks of the lake Pavin while he admitted its volcanic origin without discussion. He justified the origin and the recent age of the visible, thick lava flow located mid slope of the crater and identified the deposits on which it rests as trachytic agglomerates whose origin was to be found in the neighbouring Monts Dore.

"On n'a jamais pu refuser à ce lac [Pavin] *le nom de cratère, malgré ses dimensions et malgré la masse d'eau qui s'y est accumulée. C'est un vaste cirque dont les bords sont abruptes quoique accessibles, et se prolongent sous l'eau avec le même degré d'inclinaison. Il paraît cependant que cette pente ne continue pas longtemps, et, d'après quelques sondages qui ont été faits, on s'est assuré que le fond du lac est presque plat, comme le serait celui d'une soucoupe, malgré la grande inclinaison de ses bords.*

De la surface au fond de l'eau, la profondeur est de 280 pieds; le diamètre du lac est dix-sept fois plus considérable; l'eau s'échappe par une échancrure pratiquée dans un des bords, son niveau doit baisser continuellement à mesure qu'elle use la digue qui la retient. En faisant le tour du lac sur le bord de l'eau, on voit facilement les sources qui l'alimentent, et qui sortent d'une coulée de lave qui paraît suspendue à plus de quarante pieds au-dessus de son niveau. La nature de la roche et les scories qui l'accompagnent font présumer, avec bien de la vraisemblance, que c'est une coulée moderne, d'autant plus qu'immédiatement au-dessus s'élève le puy de Montchalme, qui domine Pavin, et qui présente encore un cratère très bien conservé. La lave est accompagnée d'amas de pouzzolane; elle offre des traces de structure prismatique, et se montre sur plusieurs points autour du lac. Partout où l'on peut observer le terrain qui supporte cette lave, on le voit formé de tufs trachytiques, de dépôts ponceux, qui contiennent des morceaux de trachyte, et l'on distingue même souvent à travers l'eau limpide de Pavin, cette roche qui paraît stratifiée, et dans laquelle, suivant toute apparence, le cratère est creusé. / We were never able to refuse to this lake [Pavin] *the name of crater despite its size and the body of water which accumulated there. It is a vast cirque whose edges are steep, although accessible, and keep on going under the water with the same slope value. It seems, however, that this slope does not continue for a long time, and, according to some soundings which were made, we ensured that the bottom of the lake is almost flat, as one of a saucer would be despite the steep slope of its edges.*

From the surface to the bottom of the water, the depth reaches 280 feet; the diameter of the lake is seventeen times as large; the water escapes by a notch cut in one of the edges, and its level must drop continuously as it wears out the dam which retains it. Around the lake on the edge of the water, we easily see the springs feeding it and stemming from a lava flow which seems hanging more than forty feet over the water level. The nature of the rock and the scoriae which accompany it lead us to assume that it is very likely a modern lava flow, especially as the volcanic puy of Montchalme rises immediately above and dominates Pavin, and presents another very well preserved crater. The lava is accompanied with an accumulation of pozzolana; it exhibits indications for jointed structure and crops out in several sites around the lake. In every site where we can observe the deposit which supports this lava, we see that it is formed by trachytic tuffs of pumiceous deposits, which contain fragments of trachyte, and we often distinguish this rock through the crystal clear water of Pavin, which seems laminated and in which, in appearance, the crater is opened." (Lecoq 1835, p 80–82).

His detailed work prevented him from missing what escaped to the previous authors: the ejecta associated with the eruption of the lake Pavin. However he did not recognize juvenile lava clasts as he considered them as formed only by fragments of the basement:

> *"Après un examen approfondi, on ne peut donc refuser à Pavin le nom de cratère qu'il mérite sous tous les rapports; mais comme le lac de Servière, c'est un cratère d'explosion. Il a été établi au milieu de ces dépôts trachytiques et ponceux qui couvrent tous les environs du Mont-Dore.*
>
> *Jusqu'ici on n'avait trouvé aucun fait géologique qui puisse indiquer l'âge de ce cratère, et on le considérait comme antérieur aux volcans modernes. En examinant le sol des environs, et surtout la déchirure par laquelle s'échappent les eaux, on voit que le terrain est formé de débris de trachyte, de fragmens de lave, de sables ponceux, et qu'il présente enfin toutes les apparences d'un sol où se trouvent amoncelés les débris de plusieurs couches superposées. Ce sol, en effet, n'est autre chose que celui qui a été chassé du cratère lors de l'explosion qui l'a produit, et qui est retombé tout autour de la bouche. / After a thorough examination, we cannot thus refuse to Pavin the name of crater which it deserves in every respect; but like the lake of Servière, it is a crater of explosion. It was established in the middle of these trachytic and pumiceous deposits which cover all the neighborhood of the Mont-Dore volcano.*
>
> *So far we had found no geological fact which can indicate the age of this crater, and we considered that it preceded the modern volcanoes. By examining the neighbouring ground, and especially the notch through which waters escape, we see that the deposit is formed by fragments of trachyte, lava, and pumiceous sand, and that it presents finally all the appearances of a soil where the fragments of several superimposed layers were accumulated. This deposit, indeed, is not different from what was driven out by the crater during the explosion which produced it, and which fell back around the mouth."* (Lecoq 1835, p 84–85).

Putting his observations in perspective, he was therefore able to propose the first and accurate chronostratigraphy of the Montchal-Pavin system:

> *"Si l'on suit, en remontant au delà de Besse, le cours de la lave [du Montchal], elle disparaît avant qu'on ne soit à Pavin, sous les débris dont nous avons déjà parlé, et sous la pelouse qui recouvre tous les environs du lac. Cette superposition de fragmens et de tufs trachytiques au-dessus d'une lave moderne, paraîtrait inexplicable, si elle ne donnait elle-même l'époque relative de l'explosion du cratère. Ce n'est qu'après l'épanchement de la coulée de Montchalme, qu'a eu lieu la formation de Pavin. / If we follow up, by going back up beyond Besse, the course of the lava [of Montchal], it disappears before we reach Pavin under the fragments we spoke about already, and under the meadows which cover all the surroundings of the lake. This overlapping of fragments and trachytic tuffs overlying a modern lava would seem inexplicable, if it did not provide the relative time of the explosion of the crater. Hence, it is after the effusion of the lava flow of Montchalme that the formation of Pavin took place."* (Lecoq 1835, p 83–84).

Lecoq copied almost word for word these descriptions and conclusions in a summary work (Lecoq 1867). At this time, Alphonse Julien (1869) highlighted in his thesis the action of the Quaternary glaciers in the area of Sancy. From inquestionable observations made on the surficial deposits, he generalized and proposed a glacial (moraine) origin for all the trachytic agglomerates, in particular those of the valley of the Couze Pavin. Just like the volcano Tartaret which, by blocking the valley of the Couze Chambon dammed the lake of the same name, Julien imagined that Montchal blocked a glaciated valley dug out in an "*erratic ground*" and thus formed the lake Pavin, while he attributed its flat bottom due the erosion of a glacier:

> "*Leur profondeur* [lac Pavin, Montcineyre, Bourdouze] *est tout juste égale à celle du terrain erratique dans lequel ils sont creusés; ainsi Pavin, qui d'après des mesures récentes à* [sic] *90 m de profondeur. Si l'on supprimait par la pensée ce vaste manteau erratique, ces lacs disparaîtraient avec lui. Il serait intéressant d'explorer le fond du lac Pavin car des sondages ont démontré que ce fond était parfaitement uni. Il ne peut avoir été dressé de cette façon qu'à l'époque où le puissant glacier qui couvrait le plateau exerçait son action sur le sous-sol. / Their depth* [lake Pavin, Montcineyre, Bourdouze] *is just equal to that of the erratic ground in which they were excavated; hence Pavin is 90 m* [sic] *deep according to recent measurements. If we discard this vast erratic soil cover, these lakes would disappear with it. It would be interesting to explore the bottom of the lake Pavin because soundings demonstrated that this bottom was perfectly uniform. It may have been shaped this way only when the powerful glacier which covered the plateau exerted its action on the basement.*"

He observed this erratic formation in the walls of the outlet and supposed that this formation kept going all the way down to the bottom of the lake.

This poorly supported – or perhaps dogmatic – explanation will be defended again by Morin in 1888 and later by Giraud in 1909 and will spread confusion among the naturalists who were not geologists at this time (e.g. Bruyant in 1894). This will not escape the mind of those who had gained enough field experience like Vimont (1874) who, in his 1889 publication, strongly opposed Julien's hypothesis and defended the "*crater-lake*" nomenclature (Vimont 1889).

However this author, who neglected the observations made by Lecoq on Pavin's ejecta, confused them with the trachytic tuffs of the Monts-Dore and invented his own radical explanation to explain their absence:

> "*Un trait caractéristique et fréquent chez les grands cratères comme celui de Pavin est l'absence complète ou à peu près complète du bourrelet saillant, ou paroi circulaire formant relief au-dessus du sol préexistant qui constitue le cône volcanique, et qui est dû à l'accumulation des déblais produits par le creusement du cratère et rejetés circulairement tout autour par la force de projection des gaz et des vapeurs. L'absence de ces déblais est ici d'autant plus réelle qu'on ne voit nulle part aux alentours de fragments de gneiss, contrairement à ce qui devrait être, Pavin s'enfonçant, comme nous venons de le démontrer, dans cette dernière roche. Voici l'explication probable de ce fait. Les cratères de ce genre paraissent avoir été formés par une série d'explosions violentes et répétées qui ont successivement brisé les roches et les ont broyées en particules assez ténues et d'assez faible poids pour que la presque-totalité ait dû être projetée et dispersée au loin, ou même entraînée par les vents, semblables en cela à ces cendres volcaniques qui, lors de certaines éruptions, sont transportées à d'immenses distances. / A characteristic and frequent feature associated to the big craters as that of Pavin is the more or less complete absence of the prominent ridge or circular wall forming relief above pre-existing ground level which constitutes the volcanic cone, and which is due to the accumulation of the debris produced by the excavation of the crater and ejected circularly around by the strength of projection of gases and vapors. The absence of these debris is all the more real here as nowhere do we see fragments of gneiss, contrary to what should be, if Pavin were sinking, as we have just demonstrated it, in this rock. Here is the likely explanation for this fact. This kind of crater appears to have been formed by a series of violent and repeated explosions which successively broke rocks and crushed them in rather tiny particles and of rather weak weight so that it was almost entirely thrown out and dispersed away, or even entrained by winds, in a similar way to this volcanic ash which, during some eruptions, is transported at great distances.*" (Vimont 1874, p. 8).

Finally, the century ends with a note written by Boule who, using the bathymetric sections of Delebecque later published (Delebecque 1898), concluded abruptly the debate:

> "*b. – Cratères d'explosion ou d'effondrement. – Ceux-ci sont les plus profonds; leurs parois latérales sont très escarpées; le fond est constitué par un grand espace plat. L'auteur les regarde comme produits par un effondrement et il critique l'expression de cratères d'explosion qu'on leur applique ordinairement. Tels sont le lac d'Issarlès (Ardèche) dont la profondeur atteint 108 m50, le lac Pavin (92 m), le lac Chauvet (63 m), le gour de Tazanat (66 m50), dans le Puy-de-Dôme. / b. – craters of explosion or collapse. – The latter are the deepest; their side walls are very steep; its bottom is a large flat space. The author considered them as resulting from a collapse and he criticized the expression of craters of explosion which one usually applies to them. Such are the lake of Issarlès (Ardèche) the depth of which reaches 108.50 m, the lake Pavin (92 m), the lake Chauvet (63 m), the gour of Tazanat (66.50 m) in Puy-de-Dôme*". (Boule 1896, p. 759).

5.4 The Twentieth Century: More Scientific Rigor and Data

The XIXth ends as no true professional geologist studied in detail the issue of the lake Pavin and all the explanations are equally dismissed. The XXth sees Philippe Glangeaud's arrival who returned to geological mapping at scale 1:80,000 of Auvergne, in particular its volcanic areas. Thanks to its field notes, rich in maps and cross sections, we can follow the progress of his very rigorous work (Bibliothèque Clermont Université 2011). This is how we learn that at the beginning of June 1911, he visited the area of the lake Pavin for the first time while he bore in mind the interpretation of Julien and Giraud, his colleagues at the university, on the presence of glacial formations. Actually, he noticed the existence, everywhere around the lake, of a deposit containing a wide range of different rocks from the Monts-Dore and the basement for which he admitted the glacial origin back then. However, making the detailed stratigraphy of the lake edges, he rejected flatly the hypothesis of a dam and supported instead that of a crater of explosion opened "cut-die" in the preexisting formations. But he was not entirely convinced by his own conclusions so that he wrote "*this needs to be verified*"on one page of his field notes.

He returned to the site in August 1914 and in this opportunity, his observations left no doubt: the formations hitherto considered as glacial are volcanic projections (Fig. 5.6). He thus arrived at the same conclusions as Lecoq, but with more precise evidence.

These results will quickly be published (Glangeaud 1916; Fig. 6.2 in Leyrit and al, chapter 6) and will be definitely accepted. However, one mistake persisted! Philippe Glangeaud did not know how to recognize the lava at the origin of the eruption of Pavin because, just as he did at Tazenat, he was still looking for bombs, ash and projections of basaltic lava. As a first step, his cross sections show that he drew a pipe filled with basalt at the bottom of Pavin. But very quickly he would explain Pavin as the result of multiple purely gaseous explosions (which was actually the original hypothesis submitted by Montlosier!) because he did not observe magma associated with the projections (Roubille 1916).

No new study documented the eruption of the lake Pavin after 1916 to a point that the observations made by P. Glangeaud will be "forgotten" by the defenders of the "*collapse caldeiras*" (Bout and Brousse 1969; Brousse 1969). Nevertheless, the geochronological studies of both

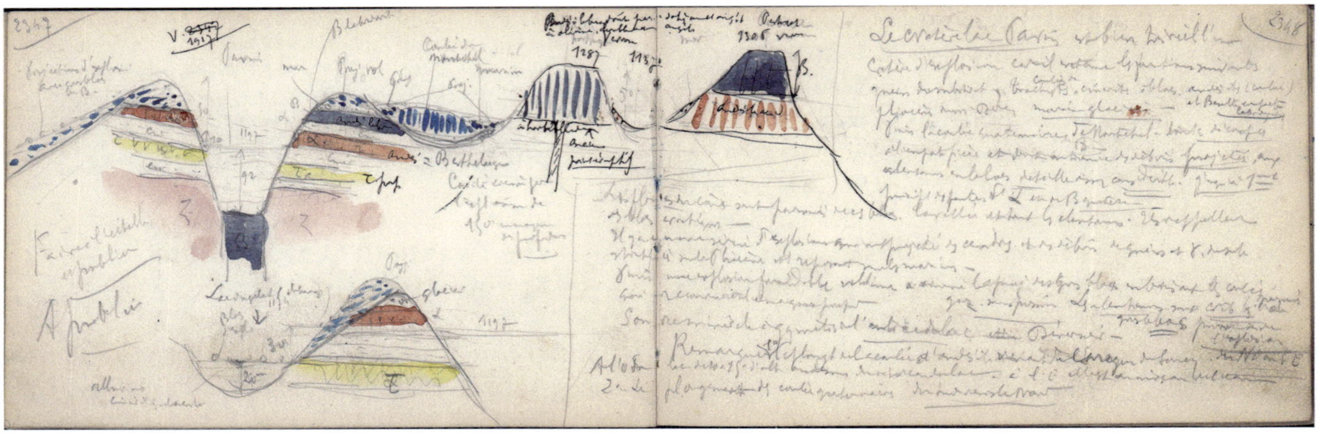

Fig. 5.6 Filed note #2300 of P. Glangeaud, pp 2347–2348 (Bibliothèque Clermont Université 2011). Transcription of the first lines: "*Le cratère-lac Pavin est bien* [illisible] *d'un cratère d'explosion car il entame les formations suivantes : gneiss du substrat, coulées de trachyte, cinérite à blocs, andésites (coulées) pliocènes monts Dore, moraines glaciaires et basalte compacte labradoritique. Puis la coulée quaternaire de Montchal. Bords découpés à l'emporte pièce et dont on trouve des débris projetés aux alentours en blocs de taille assez considérable, jusqu'à 1 m3, principalement dans les pentes N./The crater-lake Pavin is* [illegible] *of a crater of explosion because it is cut down into the following formations: gneiss of the basement, trachytic lava flow, cinerite with blocks, Pliocene andesites (lava flow) Monts-Dore, glacial moraines and massive labradoric basalt. Then the Quaternary lava flow of Montchal. Edges are die-cut and one finds their fragments thrown in the surroundings as blocks of rather considerable size up to 1 m3 mainly on north slopes*"

authors showed that a deposit of pumice coincided with the end of volcanic activity in the south of the lake Pavin (Bout 1969; Brousse et al 1969; Leyrit et al, chapter 6; Juvigné et Miallier, chapter 8). But Bout and Brousse mistakenly attributed the pumice deposit to Montcineyre.

Finally, this story ends with the work by Camus et al. (1973) who reconciled all the existing hypotheses by establishing a consistent tephrostratigraphy of the eruptions of the Pavin area, mapping the extent of the deposits of the Pavin eruption and neighbouring volcanoes, describing the petrography of the juvenile magma of the eruption ("*an amphibole-bearing leucotrachyte*"), and by showing that the magma produced either pumice or typical dense blocks owing to a dynamism affected by the arrival of superficial water. By identifying Pavin with a maar, these authors eventually agreed with Scrope and Dufresnoy. Meanwhile the model of a maar formation, which implies violent explosions to enlarge the crater by concentric collapses, was clarified and its demonstration has been long documented. The results of the most recent research on the eruption of the lake Pavin which constrain this model, are presented in the next three following chapters.

5.5 Summary

At the time of the discovery of the Auvergne volcanoes in 1751, the model which was used as the existing paradigm was a volcano with crater, built with red scoriae – '*fire print*' – and which had emitted lava flows. Pavin, with its large lake occupying a deep circular depression and without red ejecta or lava flow, would not be considered as a volcano – in the absence of any better hypothesis – until 20–30 years later. As soon as this basic premise was accepted, the mechanism of its formation was immediately the object of controversies between those who advocated for gaseous explosions without deposit, and those who defended a large-scale collapse. Very quickly, the link with the morphology of the maars of Eifel was established, but this did not offer the solution to the problem because the formation of the German maars was not better understood at this time.

No observation allowed for militating in favor of one hypothesis or another until 1835, the year when deposits of heterogeneous breccia were described. These deposits were interpreted as eruptive deposits but in the absence of petrographic analysis no juvenile magma was underlined. Shortly after, the findings of the action of the glaciers in the region raised a new hypothesis, namely the damming of glacial valleys by lava flows. Breccias that were previously described were subsequently likened to moraines. The systematic geologic mapping of France at the scale 1:80,000 with its rigorous methodology allowed geologists to discard the dammed glacial valley hypothesis and retain only that of an explosive activity with associated ejecta. But at that time nobody knew as yet how to describe and highlight the juvenile magma in this deposit that reworked the old formations with pumice and trachytes of Monts-Dore. After a brief, rather confidential return to the collapse hypothesis coinciding with the theory of the "collapse calderas", the general framework of the modern interpretation was published in 1973 with the identification of juvenile magma in the deposits, mapping of their extent and pinpointing of their position in the local tephrochronology. The juvenile magma, which appears either as pumice or dense blocks, suggested the intervention of water at the time of the eruption. This allowed the genetic maar model to be used for the Pavin case, which state that maars are created by phreatomagmatic explosions and accompanied with circular collapses enlarging their crater.

Acknowledgements The authors thank Jean Gaudant who efficiently reviewed the manuscript as well as Hervé Leyrit for his constructive remarks. We are very grateful to Jean-Claude Thouret for his relevant corrections of the English text.

References

Bibliothèque Clermont Université (2011) Carnets Géologiques de Philippe Glangeaud. http://bibliotheque.clermont-universite.fr/glangeaud/

Boule M (1896) Sur l'origine géologique des lacs de l'Auvergne et du Velay. Bulletin de la Société Géologique de France XXIV:759

Bout P (1969) Problèmes du volcanisme. XI-Scories et ponces dans la Chaîne des Puys. Revue d'Auvergne 83(1):39–67

Bout P, Brousse R (1969) Livret-guide de l'excursion C 13, Auvergne-Velay. 8th congress of the International Union for Quaternary Research. International Union for Quaternary Research Publisher, 87

Brousse R (1969) L'âge des lacs du Sud du Mont Dore. Annales de la Station Biologique de Besse en Chandesse 4:201–207

Brousse R, Delibrias G, Labeyrie J, Rudel A (1969) Eléments de chronologie des éruptions de la Chaîne des Puys. Bull Soc Geol Fr 7(XI):770–793

Camus G, Goër de Herve A, Kieffer G, Mergoil J, Vincent PM (1973) Mise au point sur le dynamisme et la chronologie des volcans holocènes de la région de Besse-en-Chandesse (Massif Central français). C R Acad Sci Paris, sér D 277:629–632

Delarbre A (1795) Flore d'Auvergne, ou recueil des plantes de cette ci-devant province; suivi de la description du lac de Pavin. B Beauvert & L Deschamps Clermont-Ferrand

Delarbre A (1805) Notice sur l'ancien royaume des Auvergnats et sur la ville de Clermont. Landriot, Clermont

Delebecque A (1898) Les Lacs Français, vol 1 in-4°. Chamerot et Renouard, Paris

Delécluze EJ (1821) Voyage en Auvergne. Album de 72 dessins et Explications des dessins faits en Auvergne en 1821. Ville de Clermont-Ferrand, musée d'art Roger-Quilliot, inv. 2010.9.1 et 2010.9.2

Desmarest N (1771) Mémoire sur l'origine et la nature du basalte à grandes colonnes polygones, déterminées par l'histoire naturelle de cette pierre observée en Auvergne. Mém Acad Roy Sci, Paris (parts 1 & 2 [published in 1774]):705–775

Desmarest N (1823) Carte topographique et minéralogique d'une partie du département du Puy-de-Dôme dans la ci-devant province d'Auvergne où sont déterminées la marche et les limites des matières fondues et rejetées par les volcans ainsi que les courants anciens et modernes pour servir aux recherches sur l'histoire naturelle des volcans. Desmarest Fils edn. Charles PICQUET, Géographe ordinaire du Roi et de S.A. S. Mgr le Duc d'Orléans, Quai de Conti, n 17, entre l'Hôtel des Monnaies et le Pont des Arts. Engraved by Guill.de la Haye and Perrier. 6 sheets and table of assembly, Paris

Desmarest N, Bory de Saint-Vincent C (1827) Lacs volcaniques de la Iere, IIeme. et IIIeme. époques. Planche 44, In: Atlas encyclopédique contenant les cartes et les planches relatives à la géographie physique. Engraved by Ambroise Tardieu. Chez Mme veuve Agasse, Imprimeur-Libraire, rue des Poitevins, n 6, Paris

Dolomieu D (1797) Sur les volcans d'Auvergne et sur la volcanisation en général. J des Mines 7:393–420

Dufrénoy A, Elie de Beaumont L (1833) Mémoire sur les groupes du Cantal, du Mont-Dore, et sur les soulèvements auxquels ces montagnes doivent leur relief actuel. Ann des Mines 3(3):531–631

Dulieu L (1955) Apothicaires, académiciens de Montpellier. Rev Hist Pharm 145:86–91

Ellenberger F (1978) Précisions nouvelles sur la découverte des volcans de France: Guettard, ses prédécesseurs, ses émules clermontois. Histoire et nature 12–13:3–42

Giraud J (1909) Notes géologiques sur la région du Mont-Dore. Annales de la Station Limnologique de Besse 1:147–161

Glangeaud P (1916) Le cratère-lac Pavin et le volcan de Montchalm (Puy-de-Dôme). C R Acad Sci Paris 162:428–430

Guettard J-E (1752a) Mémoire sur quelques montagnes de la France qui ont été des volcans. Mém Acad Roy Sci Paris:27–59

Guettard J-E (1752b) Mémoire sur quelques montagnes de la France qui ont été des volcans. Histoire de l'Académie royale des sciences (published in 1756):1–8

Hamilton WC (1767) Two Letters from the Hon. William Hamilton, His Majesty's Envoy Extraordinary at Naples, to the Earl of Morton, President of the Royal Society, Containing an Account of the Last Eruption of Mount Vesuvius. Phil Trans Roy Soc London 57:192–200. doi:10.1098/rstl.1767.0021

Hamilton WC (1776) Campi Phlegraei: observations sur les volcans des Deux Siciles. Pierre Fabris, Naples

Huot JJN (1828) Volcans in Desmarest, Nicolas: encyclopédie méthodique. Géographie-physique, vol V. Chez Mme veuve Agasse, an III, Paris

Julien A (1869) Des phénomènes glaciaires dans le Plateau Central de la France et, en particulier, dans le Puy-de-Dôme et le Cantal. Paris

Lacoste P-F (1802–1803) Observations sur les volcans de l'Auvergne suivies de Notes sur divers objets; recueillies dans une course minéralogique faite l'année dernière, an X. Delcros, Granier et Froin, Clermont-Ferrand, An XI

Lacoste P-F (1805) Lettres Minéralogiques et géologiques sur les volcans de l'Auvergne, écrites dans un voyage fait en 1804. Imprimerie de Landriot, Clermont An XIII

Larouzière F-DD (2003) Le comte de Montlosier: une vision originale des volcans d'Auvergne à la fin du XVIIIe siècle. Travaux du Comité Français de la Géologie 3 (17):99–120 publié in Gaudant J. (Coord.) (2008). Géologues et paléontologues – De la passion à la profession. Presses des Mines, Paris, pp 53–81

Lecoq H (1835) Description pittoresque de l'Auvergne. II. Le Mont-Dore et ses environs. Baillière ed. Paris, Paris

Lecoq H (1861) Carte géologique du département du Puy-de-Dôme au 1/40.000e, 24 sheets. Clermont-Ferrand, Paris

Lecoq H (1867) Les époques géologiques de l'Auvergne, vol in-8 with maps. Baillière & Fils, Paris

Legrand d'Aussy PJB (1788) Voyage en Auvergne. Edn. E. Onfroy, Paris

Legrand d'Aussy, PJB (1794–1795) Voyage fait en 1787 et 1788, dans la ci-devant haute et basse Auvergne, aujourd'hui départemens du Puy-de-Dôme, du Cantal et partie de celui de la Haute-Loire. Edn. Chez le Directeur de l'Imprimerie des Sciences et arts, Rue Thérèse, près la rue Helvétius, Paris. 3 volumes:1491

Monnet A-G (1787–1887) Les bains du Mont-Dore en 1786: Voyage en Auvergne de Monnet. Publié et annoté par Henry Mosnier en 1887. Mémoires de l'Académie des Sciences, Belles-Lettres et Arts de Clermont-Ferrand Edt Ribau-Collay Clermont-Fd 29:71–174

Montet J (1760) Mémoire sur un grand nombre de volcans éteints qu'on a trouvé dans le Bas-Languedoc. Histoire Académie Royale des Sciences 1760:466–476

Montlosier FDdRcd (1788) Essai sur la théorie des volcans d'Auvergne, vol in-8. 2nd edit Landriot, 1802 edn., Riom

Poulett-Scrope G (1827) The geology and extinct volcanoes of central France, vol 8, Pl. and maps, 2nd edn, 1858 edn. Murray, London

Roubille J (1916) Compte rendu d'une communication de Ph. Glangeaud sur les cratères-lacs du Massif Central. Revue d'Auvergne XXXIII:210–212

Taylor KL (2007) Geological travellers in Auvergne, 1751–1800. Geological Society Special Publications 287:73–96@71sheet. doi:10.1144/SP287.7

Vimont E (1874) Les lacs Pavin, de la Montsineyre et de la Godivelle (Auvergne). Annuaire du Club Alpin:337–349 with map HT

Vimont E (1889) Le lac Pavin est-il un cratère-lac? Revue d'Auvergne VI:8–14

Characterization of Phreatomagmatic Deposits from the Eruption of the Pavin Maar (France)

6

Hervé Leyrit, William Zylberman, Pascale Lutz, Alexis Jaillard, and Pierre Lavina

Abstract

The crater-lake Pavin, located in Auvergne and dated 6740 years BP, is the youngest volcano and one of the two acidic maars in metropolitan France. Field missions to the Pavin area were conducted for the past 5 years, leading to a better understanding of the Pavin tephrostratigraphy and geological history. Based on field textures, componentry and SEM morphoscopy of juvenile ash particles, a new complete tephrostratigraphy of the Pavin volcanic deposit is defined with a new reference section named Clidères. The 26 tephra beds and bed sets correspond to 4 volcanic units. The deposits are composed of high energy basal surges, lapilli fall and mixed dynamisms. The vertical variations of the maar deposits provide a way to access the fluctuating eruptive conditions related to changing magma-water interactions of the 4 main phases. The changes are associated to simultaneous variations of three factors: the pulsating mass eruption rates, the depth of fragmentation and the aquifer yield.

Based on the combination of two geophysical methods, ground penetrating radar and electrical resistivity surveys, the boundaries of the volcanoclastic deposits are visualized and the average thickness of the formation is followed from proximal to intermediate locations. Including the combination of field observations of near 50 trenches of 1–2 m depth, a core drilling and geophysical profiles, the total volume of deposits is now estimated at 5.2×10^7 m^3 which is 31 % less than previously estimated.

Keywords

France • Massif Central • Volcanoes – Pavin • Tephrostratigraphy • Maar

6.1 Introduction

Depending on the chemical and physical nature of the erupting magma and the environment around the vent (dry or wet), the mode of eruption varies widely. In general, formation of pyroclasts derived from the fragmentation of magma which results in the exsolution of volatile phases during decompression when it rises toward the Earth's surface. The fragmentation depends essentially on compositional and physical characteristics of magma, such as viscosity and volatile content. In this case, explosive eruptions are called magmatic eruptions, and lead to formation of vesicular pyroclasts. In contrast, phreatomagmatic explosive volcanism results from the interaction of magma with external water, that is groundwater or surface water, close to or at the Earth's surface. The gas phase is mainly steam derived from ground or surface water (Lorenz 1973, 1987; Wohletz and Sheridan 1983). Different interaction sites and variations of the magma/water mass ratio may result in a spectrum of eruption efficiency, eruptive styles and emplacement mechanisms (Wolhetz 1986; Houghton et al. 2000; White and Houghton 2000; Carrasco-Nunez et al. 2007).

H. Leyrit (✉) • W. Zylberman • P. Lutz • A. Jaillard • P. Lavina
Institut Polytechnique LaSalle Beauvais,
19 rue Pierre Waguet, 60026 BEAUVAIS cedex, France
e-mail: herve.leyrit@lasalle-beauvais.fr

T. Sime-Ngando et al. (eds.), *Lake Pavin*, DOI 10.1007/978-3-319-39961-4_6

Maars are one of the landforms caused by phreatomagmatic explosive volcanism. Maars consist of a crater, which reaches or extends below general ground level and is considerably wider than deep, and in a surrounding rim constructed of material ejected from the crater (Ollier 1967, p.66). The resulting predominant base surge and subordinate ballistic and pyroclastic fall deposits of phreatomagmatic eruptions consist of both juvenile and comminuted country-rock material (Lorenz 1973). Generally, such eruptions are characterized by a basaltic magma. Explosive interaction between magma and external water may occur during vesiculation (thus juvenile clasts may have a low degree of vesiculation) or prior to it (thus juvenile clasts are dense) (Lorenz 1986).

Two conceptual models have been proposed to explain the formation and evolution of magma-water explosions: the Lorenz's model (Lorenz 1986) and the Valentine and White's model (Valentine and White 2012). In the Lorenz model, the magma interacts explosively with groundwater via molten fuel-coolant interaction (MFCI; Büttner and Zimanowski 1998). Such thermohydraulic explosions start only at shallow initial depth below preeruptive ground because a hydrostatic pressure barrier limits the occurrence of MCFI at low pressure (2–3 MPa). Repeated explosions induce the ejection of hydroclasts and in part the evaporation of groundwater. This mechanism induces the water table drawdown and the associated downward migration of explosion chambers. Deep-seated country rock lithic clasts are derived from direct ejection of the deepest explosions. This explains the presence of progressively deeper-seated country rock lithics in the upper parts of the tephra ring stratigraphy. The repeated ejection of the fragmented country rocks causes the instability of the walls and roof of the root zone. Moreover, the overlying rocks collapse into the partially evacuated root zone, forming a cone of subsidence: the diatreme. At the surface, the maar crater is the result of the subsidence (Lorenz 1986, 2007). In the Valentine and White's model, the deepening cone of depression in the water table is not a necessary condition because explosions can happen at any depth where hydrostatic pressure is less than critical pressure. This second model suggests the water table remains relatively constant because permeability limitations prevent the rapid draining in the diatreme and because the diatreme material is water-saturated. Analog experiments show that shallow explosions are more likely to erupt and are more effective at depth < 100 m. Also, clasts from shallow country rock should be more frequent in the tephra deposits. The deep-seated country rocks can be present in the tephra ring if repeated deep explosions have mixed subcrater deposits (upward mixing) and later shallow explosions have ejected these lithics (Valentine and White 2012).

The Pavin maar offers the opportunity to study one of the rare and best examples of acidic maar. After the geological setting presentation, we describe new field data with two reference cross-sections to establish the stratigraphy of the deposit. Then, in order to perform subsurface imaging between outcrops and realize three-dimension visualization, we establish long geophysical profiles thanks to two methods: Electrical Resistivity Tomography (ERT) and Ground Penetrating Radar (GPR). The results are used to estimate the eruption volume and to understand the evolution of the eruption with the two conceptual models.

6.2 Geological Setting

The Pavin system is the youngest volcanic group in metropolitan France. It was formed around 6740 years BP (Juvigné and Miallier 2016) with the eruption of four volcanoes (Fig. 6.1): two strombolian cones (Montchal and Montcineyre), a basaltic maar (Estivadoux) and the Pavin lake, a trachy-andesitic maar that is one of the rare examples of acidic maar within French Holocene volcanoes (Camus et al. 1973).

The nature of Pavin remained unclear until the publications of Henry Lecoq (1835) and Philippe Glangeaud (1916), who established the relations between the volcanoes and the bedrock (see Boivin and Jouhannel 2016; Fig. 6.2). According to Thonat et al. (2015), the bedrock of the Holocene volcanism period consist of three main geological units, from the base to the top:

- the crystalline basement, formed mostly of cordierite-rich migmatite, muscovite-biotite leucogranite and biotite-sillimanite-garnet gneiss,
- the Cézalier Pliocene volcanic formations (e.g. Cocudoux, Jansenet),
- the Guéry and Sancy plio-pleistocène volcanic formations (e.g. Pertuysat, Fig. 6.1).

In the area of puy de Pertuysat, the Holocene stratigraphic column is as follows (from oldest to youngest; Bourdier 1980):

- glacial moraine,
- white clay-altered trachytic tephra bed of debated origin,
- Montcineyre black lapilli,
- Estivadoux basaltic bedded tuffs,
- Montchal red lapilli,
- Pavin deposit,
- Topsoil

The Pavin crater is roughly subcircular with a rim diameter between 900 m and 1000 m and with a mean slope of 45° toward the center which is filled by a lake located 60 m below the summit of the rim. This lake has a diameter of 750 m and a maximum depth of 92 m. The lake represents

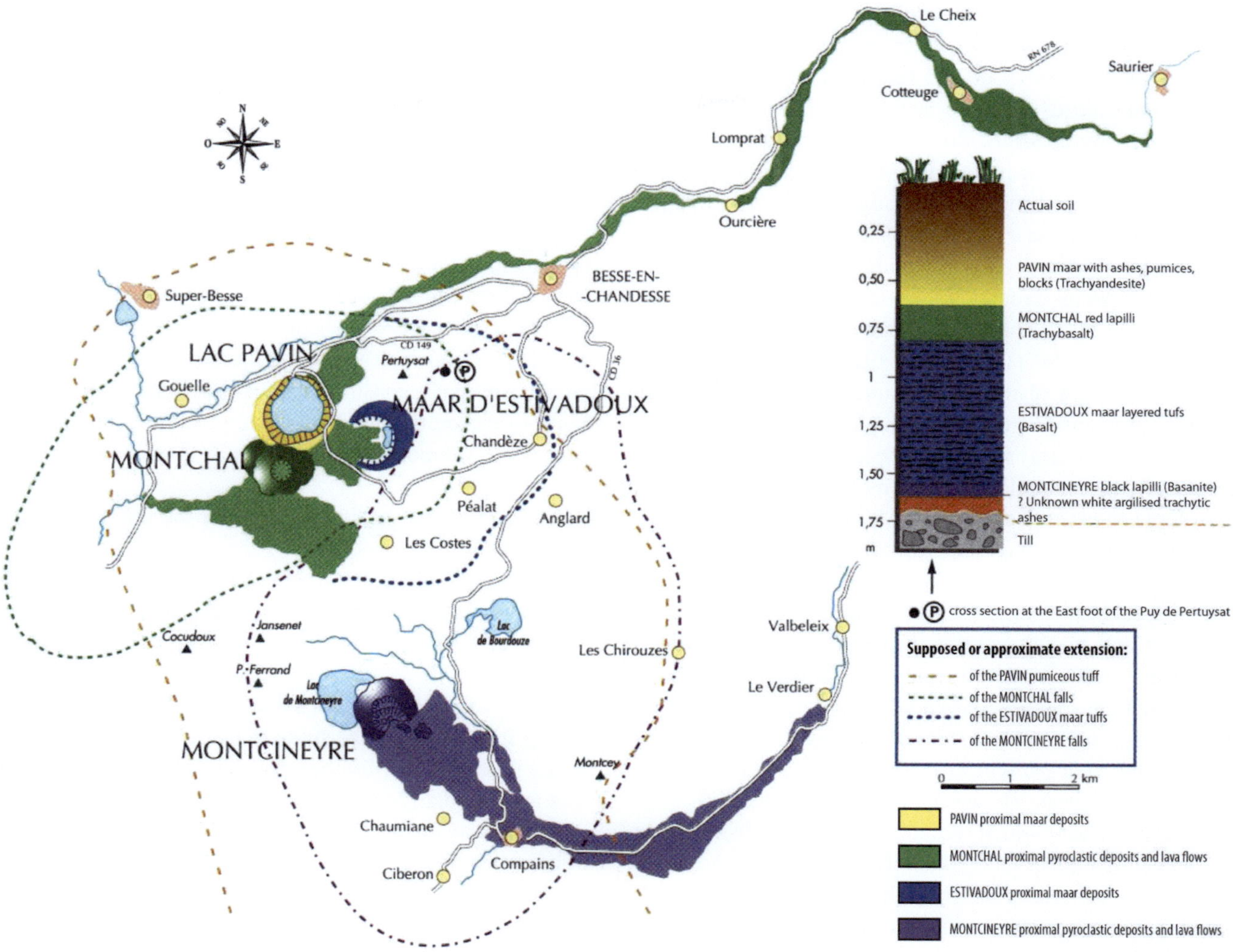

Fig. 6.1 Simplified geological map of the Pavin volcanic system (Goër de Herve 1997)

Fig. 6.2 Geological cross-section showing relations between the basement, the Montchal volcanic formations and the Pavin maar (with an actualized traduction, after Glangeaud 1916)

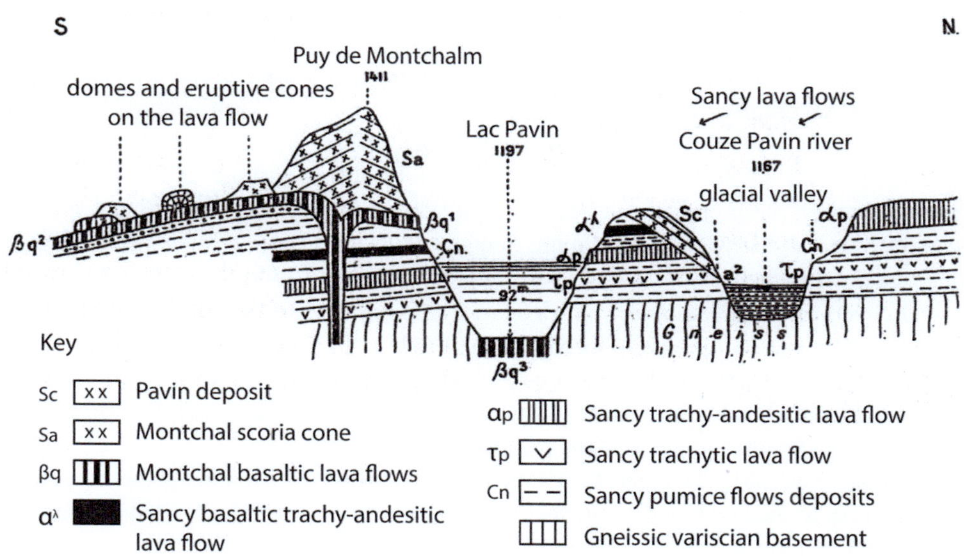

the surface expression of the regional groundwater table in the volcanic terrains. The Pavin crater cut the variscan granitic-gneissic rocks of the basement, the Sancy basic tranchyandesitic to trachytic lava flows, a Sancy trachytic pumiceous pyroclastic flow (named "Rioubes-Haut", Ménard 1979) and the base of the Montchal scoria cone with its trachybasaltic lava flow (Glangeaud 1916; Bourdier 1980). The Pavin maar ("*Lac Pavin*") is surrounded by gently outward-sloping tephra ring with beds of mostly unconsolidated deposits, except on the southwest where the crater is dominated by the older Montchal scoria cone (formerly "*Puy de Montchalm*") (Glangeaud 1916; Bourdier 1980).

The formation of the crater is due to an eruption with complex dynamism which has both plinian and phreatomagmatic characteristics. Ejecta have been dispersed by aerian clouds giving air fall deposits and by basal blasts producing base surge deposits (Bourdier and Vincent 1980). The deposits have a phreatomagmatic origin with polylithological composition characterized by association of basement clasts (granitic-gneissic rocks, basaltic to trachy-andesitic lavas, scoria…) and two types of K-benmoreitic juvenile clasts: commonly pumice lapilli and less frequently poorly-vesiculated dense glassy lapilli (Camus et al. 1973). When its thickness is greater than 50 cm, the deposit has a regular stratification near-parallel to the surface of the substratum, suggesting fall beds. However, some layers have low-angle cross stratifications which characterize surge deposits. When the deposit is thinner than 50 cm, it is generally an unstratified homogeneous ashy matrix with some centimetric to millimetric pumice fragments, interpreted as pyroclastic flow or surge deposits (Bourdier 1980; Boivin et al. 1982).

500 m from the crater, a drilling called "Drilling Pavin 1979" intersected 10.9 m of Pavin trachy-andesitic pumice deposit (PD) based on 1.1 m of Montchal strombolian pyroclastic fall and 18 m of Montchal lava flow. Within the drilling the deposits are characterized by the absence of lithic with great size (>5 cm). Furthermore, the main features are a large variation in the proportion of xenoliths (20–75 %) depending on the level, the initial products being particularly lithic-poor. An abrupt compositional change is localized near 4.9 m from the bottom, with an increase of lithics in the ejecta from 20 to 60 % (mostly from the basement). Simultaneously, the proportion of pumice lapilli decreases. Moreover dense glassy lapilli seem more frequent at the end of eruption (Bourdier 1980).

PD covers an elliptic area of 17 km × 6 km with a NNW-SSE major axis (Bourdier and Vincent 1980; Boivin et al. 1982) The total area was estimated to be 95 km^2. Its extension is limited by Super-Besse to the north, Chirouzes to the east, Lake Chambedaze to the west and exceeds the Godivelle to the south (Fig. 6.3). The deposit asymmetric area appears to extend far to the south and southeast (up to 14 km from Pavin) and slightly to the north (only 3 km). The deposit close to the crater is 15 m thick and the total eruption volume was estimated to be 75×10^6 m^3 (Bourdier 1980; Bourdier and Vincent 1980; Boivin et al. 1982).

6.3 General Methodology

In this volcanic province, natural outcrops are very rare due to the relatively recent age and nature of the eruptions. The volcanic ash deposits soften the reliefs, creating smooth shaped hills that are typical of Auvergne countryside. The lack of uncovered volcanic deposits, due mostly to vegetation or man-made constructions, is a serious brake on the studies of these deposits. For studying and mapping the PD boundaries, two methods are used:

- Near 50 trenches of 1–2 m depth in the intermediate to distal area,
- Combination of core drilling (up to 50 m) and geophysical sections in the proximal to intermediate area of the deposit (Fig. 6.4).

Among the common geophysical methods, two are particularly interesting for the study of pyroclastic deposits: the Electrical Resistivity Tomography (ERT) and the Ground Penetrating Radar (GPR).

For instance Russel and Stasiuk (1997) showed that GPR can be extremely effective in defining the bases of volcanic deposits and has tremendous potential for quantifying distributions, thicknesses, and volumes of volcanic deposits. Cagnoli and Ulrych (2001a, b) point out that an amplitude decrease in the GPR signal probably reflects lateral facies variation of base surge deposits (decrease in grain size). Gómez-Ortiz et al. (2007), in a joint application of GPR and ERT in Tenerife, showed that ERT usually provides a good definition of the boundaries between the volcanic units and allows to locate structures such as lava tubes and faults, whereas GPR is most effective to characterize the internal structure of the volcanic deposits.

In the present study, the ERT is mostly used to estimate the width of the different units (greater penetration depth) and the GPR allows to image sedimentary features (better resolution). Note that a compromise has been made between the profile localization, length, the type of antennae (500 MHz shielded, 100 MHz unshielded) and the duration of the measurements.

6.3.1 Field Observation and Description

First, a new reference section named "Clidères" is described in order to define a detailed lithostratigraphic column for the PD, which has never been done in previous works. A second

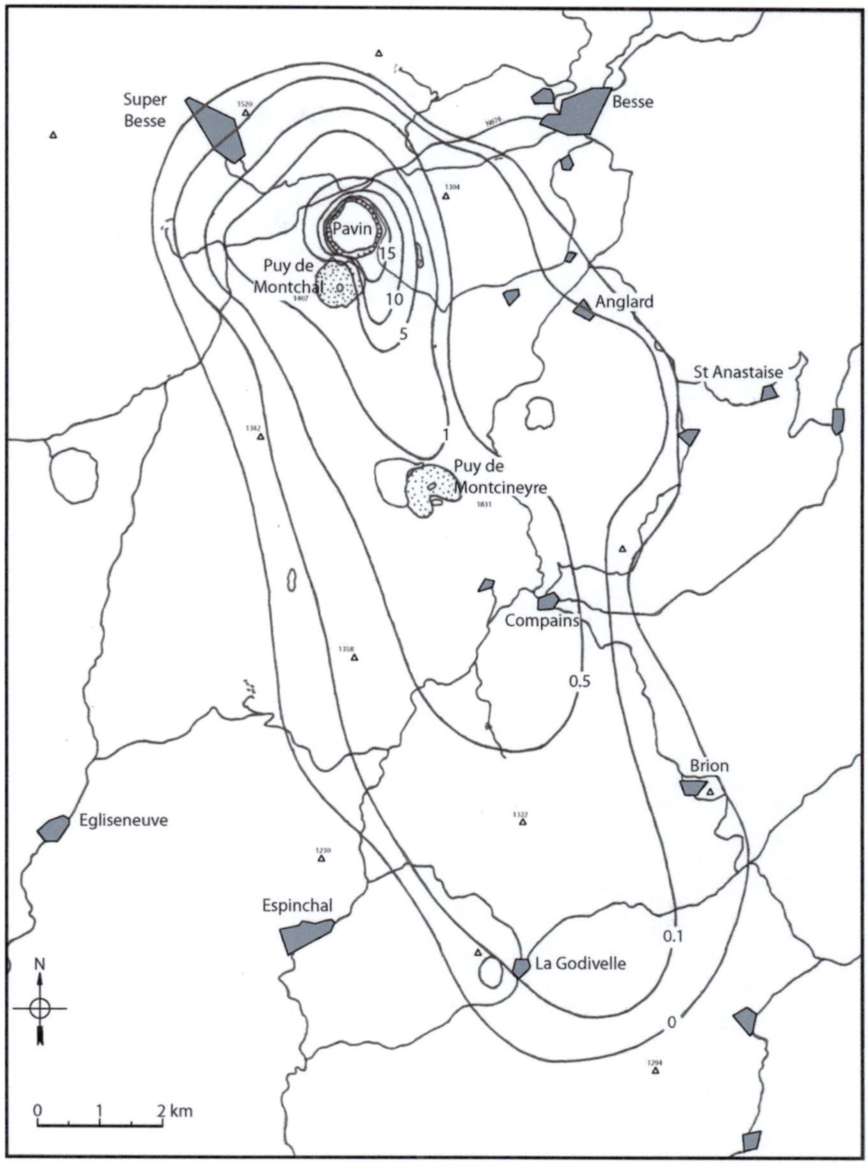

Fig. 6.3 Isopach map of the trachy-andesitic pumiceous flow from the Pavin (thickness in meters) (Bourdier 1980)

reference section, named "La Liste" is compared and correlated. These two reference sections were chosen for the good quality of the outcrops, their very interesting geographical locations and the possibility of studying the deposit in different orientations with a 7–10 m extended side. Clidères and La Liste outcrops are located respectively 1.5 km south-east and 2.4 km south from the vent (Fig. 6.4) in the intermediate area of PD. They are located at the appropriate distance from the vent to observe the bulk of sedimentary structures, that is to say neither too proximal (thus the deposit is unstratified and too thick to reach the basement) or too distal (where energy is too low to observe any sedimentary structure). By describing structural, petrological and geometrical features of each tephra bed, the two sections units can be correlated and the main evolution of eruptive dynamics defined. We estimate the roundness and sphericity of grains using the visual chart of Krumbein and Sloss (1956). These outcrops, associated with the 1979 Pavin drilling, are used to calibrate the geophysical data: Electrical Resistivity Tomography (ERT) and Ground Penetrating Radar (GPR).

6.3.2 Electrical Resistivity Tomography

The ERT is an electrical method based on the measurement of the apparent resistivity ρ_a (Ω.m) in the ground after the injection of a direct current I_{AB} using two electrodes (A for the injection and B for the reception). The voltage difference ΔV_{MN} is measured with two others electrodes (M and N) along a line. The apparent resistivity is then calculated with the following equation: $\rho_a = k\frac{\Delta V_{MN}}{I_{AB}}$

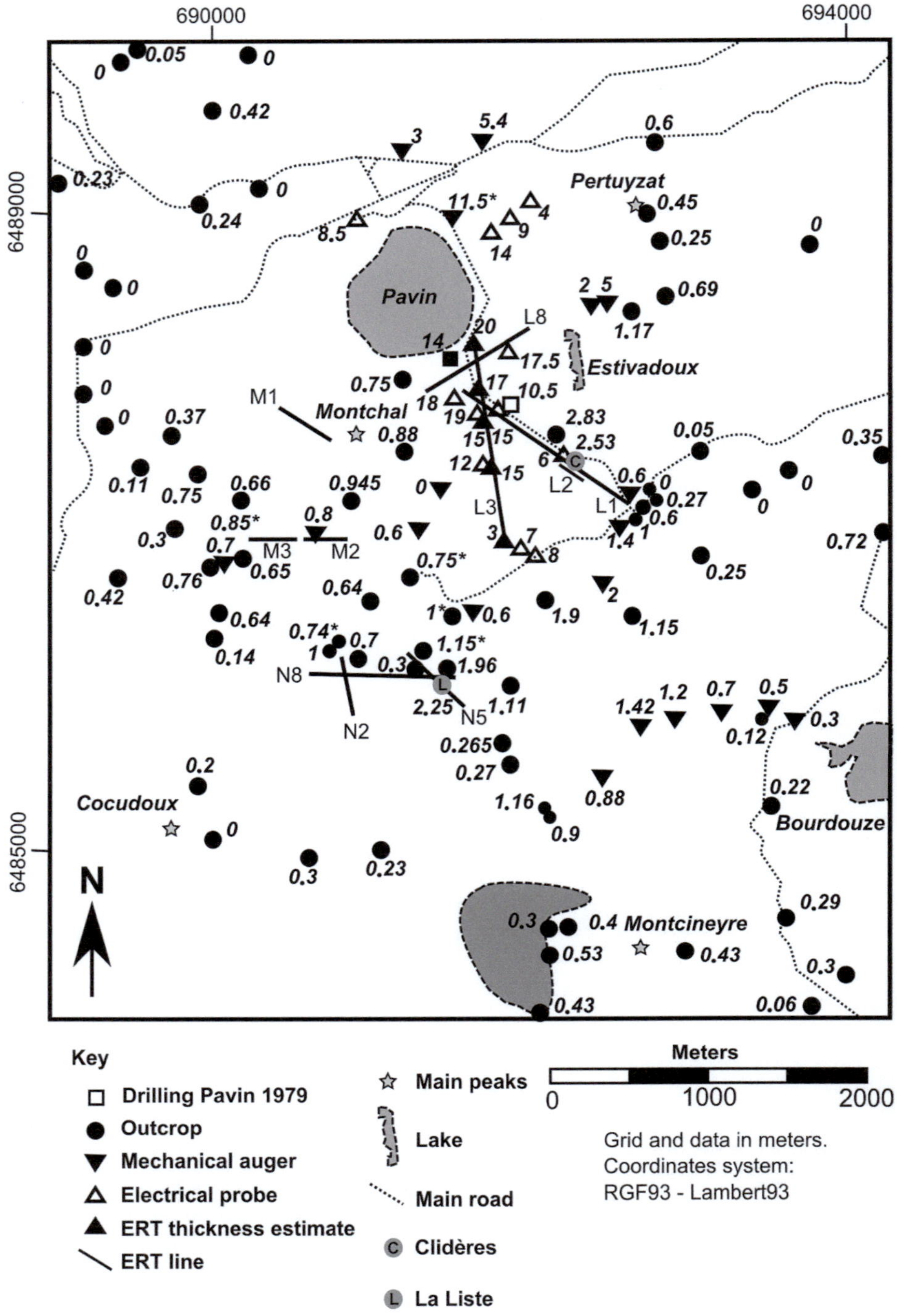

Fig. 6.4 Synthetic map of the geological and geophysical data on the Pavin proximal and intermediate volcanic deposits

where $k = 2\pi\left[\frac{1}{AM} - \frac{1}{MB} - \frac{1}{AN} + \frac{1}{NB}\right]^{-1}$ is the geometric factor.

Electrical Resistivity Tomography (ERT) measurements were undertaken with a multi-electrode 2D device (ABEM TERRAMETER SAS 4000), using 64 electrodes with spacing depending on the desired resolution and depth. We used a spacing of 0.5 m above the outcrops (high resolution but low depth) and a spacing of 5 m for the prospective lines. We usually used a gradient array to obtain both high resolution and satisfactory depth (1252 points of measure, 45 m depth) and occasionally pole-dipole array in order to increase the investigation depth (588 points, 70 m depth). Each prospective line was acquired using several 315 m ERT sections, according to a roll-along procedure, with a half-covering. Prior to the acquisition, a check-up of the electrodes has been realized consisting in an injection of 20 mA (ABEM Instrument 2006).

The injection current intensity was adjusted automatically according to the signal-to-noise ratio between 10 mA and 200 mA. Reciprocal measurements (Electrode A is changed in electrode B and vice versa) were taken to check for reproducibility of data. For average reciprocal errors above 1 %, four reciprocal data were measured. The topographic variations were also measured using a level and a rod ruler, and the lines were positioned using a GARMIN GPS with a 3 m precision and remarkable places (e.g. cross-roads, altimetric marks, topographic peaks, canyon's entry or working face).

The data, once recorded, were transferred to and presented in the form of pseudo-sections, which show the apparent resistivity according to levels. These sections were then processed on a computer. Inversion of the apparent resistivity data is indeed required to obtain a model of the subsurface resistivity in 2D vertical section. The sections were inverted with the standard Gauss-Newton code Res2dinv described by Loke and Barker (1996) and Loke (2003), taking into account topographic variations. A standard constraint was chosen as the dataset wasn't particularly noisy.

The code divides the 2D model into a number of rectangular blocks. The purpose of the data inversion is to determine the resistivity of the rectangular blocks that will produce an apparent resistivity pseudo-section that agrees with the actual measurements. The optimization tries to reduce the difference between the calculated and measured apparent resistivity values by adjusting the resistivity of the model blocks. The root-mean-squared (RMS) error allows measuring this difference. To obtain a resistivity model in agreement with the geology, the most prudent approach is to choose the model at the iteration after which the RMS error does not change significantly (generally between the 3rd and 5th iterations). If the RMS error is high (typically more than 10 %), some points corresponding to a high error values can be deleted, and the inversion has to be computed with this new data set (Loke and Barker 1996).

6.3.3 Ground Penetrating Radar (GPR)

The GPR is a geophysical method based on the propagation and reflection of electromagnetic waves in the ground. The waves are transmitted into the ground by an antenna. The wave propagates until reaching a contrast in the dielectric parameters (permittivity and conductivity), then a part of the wave energy is refracted whereas the other part is either reflected towards the surface and captured by the receiver antenna or absorbed by the medium. This method uses the contrasts of dielectric permittivity ε (F/m) and electric conductivity σ (mS/m) to characterize the subsurface (Fauchard and Mériaux 2004).

The primary goal of investigation is to differentiate subsurface interfaces. The reflectivity of radar energy at a boundary between volcanic formations is a function of the magnitude of the difference in relative dielectric permittivity and electric conductivity between the two deposits. The greater the contrast in dielectric properties, the greater the velocity change, and the stronger the reflected signal (Sellmann et al. 1983). In volcanic deposits, velocities are generally between 0.04 and 0.14 m.ns^{-1}. They are controlled principally by porosity, water content and matrix composition (presence of organics, clays, etc.) (Olhoeft 1984; Gómez-Ortiz et al. 2007).

The field work was performed using Ramac GPR system (Malå Geoscience). Two kinds of GPR' sections were realized:

- Common-offset profiles (COP) in order to obtain 2D time sections showing the different reflectors;
- Common-midpoint profiles (CMP) to estimate the velocity of the electromagnetic waves underground, and then convert the 2D time sections into 2D depth sections.

In COP surveys, both antennae are regularly moved along a line, while maintaining constant antennae offset. The midpoint is also regularly moved forward. In CMP surveys, the receiver and the transmitter are moved in opposite directions to regularly increase the antennae offset.

Two different types of antennae were used in this study: 100 MHz frequency unshielded antennae and 500 MHz frequency shielded antennae. The shielding avoids air wave emission and associated noise and allows achieving a greater acquisition speed (shielded antennae have a wheel coder and are pulled whereas unshielded antennae are regularly moved along a decameter). The antennae frequency is chosen according to the required investigation depth and resolution, and the other acquisition parameters depend on the frequency (Tables 6.1 and 6.2).

Ground penetrating radar signal attenuation occurs as waves pass through the earth. The maximum effective depth of radar wave penetration depends mainly on the frequency of the radar waves and the physical characteristics of the volcanic deposits. A higher groundwater and/or clay content (high dielectric constant ε, high electrical conductivity σ) leads to a stronger attenuation and, therefore, a markedly reduced penetration depth. In ashy tephra deposits, the 100 MHz frequency antenna was used to obtain an investigation depth of approximately 15 and 30 m while the 500 MHz frequency antenna was used to investigate the first 1–4 m (Smith and Jol 1992; Schrott and Sass 2008).

The raw data is expressed in terms of propagation time to amplitude of the wave. To obtain a radar section in terms of depth, the raw data must be processed and the wave velocity in the ground has to be estimated using "Common Mid-Point" (CMP) data using unshielded 100 MHz antennae. The processing was realized using REFLEX W Software

Table 6.1 Main acquisition parameters of Ground Penetrating Radar (GPR) for each antennae frequency

Frequency (MHz)	Antennae offset (m)	Step size (m)	Stack	Time window (ns)
100	1	0.2	16	400 or 500
500	0.18	0.02	16	100

Table 6.2 Frequency, velocity and resolution of the GPR sections

Section	Frequency (MHz)	Velocity (m/ns)	Resolution (m)
L2 – Clidères	100	0.0582	0.15
N5 – La Liste	100	0.059	0.15
D1 – Drilling Pavin 1979	100	0.0736	0.18
L1– Along Costes-Pavin road	500	0.0736 from 0 to 540 m	0.04
		0.0582 from 540 to 1350 m	0.03

Table 6.3 GPR processing for the common-offset profiles

Processing	Parameters values	Aim
1D Filter – substract mean (dewow)	Time window = 10 ns (100 MHz)	Correct the zero in amplitude
	Time window = 2 ns (500 MHz)	
Static correction – move start time	According to the antennae offset	Correct the zero in time
Static correction – topography corrections	According to the topography	Introduce the topography if necessary
Gain – Energy decay	According to the envelope of the amplitude spectrum	Correct the amplitude energy decay
1D-Filter – bandpassfrequency	For 100 MHz: [12,5; 50; 150; 300] MHz or [10; 40; 120; 240] MHz (depending on the central frequency)	Suppress the low and the high frequencies
	For 500 MHz [47.5; 190; 570; 1140] MHz	
2D-filter – running average	Traces number = 4	Suppress the average noise
Kirchoff migration	Filter parameter summation width = 25 and a constant velocity (average)	Move dipping reflections to their correct position, unravel crossing events, and collapse diffractions

Table 6.4 GPR processing for the common-midpoint profiles

Processing	Parameters values	Aim
1D Filter – substract mean (dewow)	Time window = 10 ns (100 MHz)	Correct the zero in amplitude
Static correction – move start time	According to the antennae offset	Correct the zero in time
Gain – energy decay	According to the envelope of the amplitude spectrum	correct the amplitude energy decay
1D-Filter – bandpassfrequency	[12,5; 50; 150; 300] MHz	Suppress the low and the high frequencies
Velocity analysis – reflection hyperbolae method	In order to overlay hyperbolae on the signal	To evaluate the velocities

(Sandmeier Software) and consists in several stages described in the tables (Tables 6.3 and 6.4).

6.4 The Pavin Deposits

6.4.1 The Components

According to Bourdier (1980), the Pavin deposit (PD) is typically composed of three main types of particles: glass, lithics and minerals (Fig. 6.5). The glass corresponds to juvenile pyroclasts, which include pumices and dense glassy lapilli. The lithics consist of particles from the crystalline basement and old lavas (dense or vesiculated).

6.4.1.1 Juvenile Pyroclasts (JP)

The PD is characterized by the presence of pumices and glassy lapilli in all units. Juvenile pyroclasts are found at many scales from the fine fraction (ash) to blocks. Pumices are grey but frequently take a yellow color due to alteration. On hand specimens, phenocrysts are amphibole, pyroxene, plagioclase and opaque.

Mean mineral content estimates from 17 thin sections give 26 % of phenocrysts: 7 % brown hornblende (Fig. 6.5),

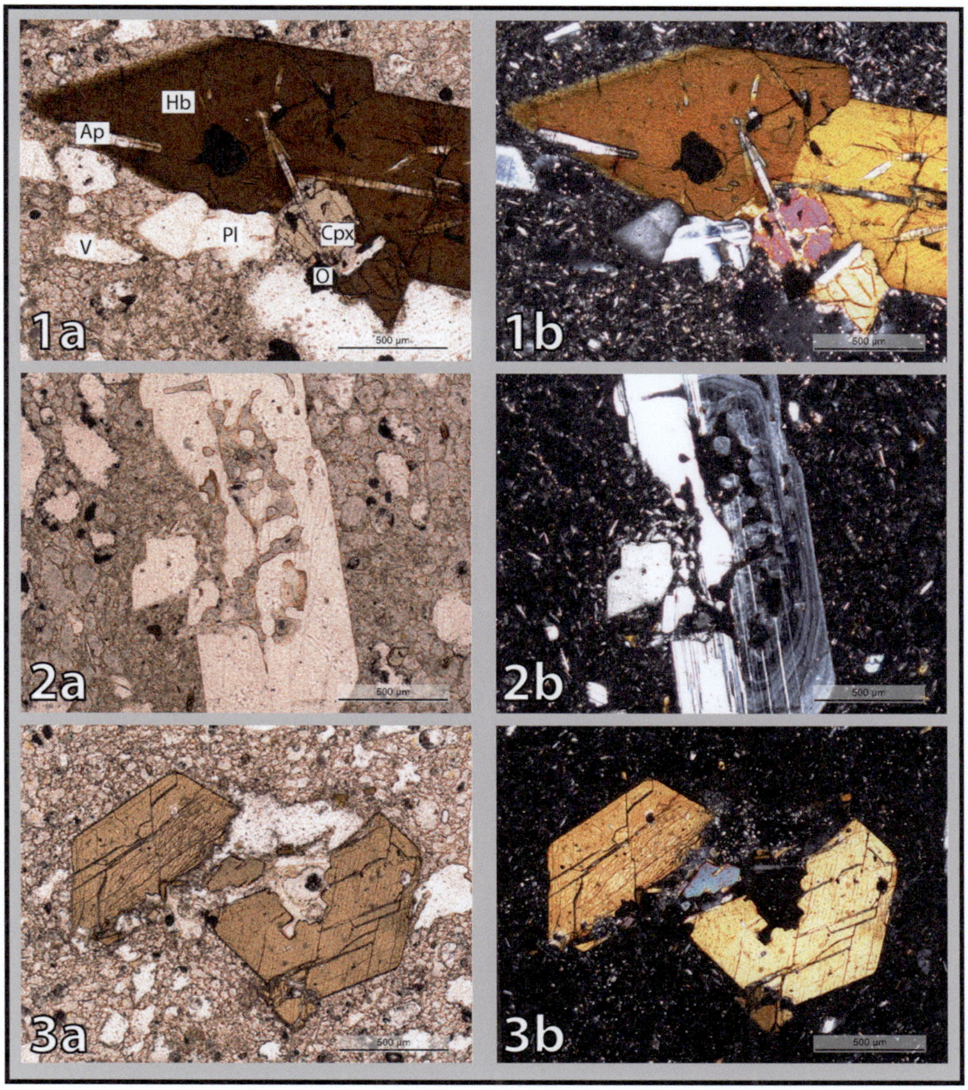

Fig. 6.5 Photomicrographs of the Pavin's juvenile melt. (**a**) is the PPL photomicrograph and (**b**) is the same photograph in XPL. (1) Typical paragenesis showing brown hornblende (Hb), clinopyroxenes (Cpx), plagioclase (Pl), apatite inclusions (Ap) and opaques (O). (2) Destabilization structure in a twinned and zoned plagioclase. (3) Shearing of an amphibole during the phreatomagmatic eruption

7 % clinopyroxenes (including augite and aegyrinic augite), 6 % plagioclases (andesine) and 3 % opaques. The groundmass includes 52 % glass and 22 % microcrysts with 15 % feldspars, 3 % hornblende, 3 % clinopyroxene, 2 % apatite, 1 % sphene and 1 % opaque (Melet 2009; Lêvèque and Vaillant 2010; Jaillard and Zylberman 2012). Microscopic textural criteria indicate two plagioclase populations: the first typically show disequilibrium textures such as sieved, spongy and different types of oscillatory zoning (2a and 2b, Fig. 6.5). Abundant melt inclusions varying from elongate to circular shapes result in coarse sieved texture. The second population represents an equilibrium texture with normal oscillatory zoning. Pumices can contain two main forms of vesicles: tubular microvesicles that can impart a fibrous fabric, and spherical to subspherical vesicles resulting from high vapor pressure during eruption. Sometimes glass splinters can be observed within the matrix, in particular on the periphery of vesicles. It is also important to notice that the vesicle content can vary widely. This was already noted by Bourdier (1980) who identified dense juvenile pyroclasts in some parts of the deposit.

6.4.1.2 Lithic Pyroclasts

The lithic pyroclasts consist of particles from the crystalline basement and old lavas. Three types of basement clasts (noted B) are identified: (1) cordierite-bearing migmatite (2) muscovite leucogranite with an assemblage of muscovite + biotite + perthitic microcline + oligoclase + quartz and (3) sillimanite-biotite-garnet gneiss with foliated texture made up of granular mineral grains, with an assemblage composed of rare cordierite + late muscovite + oxides + garnet + sillimanite + biotite + perthitic microcline + oligoclase + quartz (Bourdier 1980; Lêvèque and Vaillant 2010). The old lavas are subdivided in dense lavas (noted DLP for Dense Lithic Pyroclasts) and black to red scoria (noted VLP for Vesiculated Lithic Pyroclasts).

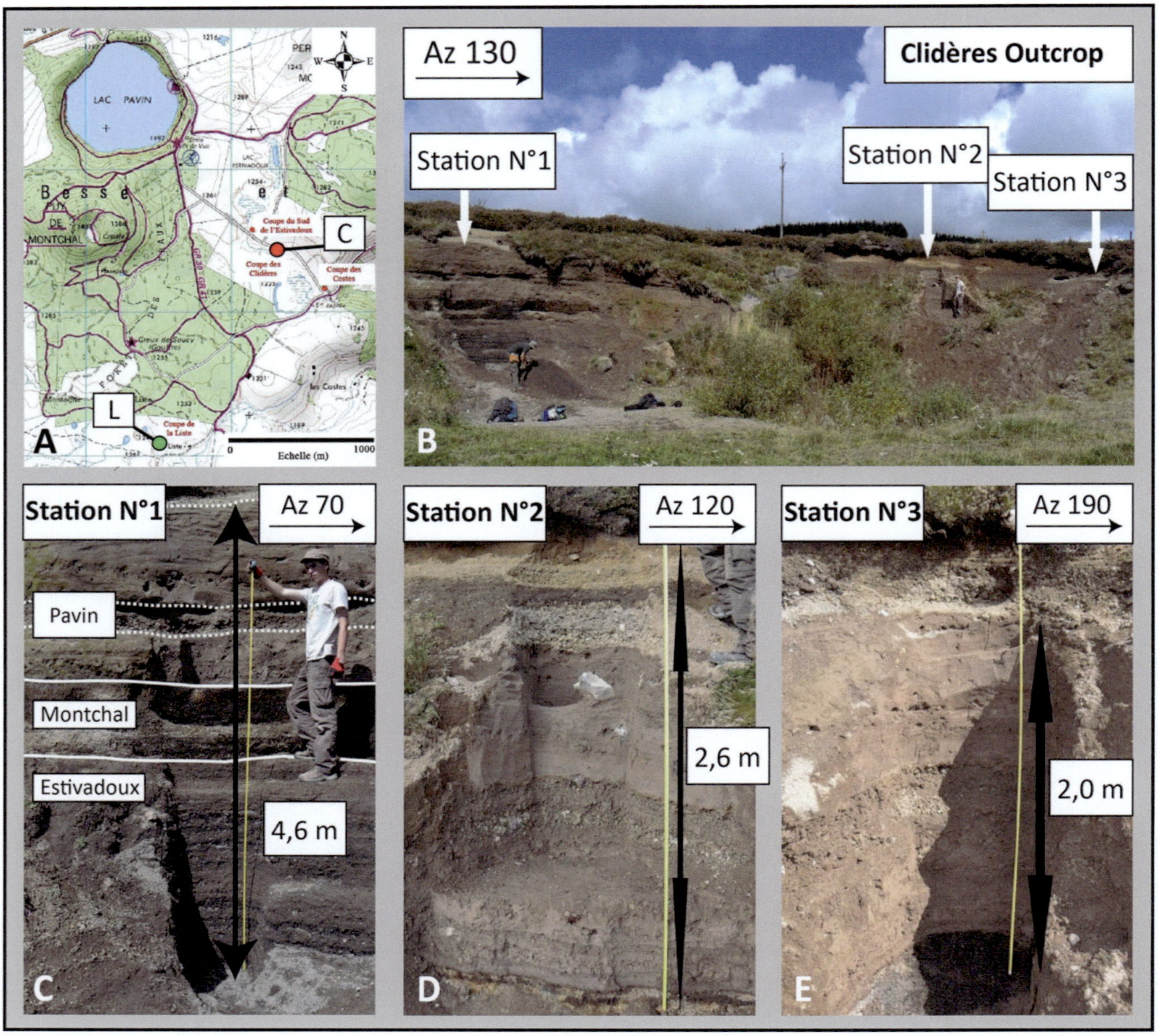

Fig. 6.6 The Clidères outcrop. (**a**) Location of the Pavin's reference outcrops; C – Clidères, L – La Liste, The topographic map is from IGN, 2006. (**b**) Photography of the Clidères outcrop and localization of the 3 study stations. (**c**) Station 1. (**d**) Station 2. (**e**) Station 3

6.4.1.3 Minerals (Mnls)

Free minerals in the PD are diverse in nature. Two categories of free minerals have to be distinguished: xenocrysts from the basement (quartz, some feldspars and micas) and free minerals from the juvenile magma (feldspar, amphibole and pyroxene). Minerals are often found broken. The amounts vary widely depending on the fractions' size.

6.4.2 The Clidères Section

The Clidères section is located 1.5 km south-east from the vent. By creating three study stations with different orientations, the Clidères outcrop allows an optimal characterization of the sedimentary structures in the PD (Fig. 6.6). Station n°1 shows the regional tephrostratigraphy of the Pavin group whereas the two other stations include only the Pavin deposit and a part of its direct basement layer as a reference level (*i.e.* Montchal trachy-basaltic lapilli deposit, noted BML). The Montcineyre basaltic lapilli are not present at this distance from the vent.

Stratigraphic subdivisions were performed according to petrographical, sedimentological and grain size data. 26 tephra beds are grouped in four main units in the PD, which has a total thickness of 2.53 m at the Clidères outcrop (Fig. 6.7). The nomenclature used to describe the geological formations is as follows: P refers to the Pavin deposit, which is then subdivided in 4 units numbered from the oldest to the

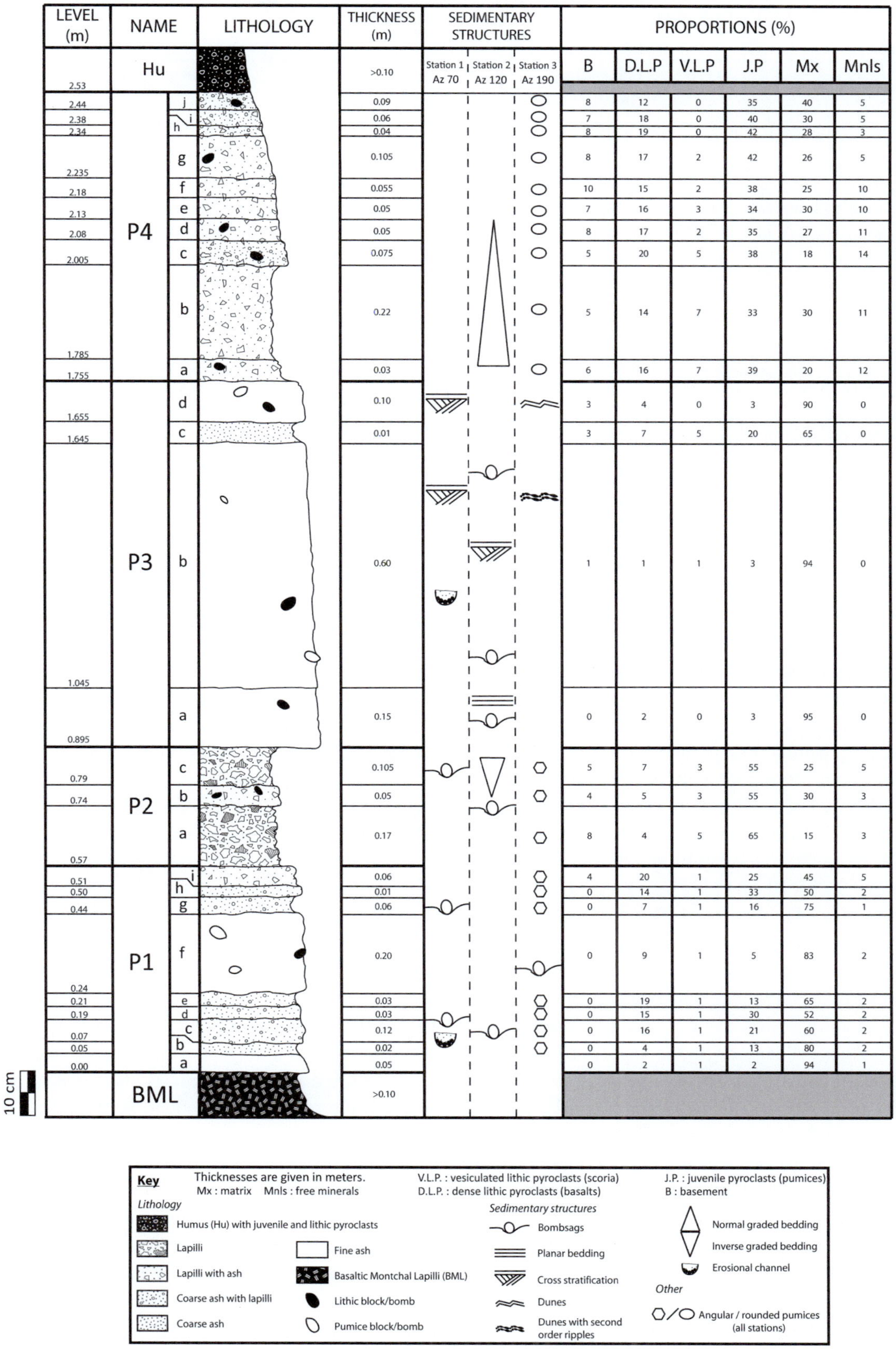

Fig. 6.7 Lithostratigraphic column of the Pavin deposit

Table 6.5 Characteristics of the Pavin deposit

Clidères outcrop	P1	P2	P3	P4
	Fine-grained	Coarse-grained	Very fine-grained	Coarse-grained
Color	Grey	Grey	Grey	Grey to yellow
Thickness mean (mini-maxi) (cm)	58 (30–62)	32 (20–37)	86 (82–95)	77 (55–92)
Number of beds	9	3	4 (minimum)	10
Matrix %	83 % (59–99 %)	22 % (19–39 %)	91 % (65–99 %)	34 % (18–65 %)
Juvenile / (juvenile + lithics) %	40 % (10–70 %)	80 %	50 % (30–60 %)	60 % (55–75 %)
Granitic-gneissic lithics / total lithics %	2 % (0–5 %)	33 % (0–47 %)	20 % (0–33 %)	30 % (15–40 %)
Bigger pumice block (cm)	20 × 10	18 × 14	15 × 11	6 × 4
Bigger lithic block (cm)	14 × 11	25 × 13	45 × 35	11 × 5
Pumice roundness mean (mini-maxi)	0.4 (0.4–0.6)	0.6 (0.5–0.7)	0.4 (0.3–0.5)	0.5
Pumice sphericity mean (mini-maxi)	0.5 (0.5–0.7)	0.7 (0.7–0.8)	0.5 (0.4–0.7)	0.7
Lithic roundness mean (mini-maxi)	0.5 (0.3–0.6)	0.3 (0.3–0.4)	0.2 (0.2–0.3)	0.4 (0.3–0.4)
Lithic sphericity mean (mini-maxi)	0.7 (0.5–0.7)	0.7 (0.5–0.8)	0.5 (0.5–0.6)	0.6 (0.5–0.7)

youngest (e.g. P1 is the first unit formed and P4 the last). Then, each of these units is detailed in a set of tephra beds named with letters where "a" is the oldest bed.

Some key features identified for each unit are described below (Table 6.5):

P1 unit is generally fine-grained and grey. The mean thickness is 58 cm varying laterally between 30 and 62 cm. It is subdivided in 9 beds with sharp contacts and horizontal bedding (Fig. 6.7). Beds with coarser grain-size have a red color on the top and around lapilli and blocks. The largest pumice measures 20 × 10 cm (bed P1g) and the largest basaltic bomb measures 14 × 11 cm (bed P1d). Sedimentary structures are rare. U-type erosional channels are found in P1e and are associated with a filling characterized by a normal graded deposit in P1c (Fig. 6.8, 3a and 3b). Moreover, some beds bevel laterally which explains the difference of bed number with other colleagues (Boivin et al. 2010). The beds have a bimodal grain-size distribution, except the bed P1b. For example, in P2e the two pics correspond to −2 ϕ and 2 ϕ. The lithological composition varies with the size fraction; the coarser is a pumice-rich fraction and the finer is an ash and mineral-rich fraction. Based on the lapilli + blocs fractions (> 1 mm), the relative proportions between juvenile clasts and lithics are estimated to 67–33 % in the coarser beds (1d, 1j) and near 33–67 % in finer beds (1a, 1f). We interpret this difference due to the fragmentation mechanisms. The poor-sorting associated with bimodal grain-size distribution and the sedimentary structures are typical of surge deposits. Furthermore, sedimentary structures such as bombsags (beds 1c, 1d, 1f and 1 g) are typical of ballistic deposits. Using the visual chart (Krumbein and Sloss 1956), the pumices have an average roundness/sphericity ratio of 0.4/0.5. In this unit, no accretionary lapilli are detected on the field but Boivin et al. (2010) have described proto-accretionary lapilli at a microscopic scale in beds 1a and 1 f.

P2 unit is grey, coarse-grained and has a 32 cm mean thickness (between 20 and 36 cm). It is composed of three beds (Fig. 6.7). This unit has an intermediate thin bed consisting of finer clasts (P2b) separating two thicker beds with reverse graded-bedding composed of lapilli. The beds P2a and P2c are well-sorted (pic: -3 ϕ), composed of 85 % lapilli clasts with a rare ashy matrix (15 %). The largest clast is a basaltic bloc measuring 23 × 13 cm (bed P2c). The pumices have an average roundness/sphericity ratio of 0.6/0.7. Based on the lapilli and block fractions (> 1 mm), the relative proportions between juvenile and lithic clasts are estimated to 80–20 % in the P2c bed. The lithic fraction is composed of dense basaltic to trachyandesitic lava, red and black scoriaceous lapilli and variscan gneiss and granites. In the lithic fraction, the ratio lavas/basement is near 3/2. This unit has typical characteristics of a volcanic fall deposit.

P3 unit is partially indurated and very fine-grained. It has a mean thickness of 86 cm (variation between 82 and 95 cm). Due to the presence of many cross beddings and discontinuities, we have distinguished only four main beds (or bedsets). P3 is mainly grey with some pinkish levels. This unit is well sorted, because the matrix represents more than 90 % of the deposit, but the grain-size distribution is bimodal because some lapilli and blocks are isolated in the ashy matrix (Fig. 6.7). The pic of the ash population is between 3 ϕ (bed P3b) and 2 ϕ (bed P3c) while the pic of the coarse population is −3 ϕ. The ratio juvenile/(juvenile + lithics) on the

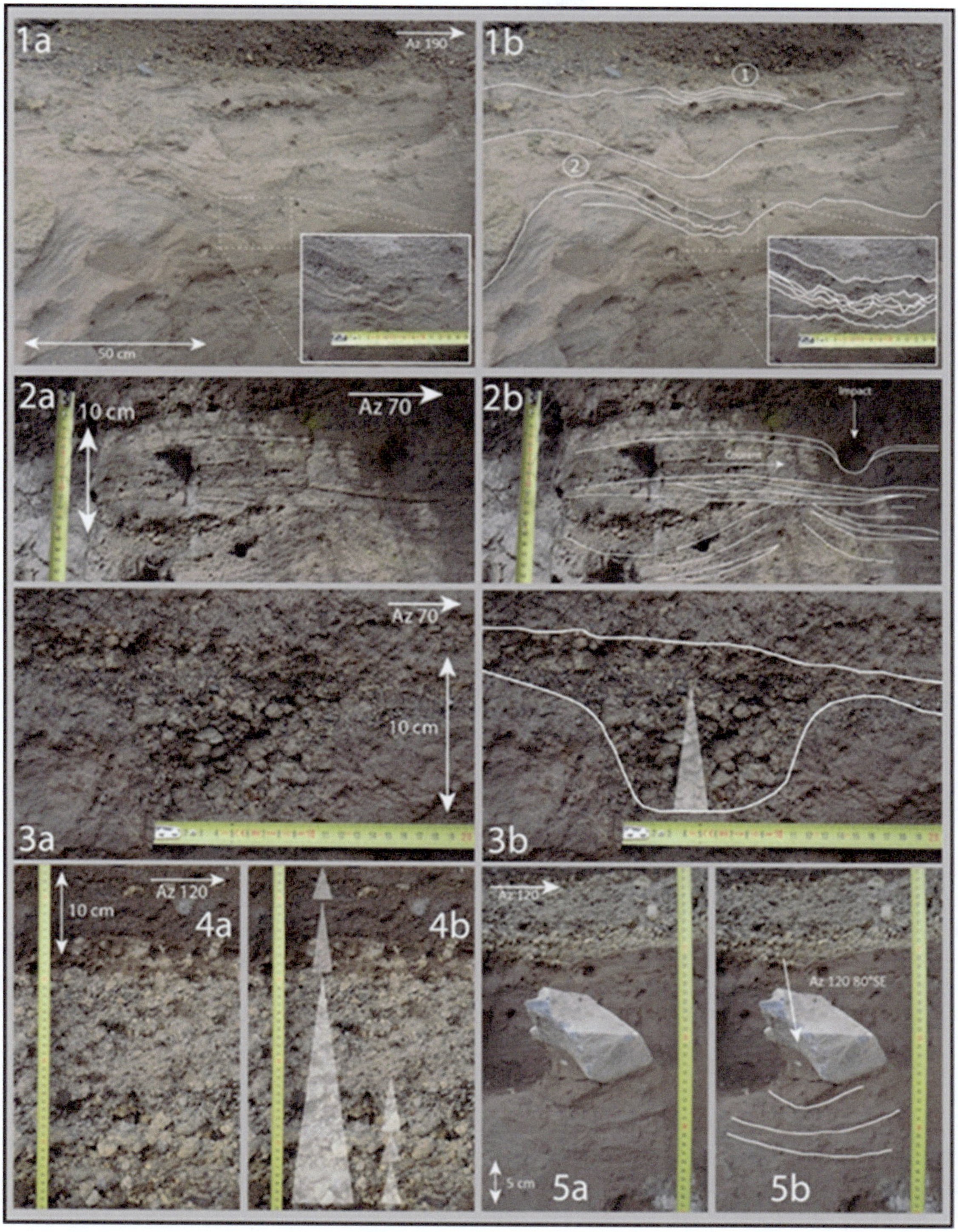

Fig. 6.8 The five main types of sedimentary structures of the Pavin deposit. Each number is accounting for a type of structure. (a) is the sedimentary structure as observed on the outcrop. (b) is the interpretation. (1) Anti-dunes and associated second order ripples. (2) Cross-stratifications. (3) U-type erosional channels. (4) Normal grading. (5) Impact-sags

lapilli + bloc fraction is difficult to estimate, varying between 30 % (bed P3d) and 60 % (bed P3c). Sedimentary structures typical of surge deposits are more frequent than in other layers. Some channels, filled by coarser ash and characterized by diminishing grain-size towards the margin, are interpreted as erosional features of the pyroclastic surges. Cross-stratifications are observed in the P3b bed (Fig. 6.8, 2a and 2b) as well as a dune prograding to azimuth 70° overlapping oblique stratifications in which coarser beds are found (bottom left of the photography). This structure indicates that the

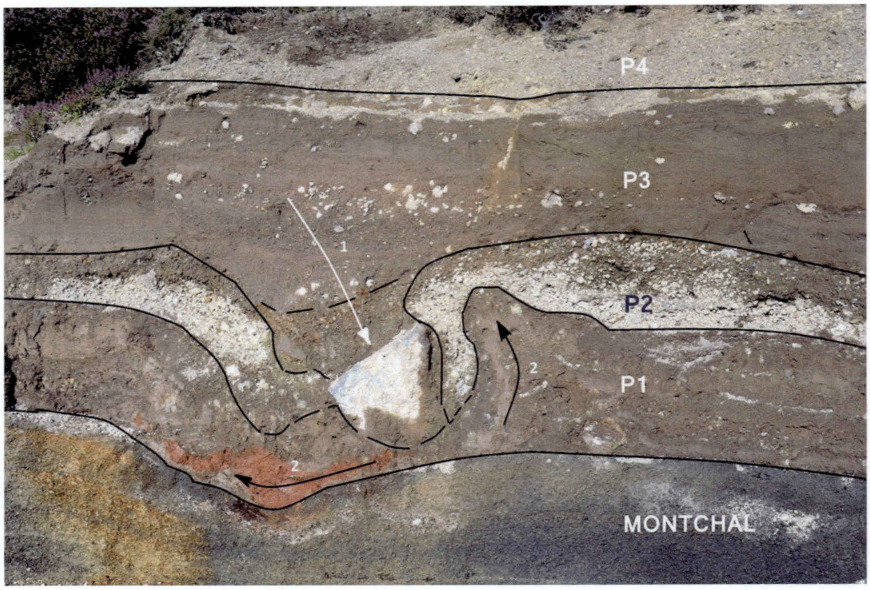

Fig. 6.9 Bombsag block at the station n°3 on the Clidères outcrop

main flow direction is near 160° azimuth (perpendicular to the photography). Antidunes and associated ripples (Fig. 6.8, 1a and 1b) were observed at study station n°2 with a 190° azimuth orientation in the P3b and P3c beds. These high-energy structures are characterized by the shortest side closer to the vent. Based on a wavelength (λ) and an amplitude (A), two main types of antidunes are distinguished: first, simple antidunes with low amplitude and short wavelength (A ≈ 5–10 cm; λ ≈ 20 cm) located in the P3d bed (labeled 1 in Fig. 6.8, 1b) and, second, complex antidunes with large amplitude and long wavelength (A ≈ 20–40 cm; λ ≈ 50 cm) with second order ripples (A ≈ 1.5 cm; λ ≈ 5 cm) themselves affected by third order ripples (A ≈ 0.5 cm; λ ≈ 1 cm) located in the P3b bed (labeled 2 in Fig. 6.8, 1b). Furthermore, sedimentary structures such as impact sags under blocks are typical of fall deposits. The largest basaltic lithic bomb (45 cm × 35 cm) cuts the P2 and P1 units and induces deformation of the Montchal trachy-basaltic lapilli deposit; the bombsag, filled with P3 material, has 0.8 m depth and 1.3 m large (Fig. 6.9). Bombsag deformation within the P3b bed (station n°2, Fig. 6.8, 5a and 5b) indicates a movement towards the southeast (N120). In this unit, no accretionary lapilli are detected.

P4 unit has a grey coarse-grained base (P4a to P4g) and an upper yellow fine-grained part (P4h to P4j). The thickness varies between 55 and 92 cm with a mean value of 77 cm (Fig. 6.7). At station n°3, it is subdivided into 10 beds that are less individualized than in P1. The beds have a poor sorting and, generally a bimodal grain-size distribution. For example, the P4d has a population centered on -3 ϕ and a second population around 1 ϕ. The lithological composition varies with the size fraction; the coarser is a pumice-rich fraction and the finer is an ash and mineral-rich fraction. Based on the lapilli + bloc fractions (> 1 mm), the ratio juvenile/(lithic + juvenile) varies around 60 % (55–75 %). In the lithic fraction, granitic gneissic xenoliths proportions are higher than in the P3 layer. P4 pumices also appear to be denser and less vesiculated than pumices of other levels. Moreover, they have a more yellowish color due to alteration. The pumices have an average roundness/sphericity of 0.5/0.7. Normal graded-beddings were observed in P4b (Fig. 6.8, 4b). We do not identify the presence of another unit on the top of the section (unit 5 in Boivin et al. 2010): due to the much higher clay and iron oxide-hydroxide content, we consider that the last P4 bed has undergone pedogenetic processes leading to weathering of primary volcanic materials (especially volcanic glass) and clay illuviation. In the P4 unit, the poor-sorting associated with bimodal grain-size distribution and the sedimentary structures are typical of surges deposits. In this unit, no accretionary lapilli are detected.

Contacts between the four units are sharp and clearly visible, relatively horizontal even if irregular at a smaller scale. No paleosol is detected between the units and with the Montchal lapilli fall.

6.4.3 The La Liste Section

The second reference section is located 2.4 km south of the Pavin maar in the La Liste area, beyond the west side of a small stream (Fig. 6.6, A, noted L). The PD covers the Montchal deposit with no paleosol. The total thickness of the section is 250 cm. Based on the sharp grain-size variations (or proportion of matrix) and the material composition, the section can be subdivided in 4 units similar with the Clidères section. The common characteristics are the coarser grain size of P2 and P4 units, the greater abundances of pumices in P2 and basement lithics in P4, as well as the lower abundance of scoria in P1. Moreover, the relative proportions between juvenile clasts and lithics in the lapilli+block fractions (> 1 mm) are similar to the Clidères section in the two coarser grain units (J% = 81–100 % in P2 and 59–81 % in P4).

However, some differences are observed. The first difference is the variation of the thickness of each unit: P1 and P3 units are thinner and P2 and P4 units are thicker. The second difference is the finer average grain-size distribution of the 4 units as indicated by the matrix proportion. The third difference is the size of the largest clasts: no pumice measures more than 16 × 10 cm and no lithic exceeds 10 × 7 cm. The fourth difference is the greater juvenile proportions in P1 (J% = 90–100 %) and P3 units (77–80 %) compared to the Clidères section, which comprises respectively 33–61 % and 30–57 %. The fifth difference is the lesser proportions of gneissic rocks in the lithic fraction (mostly 0–10 %). The last difference is the quasi-absence of sedimentary structures in the deposit, which is essentially planar.

These differences can be explained by the distance between the Pavin crater and the localization of the two sections: 2400 m for La Liste and 1500 m for Clidères. With distance, the average grain size, the size of the largest clasts and the thickness of units decrease while the pumice proportion increases in the coarser fraction due to their lower density. The greater thickness of P2 and P4 units could be linked to the greater southward extension of the dispersion area. The maar deposits range from thick, structureless and commonly block-rich beds near the vent to well-developed intermediate cross-bedding and duneform beds and thin distal planar beds.

6.5 Geophysical Study of the Pavin Deposit

6.5.1 Calibration of Geophysical Sections

First, we calibrate ERT and GPR results using the profiles above the two reference sections, allowing correlations between geophysical and geological sections (Fig. 6.4). We also used the results of the Pavin Drilling, located near the section L1 (square in Fig. 6.4). All the figures of the ERT and GPR surveys have vertical exaggeration.

The velocity value used for each GPR section is estimated from the CMP located nearby (Table 6.2). The method cannot resolve the thickness of beds thinner than approximately onefourth of the radar wavelength in the best case. For a 0.06 m.ns^{-1} velocity, the vertical resolution limits in volcanic deposits range between 15 cm for 100 MHz antenna and 3 cm for 500 MHz antenna. Where strata are thinner than this resolution limits, the GPR sections do not yield to an image showing the correct geometry of beds.

Above the Clidères outcrop (Line L2), the ERT maximum investigation depth is about 5 m with 0.5 m spacing between each electrode (Fig. 6.10). The model discretization yields to blocks with a width of 25 cm and a height between 18 and 62 cm (equivalent to the horizontal and vertical resolution).

On the ERT section L2 achieved above the Clidères outcrop, the PD has three sectors with lateral extension: an intermediate part with high resistivity values, between 3000 and 13,000 Ω.m framed by lower and upper parts with low resistivity values, between 300 and 3000 Ω.m (Fig. 6.10). At this scale of investigation, the P1-P2-P3 units are a set characterized by high resistivity values according to the total thickness of the very fine grained P1 and P3 units (near 160 cm) relative to the thinner P2 coarse-grained unit (about 25 cm). In this case, the low-resistivity values of the upper part are characteristic of the P4 unit and the low-resistivity values of the lower part are associated with Montchal and Estivadoux coarser deposits.

GPR section shows some reflectors that are well correlated with the depth and the thickness of deposits (Fig. 6.11). First, the trace clearly shows the air and surface waves with perfect horizontal reflections up to a time of 20 ns. The main reflector near 2.6 m depth represents the limit between the Montchal trachy-basaltic lapilli fall deposit and the PD (red arrows). The PD units form a stratified deposit near parallel to the surface. Moreover, near a depth of 2 m, two reflectors are near parallel to the surface (black arrows) and separate

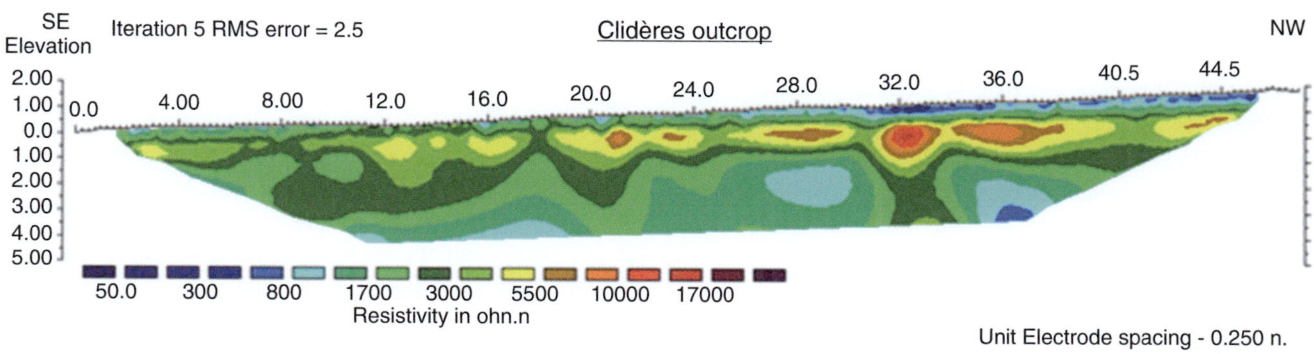

Fig. 6.10 Southeast-Northwest L2 ERT section above Clidères (Depth in meters)

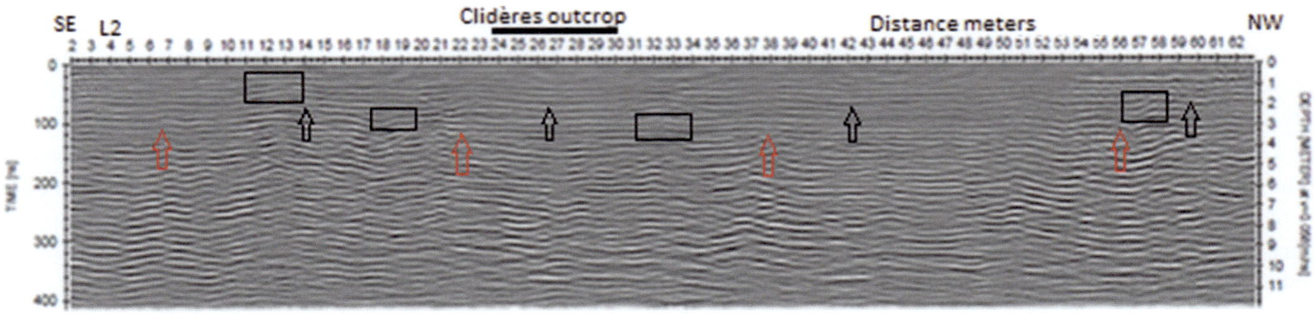

Fig. 6.11 Southeast-Northwest L2 GPR section above Clidères (Depth in meters)

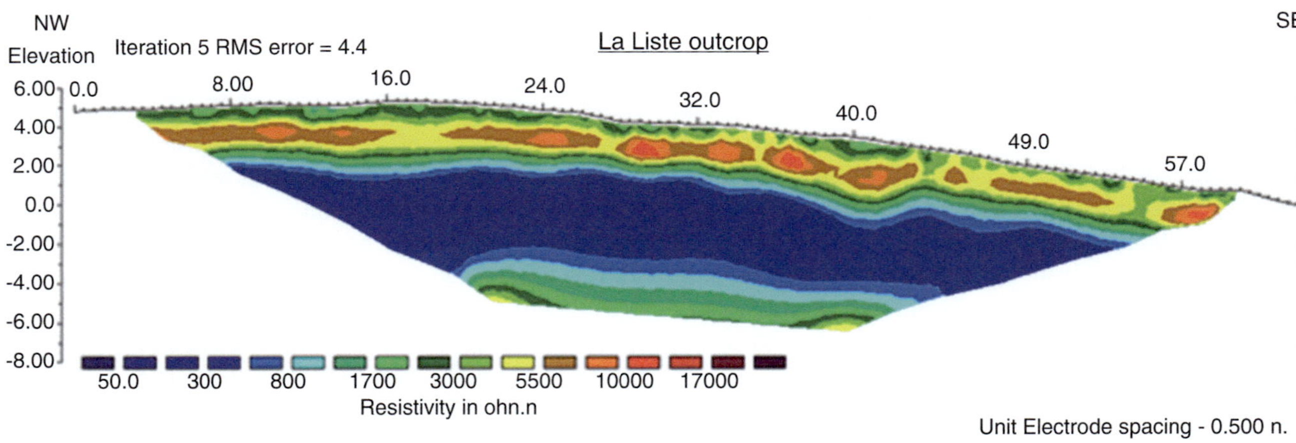

Fig. 6.12 Northwest – Southeast N5 ERT Section above La Liste (Depth in meters)

units characterized by oblique reflectors interpreted as cross bedding of base surge deposits (surrounded areas). The two reflectors could be interpreted as P2 lower and upper limits. The reflector near the depth of 6 m can be interpreted as the base of Estivadoux surges deposits.

For the La Liste outcrop (line N5 on Fig. 6.4, oriented NW-SE), the ERT investigation depth is about 10 m with a 1 m spacing between each electrode. The resolution values are also twice less than those of the Clidères section. At this scale of investigation, the P2 unit is undistinguishable and the higher resistivity (3000–13,000 Ω.m) is associated with the finest-grained deposits (P1-P3). The P4 layer is characterized by low resistivity values, between 1300 and 3000 Ω.m (Fig. 6.12).

The GPR section is oriented NW-SE (Fig. 6.13). Using the 100 MHz antenna, the depth of investigation is a dozen meters. The main reflector has a depth between 2 and 2.2 m. Above this limit, reflectors tend to be fairly continuous and horizontal while just below there are many diffraction hyperbolas.

On the first two meters deep, there are in average six sub-horizontal and sub-parallel reflectors. Each reflector is near 35 cm thick. The two first reflectors and the base of the top reflector are continuous over the entire length of the section. The other three reflectors have some lateral discontinuities, marked by oblique reflectors; for example, at the points 10 m, 17 m, 33 m and 49 m (black ovals). The above-described reference geological cross-section is located between 28 and 38 m.

The boundary between subhorizontal reflectors domain and the hyperbolae domain corresponds to the contact between the Holocene volcanic deposits and the bedrock which corresponds to glacial moraine deposits. The hyperbolae therefore represents the blocks within the moraine, whose size is greater than 15 cm to be detectable as a point of diffraction with the conditions used (100 MHz, unshielded, v = 0.059 m/ns). This interpretation is consistent with field observations and bibliographic data (Van Overmeeren 1998).

Among the Holocene volcanic deposits (upper part), the outstanding features are the absence of diffraction with hyperbolae which indicates the quasi-absence of ballistic blocks with size larger than 15 cm in these deposits according to the lithological data. There are also oblique or undulating reflectors diagnostic of surge deposits (Cagnoli and Ulrych 2001a, b). These reflectors correspond to the P3 deposit (fine material with dunes and cross-bedding). However, we must remain cautious because the device used is near the limit of detection of objects (small dunes measuring a few decimeters).

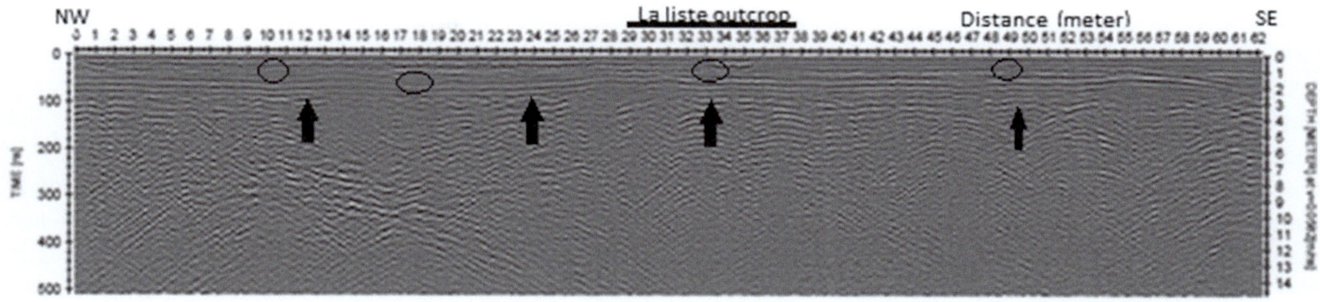

Fig. 6.13 Southeast–Northwest N5 GPR section above La Liste (Elevation in meters) Northwest–Southeast GPR section above La Liste (Depth in meters). Without Kirchoff migration

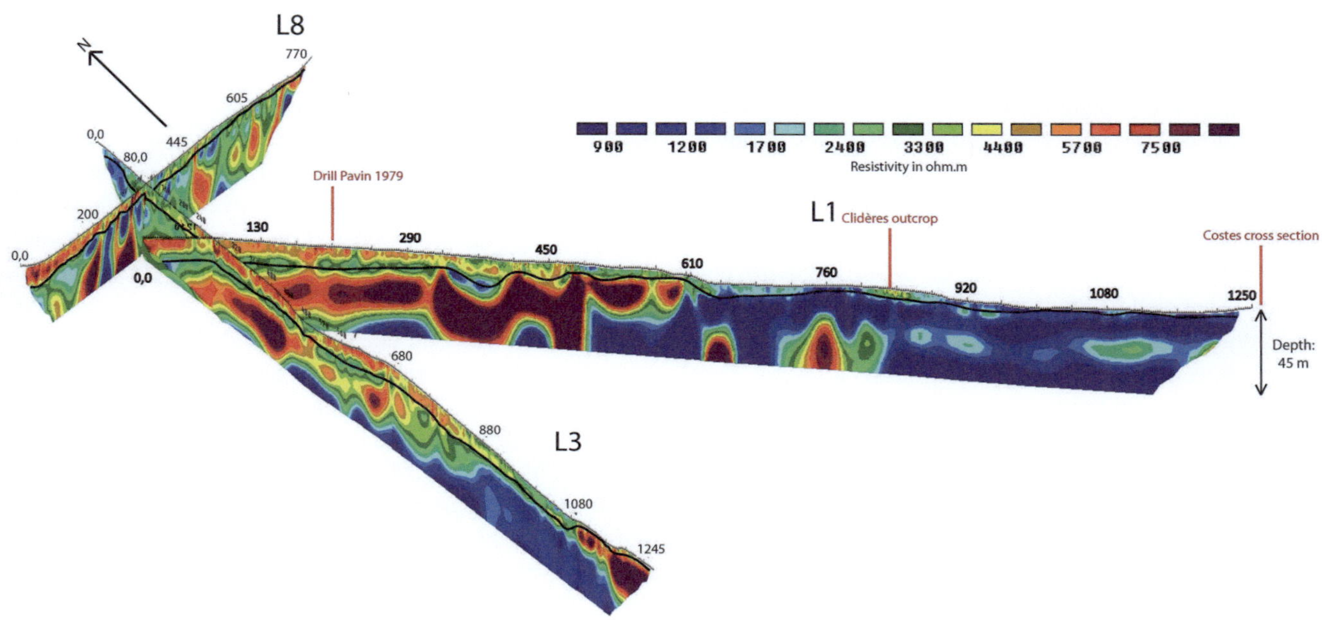

Fig. 6.14 Pseudo-3D representation of the L1, L3 and L8 resistivity sections

6.5.2 Long Geophysical Transverses from Proximal to Intermediate Area

ERT data was collected from three transverses (L1, L3, L8) over long unexposed fields to estimate thickness variations of the PD from proximal to intermediate area. The ERT sections were carried out with a 5 m spacing between electrodes leading to an investigation depth of about 45 m (Fig. 6.14). Section L1 is a geophysical line of 1275 m oriented towards the SE along the road between the Pavin crater and the Costes outcrop. This section is calibrated with the Clidères outcrop (this study) and the 1979 Pavin Drilling. The L3 section follows the GR30-GR41 trail with a N-S orientation over a distance of 1275 m. Finally, the section L8 extends 795 m in a SW-NE direction.

At this scale of investigation, the PD has resistivity values compatible with the reference sections (3000–13,000 Ω.m) in the proximal area. On ERT sections, the PD has a thickness of about 15 m in the proximal zone which decreases less rapidly to the south than to the southeast. In fact, at 1.2 km from the crater rim, the thickness of the deposit reaches 7–10 m southwards whereas it is only 1 m to the SE. On distal area, the decrease in thickness explains the low resistivity values because the horizontal resolution is 2.5 m and the vertical one varies between 1.8 m and 6 m. Moreover, in the proximal area, the PD covers the Montchal lava flow which is characterized by a high resistivity (> 3000 Ω.m). In this sector, the Montchal lava flows northward from the point 700 m on the L3 section. A second Montchal lava flow appears at the southern end of the L3 section, between 1175 and 1245 m. On the section L1, the Montchal lava flow is located between 1250 m (proximal) and 645 m. The abnormal change in the flow geometry between 900 m and 600 m could be attributed to the edge of the Estivadoux maar.

A GPR survey was conducted on the road with a direction parallel to ERT L1 sections (except for the 200 first meters with a direction parallel to ERT L3). The beginning of the GPR survey is the edge of the crater. Data collected with the 500 MHz antennae allow high resolution images of the subsurface and highlight bedding features from the surface to about 3.6 m (Fig. 6.15). In the proximal area (0–700 m), the topographic surface has a gentle slope near 2°. The continuous radar section shows the upper part of the P4 unit, according to the Pavin drilling located at 325 m. The deposits are characterized by subhorizontal reflectors (except near 200 m) and by the presence of diffraction with hyperbolae and attenuated area just below which indicates the presence of ballistic blocks. Of particular interest, the area between 400 and 500 m is littered with metric blocks which represent the last explosion (one example is surrounded). This is consistent with our field observations along the profile and the localization of the largest ejected blocks with a diameter of 4.5 m (Lorenz 2007). Between 550 and 750 m, the northwest dip is an artefact because no topographic correction is done here; this area corresponds to a natural higher slope (10–15°) associated with the eastern boundary of the Montchal lava flow. In the intermediate and distal area (700–1350 m), three reflectors are well marked: at a depth of 1 m, 2 m and 2.7 m. According to the Clidères section (point 950 m), they correspond respectively to the boundaries between P4-P3, P3-P2 and P1-Montchal. Surprisingly, the P4 deposit partially molds and fills an undulating topography composed of subparallel reflectors of P1-P2-P3 deposits. The attenuated reflection area near the base of P4 is explained by the water content of this permeable unit and the impermeable behavior of the P3 fine ash unit. The P4 unit has characteristics of pyroclastic surges because it fills the topographic lows, have oblique reflectors corresponding to cross-bedding (surrounded area near 1090 m and 1200 m) and some little blocks with hyperbola reflections.

6.6 Discussion

6.6.1 The Pavin Eruption Dynamics

The Pavin layers are characterized by alternating base surge associated with ballistic blocks and ash fall beds which represent instantaneous pulses of explosive activity during the formation of the crater. Consequently, a detailed study of the vertical variations of the maar deposits, based on field textures, componentry, SEM morphoscopy of juvenile ash particles and geometry of beds provides a way to access the changes in dominant eruptive styles. During the Pavin eruption, the 4 main units occurred in a relatively short period of time, almost simultaneously at the geological scale (no paleosol).

According to the Lorenz' model (1986), the Pavin maar could result from fluctuating eruptive conditions related to changing magma-water interactions.

P1 unit is a typical period of alternating base surge and ash fall. In this case, the ascending trachy-andesitic magma enters in contact with water. During the P2 phase, the magmatic component is more expressed, with fragmentation and eruption driven by exsolved magmatic volatiles, resulting in the generation of a sustained plume with a subplinian intensity. The fallout products mostly consist of pumice lapilli-rich deposits, relatively lithic-poor. During the P3 sequence, depending on the Lorenz's model, the largest water availability in the aquifer just above the basement could lead to the best conversion rate of energy into fragmentation, thus resulting to fine particles and to a stratified pyroclastic surges sequence. Then, a dramatic change occurs: within P4, fragments from the basement are more abundant, indicating deeper magma-water interactions. Lithic clasts data show that the depth of magma-water explosive interaction was in the order of a few 100 m based on the presence of granitic gneissic lithic rocks in the deposit. According to the Pavin drilling, the depth of the Variscan basement is inferred near an altitude of 1050 m, i.e. 170–200 m below the top of the tuff ring (Bourdier 1980). In the basement, there is a quasi-absence of aquifer. The evolution could confirm the enlargement of the diatreme, and its deeper evolution with time, as indicated by Lorenz (2007).

However, interacting with the depth of fragmentation, pulsating mass eruption rates control the fluctuating magmatic-hydromagmatic activity (Sheridan and Wohletz 1983; White and Houghton 2000). So, the P2 phase coincides with the highest magmatic eruption rate. Moreover, the evolution towards increased fragmentation and a more hydromagmatic character during P3 may reflect the progressive depletion of magmatic volatiles and a decrease in conduit pressure during the last stage of the eruption. The P4 unit could reflect a decrease of the mass eruption rate with a better crystallized magma composed of less volatiles.

The last interdependent factor that can control the fluctuating magmatic-hydromagmatic activity is the aquifer yield. Generally, maars are found in subaerial areas underlain with hard rocks of varying compositions where groundwater is usually located in hydraulically active zones of structural weakness such as the Eifel in Germany (Lorenz and Büchel 1980) or they occur in soft rock environments as grabens with synsedimentary volcanism such as the Limagne basin in France (de Goër 2000). The localization of the Pavin maar belongs to the first category. The aquifer was formed in fractured rocks of predominantly volcanic lavas and granitic gneissic basement. The maar is in exact coincidence with the ancient valley floor corresponding to the Montchal lava flow.

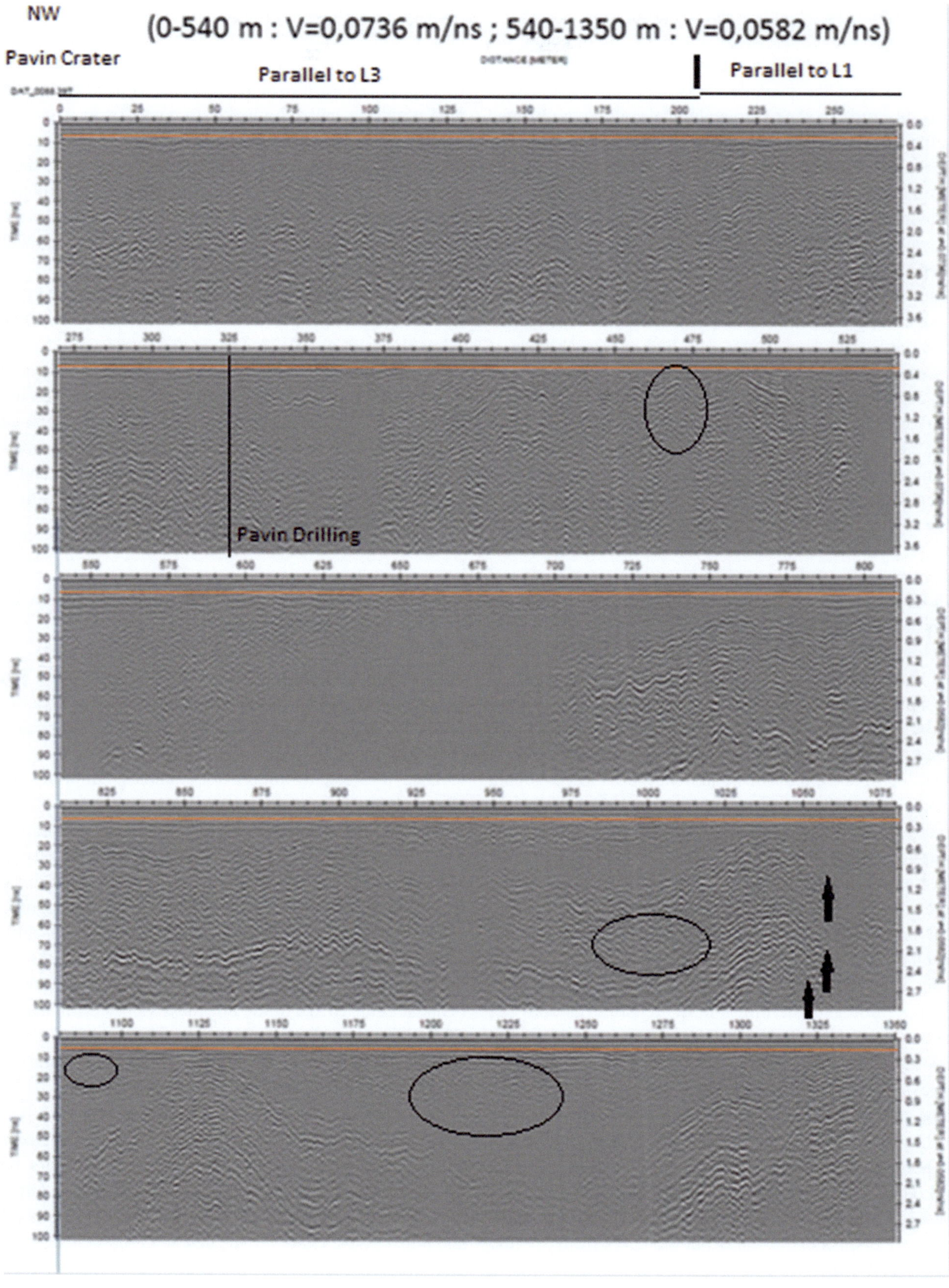

Fig. 6.15 GPR section near parallel to ERT L1 section (Zero is near Pavin crater and 1350 is near Costes outcrop). Two velocities are applied (0–540 m: 0.0736 m/ns; 540–1340 m: 0.0582 m/ns) because of lateral velocity variation revealed by CMP. As no topographic correction is made, the zero depth is the ground surface

This suggests that stream water entering from above, or ground water circulating along joints or faults below the valley floor, provided the source for the large amounts of gas required to eject the juvenile rocks and wall-rock debris. During the eruption, the hydrogeological conditions changed with the nature of the country rocks and the wall rock collapsed in the diatreme zone. Moreover, abundant basement clasts in the upper part of the deposit (P4) could indicate that the Pavin maar formed during the downward migration of foci explosions as the depression cone formed in a shallow aquifer according to the Lorenz's model.

Another interpretation of the Pavin eruption dynamics can use the results of the scaled subsurface blast experiments with variable explosion depths, and presence or absence of preexisting crater (Valentine and White 2012; Valentine et al. 2012, 2014; Graettinger et al. 2014). In recent experimental subsurface explosions (Goto et al. 2001; Valentine et al. 2014), the crater morphology and ejecta dynamics are largely determined by scaled depth, $D_{sc}=d.\ E^{-1/3}$, where d is the physical depth of the explosion site and E is the mechanical energy (J) produced by the explosion. For the optimal scaled depth ($D_{sc}\approx 0.004$ m $J^{-1/3}$), the crater has its largest value. With an increase of the scaled depth relative to the optimal depth of burial for a given explosion energy, the resulting craters are smaller until $D_{sc}\approx 0.008$, which marks the approximate transition to non-eruptive explosion (Graettinger et al. 2014; Valentine et al. 2014). Because most phreatomagmatic explosions have energies between 10^9 and 10^{13} J, the non-eruptive depth calculated from the equation is in the range between 8 and 172 m. So, the explosions mostly occur at depth < 200 m and explosions that contribute most to tephra ring deposits are likely to occur at depths < 100 m (Valentine et al. 2014). This is consistent with our field data because (1) the depth of the gneissic basement is 170–200 m below the highest top of the Pavin tuff ring and near 100 m below the lowest top of the tuff ring, (2) the gneissic lithics are rare or absent in P1 and (3) the shallow-derived lava lithics are between 60 and 80 % of the lithic fraction of P2, P3 and P4 deposits at Clidères (Fig. 6.7). However, note that the presence of granitic gneissic lithics in the tephra beds does not mean that the explosion occurred at the basement depth, but rather, according to Valentine and White (2012), that some prior explosions occurred at depth and mixed the lithics upward until they could be ejected by shallow explosions.

In the field-scale analog experiments, the explosion jets have heights and shapes that are strongly controlled by scaled depth and by the presence or absence of a crater. Jet properties influence the distribution of ejecta deposits outside the craters. As explosion depth increases from optimal depth (largest crater) to non-eruptive explosion depth, the eruption jets become increasingly vertically focused. In this case, the vertical core of the jet is associated with rapid fall and sedimentation of coarse material in the crater. The fall induces the formation of dilute fine-grained pyroclastic density currents outer the crater (Valentine et al. 2012; Graettinger et al. 2014). This mechanism could explain the P3 deposits characterized by very fine-grained clasts associated with rare ballistic blocks. In comparison with P1, the abundance of surges in P3 could result in the deepening of explosions or in the presence of a deeper crater after the subplinian P2 phase. This mechanism could also explain the greater relative abundance of granitic gneissic lithics in P4, because the ballistic blocks, vertically focused, fallback in the crater during P3 phase.

Finally, in the blast experiments, the deep-seated materials are not ejected as far as shallow-seated materials (Graettinger et al. 2014). This is consistent with the difference between the two reference sections and the Pavin drilling. The La Liste section, 2.4 km distant from the crater, contains less granitic gneissic rock in the lithic fraction (mostly 0–10 %) than the Clidères reference section, localized 1.5 km from the crater, which has between 20 and 33 % granitic gneissic lithics and the Pavin drilling (500 m from the crater) with 20–50 % granitic gneissic lithics.

6.6.2 Dispersal Area and Volume of the Pavin Volcanic Deposit

After defining the main characteristics of the deposit, a mapping campain was performed in order to map the area of dispersion and to build an isopach map for the Pavin deposit (Fig. 6.4).

Data available from previous studies (Bourdier 1980; Melet 2009; Lêvèque and Vaillant 2010; Davesne and Demoulin 2010; Batailler and Jallais 2012; Jaillard and Zylberman 2012) was collected and synthesized on a map built with the ArcGIS software (©ESRI, 2010). To this data were added some thickness measures from the significant work carried out by Lavina (Lavina 2002; Lavina and del Rosso d'Hers 2006, 2009; del Rosso d'Hers et al. 2008) for the achievement of the geological map of the Pavin area (Thonat et al. 2015). Some thickness values estimated from ERT sections or from digging were also added to the map (Lêvèque and Vaillant 2010; Davesne and Demoulin 2010; Batailler and Jallais 2012; Jaillard and Zylberman 2012).

Based on the data, an isopach map is proposed for the Pavin deposit (Fig. 6.16). The extension of the volcanic cloud to the south-east is coherent with Bourdier's result (1980). This extension axis is also consistent with information given by sedimentary structures such as cross-bedding and impact sags. However, the new map shows a more irregular and asymmetrical shape, with locally important thickness variations. Those changes are mostly reflecting the variability of the topography before eruption (Montchal scoria cone and lava flows with rootless cones) but also last

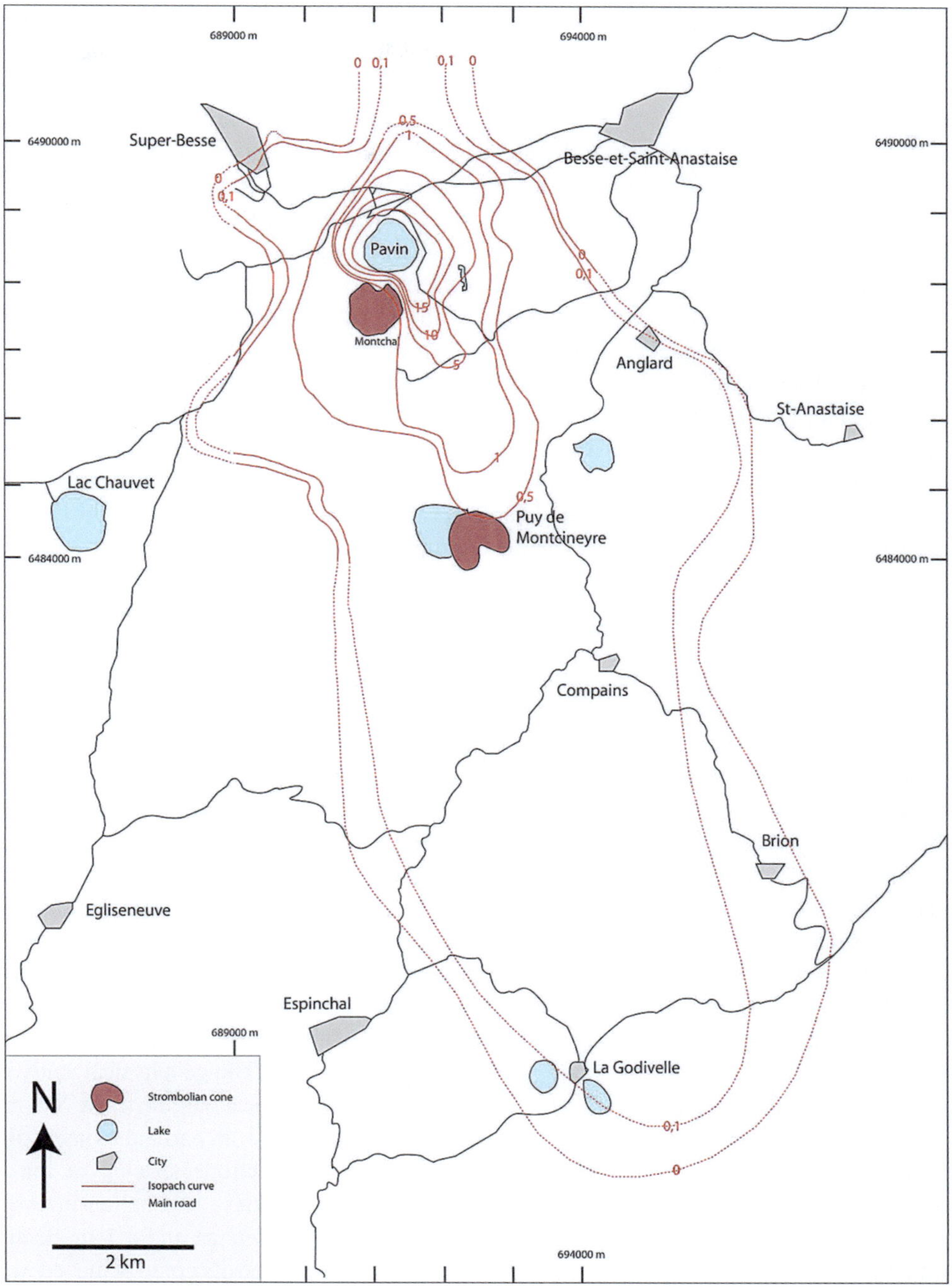

Fig. 6.16 Isopach model of the Pavin deposit (thickness and grid in meters, RGF93 – Lambert93)

eruptive dynamics dominated by surges deposits controlled by the topography, even if significant variations may be due to later erosional processes. To avoid at the maximum the effect of post-eruptive erosional processes, some anomalous points with very low thickness values were not taken into account.

Most important thickness variations result in isopachs undulations (see isopachs 0.5 m and 1 m east to the Pavin, and between the Montchal and Montcineyre cones). Such geometry cannot only result from a unique dynamism such as wind-controlled ash fall deposits. Thus, it is likely that this phenomenon expresses when some particular volcanic dynamism such as base surges flow on an irregular topography at intermediate distances from the vent. The bulk of sedimentary structures, such as cross-stratification, are also mostly found in this zone which is consistent with this observation.

On the Fig. 6.16, dashed isopach curves indicate that the drawing is not based on measured thicknesses but that it is either an interpretation from this study (to the west) or based on other observations: to the south and east, dashed isopach curves are based on Bourdier (1980), and to the north it is based on observations from Lavina who found rare outcrops of preserved fine trachytic particles attributed to the Pavin (Thonat et al. 2015).

Based on the digitalized isopachs surfaces (Surfer 8, Golden Software Inc., 2002), the calculated total area is 146 km^2 and the calculated total volume 5.2×10^7 m^3. This new area is 51 km^2 bigger (54 %) and the volume is 2.3×10^7 m^3 lesser (31 %) than previously estimated with 95 km^2 and 7.5×10^7 m^3 (Bourdier 1980), which is mainly due to the estimation of a thinner proximal deposit. However, it is today known that fine particles from the Pavin eruption extended far beyond what was previously thought: to the south, up to 60 km on the Cantal and Cézallier reliefs (Juvigné and Gilot 1986; Gewelt and Juvigné 1988) and to the north, up to 30 km on the Chaîne des Puys *s.s.* (Juvigné et al. 1988; Juvigné and Stach-Czerniak 1998). The amount of ash loss during maar eruption is difficult to estimate; in Italian examples, it was estimated to 50 % (Giaccio et al. 2007; Sotilli et al. 2012), then the estimated total volume could be near 7.8×10^7 m^3. The Pavin maar has a volume of erupted products in the range typical of maar and tuff-ring volcanoes $1 \times 10^5 - 1 \times 10^9$ m^3 (White and Ross 2011).

Based on the recent geomorphology section of the lake (Chapron et al. 2012), the crater volume yields a value of 3.7×10^7 m^3. For comparison, within the deposits, the proportions of lithic clasts are approximately 40 and 33 % in proximal outcrops and 33–20 % on distal outcrops. So, the excavated volume is estimated to the third of the total volume of deposits, that is approximately 1.7×10^7 m^3 (if 50 % of ash loss, 2.6×10^7 m^3). Probably the proportions of lithic clasts are underestimated in the proximal area (according to the experiments of Valentine et al. 2012), the excavated volume could be the half of the total volume of the deposits, *i.e.* 2.6×10^7 m^3 (and near 3.7×10^7 m^3 if 40 % of ash was lost). Also, the estimated volume for the lithic component in the erupted products corresponds broadly to the excavated volume if we consider 40 % of ash loss.

The asymmetry of the disposal area could be induced by laterally shifting explosion sites during the eruption. In analogy with the migration of explosions in field subsurface experiments (Valentine et al. 2015), a tilted-jet can result from the localization of the explosion site near one side of the crater. In this case, the material is mostly dispersed in the same direction than the jet and in the opposite direction of the nearest crater wall. Even if the experiments are most directly related to early phase of maar development, Valentine et al. (2015) expect that the phenomena will be similar to a mature maar. Application to the Pavin deposit could indicate that the major site of explosions is near the northwest crater wall.

6.7 Conclusion

A new complete tephrostratigraphy of the Pavin volcanic deposit is defined with a new reference section named Clidères. The 26 tephra beds and bed sets correspond to four volcanic units. The deposits are composed of high energy basal surges, ballistic blocks, lapilli fall and mixed dynamisms. The vertical variations of the maar deposits provide a way to access the fluctuating eruptive conditions related to changing magma-water interactions of the four main phases. The changes are associated to simultaneous variations of three factors: the pulsating mass eruption rates, the depth of fragmentation and the aquifer yield. However, a more comprehensive study of facies associations and their lateral distributions is needed to reconstruct the explosion conditions that occurred during the eruptive sequence.

In comparison with trachytic or benmoreitic volcanoes of the Chaîne des Puys, the Pavin maar shows the particularity that no extrusion of a gas-poor magma took place within the crater at the end of the eruption. It seems thus likely that a gas-poor part of the magma was not able to reach the surface but now rests at a shallow depth (Bourdier and Vincent 1980). The last phase, P4, has some biotite-rich juvenile fragments which could correspond to the north cloud identified by Juvigné and Miallier (2016). A more detailed mapping of proximal to intermediate products is necessary to identify the extension of deposits of each phase.

Based on the combination of two geophysical methods, ground penetrating radar and electrical resistivity surveys, the boundaries of the volcanoclastic deposits are visualized and the average thickness of the formation is followed from proximal to intermediate locations. Including the combination of field observations of near 50 trenches of 1–2 m depth, a core drilling and geophysical profiles, the total volume of deposits is now estimated at 5.2×10^7 m^3.

The geophysical study could be improved by longer profiles along the main N-S dispersal axis and orthogonal E-W axis. In particular, using 50 or 100 MHz antennae and a 5 m electrode spacing in the thicker proximal deposits could achieve to a higher investigation depth and using 500 MHz antennae and a 1 m electrode spacing in the distal thinner deposits could achieved to a better resolution.

Acknowledgements This work was funded by the Institut Polytechnique LaSalle Beauvais. We give special thanks to field technicians Antoine Auffray, Thibaut Allary and Camille Dufeu and to engineering students, without whom digging and carrying geophysical equipment would not be so easy. We thank Alexandra Batailler, Erika Döhring, Nolwenn Jallais, Caroline Melet, Jean-Marie Davesne, Mathieu Demoulin, Yowen Lévêque, and Matthieu Vaillant for useful discussions and sharing observations and opinions in the field.

We are grateful to Lahcen Zouhri and Claire Podevin for discussions on hydrogeological characteristics in volcanic fields. We wish to thank Philippe Rocher (BRGM-Auvergne, Clermont-Ferrand), Philippe Rossi and Olivier Blein (BRGM-CDG, Orléans) providing us the authorization to use the drilling results on mecanical augers and share mapping data from Pierre Lavina to the geological map Besse N°741 (Thonat et al. 2015, in press). The paper greatly benefited from the thorough revision of Jean-Louis Bourdier and Franck Donnadieu.

References

ABEM Instrument (2006) Instruction Manual – Terrameter SAS 4000 – SAS 1000

Batailler A, Jallais N (2012) Origine des dépressions subcirculaires énigmatiques de la plaine de La Liste (Massif Central français) volcanique, glaciaire ou anthropique ? Caractérisation des formations et apports de la géophysique. Mémoire d'Initiation à la Recherche n°570, Unpublished data. Institut Polytechnique LaSalle Beauvais, p 124

Boivin P, Jouhannel S (2016) The origin of lake Pavin: 200 years of controversy. Chapter 5, this book

Boivin P, Bourdier J-L, Camus G, de Goër de Herve A, Gourgaud A, Kieffer G, Mergoil J, Vincent PM, Auby R (1982) Influence de la Nature des Magmas sur l'Activité Phréatomagmatique: approche Volcanologique et Thermodynamique. Bull Volcanol 45:25–39

Boivin P, Besson JC, Ferry P, Gourgaud A, Miallier D, Thouret JC, Vernet G (2010) Le point sur l'éruption du lac Pavin il y a 7000 ans. Rev Sc Nat Auvergne 74:47–58

Bourdier JL (1980) Contribution à l'étude volcanologique de deux secteurs d'intérêt géothermique dans le Mont-Dore: le groupe holocène du Pavin et le massif du Sancy. Thèse de doctorat Sciences de la Terre : Université de Clermont-Ferrand. 196p

Bourdier JL, Vincent PM (1980) A geological and geothermal approach of the Pavin area, southern chaîne des Puys (France). Advances in European Geothermal Research, In: Proceedings of the second international seminar on the results of EC geothermal energy research, 238–249 Springer Netherlands: Amsterdam.

Büttner R, Zimanowski B (1998) Physics of thermohydraulic explosions. Phys Rev E Stat Nonlinear Soft Matter Phys 57:5726–5729

Cagnoli B, Ulrych TJ (2001a) Downflow amplitude decrease of ground penetrating radar reflections in base surge deposits. J Volcanol Geotherm Res 105:25–34

Cagnoli B, Ulrych TJ (2001b) Ground penetrating radar images of unexposed climbing dune-forms in the Ubehebe hydrovolcanic field (Death Valley, California). J Volcanol Geotherm Res 109:281–300

Camus G, (de) Goër De Herve A, Kieffer G, Mergoil J, Vincent PM (1973) Mise au point sur le dynamisme et la chronologie des volcans holocènes de la région de Besse-en- Chandesse (Massif Central français). C R Acad Sci 27:629–632

Carrasco-Nunez G, Ort MH, Romero C (2007) Evolution and hydrogeological conditions of a maar volcano (Atexcac crater, Eastern Mexico). J Volcanol Geotherm Res 159:179–197

Chapron E, Ledoux G, Simonneau A, Albéric P, St-Onge G, Lajeunesse P, Boivin P, Desmet M (2012) New evidence of holocene mass wasting events in recent Volcanic Lakes from the French Massif Central (Lakes Pavin, Moncineyre and Chauvet) and Implications for natural hazards. In: Yamada Y. et al. (ed), Submarine Mass Movements and their Consequences, pp 255–264

Davesne JM, Demoulin M (2010) Volcanologie du système Pavin: caractérisation du puy de Montchal et de ses produits. Mémoire d'Initiation à la Recherche n°49, Unpublished data. Institut Polytechnique LaSalle Beauvais, p 156

(de) Goër De Herve A (1997) Volcans d'Auvergne: la menace d'une éruption ? Editions Ouest-France, Luçon, 64 p

del Rosso d'Hers T, Lavina P, Levy-Faict TW (2008) Le complexe volcanique Montchal – Pavin – Montcineyre: caractérisation des risques naturels associés. 18ème réunion Annuelle des Sciences de la Terre, Nancy.

Fauchard C, Mériaux P (2004) Méthodes géophysiques et géotechniques pour le diagnostic des digues de protection contre les crues – Guide pour la mise en œuvre et l'interprétation. Cemagref Editions, Paris, 124p

Gewelt M, Juvigné E (1988) Téphrochronologie du tardiglaciaire et de l'holocène dans le Cantal, le Cézallier et les Monts Dore (Massif Central, France): résultats nouveaux et synthèse. Bulletin de l'Association française pour l'étude du Quaternaire 1:25–34

Giaccio B, Sposato A, Gaeta M, Marra F, Palladino DM, Taddeucci J, Barbieri M, Messina P, Rolfo MF (2007) Mid-distal occurences of the Albano Maar pyroclastic deposits and their relevance for reassessing the eruptive scenarios of the most recent activity at the Colli Albani volcanic district, central Italy. Quat Int 171–172:160–178

Glangeaud P (1916) Le cratère-lac Pavin et le volcan de Montchalm (Puy-de-Dôme). C R Acad Sci 162:428–430

Goër De Herve A (de). (2000) Peperites from the Limagne trench (Auvergne, French Massif Central): a distinctive facies of phreatomagmatism pyroclastics, history of a semantic drift. In: Leyrit H, Montenat C (eds) Volcaniclastic Rocks, from magmas to sediments. Gordon and Breach Science Publisher, Amsterdam, pp 91–110

Gómez-Ortiz D, Martín-Velázquez S, Martín-Crespo T, Márquez A, Lillo J, López I, Carreño F, Martín-González F, Herrera R, De Pablo MA (2007) Joint application of ground penetrating radar and electrical resistivity imaging to investigate volcanic materials and structures in Tenerife (Canary Islands, Spain). J Appl Geophys 62:287–300

Goto A, Taniguchi H, Yoshida M, Ohba T, Oshima H (2001) Effect of explosions energy and depth to the formation of blast wave and crater: field explosion experiment for the understanding of volcanic explosion. Geophys Res Lett 28:4287–4290

Graettinger AH, Valentine GA, Sonder I, Ross P-S, White JDL, Taddeucci J (2014) Maar-diatreme geometry and deposits: subsurface blast experiments with variable explosion depth. Geochem Geophys Geosyst 15:740–764, doi: 10.1002/2013GC005198

Houghton BF, Wilson CJN, Smith RT, Gilbert JS (2000) Phreatoplinian eruptions. In: Sigurdsson H (ed) Encyclopedia of volcanoes. Academic, San Diego, pp 513–525

Jaillard A, Zylberman W (2012) Caractérisation géophysique et pétrologique des dépôts volcaniques du cratère-lac Pavin. Nouvelle estimation du volume de la nuée. Mémoire d'Initiation à la Recherche n°591, Unpublished data. Institut Polytechnique LaSalle Beauvais, p 126

Juvigné E, Gilot E (1986) Ages et zones de dispersion de téphra émises par les volcans du Montcineyre et du lac Pavin (Massif Central, France). Zeit der deutschen geologischen Ges 137:613–623

Juvigné E, Miallier D (2016) Distribution, tephrostratigraphy and chronostratigraphy of the widespread eruptive products of Pavin volcano. Chapter 8. This book

Juvigné E, Stach-Czerniak A (1998) Etude sédimentologique et palynologique des dépôts lacustres tardiglaciaires et holocènes du Gour de Tazenat (Massif Central, France). Quaternaire 9(1):15–23

Juvigné E, Lousberg N, Gewelt M (1988) Evolution morphosédimentaire de la Narse d'Espinasse. Rev Sc nat d'Auvergne 53:7–14

Krumbein WC, Sloss LL (1956) Stratigraphy and sedimentation. Freeman and Company, San Francisco

Lavina P (2002) Rapport d'activité de l'année 2001–2002 pour le levé de la carte géologique de Besse n°741: cartographie géologique et campagne de sondages à la tarière

Lavina P, del Rosso d'Hers T (2006) Les éruptions volcaniques holocènes du système Montcineyre – Montchal – Pavin: les nouvelles découvertes. 16ème réunion Annuelle des Sciences de la Terre, Dijon.

Lavina P, del Rosso d'Hers T (2009) Le système volcanique du groupe Montchal – Pavin: nouvelle stratigraphie des formations volcaniques et volcano-sédimentaires et nouvelles datations, volcanologie dynamique et structurale. Colloque International Lac Pavin et autres lacs méromictiques, Besse-et-Saint-Anastaise

Lecoq H (1835) Description pittoresque de l'Auvergne, le Mont Dore et ses environs. Baillière ed, Paris, 437p

Lêvèque Y, Vaillant M (2010) Système Pavin: caractérisation de la dynamique éruptive du cratère-lac Pavin. Mémoire d'Aptitude à la

Géologie n°511. Unpublished data. Institut Polytechnique LaSalle Beauvais, p 121.

Loke MH (2003) RES2DINV Rapid 2-D Resistivity and IP inversion using the least squares method. Manual. Geotomo Software, Penang

Loke MH, Barker RD (1996) Rapid least-squares inversion of apparent resistivity pseudosections by a Quasi-Newton method. Geophys Prospect 44:131–152

Lorenz V (1973) On the formation of Maars. Bull Volcanol 37:183–204

Lorenz V (1986) On the growth of maars and diatremes and its relevance to the formation of tuff rings. Bull Volcanol 48:265–274

Lorenz V (1987) Phreatomagmatism and its relevance. In: Bottinga Y, Mathieu J-C (eds) Liquid Silicates, vol 62, Chem Geol., pp 149–156

Lorenz V (2007) Syn- and posteruptive hazards of maar-diatreme volcanoes. J Volcanol Geotherm Res 159:285–312

Lorenz V, Büchel G (1980) Zur Vulkanologie der Maare und Schlackenkegel der Westeifel. Mitt Pollichia 68:29–100

Melet C (2009) Montcineyre – Groupe Montchal – Pavin: Nouvelle caractérisation des formations et apport à la téphrochronologie du système Pavin. Mémoire d'Aptitude à la Géologie n°473. Unpublished data. Institut Polytechnique LaSalle Beauvais, 215 p.

Ménard JJ (1979) Contribution à l'étude pétrogénétique des nappes de ponces du massif volcanique du Mont-Dore. Thèse 3° cycle, Université Paris-Sud.

Olhoeft GR (1984) Applications and limitations of ground-penetrating radar. In: Expanded abstracts of the 54th Annual SEG International Meeting and Exposition. Atlanta, pp 147–148

Ollier CD (1967) Maars: their characteristics, varieties and definition. Bull Volcanol 31:45–75

Russel JK, Stasiuk MV (1997) Characterization of volcanic deposits with ground-penetrating radar. Bull Volcanol 58:515–527

Schrott L, Sass O (2008) Application of field geophysics in geomorphology: Advanced and limitations exemplified by case studies. Geomorphology 93:55–73

Sellmann PV, Arcone SA, Delaney AJ (1983) Radar profiling of buried reflectors and ground water table. Cold Regions Research and Engineering Laboratory Report 83-11, pp 1–10

Sheridan MF, Wohletz KH (1983) Hydrovolcanism: basic considerations and review. J Volcanol Geotherm Res 17:1–29

Smith DG, Jol HM (1992) Ground-penetrating radar investigation of a Lake Bonneville delta, Provo level, Brigham city, Utah. Geology 20:1083–1086

Sotilli G, Palladino DM, Gaeta M, Masotta M (2012) Origins and energetics of maar volcanoes: examples from the ultrapotassic Sabatini Volcanic Distric (Roman Province, Central Italy). Bull Volcanol 74:163–186

Thonat A, Blein O, Feybesse JL, Chévremont P, Morel M, Bouillier R et Lavina P (2015, in press). Carte géologique de Besse n°741. BRGM-SDG édition

Valentine GA, White JDL (2012) Revised conceptual model for maar-diatremes: Subsurface processes, energetics, and eruptive products. Geology 40:1111–1114

Valentine GA, White JDL, Ross P-S, Amin J, Taddeucci J, Sonder I, Jonhson PJ (2012) Experimental craters formed by single and multiple buried explosions and implications for volcanic craters with emphasis on maars. Geophys Res Lett 39, L20301. doi:10.1029/2012GL053716

Valentine GA, Graettinger AH, Sonder I (2014) Explosion depth for phreatomagmatic eruptions. Geophys Res Lett 41:3045–3051

Valentine GA, Graettinger AH, Macorps E, Ross P-S, White JDL, Döhring E, Sonder I (2015) Experiments with vertically- and laterally-migrating subsurface explosions with applications to the geology of phreatomagmatic and hydrothermal explosion craters and diatremes. Bull Volcanol 77:15. doi:10.1007/s00445-015-0901-7

Van Overmeeren RA (1998) Radar facies of unconsolidated sediments in The Netherlands: a radar stratigraphy interpretation method for hydrogeology. J Appl Geophys 40:1–8

White JDL, Houghton B (2000) Surtseyan and related phreatomagmatic eruptions. In: Sigurdsson H (ed) Encyclopedia of volcanoes. Academic, San Diego, pp 495–513

White JDL, Ross PS (2011) Maar-diatreme volcanoes: a review. J Volcanol Geotherm Res 201:1–29

Wohletz KH, Sheridan MF (1983) Hydrovolcanic explosions II: evolution of basaltic tuff rings and tuff cones. Am J Sci 283:385–413

Wolhetz KH (1986) Explosive magma-water interaction: thermodynamics, explosive mechanisms, and filed studies. Bull Volcanol 48:248–264

Magmatic Evolution of Pavin's Group of Volcanoes: Petrology, Geochemistry and Modeling of Differentiation Processes. A Preliminary Study

7

Benoît Villemant, Benoît Caron, Pauline Thierry, and Pierre Boivin

Abstract

The volcanoes of the Pavin group (Montcineyre, Estivadoux, Montchal and Pavin) represent the most recent (~7 ky) volcanic activity in the Chaîne des Puys s.l.) and also occupy a particular geographical and structural position in this volcanic chain. Petrology and geochemistry of the volcanic products (lava flows, scoriae, bombs and pumice clasts) of the different edifices show that they define a magma differentiation series from primary basalts (Montcineyre lava flow) to benmoreites (Pavin pumice clasts) with close similarities with that of the Chaîne des Puys (stricto sensu). It is however characterized by some significant difference in the relative behavior of K_2O and of numerous trace elements such as Sc, Co, Y, REE, Nb, Ta. These specific behaviors, in particular the relatively high bulk partition coefficients of Nb and Ta are characteristic of the fractionation of amphibole which is the only major mineral phase able to significantly fractionate these incompatible elements in basaltic to benmoreitic magmas. A quantitative modeling of the differentiation process using major and trace element partition coefficients is proposed which evidence the very early fractionation of amphibole from the most primary basaltic melts. This result is consistent with petrological observations and confirms the earlier and more efficient fractionation of amphibole at the expense of clinopyroxene in magmas of the Pavin group compared to those of the Chaîne des Puys (stricto sensu). The results are used to discuss the influence of these specific differentiation conditions involving large H_2O contents in primary melts likely contributing to the particularly high explosivity of the Pavin eruption which produced benmoreitic to trachytic magmas.

The particular features of the Pavin group within the Chaîne des Puys volcanic system, namely, the southernmost position, the younger age, the specific magmatic evolution from the most primary basalt ever erupted in Chaîne des Puys, coupled with a short activity duration (~400 years) justify a renewal of petrological and geochemical interest for these volcanoes.

Keywords

France • Massif Central • Chaîne des Puys • Lake Pavin • Geochemistry • Trace elements • Magmatology

7.1 Introduction

On the basis of the petrology, mineralogy and major and trace element composition of the lavas of the Chaîne des Puys, Maury et al. (1980) and Villemant et al. (1980) have shown that this magma series evolves following a crystal fractionation process. Amphibole fractionation plays a critical role in its evolution, at least from the intermediate magmas (Mugearite, Benmoreite). The detailed analysis of

B. Villemant (✉) • B. Caron • P. Thierry
Laboratoire de Pétrologie, Géochimie et Minéralogie Magmatiques, Sorbonne Universités, UPMC Univ Paris 06, CNRS, Institut des Sciences de la Terre de Paris (iSTeP), 4 place Jussieu, 75230 Cedex 05 Paris, France
e-mail: benoit.villemant@upmc.fr

P. Boivin
LMV, Université Clermont-Auvergne - CNRS - IRD - OPGC, Aubière, France

T. Sime-Ngando et al. (eds.), *Lake Pavin*, DOI 10.1007/978-3-319-39961-4_7

trace element distribution coefficients and modeling of the differentiation process has led Villemant et al. (1981) to the conclusion that amphibole fractionation was effective from the most primary magmas. On the basis of experimental petrology Foury (1983) reached similar conclusions. The field and petrological observations of Bourdier (1980) on eruptive products of the volcanoes of the Pavin group also evidenced the especially important role of amphibole fractionation in these magmas: large abundance of amphibole phenocrysts in lava flows (resorbed; Montchal) or in basaltic tephra (unresorbed; Montcineyre, Estivadoux), and in all Pavin products (pumice, bombs, amphibole cumulates). Furthermore, a recent petrological study of trachytic magmas of the Chaîne des Puys (Martel et al. 2013) shows that from a same initial differentiation series (basalt to intermediate magma) various P and T conditions have prevailed to produce the large variety of trachytic magmas erupted during the whole volcanic history of this volcanic chain.

This preliminary geochemical study shows that the different magmas which have been emitted by the volcanoes of the Pavin group on a relatively short interval of time (~500 years) have evolved following a single differentiation process from primary basalts to benmoreite. This magmatic activity produced the most primitive basalts (Montcineyre lavas) ever erupted since 200 ky in the Massif Central. This differentiation process differs from that observed in further north volcanoes of the Chaîne des Puys (stricto sensu) by the evolution of some specific trace elements as Nb, Ta or LREE from the most primitive basalts. Modeling the differentiation process using major and trace elements distribution coefficients highlights a very early and substantial amphibole fractionation, leading to the particular features of the Pavin benmoreitic magmas and likely indicating specific differentiation conditions in relation with its specific eruptive activity.

7.2 Samples and Analytical methods

The magmas emitted by the four volcanic edifices of the Pavin group (see geological maps in chapter 6, Leyrit et al. this issue) have been sampled through their different volcanic products. Montcineyre volcano (basaltic lava flow), Estivadoux basaltic maar (scoria fall deposits), Montchal volcano (scoria fall deposit and basaltic lava flow) and Pavin benmoreitic tephra (cauliflower bombs and pumice fall deposit). Only few geochemical data have already been published for this volcanic group (see e.g. Villemant et al. 1981 for trace elements) and mainly concern the lava flows and the Pavin bombs and blocks. In this preliminary study we have completed the available sampling (mainly Estivadoux scoria falls and Pavin pumice fall) and performed new trace element measurements.

Volcanic products of the Pavin eruption are frequently contaminated by fragments of surrounding rock material (gneiss, granite and old lava fragments of Mont Dore volcano). The granite and gneiss xenoliths are more or less modified by interaction with the Pavin magma (thermal metamorphism, degassing, sometimes partial melting). As far as possible, these xenoliths have been manually separated before fine crushing. Some volcanic fragments still containing significant fractions of various types of xenoliths and bulk rocks of some basement xenoliths have also been analysed to estimate the influence of this contamination on bulk compositions of Pavin products.

Trace element analysis of around 40 samples of Pavin group magmas has been performed by ICP-MS (see below). For comparison, a suite of 35 samples representative of the whole magma series of Chaîne des Puys corresponding to the sampling of Villemant et al. (1981) have also been re-analysed, with the same method during the same runs (unpublished data). Major elements compositions of some representative samples of Pavin group magmas have also been analysed (ICP-AES, CRPG Nancy). The compositions of uncontaminated magma fragments (~30 analyses) are reported in Table 7.1. Finally some mineral separates of Pavin products (major mineral phases and glass of pumice fragments and cumulates) have been analysed by ICP-MS and by electron probe (CAMPARIS, UPMC) to calculate mineral melt distribution for quantitative modeling. Detailed analyses will be published elsewhere. Major element compositions of glass are reported in Fig. 7.1 along with bulk rock compositions.

*Analytical Method (ICP-MS) f*or trace element analyses, between 10 mg (minerals and glass separates) and 50 mg (powdered bulk rock samples) were dissolved in 5 ml of concentrated HF-HNO_3 mixture and then evaporated to dryness. The residue was dissolved in 2 ml of HNO_3-H_3BO_3 mixture and slowly evaporated to remove possible weakly soluble fluorides. The new residue was dissolved in 2 % HNO_3 (i.e., dilution factor of ~10^4) for analysis. For mineral separates, somewhat less material was used per sample (~10–20 mg), but dilution in the final solution is similar for all analysed samples. Trace element compositions of prepared samples were measured by inductively coupled plasma mass spectrometry (ICP-MS) using a X Series II instrument of Thermo Scientific at the University P& M Curie (LAGE-ISTEP). For each set of measurements, two aliquots of the geochemical reference material BHVO-2 were prepared according to the same procedure as for the samples and analysed every five samples to correct instrumental drift during analysis and to calculate unknown sample compositions. The mean reproducibility of the analytical procedure was estimated from a series of more than 20 independent measurements of BHVO-2 and reported in Table 7.1. The mean analytical error is ≤5 % for most trace elements, except Rb, Cs, Sc and Pb (error of ca. 10 %).

7.3 Results

7.3.1 The Pavin Group Magmas: Comparison with Chaîne des Puys (stricto sensu)

7.3.1.1 Major Elements

The different edifices and eruptive units of the Pavin group which have been analysed correspond to a large composition domain: basalt for Montcineyre lava flows, hawaïite for Montchal lava flow ('Couze Pavin' flow) and Estivadoux scoriae, benmoreite/trachyte for Pavin products (bombs and plinian clasts).The evolution of the major element compositions is closely similar to that of the Chaîne des Puys (stricto sensu) magmas (Fig. 7.1) with however two significant differences: the Pavin group magmas are slightly more K-rich as a whole, difference which increases with magma differentiation process and Montcineyre basalts are the most primitive basaltic lavas erupted since ~200 ky. The composition of Montcineyre basalts is compared to that of primitive melt inclusions of olivines from Puy Beaunit volcano, which are also considered as the most primitive melts analysed in Chaîne des Puys products (Jannot et al. 2005). Both compositions are very close (except in Fe and Ca), but with slightly more primitive characters for Montcineyre basalts. Post-entrapment evolution and analytical difficulties inherent in melt inclusions analysis may explain the discrepancies in Ca and Fe, and they require further investigations to address these uncertainties.

The benmoreite of Pavin and the trachyte of Clierzou – emitted by a volcano north of Chaîne des Puys have a very similar mineralogy (alkali-feldspar, abundant amphibole – Hb-, minor Cpx, Fe-Ti oxides) and are very close in major element composition. The Pavin benmoreite is however sigificantly more K-rich and less Na-rich than the Clierzou trachyte. Major element evolutions from basalts to benmoreite-trachyte are thus very similar for both volcanic systems and the differentiation process is mainly driven by crystal fractionation (Maury et al. 1980; Villemant et al. 1980, 1981).

Electron-probe analyses of the interstitial glass of Pavin benmoreite in cauliflower bombs and in amphibole cumulates are also reported in Fig. 7.1 along with glass separates of Chaîne des Puys (stricto sensu) lavas (Villemant et al. 1981; Lemarchand et al. 1987). These data plot in the trachyte domain and are wholly included within the main differentiation trend of Chaîne des Puys magmas (defined by both whole rock and glass compositions of benmoreites to evolved trachytes), except for K_2O : the glass of Pavin benmoreite displays a large enrichment in K_2O from ~5 % to ~6.5 %, at almost constant SiO_2 (~65 %). The residual glass of amphibole cumulates is the most evolved glass analysed in Pavin magmas.

7.3.1.2 Trace Elements

Trace element composition evolutions in function of Th content are reported in Fig. 7.2. Th is a highly incompatible element and used as a reference because of its very low bulk solid/melt partition coefficient in these magmas. Its relative variations are almost exactly proportional to the differentiation degree of the melt (or fraction of residual melt *f*; Villemant et al. 1981). REE compositions normalized to Chondrites are reported in Fig. 7.3.

Contrary to major elements, compositions and evolution trends in most trace elements are clearly distinct between magmas of the Chaîne des Puys (stricto sensu) and of the Pavin group. Though trace element compositions of initial basaltic magmas are very close the evolution trends differ markedly from the hawaïtes to the benmoreites, indicating significant differences in the differentiation processes.

Basalts Montcineyre basalts (in fact, basanites in classification IUGS) have particularly high contents in compatible elements (3d transition series trace elements: Ni ~240 ppm, Cr ~350 ppm and Co ~ 56 ppm and major elements: TiO_2 ~ 2.8 %, FeO_t ~ 11.2 %, MgO ~ 12.8 %) which indicates their highly primitive character. Significant effects of olivine accumulation which could explain the high MgO and Ni contenst in these lavas can however be excluded because they are also enriched in Ti Sc and Co which are not compatible in olivine.As for major elements, comparison with measurements in melt inclusions of Puy Beaunit olivine (Jannot et al. 2005) often show close similarities between both magma compositions. However, it is not clear if the lowest contents in incompatible elements (Th, LREE, Nb, Ta, Hf) reported in some melt inclusions are real or related to analytical difficulties (apparent 'dilution effect' of LA-ICP-MS relative to EPMA) because they are inconsistent with their relatively low contents in Fe, Mg or Ti (compare melt inclusion data in Figs. 7.1 and 7.2). Further studies are needed to explain the apparent discrepancies between trace and major element compositions of basaltic melt inclusions and bulk rocks. The bulk composition of Montcineyre basalt thus represents the best available representative of the primary melts generated by partial melting at depth during the recent volcanic activity of Chaîne des Puys and Pavin group volcanoes. It likely represents the common initial melt from which different differentiation processes have operated to produce the large diversity of trachytic magmas produced in this area. The trace element characteristics of these primary melts have been discussed in details by Jannot et al. (2005) using melt

Table 7.1 Major and trace element composition of Pavin group magmas

Sample	Montcineyre		Estivadoux	Montchal						Pavin			
	P13-8	P7699	P13-7TE	P13-7Sa	P13-7Sb	P7698	P13-4	P13-5	P13-9	KBP2b*	KBP3b*	PAV-0	PAV1-1
	Lava Flow			Scoria fall deposit		Lava Flow (couze Pavin)				Cauliflower bombs		Plinian Fall	
				Red (Top) Black (bottom)		*Besse en Chandesse*			*Sanner*	(1981, unpub.)	in Bourdier (1980)		
SiO_2	43.80						49.20			57.71	58.45		
TiO_2	2.80						2.60			1.37	1.30		
Al_2O_3	14.10						16.30			18.01	18.05		
Fe_2O_3	5.96						5.02						
FeO	5.80						4.93			5.99	6.11		
MnO	0.17						0.17			0.15	0.15		
MgO	12.70						7.25			1.78	2.11		
CaO	10.00						9.00			4.80	4.71		
Na_2O	3.20						3.70			5.50	5.01		
K_2O	1.45						2.20			4.50	4.11		
P_2O_5										0.37			
PF	0.06												
Total	100.04						100.37			99.73	100.00		
Trace elements ppm													
Rb	39.0	33.5	60.6	62.4	71.2	56.8	62.2	61.4	55.2	134.5	141.4	132.5	132.8
Cs	0.86	0.66	1.47	1.38	1.58	0.76	0.92	1.05	0.88	0.96	2.34	2.83	2.12
Sr	685	722	627	610	640	701	422	623	649	537	731	788	729
Ba	480	466	530	545	593	559	579	589	549	881	909	901	1006
Y	21.8	21.4	23.6	22.2	23.9	23.8	24.3	24.4	22.9	21.0	24.3	27.4	25.6
La	43.0	42.0	46.3	47.5	52.3	45.8	50.3	51.1	48.3	65.6	74.5	74.3	79.6
Ce	85.6	82.3	91.9	92.7	102.0	88.6	97.9	99.9	94.6	115.0	134.7	134.3	141.5
Pr	10.1	9.6	10.8	10.7	11.7	10.0	11.2	11.5	10.9	12.1	14.0	14.5	14.7
Nd	39.2	37.1	41.3	40.4	43.8	37.8	42.2	43.0	41.3	41.8	47.8	50.1	50.0
Sm	7.70	6.84	7.91	7.72	8.32	6.92	7.93	8.16	7.89	7.09	7.95	8.45	8.19
Eu	2.47	2.36	2.50	2.44	2.60	2.36	2.54	2.55	2.48	2.39	2.44	2.77	2.81
Gd	7.56	7.79	7.85	7.58	8.13	8.11	7.96	8.06	7.83	7.05	7.85	10.13	10.10
Tb	0.98	0.95	1.02	0.99	1.05	1.00	1.04	1.05	1.02	0.87	0.96	1.14	1.12
Dy	5.01	4.54	5.39	5.16	5.51	4.88	5.41	5.42	5.29	4.59	5.03	5.39	5.08
Ho	0.92	0.83	1.01	0.97	1.03	0.91	1.01	1.02	1.01	0.89	0.97	1.04	1.00
Er	2.34	2.22	2.63	2.55	2.72	2.51	2.62	2.66	2.58	2.45	2.68	3.03	2.96
Tm	0.30	0.28	0.36	0.35	0.36	0.33	0.36	0.37	0.35	0.35	0.38	0.41	0.39
Yb	1.80	1.64	2.15	2.12	2.22	1.94	2.17	2.21	2.16	2.29	2.44	2.65	2.52
Lu	0.26	0.24	0.31	0.30	0.32	0.28	0.31	0.31	0.30	0.34	0.36	0.38	0.37
Zr	235	219	270	278	298	280	300	304	281	448	497	474	507
Hf	5.18	4.61	6.06	6.18	6.44	5.55	6.39	6.47	6.18	8.63	9.48	8.84	9.41
Nb	59.9	54.3	61.9	63.3	68.8	61.0	69.3	69.2	64.6	106.4	110.2	100.8	110.3
Ta	3.41	3.02	3.70	3.79	4.08	3.34	3.92	3.97	3.79	5.17	5.96	5.39	5.77
Pb	2.28	3.67	4.43	4	5	3	4	4	4	8	10	9	9
Th	4.74	4.73	6.87	7.13	7.74	6.94	7.29	7.35	6.92	15.98	17.12	17.51	18.04
U	1.30	1.26	1.96	2.04	2.26	1.75	1.96	2.10	1.87	4.60	4.71	4.71	4.89
Sc	27	26	23	19	18	18	22	21	21	10	10	13	12
V	280	275	223	189	202	230	232	226	216	103	103	110	96
Cr	345	377	146	108	83	148	180	145	181	10	11	16	14
Co	56.4	57.1	33.5	30.4	32.0	35.8	37.9	35.3	35.3	10.9	11.8	11.5	10.5
Ni	237	221	57	51	46	73	78	64	75	6	8	9	7
Cu	51	67	38	42	40	54	52	45	44	14	12	36	47
Zn	83	89	90	70	78	85	80	79	73	69	70	67	65

* indicates that xenoliths fragments have been removed by hand

Trace elements in ppm. BHVO-2 is used as the reference for ICP-MS measurements. Reproducibility (in %) is calculated on 25 repeated measurements

															Reproducibility	
PAV1-11	PAV1-12	PAV-1-15	PAV1-2	PAV-1-25	PAV1-3	PAV1-4	PAV2-1	PAV2-3	PAV2-5	PAV2-6	PAV2-7	PAV2-8	PAV2-9	PAV1-7	BHVO-2 Ref	%
121.6	115.1	123.6	131.9	123.9	130.0	121.1	121.1	115.2	109.1	102.8	115.0	131.9	128.3	124.3	9.60	5.0
2.26	2.10	3.05	2.65	2.70	2.37	2.24	1.98	2.05	2.05	3.04	2.02	2.65	2.38	2.29	0.03	5.0
735	746	661	725	745	719	723	696	743	779	681	723	741	700	712	381	1.2
871	840	843	884	889	908	858	910	847	872	929	844	903	886	879	133	4.7
22.8	24.7	21.7	24.5	24.8	25.4	25.2	21.2	23.1	23.2	23.9	24.3	24.1	24.4	23.6	23.1	5.0
73.0	75.0	69.5	72.4	77.4	76.3	74.0	69.3	69.5	68.4	69.4	73.2	76.6	84.1	73.9	15.2	5.5
13.03	134.2	125.6	131.9	142.1	137.3	134.4	125.6	125.9	126.2	130.2	134.5	137.0	149.7	131.7	37.9	2.3
13.7	14.1	13.1	13.7	14.9	14.3	14.2	12.9	13.4	13.6	14.0	14.2	14.3	15.3	13.8	5.38	0.1
45.8	48.2	44.5	47.1	51.3	48.8	48.1	44.2	46.0	46.4	48.3	49.0	47.9	51.6	46.5	24.4	5.1
7.77	7.81	7.36	7.64	8.45	7.94	8.03	7.14	7.59	8.04	8.37	8.31	7.69	8.45	7.57	6.06	8.0
2.52	2.65	2.25	2.62	2.56	2.60	2.63	2.57	2.59	2.67	2.76	2.69	2.61	2.64	2.55	2.05	1.7
9.20	9.66	7.34	9.41	8.33	9.78	9.75	8.94	9.14	9.44	10.07	9.98	9.63	10.37	9.29	6.22	0.1
1.05	1.10	0.90	1.05	1.02	1. 10	1.10	0.98	1.07	1.08	1.14	1.11	1.07	1.11	1.06	0.93	3.8
4.77	5.08	4.66	4.90	5.22	5.16	5.20	4.45	4.90	5.08	5.27	5.24	4.97	5.30	4.80	5.25	0.2
0.92	0.98	0.90	0.93	1.01	0.99	1.01	0.87	0.96	0.96	1.01	1.02	0.97	1.03	0.92	1.00	3.7
2.72	2.82	2.50	2.74	2.76	2.93	2.92	2.62	2.74	2.85	2.96	2.95	2.82	3.03	2.81	2.51	1.0
0.38	0.38	0.36	0.36	0.40	0.41	0.40	0.36	0.38	0.40	0.41	0.41	0.39	0.41	0.39	0.34	8.0
2.33	2.46	2.34	2.35	2.52	2.51	2.57	2.02	2.44	2.50	2.71	2.58	2.51	2.67	2.56	1.99	0.4
0.34	0.36	0.34	0.35	0.37	0.36	0.36	0.31	0.34	0.34	0.37	0.37	0.34	0.38	0.36	0.27	0.4
475	483	440	469	473	503	496	467	457	474	465	465	499	493	485	174	4.0
9.05	9.14	8.81	8.49	9.24	9.44	9.42	8.90	8.70	9.11	8.93	8.95	9.24	9.60	9.14	4.28	0.4
105.3	106.3	99.3	103.1	111.8	110.8	109.8	103.4	101.6	107.1	107.2	104.6	111.3	111.8	107.4	17.2	6.0
5.59	5.65	5.71	5.33	6.16	5.87	4.63	5.53	5.29	5.66	5.67	5.52	5.90	5.60	5.72	1.06	5.0
9	8	11	9	10	9	8	8	9	9	10	9	9	9	29	2	8
17.72	17.11	16.70	16.89	17.32	18.30	17.41	17.31	16.68	17.74	18.57	17.44	17.97	18.84	17.95	1.12	4.5
4.96	4.71	4.68	4.56	4.73	5.01	4.82	4.08	4.60	5.17	5.35	4.76	5.10	5.20	5.05	0.39	4.0
12	11	10	15	9	13	12	11	13	12	12	12	10	11	11	29	8.4
83	90	81	96	94	96	105	81	105	101	113	114	86	84	83	310	3.0
9	10	11	14	10	11	16	9	16	13	18	29	7	7	8	283	2.7
8.6	9.2	8.8	10.4	10.4	9.8	10.8	8.8	11.8	10.5	13.1	12.6	8.8	8.9	8.7	46.5	4.0
6	6	7	7	5	6	7	6	9	8	12	14	5	6	5	121	2.6
55	61	70	16	16	78	135	14	37	34	37	47	59	76	56	117	5.5
61	74	69	76	76	67	67	64	64	61	68	63	67	88	61	95	7.0

of this standard

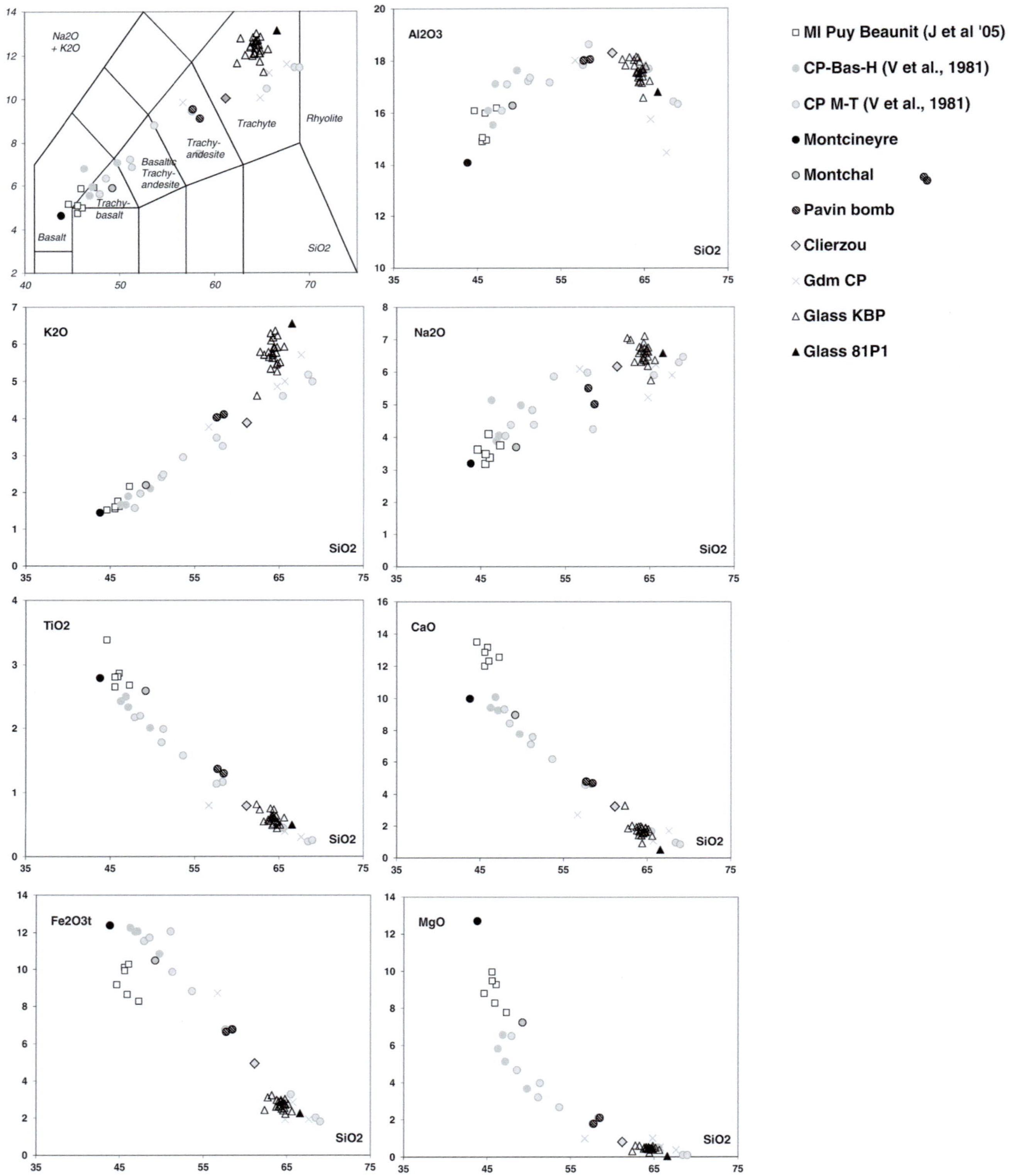

Fig. 7.1 Evolution of major element compositions (wt%) of Pavin group magmas. *Black dot*: Montcineyre basalt (lava flow); *circled grey dot*: Montchal hawaïite (lava flow); *dashed circles*: Pavin benmoreites (bombs). *Triangles*: glass compositions of Pavin benmoreite (*open triangles bombs* (KBP), *solid triangle*: residual glass of amphibole cumulates(81P1)). *Light grey dots*: compositions of Chaîne des Puys (stricto sensu) magmas (Bas-H: basalts- hawaïites, M-T: mugearites-trachytes; Villemant et al. 1981 and unpublished data). *Light grey crosses*: glass compositions of Chaîne des Puys (stricto sensu) samples (data from Villemant et al. 1981; Lemarchand et al. 1987). *Grey diamond*: amphibole-bearing trachyte of Clierzou volcano (north Chaîne des Puys). *Open squares*: compositions of melt inclusion in olivine of Puy Beaunit (north Chaîne des Puys; data from Jannot et al. 2005)

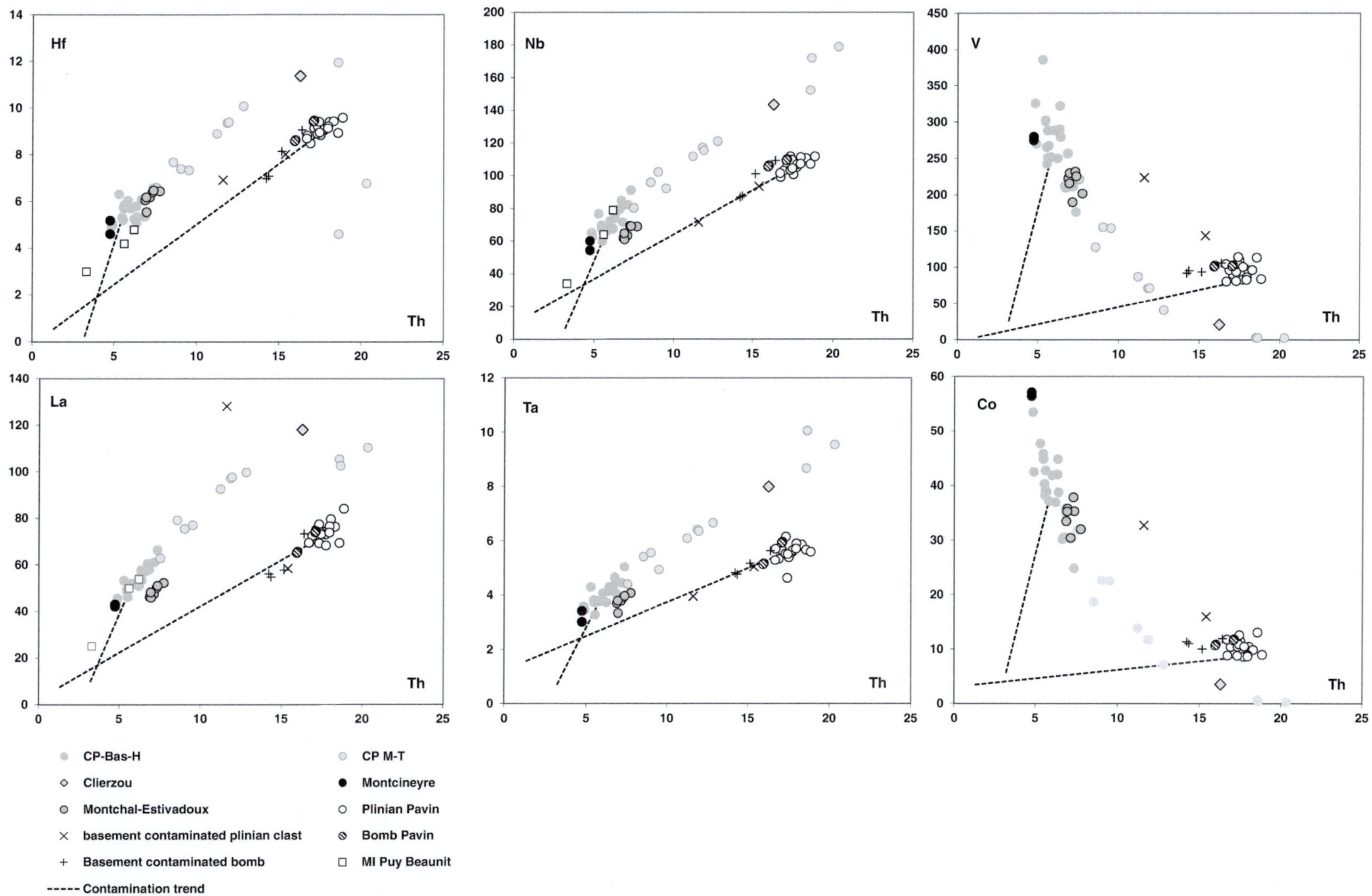

Fig. 7.2 Evolution of trace element compositions (ppm) of Pavin group magmas. *Black dots*: Montcineyre basalts (lava flows); *circled grey dots*: hawaïtes of Montchal (lava flows and scoria fall) and Estivadoux maar (scoria fall); Pavin benmoreites: *dashed circles* (cauliflower bombs), *open circles* (plinian clasts). *Crosses* : basement contaminated magmas (+ bombs, × plinian clast). Chaîne des Puys magma compositions (unpublished data): symbols as in Fig. 7.1. *Dotted line*: possible contamination trends of Pavin clasts by xenoliths (basement xenoliths or older basaltic fragments –unpublished data; not represented-) and emitted during the phreato-magmatic activity

inclusion compositions. Though some analytical limitations (see above) may slightly blur the geochemical signature of the primary magmas, the Montcineyre basalt and Puy Beaunit olivine melt inclusions are close enough to assume that the same mantle source produced the primary basalts of Chaîne des Puys (stricto sensu) and of Montcineyre at the origin of distinct differentiation suites.

Hawaïites they are represented by both the Puy Montchal lava flows (in particular the 'Couze Pavin' flow) and strombolian scoriae and by the tephra generated by the Estivadoux maar. The trace element compositions of Pavin group hawaïites are relatively close to those of Chaîne des Puys (stricto sensu) but are slightly more rich in Th, U, and significantly depleted in REE, Nb, Ta, Y and Ba. They have similar contents in compatible elements (3d transition series elements) except enrichment in Ni (80–180 ppm) against 20–100 ppm in hawaïites of Chaîne des Puys (stricto sensu). Clear evolution trends are defined by the Montcineyre basalts and the Montchal-Estivadoux hawaïites. They are clearly distinct from that of Chaîne des Puys (stricto sensu), except for some highly incompatible elements: U, Th, Rb, Zr and Hf. For all other incompatible (such as Nb, Ta or REE) and compatible elements (as 3d transition series elements) the trends defined in binary diagrams using Th as a reference (Fig. 7.2) are significantly distinct from those of the Chaîne des Puys (stricto sensu)

Pavin Benmoreites differentiated magmas are only represented in the phreato- magmatic and plinian products of Pavin eruption. Cauliflower bombs are magma fragments more or less rapidly quenched by interaction with rock and water of the surrounding basement. Due to this explosive interaction they are very frequently contaminated by solid fragments of basement rock material (mainly granite and gneiss) which is also sometimes (mainly for small < cm fragments) thermally metamorphosed and sometimes partially melted and/or degassed. Plinian clasts collected in the pumice fall deposit are less frequently contaminated. Though careful magma fragment selection was performed before crushing, accidental contamination by solid rock fragments or contamination by fluids and partial melts of interacting rocks may not be a priori discarded. In order to estimate the possible effects of these interactions, different xenoliths fragments and magmatic clasts with evidence of contamination have been analysed together with 'fresh' material. A detailed study of the interaction of surrounding material with Pavin magma is beyond the scope of this study. To test however the possible role of such contamination processes we have also reported in Fig. 7.2 the compositions of 'basement-bearing' magma fragments and two 'tie-lines' between xenolith fragments and their host magma (a Pavin cauliflower bomb and a basalt of Tartaret volcano, north the Pavin group). The compositions of contaminated benmoreites display linear trends which are close to the benmoreite-xenolith tie line. These common trends are clearly distinct from the basalt – hawaïite – benmoreite trends (Pavin group or Chaîne des Puys). In addition the basalt-xenolith tie line does not correspond to any magmatic trend. Finally, some contaminated clasts from plinian deposit display a distinct trend to the main contamination trend for compatible elements (see e.g. Cr, Co, Ni, Sc, V; Fig. 7.2) suggesting an additional contamination by basaltic clasts. These observations indicate that the late interactions of differentiated magmas and the surrounding rock material during their explosive eruption, do not significantly affect the melts compositions and that their effect are restricted to a pure mechanical contamination of erupted products. In addition, most plinian clasts of Pavin are uncontaminated and define a narrow composition range for all trace elements. Finally the basaltic and hawaïitic magmas are clearly unaffected by these processes.

Though close in major element composition (except K_2O) to the Clierzou trachyte, the Pavin benmoreite has very different trace element contents and specifically much lower contents in numerous incompatible elements such as REE, Nb, Ta, Zr, Hf and Ba. Pavin benmoreites are also significantly richer in 3d transition series elements compared to their benmoreitic and trachytic equivalents (i.e. at similar Th or SiO_2 contents) of Chaîne des Puys (stricto sensu). They are particularly rich in Co and V and in Ti and Fe.

The compositions in REE of Pavin group magmas, normalized to Chondrites are reported in Fig. 7.3 and compared to Chaîne des Puys (stricto sensu) compositions. It displays a typical REE pattern of alkaline series with a large enrichment in LREE relative to HREE ($(La/Yb)_N$ ~20) and a bulk enrichment of REE with differentiation degree. It is in the same range as other Chaîne des Puys magmas, with a particular feature for Clierzou trachyte which is significantly more enriched than all other magmas. A specific feature of Pavin group magma is that differentiation leads to enrichment in LREE and HREE, but more limited enrichment in MREE. This particular evolution is typically due to the fractionation of accessory minerals as apatite or titanite which is usually only observed in most differentiated alkaline magmas (Villemant et al. 1979). This is consistent with petrological observations which point out the significant fraction of apatite inclusions in clinopyroxenes of Pavin group magmas (Bourdier 1980).

The trends defined by the three main magma types (basalts, hawaïites, benmoreites) of the Pavin group represent an actual magma differentiation process which is significantly distinct from that of the main Chaîne des Puys series.

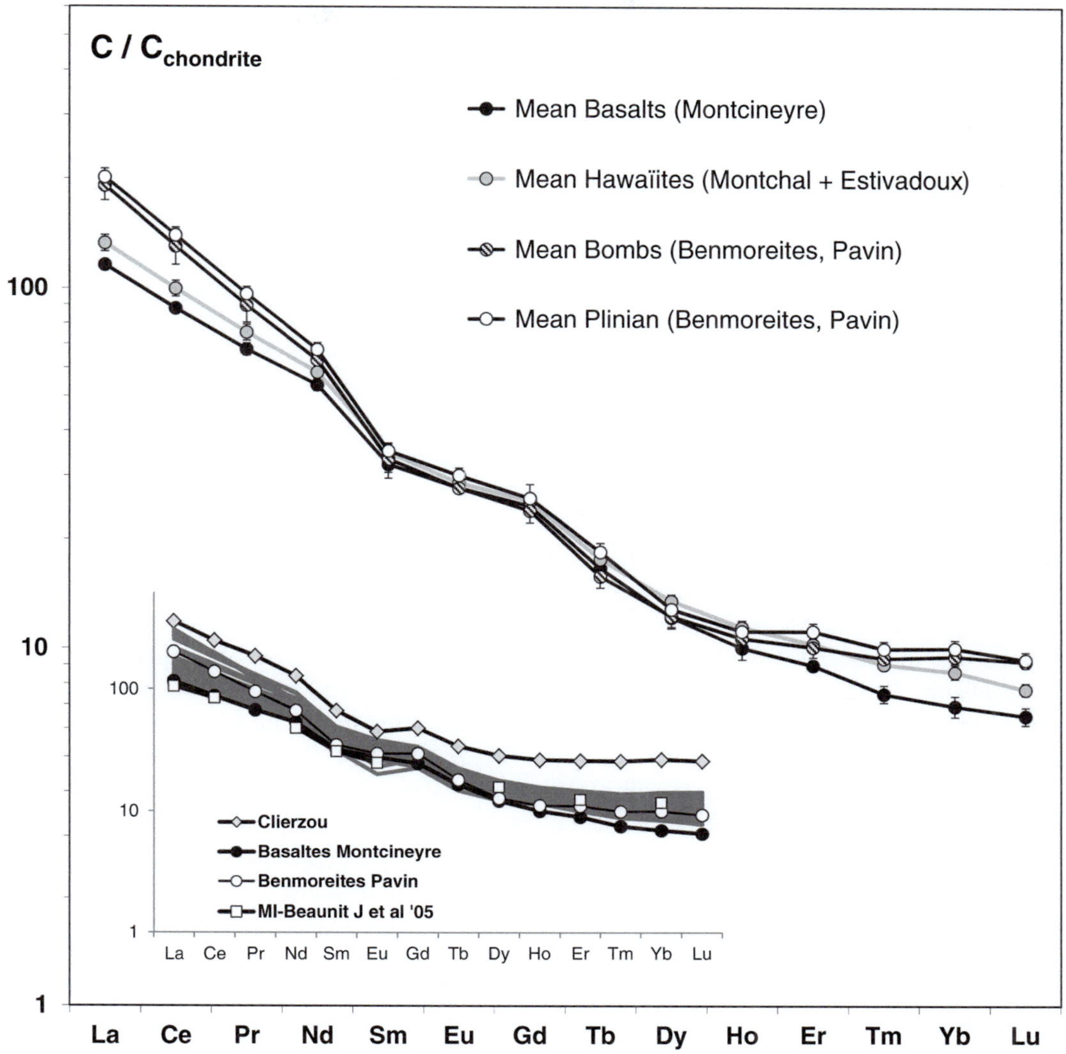

Fig. 7.3 Chondrite normalised REE diagrams of Pavin group magmas. Symbols as in Fig. 7.2. *Inset*: comparison with Chaîne des Puys magmas (*grey lines*)

7.3.2 The Pavin Differentiation Series: A Crystal Fractionation Trend Dominated by Amphibole Fractionation

The compositions of Pavin group magmas define variation trends that have the same main characteristics as those observed for the Chaîne des Puys (stricto sensu) series: from basalts to benmoreites and trachytes, early depletion in 3d transition series elements (Sc, Ti, V, Cr, Fe, Co, Ni) correlated to a large enrichment in highly incompatible elements (Th, U, Rb, LREE among others) that is characteristic of a dominant evolution by crystal fractionation (Villemant et al. 1980, 1981). In addition, for all the northern edifices of Chaîne des Puys (stricto sensu), there is a single mode of differentiation from basalts to benmoreites, and it diverges in several distinct paths only at the stage of trachytes differentiation (Villemant et al. 1981; Martel et al. 2013). For the Pavin group the differentiation process differs significantly from that of Chaîne des Puys (stricto sensu) as early as at the stage of evolved basalts indicating significant differences in deep differentiation conditions between both volcanic systems.

The differentiation steps from basanite to hawaïite and from hawaïite to benmoreites in Pavin group series are not represented by erupted magmas contrary to what is observed for the whole Chaîne des Puys series. These apparent discontinuities are however not surprising because the Pavin group represents only a very short activity period (~400 years). If Pavin benmoreitic magmas is actually related by a continuous crystal fractionation process to Montcineyre basalts through Montchal and Estivadoux hawaïites, it is therefore possible to quantify this process using the method developed by Villemant et al. (1981).

The basic ideas of this classical modeling method are summarised thereafter. The Rayleigh law describing the crystal fractionation process writes $C_i = C_i^\circ f^{(Di-1)}$, where C_i and C_i° represent the concentration of element *i* in the melt and in the initial melt respectively. *f* is the residual melt fraction ($f = M_L/M_L^\circ$, where M_L and M_L° are the mass of residual melt and of initial melt respectively). *Di* is the bulk solid/melt partition coefficient of element *i* ($D_i = C^s_i/C_i$, where C^s_i is the concentration of element *i* in the solid in equilibrium with the melt). If a highly incompatible element as Th is used as a reference, it can be shown that in log(C_i) – log (Th) diagrams, for composition domains where D_i is constant to first order the crystal fractionation trend is represented by a straight line with a slope $\Delta_i \sim 1 - D_i$. It has been shown that the Chaîne des Puys series evolved by steps over which the D_i values are approximately constant (Villemant et al. 1980, 1981; see Fig. 7.4a). During each differentiation step the modal composition of the fractionating minerals is in first approximation constant, and the abrupt changes in D_i values are clearly related to the changes in the mineralogy of the fractionating solid. The bulk solid/melt partition coefficients D_i are directly related to the mineral (j)/melt partition coefficients d^j_i by the relation: $D_i = \Sigma\ \alpha^j\ d^j_i$, where the α^j are the weight fraction of minerals j in the fractionating solid. If the d^j_i are known or measured independently (see for example Lemarchand et al. 1987, for determination in alkali basalt series or Blundy and Wood 2003, for theoretical calculation of mineral/melt partition coefficients) it is thus possible to calculate the modal composition of the fractionating solid (see Villemant et al. 1981 for resolution methods and a detailed application to the Chaîne des Puys model) and to compare it to petrological observations (determination of modal composition using point counting) or mass balance calculations on major elements (see e.g. Wright and Doherty (1970) and derived methods).

We have reproduced this model for the Pavin group magmas assuming a single continuous differentiation process and calculated the composition of the fractionating solid consistent with this interpretation. The data are compared to the modeling already performed for the Chaîne des Puys (Villemant et al. 1981) and to the mass balance calculations performed on major elements by Bourdier (1980). In Table 7.2 and Fig. 7.4 are reported the bulk solid/melt partition coefficients of trace elements calculated for the Pavin group and the Chaîne des Puys (stricto sensu) for the first two differentiation steps: Basalts-Hawaïites and Hawaïites-Benmoreites.

The bulk partition coefficients calculated in log diagrams are very different for Pavin group and for Chaîne des Puys (stricto sensu) for both differentiation steps, especially for elements usually considered incompatible. The D values of REE, Nb, Ta are unusually large (Fig. 7.4b) in both series which has been attributed to the fractionation of amphibole at the expense of clinopyroxene in crystallising solid (Villemant et al. 1981), in agreement with petrological observations (see e.g. Foury 1983). This effect is even more pronounced for Pavin group magmas and consistent with the ubiquitous presence of amphibole from basalts to benmoreites (Bourdier 1980). The mineral/melt partition coefficients of highly incompatible elements for amphibole and clinopyroxene are different enough to induce a significant effect on corresponding bulk D values (Fig. 7.4b). Indeed, for most elements mineral/melt partition coefficients of these two minerals are very close (Lemarchand et al. 1987) except in Nb and Ta and LREE. For these elements, other major mineral phases that may crystallise in basic and intermediate magmas (olivine, feldspars, Fe-Ti oxides) have all very low mineral/melt partition coefficients. Thus, the bulk D value patterns directly reflect the fraction of amphibole and/or clinopyroxene that are involved in crystallising solid. Fig. 7.4b provides a clear evidence for the predominance of amphibole as early as in basaltic melts in Pavin group differentiation series. However, this comparison also shows that REE behavior cannot only be explained by amphibole fractionation and requires also the fractionation of mineral phases having higher partition coefficients for MREE than HREE and LREE. This qualitative result also suggests that an accessory mineral phase (such as apatite) may also play a significant role in REE fractionation.

Mass balance calculations using the method described above and mineral/melt partition coefficients measured in alkaline series (Lemarchand et al. 1987) allow estimation of the modal compositions of the crystallizing solids during the two differentiation steps (Table 7.3). They are consistent with modal compositions calculated using other mass balance methods (major elements) and with petrological data (point counting). In agreement with the above discussion the weight fraction of amphibole and the total fraction of clinopyroxene + amphibole that crystallise in basaltic melts are larger in Pavin series than in Chaîne de Puys series. The mean calculated amphibole/clinopyroxene ratio is ~1:2 in Chaîne des Puys basalts and ~2:1 in Pavin group basalts. Amphibole is generally absent in Chaîne des Puys basalts and hawaïite lava flows but it sometimes occurs in basaltic tephra of Chaîne des Puys as resorbed phases, and its presence is ubiquitous in Montcineyre basaltic scoriae (Bourdier 1980). Notice that only modeling using trace elements is able to quantify the amphibole fractionation in Chaîne des Puys basic magmas. Conversely, during the second differentiation step (hawaïite-benmoreite) amphibole fraction is larger in Chaîne des Puys series. More generally, the crystallisation of ~75% of basaltic melt are necessary to produce benmoreite magmas of which approximately one half consists in amphibole in Pavin series against only one third in Chaîne des Puys series.

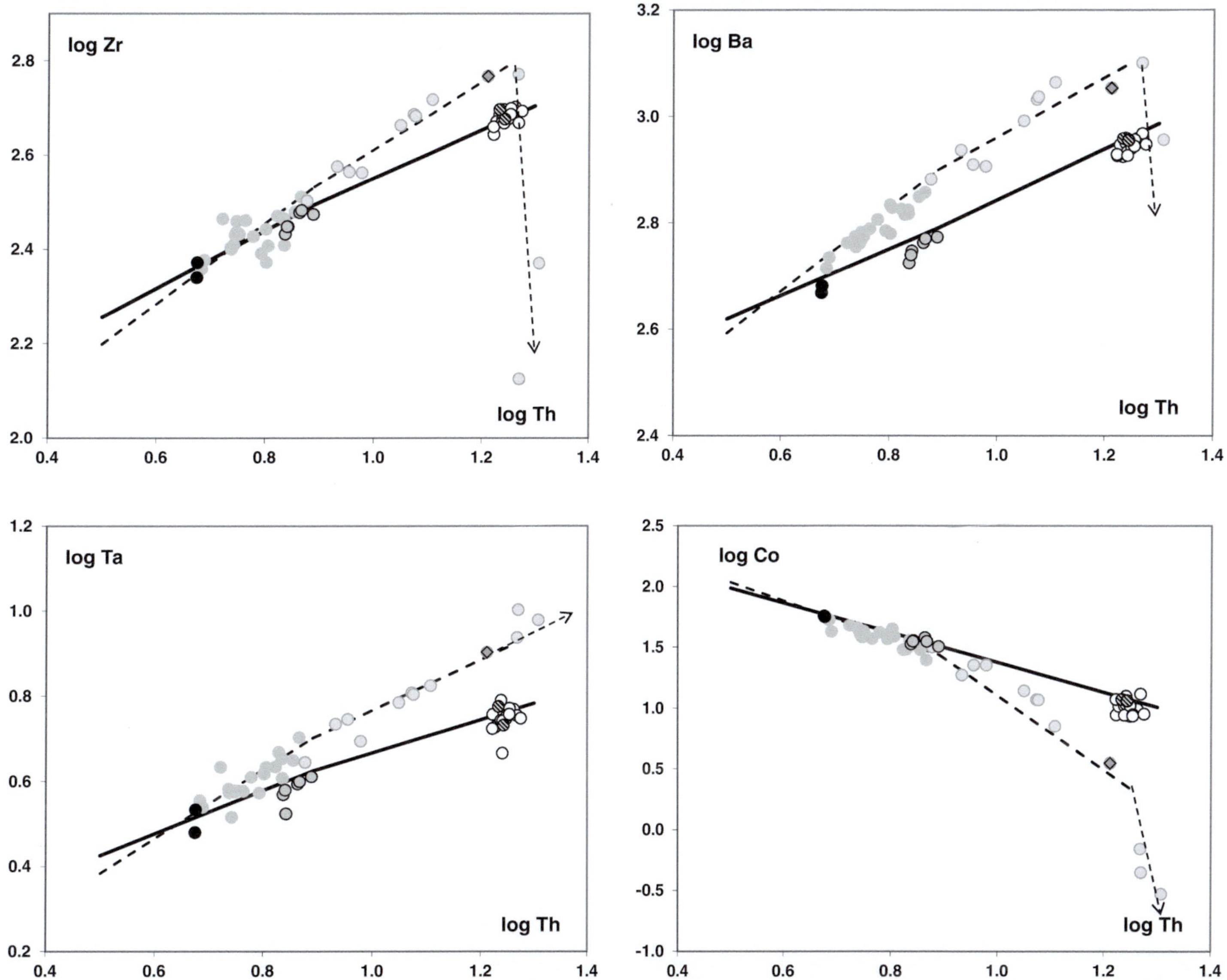

Fig. 7.4 Calculation of bulk D values in logarithmic diagrams. (**a**) Example of logarithmic diagrams. Symbols as in Figs. 7.1 and 7.2. Lines are regression lines of slopes Δ_i with Δ_i ~1–Di. Comparison between Chaîne des Puys (stricto sensu) and Pavin group magmas evidence a stepwise differentiation process, each step being characterized by approximately constant D_i values. (**b**) Bulk D values of some incompatible elements, comparison with amphibole and clinopyroxene/melt partition coefficients. Zr, Hf, Nb, Ta and REE partition coefficients calculated for Basalt-Hawaïte and Hawaïte-Benmoreite differentiation steps in Pavin group (Dg Pav) and Chaîne des Puys (stricto sensu; Dg CP) differentiation series. Mineral/melt partition coefficients for clinopyroxene (d CPx) and amphibole (d Amph) in alkaline series are from Lemarchand et al. (1987). For Basalt-Hawaïite differentiation step, mineral/melt partition coefficients vary in a relatively large range: extreme values are represented). Notice the large differences in bulk D values of LREE in basalt-hawaïite differentiation steps in Pavin and Chaîne des Puys series. The mineral/melt partition coefficients of amphibole and clinopyroxene are generally close for all elements ('parallel D values patterns'), except Nb and Ta which are much more efficiently partitioned into amphibole

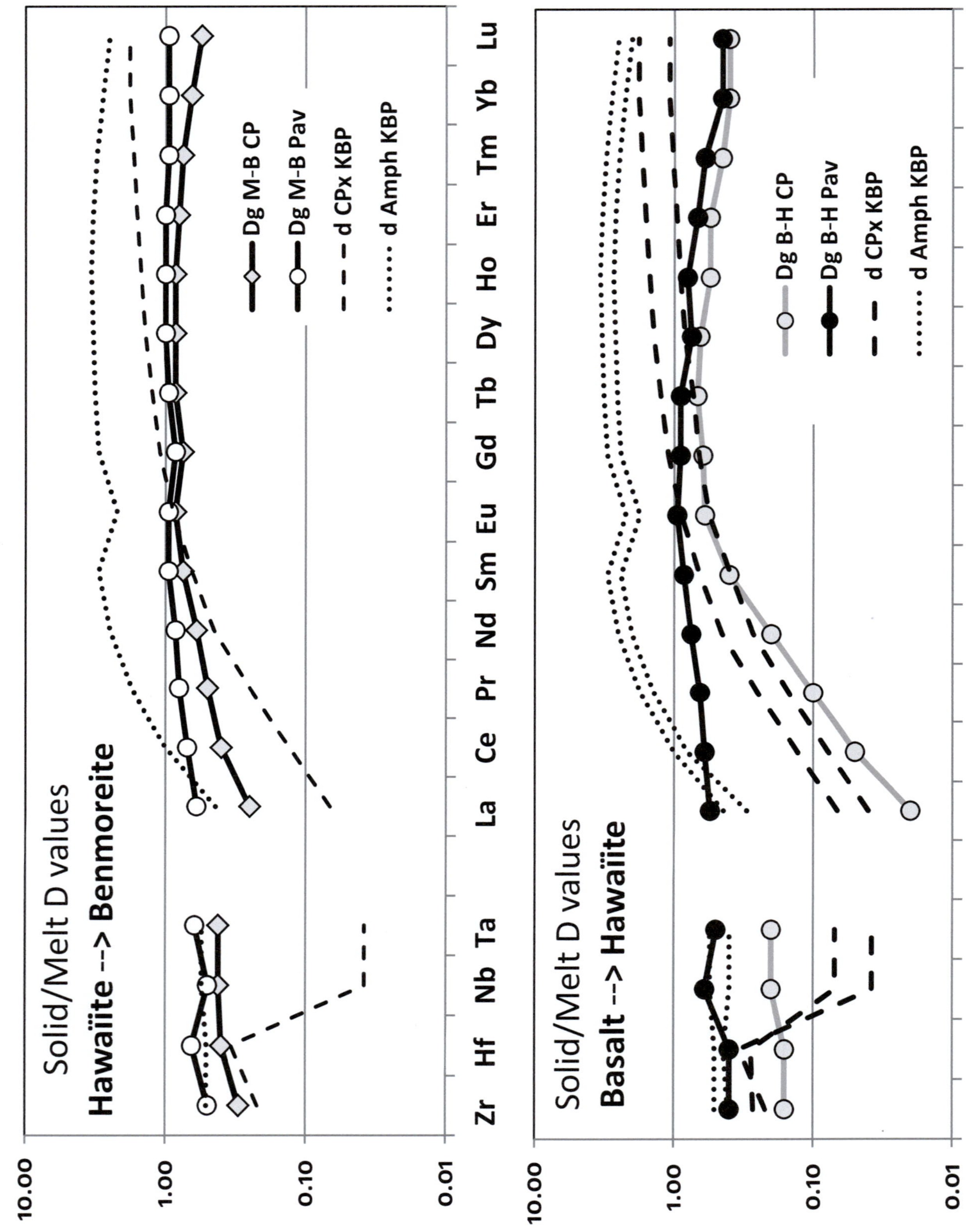

Fig. 7.4 (continued)

Table 7.2 Bulk solid/melt partition coefficients of Pavin and Chaîne des Puys series

	Pavin group		CP	
	β–H	M–B	β–H	M–B
SiO_2	0.72	0.82	0.75	0.70
TiO_2	1.20	1.80	1.20	2.50
Al_2O_3	0.70	0.90	0.70	1.10
Fe_2O_3 t	1.30	1.60	1.10	2.50
MgO	2.50	2.50	2.00	3.50
CaO	1.28	1.80	1.30	2.20
Na_2O	0.35	0.60	0.35	0.60
K_2O	0.10	0.25	0.10	0.25
Th (reference)	0.02	0.02	0.02	0.02
Zr	0.40	0.50	0.16	0.30
Hf	0.40	0.65	0.16	0.40
Ta	0.50	0.62	0.20	0.42
Ba	0.57	0.53	0.23	0.45
Sr	1.21	0.85	1.21	1.06
Rb	0.02	0.20	0.02	0.02
Cr	3.35	3.80	5.50	4.50
Co	2.20	2.20	2.50	4.00
Ni	4.50	3.00	3.60	6.00
Sc	1.69	1.50	2.50	2.30
La	0.55	0.60	0.02	0.25
Eu	0.95	0.95	0.60	0.85
Tb	0.90	0.95	0.68	0.85

The D values are calculated from logarithmic diagrams (see Fig. 7.4) for the first two differentiation steps (Basalts-Hawaïites and Mugearites-Benmoreites)

Table 7.3 Modal composition of fractionating solids (mass balance calculations)

Differentiation	Minerals weight fractions (wt %)								
Step		$\Delta\phi$[a]	Ol	Cpx	Hb	Mt	Fp	Total	MSRRT
CP $\beta-H$	1	50 %	10 %	50 %		10 %	30 %	100 %	
	2	50 %	12 %	57 %		10 %	21 %	100 %	
	3		10 %	40 %	20 %	5 %	25 %	100 %	
Pavin $\beta-$H	***This work***	**39 %**	**13 %**	**25 %**	**45 %**	**7 %**	**11 %**	**100 %**	**4.7**
	Bourdier (1980)	**52 %**	**18 %**	**26 %**	**48 %**	**8 %**	**-**	**100 %**	
CP M – B	*1*	20 %		30 %	25 %	10 %	35 %	100 %	
	2	33 %		10 %	43 %	10 %	36 %	100 %	
	3			10 %	60 %	5 %	25 %	100 %	
Pavin M - B	***This work***	**63 %**		**10 %**	**31 %**	**0.2 %**	**59 %**	**100 %**	**6.0**
	Bourdier (1980)	**40 %**	**6 %**	**18 %**	**37 %**	**6.4 %**	**33 %**	**100 %**	

Modal compositions for each differentiation step (Basalt-Hawaïte and differentiated Hawaïte-Benmoreite) have been calculated using the method of Villemant et al. (1981) and this work. $\Delta\phi$ is the weight fraction of crystallised melt and is calculated from Th contents: $\Delta\phi$ (1−>2) is the fraction of melt crystallised between steps 1 and 2: $\Delta\phi \sim 1 - [Th]_1/[Th]_2$. MSRR values (mean square reduced residuals) give estimates of the quality of fit; they should be close to 1. The relatively high MSRR values indicate a moderate fit quality (see text for further explanations). Results are compared to petrological estimations (Foury 1983 for Chaîne des Puys) and mass balance calculations based on major elements (Wright and Doherty 1970) performed by Bourdier (1980) for Pavin group and Maury et al. (1980) for Chaîne des Puys. Agreement between the different methods is relatively good

1 Maury et al. 1980; point counting (lavas)
2 Foury 1983, point counting (scoriae)
3 Villemant et al. 1981; mass balance calculations (major and trace elements)
4 Bourdier 1980; mass balance calculations on major elements
[a] $\Delta\phi$ measured from Th contents

7.4 Conclusions

This preliminary study of the geochemistry of the eruptive products of Pavin group magmas shows that they define a single differentiation series from basalts to benmoreïtes. Though having very close major element compositions both magma series (Pavin and Chaîne des Puys (stricto sensu)) have specific trace element characteristics that distinguish them from the earliest differentiation steps. The basalts erupted at Montcineyre volcano are the most primary melts ever erupted in Chaîne des Puys and likely represent the primary melt at the origin of the two magma series. The Pavin series is characterized by the very large fractionation of amphibole from the most primitive basalts, whereas amphibole becomes the dominant mineral phase in Chaîne des Puys (stricto sensu) only at the step of benmoreite differentiation. This indicates the differentiation conditions of the Pavin group basalts are significantly different from that of the northern volcanoes with likely much higher initial H_2O content to favor amphibole crystallization at the expense of clinopyroxene. Comparison of the trace element composition and mineralogy of Clierzou and Pavin magmas (both close to the benmoreite-trachyte limit) suggests that they have been likely produced in significantly distinct P (or P_{H2O}) and T conditions. Martel et al. (2013) have shown that Clierzou trachyte was produced at a well constrained T of around 800°c, but the presence of amphibole as the only Fe-Mg silicate did not allow to constrain precisely the P_{H2O} (= P total). The Pavin benmoreite is on the contrary characterised by the coexistence of both amphibole and clinopyroxene at liquidus (and likely low Fo olivine, as described in some amphibole cumulates). The petrological grid of Martel et al. (2013) suggests that the mineral paragenesis of Pavin benmoreite could have equilibrated at 750–800 °C and possibly P_{H2O} ~250–300 MPa, but new experiments and comparison with the Pavin phenocrysts compositions are needed to refine these conditions.

In addition this study shows that there is no evidence that the interaction of the benmoreitic magma with the basement material during the phreato-magmatic and plinian activity of Pavin have significantly modified the melt compositions of the erupted products (cauliflower bombs and pumice clasts). However, since some highly incompatible elements as REE, Nb and Ta display unusual behaviors (unusually high bulk solid/melt partition coefficients) and since Bourdier (1980) reported increasing $^{87}Sr/^{86}Sr$ from Montcineyre basalts to Pavin benmoreite (0.7033–0.7040) the possible role of an assimilation-crystal fractionation process (ACF) involving surrounding rock assimilation cannot be discarded.

The volcanoes of Montcineyre, Montchal, Estivadoux and Pavin, form the most southern and the most recent volcanic area of the Chaîne des Puys (s.l.) and that evolve following a very specific magma differentiation path, although initiated from a primary basaltic melt identical to those of the northern edifices. The short duration of this volcanic activity (~400 years) is also an opportunity to estimate the duration of differentiation processes. The highly original features of Pavin group magmas justify the development of more complete mineralogical, petrological and geochemical studies to establish the particular emplacement and differentiation conditions of this magma series and their possible role in controlling, at least *pro-parte*, the explosive activity of Pavin volcano.

Acknowledgments We thank G. Grassi for her participation in preliminary analyzes. One of us (BV) is particularly grateful to the late P. Vincent and his student JL. Bourdier for their initiation, 35 years ago, to the tephrostratigraphy of Pavin eruptions. We thank C. Martel and I. Vlastelic for their constructive reviews.

References

Blundy J, Wood BJ (2003) Partitioning of trace elements between crystals and melts. Earth Planet Sci Lett 210:383–397

Bourdier JL (1980) Contribution à l'étude volcanologique de deux secteurs d'intérêt géothermique dans le Mont Dore : le groupe holocène du pavin et le Massif du Sancy. Thesis, University Blaise Pascal, Clermont-Ferrand II,185 p

Foury P (1983) Etudes pétrologiques et expérimentales (à une atmosphère) d'une série alcaline continentale de la Chaîne des Puys (M.C.F.). Thesis, University Blaise Pascal, Clermont-Ferrand II,150 p

Jannot S, Schiano P, Boivin P (2005) Melt inclusions in scoria and associated mantle xenoliths of Puy Beaunit Volcano, Chaine des Puys, Massif Central, France. Contrib Mineral Petrol 149:600–612. doi:10.1007/s00410-005-0670-y

Lemarchand F, Villemant B, Calas G (1987) Trace element distribution coefficients in alkaline series. Geochim Cosmochim Acta 51:1071–1081

Martel C, Champallier M, Prouteau G, Pichavant M, Arbaret L, Balcone-Boissard H, Boudon G, Boivin P, Bourdier JL, Scaillet B (2013) Trachyte phase relations and implication for magma storage conditions in the Chaîne des Puys (French Massif Central). J Petrol 54(6):1071–1107. doi:10.1093/petrology/egt006

Maury R, Brousse R, Villemant B, Jaffrezic H, Joron JL, Treuil M (1980) Cristallisation fractionnée d'un magma basaltique alcalin: la série de la Chaîne des Puys (Massif Central, France). I Minéralogie et Pétrologie. Bull Minéral 103:250–266

Villemant B, Joron JL, Treuil M (1979) Origine des Quartz des nappes de ponces du Mont Dore (Massif Central, France) : arguments géochimiques. C R Acad Sci Paris 290(D):687–690

Villemant B, Joron JL, Jaffrezic H, Treuil M, Maury R, Brousse R (1980) Cristallisation fractionnée d'un magma basaltique alcalin: la série de la Chaîne des Puys (Massif Central, France). II Géochimie. Bull Minéral 103:267–286

Villemant B, Jaffrezic H, Joron JL, Treuil M (1981) Distribution coefficients of major and trace elements; fractional cristallization in the alkali basalt series of Chaîne des Puys (Massif Central, France). Geochim Cosmochim Acta 45:1997–2016

Wright TL, Doherty PC (1970) A linear programming and least squares computer method for solving petrologic mixing problems. Geol Soc Am Bull 81:1995–2008

Distribution, Tephrostratigraphy and Chronostratigraphy of the Widespread Eruptive Products of Pavin Volcano

8

Etienne Juvigné and Didier Miallier

Abstract

A tephra erupted by Pavin volcano was found as millimetre- to centimetre-thick tephra beds in numerous peat-bogs and lake deposits of the Massif Central (France). At least two distinct Plinian pulses are attested by differences in chemical compositions of glass shards. A southern lobe was clearly delimited; it is widespread at least to Cantal (30 km from the volcano). At three northern localities as far as the Gour de Tazenat (50 km from the volcano), two populations of glass shards present in the tephra deposit indicate either overlapping of two compositionally-distinct lobes or a single lobe characterized by bimodal differences arising from magma mixing in the source volcano in the Chaîne des Puys. The Bayesian application, RenDateModel, to a sequence of radiocarbon and TL dates enabled us, at the 95 % confidence level, to constrain the age of Pavin eruption to 4720 ± 170 B.C., i.e. a current age of 6,730 ± 170 year, and to assess that between 100 and 700 years lapsed between the eruptions of Montcineyre and Pavin. Since Pavin Tephra was identified in sediments containing known pollen sequences pertaining to each lobe, the volcanic activity can be placed within the Atlantic palynozone between the spreading of *Fraxinus* and the synchronous palynozones of *Abies* and *Fagus* in that part of the Massif Central. The Pavin volcano generated the latest cataclysmic eruption of Plinian type in the Massif Central.

Keywords

France • Massif Central • Volcanoes • Pavin • Montcineyre • Tephra • Pollen diagram

8.1 Introduction

Brousse et al. (1969) and Brousse and Rudel (1973) interpreted sandy beds as tephras which were plotted in pollen diagrams, but no mineralogical or geochemical control was undertaken. Nevertheless, the former two works initiated Holocene tephrostratigraphy of the Massif Central (Fig. 8.1). Camus et al. (1973) and Bourdier (1980) described outcropping proximal products of the volcanoes of the group of Besse-en-Chandesse as well as Montcineyre volcano (Compains), so that it was demonstrated that the relevant eruptions had taken place in the following order: Montcineyre, Estivadoux, Montchal, and Pavin. Brousse and Bardintzeff (1987) proposed a tephrochronologic model based on the previous works of Brousse et al. (1969) and Brousse and Rudel (1973) (see above). Later on, cores were taken from peat-bogs and lakes in the Cézalier that enabled Gewelt and Juvigné (1988) to confirm and develop the tephrostratigraphic model of Bourdier (1980) (see above).

Then, more cores were taken from some forty peat-bogs and lake deposits of the Massif Central from the northern Chaîne des Puys to the Planèze de Saint Flour/Cantal

E. Juvigné (✉)
Laboratoire de Géomorphologie et de Géologie du Quaternaire, Sart-Tilman, Bât. 11, Université de Liège, B 4000 Liège, Belgium
e-mail: ejuvigne@skynet.be

D. Miallier
Laboratoire de Physique Corpusculaire, BP 10448, Clermont Université, Université Blaise Pascal, CNRS/IN2P3, F-63000 Clermont-Ferrand, France

T. Sime-Ngando et al. (eds.), *Lake Pavin*, DOI 10.1007/978-3-319-39961-4_8

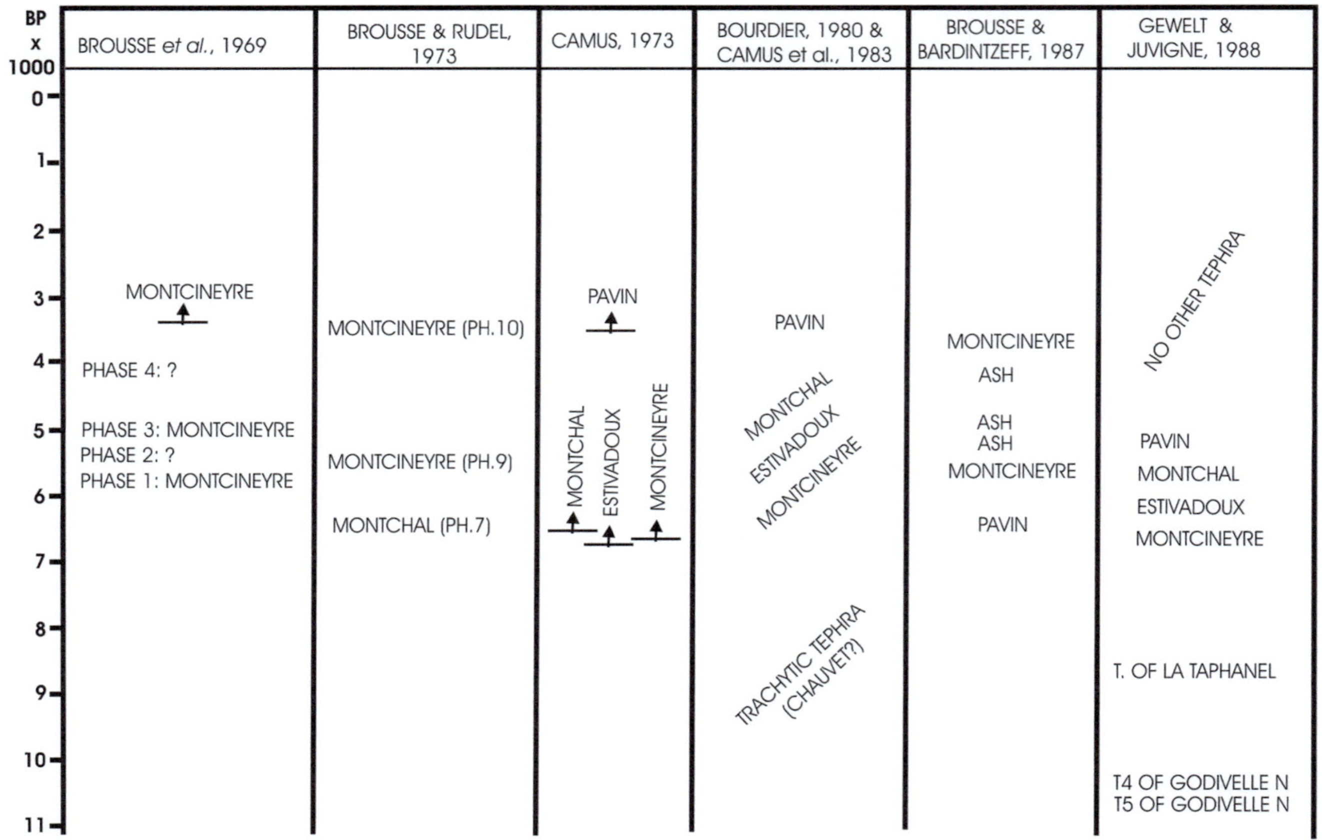

Fig. 8.1 The evolution of the Late Glacial to Holocene tephrostratigraphical model in the Massif Central (After Gewelt and Juvigné 1988, p.26, Fig. 1)

(Fig. 8.2). Altogether, 10 distinct tephras were found in stratigraphic positions distributed from the Late Glacial (Allerød) through the mid-Holocene (Atlantic) (Juvigné 1992a). The tephras of the volcanoes Montcineyre-Estivadoux- Montchal-Pavin (MEMP group hereafter) are part of them: Juvigné and Gilot (1986), Juvigné et al. (1986, 1988b, 1994), Etlicher et al. (1987), Gewelt and Juvigné (1988), Juvigné (1992b, 1993), Juvigné and Bastin (1995a, 1996), Juvigné and Stach-Czerniak (1998). The application of ^{14}C-dating method on samples from several sites has allowed Juvigné and Gillot (1986) to show that the series of eruptions have occurred within a time range of about five centuries around 6000 B.P. (we use the standard notation B.P. for non-calibrated ^{14}C dates).

8.2 Description of Tephras from MEMP Volcanoes

The current paper focuses on the tephras of the MEMP volcanoes, and more specifically on the Pavin tephra. Bourdier (1980) realized petrographic and geochemical analyses of these tephras and he produced a map of the distribution of relevant lobes based on outcrops in road-sides and various excavations (Fig. 8.3a). Later on, the investigation of the cores taken from lake and peat-bogs (see above) has enabled maps of each lobe to be extended (Fig. 8.3b) since millimetre-thick tephra beds can be easily identified in lacustro-palustrine sediments where they have not been disturbed nor altered by pedological processes.

8.2.1 Pavin Tephra

Various activities during the eruptive phases were described by Bourdier (1980) and later refined by Boivin et al. (2010) and Leyrit et al. (2016). Base surges linked to phreatomagmatic explosions have accumulated products in the vicinity of the volcano, while Plinian activity was responsible for the widespread distribution of pumice. Due to the high amount of pale grey to yellowish pumices, a tephra bed is easily recognizable to the naked eye up to tens of kilometres far from the volcano as decimetre- to millimetre-thick beds in dark organic material of peat bogs and lakes. The spatial distribution of pumices in the tephra beds in peat bogs is strongly asymmetrical due to floating and drift on lacustrine areas at the time of the ash-fall, so that the ratio 'pumices/(minerals + lithics)' does not confirm with normal patterns in such

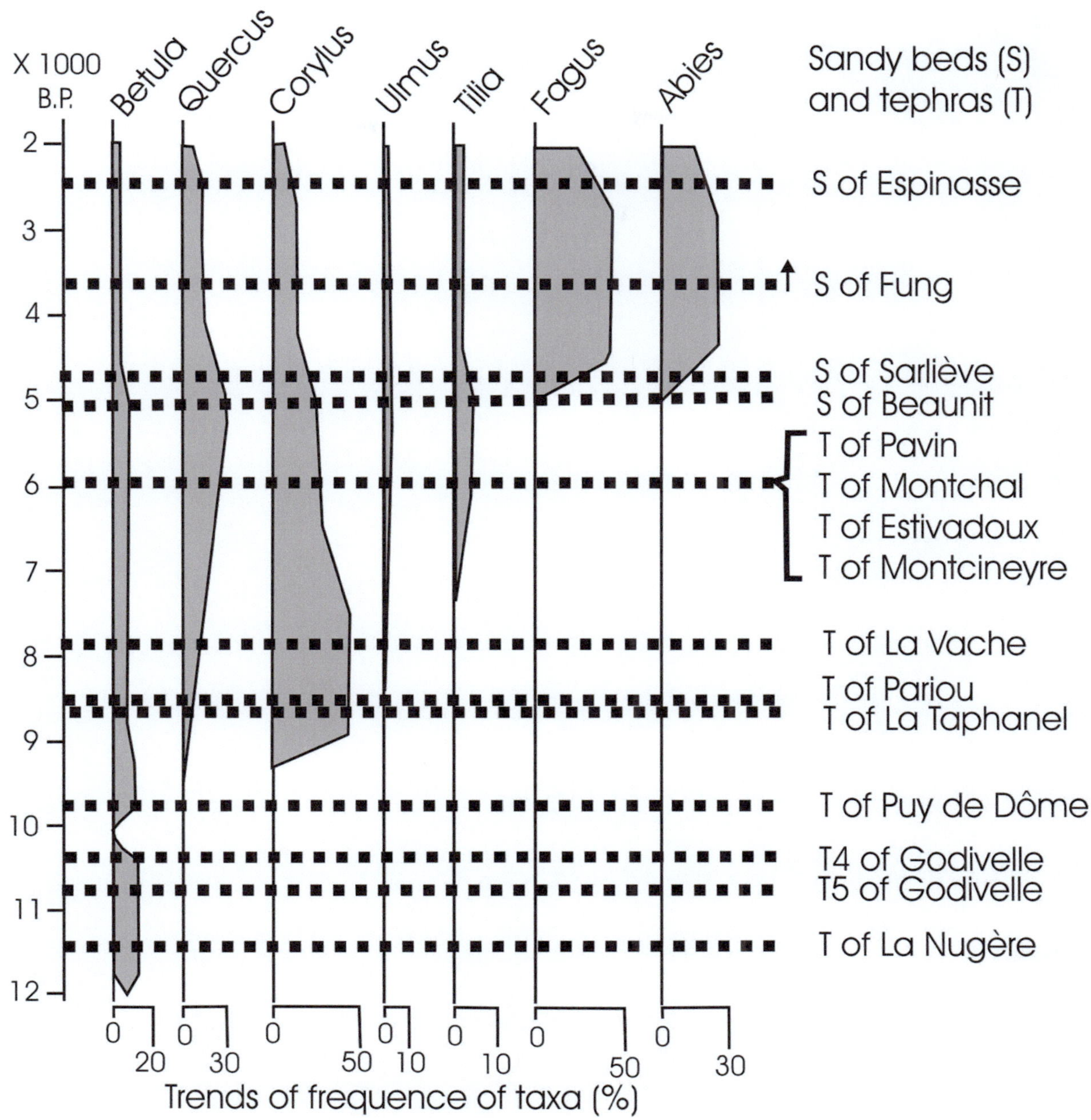

Fig. 8.2 Tephra age model after investigation of cores taken from peatbogs and lakes of the Massif Central (after Juvigné 1991, p. 166, Fig. 1, updated). *T* = recognized tephra; *S* = sandy layer (detrital or tephra; see below)

sites. The tephra is distributed in vast lobes: (1) the southern lobe is oriented southwards and it was delimited at La Godivelle by Bourdier (1980). The identification of the tephra layer in lakes and peat bogs outside that lobe has enabled its known distribution to be extended at least as far as the Planèze de Saint-Flour (Cantal), where the tephra bed is still 2 cm thick; (2) the northern lobe of the tephra was only identified by thin tephra beds in two peat bogs and one lake as far as the Gour de Tazenat in the northernmost part of the Chaîne des Puys. For both lobes, the tephra materials correspond to Plinian fall-out deposits.

A. Southern lobe

Tephra occurrences were described in the vicinity of the volcano by Bourdier (1980) and in distal lacustro-palustrine sites by Juvigné and Gilot (1986). Juvenile magma is represented by highly vesiculated pumice clasts which are essentially yellowish in dry terrains, but light grey in peat bogs. Their size reaches 20 cm across on the edge of the volcano, and decreases to ash size towards the western and southern margins of the lobe, while centimetre-thick lapilli are still present in the easternmost peat bogs. The trend is similar for

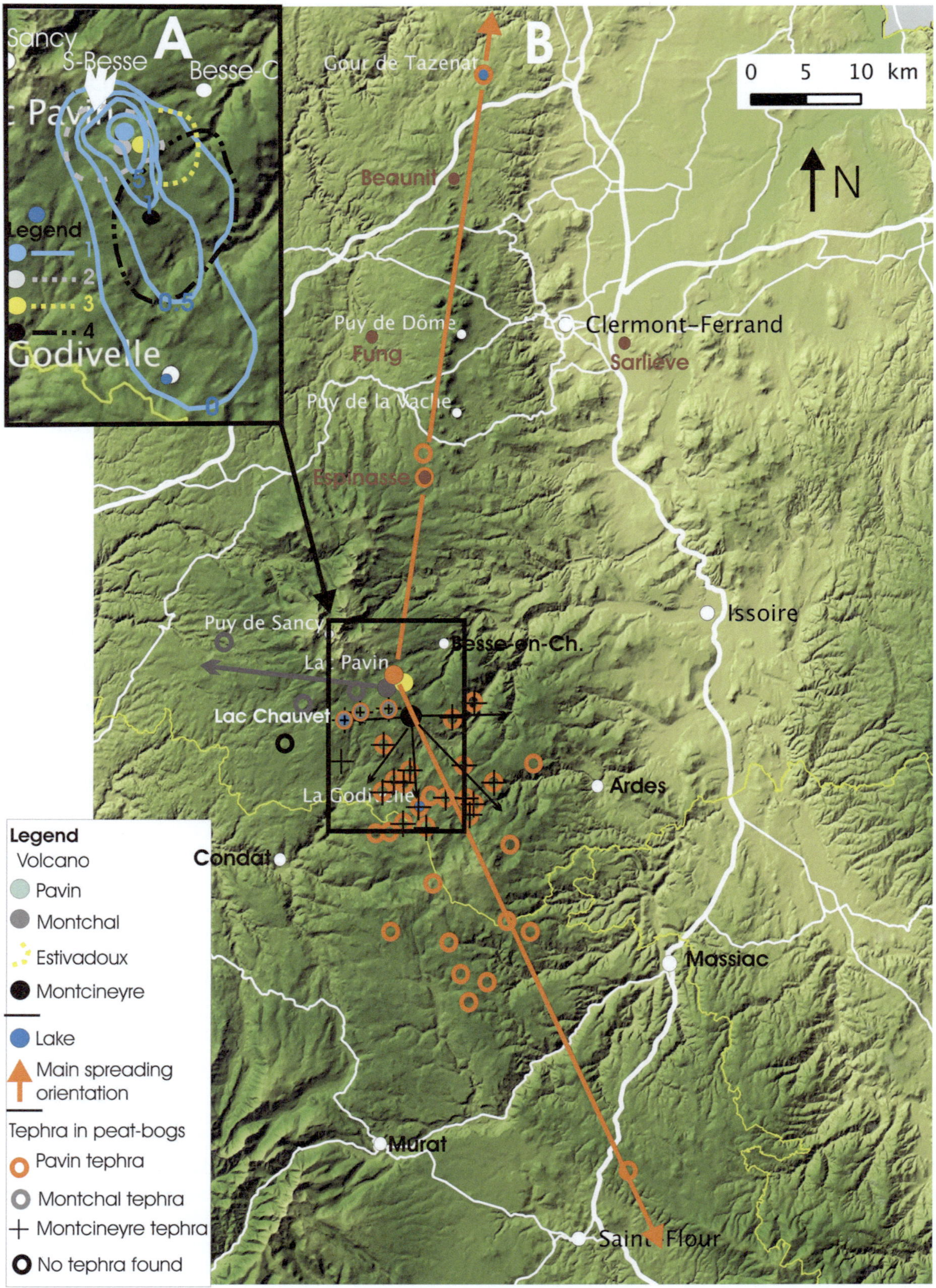

Fig. 8.3 Localization of the MEMP volcanoes and the relevant tephra lobes. (**a**), after Bourdier (1980): #1, Pavin; #2, Estivadoux; #3, Montchal; #4, Pavin (isopach in cm). (**b**), peat bogs and lakes where tephra layers were found, as well as localities mentioned in the text

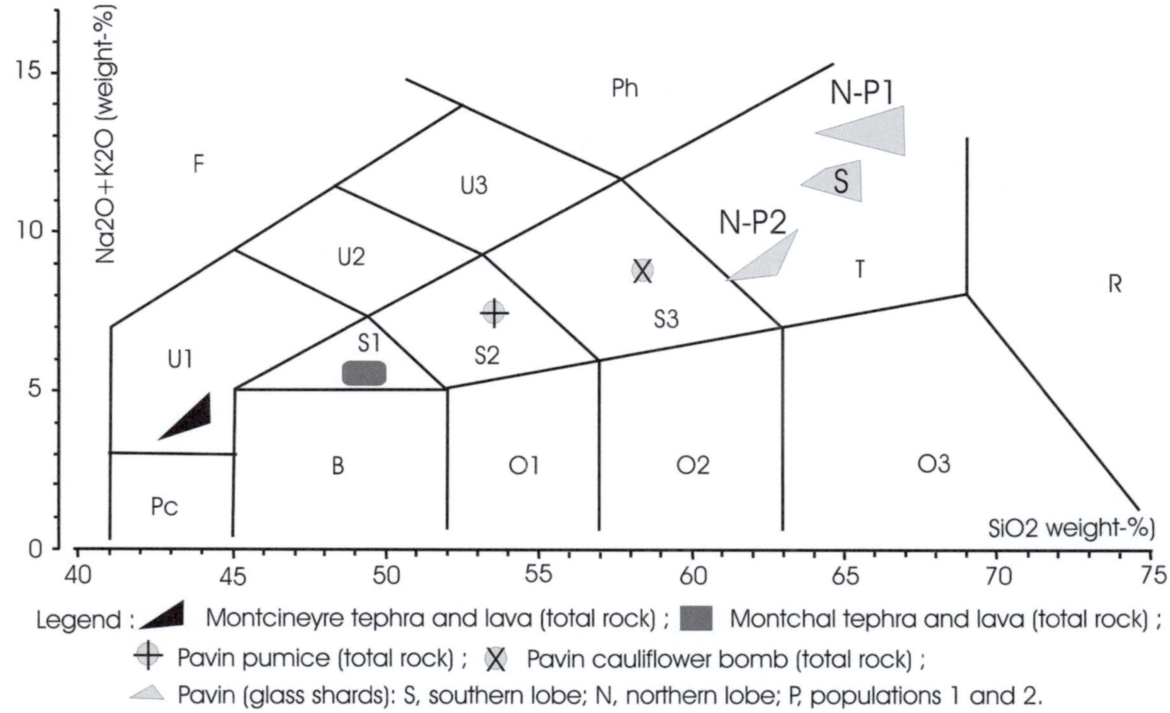

Fig. 8.4 Determination of the magma composition of the Montcineyre*, Montchal* and Pavin** volcanoes after the TAS classification (Le Bas et al. 1986). (*) Total rock analyses of ash after Juvigné and Bastin (1995a), and lava flows after Bourdier (1980); (**) glass analyses of ash after Juvigné and Stach (1998) and Gourgaud for glass shards of the southern lobe (unpublished), and total rock of one pumice and one cauliflower bomb after Bourdier (1980) Legend of TAS-classification (Le Bas et al. 1986). *F* foïdite, *Pc* picro-basalte, *U1* basanite, *U2* phonotephrite, *U3* tephriphonolite, *Ph* phonolite, *B* basalt, *S1* trachybasalte, *S2* basaltic trachyandesite, *S3* trachyandesite, *T* trachyte, *O1* basaltic andesite, *O2* andésite, *O3* dacite, *R* rhyolite

the thickness of the tephra bed that attains 10–30 cm at the easternmost sites, and 3 cm at the southernmost site (Planèze de Saint Flour). Due to the lack of peat bogs outside the investigated zone, and thus no opportunity to assess its occurrence and thickness beyond this zone, it is nevertheless reasonable to assume that the lobe must extend much further eastwards and southwards. The minerals consist essentially, in order of decreasing frequency, feldspars, brown amphiboles, green clinopyroxenes, and opaque Fe-Ti oxides. Various xenoliths from the basement and volcanic products of the Monts Dore are also present (lithics).

Microprobe analyses were performed to constrain the geochemical composition of glass shards in a sample of the tephra from the La Godivelle area. That enabled its determination in the TAS classification (glass basis) as a trachyte (Fig. 8.4). Total rock analyses of proximal products were performed by Bourdier (1980); since minerals were part of the sample, the magma is less differentiated than the glass and it was determined to be (whole-sample basis) trachyandesite and basaltic trachyandesite.

B. Northern lobe

So far the tephra layer is known only at three localities (Fig. 8.3b): the Narse d'Ampoix (Juvigné and Gewelt 1987), the Narse d'Espinasse (Juvigné et al. 1988b), and the lake of the Gour deTazenat (Juvigné and Stach 1998). Juvenile magma is represented by light grey pumice clasts up to 1.6 mm in diameter at Ampoix and Espinasse, and by bubble-wall shaped glass shards smaller than 0.3 mm at Tazenat. That decreasing size trend (from Ampoix and Espinasse to Tazenat) is matched by a trend of decreasing thickness, the layer being 1 mm thick at Ampoix and 0.5 mm at Tazenat. In the mafic mineral suite, brown amphiboles exceed green clinopyroxenes by a ratio of approximately 4:1. A few titanites, apatites, and brown micas are also present. Beyond the investigated area it can be assumed that the lobe should overlap more regions farther eastwards and/or westwards while farther northwards the tephra is close to the limit of detection.

Microprobe analyses of glass shards were performed on samples of tephra from the Narses d'Ampoix and d'Espinasse bogs as well as from the lake sediments at Gour de Tazenat. Those analyses enabled the glass shards to be determined in the TAS classification as trachyte; nevertheless two populations can be clearly distinguished based on silica and alkali content (Fig. 8.4). The joint presence of those populations can be explained either by a single pulse involving magma mixing, as shown by banded pumices in proximal sites of the volcano (Leyrit, personal communication), or the mixing of two separate Plinian-derived eruptives within a short time range (days to weeks).

8.2.2 Montchal Tephra

Bourdier (1980) described the products and the activity of Montchal volcano. Short lava flows are present at the southern foot of the volcano, but another one that is 16 km long followed the Clamouze valley where Besse-en-Chandesse is located. The tephra, composed essentially of scoriaceous lapilli and ash, was not identified in outcrops beyond 3 km from the volcano (Fig. 8.3a). Its phenocrysts are, in order of frequency, clinopyroxenes, olivines, feldspars (plagioclases), and opaque Fe-Ti oxides. Juvigné and Bastin (1995a) identified Montchal tephra in three peat bogs situated west of the fallout lobe as delimited by Bourdier (1980) (Fig. 8.3a, b). Since the macroscopic features of that tephra are similar to those of an immediately older tephra deposit (see below), its identification was only possible after geochemical analyses of bulk magma, which is (whole sample basis) a trachybasalt (box S1 in Fig. 8.4).

8.2.3 Montcineyre Tephra

Bourdier (1980) described the products of Montcineyre volcano. A 7-km-long lava flow followed southwards the valley of Couze de Valbeleix where the village Compains is located. The tephra, essentially composed of black scoriaceous lapilli and ash, was identified in various outcrops as far as 5 km away from the volcano towards the N-NE and S-SW (Fig. 8.3a). The tephra consists of an accumulation of centimetre-thick beds without internal erosion features, so that it corresponds to a series of fall deposits related to as many pulses (Strombolian activity) during a very short time span (weeks to months). The occurrence of Montcineyre tephra in peat bogs and lakes enabled it to be mapped as delimited by Bourdier (1980) (Fig. 8.3a, b). By naked eye, the Montcineyre tephra can be confused with the Montchal tephra in the area where both lobes overlap west of Pavin volcano. Otherwise, the tephra was not found in peat-bogs south of La Godivelle and northwards in the Sancy mountainous area, while the bed is still centimetre to decimeter in thickness in the most western and eastern investigated peat bogs. Unfortunately beyond those locations, the lack of peat bogs means it cannot be mapped further.

In the peat bogs near the Pavin-Montchal-Estivadoux group, the Montcineyre tephra is directly overlain by coarse volcanic products that can be linked to ash-falls from Montchal and/or Estivadoux, but these younger, coarse ash-fall deposits were not investigated.

8.3 Pollen Sequences and Tephrochronology

In all pollen diagrams where both Pavin and Montcineyre tephras were depicted, the tephras are shown to occur between the beginning of the continuous curves of *Fraxinus* (earlier) and those of *Abies* and *Fagus* (later) (see above, Fig. 8.2), i.e. in the Atlantic palynozone (Reille et al. 1985). Later on, referring to the age of the 2 %-point of the increase of individual pollen curves, Juvigné et al. (1996) obtained the following mean ^{14}C ages: for *Fraxinus*, 6702 ± 72 B.P.; for *Abies*, 4994 ± 50 B.P.; and for *Fagus* 4994 ± 47 B.P. Hence, the ^{14}C ages derived from pollen-bearing sediment and bracketing the eruptive epoch of the MEMP volcanoes would be 6702 ± 72 B.P. and 4994 ± 50 B.P. (see Juvigné et al. 1988a, p. 38), that is about 5620 B.C. and 3800 B.C. respectively, after calibration. In this calculation, uncalibrated ^{14}C data were processed for evaluating the age of the 2 % point of the beginning of the continuous curve of both taxa in various logs and finally averaged. Indeed, it would be more rigorous to process initially calibrated data since the ^{14}C B.P. time scale is not linear. This new calculation, which needs a devoted development of the Bayesian model, has not yet been achieved; however, preliminary results suggested that the bracketing limits will not be significantly modified.

In each peat bog where both the Montcineyre and Pavin tephras are present, they are separated by a few centimetres of peat attesting that some time has elapsed between the relevant eruptions. Based on the rate of peat formation of the interbedded segment, the duration of the hiatus was estimated at between 93 and 456 years; the large amplitude of the interval can be explained by various factors including changing rate of peat formation along the relevant segments and in the different peat bogs, disturbance of the original thickness of the interbedded segment while coring, and deviation of ^{14}C dates of the dated segments below and above the pair of tephra. This hiatus between the eruption of the Montcineyre and Pavin tephras is shown also by weak soil development in the upper part of the Montcineyre tephra (Camus et al. 1973).

8.4 Numerical Dating of the MEMP Volcanoes

Current numerical ages for the crater of Pavin and its three companion volcanoes (the MEMP group) mainly rely on the radiocarbon method and, to a lesser extent, on the thermoluminescence method (Table 8.1).

Table 8.1 Age data used in the present work

References (Articles)	Volcano and site	Material	Age reference (Lab. No)	^{14}C result (B.P.)	Calibrated date interval (B.C., 95 %)		Comments
	Pavin						
Juvigné et Gillot (1986)	Col de Chamoune (peat bog)	Peat. Tephra centred in the sample	Lv 1491	=5680±100	4340	4230	
Juvigné et Gillot (1986)	Col de Chamoune (peat bog)	Peat. Tephra centred in the sample	Lv 1492	=5990±80	5110	4690	
Juvigné et Gillot (1986)	Redondel (peat bog)	Peat. Tephra centred in the sample	Lv 1493	=5710±90	4720	4360	
Gewelt et Juvigné (1988)	Narse d'Ampoix, (peat bog)	Peat. Tephra centred in the sample	MBN 327	=5990±140	5220	4550	
Gewelt et Juvigné (1988)	Landeyrat (peat bog)	Peat. Tephra centred in the sample	MBN 382	=6000±110	5210	4650	
Juvigné et Stach-Czerniak (1998)	Tazenat (crater lake)	Very organic mud. Tephra centred in the sample (30 cm thick)	LV-2174	<6770±80			Partially reworked older sediments ?
Chapron et al. (2012)	Bottom of lake Pavin (sediment core)	Leaves	Poz- 27050	>4995±35			
Chapron et al. (2012)	Bottom of lake Pavin (sediment core)	Leaves	Poz- 27051	>4820±40			
	Montcineyre						
Juvigné et Gillot (1986)	Peat bog # 16	Peat beneath the tephra	Lv 1522	<6530±100	5480	5060	
Juvigné et Gillot (1986)	Peat bog # 10	Peat beneath the tephra	Lv 1523	<6520±100	5630	5310	
Gewelt et Juvigné (1988)	Redondel (peat bog)	Peat. Tephra centred in the sample	MBN 383	=5750±110	4830	4360	
Gewelt et Juvigné (1988)	Auzolle (peat bog)	Peat. Tephra centred in the sample	MBN 389	=6180±125	5380	4800	
				TL (real age)	**B.C. Interval (95 %)**		
	Montcineyre						
Guérin (1983)	Compains	Feldspars (lava)	MD 73	=8500±400	7290	5690	
	Montchal						
Guérin (1983)	Road CD # 149	Feldspars (lava)	MD 127	=7560±385	6320	4780	

Explanations: '=' means that the tephra layer was centered in the dated peat sample (same thickness above and beneath the tephra); it can then be assumed that the measured age is very close to the age of the tephra; '<' means that the age is a maximum one for the tephra because the peat was sampled beneath the tephra; '>' means that the age is a minimal one because the measured peat sample is younger than the tephra. For homogenizing the display (column # 5), TL error limits are given as for ^{14}C at the 68 % confidence level. The assessed relative chronology, beginning by the oldest eruption is as follows: Montcineyre, Estivadoux, Montchal, Pavin. E is the thickness of the sampled peat

The ages of the four volcanoes, namely crater Pavin, Montchal, maar d'Estivadoux, and Montcineyre, were estimated by means of a Bayesian approach developed in the program RenDateModel by Lanos and Dufresne (2012). This method allows combining together ages obtained by various techniques and it can take stratigraphic constraints into account (e.g., see Blockley et al. 2008; Lowe et al. 2013). Actually, such constraints are very strong in the present case as outlined earlier. For sake of processing real years, the RenDateModel soft will begin by calibrating the radiocarbon conventional results (B.P.); for that purpose, the data from Intcal09 (Hughen et al. 2009) were used in the present work.

Table 8.1 shows the data assumed as reliable and put in the calculations. The five results for Pavin were measured on

Table 8.2 Published age data, not used in the present work

References (Articles)	Volcano and site	Material	Age reference (Lab. No)	^{14}C result (B.P.)	Calibrated age B.C. interval (95 %)		Comments
	Pavin						
Brousse et Delibrias (1969)	Ciberons, near Chaumiane	palaeosoil	Gif 1496	3450 ± 110			
	Montchal						
Brousse et Delibrias (1969); Brousse et al. (1969)	Banks of Couze Pavin river, upstream to Gouelle bridge	Charred plant remains	Gif 1164	6670 ± 160	5890	5320	Attribution to Montchal must be confirmed
Brousse et Delibrias (1969); Brousse et al. (1969)	Banks of Couze Pavin river, upstream to Gouelle bridge	Charred plant remains	Gif 1191	6650 ± 160	5830	5320	Attribution to Montchal must be confirmed
Brousse et Delibrias (1969)	Banks of Couze Pavin river, upstream to Gouelle bridge	Charred plant remains	?	6660 ± 140			Attribution to Montchal must be confirmed
	Estivadoux						
Camus et al. (1973)	1 km North of Les Costes	palaeosoil	Gif 2350	6760 ± 130			Ex. "maar des Costes"
	Montcineyre						
Brousse et Horgues (1969)	CD 36 Road, near Anglards	palaeosoil	Gif 1278	5750 ± 150	4960	4330	

For Montchal, the 6660 B.P. result seems to be averaged from two distinct measurements, 6650 B.P. and 6670 B.P. (Brousse et Delibrias 1969)

peat embedding tephra attributed to this volcano; their mean age was considered as denoting the Pavin eruption. In the local context, it was estimated that one centimetre of Holocene peat represents 25–50 years. Therefore, when about 10 cm of peat was sampled beneath the tephra only, instead of beneath and above the tephra, the measured ^{14}C age predated the tephra by about 125–250 years as was the case for Montcineyre.

Table 8.2 gives for information the age results which were not used in the process. Most of them were measured many years ago for studies on palaeosols. Such ^{14}C dating applied to sediments and palaeosols was possibly biased (Goh et al. 1977; Geyh 1983; Hétier et al. 1983); the authors of the initial application of ^{14}C dating method to the Pavin confirmed such difficulties (Delibrias 1979). Also, ^{14}C measurements for palaeosols sampled at other sites during the same campaign as for the MEMP group and published in the same articles, were later invalidated (for example, for the Gravenoire lava flow in the Royat valley: Huxtable et al. 1978; Boivin et al. 2009). The date for the Pont de Gouelle (Montchal, Table 8.2) was rejected although it seems to really concern carbonized plant remains and although it has been repeated twice, because the authors of the work were not sure of the origin of the correlated layer of tephra. Since then, the corresponding outcrop has been identified and a new study is now planned.

New ^{14}C dates were obtained a few years ago for samples extracted out of sediments drilled in the bottom of Lake Pavin (Chapron et al. 2010). Of them we considered only two results (Table 8.1) related to tree-leaves because the sampling levels were sufficiently close to the base of the lake sediments for producing useful ages. A ^{14}C age measured in the same work for undetermined organic material at the very base of the log was not used despite the fact that it was in agreement with the other data (Poz-27052: 6090 ± 40 B.P.) because the particular conditions of the environment induced a "reservoir effect" (Albéric et al. 2013). Sampling 'terrestrial' plant macrofossils, such as leaves or pollen, obviates such effects (e.g., Newnham et al. 2007; Staff et al. 2013).

Indirect age evaluation by palynostratigraphic correlation was not accounted for in the calculations because it necessitates further theoretical investigation and because it has been demonstrated that pollen features are diachronous even at the regional scale (Juvigné et al. 1988a; 1996).

Strictly speaking, calibrating conventional ^{14}C ages will result in probability distributions (Fig. 8.5) but for sake of simplifying their expression in the text, we opted for quoting the median ages (Table 8.3). The proposed age for the Pavin crater is finally 4720 ± 170 B.C., i.e. a current age of 6730 ± 170 year. Thanks to the stratigraphic constraints, the RenDateModel numerical method allowed us to estimate the age of the maar of Estivadoux without any direct measure-

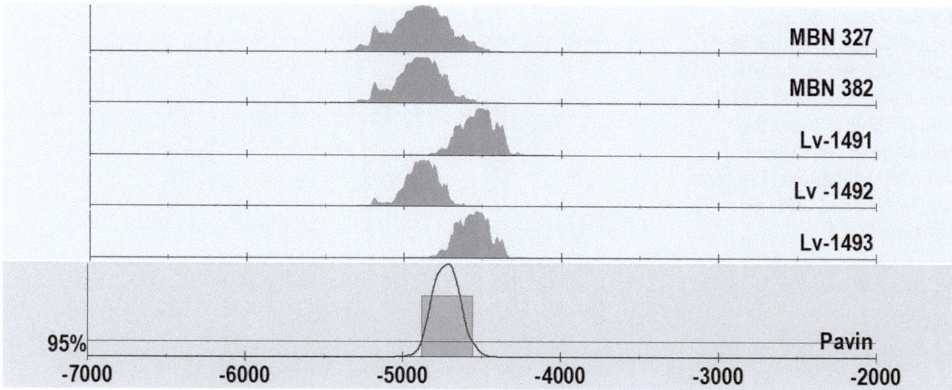

Fig. 8.5 Crater Pavin: smoothed probability distributions for the calibrated ages (years B.C.) for individual ^{14}C measurements and resulting distribution (*lower diagram*), based on the assumption that the 5 age results concern a unique event, the Pavin eruption (Program RenDateModel, Lanos et Dufresne 2012). Data from Table 8.1

Table 8.3 Results of the compilation of the age data given in Table xxx, in years, obtained by means of the program RenDateModel (Lanos et Dufresne 2012) along with the calibration data from IntCal2009 (Hughen et al. 2009)

Volcano	B.C. interval	B.C. median age	Present age (in 2014)	+/–
Pavin	4880–4560	4720	6730	170
Montchal	5190–4640	4920	6930	270
Estivadoux	5310–4730	5020	7030	290
Montcineyre	5410–4830	5120	7130	290

The quoted age is the median one, given as B.C. (column # 3) and as present age (column # 4; "present age" = B.C. age + 2010). Uncertainties are given at the 95 % confidence level. Stratigraphic constraints reminded in the text were accounted for in the computing process

ment for this volcano (Table 8.3). The RenDateModel also enabled us to assess that the interval between the eruptions of Montcineyre and Pavin was about 100–700 years in duration at the 95 % confidence level.

The present data collection of the available chronological information on the MEMP group shows the potential for improvement. The poor precision of half of the measured ages is also shown by the large width of the distribution curves (Fig. 8.6). In fact, the age corpus is still very weak for the three companion volcanoes of the Pavin, whereas the region offers various possibilities to find, in palaeosols principally, charred vegetable remains which could be dated by means of A.M.S.; it was not the case at the time of the publication of the majority of the dates available today.

8.5 Was Pavin Eruption the Latest Eruption in the Chaîne des Puys?

The literature refers to possible volcanic activity in the Chaîne des Puys after the Pavin eruption. At four localities, sandy beds are present in stratigraphic positions postdating the Pavin eruption.

N.B. For pollen features cited below, see above the sketch pollen diagram of Fig. 8.2.

Camus (1975) pointed out the presence of *a sandy layer interbedded in swamp deposits of the Etang de Fung* in the Sioule valley. Underlying non-charred wood remains were dated at 3890 ± 110 B.P. (GIF-2349), which is a maximum age for the sandy layer and which places it in the protohistoric period.

Juvigné et al. (1986) described a millimeter-thick sandy layer in the pure peat body of the Maar of Beaunit. The layer is composed of a large variety of minerals so that the material cannot be linked with any eruptives of previously investigated volcanoes. A peat segment containing the sandy bed was dated at 5020 ± 120 B.P. Nevertheless, in the relevant pollen sequence the bed is situated at the 2 %-point of the continuous curve of *Fagus* (mean date in the Massif Central: 4994 ± 47 B.P. after Juvigné et al. 1988a) and much deeper than the Neolithic regression of the pollen trees (Juvigné et al. 1986: p. 34, Fig. 2; see above Fig. 8.2). The provisional name of *Beaunit tephra* was applied to the sandy layer while waiting for further identical findings elsewhere.

Juvigné et al. (1988b) described *two cm-thick sandy beds at depth 6.5 m in the 13.5 m thick pure peat body of the Narse d'Espinasse* (Juvigné et al. 1986: p. 12, Fig. 3; see above Fig. 8.2). Mineralogical composition and texture were different from each other. Both layers, and the lowest depletion of pollen trees corresponding to the Neolithic epoch in the pollen diagram, are synchronous. Since these layers are not present in the nearby peat bog of the Narse d'Ampoix, their origin is attributed to the accelerated erosion due to drastic Neolithic deforestation. The distribution area of the sandy beds has not been verified. The site, being off a delta front, indicates that scattered lenses can be expected in the peat.

Fourmont et al. (2006) found in the Marais de Sarliève (Grande Limagne plain, near Clermont-Ferrand) a thin sandy

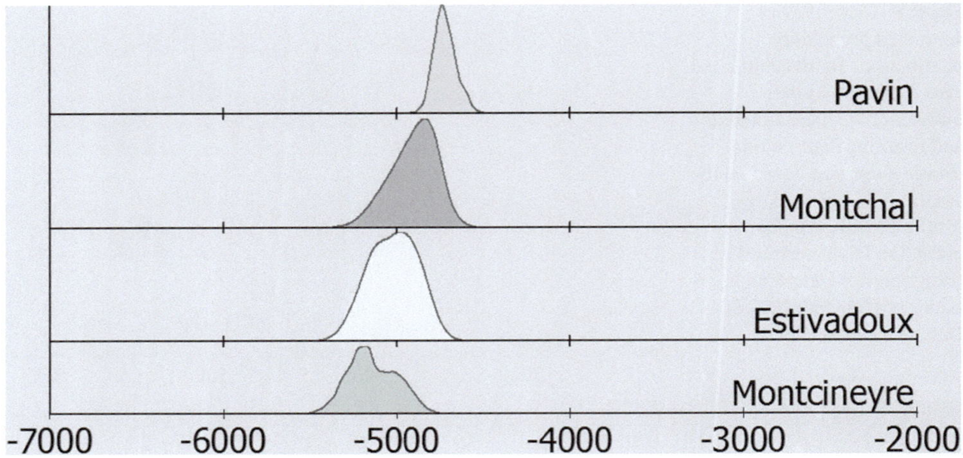

Fig. 8.6 Smoothed age distributions (B.C.) for the four volcanoes of the MEMP volcanoes, computed with the data of Table 8.1 and the stratigraphic constraints reminded in the text (Program RenDatmodel, Lanos and Dufresne 2012)

bed named *Sarliève tephra*, but it cannot yet be linked with any known volcano. Chronological data for *Sarliève tephra* are puzzling: the TL-method has given an age of 16 ± 4 kyr (i.e., emergence from, or final part of, the last glaciation). In a pollen sequence it is clearly situated in the Atlantic palynozone (i.e., Holocene; Fourmont et al. 2006: p. 1146, Fig. 3; see above Fig. 8.2). Its position within the increase of *Fagus* curve provides an approximate age of 4690 ± 40 B.P. (after Juvigné et al. 1986), and necessarily more recent than 4994 ± 47 B.P. which is the age of the underlying 2 %-point of the beginning of the continuous curve of *Fagus* (after Juvigné et al. 1988a). Mineralogical composition does not support any correlation with the other sandy beds described above. Otherwise, the site is far from the other three ones and on the opposite side of the Chaîne des Puys in Limagne. The bed of sandy volcanogenic materials is also clearly younger than the most recent (primary) tephra found previously in Limagne including the CF7 trachytic tephra (about 7200 B.C.), the Marsat trachytic tephra (about 7600 B.C.; Vernet and Raynal 2000, 2008), and the Pariou tephra (about 8000 B.C.; Juvigné et al. 1992; Vernet and Raynal 2000).

In all sites described above, each sandy bed corresponds to a single local event. No similar bed has been found even in a nearby site, so that a volcanic origin is doubtful for most of them. The sandy beds of Fung and Espinasse were formed during the vast deforestation of the protohistorical epoch that has enabled increasing soil erosion by run-off and aeolian deflation. Hence the detrital origin of all the sandy beds is very likely. The tephras of Beaunit and Sarliève predate the protohistoric period of extensive deforestation in the Massif Central. In spite of their similar age, around 5000 B.P., they cannot be linked with any volcanic activity known in the Massif Central. Otherwise they cannot be correlated with one another, since neither the mineralogical composition, nor the palynostratigraphical positions are compatible. Hence, they cannot be definitively assessed as the youngest tephras of the French Massif Central as long as similar occurrences are not found in nearby sites. Currently, therefore, the eruption of Pavin volcano can be considered as the latest one in the Massif Central.

8.6 Conclusion

Pavin volcano is responsible for the most recent cataclysmic eruptions in the Massif Central. Among various types of activity, products of at least two Plinian columns were injected in the high troposphere then spread northwards and southwards by the wind. That eruption took place at 4720 ± 170 B.C., i.e. a current age of 6730 ± 170 year, and activity occurred within a very short time range (days to weeks). It is the final one of an eruptive epoch which lasted between 100 and 700 years after the initial Strombolian eruption of Montcineyre volcano, followed by the phreatomagmatic activity of the Estivadoux Maar and the Strombolian eruption of Montchal volcano. Sandy beds attributed as tephras in some earlier work are likely to be erosion products relating to deforestation.

References

Albéric P, Jézéquel D, Bergonzini L, Chapron E, Viollier E, Massault M, Michard G (2013) Carbon cycling and organic radiocarbon reservoir effect in a Meromictic Crater Lake (Lac Pavin, Puy-de-Dôme, France). Radiocarbon 55:3–4

Blockley SPE, Bronk Ramsey C, Lane CS, Lotter AF (2008) Improved agemodelling approaches as exemplified by the revised chronology for the central European varved lake Soppensee. Quat Sci Rev 27:61–71

Boivin P, Besson JC, Briot D, Camus G, de Goër de Herve A, Gourgaud A, Labazuy P, Langlois EP, de Larouzière FD, Livet M, Mergoil J, Miallier D, Morel JM, Vernet G, Vincent P (2009) Volcanologie de la Chaîne des Puys. Parc Naturel Régional de la Chaîne des Puys (Ed.), Carte et fascicule, 5e édition, 200 p

Boivin P, Besson J-C, Ferry P, Gourgaud A, Miaillier D, Thouret J-C, Vernet G (2010) Le point sur l'éruption du lac Pavin il y a 7 000 ans. Revue des Sciences Naturelles d'Auvergne 74–75:45–56

Bourdier JL (1980) Contribution à l'étude volcanologique de deux secteurs d'intérêt géothermique dans le Mont Dore: le groupe du

Pavin et le Massif du Sancy. Thèse, Université de Clermont-Ferrand II, Clermont-Ferrand, 180 p

Brousse R, Bardintzeff JM (1987) Le volcanisme quaternaire, influence sur le cadre de vie. In: Miskowski JC (ed) Géologie de la Préhistoire: Méthodes, techniques, applications. GEOPRE, Paris, pp 241–249

Brousse R, Delibrias G (1969) Une éruption vieille de 6660 ans en Auvergne. C R Acad Sci Paris 268:1175–1177

Brousse R, Horgues M (1969) Le volcan de Montcineyre, l'un des plus récents d'Auvergne. C R Acad Sci Paris 269:1835–1838

Brousse R, Rudel A (1973) Activité paroxysmale de la Chaîne des Puys à 10000 ans B.P. Bull Hist Sci Auv 86:1–11

Brousse R, Delibrias G, Labeyrie J, Rudel A (1969) Eléments de chronologie des éruptions de la chaîne des Puys. Bull Soc Géol Fr 7(XI):770–793

Camus G (1975) La Chaîne des Puys. Etude structurale et volcanologique. Thèse Doct. Etat, Univ. Clermont-Ferrand II, Clermont-Ferrand, 321 p

Camus G, de Goër de Herve A, Kieffer G, Mergoil J, Vincent PM (1973) Mise au point sur le dynamisme et la chronologie des volcans holocènes de la région de Besse-en-Chandesse. C R Acad Sci Paris 277:629–632

Chapron E, Albéric P, Jézéquel D, Versteeg W, Bourdier JL, Sitbon J (2010) Multidisciplinary characterization of sedimentary processes in a recent maar lake (Lake Pavin, French Massif Central) and implication for natural hazards. Nat Hazards Earth Syst Sci 10:1815–1827, 9.5194/nhess-10-1815-209

Chapron E, Ledoux G, Simonneau A, Albéric P, St-Onge G, Lajeunesse P, Boivin P, Desmet M (2012) New evidence of Holocene mass wasting events in recent volcanic lakes from the French Massif Central (Lakes Pavin, Montcineyre and Chauvet) and implications for natural hazards. In: Volcanic lakes ISSMMTC Japan proceedings 2011. In: Yamada Y et al (eds) Submarine mass movements and their consequences, 255 advances in natural and technological hazards research 31. Springer, pp 255–264. doi:10.1007/978-94-007-2162-3_23

Delibrias G (1979) Problèmes liés à la datation des sols fossiles par le carbone 14. Bulletin de l'Association française pour l'étude du quaternaire 16(12):35–42

Etlicher B, Janssen C, Juvigné E, Van Leeuwen J (1987) Le Haut Forez (Massif Central, France) après le pléniglaciaire würmien: environnement et téphra du volcan de la Nugère. Bull Ass Fr Et Quat 32:229–239

Fourmont A, Macaire JJ, Bréhéret JG, Argant J, Prat B, Vernet G (2006) Tephras in lacustrine sediments of the Sarliève marsh (French Massif Central): age and preservation. C R Geosci 338:1141–1149

Gewelt M, Juvigné E (1988) Téphrochronologie du Tardiglaciaire et de l'Holocène dans le Cantal, le Cézalier et les Monts Dore (Massif Central, France): résultats nouveaux et synthèse. Bull Ass Fr Et Quat 1:23–32

Geyh MA (1983) The unreliability of ^{14}C dates obtained from buried sandy podzols. Radiocarbon 25:409–416

Goh KM, Molloy BPJ, Rafter TA (1977) Radiocarbon dating of quaternary loess deposits, Banks peninsula, Canterbury, New Zealand. Quat Res 7:177–196

Guérin G (1983) La thermoluminescence des plagioclases, méthode de datation du volcanisme. Applications au domaine volcanique français : Chaîne des Puys, Mont-Dore et Cézallier, Bas-Vivarais. Doctorat d'État, Université Pierre et Marie Curie, Paris VI, Paris, 253 p

Hétier JM, Guillet B, Brousse R, Delibrias G, Maury RC (1983) ^{14}C dating of buried soils in the volcanic Chaine des Puys (France). Bull Volcanol 46(2):193–201

Hughen KA, Kaiser KB, McCormac FG, Mannings W, Reimer RW, Richards DA, Southon JR, Talamo S, Turney CSM, Van der Plicht J, Weyhenmeyer CE (2009) IntCal09 and Marine09 calibration curves. Radiocarbon 51:1111–1150

Huxtable J, Aitken MJ, Bonhommet N (1978) Thermoluminescence dating of sediments baked by lava flows of the Chaîne des Puys. Nature 275:207–209

Juvigné E (1991) Distribution de vastes retombées volcaniques originaires de l'Eifel et du Massif Central aux temps post-glaciaires dans le NE de la France et les régions voisines. C R Acad Sci Paris 312, Série II:415–420

Juvigné E (1992a) Distribution of Late Glacial and Holocene tephra beds in the French Central Massif. Quat Int 13/14:181–185

Juvigné E (1992b) Approche de l'âge de deux cratères volcaniques lacustres d'Auvergne (France). C R Acad Sci Paris 314, Série II:401–404

Juvigné E (1993) Contribution à la téphrostratigraphie du Quaternaire et son application à la géomorphologie. Mémoires pour servir à l'Explication des Cartes Géologiques et Minières de la Belgique 36, Service géologique de Belgique, Bruxelles, 66 p

Juvigné E, Bastin B (1995a) Téphrostratigraphie et palynologie de tourbes du Boréal et de l'Atlantique dans le Massif Central (France). Géog Phys Quatern 49:207–216

Juvigné E, Bastin B (1995b) Le Gour de Tazenat: évolution de l'environnement du Tardiglaciaire au Boréal dans la Chaîne des Puys septentrionale (Massif Central, France)

Juvigné E, Bastin B (1996) Le Gour de Tazenat: évolution de l'environnement du Tardiglaciaire au Boréal dans la Chaîne des Puys septentrionale (Massif Central, France). C R Acad Sci Paris 321, série IIa:993–999

Juvigné E, Gewelt M (1987) La Narse d'Ampoix comme téphrostratotype de la Chaîne des Puys méridionale (France). Bull Ass Fr Quat 29:37–48

Juvigné E, Gilot E (1986) Ages et zones de dispersion de téphra émises par les volcans du Montcineyre et du lac Pavin (Massif Central, France). Zool Geol Ges 137:613–623

Juvigné E, Stach-Czerniak A (1998) Etude sédimentologique et palynologique des dépôts lacustres tardiglaciaires et holocènes du Gour de Tazenat (Massif Central, France). Quaternaire 9(1):15–23

Juvigné E, Bastin B, Gewelt M (1986) Découverte de retombées volcaniques d'âge holocène dans la Chaîne des Puys septentrionale (Massif Central, France). Rev Sc Nat D'Auvergne 52:33–36

Juvigné E, Milcamp V, Delibrias G, Evin J (1988a) Ages de traits polliniques et chronozonation du Tardiglaciaire et de l'Holocène dans le Massif Central (France). Med Rijks Geol Dienst 4(4):33–50

Juvigné E, Lousberg N, Gewelt M (1988b) Evolution morphosédimentaire de la Narse d'Espinasse. Rev Sci Nat Auvergne 53:7–14

Juvigné E, Kroonenberg S, Veldkamp A, Arabi E, Vernet G (1992) Widespread Allerød and boreal trachytic tephra layers as stratigraphical markers in the Massi Central, France. Quaternaire 3(3–4):137–146

Juvigné E, Bastin B, de Goër de Herve A (1994) Nouvelles données sur les téphras et l'environnement tardiglaciaires du Cézalier et de l'Artense (Massif Central, France). Ann Soc Géol Belg 117(2):321–332

Juvigné E, Bastin B, Delibrias G, Evin J, Gewelt M, Gilot E, Streel M (1996) A comprehensive pollen and tephra-based chronostratigraphic model for the Late glacial and Holocene period in the French Massif Central. Quat Int 35/36:113–120

Lanos P, Dufresne P (2012) Modélisation statistique bayésienne des données chronologiques. In: Sophie A (ed) de Beaune et Henri-Paul Francfort (dir.). L'Archéologie à découvert, CNRS Editions, Paris, pp 238–248

Le Bas MJ, Le Maître RW, Streckheisen A, Zanettin B (1986) A chemical classification of volcanic rocks based on the total alkali-silica diagram. J Petrol 27:745–750

Leyrit H, Zylberman W, Lutz P, Jaillard A, Lavina P (2016) Characterization of phreatomagmatic deposits from the eruption of the Pavin maar (France). Chap 6, this volume

Lowe DJ, Blaauw M, Hogg AG, Newnham RM (2013) Ages of 24 widespread tephras erupted since 30,000 years ago in New Zealand, with re-evaluation of the timing and palaeoclimatic implications of the Lateglacial cool episode recorded at Kaipo bog. Quat Sci Rev 74:170–194

Newnham RM, Vandergoes MJ, Garnett MH, Lowe DJ, Prior C, Almond PC (2007) Test of AMS 14C dating of pollen concentrates using tephrochronology. J Quat Sci 22:37–51

Reille M, de Beaulieu JL, Pons A (1985) Recherches pollenanalytiques sur l'histoire tardiglaciaire et holocène de la végétation du Cézalier, de la Planèze de St Flour et de la Margeride. Pollen Spores 27(2):209–270

Staff RA, Nakagawa T, Schlolaut G, Marshall MH, Brauer A et al (2013) The multiple chronological techniques applied to the Lake Suigetsu SG06 sediment core, central Japan. Boreas 42:259–266

Vernet G, Raynal JP (2000) Un cadre téphrostratigraphique réactualisé pour la préhistoire tardiglaciaire et holocène de Limagne (Massif Central, France). C R Acad Sci II A Paris 330:399–405

Vernet G, Raynal JP (2008) La formation de Marsat et le téphra CF7, marqueurs distaux d'éruptions trachytiques violentes de la Chaîne des Puys au Boréal. Quaternaire 19(2):97–106

9 Geology, Geomorphology and Slope Instability of the Maar Lake Pavin (Auvergne, French Massif Central)

Jean-Claude Thouret, Pierre Boivin, Philippe Labazuy, and Alberic Leclerc

Abstract

Maars are craters created by violent phreatomagmatic eruptions. The crater shape involves roughly circular rims whose asymmetric slopes may be unstable long after the initial eruption. Lakes that occupy many maars are natural receptacles that enclose geologic archives such as slope deposits. Here we describe the geologic and geomorphic characteristics of the maar and Lake Pavin in Auvergne with emphasis on recent and current slope instability. This is due to its geometry (the 800 × 92 m lake occupies a wide depression cut deep in pre-existing lava flows and Montchal cone), loose surficial formations on steep subaerial slopes and fractured lava scarps, and loose and gas-rich sediment on sub-lacustrine steep slopes.

Our study of the outer rim slopes (<20°) of the maar shows that current geomorphic processes apparently act slowly, but mapping of the steepest (>31°), inner rim slopes suggests that instability is now related to runoff, *solifluction* and perhaps rockslides or deep-seated landslides. The slow and often small-sized mass movements occur on steep slopes >31° and fractured lava flow scarps while solifluction is favored by loose and thick, surficial maar deposits and a 4–5 month-long snow cover. Geomorphic anomalies on top of the north and NE maar rims suggest deep-seated, (slow moving?) rotational landslides that may record a long-lasting post-eruptive response to maar collapse. One of the large, recent rock fall scree on the NNE edge of the lake is apparently connected to the submerged platform capping a syn-eruptive collapse mass. The quasi-vertical edge of this platform may act as a source of debris transfer towards the deep lake bottom. Long subaerial slopes south and SE of the maar point out to the most unstable sector: fractured, thick lava flow scarps topple and produce scree, and permanent springs feed runoff and streams above the underlying clay-rich pyroclastic deposit. The south slope overhangs subaquatic lava cliffs which can transfer debris directly to the lake bottom 90 m below. In contrast, mapping of the recent fan of the lake outlet and the adjacent Gelat valley to the north, in which the outlet stream is incised, show no evidence for debris-flow deposits that were claimed to be emplaced by a historical catastrophic event triggered by a lake breakout.

Keywords

Lake Pavin • Maar • Slope • Instability • Lava • Pyroclastic deposit • Fan • Hazard

J.-C. Thouret (✉) • P. Boivin • P. Labazuy
LMV, Université Clermont-Auvergne - CNRS - IRD - OPGC, Aubière, France
e-mail: J.C.Thouret@opgc.univ-bpclermont.fr

A. Leclerc
Ecole Supérieure des Géomètres et Topographes, 1 Bd. Pythagore, 72000 Le Mans, France

T. Sime-Ngando et al. (eds.), *Lake Pavin*, DOI 10.1007/978-3-319-39961-4_9

9.1 Introduction

9.1.1 Maar, Landform and Process Definition

Hydrovolcanic landforms created by phreatomagmatic eruptions encompass tuff rings, tuff cones and maars. Like calderas but smaller sized, maars are negative landforms excavated in the pre-eruption surface, in volcanic and non-volcanic bedrock. Unlike calderas maars characterize small-volume and short lived volcanoes commonly defined as monogenetic volcanoes with a central crater cut deeply into the pre-eruptive country rocks, surrounded by a low ring wall of pyroclastic material (tephra/tuff ring), and underlain by a diatreme (Ollier 1967; Lorenz 1973, 1986, 1987, 2007; Cas and Wright 1987; Vespermann and Schmincke 2000). Originally the term maar described a topographic feature, consisting of a crater and a tephra/tuff rim, but this term now incorporates the ring wall, the crater sediments, the diatreme and the feeder dyke. A maar is the crater cut into the ground below the syn-eruptive surface and surrounded by an ejecta ring, while the diatreme structure continues downward and encloses diatreme and root zone deposits (White and Ross 2011; Németh and Kereszturi 2015).

The syn-eruptive processes are driven by magma/groundwater, Molten Fuel-(Impure) Coolant explosive Interaction (MFCI) and produce mostly base *surge* and phreatomagmatic fallout beds up to a few tens of meters thick around the excavated maar crater (Wohletz 1986; Buettner et al. 2002; Lorenz 2007). During the post-eruptive processes subsequent to the formation of the maar basin, the intersection of the groundwater level leads to formation of a lake (Büchel and Lorenz 1993; Christenson et al. *in* Rouwet et al. 2015). The lake formation in the maar crater is usually a fast process but it depends on the permeability of the disrupted country rock, the geometry of the water storage system, and the availability of groundwater (Pirrung et al. 2008). The long-term sedimentary filling of the lake is controlled by (i) mass movements (mass flows of any type from the inner crater wall), (ii) delta deposits, (iii) atmospheric loads, mostly ash fall from nearby eruptive sites, (iv) production of organic matter in the lake, and (v) intensive mineral-rich spring activity (Pirrung et al. 2007, 2008; Németh et al. 2008; White and Ross 2011; Fox et al. 2015). Other processes can be called in for the relatively rapid sediment infill such as the post-eruptive subsidence of the maar, which may contribute to the maar lake deposit compaction and resettlement (Suhr et al. 2006; Lorenz 2007).

9.1.2 Scope and Objectives

We describe the geologic and geomorphic context of the maar and lake in order to examine potential factors of instability linked to both maar slopes and/or the lake outlet. The motivation of this study lies on the claim related to recent (post 6700 years) eruptive activity and Middle Ages catastrophic mudflow triggered by lake breakout subsequent to subaquatic landslide(s) (Lavina and Del Rosso-D'Hers 2008, 2009). This claim originated from geologic mapping of the area by the authors following earlier studies on volcanic lakes in Auvergne (e.g. Glangeaud 1916). As limnic eruptions occurred in crater lakes such as the lethal Lake Nyos event in Cameroon in 1986 (Lockwood et al. 1988; Tassi and Rouwet 2014), investigations have re-considered the volcanic lakes in Auvergne and Pavin in particular (Fig. 9.1). Camus et al. (1993) and more recently Olive and Boulègue (2004) concluded that despite its meromicticity, Lake Pavin does not present an imminent threat of violent CO_2 degassing due to lake overturn induced by a volcanic or limnic eruption. Meromicticiy defines lakes that have layers of water that do not intermix (Touchard 2000; Jézéquel et al. 2008; Rouwet et al. 2015). Various hydrothermal venues and carbogaseous springs as well as CO_2 emissions are known in the area at Escarot 3 km WSW and Fontaine Goyon 2 km ENE of Lake Pavin (Gal and Gadalia 2011; Fig. 9.1). The regionally known CO_2 has carbon isotope ratios that bind it to a magmatic origin from the upper mantle, linked to attenuation and fracturing of the continental crust beneath the Massif Central (Olive and Boulègue 2004; see also Alberic et al. 2013 for a more recent carbon reservoir analysis).

The objectives of the paper are twofold: (1) Link evidence of instability on the maar rims with subaquatic landforms such as the syn-eruptive, subaquatic platform mantled by sediment cover in the NNE sector of the lake described by Chapron et al. (2010a,b); (2) Unravel whether or not the fan created by the lake outlet can be related to recent events that have been postulated on the basis of historical landslide deposits dated from lake sediment (Fig. 9.1). Besides the geomorphological mapping of maar rims and the adjacent Couze Pavin Valley, we seek to detect geomorphic signs which may suggest current or recent instability above the lake shore. Slope instability may be related either to the steep maar rims (rock fall from lava flow scarp, slump and runoff in *tephra*) and to mass movements (subaquatic slumps and *turbidites*) that occurred in the recent past in the lake. Chapron et al. (2010a,b, 2012) have summarized natural hazards linked to a group of lakes in the French Massif Central and the Pavin maar lake in particular. Geophysical investigations, sedimentary cores and C^{14} ages have allowed the authors to reveal that landslides occurred twice inside the lake between AD 580–640 and as recently as AD 1200 and 1300. Lavina and Del Rosso-D'Hers (2009) and Del Rosso-D'Hers et al. (2009) claimed that debris-flow deposits they observed in the adjacent Gelat valley to the north resulted from the most recent of these events (see Sect. 9.4.4 below). We will report geomorphic indicators for current slope instability of the maar rims and we discuss the claim on debris-flow deposits related to the postulated lake breakout.

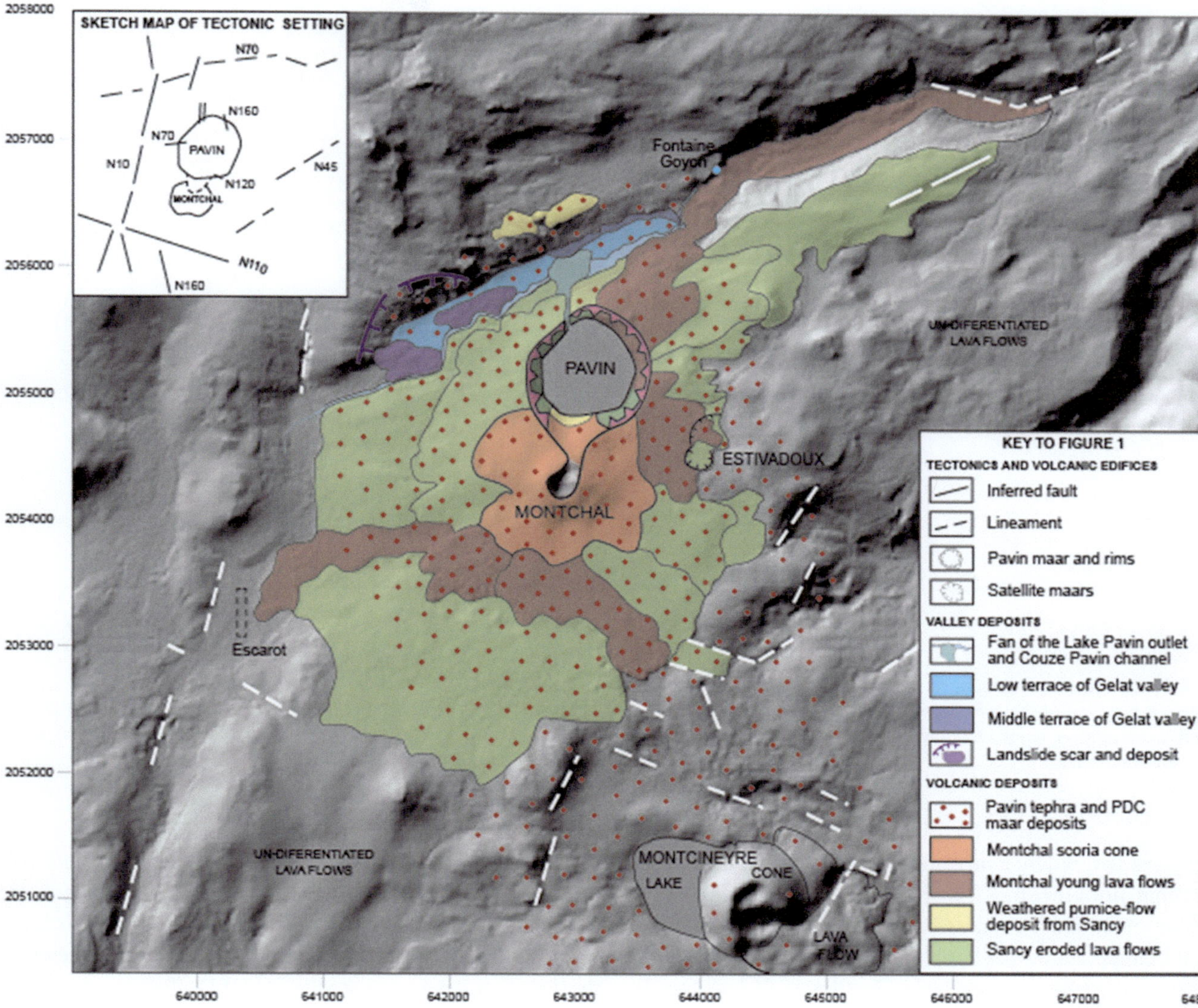

Fig. 9.1 Geologic setting of the Lake Pavin maar and surroundings. The area of carbogazeous springs and mofettes near Escarot and the CO_2-rich spring of La Fontaine Goyon are indicated

9.2 Study Site: Pavin Maar, Lake and Surroundings

Pavin is the only meromictic maar lake in France (Chaps. 1, 5 and 10, in this volume), which is surrounded by several older, small lakes of volcanic origin as shown in Chap. 23, Fig. 23.1. Lake Pavin (45° 55′ N; 2°54′ E) is located at 1197 m asl. in a remote area of the department of Puy-de-Dôme on the eastern slope of the Monts-Dore-Sancy strato-volcano in the French Massif Central. The almost circular lake has a diameter of about 800 m, a surface area of 44.5 ha and a maximum water depth of 92 m (Fig. 9.1). The Montchal lavas have then been partly removed by the Pavin phreatomagmatic eruption, first ^{14}C dated at 6900 ± 50 year ago (Gewelt and Juvigné 1988; Juvigné 1992a,b), but statistically revised as 6730 ± 130 years ago using new calibration curve and new data (Juvigné and Miallier; Chap. 8, in this volume). Primary deposits of eruption have been recognized on a large area around the lake (Bourdier 1980; Leyrit et al., Chap. 6, in this volume). This recent volcanic event formed Lake Pavin: a roughly circular and deep maar lake draining a steep and well preserved crater rim reaching an altitude of 1253 m asl. Nowadays, the maar rims wall rise, for the most part, 50–80 m above the lake surface but the maar intersects the north flank of the Montchal strombolian cone that rises as high as 210 m above the south lake edge (Figs. 9.1, 9.2, and 9.3).

As shown in Fig. 9.3, the edge of the Pavin maar tephra ring matches the limit of the topographic drainage basin of Lake Pavin except on the south flank of Montchal. It is however important to keep in mind that this topographic drainage basin is smaller than the still poorly defined watershed of Lake Pavin, approximately 2 km^2, draining several subaerial and subaquatic springs (Bonhomme et al. 2011). Pavin's

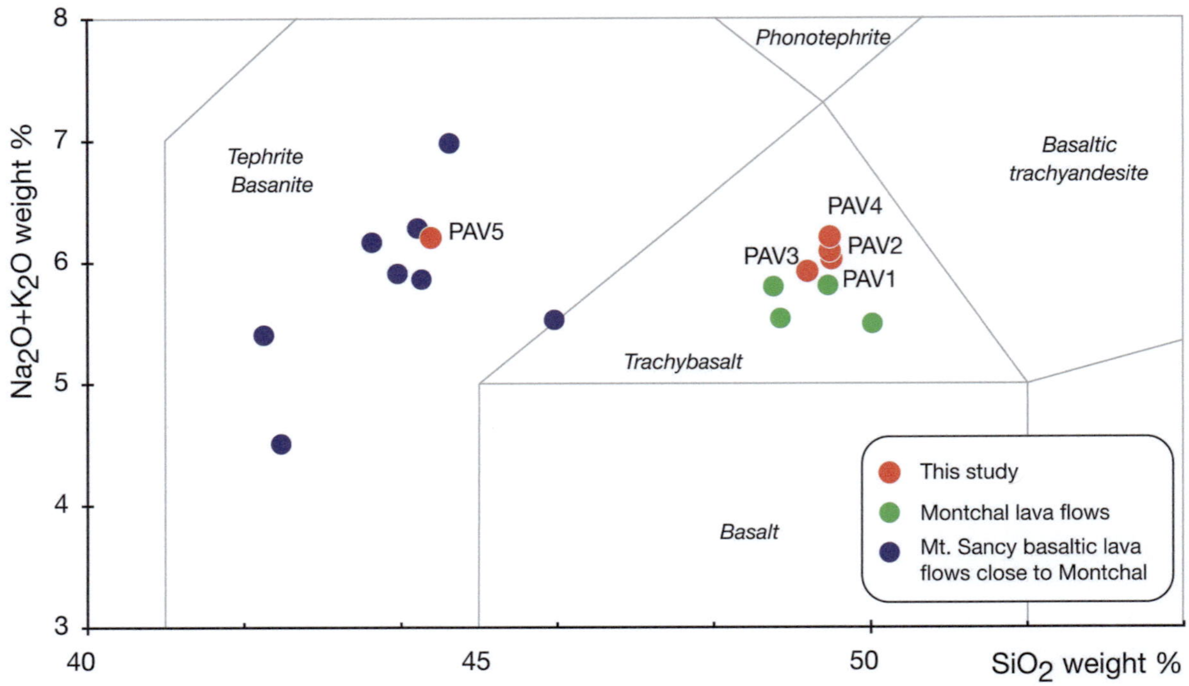

Fig. 9.2 TAS diagram showing five lava samples (no.1–5 *red* dots, located in Fig. 9.3) of the lava flows cut by the maar and surrounding the lake: one quasi-aphanitic tephrite sample (PAV5) with a few phenocrysts of leucite and olivine from the NE cliff, and four porphyritic trachybasalt samples (PAV1-4) with phenocrysts of olivine, clinopyroxene and amphibole from the NW, west, south and east cliffs. Other Montchal (*green*) and Mt. Sancy-type (*blue*) lava flows are shown for the purpose of comparison (after Brousse 1961; Bourdier 1980; Boumehdi 1988)

slopes are covered by a mixture of deciduous and coniferous trees that limit erosion. The region is characterized by an oceanic-mountain climate (Stebich et al. 2005). A mean annual temperature of 6.5 °C (between 1946 and 1974) was observed at Besse-en-Chandesse, the closest (4 km) meteorological station to the lake, located at an elevation of 1050 m asl. The annual thermal amplitude ranges between −5 and 20 °C (Stebich et al. 2005). Annual precipitation in the catchment averages between 1600 and 1700 mm. The climate diagram from this station indicates precipitation maxima in May and also between November and January. The lake surface usually freezes during the winter months while daily frost and thaw conditions may prevail between November and March (≥100 days/year in Besse's weather station located 150 m lower in elevation).

9.2.1 Geological Context Including Tectonic and Volcanic Features

Auvergne maars are typical landforms in the Chaîne des Puys (e.g. Beaunit, Gour de Tazenat), on the Mont-Dore stratovolcano (e.g. Chauvet) and in the Cézalier massif (e.g. Godivelle d'en haut). Pavin maar is the youngest volcano in continental France, which stands among a group of four volcanoes which erupted within a period of five centuries (Juvigné 1992a,b; Juvigné & Miallier, Chap. 9, in this volume). In chronological order (Camus et al. 1973), these are: Montcineyre strombolian cone 4 km south, Estivadoux maar 1 km southeast and Montchal strombolian cone adjacent to the Pavin crater to the south (Fig. 9.1). The regional map based on the IGN DEM (10 m pixel) shows the geologic context of the Pavin maar on the western flank of the Sancy massif, which is the youngest edifice of the Monts-Dore stratovolcano (Bourdier 1980; Lavina 1985). The maar cut away the high (210 m) north flank of the Montchal cone and its lava flows to the ENE (Figs. 9.1, 9.2, and 9.3). Montchal lava flows cut by the maar are older than the strombolian scoria-fall from the Montchal cone, which directly underlies (without any soil) the 6730 ± 130 year Pavin maar deposits. Porphyritic, trachybasaltic lava flows of Montchal, which were cut by the maar, surround the lake except on the NNE rim slope. There, the leucite-bearing tephritic lava flow that makes the fractured cliff shares mineralogical and geochemical characteristics with Mont-Dore Sancy lava flows as shown by the TAS diagram (Fig. 9.2).

The group of four volcanoes provides no evidence for a north-south graben. Lavina (1985) locates the Pavin-Montchal group at the southern end of a N105-125°E dyke group linked to the Sancy stratovolcano. Instead, the Pavin maar together with the Montchal and Montcineyre cones are roughly aligned N160° although no fault is observed at

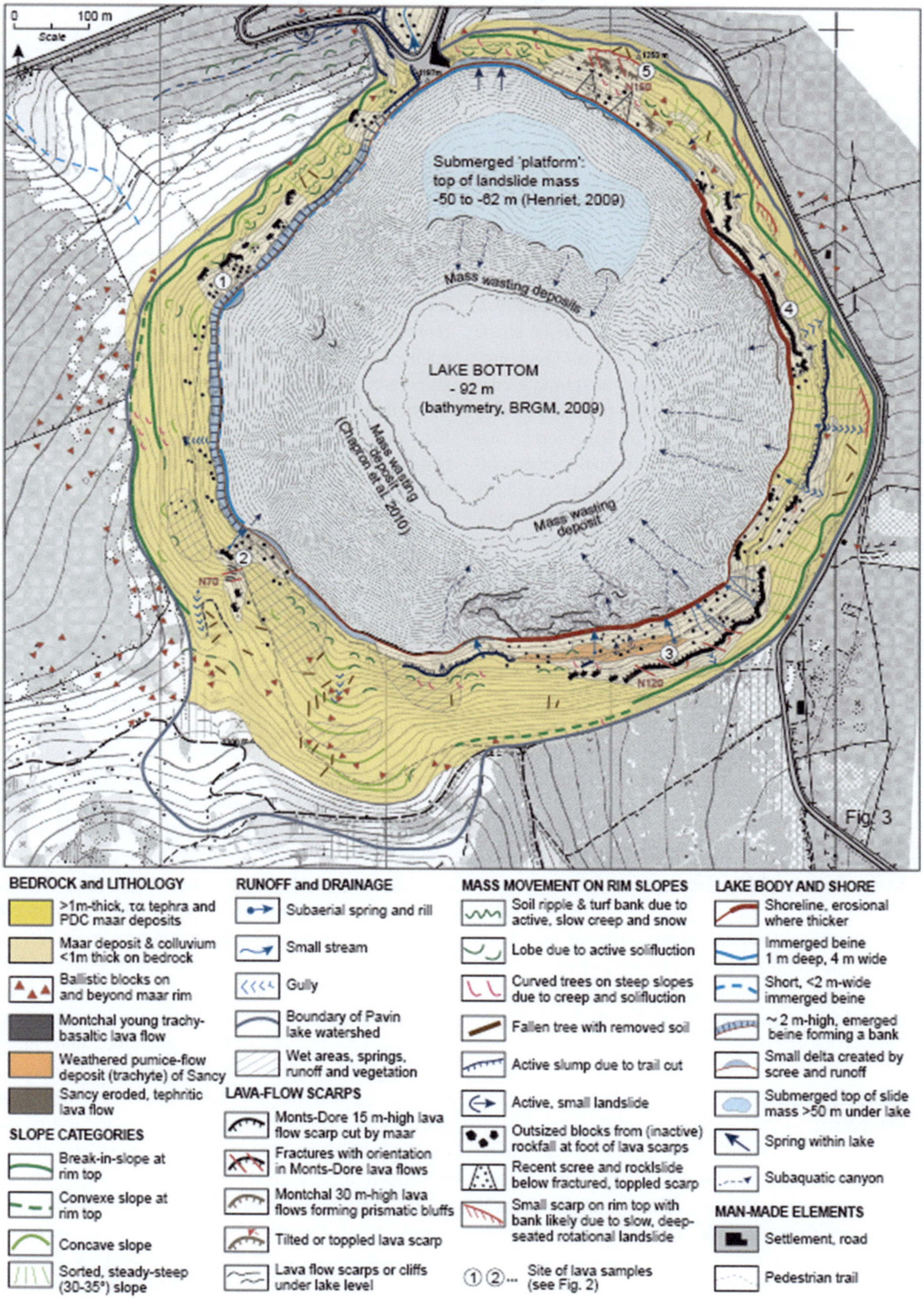

Fig. 9.3 Geologic and geomorphological map of the maar rim slopes with emphasis on slope landforms and inferred processes

ground surface. Fig. 9.1 portrays four groups of lineaments at the intersection of which the four volcanoes are located: (1) N10°E fractures appear west of Pavin (e.g. the Escarot mofette 3 km west of the lake extends 0.4 km N-S) and SSE of Montchal (Fig. 9.1), (2) N70°E lineaments guide large valleys in the region, (3) N110°E fractures which may offset N10° lineaments, and (4) N160°E fractures. These lineaments likely reflect deeper faults in the basement and the Pleistocene Sancy volcano.

The polygonal shape of the Pavin maar displays six sides, the most rectilinear of them (N10-20, N70, N110) paralleling lineaments shown in the tectonic sketch of Fig. 9.1. After almost 6700 years of existence, the non-circular shape of the maar rims suggests that the post-eruption slope erosion has been relatively slow under temperate climate conditions as vegetation probably covered the slopes soon after the eruption ended (e.g. Schwab et al. 2009). Another reason for the irregular rim is that the deep, funnel-shaped crater nested in lava rock together with resistant lava cliffs around the lake have preserved a large amount of the original maar form except the north flank of Montchal cone. Tops of maar rims still show a strong asymmetry (gentle outer slope versus steep inner slope) around three quarters of the maar except the wide, gently dipping and concave, north Montchal's flank over which maar tephra and cone scoriae have been redistributed since the maar eruption. The asymmetry of the crater inside the maar lake, as inferred from the bathymetry (*in* BRGM 2009), is due to the quasi vertical SE and east slopes cut in thick lava piles, which contrast with the subaquatic plateau submerged in the NNE part of the lake (Fig. 9.3).

9.3 Materials and Methods

Three categories of field and laboratory methods have been undertaken to map the maar and analyze its slope stability: geologic and geomorphologic mapping, very high resolution DEMs, and analysis of geomorphic parameters.

9.3.1 Geologic and Geomorphologic Mapping

Mapping of the maar rims has been conducted with students over the past 4 years in order to delineate surficial deposits and highlight landforms and processes that may indicate slope instability around the lake. Analysis of DEMs and delineating from aerial photographs taken by IGN in 1948, 1956 and 1962 were carried out with a stereoscope and completed by field observations. Deposits have been studied in the field and lava and tephra samples (thin sections, geochemistry and grain size distribution) were analyzed in the LMV laboratory (Boivin et al. 2012; Fig. 9.2). We had access to the old geologic map of Brioude (1964, no.175, scale 1:80,000) and the unpublished sketch map of Besse (after the name of the main town 3 km ENE of Lake Pavin) carried out by P. Lavina (BRGM Puy-de-Dôme).

The rim slopes map (Fig. 9.3) was drawn on a 1/10,000 scale topographic map with 10 m contours (Camus et al. 1993) outside the lake, to which a more precise, 1 m-contour bathymetry derived from a DEM was added (MESURIS Bathymetrie in: BRGM 2009). Two maps of the Gelat and Couze Pavin River valley were drawn on a 50 cm-pixel surface DSM (i.e. without filtering vegetation above soil), which represents a new basis for mapping landforms with unprecedented detail and allowing for a better interpretation of slope and valley processes. Thus, Figs. 9.4 and 9.5 display the bedrock, surficial deposits and mass movements along the Gelat – Couze Pavin valley adjacent to the north maar rim. We have focused on lithology, deposit types and landforms indicating mass wasting and mass movements on slopes at all scales.

9.3.2 Very High Resolution DEM Acquisition and Calibration

Over the past 10 years, Unmanned Aerial Vehicles (UAVs) have become relevant tools for high resolution Digital Elevation Model computation using stereophotogrammetry methods. A survey of the Plaine du Gelat area adjacent to the north maar rim was carried out in April 2015 to acquire a set of low-altitude photographs that can be used to compute high-spatial resolution DEMs. A fixed-wing drone (Lehmann Aviation LA300) was operated by Technivue society and equipped with a 12 Mpx Canon PowerShot S110 camera (focal length: 5.2 mm). The plane flew at about 200 m in elevation above ground using preprogrammed GPS positioning. A simple constructed two dimensional stabilization plant is designed to keep the image sensor's optical axis directed at vertical. Due to the limited battery capacity the grid based flight plan had to be subdivided into eight flights, allowing us to acquire a complete set of 2608 images, with a total coverage area of 4.17 km^2. As many as 908 images were selected to ensure an image overlap of about 80 % and side overlap of 30 % over the entire study area with a mean ground resolution of 0.068 m per pixel.

The photogrammetric workflow has been carried out using Agisoft PhotoScan software. Before the photogrammetric survey, 52 targets (50 cm-wide white squares) were installed in the field. Targets coordinates were measured using two Differential GPS (DGNSS) receivers (one being a fixed base) with a positioning accuracy of c. 2 cm. The entire set of targets was defined as Ground Control Points GCPs and manually positioned on the images. The resid-

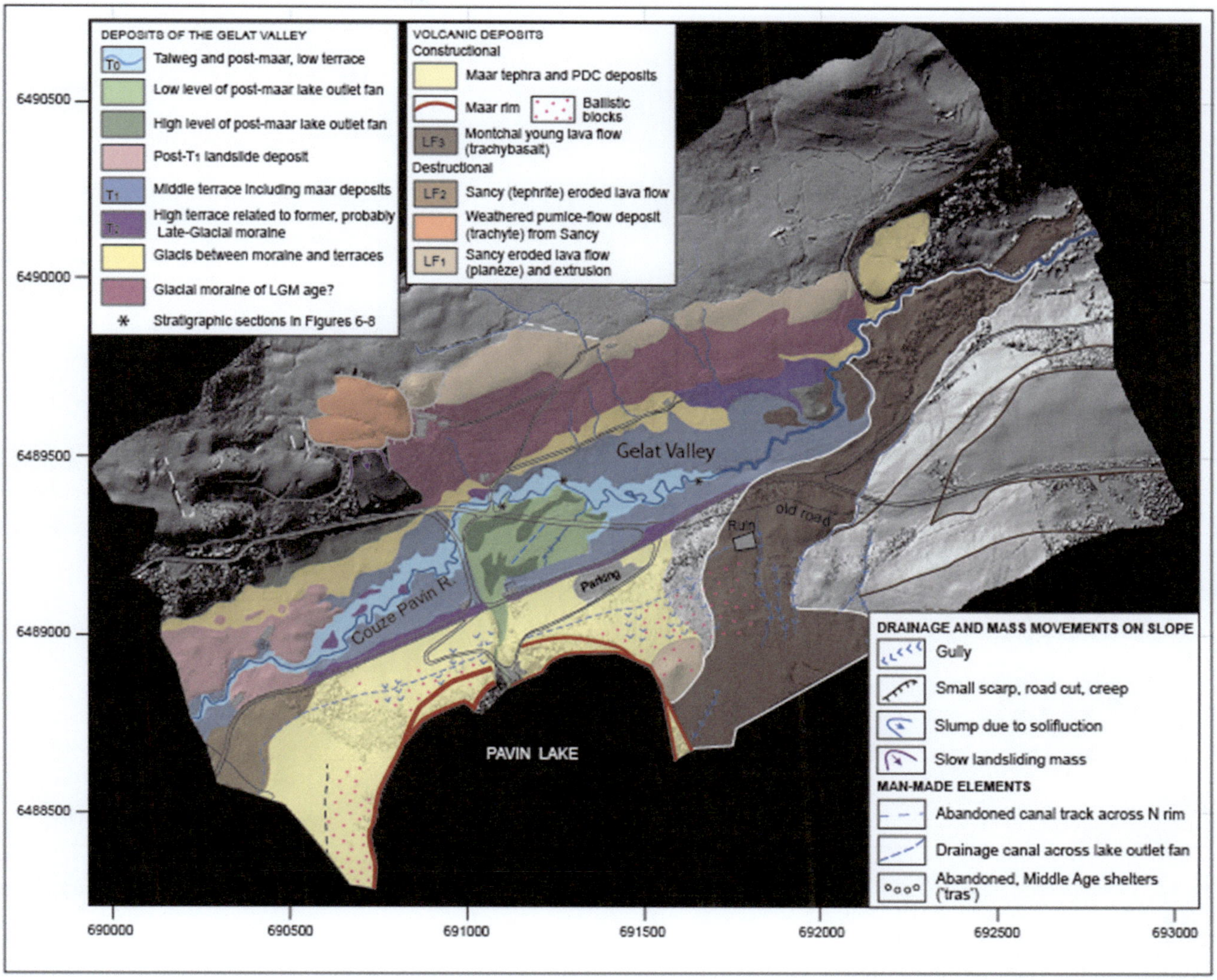

Fig. 9.4 Geologic map draped on very high resolution DSM (here cropped at 50 cm) showing a stratigraphic record of landforms in the Plaine du Gelat, middle course of Couze Pavin valley

ual difference between the models and the GCPs's coordinates was estimated by Photoscan to be more precise than 4 cm. The next step was the automatic matching of tie points using a feature based approach. The resulting tie point coordinates ($3.3 \cdot 10^6$ points) were used as observations in a robust bundle adjustment. The final processing step was the densification of the scattered 3D point cloud of tie points by a dense matching algorithm of the automated workflow in Photoscan. The resulting dense point cloud (density of 13.4 points per m^2) was used to derive the high resolution DEM, generated in the RGF93/Lambert93 projection (NGF IGN69 elevation), and the associated orthophoto of the entire scene. This drone-based multiview photogrammetry acquisition allowed us to compute a 25 cm-spatial resolution DSM (Digital Surface Model, including vegetation cover) with planimetric and altimetric accuracy of c. 5 cm.

9.3.3 Analysis of Geomorphic Parameters of the Maar and Lake

The Pavin maar belongs to the first category of Seib et al.'s (2013) geomorphological classification on the basis of five parameters (Table 9.1): rim diameter, rim height, rim height/diameter ratio, excavation depth and bedding dip (Vespermann and Schmincke 2000). Other parameters are tephra ring, angle of the inner rim slope, angle of the outer rim slope, diatreme depth and depth coefficient (e.g. Büchel et al. 1993).

Pavin maar large diameter (900 m) lies in the average values for maar crater diameters (0.7–0.8 km) albeit their entire range varies widely between 0.2 and 3 km. Pavin maar is wide to the point that the mean height / rim diameter ratio of 0.15 defines a high-relief depression at the upper end of the 0.13–0.05 range for usual maars. If we take into account the

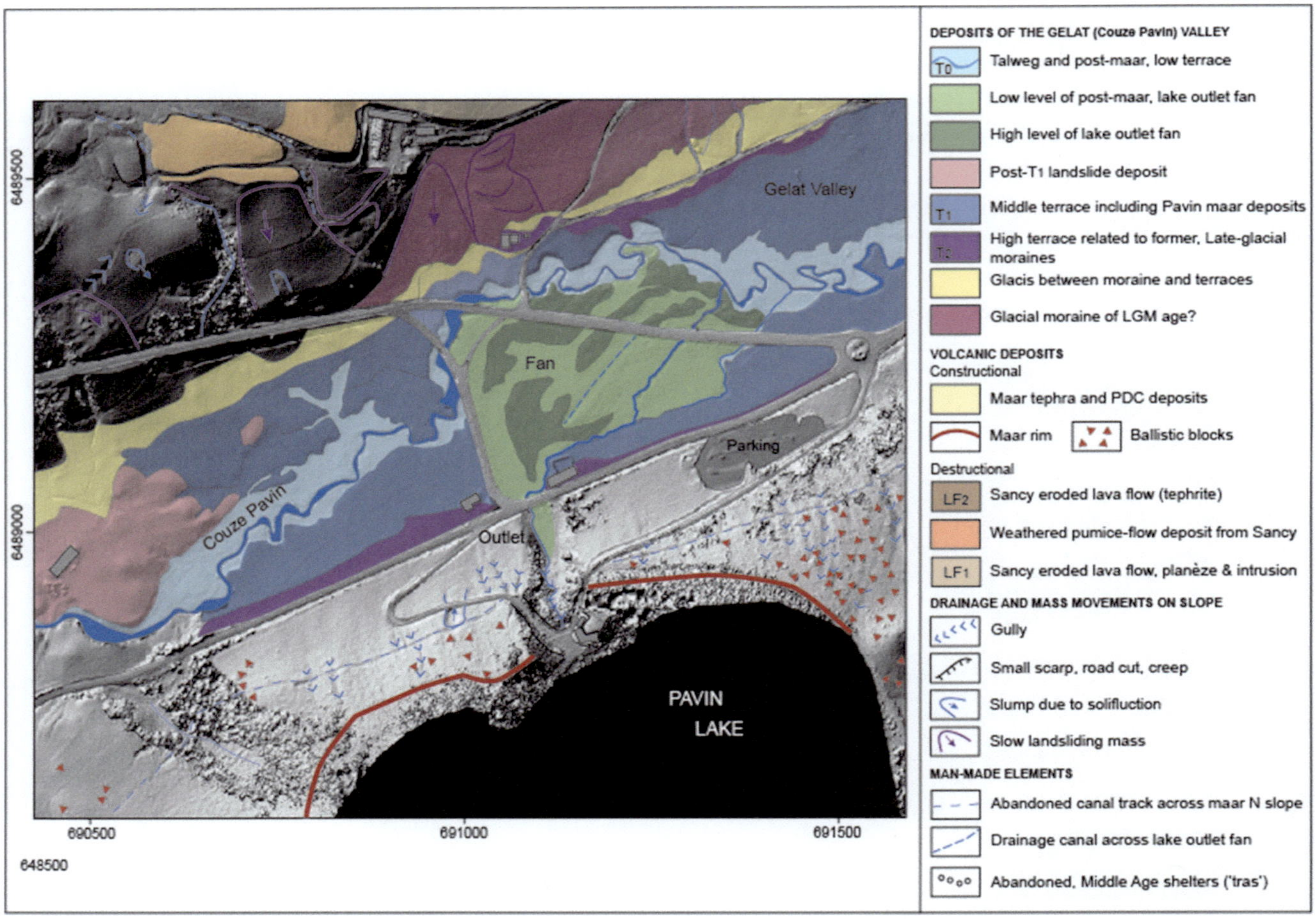

Fig. 9.5 Close-up geologic map draped on very high resolution DSM showing the alluvial fan created by the lake outlet in the alluvial terrace setting of the middle course of the Gelat valley

Table 9.1 Geometric and morphometric characteristics of the maar and comparison with average parameters of maars worldwide

Morphometric parameters	Rim diame-ter (km)	Mean/maximum rim height (m)	Rim height/diam. ratio	Excavation depth (m)	Mean/Max bedding dip (°)
Pavin maar	0.9	130/145	0.15	average 140	inner 30–45
		(56 above lake)		max 290	outer 10–25
Pavin	0.8	Na	0.05	92 max 102	deposits >30
Meromictic lake				incl. bottom turbidites	in lake
Min/Max size	0.2–3; aver. 0.75	Average 50	0.05–0.13	50–100	5–20

92 m crater depth under the lake surface and the highest rim elevation above lake (210 m to the SE), the latter ratio increases to 0.33. This contributes to the original and unusual depth coefficient of 0.18 accounting for the funnel-shaped 92 m crater depth (i.e. average maximum depth 145 m/√surface area (0.64 km^2). This relatively high depth coefficient together with the steep rim elevation facing prevailing winds from the WNW explains the limited wind and wave action inside the maar. This in turn favors the meromictic characteristic of the Pavin lake. The high relief of Pavin maar rims is enhanced if we account for the strong asymmetry between the short (≤0.1 km) inner slopes (30–45° on average, with vertical cliffs on the east and south flanks) and the wide (≥0.70 km), gently dipping outer slopes (<20° to the west). These become flattish eastward where the maar ring abuts low relief, young Montchal lava flows (Fig. 9.1). As a result, the shape of the maar Pavin is closer to a high and large bowl than the low-relief maar depressions in the Chaîne des Puys and Cézalier (e.g. Beaunit, La Godivelle d'en Haut). Further north-south asymmetry stems from the contrast between the highest (210 m) rim to the south, which coincides with the Montchal cone flank, and the low-relief

NNE rim <60 m above the lake level, with a notch (70 m wide) open in the rim towards the north.

Given the relatively young age of the maar (c. 6700 years) and the moderate geomorphic processes acting under a temperate mountain climate, the original shape of the subaerial depression has not degraded much over time; hence any post-erosion maar basin enlargement (Jordan et al. 2013) and shift of the geometrical parameter ratios can be neglected. In contrast, geophysical investigations under the lake have shown that the crater floor and rim have considerably changed with a flat central bottom covered by 5 m-thick turbidites, a submerged plateau with eroded edges and sectors between the bottom and the walls that are mantled by thick debris fans (Chapron et al. 2010a, b; 2012). The main processes having formed the shoreline shortly after maar eruptions are collapse events followed by the propagation and interfingering of debris fans. The largest debris fan in the lake is on the south but a submerged platform on the NE side of the lake is the top of a collapsed mass (Henriet 2009). If we consider the subaerial maar, the stage of erosion compared to other maars worldwide can be ranked two out of five after Büchel (1993) and one out of three after Németh (2001). If the maar-crater lake system is considered, the stage of erosion after c. 6700 years of existence does not fit the "filled stage" of Büchel (1993) or the "moderately eroded" stage of Németh (2001).

9.4 Geologic and Geomorphologic Mapping to Determine Slope Instability

9.4.1 The Maar Rim and Slopes

Figure 9.3 displays geologic and geomorphologic mapping of the rim slopes: bedrock and lithology, slope categories, runoff and drainage, lava flow scarps, mass movements on rim slopes, lake shore and man-made elements. Bedrock is shown only where maar deposit is less than 1 m thick. Bedrock lithology includes 'old' lava flows of Monts-Dore-Sancy, limited outcrops of weathered pumice-rich deposit from Sancy volcano on the south edge, and recent Montchal lava flows forming >30 m-thick cliffs cut by the maar on the eastern side.

The 15–30 m thickness of the rim sequence (BRGM 2009) is an average value for maars usually <50 m thick. Maar deposits covering the inner slopes at Pavin are massive and much less well bedded than the outer slope deposits 0.75 km ESE away from the maar (Boivin et al. 2012). Deposit is typically coarse sand and gravel, pumice- and clast-rich, trachyandesitic tephra 1- to 5-m thick on the steep inner slopes of the maar rims. The maar deposits are coarser than the average fine grained deposits that dip gently to less than a few degrees outward. Maar deposits encompass fine grained, bedded surge deposits, pumice-fall layers attributed to a sub-Plinian episode (Boivin et al. 2012) and massive pyroclastic-flow deposits totaling 2–5 m on the outer slopes of the rim. Accidental clasts are less abundant at Pavin than common maar deposits in which they may reach up to 90% by volume. Maximal diameter of country rock fragments may exceed 2 m, which is not at odds with the 3 m maximal size of maar ejecta elsewhere. Strewn on the rim and outer slopes, ballistic blocks (commonly 1 m^3 and up to 4 m^3), now observed towards the west and WSW and towards the E and NE, suggest two prevailing angles for the ejecta dispersal (Fig. 9.5). Violent ballistic explosions expelled blocks exceeding 2 m across to a distance of 700 m while ballistics 1 m across reached 1.2 km from the crater (see Lorenz 2007, for a comparison of size/travel distance of maar ballistics).

Slope categories are based on slope aspect (exposure, angle) and the shape of the top rim slope. We emphasize the persisting break-in-slope at the top of the rim around three quarters of the maar circumference. Some maar rim slopes remain vertical as lava cliffs crop out on the eastern and southeast edges. Other maar rims on the NW, W, E and NE edges evolved from cliffs to short, rectilinear, sorted slopes close to angle of rest (31–35°) covered by only thin and sorted tephra or colluvium. Lava flow scarps drawn on the map (Figs. 9.3 and 9.6) include three cliff segments showing open fractures and toppling over with debris feeding screes below. Lava flows cut by the maar include: the 'older' Mont-Dore (Sancy) cliffs of basalt higher up on the maar rims and the 'younger' Montchal, trachybasalt lava flows which form 20–30 m high cliffs armoring the east rim. The thick Montchal lava flows were channeled in a valley cut perpendicularly by the eastern maar rim. One channeled lava flow has blocked the Gelat plain in the Couze Pavin Valley to the NE (Fig. 9.4) and continues beyond the town of Besse at a distance of 14 km down valley (Bourdier 1980; Boivin et al., 2009, pp. 118–119).

Runoff and drainage are fed by springs above and below the lake level. Springs are more abundant on the south and SE rims of the maar at the base of the Montchal cone as permeable boundaries exist between lava flows and the underlying, clay-rich, weathered pumice deposit of Sancy. Permanent springs and rills on the north Montchal flank supply water that favors runoff and slumps in the sector. Mass movements are observed on all rim slopes but prevail on steep slopes to the north, NE, NW and SE. We distinguish three types of mass movements based on size and processes (Figs. 9.3, 9.4, 9.5, 9.6 and Table 9.2).

1. Small (tens of m^2) forms such as soil ripples, turf bank terracettes are due to creep and solifluction. Slow slope processes are also indicated by curved trees and many fallen trees that remove soil and colluvium. Wet areas around springs with hygrophilous vegetation (Fig. 9.3) and a 3–5

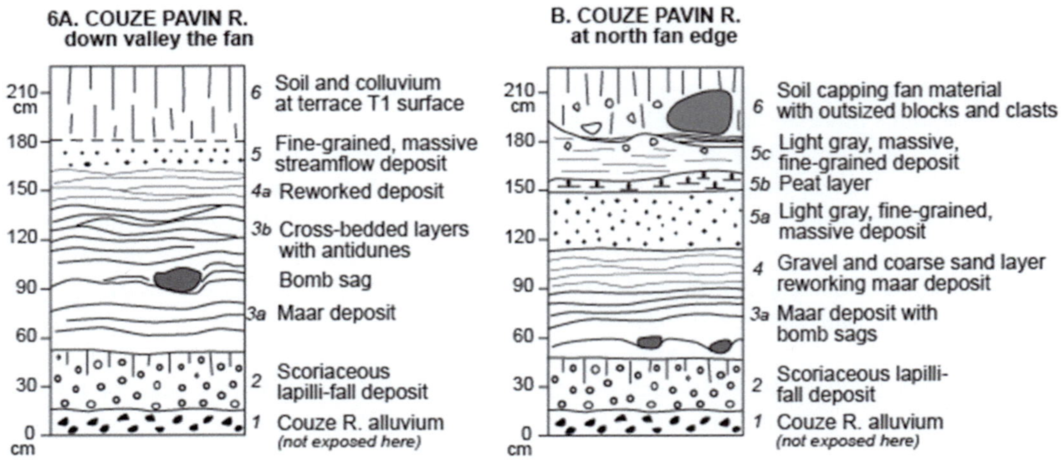

Fig. 9.6 Two stratigraphic sections located along the Couze River in the Gelat Valley north of the Pavin maar. A. Deposits observed in the terrace T_1 section on the Couze River south bank about 0.25 km downstream of the lake outlet fan. B. Deposits observed in the Couze River north and south banks at the northern end of the fan of the lake outlet

Table 9.2 Parameters used to assess and delineate hazard-prone areas around the lake Pavin and on the maar rims (including previous data from subaquatic investigations led by Chapron et al. 2010a, b)

Geomorphic areas	Slope categories	Landforms: Erosional	Landforms: Depositional	Mass movement	Criteria for mass movements and slope evolution
Maar rim slopes	Break in slope at top, steady, rectilinear below	Cliffs	Tephra	Landslide	Top scars and banks, open fractures, failure plane
		Steep slopes	None	Rock fall rockslide	Chaos / scree
	Convex or multi convex-concave		Tephra cover	Solifluction	Concave scar and convex lobe, curved trees, soil deformation
	Irregular, undulated		Colluvium, soil	Creep	Small turf banks, soil ripples (snow cover)
	Concave, incised		Alternate layers	Gullying	Drainage: sources, streams, heavy rainfall
Lake shore	Above line	Emerged Subaerial	Immerged	Lake oscillation?	Riverine bank >2 m high above N,W&SW shoreline
	Under line	beine?	beine	Artificial level rise	Coarse deposit >1 m deep below shoreline
Lake outlet and adjacent Gelat Valley		Outlet fan			Two levels of deposits
					Canal drainage
		Terraces T_1-T_3			Inset channel deposits

month-long snow cover favor solifluction as do persisting firn on north aspect slopes.

2. Relatively small landforms tens to hundreds of m^2 in size are rock fall blocks below small or thin lava scarp (SW, NW) or recent scree slopes below fractured lava flow scarps to the NE. Rock fall is not a continuous process as witnessed to by moss and lichen covering boulders. Stacked, outsized blocks (up to 10 m across) form chaos below small fractured cliffs, e.g., on the NW and SW rims and the longest scarp of Mont Dore-Sancy lava flow on the SSE rim. This cliff <200 m long and 10–20 m thick shows open fractures and many blocks below (Fig. 9.6). Rock fall is a discontinuous process as several blocks meter sized, not covered by moss or lichen, are sliding on the steep and wet slope cut on clay-rich material below the cliff foot. These lava flows show pervasive fractures oriented N160 on the NNE rim, N120 on the south rim and N70 on the SW rim. They parallel regional lineaments portrayed in the tectonic sketch of Fig. 9.1. Recent scree may result partly from cliff rock shuttering due to frost/thaw effects that produced debris accentuating scree and rock falls (NE and SSE slopes) during the Little Ice Age. Cold spells occurred during the sixteenth and seventeenth centuries on the basis of pollen and diatom fluctuations recorded from cores collected in the lake sediment (Stebich et al. (2005), Schettler et al. (2007) and Schwab et al. (2009). Yet, scattered blocks still fall from the fractured cliff as they

remain lichen-free and blocked by young trees at the foot of the lava scarps.

3. The third landforms are more subtle such as small (2–4 m high) scarps and adjacent banks near the rim top on the NE and east side of the maar (Fig. 9.3). On the NNE rim top, a scarp overhanging the adjacent bank associated with a shallow gully forming a scar, together with curved and fallen trees around the scar, suggest a (slow?) process of deep-seated rotational landslide. This coincides with the top of a landslide mass (attributed to the maar eruption based on the ^{14}C age of the base of the overlying sediment) forming a platform at −42 m in the lake at the base of the NNE rim slope. Failure plane may correspond to the contact between the maar tephra deposit and bedrock or along impermeable layers within the tephra deposit. Other scarps and banks on the east rim top show no scar or curved trees.

9.4.2 The Lake and Riverine Maar Rim Contact

The lake occupies 0,445 km^2 in area and hosts a water volume of about 24.1 million m^3. Pavin is known to be the only meromictic lake in Auvergne, which consists of two superimposed and geochemically different water bodies, termed a mixolimnion above a monimolimnion (Jézéquel et al. 2008). This characteristic is related to the unusual lake depth coefficient as a result of the funnel-shaped, deep crater inside a relatively high-relief maar ring.

The riverine lake contact can be divided in two areas, one immerged and another subaerial, separated by a bank a few meters high. Half of the lake shoreline is erosional (near lava cliffs) while the other half is depositional (concave slopes covered by screes) (Fig. 9.3). The erosional shore, which prevails on the east and south edges of the funnel-shaped crater lake, is characterized by a narrow shoreline above deep water without a *beine* (subaquatic terrace). This is the usual case of lakes in erosional context where the riparian bank directly overhangs a submerged beine at the water contact (Provencher and Dubois 2008; Touchard 2000). Lake Pavin shows no high water level due to heavy rain inflow so the present-day shoreline corresponds to the stable high water lake despite limited seasonal fluctuations. Accumulation shoreline includes a 1 m-deep, immerged beine along the western and NE edges, which can reach 4 m wide above deep water steep slopes. Locally (SW, NW and at the foot of the Montchal lava bluffs to the east), a narrow beine has been formed by screes or rock falls from adjacent cliffs (Fig. 9.3). On the western lake edge the beine is slightly larger due to runoff, tephra redistribution and scree or rock fall. Screes and runoff have formed small deltas in the NW and SW corners of the lake. The erosional and depositional pattern of the lake shore is also a result of wave action, which is more important on the eastern edge than the western side. Primary wind direction is from west-northwest. The lower north, NNW and NNE crater rims provide less shelter against the wind and the longest lake diameter from NNW to SSE with the highest fetch is oriented in a SE direction. However, wind action on waves is limited by the high maar rims and the funnel shape of the maar-lake system.

The immerged beine observed near the west and north shoreline is largely due to recent anthropogenic impact. The level of the lake was artificially raised in 1859 when works (allowing noble fish introduction) heightened the lava threshold and covered the outlet. Today a large platform exists at the northern entrance of the maar for touristic and access purpose where the maar rim was the lowest originally.

9.4.3 Possible Links Between Subaerial Slopes and Subaquatic Landforms

When slope angles in Lake Pavin are above 31°, they are free of any sediment and characterized by the development of numerous steep canyons clearly visible on *multibeam bathymetric* data (Figs. 23.4, 23.7 and 23.10; Fig. 9.3). As sub-bottom profiles along steep slopes only illustrate the morphology of the acoustic substratum, Chapron et al. (2010a, 5e; Figs. 23.5a, 5d) claimed that these canyons draining the steep slopes of Lake Pavin crater are still active path for sporadically bypassing sediment towards the deep central basin. In such a context it is highly possible that sediment from subaquatic littoral environments, lake shores and subaerial slopes from the crater ring draining into the lake can be exported directly to the deep central basin (Fig. 23.10). Abundant sources near the lava cliff and streams cutting the south and SE subaerial rim slopes probably continue in the lake as gullies and 'subaquatic canyons' described across the south flank of the crater (Fig. 9.3; Chapron et al. 2010a,b).

Mapping suggests probable relations between topographic anomalies reflecting mass movements on the NE rim slope with the submerged landslide mass in the NE lake sector (Figs. 9.3 and 9.6). Chapron et al. (2010a) related the slump deposit ^{14}C dated and calibrated between 580 and 640 AD (PAV08 site) above the subaquatic platform and the most recent scree slopes on the NE maar rim. The fractured 'old lava' cliffs on the north flank of Montchal are only the subaerial part of a pile of immerged lava flows that are drawn along the south rim edge (Fig. 9.3). Locally, >45° slopes at the eastern and southern edges of Lake Pavin coincide with outcropping, unstable lava cliffs within the inner slopes of the crater ring (Figs. 9.3 and 9.6; Chapron et al. 2010a). Blocks along the shoreline and reported on the steep slopes near these lava cliffs highlight the occurrence of relatively small sized but recurrent rock falls.

Fig. 9.7 Pictures (©J.-C. Thouret and D. Miallier) showing relevant landforms, deposits and mass movements around the maar lake and in the Gelat Valley. (**a**) Cliff of trachybasalt lava flow below the Montchal scoria cone on the SSE edge of the maar. (**b**) Note fractures and topple across the 12 m-thick, trachybasaltic Montchal lava flow scarp. (**c**) Scar of slip, spring and runoff on weathered, Sancy pumice-rich pyroclastic deposits overlain by the Montchal lava flow. (**d**) Fractured cliff of Mt. Sancy tephrite lava flow, NNE slope of the maar. (**e**) Fallen blocks retained by young trees below the previous cliff are not covered by moss or lichen. (**f**) Open fracture in the Montchal trachybasalt lava flow, SW slope of the maar. (**g**) Chaos of outsized boulders below an eroded cliff of Montchal lava flow at the foot of the NW rim cliff.

9.4.4 The Lake Outlet and Fan

We have mapped the landforms of the Couze Pavin Valley adjacent to the maar, termed Plaine du Gelat, including a geologic and geomorphologic map draped over a very high-resolution DEM (see 9.3.2) together with a close-up map of fan deposits associated to the lake outlet in the middle reach of the Plaine du Gelat.

The first map (Fig. 9.4) shows the alluvial and glacial deposits of the Gelat plain, the volcanic deposits of the Monts Dore east flank, the Montchal lava flows and Pavin maar deposits, and finally mass movements on slopes. Pre-Pavin maar deposits are represented by Monts Dore (Sancy) basalt lava flows and trachytic pumice-flow deposits from the Sancy massif (associated to the Sancy caldera by Lavina 1985). Glacial moraines cover the flanks of the valley particularly the north one. On the south flank, the lava flow from Montchal, which was channeled in a valley and flowed towards the Couze valley, dammed the Plaine du Gelat. The north outer rim of the maar shows solifluction lobes and shallow gullies cutting down the thick maar deposit. Most gullies seem inactive or subdued. A sub-horizontal, subdued track on the slope marks out an abandoned, historic drainage canal that extends from the NW slope and crosses the outlet ravine towards the east along the south slope of the Gelat valley (Fig. 9.4).

Figure 9.5 is a close-up map of the central part of the Gelat Valley just north of the maar lake in which the lake outlet formed a fan. The Pavin lake small outlet (mean discharge 50 l/s) feeds a stream draining a ravine open to the north, 50 m wide down valley and 40 m deep below the lava scarp threshold that dammed the lake (visible in the historic Lecoq painting of 1867). Adjacent to the outlet tunnel, an iron-bearing source has been shown to result from outward drainage of the bottom lake (monimolimnion) (Jézéquel et al. 2008). The stream has formed an alluvial fan starting at the foot of the rim slope of the maar at 1175 m asl. The subdued fan that extends 250 m towards the north and 300 m NE across the Gelat valley, pushed the Couze River against the north valley bank (Fig. 9.5). In turn the Couze River became meandering as the fan blocked the valley and lowered its topographic gradient to as little as 1.1 %. We distinguish two levels in the fan formation based on close inspection of the very high resolution DEM (Fig. 9.5). High level, subdued deposits have been left by successive ridge axes of alluvial deposits when the fan was formed by repeated outlet floods. Low level, flattish deposits of the fan between bar bumps have been abraded by floods and smoothed out by runoff. The present-day topography compared with aerial photographs of 1956 reveals that the fan surface has also been modified by drainage canals and smoothed out by cultivation techniques.

9.4.5 Deposits of the Gelat Valley and Lake Fan

The Gelat Valley bottom and banks contains four types of terraces as follows (Figs. 9.4, 9.5 and 9.6): (1) The lowest T_0 lies near the bottom of the river channel that meanders (slope <1.5 %) upstream of a rockbar. This rockbar corresponds to Mont-Dore lava flows overlain by the Montchal channeled lava flow. (2) The middle T_1 terrace in which the Couze Pavin River channel is cut down about 3 m, contains the maar deposits, which have been reworked by streamflow processes (Figs. 9.7 and 9.8). (3) Remnants of the middle T_2 terrace are observed on both sides of the valley about 4 m above the channel. This terrace can be linked to a valley bottom enclosed by moraines of Late or Full Glacial age or to alluvium released by former glaciers upstream of the valley. (4) Landslide deposits have covered the T_1 terrace, triggered by landslides the long and deep scar of which is visible on the north flank of the valley.

Figure 9.6 shows the stratigraphic succession of six deposits associated to the terraces of the Couze River and the fan formed by the lake outlet in the Gelat Valley. Deposits no.1–6 are illustrated in Figs. 9.7 and 9.8. On the south bank of the Couze Pavin River ~0.35 km down valley from the fan, the 3 m-thick T_1 terrace is 2.5 m lower than T_2 (Fig. 9.7j). Overlying the coarse, reddish, oxidized alluvium deposit 1 and the Montchal scoria-fall deposit 2, the 1-m thick, massive and fine-grained maar deposit no.3a is in turn overlain by a grey, cross bedded surge deposit 3b showing 2 m wavelength antidunes (Fig. 9.7k). Bomb sags have deformed the muddy, silt and sand-sized, stratified layers of the maar surge deposit 3a from left (SE) to right (NW) (Figs. 9.7l and 9.8a,b).

On the Couze River north bank section (Fig. 9.6a), the surge deposit 3b of the maar grades up in a massive, fine grained deposit (Figs. 9.7j, 9.8c, d). Lenticular, 50 cm-thick, sand and gravel layers filled the undulated, eroded top surface of the maar deposit pointing to a streamflow deposit 4a. (Fig. 9.8c, e). The layers grade up into stratified layers of sorted sand deposit 4b, which indicate that sand was

Fig. 9.7 (continued) (**h**) Gentle (<20 %) north outer slope of the maar showing undulated landforms due to solifluction and a network of inherited gullies. (**i**) View of the alluvial fan (note two *top* surfaces) at the mouth of the ravine formed by the lake outlet that has cut deeply down the north maar rim. (**j**) View of the T_1 terrace on the Gelat valley south side cut down c. 3 m by the Couze River channel, and fan surface in the background. T_2 terrace, 2.5 m higher than T_1, is visible in the background. (**k**) The T_1 deposits (geologist for scale) on the south bank of the Couze Pavin River ~0.35 km down valley from the lake outlet fan are described in Fig. 9.6a. (**k**) T_1 terrace shows the maar deposit 2 overlying the oxidized alluvium and Montchal scoria-fall deposit 1. The massive, fine-grained maar deposit is overlain by a *grey*, cross-bedded surge deposit showing 2 m wavelength antidunes. (**l**) The Pavin surge deposit 3a including bomb sags overlies the coarse, reddish Montchal scoria-fall deposit 2 mixed with alluvium

Fig. 9.8 Pictures (©P. Boivin, D. Miallier) showing the stratigraphic succession of six deposits associated to the terraces of the Couze River and the fan formed by the lake outlet in the Gelat Valley (see Figs. 9.6a, b). (**a**) Close up view of two bombs sags in the muddy, silt and sand sized maar deposit 3a. (**b**) Stratified, cross-bedded layers of maar surge deposit 3a impacted by bomb sags above the oxidized lapilli-fall deposit of Monchal. (**c** and **d**) Gravel and coarse sand layers 4a about 0.5 m thick fill shallow channel cut down in the undulated top of the maar

emplaced in a low energy alluvial channel. The uppermost, massive bed of coarse sand including small gravel suggests a flood deposit 4b associated to river overbank in a flat-lying alluvial plain (Fig. 9.8e). Although the section of the north bank of the Couze terrace T_1 is located near the northern toe of the fan, the sand and gravel, streamflow layers 4a show no evidence for catastrophic erosion above the contact with the maar deposit. Outside the fan and along the Couze River banks, 0.50–1 m-thick colluvium and soil (no. 5 in Fig. 9.6a) cap the T_1 terrace.

At the northern toe of the fan, the south bank section shows two additional deposits (Fig. 9.6b). First, a peat layer deposit 5 (dating in progress) overlies the sorted, sand layer 4b (Fig. 9.8e, f). The peat layer suggests that the flat, alluvial Gelat plain was impounded due to damming downstream. Brief impoundment is further suggested by the peat layer being intercalated between (a) a whitish, laminated, fine-grained deposit 5a beneath the peat layer and stratigraphically equivalent to the deposit 4b of Fig. 9.6a, and (b) a gray, massive and fine-grained deposit 5c above the peat layer, pointing to overbank deposit. Second, angular boulders and small clasts in thick, coarse fan material deposit 6 overlie the deposits 5a-c (Fig. 9.8g, h). The outsized blocks and angular clasts may have been emplaced by a combination of processes: (1) Glangeaud (1916) described scattered ballistic blocks strewn at the surface of the fan north of the maar. They no longer appear at the fan cultivated surface but the largest blocks may have been tossed aside by local farmers at the field boundary near the Couze Pavin channel; (2) clasts may have been transported from the maar north rim when the (pre-?)historical fan was formed by lake outlet floods, and; (3) some boulders may have been washed out by the Couze River floods from the moraines that mantle the valley slopes and from landslide deposits that covered the T_1 terrace upstream (Fig. 9.4).

9.5 Discussion

9.5.1 Observations, Criteria and Uncertainties on Lake and Slope Instability

We discuss relationships between slope processes and the evolution of the maar, potential links with subaquatic landforms described by Chapron et al. (2010a,b; 2012), and past climatic conditions. Slope erosion usually takes place right after the maar eruption and ceases when the tephra ring is stabilized by a cover of vegetation or becomes indurated diagenetically. Studies of young maars such as Ukinrek 1977 in Alaska (Pirrung et al. 2007) and young (1913) tuff ring in Vanuatu (Németh and Cronin 2007) have both pointed out that gullies and sheet erosion occurred shortly after the eruption and the following few weeks to months at most. The denudation of tephra proceeds at significantly lower rates compared with the retrogressive erosion of the inner crater walls. Height of crater rim decreased –whereas mean crater diameter increased as the horizontal central area of the crater floor is formed during first years of a maar. Re-sedimentation is due to collapse of nearly vertical parts of crater wall and undercutting of debris fans by waves. Water erosion lowers the tephra ring, widens the crater and deposits eroded material both inside the crater and also outside the tephra ring. As a result, the long term (at least tens of thousands of years), erosional enlargement of the maar leads to an unusually large basin with a relatively low surrounding rim (Németh et al. 2012; Jordan et al. 2013).

Evolution at Pavin consisted in redistribution of tephra mostly on the outer rim slopes which are much gentler than the inner slopes. Sheet runoff on outer slopes and some small gullies are apparently inactive today--except in cases of heavy rainstorm or intense forest clearing and cattle grazing during the Middle Age and the eighteenth–nineteenth century (see shelters locally termed 'tras' of that age in Figs. 9.4 and 9.5). Pavin maar inner slopes show limited tephra re-deposition as the NW, NE and East sides cut cliffs or bedrock covered by thin soil and colluvium. On the SSE edge of the lake, runoff and rainwater streams, slumping and rock falls have removed and are slowly redistributing tephra on slopes. In contrast, maar tephra together with scoria material from Montchal was widely re-distributed on the north slope of the cone towards the lake.

9.5.2 Implications for Natural Hazards

Figure 9.8 displays six types of rim slopes according to degrees of instability. Instability has been determined based on a series of geomorphological criteria determined in Table 9.2. We also took into account the unpublished survey carried out by the regional BRGM office (BRGM 2009). Based on a set of parameters, we distinguish five types of unstable areas on inner slopes around the maar rim. The six categories are as follows:

Fig. 9.8 (continued) deposit on the north bank of Couze River. (**e**) Close up of the 50 cm-thick gravel and sand sized layers (deposit 5a) beneath the colluvium and soil six capping the T_1 terrace. (**f**) Angular blocks in the coarse fan deposit 6 overlie a light gray, massive and fine-grained (overbank?) deposit 5c and a peat layer 5b at the northern edge of the fan (Fig. 9.6b). (**g**) Close up of the peat layer 5b (dating in progress) interbedded between deposits 5a and 5c. (**h**) Large (50 cm) boulder and clasts at the base of the fan deposit 6. See text for possible processes that have emplaced the blocks

(1) Unstable, quasi vertical slopes on fractured cliffs with rock fall or steep landslide mass; (2) Potentially unstable, steep (31°) slopes on non-fractured cliffs, small mass movements in wet areas and/or gullying; (3) Relatively slow and surficial slope processes due to creep, solifluction and runoff; (4) geomorphological anomalies such as scarp-bank, scar suggesting top of deep-seated rotational landslide or collapsing mass; (5) Potentially unstable, subaquatic areas, and (6) rim areas with very little or surficial signs of erosion and mass movement. Figure 9.8 also suggests possible links between subaerial and subaquatic unstable slopes (based on published data for subaquatic landforms). Potentially unstable areas are the edge of the subaquatic plateau to the north and the pile of cliffs under the lake in front of the unstable cliff on the SE edge of the maar. The >30° slopes below the Montchal cliffs on the east side of the lake may be another potentially unstable area. Subaquatic plateaus, which are common in maar lakes (Chapron et al. 2012), represent the most probable potential sedimentation source area for mass wasting events along their sub-vertical edges. The authors reported a slump deposit 580–640 AD on site PAV08 on the platform not far from the most recent and wide scree slopes on the NNE rim. A large slide scar at 50 m water depth coinciding with the platform edge (Fig. 9.3) has been attributed to an event 1200–1300 AD. Chapron et al. (2010a,b) assigned these events to time periods when contrasted climatic conditions and lake level fluctuations occurred elsewhere in France, although strong evidence in Lake Pavin is lacking. Additional mass wasting deposits have also been reported on the SW and SE edges of the lake bottom (Fig. 9.5). Instability factors including the gas content of the lake sediments in the monolimnion have been pointed out by Chapron et al. (2010a, b)) & Meybeck (Chap. 3 in this volume). The occurrence of past limnic eruptions, however, has yet to be demonstrated (Fig. 9.9).

Has a higher, thus older, lake shore ever existed? The fact that a large (2–4 m) bank extends at least 2 m above the subaerial lake shore bank around the north, northwest and west edges, now used as a large trail, suggests the existence of a higher, older lake shore (emerged beine). We found no lake sand or silt layers as evidence for a lake shore on this bank, although the absence of lake sediment is not necessarily a criteria: this may be due to the limited wave action and lake level fluctuations. As the Pavin lake is not stirred by wind and its level does not fluctuate due to heavy rainfall today, it is difficult to find any evidence for lake sediment above the immerged beine and shore bank. However, pointing to a former high lake stage, Bruyant (1909) reported tracks of a previous weir of the lake. Bruyant mentioned a stair with a waterfall cut 4.84 m high above the 1909 lake level into the east slope of the present-day entrance. The platform now seen with a fence above the outlet 1 m above the lake level was erected in 1859 by Lecoq and Rico for fish breeding purpose. Bruyant (1909) also mentioned the existence of a canal whose intake was 0.21 m below the present platform and that provided water at least until about 250 years ago over 3.5 km east of Lake Pavin to the village of Olpilière 0.5 km south of Besse. We now observe the track of another canal several meters below the lake level on the north side of the maar rim running west to east (Figs. 9.4 and 9.5). Scars of slumps are located at the downstream side of the abandoned canal on the north rim slope now showing active solifluction landforms. We note that human induced erosion on the maar slopes (in particular during the 14–15th and nineteenth Centuries; Schwab et al. 2009) has been deduced from pollen studies, but historic impact has not triggered a large sediment flux comparable to that computed from other lake sediments in more populated Auvergne areas over the past 2000 years (Sarliève marsh: Macaire et al. 2010, and Aydat lake: Lavrieux et al. 2013).

Is there any evidence for lake overbank and outburst? On the one hand, the bank on the north and west side of the lake suggest a high, emerged beine while the perched weir 4.84 m high at the entrance points to a high lake standstill perhaps a few centuries ago. On the other hand, the deep and narrow ravine of the lake outlet shows no sign of catastrophic lake outburst. The boulders that pave its channel 0.2 km down valley probably slid on the slopes of the ravine from which runoff has washed them out of the rim deposit. In addition, no debris-flow deposit was found in the fan deposit created by the lake outlet except reworked maar deposits forming alluvial, gravel lenses towards the top of T_1 terrace. The platform has reportedly been slightly eroded (0.21 m?) since 1859 (Bruyant 1909) and Lecoq drawings (1867). The lava threshold is not going to collapse anytime soon but retrogressive erosion at the base of the outlet gorge will act in a long-term. Within the lake slumping on the edge of the subaquatic platform may occur and eventually destabilize the gas-rich turbidite cover on slopes and at the bottom of the lake, although hydrostatic pressure will prevent violent degassing. In addition, methane is present in the monimolimnion whereas its CO_2 content remains low (Camus 1993; Jézéquel et al. 2008).

Effects of tsunamis or lake seiches are beyond the scope of this chapter, but the reader is referred to Freundt et al.'s (2007) review of physical conditions for triggering and transferring waves in a lake. Seismic shaking, which would fluidize buried mass-wasting deposits in lake sediments, has been invoked by Chapron et al. (2012), but epicenters of reported earthquakes occurred far away and no sizable seismic activity has been recorded since instruments were deployed in Auvergne in the 1960s and around the lake by OPGC in 2009. More probable are oscillations in lake level such as those reported from Lake Albano, Colli Albano volcano, Italy, where overflows have repeatedly triggered lahars and caused damage down valley (Anzidei et al. 2008). Oscillations may be due to heavy rainfall storms and trigger lake waves

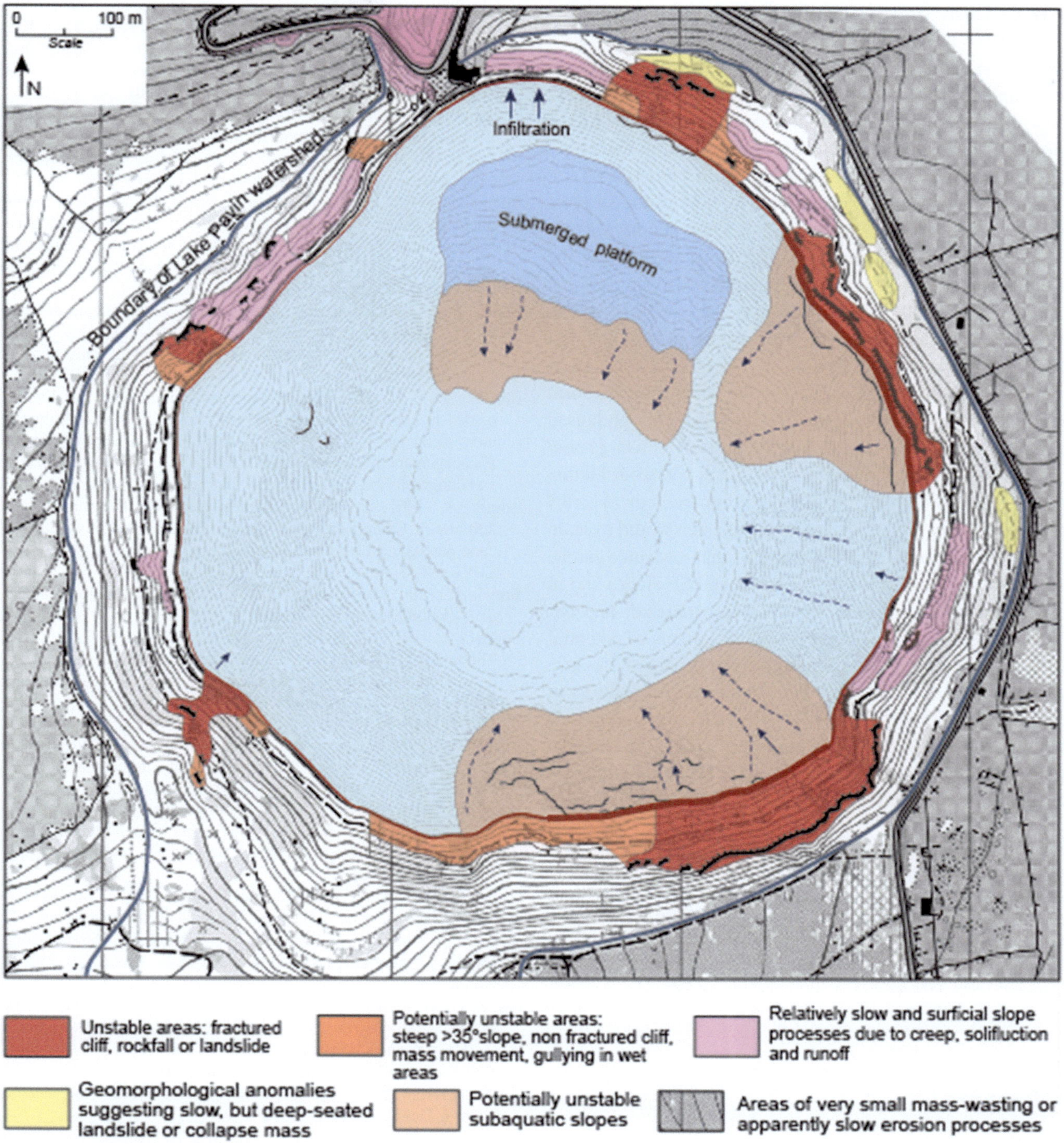

Fig. 9.9 Map of the Lake Pavin maar rim showing six categories of unstable areas: (1) Unstable slopes, (2) Potentially unstable slopes; (3) Relatively slow and small-sized slope processes, (4) Geomorphic anomalies suggesting slow but deep seated rotational landslide and/or collapse mass; (5) Potentially unstable subaquatic areas (Chapron et al 2010a, Henriet 2009); (6) Rims showing apparently very small or very slow slope processes

and seiche by resonance on lake edges. One of these exceptional storms and flooding events was reported in Auvergne around 580 AD (Vernet 2013), which may coincide with the slump of this age reported by Chapron et al. (2010a). Recurrent flooding events and the doubling of the background sedimentation rate since the early Middle Ages have been reported from the lake Aydat 27 km NE of Pavin (Lavrieux et al. 2012).

9.6 Conclusion

In contrast with postulated lake outburst catastrophic events, we found no evidence outside the lake for major lake breakouts or debris-flow deposit in the adjacent Gelat valley. However, multiple indicators for relatively slow and small sized, but numerous mass movements are observed on the steep inner slopes of the maar. Some sectors of the steepest rocky rims may pose a short-term hazard for people walking on the round-the-lake trail in particular at the foot of the south, SW and SE sectors of the maar. We also suggest that deep-seated rotational landslides may occur slowly on the NE rim, although a coupled *in situ* and laboratory geotechnical approach is needed to unravel whether this hazard is recurrent. The NE slope overhangs the subaquatic platform located on top of the syn-eruptive landmass that collapsed east of the entrance of the lake. In the absence of a large and/or near epicenter earthquake, however, it is not expected that a small (tens of m^3) to medium size (hundreds of m^3) cliff rockslide would trigger a wave in the lake that would overrun the lowest rim to the north. It is hoped that measures can be taken around the lake trail in the south and SE slopes and in two restricted areas to the SW (trail to Montchal) and NE (near the restaurant) in order to avoid the effects of potential rock falls. Drainage should be implemented below the lava cliff above the SE lake shore in order to avoid earthflows in case of exceptionally heavy rainfall. If the lake oscillations become sizeable in the near future, the drainage of the lake can be ensured by resizing the diameter of the outlet tunnel below the entrance platform.

Acknowledgments J.C. Thouret acknowledges all undergraduate students of the Department of Geology who have been mapping the maar surroundings under his supervision over the past 4 years. Thin sections and analysis of major elements have been carried out by C. Constantin, M. Benbakkar (LMV), and J. Marin (CRPG). Artwork has been achieved by L. Thouret. The content of this chapter has benefited from careful reviews and comments from our colleagues J.L. Bourdier, A. Gourgaud, E. Juvigné, D. Miallier, K. Nemeth, and C. Ollier.

References

Albéric P, Jézéquel D, Bergonzini L, Chapron E, Viollier E, Massault M, Michard G (2013) Carbon cycling and organic radiocarbon reservoir effect in a meromictic crater lake (Lac Pavin, Puy de Dôme, France). Radiocarbon 55(2):1029–1042

Anzidei M, Carapezza ML, Esposito A, Giordano G, Lelli M, Tarchini L (2008) The Albano Maar Lake high resolution bathymetry and dissolved CO_2 budget (Colli Albani volcano, Italy): constrains to hazard evaluation. J Volcanol Geotherm Res 171(3–4):258–268

Boivin P, Besson J-C, Briot D, Camus G, de Goër de Herve A, Gougaud A, Labazuy P, de Larouzière F-D, Livet M, Mergoil J, Miallier D, Morel J-M, Vernet G, Vincent P-M (2009) Volcanologie de la Chaîne des Puys, 4è édition bilingue, Editions du Parc Régional des Volcans d'Auvergne, Aydat, p 179

Boivin P, Besson JC, Ferry P, Gourgaud A, Miallier D, Thouret J-C, Vernet G (2012) Le point sur l'éruption du Lac Pavin, il y a 7000 ans. Actes du colloque international « Lac Pavin et autres lacs méromictiques », Besse-St-Anastaise, 14–16 mai 2009. Revue des Sciences Naturelles d'Auvergne 74–75, 2010–2011, 45–55

Bonhomme C, Poulin M, Vinçon-Leite B, Saad M, Groleau A, Jézéquel D, Tassin B (2011) Maintaining meromixis in Lake Pavin (Auvergne, France): the key role of a sublacustrine spring. Comptes Rendus Géosciences, Académie des Sciences 343(11):749–759

Boumehdi A (1988) Les clinopyroxènes dans les basaltes alcalins continentaux (Massif Central, France). Implications pétrogénétiques, barométriques et caractérisation de la profondeur des réservoirs magmatiques. Doctorat d'Université, Université Blaise Pascal - Clermont II, Clermont-Ferrand, 174 p

Bourdier J-L (1980) Contribution à l'étude volcanologique de deux secteurs d'intérêt géothermique dans le Mont Dore: le groupe holocène du Pavin et le massif du Sancy. Thèse, Université de Clermont II, 180 p

BRGM (2009) Evaluation préliminaire des conditions de stabilité des pentes du Lac Pavin. International Meeting "Lake Pavin and other meromictic lakes", Besse-St-Anastaise, 14–16 mai 2009. Service Aménagement et Risques naturels, DDE Puy de Dôme (Abstract p. x and unpublished presentation by O. Renaud 26 slides)

Brousse R (1961) Analyses chimiques des roches volcaniques du Tertiaire et Quaternaire de la France. Bulletin des Services de la carte géologique de la France et des topographies souterraines, 263, LVIII, 1–137

Bruyant C (1909) Le niveau du Pavin. In: Mélanges, 3è partie, Annales de la station limnologique de Besse, 203–205

Buettner R, Dellino P, La Volpe L, Lorenz V, Zimanowski B (2002) Thermohydraulic explosions in phreatomagmatic eruptions as evidenced by the comparison between pyroclasts and products from Molten Fuel Coolant Interaction experiments. J Geophys Res 107(B11):14–14

Büchel G (1993) Maars of the westeifel, Germany. In: Negendank, JFW, Zolitschka, B (eds) Paleolimnology of European maar lakes. Lecture notes in earth sciences, vol 49. Springer, pp 1–13

Büchel G, Lorenz V (1993) Syn-and post-eruptive mechanism of the Alaskan Ukinrek maars in 1977. In: Zolitschka B, Negendank JFW (eds) Paleolimnology of European Maar Lakes. Springer-Verlag, Berlin, pp 15–60

Camus G, Goër de Herve A de, Kieffer G, Mergoil J Vincent P-M (1973) Mise au point sur le dynamisme et la chronologie des volcans holocènes de la région de Besse-en-Chandesse (Massif Central francais). Copmptes-Rendus Académie des Sciences, Paris, 277(7): 629–632

Camus G, Michard G, Olive P, Boivin P, Desgranges P, Jézéquel D, Meybeck M, Peyrus J-C, Vincon JM, Viollier E, Kornprobst J (1993) Risques d'éruptions gazeuses carbonique en Auvergne. Bulletin Société géologique de France 164:767–781

Cas RAF, Wright JV (1987) Volcanic successions. Modern and ancient. Allen & Unwyn, London, p 519

Chapron E, Albéric P, Jézéquel D, Versteeg W, Bourdier J-L, Sitbon J (2010a) Multidisciplinary characterization of sedimentary processes in a recent maar lake (Lake Pavin, French Massif Central) and implication for natural hazards. Nat Hazards Earth Syst Sci 10:1–13

Chapron E, Alberic P, Jezequel D, Ledoux G and Massault M (2010b) Les archives sédimentaires de l'histoire du lac Pavin. Revue des Sciences Naturelles d'Auvergne, 74–75 (numéro spécial: Le lac Pavin), pp 57–66

Chapron E, Ledoux G, Simonneau A, Albéric P, St-Onge G, Lajeunesse P, Boivin P and Desmet M (2012) New evidence of Holocene mass wasting events in recent volcanic lakes from the French Massif Central (Lakes Pavin, Montcineyre and Chauvet) and implications for Natural Hazards. In: Yamada et al. (eds) Submarine mass move-

ments and their consequences, Advances in Natural and Technological Hazards Research 31, pp 255–264
Del Rosso-d'Hers T, Lavina P, Levi-Faict TW (2009) Risques naturels et péri-volcaniques du système volcanique Montchal-Pavin-Montcineyre. International meeting « Lake Pavin and other meromictic lakes », Besse en Chandesse, mai 2009, abstract p 72
Fox BRS, Wartho J, Wilson GS, Lee DE, Nelson FE, Kaulfuss U (2015) Long-term evolution of an Oligocene/Miocene maar lake from Otago, New Zealand. Geochem Geophys Geosyst 16(1):59–76
Freundt A, Strauch W, Kutterolf S, Schmincke HU (2007) Volcanogenic tsunamis in lakes: examples from Nicaragua and general implications. Pure Appl Geophys 164(2–3):527–545
Gal F, Gadalia A (2011) Soil gas measurements around the most recent volcanic system of metropolitan France (Lake Pavin, Massif central). Comptes Rendus Géosciences, Académie des Sciences 343:43–54
Gewelt M, Juvigné E (1988) Téphrochronologie du Tardiglaciaire et de l'Holocène dans le Cantal, le Cézallier et les Monts Dore (Massif Central, France): résultats nouveaux et synthèse. Bulletin de l'Association française pour l'étude du Quaternaire 25:25–34
Glangeaud P (1916) Le cratère-lac Pavin et le volcan de Montchalm (Puy-de-Dôme). Comptes Rendus Académie des Sciences Paris 162:428–430
Henriet JP (2009) Geophysical reconnaissance of Lac Pavin 2002. International meeting « Lake Pavin and other meromictic lakes », Besse en Chandesse, mai 2009, abstract p 29
Jézéquel D, Sarazin G, Fonty G, Tassin B (2008) Le Lac Pavin: le volcan, l'eau et la vie. Pour La Science, dossier n°58 « L'eau Attention fragile ! », janvier-mars 2008, pp 52–59
Jordan SC, Cas RAF, Hayman PC (2013) The origin of a large (>3 km) maar volcano by coalescence of multiple shallow craters: Lake Purrumbete maar, southeastern Australia. J Volcanol Geotherm Res 254:5–22
Juvigné E (1992a) Distribution of widespread late glacial and holocene tephra beds in the French Central Massif. Quat Int 14:181–185
Juvigné E (1992b) Approche de l'âge de deux cratères volcaniques lacustres d'Auvergne (France). Comptes Rendus Académie des Sciences, Paris 314:401–404
Lavina P (1985) Le volcan du Sancy et le « Massif Adventif » (Massif des Monts Dore, Massif Central Français). Etudes Volcanologiques et structurales. Thèse 3e cycle, Université de Clermont-Ferrand, 197 p
Lavina P and Del Rosso-d'Hers T (2008) Le complexe volcanique Montchal-Pavin-Montcineyre: nouvelles stratigraphie, teprochronologies et datations, vers une ré-évaluation de l'aléa volcano-tectonique en Auvergne. 22e Réunion des Sciences de la Terre, Nancy, France
Lavina P and Del Rosso-d'Hers T (2009) Le système volcanique du groupe Montchal-Pavin: nouvelle stratigraphie des formations volcano-sédimentaires et nouvelles datations, volcanologie dynamique, conséquences pour une évaluation des risques naturels. International meeting « Lake Pavin and other meromictic lakes », Besse en Chandesse, mai 2009, abstract p 15
Lavrieux M, Disnard J-R, Chapron E, Bréheret J-G, Jacob J, Miras Y, Reyss J-L, Andrieu-Pone V, Arnaud F (2013) 6700 yr sedimentary record of climatic and anthropogenic signals in Lake Aydat (French Massif Central). The Holocene 23(9):1317–1328
Lockwood JP, Costa JE, Tuttle ML, Nni J, Tebor SG (1988) The Potential For Catastrophic Dam Failure At Lake Nyos Maar, Cameroon. Bull Volcanol 50(5):340–349
Lorenz V (1973) On the formation of Maars. Bull Volcanol 37(2):183–204
Lorenz V (1986) On the growth of maars and diatremes and its relevance to the formation of tuff rings. Bull Volcanol 48:265–274
Lorenz V (1987) Phreatomagmatism and its relevance. Chem Geol 62(1–2):149–156
Lorenz V (2007) Syn- and posteruptive hazards of maar–diatreme volcanoes. J Volcanol Geotherm Res 159:285–312
Macaire J-J, Fourmont A, Argant J et al (2010) Quantitative analysis of climate versus human impact on sediment yield since the Lateglacial: the Sarliève palaeolake catchment (France). The Holocene 20(4):497–516
Németh K (2001) Long-term erosion-rate calculation from the Waipiata Volcanic Field (New Zealand) based on erosion remnants of scoria cones, tuff rings and maars. Géomorphologie: relief, processus, environnement 2:137–152
Németh K, Cronin SJ (2007) Syn- and post-eruptive erosion, gully formation, and morphological evolution of a tephra ring in a tropical climate erupted in 1913 in West Ambrym, Vanuatu. Geomorphology 86(1–2):115–130
Németh K, Kereszturi G (2015) Monogenetic volcanism: personal views and discussion. Int J Earth Sci 104(8):2131–2146
Németh K, Goth K, Martin U, Csillag G, Suhr P (2008) Reconstructing paleoenvironment, eruption mechanism and paleomorphology of the Pliocene Pula maar, (Hungary). J Volcanol Geotherm Res 177(2):441–456
Németh K, Cronin SJ, Smith IEM, Flores JA (2012) Amplified hazard of small-volume monogenetic eruptions due to environmental controls, Orakei Basin, Auckland Volcanic Field, New Zealand. Bull Volcanol 74(9):2121–2137
Olive P, Boulègue J (2004) Étude biogéochimique d'un lac méromictique: le lac Pavin, France. Géomorphologie: processus, relief, environnement 10(4):305–316
Ollier CD (1967) Maars. Their characteristics, varieties and definition. Bull Volcanol 31:45–73
Pirrung M, Büchel G, Lorenz V, Treutler H-C (2007) Post-eruptive development of the Ukinrek East Maar since its eruption in 1977 A.D. in the periglacial area of south-west Alaska. Sedimentology 55(2):305–334
Pirrung M, Buechel G, Merten D, Assing H, Schulte-Vieting U, Heublein S, Theune-Hobbs M, Boehrer B (2008) Morphometry, limnology, hydrology and sedimentology of maar lakes in East Java, Indonesia. Studia Quaternaria 21:139–152
Provencher L and Dubois J-M (2008) Proposition d'une nomenclature géomorphologique du ravage lacustre et comparaison avec les rivages côtiers et fluviaux. La société Provancher d'Histoire Naturelle du Canada, Université de Sherbrooke, pp 90–96
Rouwet D, Christenson B, Tassi F, Vandemeulebrouck J (eds) (2015) Volcanic lakes. advances in volcanology (Springer) ISBN 978-3-642-36832-5, 1–20
Schettler G, Schwab MJ, Stebich M (2007) A 700-year record of climate change based on geochemical and palynological data from varved sediments (Lac Pavin, France). Chem Geol 240:11–35
Schwab MJ, Schettler G, Bruchmann C, Acksel D, Negendank JFW, Brauer A (2009) Stratigraphy, chronology and paleoenvironment information of the sediment record from Lac Pavin, Massif Central (France). International Meeting-Lake Pavin and Other Meromictic Lakes, May 14–16, Besse et St-Anastaise, France, Abstract, p 30
Seib N, Kley J, Büchel G (2013) Identification of maars and similar volcanic landforms in the West Eifel Volcanic Field through image processing of DTM data: efficiency of different methods depending on preservation state. Int J Earth Sci 102(3):875–901
Stebich M, Brüchmann C, Kulbe T et al (2005) Vegetation history, human impact and climate change during the last 700 years recorded in annually laminated sediments of Lac Pavin, France. Rev Palaeobot Palynol 133:115–133
Suhr P, Goth K, Lorenz V, Suhr S (2006) Long lasting subsidence and deformation in and above maar-diatreme volcanoes – a never ending story. Z Dtsch Ges Geowiss 157(3):491–511
Tassi F, Rouwet D (2014) An overview of the structure, hazards, and methods of investigation of Nyos-type lakes from the geochemical perspective. J Limnol 73(1):55–70

Thouret J-C, Boivin P, Labazuy P, Leclerc A (2016) Geology, geomorphology and slope instability of the maar Lake Pavin (Auvergne, French Massif central). In: Lake Pavin. Springer, Cham

Touchart L (2000) Les lacs: origine et morphologie. L'Harmattan, Paris, 202 p

Vernet G (2013) La séquence sédimentaire des Gravanches/Gerzat: enregistrement d'événements catastrophiques à valeur chronologique en Limagne d'Auvergne (Massif Central, France). Quaternaire 24(2):109.127

Vespermann D, Schmincke HU (2000) Scoria cones and tuff rings. In: Sigurdsson et al (eds) Encyclopedia of volcanoes. Academic Press, pp 683–694

White J, Ross P-S (2011) Maar-diatreme volcanoes: a review. J Volcanol Geotherm Res 201:1–29

Wohletz K (1986) Explosive magma- water interactions: thermodynamics, explosions mechanisms, and field studies. Bull Volcanol 48:245–264

Part III

Geochemistry and Biogeochemical Cycles

Didier Jezequel

Lake Pavin Mixing: New Insights from High Resolution Continuous Measurements

10

Céline Bonhomme, Didier Jézéquel, Michel Poulin, Mohamed Saad, Brigitte Vinçon-Leite, and Bruno Tassin

Abstract

As a meromictic lake, Lake Pavin mixing is very specific. The chemocline located at about 60 m depth separates the mixolimnion (fully or partially mixed according to the year) and the monimolimnion. Deep layers are geothermally heated and stability is ensured at the bottom of the lake by the increasing dissolved substance concentration. The monimolimnion forms a compartment which has its own specific dynamics but that may interact with the mixolimnion at large time scales. Understanding of physical mixing processes is crucial to study further geochemical processes.

Temperature and turbulence were investigated in 2006 and 2007 using continuous measurements, a CTD and a high resolution temperature microstructure profiler (SCAMP). Continuous measurements give the evidence of a sublacustrine spring discharging intermittently into the mixolimnion around 55 m depth. This cold water input was observed using thermistor chains at different depths in 2007. Because of its low saline content, the spring water input rises in the water column by saline convection. The use of a simple conceptual model, representing turbulent diapycnal diffusivity and convection shows its role in maintaining the meromixis characteristic of the lake on the intra-annual time scale. The spring also influences seasonal overturns and thus contributes to establish the depth of the mixolimnion–monimolimnion interface on the interannual time scale.

Using SCAMP, vertical dispersion coefficients are estimated by different methods. Vertical dispersion coefficients show a high space and time variability. The use of these data in the geochemical model AQUASIM applied to Lake Pavin shows a variability of model outputs directly depending on mixing inputs and their variability.

Keywords

Meromixis • Water mixing • K_z • Turbulence • Temperature

This chapter is devoted to revisiting the mixing patterns of Lake Pavin thanks to new insights gained from high-resolution continuous measurements. In the first section, the physical regime of Lake Pavin is presented. Then the factors causing the meromixis are explained, namely the role of a sublacustrine spring in the mixolimnion and the role of the winter meteorological conditions. In a third section, high frequency measurements performed in order to assess the vertical diffusivity in the water column are described. The impact of the intermittency of the diffusion coefficients on the fate

C. Bonhomme (✉) • M. Saad • B. Vinçon-Leite • B. Tassin
LEESU, UMR MA 102, École des Ponts ParisTech, Marne-la-Vallée, France
e-mail: celine.bonhomme@leesu.enpc.fr

D. Jézéquel
IPGP (Institut de Physique du Globe de Paris), Sorbonne Paris Cité, Univ. Paris Diderot, UMR 7154 CNRS, 1 rue Jussieu, 75005 Paris, France

M. Poulin
Centre de Géosciences, Ecole des Mines ParisTech, Paris, France

T. Sime-Ngando et al. (eds.), *Lake Pavin*, DOI 10.1007/978-3-319-39961-4_10

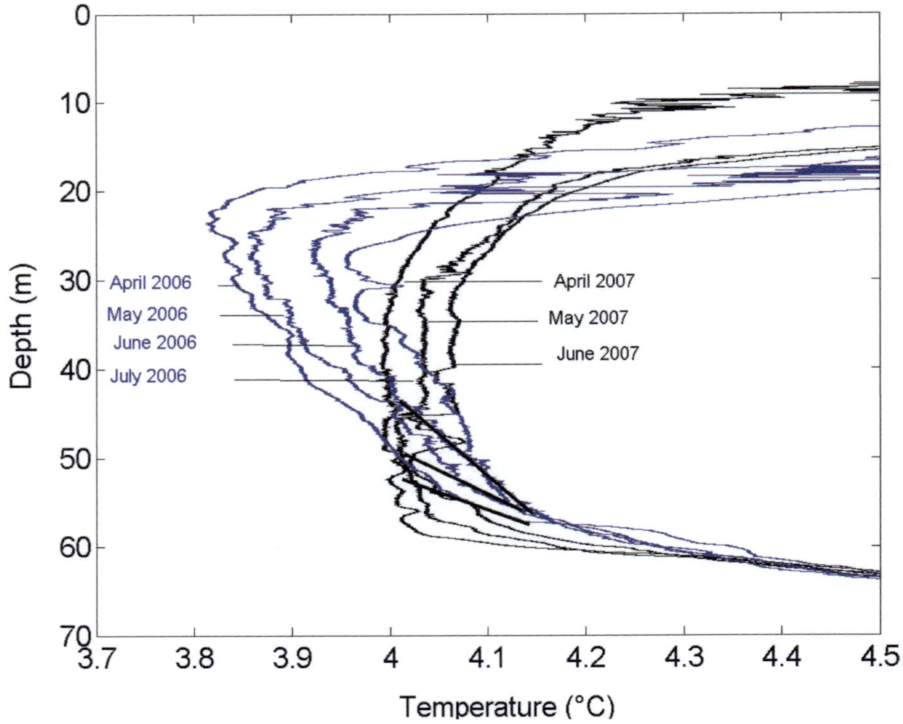

Fig. 10.1 Comparison of the evolution of temperature profiles (in °C) between 2006 (*blue*) and 2007 (*black*). Each profile is separated from the next by 1 or 2 months

of dissolved compounds is considered using a one-dimensional vertical model.

10.1 Physical Presentation of the Lake

In Lake Pavin, only the mixolimnion (from 0 to 60 m) is affected by seasonal overturns, as opposed to the monimolimnion (layer in the 60–92 m depth range) (Martin 1985). As well as many other crater lakes, the lake bottom is exposed to a heat flux of geothermal origin. Therefore the water temperature of the monimolimnion exceeds by about 1.8 °C that of the lower part of the mixolimnion. Warm water being less dense than cold water, this heating may lead to instability of the water column. But the concentration of dissolved substances in the monimolimnion is substantially greater than the concentration found in the mixolimnion (the specific conductivity at 25 °C between these two compartments varies from 50 to nearly 500 μS cm^{-1}), thus ensuring monimolimnion stability. As a consequence, the mixolimnion – monimolimnion interface is characterized by steep gradients of dissolved substances, dissolved gases (especially CH_4 and CO_2) and by complete and permanent anoxia (Michard et al. 1994).

Both the heat and dissolved compounds of the monimolimnion actually diffuse in the direction of the mixolimnion over time. The combined diffusion of matter and heat from the monimolimnion, although relatively weak, ensures the stability of the water column below depths of 30 m (Assayag et al. 2008). This stability differs from 1 year to the next. In 2006 for example, heat and dissolved substances diffuse more from the monimolimnion to the mixolimnion, the diffuse front reaching 30 m depth, whereas in 2007, the gradients are steeper between the two lake parts. The position and intensity of temperature gradient and dissolved species gradient (chemocline) may vary slightly along the year, depending on mixing variability in the mixolimnion. Therefore, stability of the water column in the lower part of the mixolimnion varies at the inter- and intra- annual time scale, but the meromictic characteristics of the lake remains (see Fig. 10.1).

10.2 Factors Leading to and Maintaining Meromixis

Understanding meromixis and the factors that lead to this particular regime is a major challenge at present because many inland water bodies are moving towards this state in the current context of climate change and increasing anthropogenic inputs of nutrients (Hakala 2004). In fact, the warming of the surface of continental waters causes stronger density gradients which make it difficult for the lake bottom to be mixed.

Both the origins and mechanisms involved in the stability of the Lake Pavin meromixis are poorly known. Hutchinson (1957) already mentioned different origins for meromixis in general: (i) ectogenic: in the case of a low salinity water inflow into the mixolimnion; (ii) crenogenic: inflow of highly-saline water into the monimolimnion due to volcanic activity; or (iii) biogenic: biological activity increases dissolved matter concentration on the lake bottom. In Lake Pavin, each of these three individual origins could play a role: (i) a low-salinity spring in the mixolimnion may stabilize the interface between mixo- and monimolimnion by increasing the salinity gradient; (ii) a mineral spring in the monimolimnion may similarly strengthen the gradient at the interface; and (iii) biological activity may also increase the particle content, and therefore the density, in the monimolimnion.

10.2.1 Presence of a Sublacustrine Spring in the Lower Part of the Mixolimnion

The hypothese of a sublacustrine spring was formulated a long time ago by geochemical studies. The presence of a mineralized spring within the monimolimnion was first predicted in 1975 as an explanation of the observed gradients at the mixo–monimolimnion interface (Meybeck et al. 1975). 10 years later, Martin (1985) used a box model to explain both the hydrologic balance and tritium measurements by assuming that a unique spring existed within the monimolimnion. Assayag et al. (2008) tested various scenarios by applying the one-dimensional vertical AQUASIM model (Reichert 1994), which incorporates, in particular, the hypothesis of a sublacustrine spring within the mixolimnion. This hypothesis was corroborated by the observation of a cold temperature anomaly on the vertical profiles measured with a multiparameter (conductivity-temperature-depth) probe in September 1996, which was not apparent on the September 1994 profiles (Aeschbach-Hertig et al. 2002). This approximately –0.06 °C anomaly was noticed at depths between 45 and 50 m. The authors also identified a slight drop in dissolved oxygen at these same depths, with no recorded conductivity anomaly.

Recently, the sublacustrine spring hypothesis was confirmed (Bonhomme et al. 2011a), on the basis of continuous physical monitoring of Lake Pavin carried out during years 2006 and 2007. The sublacustrine spring was identified between 53 and 56 m depth. The temperature of the entering water was shown to be about 1.3 °C lower than the temperature of the lower part of the mixolimnion during the measurement campaign. The oxygen and saline contents of the spring were low in comparison with the mixolimnion. Moreover, spring puffs going up in the water column by convection were observed, suggesting that the sublacustrine spring discharged to the lake intermittently.

In the vicinity of Lake Pavin, the hole "Creux du Soucy", located at about 2 km from the lake, is a good candidate to explain such cool water entries in the lake. The surface water inside the hole has an altitude of 1235 m (35 m higher than the surface elevation of the lake). The hole extends into a tube that drips somewhere else at a lower altitude. The water temperature and conductivity were measured in July 2007 and were respectively 2.1 °C and 23 mS/cm (conductivity at 25 °C is 40 mS/cm), which is consistent with the observations cited above. Old legends report that this hole may be directly connected to Lake Pavin by a kind of siphon (Bakalowicz 1971) which may explain the intermittent entering of water into the lake. This assumption is confirmed by recent diving explorations of the hole, which revealed that the hole is connected to a 50 m long quasi-vertical tunnel which may directly discharge into Lake Pavin.

10.2.2 Role of the Spring in the Maintaining of Meromixis

As the spring is very close to the mixolimnion-monimolimnion interface, the spring probably influences on the diffusion of solutes at this interface and therefore on meromixis. To understand the impact of the sublacustrine spring on mass transfer at the mixolimnion–monimolimnion interface, a modelling tool was developped to investigate the role of the sublacustrine spring role in maintaining meromixis (Bonhomme et al. 2011a, b). The model is based on the diffusion equation for solutes. The influence of the spring is evaluated only during the ice-free period of the lake and compared to the diffusion of solutes that would have happened if the spring was not present.

As described before, if no spring discharges into the lake, heat and dissolved substances naturally diffuse in the monimolimnion-to-mixolimnion direction. The decrease in heat and salinity gradients through diffusion at the mixo-monimolimnion interface causes a reduction of the stability at the interface.

On the contrary, a water inflow able to strengthen a low conductivity value near the interface raises the interface stability by sharpening the gradients, while stimulating mixing with the layers of the mixolimnion shallower than 50 m depth. This additional water therefore helps insulate the mixolimnion from the monimolimnion by reducing vertical diffusion near the interface.

The flux evaluation was performed with a fictitious independent passive tracer presenting a gradient analogous to conductivity between the mixo and the monimolimnion over a 1-year period. This fictitious passive tracer has a concentration equal to 1 at 80 m depth and to 0 at 30 m depth. It is only transported, it does not react with any other tracer in the environment and its concentration does not influence the water density. The results are summarized in Table 10.1. The presence of a spring would thus help reduce fluxes between the monimo- and mixolimnion by some 20 %, with this effect tending to become more pronounced over time. This flux decrease is significant at the annual scale and serves to limit diffusion above the monimolimnion. On the other hand, the absence of a spring causes a gradual stability decline by diffusion.

Table 10.1 Simulated tracer flux (arbitrary unit) in the presence or absence of the spring above the monimolimnion

		After 1-year simulation	
	Starting time	No spring	With a spring
Tracer mass (arbitrary unit) between 30 and 80 m	59.04	55.69	56.38
Tracer mass (arbitrary unit) between 30 and 60 m	0.67	3.35	2.66

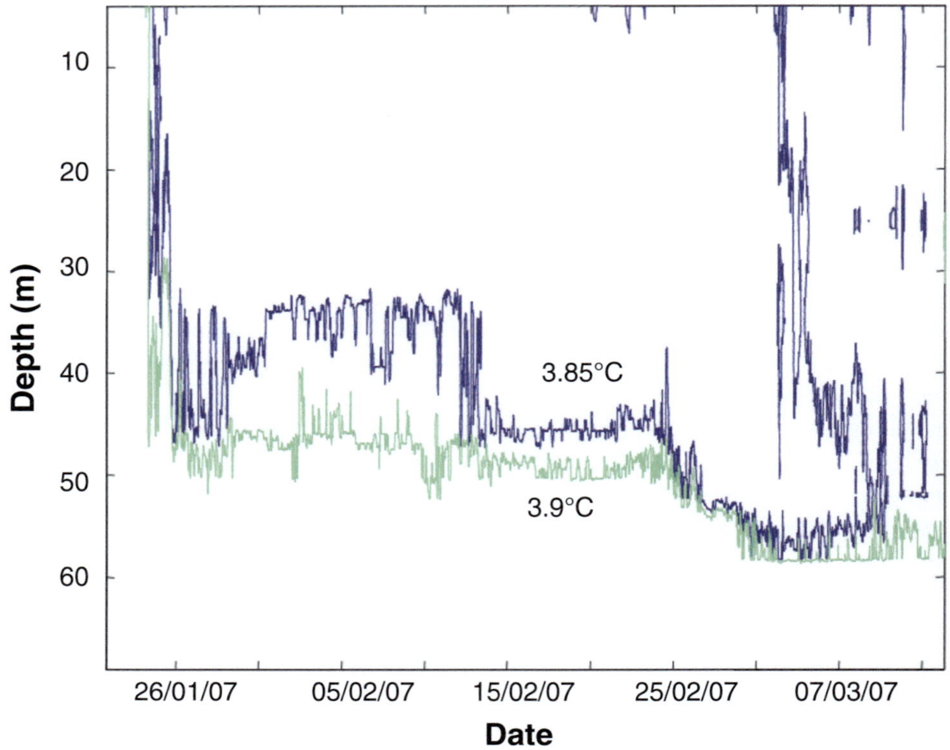

Fig. 10.2 Observation of isotherms dipping between January 25 and March 10, 2007, until a depth of 58 m. Only those isotherms in the vicinity of the maximum density temperature are shown

10.2.3 Winter Influence, Interannual Variability and Meromixis

The winter overturn has undeniably a great influence on the state of the water column during the following year. It seems that the ability of the seasonal overturn to keep the interface mixolimnion-monimolimnion around 60 m depth is linked both to the meteorological conditions of the winter and to the presence of the spring. A seasonal overturn was monitored during the 2006–2007 winter thanks to a thermistor chain recording continuously the temperature of the lake, even underneath the ice. Figure 10.2 shows that the seasonal overturn of the mixolimnion occurred in two stages: in the initial stage, water adjacent to the density maximum at these depths (3.9 °C) (Eklund 1965) dipped at the end of January to a depth in the range of 50 m. The 10 m layer above the chemocline (between 50 and 60 m) was therefore not mixed. As the temperature of the surface falls below 4 °C during this period, an inverse stratification then occurred between the surface and 50 m depth in the mixolimnion (data not provided).

In a second time, the sudden inflow of a large quantity of low-salinity water around February 25, 2007 (see Fig. 10.2) eventually flattened the salinity gradient, which enabled the seasonal overturn to reach the chemocline (at 60 m). A quick calculation of heat balance between 40 and 60 m suggests to attribute the temperature drop at these depths to a cold external water inflow and thus definitely to the presence of a spring.

During the 2006–2007 winter, it thus seems likely that the sublacustrine spring helped fix the mixing depth at the time of spring mixing as well as diffusion at the mixolimnion–monimolimnion interface. Therefore, the difference between the 2006 and 2007 temperature profiles (shown in Fig. 10.1) may be explained by winter overturn differences not only due to changing meteorological conditions, but to varying hydrological processes, favouring either the occurrence of water inflow from the spring or an absence of entering sublacustrine water.

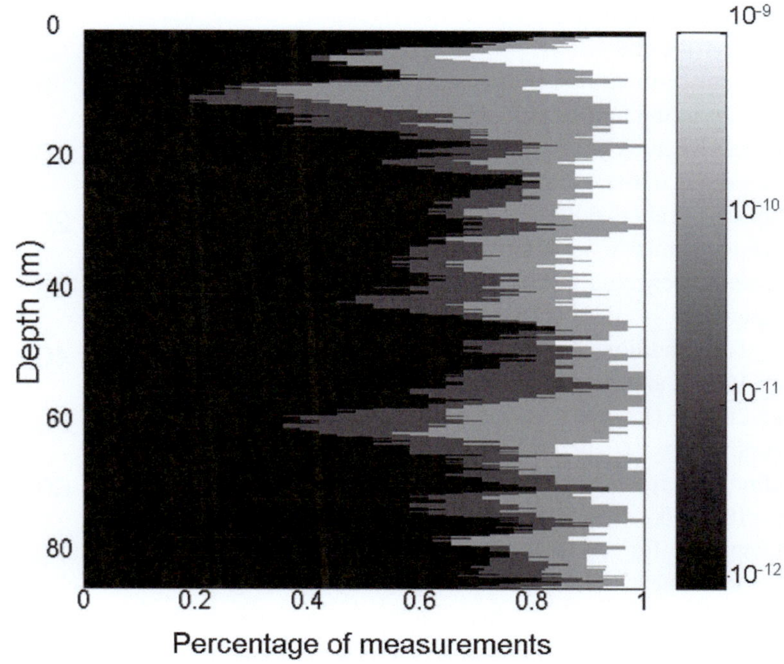

Fig. 10.3 Histograms representing the intermittency of the dissipation rate (in W/kg) in the whole water column in July–September–November 2006

10.2.4 High Frequency Measurements and Quantification of Dissipation and Vertical Diffusivity in the Water Column of Lake Pavin

When the water column of a lake is turbulent, the turbulent kinetic energy is transmitted from big scales to smaller structures and scales. When the scale reaches the one of viscous forces, the kinetic energy is transformed into heat and the temperature of the fluid rises. It is possible to calculate the dissipation in the water column from the fluctuations of temperature profiles with the Batchelor method (Batchelor 1959). This method relies on the fit between a theoretical and an experimental spectra around the length scale of viscous forces.

Nevertheless, both turbulent and diffusive processes cause micro-fluctuations of temperature in lakes. Time-series of a depth averaged "effective dissipation rate" (ε_μ) can be computed by assuming that all fluctuations are due to mechanical turbulence. This approximation is only valid when the turbulent diffusion coefficient exceeds molecular diffusivity. When the dissipation falls below or close to molecular diffusivity, this approach under-evaluates the diffusivity. It happens in Lake Pavin around the chemocline and therefore the vertical diffusivity has a lower bound which is molecular diffusivity.

For steady conditions and neglecting the divergence terms, the production of turbulent kinetic energy (by the Reynolds stress working on the mean shear) is balanced by the dissipation rate ε and the buoyancy flux b. The turbulent kinetic energy balance becomes:

$$-\left\langle u_i u_j \right\rangle \frac{\partial U_i}{\partial x_j} = b + \varepsilon \qquad (10.1)$$

where U_j is the mean velocity and u_j is the fluctuating velocity.

Temperature fluctuations are measured through temperature microstructure surveys (SCAMP – Self-Contained Autonomous Micro-Profiler designed by PME) in 2006 and 2007 completed by CTD profiles (Seabird, SBE 19). The microprofiler is used in falling mode, at a speed of ca. 0.1 m/s, with a 100 Hz sampling rate for temperature and 40 Hz for conductivity. The resolution of the device is 0.005 °C and the relative error for conductivity is 5 %.

10.2.5 Determination of the Dissipation Rate ε from Microstructure Measurements

Several profiles were performed at a central position in Lake Pavin from April to June 2007, with a time step of about 15' (this time step is the minimum delay to acquire successive measurements with the SCAMP because the falling speed is set). Based on SCAMP profiles, histograms of measured dissipation rates with different boundaries (from 10^{-12} W/kg to 10^{-9} W/kg, from black to white) as a function of depth are plotted on Fig. 10.3. From Fig. 10.3, it can be observed that the dissipation rate ε is highly intermittent in space and time in the water column of Lake Pavin.

The water column of Lake Pavin spends most of time at low dissipation rates (below the detection threshold of the instrument). Making the assumption that the measured profiles are representative of the state of intermittency in the water column, the dissipation is below 10^{-11} W/kg more than 50 % of the time. Similar observations were performed in the ocean interior in the California Current in the eastern North Pacific Ocean where (Gregg et al. 2003) observed that over a large proportion of the water column, ε is less than the instrumental noise level about 60 % of the time. Strikingly, lakes and ocean behave similarly on this point and no significant difference appears concerning the turbulent weather of the water column.

The conditions for a significant buoyancy flux to occur is $\varepsilon > 19\upsilon N^2$ and for isotropic condition to be fulfilled is $\varepsilon > 200\upsilon N^2$ (Itsweire, 1993). In the hypolimnion in 2006, the condition is satisfied when $\varepsilon > 10^{-11}$ W/kg. At the chemocline, when $\varepsilon > 10^{-9}$ W/kg and at the thermocline, when $\varepsilon > 10^{-8}$ W/kg. The conditions are increased by one order of magnitude for the isotropic condition. In the case of Lake Pavin, it means that significant buoyancy flux is generated 30–40 % of the time in the hypolimnion and less than 10 % of the time at the thermocline and chemocline. As far as the isotropy condition is concerned, rarely the turbulence is isotropic at the density interface and about 10 % of the time in the hypolimnion. Comparing to the ocean, the data set available here on Lake Pavin shows that the mixed layer is far less mixed than in the ocean. Because of small hills around the lake surface, Lake Pavin surface is protected from strong winds.

General trend of the dissipation rate shows more intense dissipation where the generated shear by internal wave motions is the highest. Internal waves occur in stratified water bodies after the wind stops. In fact, the wind tilts the lake surface when it blows. When the wind stops, the gravity brings the surface back to its equilibrium position and the lake surface oscillates several times. At the thermocline, this observation was obvious from profiles collected in 2006 (data not shown) and 2007 (see Fig. 10.3). At the chemocline, an enhanced dissipation is only noticeable in 2007 (see Fig. 10.3), which is marked by a higher density gradient around 60 m depth as shown in Fig. 10.1.

In 2007, the dissipation rate at the thermocline is higher than 5×10^{-10} W/kg about 70 % of the time. On the contrary, dissipation in the hypolimnion is very low: below the mixed layer, the water column presented turbulence levels below 1×10^{-11} m^2 s^{-1} more than 50 % of the time. But high values of dissipation (above 1×10^{-9} m^2 s^{-1}) may occur about 15 % of the time. At the chemocline, dissipation levels are close to dissipation levels at the thermocline.

10.2.6 Estimates of Vertical Diffusivity (*Kz*)

Kz is an important parameter used by different geochemical models to simulate the transport of solutes in a water body.

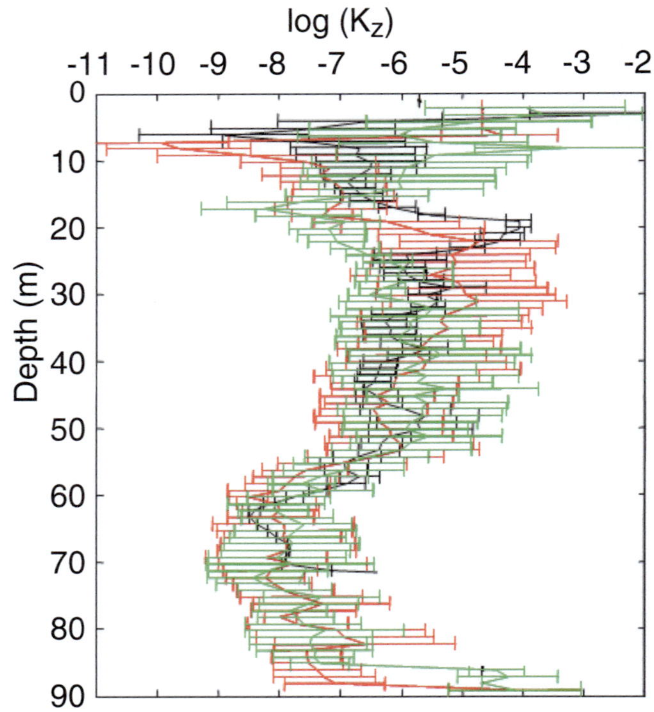

Fig. 10.4 K_z (in m^2/s) with error bars calculated from microstructure measurements by the Batchelor fitting method on the whole water column. *Black*: May 2006, *Red*: July 2006, *Green*: November 2006

An example applied to Lake Pavin is given in the next section.

Knowing ε, the vertical diffusivity (*Kz*) can be calculated by the Osborn method (Osborn 1980).

$$K_z = \frac{\gamma_{mix}\varepsilon}{N^2} \tag{10.2}$$

with γ_{mix} the mixing efficiency,and *N2* the buyoancy or Brunt-Väisälä frequency. $\gamma_{mix} = 0.2$ is the usual value given for the mixing efficiency (Ellison 1957) that was used in this study.

Computed *Kz* are geometrically averaged (Baker and Gibson 1987) to calculate the averaged value of *Kz*. The error bars (Fig. 10.4) represent the standard deviation of log(*Kz*) on each monthly series of *Kz*, at each meter. They relate the intermittency of the mixing in the water column. The intermittency of turbulence is caused by the intermittency of forcings (wind for example) and by the sporadic and statistic occurrence of turbulent bursts.

Averaged profiles are reported in Fig. 10.4. In black, averaged *Kz* of May 2006 lays slightly below averaged *Kz* of July (in green) and November 2006 (in red). The vertical diffusivity presents two minima at the thermocline and at the chemocline except in July: at these locations, the diffusivity is below heat molecular diffusivity. Outside of strong density gradient regions, vertical diffusivity is higher in the hypolimnion and in the monimolimnion. The stratification is higher in the monimolimnion than in the lower part of the mixolimnion due to salt gradients. Moreover, the standard deviation of *Kz* is higher in the hypolimnion than at the thermocline or at the chemocline. This standard deviation can reach 3–4 orders of magnitude in weak stratified parts of the water body, whereas it is closer to 2 orders of magnitude at the thermocline and at the chemocline.

In May, values of *Kz* between 30 and 50 m depth are in average equal to 5×10^{-7} m^2/s whereas in July and November, *Kz* can reach values of 5×10^{-6} m^2/s. Simultaneously an increase in the standard deviation is observed indicating that increase of mean *Kz* and intermittency are related.

Moreover, trends are observed on the evolution of vertical diffusivity along year 2006. In fact, stabilization is increased below 50 m depth and destabilized in the hypolimnion. The stabilization just above the chemocline comes from the diffusion of the saline interface between mixoliminion and monimolimnion. The destabilization above 50 m is certainly caused by a more active turbulence weather in the second part of year 2006.

Kz in 2007 are in average one order of magnitude higher in the hypolimnion than in the second part of year 2006 (data not shown). This is in agreement with the decrease of the static stability (N^2) of the order of one order of magnitude too.

10.2.7 Consequences of *Kz* Intermittency on the Estimate of Solute Transport

The estimate of *Kz* is very important to predict the vertical diffusion of solutes and finally the reactions between chemical compounds in the lake. In previous studies, *Kz* was evaluated from tracer experiments and only the mean value of *Kz* throughout the year was considered (Aeschbach-Hertig et al. 2002). But this average value is unable to reproduce the impact of the real small fluctuations of *Kz* value that cause the transport in the water column of Lake Pavin.

Lake physical models usually link *Kz* fluctuations to wind forcing fluctuations (mettre ref, par exemple l'article sur le modèle thermique du lac du Bourget, Hydrobiologia 2014). This kind of approach is usually effective to simulate *Kz* close to the lake surface but lacks precision to simulate *Kz* in the deep part of the lake, close to the chemocline.

Therefore, a direct observation of the variability of *Kz* in depth is useful and it can be used as an input data in a model in order to evaluate the influence on tracer concentrations in the monimolimnion and in the mixolimnion of Lake Pavin. Therefore, the impact of using *Kz* from SCAMP measurements on the prediction of solute concentrations was compared with using rough and averaged estimates of *Kz*, as used in previous studies.

To achieve this goal, the Aquasim model (Reichert 1994) is used. Aquasim model is a one-dimensional vertical numerical model based on the advection–diffusion equation. The base equation is the following :

$$\frac{\partial C}{\partial t} + \frac{\partial (Cw)}{\partial z} = \frac{\partial^2 K_z C}{\partial z^2} \tag{10.3}$$

with *C* the solute concentration, *w* the vertical speed and *Kz* the vertical turbulent dispersion coefficient.

The AQUASIM configuration adapted to Lake Pavin is described in Lopez et al. (2011). Surface processes linked to daily meteorological conditions are neglected as the study focuses on in-depth processes. The model simulates O_2, Fe particle concentration and NO_3 between two field surveys of July 2007 and August 2007. In this configuration, O_2 concentration conditions the whole reaction chain. The initial conditions are based on CTD measurements and *in situ* chemical analysis (Fe particle concentration, dissolved O_2 concentration, NO_3 concentration). Validation is done on measured O_2 and NO_3 concentration in August 2007 because no data for Fe particle concentration were available at the end of simulation period. The *Kz* variability is investigated by creating five sets of different *Kz* values. From the variability of *Kz* observed during the July survey where *Kz* was calculated from different SCAMP profiles performed at the same location in the lake (see Sect. 11.3.2), *Kz* values are allocated at random every 10 min using the values available at each depth in the data set. Therefore, 1 month long time-series of *Kz* are generated and used as input for the Aquasim model between July and August 2007. Three independent random draws of *Kz* are then compared with constant values of *Kz* for the month: arithmetic and logarithmic average of measured *Kz* from SCAMP profiles and constant formulation of *Kz* from CTD measurements during the whole duration of the simulation (Bonhomme et al. 2011a, b). In fact, CTD measurements can give a rough evaluation of *Kz* by calculating the buoyancy frequency and simply finding a proportional relationship between the inverse of the buyoancy frequency and *Kz*.

The outputs of the model are evaluated by calculating the cumulative error for O_2 and NO_3 concentrations at different depths in August 2007.

Cumulative error is defined as:

$$E = \sum_{50,53,56,59m} \frac{C_{August,model} - C_{August,observed}}{C_{August,observed}} \tag{10.4}$$

With C, the concentration of the chemical concentration of the followed species.

With randomly *Kz* values calculated from SCAMP measurements, the concentration of O_2 is predicted with a cumulative error of 0.32 whereas the NO_3 concentration is even better with a cumulative error of 0.048. With rough estimates of *Kz* from CTD measurements, the cumulative error becomes worse, about 10 times higher for both compounds: it becomes 3.41 for O_2 and 0.43 for NO_3 concentrations.

Therefore, it seems very important to estimate *Kz* with the highest precision as possible. In particular, taking into account the variability in time of *Kz* seems important to obtain more reliable modelling results.

10.3 Conclusion

Temperature and conductivity microstructure data completed by continuous measurements collected in Lake Pavin from May 2006 to June 2007 constitute a unique data set. This database is very useful to understand the reasons leading to and contibuting to keep meroximis. Among these reasons, the role of a sublacustrine spring and of winter meteorological conditions seem to be crucial to understand the maintaining of meromixis over time.

During the 2 years observed, the mixolimnion of Lake Pavin was more stable in 2006 and progressively destabilized in 2007. Gradients are steeper at the interface mixolimnion-monimolimnion in 2007. Moreover, in 2007, the hypolimnion is more unstable and clear convective processes due to a sublacustrine inlet at the bottom of the mixolimnion are observable.

Finally, temperature and conductivity microstructure data enable the estimate of dispersion coefficient in the water body. The use of the calculated dispersion coefficient and taking into account its variability in the modelling of the biogeochemical behaviour of the lake is of primary importance in the assessment of the fate of both major and trace elements.

The biogeochemical behaviour of Lake Pavin as well as its rather small dimensions makes it a life-size laboratory reactor.

Acknowledgments The authors would like to thank the French National Agency for Research which founded the Metanox program (2006–2007). The different physical observations detailed in this chapter are related to this program and were performed in this framework.

References

Aeschbach-Hertig W, Hofer M, Schmid M, Kipfer R, Imboden DM (2002) The physical structure and dynamics of a deep, meromictic crater Lake (Lac Pavin, France). Hydrobiologia 487:111–136

Assayag N, Jézéquel D, Ader M, Viollier E, Michard G, Prevot F, Agrinier P (2008) Hydrological budget, carbon sources and biogeochemical processes in Lac Pavin (France): Constraints from $d_{18}O$ of water and $d_{13}C$ of dissolved inorganic carbon. Appl Geochem 23(10):2800–2816

Bakalowicz M (1971) Le Creux de Soucy (Besse en Chandesse, Puy-de-Dome). Ann Speleologie 26(2):387–406

Baker MA, Gibson CG (1987) Sampling turbulence in the stratified ocean: statistical consequences of strong intermittency. J Phys Oceanogr 17:1817–1837

Batchelor GK (1959) Small scale variations in convected quantities like temperature in a turbulent fluid. Part I. General discussion and the case of small conductivity. J Fluid Mech 5:113–139

Bonhomme C, Poulin M, Vinçon-Leite B, Saad M, Groleau A, Jézéquel D, Tassin B (2011) Maintaining meromixis in Lake Pavin (Auvergne, France): the key role of a sublacustrine spring. Compt Rendus Geosci 343(11–12):749–759

Bonhomme C, Lopez F, Viollier E, Vinçon-Leite B, Tassin B (2011) Should intermittency of small-scale turbulence be taken into account for large-scale biochemical modelling in lakes? Application to Lake Pavin, France. Paper presented at IWA symposium on Lake and Reservoir Management : sustainable strategies to enhance water quality, Granada, Spain, 13–17 June 2011

Eklund H (1965) Fresh water: temperature of maximum density calculated from compressibility. Science 142:1457–1458

Ellison TH (1957) Turbulent transport of heat and momentum from an infinite rough plane. J Fluid Mech 2:456–466

Gregg MC, Sanford TB, Winkel DP (2003) Reduced mixing from the breaking of internal waves in equatorial waters. Nature 422:513–515

Hakala A (2004) Meromixis as a part of lake evolution–observations and a revised classification of true meromictic lakes. Finland Boreal Environ Res 9:37–53

Hutchinson GE (1957) A treatise on limnology, vol 1. Wiley, New York

Itsweire EL, Kosiff JR, Briggs DA, Ferziger JH (1993) Turbulence in stratified shear flows: implications for interpreting shear induced mixing in the ocean. J Phys Oceanogr 23:1508–1522

Lopez F, Viollier E, Thiam A, Michard G, Abril G, Groleau A, Prévot F, Carrias JF, Albéric P, Jézéquel D (2011) Biogeochemical modelling of anaerobic *vs.* aerobic methane oxidation in a meromictic crater lake (Lake Pavin, France). App Geochem 26(12): 1919–1932

Martin JM (1985) The Pavin Crater Lake. In: Stumm W (ed) Chemical processes in lakes. Wiley, New York, pp 169–188

Meybeck M, Martin JM, Olive P (1975) Géochimie des eaux et des sédiments de quelques lacs volcaniques du Massif Central français. Verh Int Verein Limnol 19:1150–1164

Michard G, Viollier E, Jézéquel D, Sarazin G (1994) Geochemical study of a crater lake: Pavin Lake, France–Identification, location and quantification of the chemical reactions in the lake. Chem Geol 115:103–115

Osborn TR (1980) Estimates of the local rate of vertical diffusion from dissipation measurements. J Phys Oceanogr 10:83–89

Reichert P (1994) AQUASIM: a tool for simulation and data analysis of aquatic systems. Wat Sci Tech 30:21–30

Carbon Cycle in a Meromictic Crater Lake: Lake Pavin, France

11

Didier Jézéquel, G. Michard, E. Viollier, P. Agrinier, P. Albéric, F. Lopes, G. Abril, and L. Bergonzini

Abstract

Lake Pavin is a meromictic maar lake located in the French Massif Central. Its maximum depth is 92 m and its mean diameter is 750 m (0.445 km^2). Waters below about 60 m are never mixed with overlying waters and are permanently anoxic. DIC (Dissolved Inorganic Carbon) and CH_4 concentrations in the deep monimolimnion are 14 and 4 mmol/L respectively. DIC has two origins: biogenic and volcanic.

Data on C species, ^{13}C and ^{14}C isotopes as well as auxiliary data were used to build a carbon mass balance in the lake. Main features of the carbon cycle are:

1. Between 70 and 92 m depth, DIC and methane fluxes coming from the sediment are estimated to 1.5 and 1.0 kmol d^{-1} respectively.
2. At about 65–70 m depth, a PM^{14}C (percent of modern carbon) minimum suggests the inflow of a mineral water which brought DIC = 2.6 ± 0.2 kmol d^{-1}.
3. As suggested by a minimum of $\delta^{13}C$ of DIC and very low CH_4 concentration at *ca.* 60 m depth, methane is almost quantitatively transformed mainly into DIC at the redoxcline. The methane escape toward the atmosphere is very low and negligible in the carbon mass balance.
4. The $\delta^{13}C$ DIC, DIC *vs.* O_2 plots and some thermal and dissolved oxygen anomalies suggest a fresh water inflow at about 53 m depth; the corresponding input of DIC is estimated to 1.6 ± 0.2 kmol d^{-1}.
5. Between 20 and 50 m depth, mineralization of organic matter produced in the photic zone occurs partially. From a $\delta^{13}C$ *versus* 1/DIC plot, a production of DIC of 0.3 ± 0.2 kmol d^{-1} can be derived.
6. The maximum of photosynthesis is located within the metalimnion (10–15 m depth). A mean value for the corresponding DIC uptake of 3.5 ± 0.5 kmol d^{-1} is assumed. During seasonal stratification of the mixolimnion (about April to December), water

D. Jézéquel (✉) • G. Michard • E. Viollier • P. Agrinier
IPGP (Institut de Physique du Globe de Paris), Sorbonne Paris Cité, Univ. Paris Diderot, UMR 7154 CNRS, 1 rue Jussieu, 75005 Paris, France
e-mail: jezequel@ipgp.fr

P. Albéric
ISTO (Institut des Sciences de la Terre d'Orléans), Observatoire des Sciences de l'Univers en région Centre, UMR 7327 (CNRS-Université d'Orléans-BRGM), 1A rue de la Férollerie, 45071 Orléans Cedex 2, France

F. Lopes
ECP (Ecole Centrale Paris), Grande Voie des Vignes, 92295 Châtenay-Malabry Cedex, France

G. Abril
EPOC (Laboratoire Environnements et Paléoenvironnements Océaniques et Continentaux), UMR 5805, Université de Bordeaux, Allée Geoffroy Saint-Hilaire, 33615 Pessac, France

L. Bergonzini
GEOPS (Géosciences Paris Sud), Université Paris-Saclay, Univ. Paris-Sud, UMR 8148 CNRS, rue du belvédère, Bât 504, 91405 Orsay Cedex, France

T. Sime-Ngando et al. (eds.), *Lake Pavin*, DOI 10.1007/978-3-319-39961-4_11

deeper than 10–12 m are isolated from the atmosphere and superficial inputs as shown by the low $PM^{14}C$ value of 45 % at 10 m depth.

7. In the superficial layer, an input of $1.3 \pm 0.2\,kmol\,d^{-1}$ of DIC is brought by several small brooks, and an output of $3.0 \pm 0.2\,kmol\,d^{-1}$ is calculated for the discharge.
8. CO_2 exchanges with atmosphere are highly variable seasonally; the lake is strongly under saturated in summer and oversaturated in November. An annual balance is derived from modeling the global lake functioning, leading to a CO_2 escape of $4 \pm 1\,kmol\,d^{-1}$.

Budgets for DIC and alkalinity (alk) are proposed. For this latter, we need to distinguish two kinds of dissolved species:

- "conservative" ions (*e.g.* Na^+, K^+, Ca^{2+}, Mg^{2+}, Cl^-) brought by streams and other water venues
- "reactive" species (*e.g.* Fe^{2+}, NH_4^+, $H_2PO_4^-$), essentially produced within the sediment and consumed or precipitated at *ca.* 60 m depth *i.e.* at the redox interface.

CO_2 behaviour is deduced from both DIC and Alk budgets. The dissolved concentration of CO_2 in the monimolimnion, associated with those of CH_4 and N_2, allow assessing the gas outburst natural hazard, which is very low considering the actual conditions.

Keywords
Carbon cycle • Lake modeling • Gas outburst natural hazard • CO_2 and CH_4 emissions

11.1 Introduction

Dissolved gases (carbon dioxide and methane) play an important role in the carbon cycle of lakes (Cole et al. 1994; Bastviken et al. 2004, 2008). Generally, CO_2 is consumed during photosynthesis in the upper layer and produced during mineralization of organic matter (OM) in the hypolimnion and in sediments. When the hypolimnion becomes anoxic, production of methane occurs especially when the sulfate content of waters is low. Depending on the water column properties (height, stratification, eddy diffusion…) a part of the produced methane is oxidized in CO_2 by bacteria within the lake whereas the other part may escape toward the atmosphere.

In temperate latitude, the exchange of carbon dioxide between water and air is generally a gas absorption during the spring and summer and a large degassing during lake overturn at the end of fall. Although it may be essential for the greenhouse effect when combined at regional scale, the net balance is often difficult to assess.

In crater lakes, possible inputs of gases from mantle still increase the amount of dissolved gases in the system. For meromictic lakes, such an increase involves gas concentrations in the monimolimnion close to saturation and a gas outburst may occur (for instance in Monoun and Nyos lakes, Cameroon, in 1984 and 1986 respectively; Sigurdsson et al. 1987; Freeth and Kay 1987; Schmid et al. 2006).

The possibility of such a gas outburst in Lake Pavin, France, has been already studied (Camus et al. 1993; Aeschbach-Hertig et al. 1999, 2002). However the answer to a controversial discussion on risk assessment and management in Lake Pavin vicinity (Lavina and del Rosso 2006) needs more than a simple chemical monitoring. A quantitative understanding of the carbon cycle in the lake is needed. In this paper, we associate previous data and some new analyses in order to present a carbon cycle with three objectives:

- Distinguish between biogenic and volcanic CO_2
- Predict the evolution of the saturation status of gases in the monimolimnion
- Evaluate the transfers of CH_4 and CO_2 to the atmosphere.

11.2 Presentation of the Lake

Lac Pavin shows several characteristics that have attracted the attention of numerous scientists (in the last decades: Olivier 1952; Alvinerie et al. 1966; Omaly,1968; Pelletier 1968; Meybeck et al. 1975; Devaux et al. 1983; Amblard and Restituito 1983; Restituito 1984, 1987; Martin 1985; Martin et al. 1992; Camus et al. 1993; Cossa et al. 1994; Michard et al. 1994, 2003; Schmid 1997; Viollier 1995; Viollier et al. 1995, 1997; Aeschbach-Hertig et al. 1999, 2002; Albéric et al. 2000; Olive and Boulègue 2004; Lehours et al. 2005, 2007; Schettler et al. 2007; Bonhomme 2008).

The lake is located in the youngest volcanic area of the French Massif Central (De Goër de Hervé 1974), 35 km SW of the city of Clermont-Ferrand, at the geographic location N 45°29,7400 E 2°53,2800 (lake center), and at an altitude of 1197 m a.s.l. It is set in a maar crater, mainly composed of basaltic, trachyandesitic, granitic and gneissic rocks, formed from 3450 to 7000 years cal BP (calibrated before present) according to different authors from ^{14}C datations (3450 ± 110 year BP in Brousse et al. 1969; 6600 year BP in Brousse 1969; 3450 year BP in Camus et al. 1973; 5800 year BP in Guenet 1986; 5800–5900 year BP in Juvigné and Gilot 1986; 5990 ± 140 year BP in Juvigné and Gewelt 1987; 6000 year BP in Juvigné et al. 1988; Gewelt and Juvigné 1988; Guenet and Reille 1991; 6700 ± 110 year cal BP in Boivin et al. 2009; 6090 ± 40 year BP *i.e.* 6970 ± 60 year cal BP in Chapron et al. 2010). Youngest ages seem to be discarded due artifacts (Delibrias 1979; Gewelt and Juvigné 1988) and a consensus would give an age close to 6900 years cal BP.

One of its major features, initially described by Pelletier (1968), is the presence of a stagnant anoxic deep-water layer called the "monimolimnion". The stability of this layer is favoured by the hollow shape of the basin: with a lake area of 0.445 km^2 and a maximum depth ($D_{max.}$) of 92 m, the aspect ratio ($D_{max.}/area^{0.5}$) is 0.138 (Delebecque 1898). This value is above the limit value of 0.1 that may lead to meromixis (Dussart 1966). Hereafter, the terminology of the bottom water layers is revisited, due to recent insights into this part of the lake.

The profiles of the physico-chemical parameters (dissolved O_2, conductivity and dissolved compounds) evidence a zone with a strong chemical gradient from about 60 to 70 m depth (Viollier 1995; Viollier et al. 1997; Michard et al. 1994, 2003). Thereafter, this layer is named "mesolimnion", the depth of the maximum conductivity gradient being the chemocline. The sharp increase in concentration of dissolved compounds within the mesolimnion leads to an increase in density of the bottom water layers and consequently strengthens the stability of the physical stratification, despite a temperature increase of about 1 °C (Aeschbach-Hertig et al. 2002).

The bottom layer located below 70 m depth is strictly anoxic, enriched in reduced compounds and not affected by seasonal vertical mixing. It can be considered at steady state (Viollier et al. 1997; Aeschbach-Hertig et al. 1999, 2002; Michard et al. 2003).

The overlying waters that represent the mixolimnion (from 0 to *ca.* 60 m depth) are mainly oxic and affected by seasonal vertical mixings. Due to seasonal variation in temperature, the mixolimnion waters are expected to overturn twice a year, in the periods from November to December and from March to April. However, as pointed by Restituito (1987), the autumn overturn is much less intense than in the spring, and may differ depending on the year from 60 to 70 m depth. According to Aeschbach-Hertig et al. (2002), some years, the complete seasonal mixing may only concern the first 30 m.

Ice usually covers the lake surface from late December to mid March or mid April, inhibiting the water column mixing during this period.

Two layers with very low dispersivity (K_z) are located around the metalimnion and the mesolimnion (Aeschbach-Hertig et al. 2002; Bonhomme 2008; Bonhomme et al. 2011): the first one is only present during the warm period and situated around 10 m depth, the second one being permanent at about 60–62 m depth.

In the following discussions, the lake will be divided into five layers (*cf.* Fig. 11.1):

- The epilimnion between 0 and 5–10 m depth
- The metalimnion, corresponding to the temperature gradient zone around the thermocline, is present from *ca.* April to November

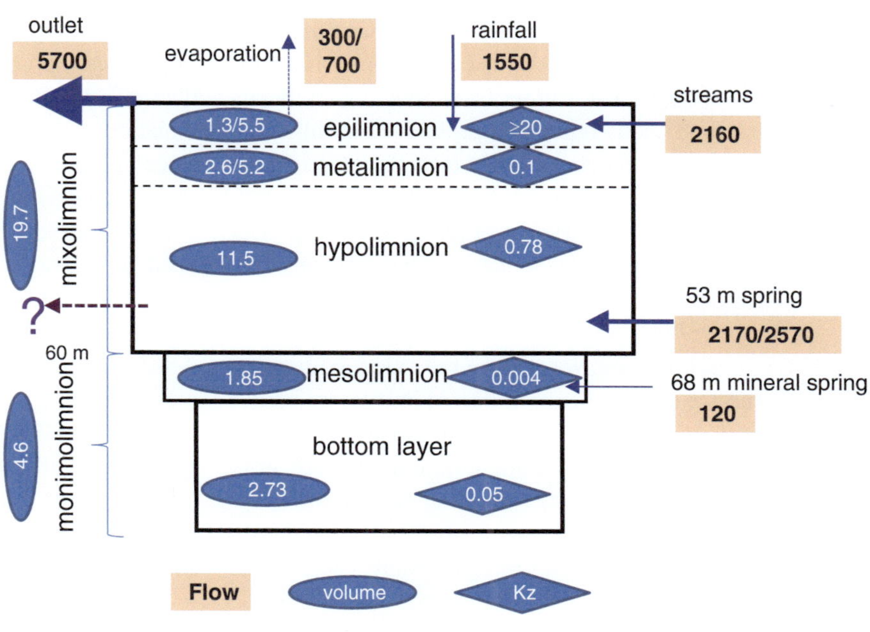

Fig. 11.1 Water balance of the lake
In *bold within rectangles*: flow ($m^3.d^{-1}$); *Within diamond*: K_z ($m^2.d^{-1}$); *Within ellipse*: volume (10^6 m^3). The *dashed arrow* with question mark may correspond to unrevealed subterranean output, for which the model is unable to give a value. The calculated inflow of the sublacustrine input at 53 m depth corresponds in fact to the difference between this inflow and the putative hidden output

- The hypolimnion between 15 and 20 m and *ca.* 60 m depth
- The mesolimnion, where is located the chemocline, between about 60 and 70 m depth
- The bottom layer between 70 m depth and the bottom.

Monimolimnion includes both mesolimnion and bottom layer.

A hydrologic budget was presented by Assayag et al. (2008), updated by constraints discussed in this paper (Fig. 11.1). New estimations lead to a water residence time of about 9 years for mixolimnion and 100 years for monimolimnion. The evaporation amount is still questionable, but as it has a low influence on the C budget, it is no further discussed in the present paper. Figure 11.1 presents also terminology and transport parameters (K_z) used in the discussion.

11.3 Methods

Temperature, dissolved oxygen, conductivity and pH were measured with a Seabird SBE 19 probe; turbidity was measured with a STBD 300 nke probe.

Water samples were obtained from a 1-l syringe actuated via a messenger. Samples were divided into different aliquots and carefully stored with classical procedures (Viollier et al. 1995):

- Major cations are measured either by AAS (before 2000) or by ICP-AES
- Anions by ionic chromatography
- Alkalinity by the spectrophotometric method of Podda and Michard (1994), or by Gran (1952) titration with hydrochloric acid, with care to avoid any oxygenation of the water.
- DIC was calculated in 1987 (Camus et al. 1993) from alkalinity and pH. pH was obtained by CEA (Commissariat à l'Energie Atomique) group using an home made pH electrode suitable for *in situ* measurement at great depth. Since that time, many other DIC profiles were either calculated with the same technique, but with commercially available pH probes (Michard et al. 1994; Viollier et al. 1995, 1997; Aeschbach-Hertig et al. 1999, 2002,…) or measured directly as DIC associated with $\delta^{13}C$ determination (Assayag et al. 2008).

For CH_4 concentration, water from the oxic mixolimnion (from the surface to 55 m) were collected into serum bottles (115 mL) and analyzed as described in Abril and Iversen (2002). In the methane rich monimolimnion (from 55 m to 90 m), 30 mL of water from the sampling syringe were rapidly transferred into sealed 115 mL serum bottles previously flushed with nitrogen gas. With this procedure, most of the depressurization occurred in the vial and the loss of CH_4 could be minimized. The methane concentrations were quantified using a gas chromatograph (GC) equipped with flame ionisation detector (FID). Sediment cores from 40 to 92 m depth were sampled by means of UWITEC corer (D9×L60 cm) in order to determine methane concentration profiles in pore waters for June 2007. For this purpose, lateral sub-sampling of the core were performed at 2 cm step, using 2.5 mL syringe with open end that was introduced in the core tube in pre-drilled holes (initially closed by tape). The small sediment cores were then transferred quickly in pre-filled serum bottles containing 10 mL of 0.5 M NaOH. Bottles were immediately capped with a septum plus aluminium capsule.

Carbone dioxide exchange rate at the air-water interface was determined in June 2010, June 2011 and December 2012 by the floating chamber method (Abril et al. 2005), measuring gaseous CO_2 in the air volume enclosed in the chamber with a LICOR LI-820 Gas Analyzer.

Sediment traps (Uwitec) have been deployed at four depths in the water column (23, 58, 70 and 88 m). At each depth were two traps each consisting of an internal tube diameter of 85.7 mm and 60 cm long completed by a removable 2 L bottle, wherein the particles accumulate. After a deployment time of about 1 month between each campaign, the particles are recovered by filtration over glass fiber filters (GF/F Whatman), dried and weighed before being analysed (C and N on analyser Thermo Scientific Flash elementary CHNS 2000, major, minor and trace elements after digestion by ICP-AES and MS). The traps of the anoxic zone were treated under nitrogen to prevent the precipitation of dissolved iron in contact with air.

Determination of the flow for both springs around the lake and the outlet was performed almost monthly by the salt (NaCl) addition and conductivity measurement method, or electrochemical gauging (May 2006–Aug. 2007). Moreover, since June 2007, the outlet was equipped with a water level piezoelectric probe (OTT Orpheus Mini) in order to monitor the outflow (after calibration of the level data *vs.* flow determination). Water level of the lake itself has been monitored since February 2009 with the same kind of probe.

$\delta^{13}C_{DIC}$ determinations were performed on gas chromatography– isotope ratio mass spectrometer; details are developed in Assayag et al. (2008). In brief, an aliquot of the water sample is injected in an Exetainer Labco tube purged with a stream of helium and charged with 1 mL of phosphoric acid. After overnight equilibration, the gas phase is sampled and introduced into a gas chromatograph coupled to a mass spectrometer (AP 2003). The calibration is made with respect to the standard of $CaCO_3$ in the PDB scale.

$\delta^{13}C_{CH4}$ and δD_{CH4}: In order to extract CH_4 dissolved in water, an aliquot of water was loaded through a septum into an evacuated volume. Water and CO_2 were then trapped at

−196 °C and non-condensable gases (CH_4, N_2, O_2, Ar) were collected with a Toepler pump. The mixture was circulated on hot CuO at 550 °C for conversion into CO_2 and H_2O. CO_2 and H_2O then were then separated from other non-condensable gas (O_2, N_2, Ar) by trapping at −196 °C. Pure CO_2, without NOx, was obtained from the trap at −140 °C. It was quantified and introduced into a Thermo Fisher Delta V dual inlet mass spectrometer to measure $\delta^{13}C$ *vs.* PDB. Precision is about ±0.5‰. Trapped water was then released and introduced into a hot uranium furnace maintained at 850 °C to convert quantitatively water-H into H_2. H_2 was then quantified. The yield of H_2 was two times that of CO_2 within less than 5 %. H_2 was collected with a Toepler pump and then measured in a Thermo Fisher Delta V dual inlet mass spectrometer to measure δD *vs.* SMOW. Precision is about ±3‰. ^{14}C analyzes were conducted on AMS (Accelerator Mass Spectrometer), ARTEMIS program, LMC14 Saclay.

11.4 Results

11.4.1 Dissolved Inorganic Carbon (DIC)

The vertical profiles (Fig. 11.2) are essentially consistent and all data is within uncertainty, which is rather low, as pH is at best obtained with an uncertainty of 0.05 unit. This implies a 10 % error on DIC calculation. For direct DIC measurements by Assayag et al. (2008), it is difficult to avoid CO_2 loss as deep waters degas spontaneously at atmospheric pressure.

DIC increases slightly but steadily in the hypolimnion, increases sharply with depth within the chemocline, and continues to increase up to the bottom. The value at 80 m depth is 14 mmol/L and the total amount in the monimolimnion is about 5.8×10^7 moles (corresponding to 2500 t of CO_2).

11.4.2 $\delta^{13}C$

$\delta^{13}C$ results on DIC were already presented by Assayag et al. (2008). The main features are the following (Fig. 11.3): $\delta^{13}C$ in epilimnion are seasonally variable and close to 0‰. It decreases steadily with depth in the hypolimnion and reaches values of about −7‰ at 58–59 m depth. It exhibits a strong increase with depth in the mesolimnion and the value at 68 m is around +1‰. The increase with depth continues in the bottom layer and the value at the bottom is between +3 and +4‰.

The value in the epilimnion is related to exchange with atmosphere, photosynthesis and inputs of streams, which have strongly negative DIC isotopic signatures. Decrease with depth in the hypolimnion corresponds to organic

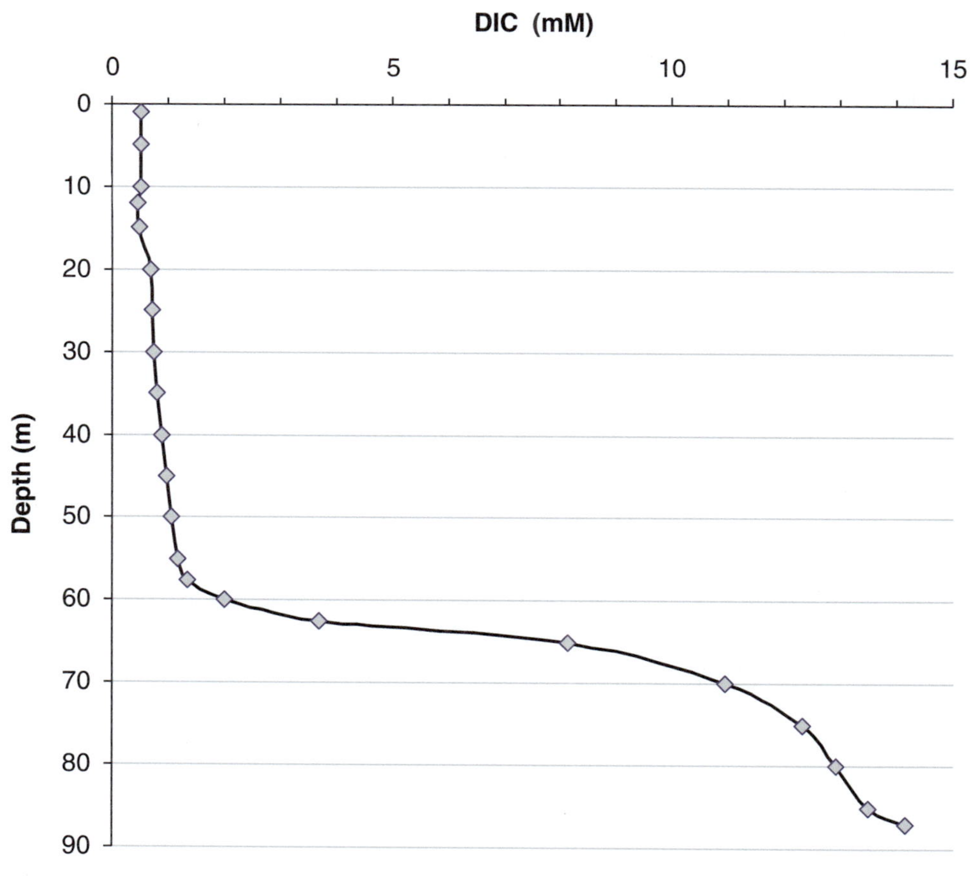

Fig. 11.2 Typical DIC profile in Lake Pavin

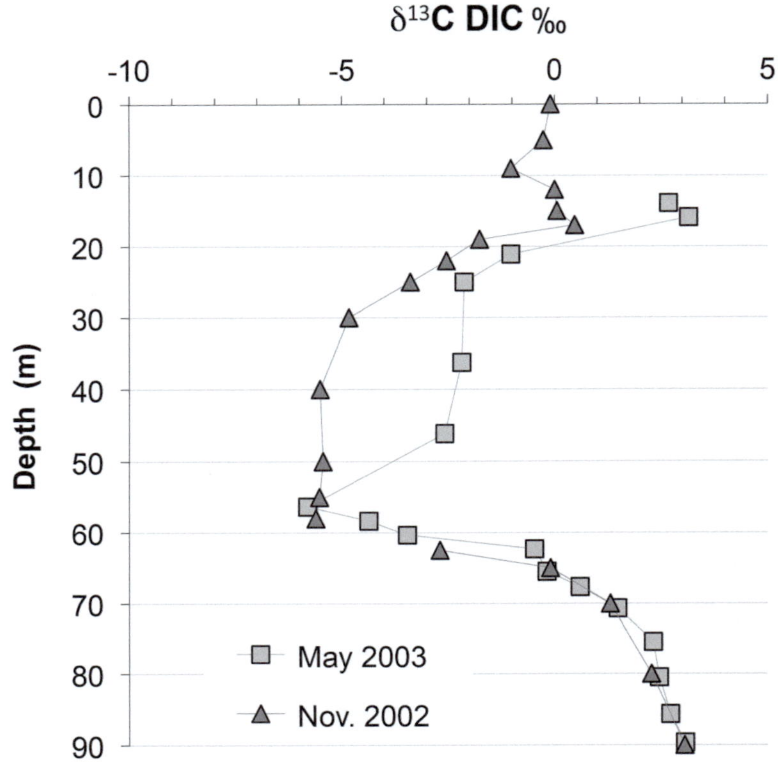

Fig. 11.3 Example of $\delta^{13}C$ of DIC profiles in Lake Pavin (Assayag et al. 2008)

matter degradation (OM with $\delta^{13}C \approx -28‰$). The increase in the monimolimnion can be associated with input of deep CO_2 and/or methanogenesis. $\delta^{13}C$ of deep C can be derived from measurements in Fontaine Goyon spring water (−4‰), which is a CO_2 – rich mineral water emerging at 1.5 km NE from the lake. Such mineral waters are very common in the French Massif Central and have $\delta^{13}C$ in the −2 to −4‰ range (Demont 1981; Matthews et al. 1987). Methanogenesis of organic matter at −28‰ yields equivalent molar amounts of methane and CO_2. $\delta^{13}C$ of methane was measured at −63‰ in 1987 (Camus et al. 1993) and in this work (about −60‰ in the monimolimnion). If we consider that acetoclastic methanogenesis is the main reaction leading to the formation of methane, we can derive from the isotopic balance that $\delta^{13}C$ of the produced CO_2 is about + 7 ‰.

From the $\delta^{13}C$ value at the bottom (+3.50‰), a simple calculation will indicate that biogenic CO_2 represents about 70 % of the total CO_2 input.

The δD of CH_4 is about −276‰ (mean value of 2 samples from 78.3 to 86.5 m depth in Oct. 2006), which excludes a hydrothermal origin for the methane (*cf. e.g.* Whiticar 1999).

11.4.3 ^{14}C of DIC

As the time scale of lake processes is less than 1 year, ^{14}C (half life ≈ 5700 years) is not suitable as a chronometer. However, it can be useful as a tracer of mixing of "volcanic" C ($^{14}C=0$) and C from superficial waters in a basaltic area ($^{14}C \approx 100$ percent of modern carbon or PMC). Actually, water from the main tributary of the lake presents a ^{14}C value of 105 PMC.

A first value in the monimolimnion in 1998 at 80 m depth was 36.0±0.8 PMC (discussed in Assayag et al. 2008). More recently, Olive and Boulègue (2004) found 31 and 33 PMC in waters from the bottom of the lake. At first glance, this suggests a proportion of volcanic C of about 65–69 %, which is in contradistinction with the result derived from the $\delta^{13}C$ value.

A recent profile in the lake (June 2007) indicates a minimum of ^{14}C at 65–68 m depth (Fig. 11.4) corresponding to an input of "old" carbon. This result suggests a direct input of volcanic carbon at intermediate and bottom depths whereas biogenic DIC would be released essentially at the bottom.

An attempt to quantify these inputs is presented in the Sect. 11.5.3.2.

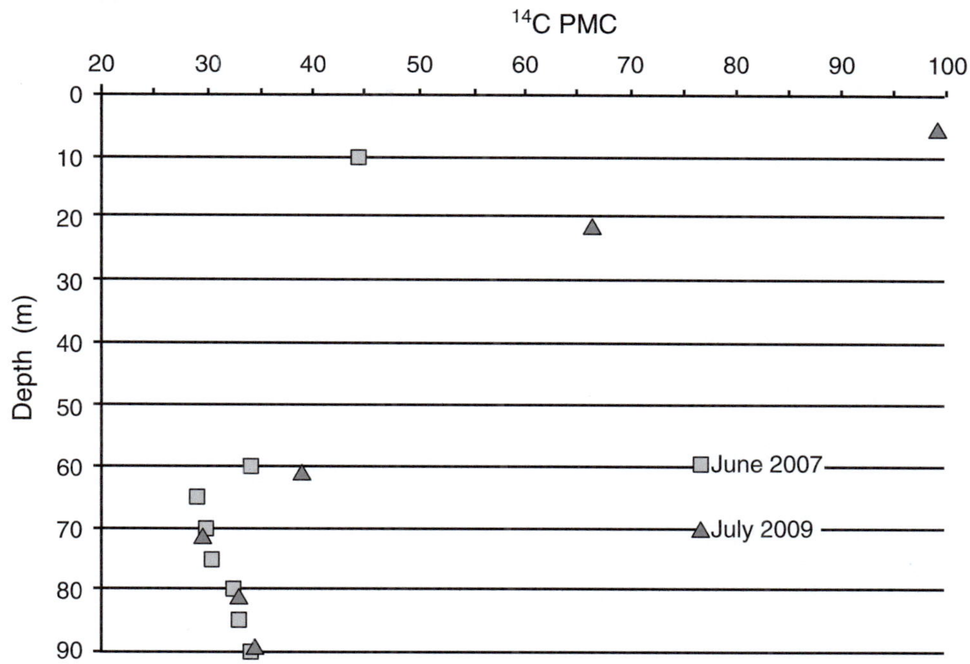

Fig. 11.4 ^{14}C of DIC values (Percentage of Modern Carbon, PMC) *vs*. depth, in June 2007 and July 2009. The minimum PMC value *ca*. 65 m depth is interpreted as a signature of a mineral water type inflow, which bring old carbon. PMC is abnormally low in the uppers layers (43.8 % at 10 m depth), reflecting exchanges between monimo- and mixolimnion. Plankton growing in surface waters is more or less ^{14}C depleted depending on the grow depth

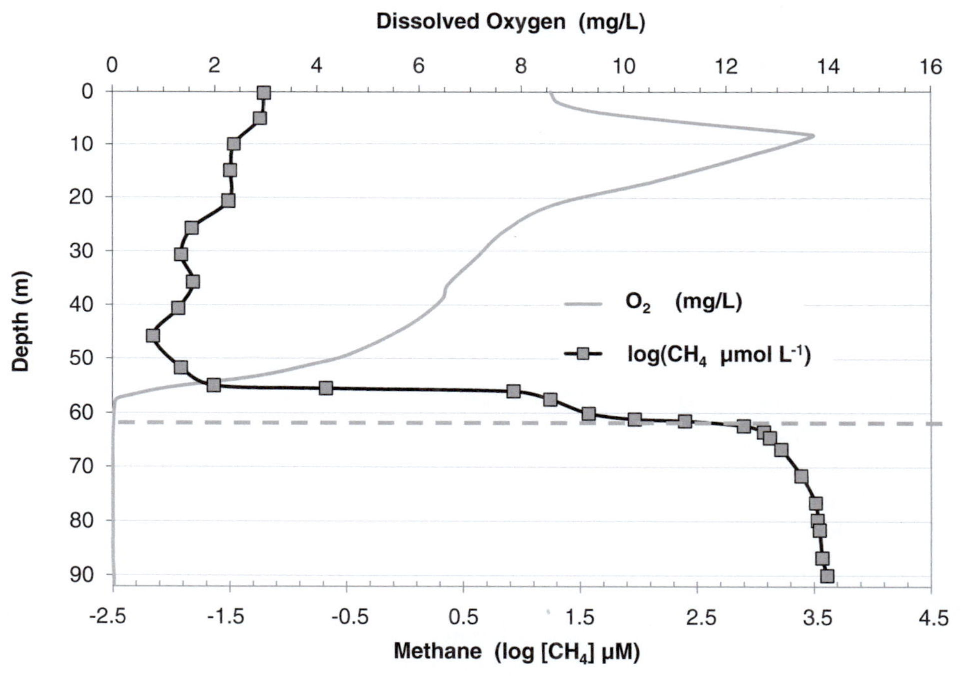

Fig. 11.5 Methane (log [CH_4] in µM) and dissolved oxygen ($mg\,L^{-1}$) profiles in Lake Pavin (June 2006). The dashed horizontal grey line corresponds to the depth of maximum of conductivity gradient (d(C25)/dz in $\mu S\,cm^{-1} m^{-1}$), *i.e.* the chemocline depth, occurring at 62 m in June 2006

11.4.4 Methane

Presence of methane in the deep waters of the lake was known for a long time, but the first precise measurements are those from Schmid (1997). In the mesolimnion, a large increase with depth is observed, from about $0\,mmol\,L^{-1}$ at 60 m to $1.5\,mmol\,L^{-1}$ at 68 m. In the bottom layer, a slower increase is observed: from less than $2\,mmol\,L^{-1}$ at 70 m to more than $4\,mmol\,L^{-1}$ at 87 m. The pattern is more scattered near the sediment interface. Thus, an amount of more than 10^7 moles of methane, *i.e.* 160 tons, is stored in the bottom of the lake. Analysis between June 2006 (Fig. 11.5) and July 2007 (not shown) agree well with the previous published data (Aeschbach-Hertig et al. 1999).

In the mixolimnion, except near the surface, the CH_4 content is lower than $5\,\mu mol\,L^{-1}$ and a minimum value close to the detection limit ($3\,nmol\,L^{-1}$) was observed at 45 m depth in June 2006.

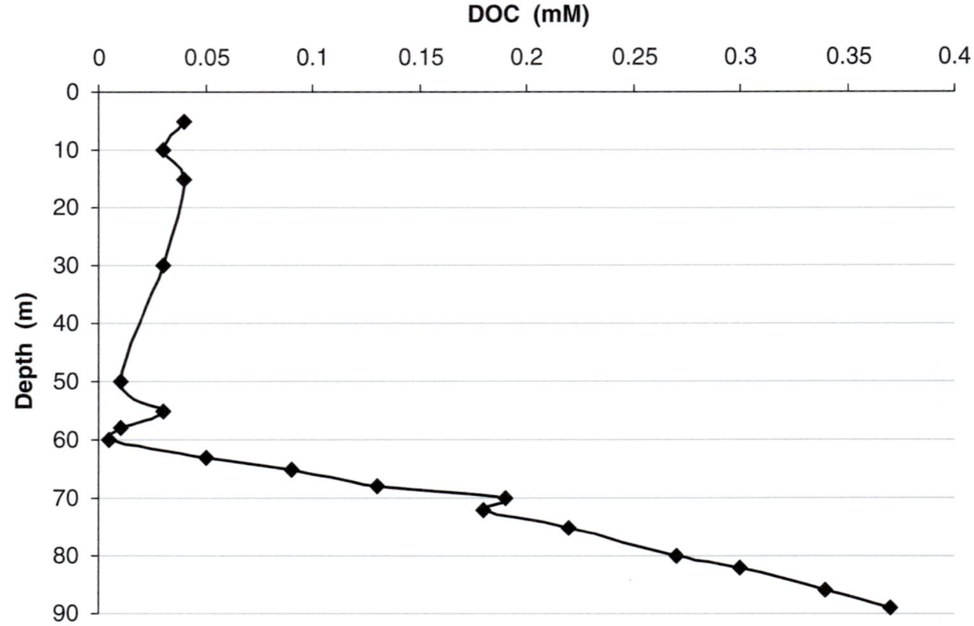

Fig. 11.6 Typical DOC profile in Lake Pavin (from Albéric et al. 2000)

11.4.5 Dissolved Organic Carbon (DOC)

In contrast with DIC and CH_4, DOC increases steadily below 65 m depth (Fig. 11.6). It has been studied in detail by Albéric et al. (2000). As its amount is rather low (<400μmol L^{-1} when DIC is about 14,000μmol L^{-1} at the same depth), it will no further be discussed in the carbon budget.

11.4.6 Particulate Organic Carbon (POC) Fluxes

In the contrary with dissolved species, POC flux is highly variable within the year. Sedimentation fluxes, as measured with sediment traps, present a maximum at the mid of spring (April-May) corresponding to the sinking of diatoms resulting from a bloom occurring some weeks before. The flux (Fig. 11.7) is about 0.11mmol.m^{-2}.day^{-1}.day^{-1} in April 1994 and only 3×10^{-4}mmol.m^{-2}.day^{-1} in September 1994 (Viollier et al. 1997). A rough annual mean value is 0.01mmol.m^{-2}.day^{-1}.

In addition, ^{14}C analysis for sediment trap particles (from July–Aug. 2006 and Nov.2006–Mar.2007 periods, at 4 depths: 23, 58, 70, 88 m) and for plankton (sampled with a 20 μm mesh net, 0–10 m depth range) gave remarkable low PMC values: 74.6±3.0PMC (mean for 6 trap samples) and 67.0±0.3PMC and 63.9±0.3PMC (plankton from April 2007 and Feb. 2008 respectively). This proves the recycling of old carbon (likely issuing from the mineral water located at 68 m depth) within the lake through photosynthesis.

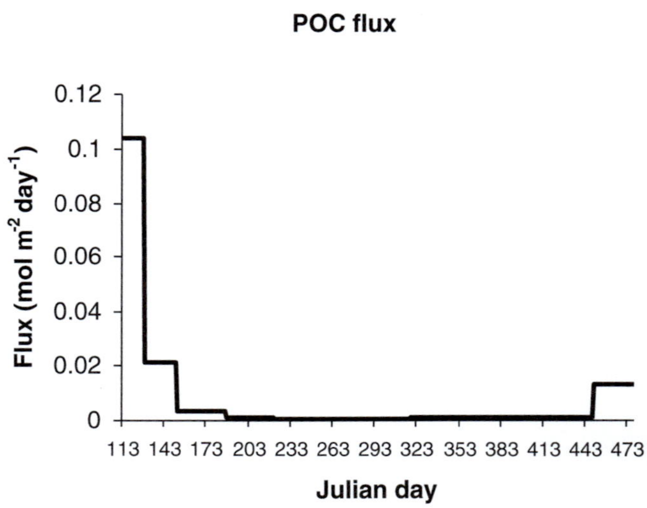

Fig. 11.7 POC flux at 85 m depth from April 20, 1994 to April 20, 1995. The period of high fluxes corresponds to the spring diatoms bloom

This variability is not in contradistinction with the steady state observed for dissolved species. For elements diffusing from the bottom of the lake resulting from POC mineralization, the very low eddy diffusion coefficient K_z (between 0.01 and 0.05m^2.day^{-1}) attenuates variations with time within a few meters.

11.4.7 Auxiliary Data

Unlike DIC and CH_4, CO_2 fluxes cannot be proposed separately in the budget for the aqueous phase, as CO_2 can be converted into HCO_3^- and CO_3^{2-}. Nevertheless CO_2 fluxes are estimated for the air-water interface, as CO_2 is the only inorganic carbon specie capable to be exchanged.

(a) In the mixolimnion, electroneutrality equation is:

$$[HCO_3^-]+2[CO_3^{2-}]=[Na^+]+[K^+]+ 2[Ca^{2+}]+2[Mg^{2+}]-[Cl^-]-2[SO_4^{2-}]$$

if only major species ($>10\mu mol/L$) are considered. This equation corresponds to the definition of alkalinity (left member) and **alkaline reserve** (right member).

$$[HCO_3^-]+2[CO_3^{2-}]=Alk$$

If DIC and Alk are known, $[CO_2]$ can be calculated by:

$$[CO_2]^2\times(1-4K')+[CO_2]\times[Alk-DIC+ 4K'(2DIC-Alk)]-K'(2DIC-Alk)^2=0$$

where K' is the apparent equilibrium constant of the reaction:

$$2HCO_3^- = CO_2 + CO_3^{2-} + H_2O$$

(b) In the monimolimnion, major species are different and the electroneutrality equation is:

$$[HCO_3^-]=[Na^+]+[K^+]+2[Ca^{2+}]+2[Mg^{2+}] -[Cl^-]+2[Fe^{2+}]+[NH_4^+]-[H_2PO_4^-]$$

In this compartment where pH ⊕6.3:

$$DIC=[H_2CO_3]+[HCO_3^-]+[CO_3^{2-}]\approx[H_2CO_3]+[HCO_3^-]$$

$$[HCO_3^-]\approx Alk$$

$$DIC >> Alk and [CO_2]=[H_2CO_3]\approx DIC-Alk$$

It is then necessary to study also the cycle of elements involved in Alk. These elements or compounds can be divided in two groups:

- Elements that are present both in the mixolimnion and the monimolimnion. They are not reactive in the lake system and inputs/outputs occurred only by brooks or springs. They will be called “conservative elements”.
- Elements or species that are present in the monimolimnion only. They can be either produced or consumed by chemical or biochemical reactions within the lake or the sediment: they will be called “reactive species”.

11.4.7.1 Conservative Elements

This group includes Na^+, K^+, Ca^{2+}, Mg^{2+} and Cl^-. All these species present the same pattern, for example for sodium (Fig. 11.8):

- Constant in the mixolimnion at about $220\mu mol\,L^{-1}$
- Large increase within the mesolimnion
- Almost constant in the bottom layer

The large increase across the chemocline is related to the low value of the eddy diffusion coefficient in this layer. The constancy in the bottom layer can be associated with an input of mineral water at *ca.* 67–70 m depth and a weak diffusion from the sediment at the bottom.

Alkali trace elements have a similar behaviour. However, the Li/Rb ratio exhibits a maximum at about 70 m depth (Fig. 11.9). It is well known (Fouillac 1983) that mineral waters in Massif Central have an unusually high Li content; the Li/Rb maximum at *ca.* 70 m depth confirms the input of a mineral water at this depth. Recently Gal et al. (2015) presented new data namely δD (H_2O) and δ^7Li that support this hypothesis.

11.4.7.2 Reactive Species

The three main species of this group are Fe^{2+}, NH_4^+ and $H_2PO_4^-$ (HPO_4^{2-} is present at concentrations lower than $30\,\mu mol\,L^{-1}$ and can be neglected). The cycle of these three species is limited to the monimolimnion and is associated with the mineralization of organic matter: N and almost P compounds result in the mineralization of organic matter, and Fe^{2+} is formed by reduction of ferric oxides by organic matter (dissimilatory reduction of iron).

Their profiles are similar:

- Essentially absent in the mixolimnion
- Large increase in the mesolimnion
- Significant increase in the bottom layer

Even if the production of these species occurs partially in the water column, the major part occurs in the sediment (or at the sediment surface) and the pattern in the monimolimnion reflects diffusion from the sediment towards the bottom waters.

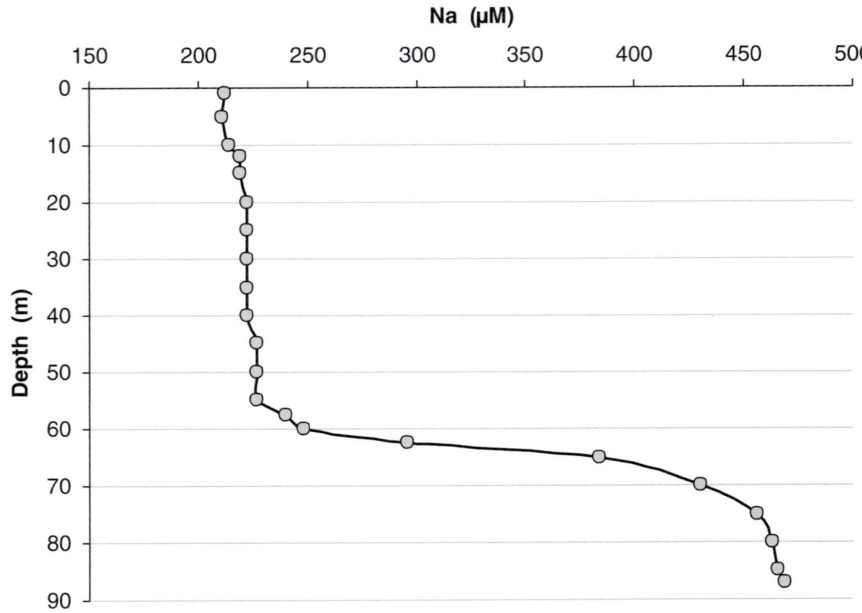

Fig. 11.8 Typical sodium profile in Lake Pavin

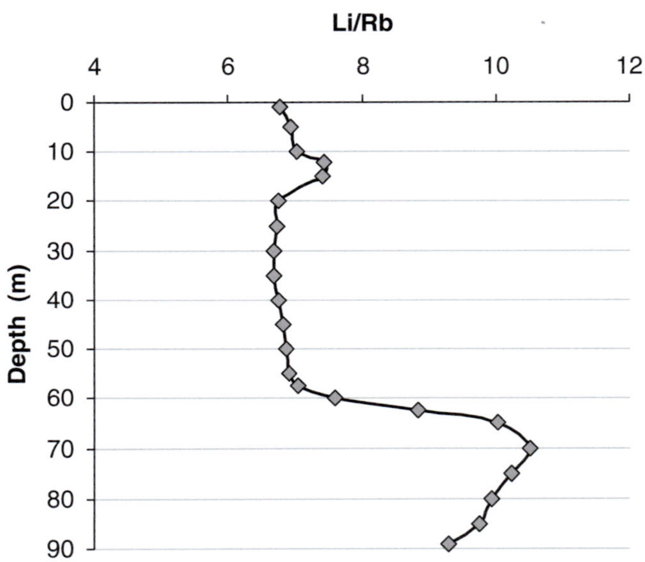

Fig. 11.9 Lithium to rubidium ratio (September 1993)

The increasing slope of the concentration *versus* depth in the mesolimnion is mainly due to the low dispersion coefficient in this layer.

At the mixolimnion-mesolimnion interface, which is also the redox interface, NH_4^+ is oxidized by O_2 into NO_3^- as suggested by the occurrence of a nitrate peak at the redoxcline (N_2 may be produced through denitrification process in the anoxic zone just below the nitrate maximum). [data not shown]

Fe^{2+} is oxidized into $Fe(OH)_3$, likely through different processes depending on the efficiency of the mixing in the mixolimnion. When O_2 is supplied in depth by a strong mixing event (*i.e.* years with favourable conditions of mixing, generally a poor ice cover during winter), dissolved oxygen reacts with ferrous iron and generates a strong turbidity peak. In most common conditions of low O_2 supply in depth, both MnO_2 and NO_3^- may oxidize Fe^{2+} in anaerobic conditions. Phosphate ions are adsorbed on authigenic $Fe(OH)_3$ (*cf.* chapter of this book about iron cycle). High efficiency phosphorus trapping with Fe(III) particles can explain the very high P content in the monimolimnion of the lake (up to 350 µM; Michard et al. 1994).

11.5 Discussion

11.5.1 Gas Saturation in the Monimolimnion

CO_2 concentrations are three times greater than CH_4 concentrations. But, as the solubility of CH_4 is much smaller than the solubility of CO_2, the partial pressure of methane is far greater and reaches about 2 bars at 88 m depth whereas the CO_2 partial pressure is less than 0.3 bar.

Another gas presents a high partial pressure close to 1 bar: nitrogen (N_2) at equilibrium with atmosphere at the surface and produced at the redoxcline by denitrification process (Anammox reaction may occur but is not considered as a significant process due to very low nitrite concentration at the redoxcline).

Finally, as total pressure of dissolved gases (Fig. 11.10) is significantly lower than the hydrostatic pressure at the same depth, in the present status, degassing by formation of gas bubbles cannot occur spontaneously in the monimolimnion. However, total pressure in the bottom waters is definitely

Fig. 11.10 Gas pressures compared to hydrostatic and atmospheric pressures

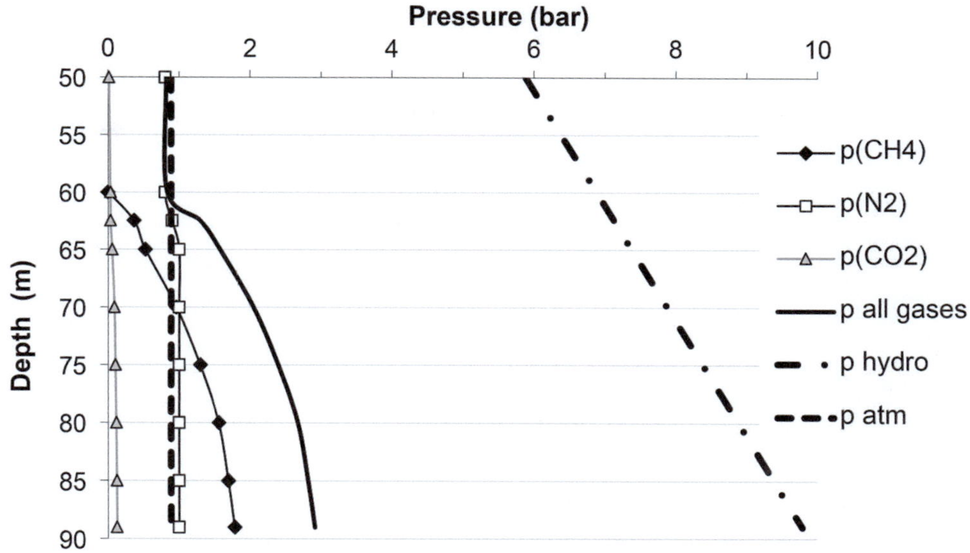

greater than 1 bar and these waters degas spontaneously with bubbles formation when they are brought to atmosphere (0.88 bar at the elevation of the lake surface), CH_4 being mainly responsible of this sparkling effect.

11.5.2 Gas Exchange with Atmosphere

Both CO_2 and CH_4 present in high concentrations in the monimolimnion are greenhouse gases and it will be interesting to check if these gases are emitted to atmosphere.

11.5.2.1 Methane

Figure 11.5 shows that methane concentrations present a minimum close to 10 nM at 45–50 m depth. Thus, the upward advective-diffusive flow of CH_4 vanishes at this depth. We shall see in the following Sect. 11.5.3.3 that CH_4 is almost completely oxidized into CO_2 at this depth.

The oversaturation of methane in the surface waters of the lake may be related to a production in the sediment in shallow areas (Bastviken et al. 2004, 2008). However, in Pavin Lake, due to the shape of the basin, epilimnetic sediments represent a weak part of the total sediment area. Alternatively, another explanation for this oversaturation, also encountered in ocean surface water and known as the methane paradox, involves CH_4 production in oxygenated superficial water (Carini et al. 2014). In this article, methane production may proceed from bacterial degradation of methylphosphonic acid, which is synthetized by marine archaea. Oxic methanogenesis is also reported in lakes such as Lake Stechlin (Germany), where the CH_4 production is performed through an interspecific transfer of H_2 and/or acetate from photoautotrophs such as cyanobacteria to attached archaea (Grossart et al. 2011). Such syntrophic associations between archaea and bacteria may be responsible of an important methane emission from aquatic systems (Bogard et al. 2014).

In June 2010 a weak emission (about 58 micromoles. $m^{-2}.d^{-1}$ on average, *i.e.* 0.026 $kmol.d^{-1}$ for the whole surface) was observed at the lake surface by direct measurements using a floating chamber and laser sensor (Guimbaud et al. 2011). Alternatively, a mean CH_4 escape ~ 10^{-3} $kmol.d^{-1}$ can be derived from the model of the lake presented by Lopes et al. (2011). This value is much lower than those obtained in shallow lakes (Bastviken et al. 2008) for which values range from 200 to 1000 micromoles. $m^{-2}.d^{-1}$, but is very similar to those from others meromictic lakes (Borges et al. 2011). Despite their usually high methane content in the bottom layer, meromictic lakes are then weak CH_4 emitters compared to classical lakes. This may be related to the permanently stratified water column that limits the diffusion (through very low eddy diffusion coefficients in the chemocline, close to molecular diffusion) and well established bacteria consortia in the redoxcline that oxidize almost all the methane diffusing from the deep compartment.

11.5.2.2 Carbon Dioxide

To quantify the gas exchange at the air water interface, it was necessary to know (i) the partial pressure of the gas in the surface water which depends on the photosynthetic activity, (ii) the kinetic constant of the gas exchange with depends on the wind velocity. A quantitative treatment of this question needs a continuous determination of these two parameters which is out the scope of the present paper and necessitates the development of specific sensors (Lefèvre et al. 1993; Prévot et al. 1998). Only some indicative results are given here.

$p(CO_2)$ was determined by calculations from Alk and pH measurements. pH measurements in natural, low ionic strength waters have a limited precision of about ±0.1 unit and accuracy on $p(CO_2)$ is then about 25 %. Nevertheless, as

Table 11.1 Fluxes of Alk, DIC, CH_4 and C burial in the global carbon budget of the Lake Pavin

Depth (m)	Process	DIC fluxes ($kmol\,d^{-1}$)	Alkalinity fluxes ($kmol\,d^{-1}$)
0 m	Output (surface outlet)	-3.0 ± 0.3	-3.0 ± 0.3
0 m	Exchange with atmosphere	-4 ± 1	0
0 m	Input by rain	0.10 ± 0.04	0.020 ± 0.004
8–10 m	Input by brooks	1.3 ± 0.2	1.3 ± 0.2
10–15 m	Photosynthesis	-3.5 ± 0.5	0.15 ± 0.1
20–50 m	Aerobic respiration	0.3 ± 0.2	-0.02
53 m	Fresh water input	1.6 ± 0.2	1.4 ± 0.2
58–60 m	Fe(II), NH_4^+ and CH_4 oxidation	1.0 ± 0.2	-0.3 ± 0.2
68 m	Mineral water input	2.6 ± 0.3	0.6 ± 0.2
75–92 m	DIC flux from the sediment	1.5 ± 0.2	0.5 ± 0.1
	CH_4 flux from the sediment	1.0 ± 0.2	
>92m	C burial in the sediment	-0.7 ± 0.2	-

Negative values correspond either to a subtraction of C species from the dissolved phase or to a carbon escape from the lake (water column)

surface waters are either very under saturated or highly over saturated, the results presented on Table 11.1 are significant.

In addition, exchange of CO_2 was directly measured in June 2010, June 2011 and December 2012 by the floating chamber method. In June 2010, flux mean values were $-13.0 \pm 1.6\,mmol\,m^{-2}d^{-1}$ in the central zone, and about $-8\,mmol\,m^{-2}d^{-1}$ near the shore (negative values are for a CO_2 uptake by superficial waters). In June 2011 fluxes were almost zero but in December 2012 the mean value in the central zone was $76.8 \pm 2.5\,mmol\,m^{-2}d^{-1}$. These values can be compared to calculated values from pH and alkalinity measurements in the surface water (*in situ* measurements for pH, Gran titration for alk after sampling) with the following relation:

$$F_{CO2} = k \times (CO_{2\_solubility} - CO_{2\_Z=0})$$

where F_{CO2} is the carbon dioxide flux ($mol\,m^{-2}d^{-1}$), k (md^{-1}) is the gas exchange velocity. $CO_2\_Z = 0$ and $CO_2\_solubility$ are given in $mol\,m^{-3}$. $CO_{2\_Z=0}$ is calculated from alkalinity (444–460 μM), the water pH (8.05–8.15 range) and the K_{a1} and K_{a2} dissociation constants at 14 – 15 °C (Alk, pH and T data from June 2010).

Calculated fluxes are in good agreement with measured ones (June 2010) if $k = 1\,m\,d^{-1}$ for the stations near the shore, and $k = 1.7\,m\,d^{-1}$ for the stations in the central zone. These k values correspond to classical values for lakes (Cole and Caraco 1998).

The surface waters are strongly under saturated with respect to atmospheric CO_2 at least from mid May to the end of September, due to the photosynthesis uptake. On the contrary, they are strongly over saturated after the mixolimnion overturn, which occurs partially in November to December, and more after the end of ice melting. With only three measured flux values the annual budget is impossible to calculate. However from the model of the lake presented by Lopes et al. (2011), a value of $4 \pm 1\,kmol.d^{-1}$ can be derived, corresponding to the main carbon output from the lake. Similar values were obtained in Lake Kivu (Borges et al. 2014).

11.5.3 Tentative Quantitative C Cycle

The quantitative carbon behaviour in the lake will be described from the bottom to the surface.

11.5.3.1 The Bottom Layer (70–92 m Depth)

In this layer DIC, CH_4 and Alk present concentration gradients of respectively 0.15, 0.10 and $0.05\,mol\,m^{-4}$. The vertical dispersion coefficient (Kz) was estimated to about $0.05\,m^2 day^{-1}$ (Aeschbach-Hertig et al. 1999). Thus the flux of these species can be estimated to 7.5, 5 and $1\,mmol\,m^{-2}day^{-1}$ respectively, *i.e.* 1500, 1000 and $\approx 500\,mol\,day^{-1}$ for the whole monimolimnion.

These values are in rough agreement with the POC total input to the bottom of the lake ($\approx 3200\,mol.d^{-1}$ for the whole monimolimnion). Only a small part of POC is buried. Uncertainties on both eddy diffusion coefficient and sediment traps efficiency may explain the difference.

11.5.3.2 Input of Mineral Water at 65–70 m Depth

The minimum of PMC and the maximum of the Li/Rb ratio at 65–70 m depth suggest the inflow of a ^{14}C depleted and lithium rich mineral water. A simple model considering a

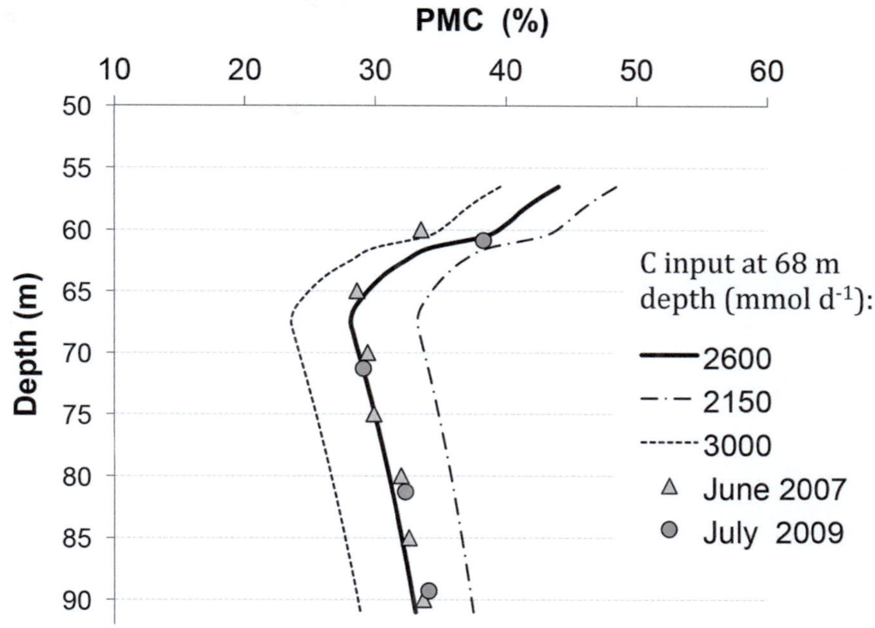

Fig. 11.11 Model and data for ^{14}C measurements in the monimolimnion The different curves correspond to different C inputs in $mmol\,d^{-1}$ in order to check the sensibility of the model for this parameter

flux of DIC with a given PMC and an input of mineral water at $68 \pm 2\,m$ depth was built using the AQUASIM code (Reichert 1994). The flow of the mineral water ($138\,m^3d^{-1}$) has been estimated by Assayag et al. (2008) from ^{18}O data, revisited here at $120\,m^3d^{-1}$. Observed results are consistent with a flux of $7.5\,mmol\,m^{-2}d^{-1}$ of DIC (as mentioned above Sect. 11.5.3.1) with a settling POC with 75 PMC at the bottom sediment-water interface (as mentioned above Sect. 11.5.3.1) and an input of $2600\,mol\,day^{-1}$ of 0 PMC DIC by the mineral water (Fig. 11.2). Therefore, the ratio of biogenic DIC to "volcanic" carbon is $1500/2600 (= 0.6)$. The disagreement between values previously obtained either by $^{13}C(\approx 2)$ and $^{14}C(\approx 0.5)$ is related: (i) to the different depths of biogenic C inputs (OM deposition onto the bottom and subsequent remineralization) and volcanic C (*ca.* 68 m depth); (ii) to the recycling of dead ^{14}C in the lake, leading to a relative low PMC for the organic matter produced into the lake. Such a large ^{14}C impoverishment of the lake autochthonous organic production compare to aerial allochthonous sources (tree leaves) may be inferred from sedimentary archives since at least 1000 years ago (Albéric et al. 2013).

The same model allows the determination of conservative species such as alkalinity to $500\,eq\,d^{-1}$. Estimation of reactive species is more difficult. However, CO_2 rich mineral waters present low concentrations of both NH_4^+ and soluble phosphates (relatively to monimolimnion concentrations). A tentative value for Fe^{2+} derived from the model is about $200\,eq\,day^{-1}$.

The model shows also a significant difference of $\delta^{13}C$ between 88 and 68 m depth (about 1‰) in agreement with the experimental data (Fig. 11.11).

11.5.3.3 The Chemocline and the Redox Boundary

The large gradients observed in this layer correspond to the low dispersion coefficient estimated at $0.004\,m^2d^{-1}$.

Redox boundary occurs at mixolimnion-mesolimnion interface. Fe(II), NH_4^+ and CH_4 are completely oxidized in this zone. The almost total methane oxidization is confirmed by the drastic decrease of $\delta^{13}C$ of DIC at about 60 m depth with respect to values at higher depths. However, $\delta^{13}C_{DIC}$ profile is very sensitive to the relative depths of the oxidation zone and of the top of the low dispersion layer. The best results are obtained with a complete oxidization of CH_4 between 59.5 and 61.5 m depth and a top of the low dispersion coefficient zone at 60 m in agreement with Lopes et al. results (2011) (Fig. 11.12).

11.5.3.4 Fresh Water Input at *ca.* 53 m Depth

Sublacustrine inputs were inferred in ancient papers (*e.g.* Perreau 1948), but the first attempt to calculate an hydrological budget was proposed by Meybeck et al. (1975), based on tritium data, rainfall budget on the catchment area and only four outflow values (pygmy current meter method). Due to incoherence between ^3H model (which gave $25\,L\,s^{-1}$) and observed outflow data (from 65 to 360 $L\,s^{-1}$), the flow of sublacustrine input was not clarified. They just inferred, from chemical composition of the monimolimnion, that missing input might correspond to mineral water feeding the bottom water. Martin (1985) proposed a box model of the lake from tritium data. The apparent water deficit was estimated to $40\,L\,s^{-1}$, corresponding to a mineral water inflow located in the monimolimnion. Camus et al. (1993) barely revisited this box-model and deduced a sublacustrine input of about $35\,L\,s^{-1}$ also located in

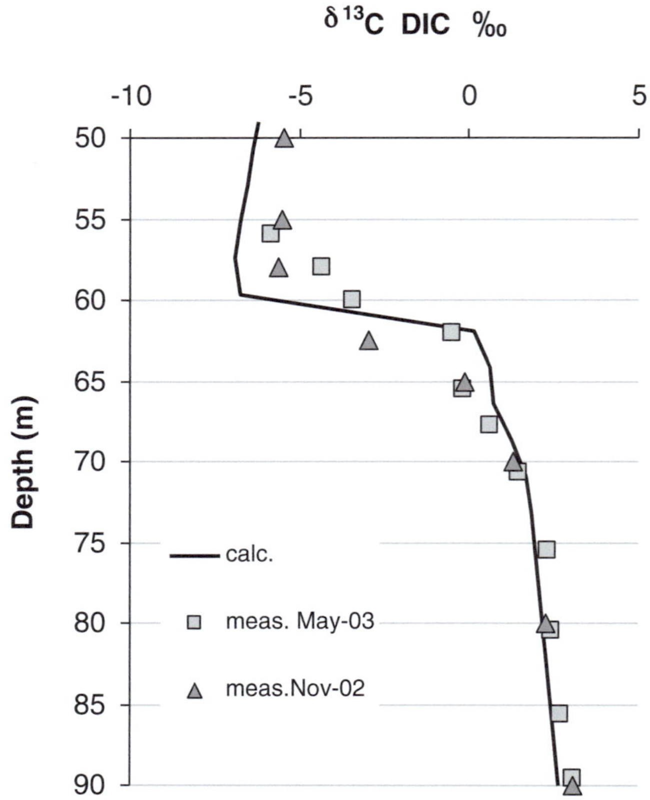

Fig. 11.12 Observed and modeled profiles of $\delta^{13}C$ of DIC

the monimolimnion. This hypothesis has been discarded in further works, as this deep input should imply a rather strong advection of water from the monimolimnion toward the mixolimnion, inducing for example a strong iron recycling at the redox interface (*i.e.* a strong iron rich particles sedimentation rate). Based on sediment trap data and concentration gradients, Viollier et al. (1997) located the sublacustrine inflow within the mixolimnion, and deduced a fresh water input of $22\,L\,s^{-1}$. Aeschbach-Hertig et al. (2002) concluded the same thanks to high-resolution CDT profiles and geochemical tracers profiles analysis and modeling (He, ^{3}H, CFCs), but with a lower inflow of $10\,L\,s^{-1}$. From temperature profiles, these last authors inferred an intrusion of cold water at about 45 m depth in the northeast part of the lake (Fig. 11.13).

Works from recent research programs have given new constrains on the hydrological budget: (i) almost monthly flow determination of up to 12 streams around the lake from May 2006 to August 2007, (ii) continuous measurement of the outlet flow (June 2007–Dec. 2010), (iii) monitoring of the lake level (Feb. 2009–Dec. 2010); (iv) high resolution CTD-O_2 profiles and (v) temperature sensors line (precision 0.001 °C; July 2006–Dec. 2006 and April 2007–Dec. 2007).

The lake level is stable within 30 cm of amplitude (level OTT probe, Feb. 2009–Dec. 2010).

Direct evidences of the sublacustrine input are based on temperature measurements, either from CDT profiles or from the

Fig. 11.13 $\delta^{13}C$ *vs.* 1/DIC in the hypolimnion

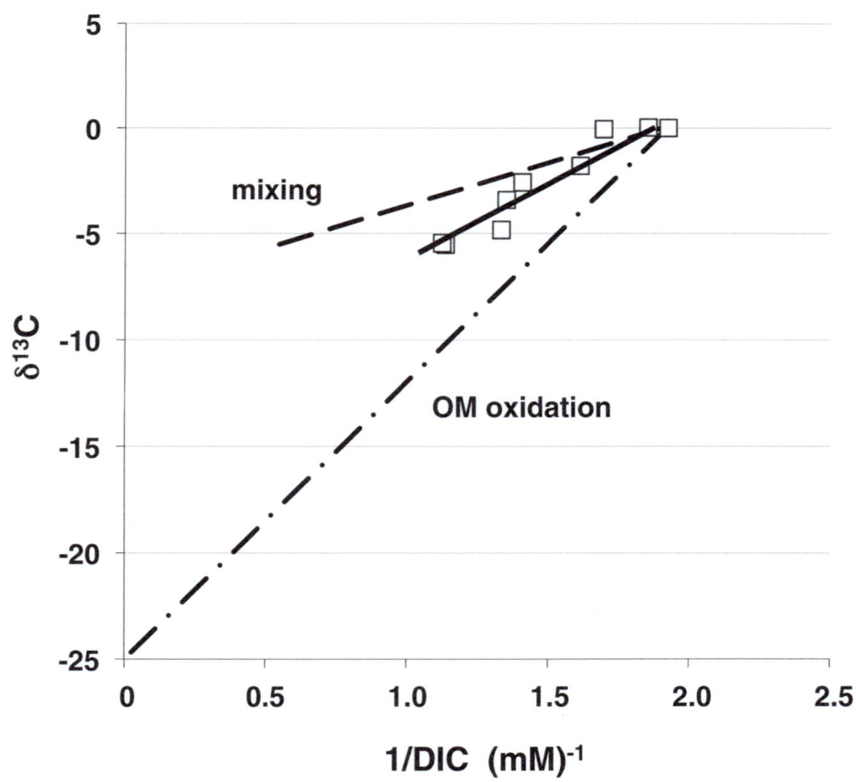

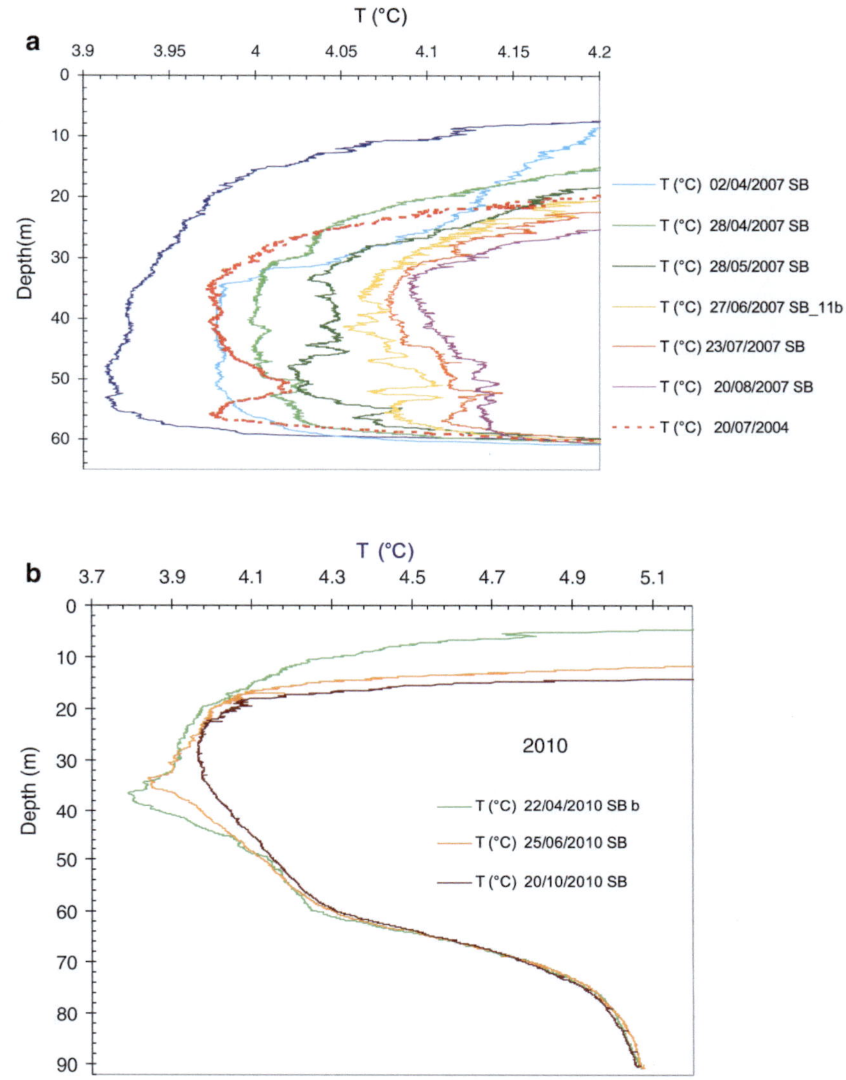

Fig. 11.14 (**a**) and (**b**): Temperature profiles in 2007 (0–65 m depth) and 2010

static thermometers line (Bonhomme et al. 2011). Temperature profiles obtained with the Seabird profiler (Fig. 11.14) revealed some anomalies in the hypolimnion but at various depths according to the period: they were located more often in the 50–55 m depth zone (July 2004, April-July 2007) but also *ca.* 34–39 m in April 2010 or 33–36 m in June 2010. For these last periods, T anomalies where well marked and reached up to −0.2 °C (April 2010). A cold intrusion where also deduced from thermometers line data at *ca.* 53 m depth in 2007 (Bonhomme et al. 2011). On the other hand, any T anomalies where observed at other periods (*e.g.* Oct. 2010 – Fig. 11.14b -, almost none in 2006 and 2009), suggesting that this inflow may be intermittent. Concerning the depth of these anomalies, it is important to remember that measurements where usually performed in the central zone of the lake, and that observed depth should be considered as the resulting effect of the isodensity and plume propagation rather than the real depth of the input, which is still to precise both in depth and direction.

Any anomalies were visible from conductivity measurements at the resolution of the probe ($2.5\,\mu S\,cm^{-1}$). This could be explained as this water, supposed chemically very near from surface streams, represents an important supply for the lake: waters from water column and this inflow are then very difficult to distinguish with this parameter.

In the absence of any identified geochemical tracer for the sublacustrine input in the mixolimnion, one can only deduce the corresponding inflow from other inputs/outputs values, leading to a less accurate determination. Moreover, the calculated value may correspond to the difference between the real input and a hypothetical sublacustrine output. Water budget is equilibrated as following:

$$Q_{53} = Q_{output} + Q_{evap.} - Q_{brooks} - Q_{mineral} - Q_{rain}$$

Q_{output} is estimated to about $67\,L\,s^{-1}$ from the monitoring of the outlet;

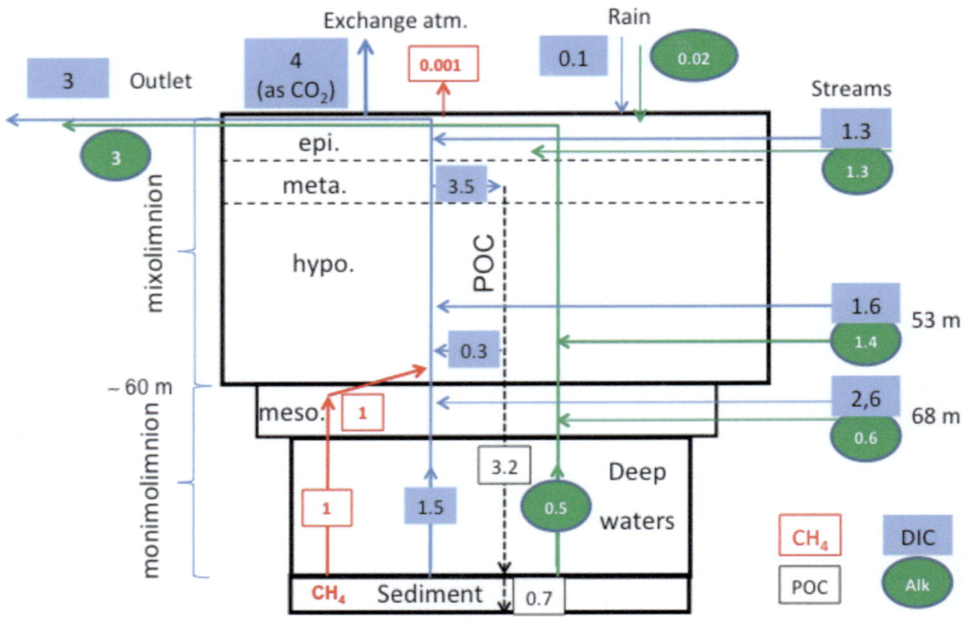

Fig. 11.15 Tentative Carbon cycle. Fluxes are in kmol d^{-1} (see Table 11.1 for associated processes and uncertainties). *Blue*: DIC; *red*: CH_4; *black*: POC; *green*: Alk

Q_{evap}. is estimated to 8.7 L s^{-1} from $\delta^{18}O$ (Assayag et al. 2008)

Q_{brooks} determination was about 25 L s^{-1}

$Q_{mineral}$ is estimated to 1.4 L s^{-1} from $\delta^{18}O$ data (revisited from Assayag et al. 2008)

Q_{rain} is estimated to 18 L s^{-1} from mean meteorological data and lake surface

As the lake level is stable, these data lead to $Q_{53} = 31.4 \pm 5.78$ L s^{-1} (2700 ± 500 m^3d^{-1}), close to the previous estimation from Camus et al. (1993).

11.5.3.5 Mineralization of POC in the Hypolimnion

In opposition to deeper layers, the layers above 50 m depth present element concentrations that are variable with the seasons and a steady state cannot be assumed. However, chemical composition is about the same at a given period of the year and an average budget can be estimated.

Between 20 and 50 m depth, DIC increases and [O_2] decreases; this partly corresponds to oxidation of POC produced by photosynthesis in superficial layers.

On the contrary of O_2 that varies with the season, the profile of DIC is almost constant during the year. Actually, DIC increase is related with POC oxidation and with an upward flux.

In a $\delta^{13}C_{DIC}$ *vs.* (1/DIC) plot (Fig. 11.13), diffusive mixing and oxidation reaction are shown as straight lines. Actual data points are between these 2 lines. Owing to the estimated flux at 55 m, we can estimate the production of DIC from POC at 200 – 500 mol d^{-1}.

11.5.3.6 The Photic Layer

The photic layer in Lake Pavin extends from surface to about 20 m depth (Secchi values *ca.* 4–10 m depth), and sharp peaks of [O_2] are observed near the thermocline. The diffusion of produced oxygen is limited by the low K_z value around the thermocline, *i.e.* near 10 m depth. Thus photosynthesis can occur in layers rather isolated from the surface where PMC can be as low as 45 %.

The amount of produced POC is rather difficult to derive from our data. From the amount of POC mineralized in the deeper layers, it cannot be less than about 3 kmol d^{-1}.

11.5.3.7 Input and Output of Rivers at the Surface

Small brooks are the tributaries of the lake. Their discharges have been measured all along the year and an average total discharge can be estimated as 2160 m^3d^{-1} (May 2006–Aug. 2007). The mean concentration of DIC and alkalinity is 0.58 mol m^{-3}. The net input of both DIC and Alk is therefore 1300 mol d^{-1}.

From a metrological station installed on the lake, the rainwater input is 1550 m^3d^{-1}. The mean concentration of DIC and alkalinity are respectively *ca.* 0.05 and 0.012 mol m^{-3}. The net input of both DIC and Alk are respectively ~ 100 and 20 mol d^{-1}.

The superficial lake outlet carries away *ca.* 6000 m^3d^{-1} of water, *i.e.* 3000 mol d^{-1} of DIC and alkalinity. The main species of DIC is HCO_3^-.

The budget of DIC and Alk is summarized in Fig. 11.15.

11.6 Conclusions

The monimolimnion of the lake Pavin contains high concentrations of greenhouse carbon gases (CO_2 and CH_4), up to respectively 14 and 4 mmol L^{-1}. The total pressure of gases is about 3 bars at 80 m depth. This is significantly lower than the hydrostatic pressure and for the present time spontaneous gas outburst cannot occur in the water column.

It is difficult to estimate the gas exchange at the air-water interface. It needs a continuous monitoring which is not available in Lake Pavin up to now. Nevertheless, CO_2 and CH_4 fluxes were estimated by modeling the whole water column functioning, as presented in Lopes et al. (2011). A mean value of $9\,mmol.m^{-2}.d^{-1}$ and $\sim 2\times10^{-3}\,mmol.m^{-2}.d^{-1}$ escaping from the lake were derived from the model for CO_2 and CH_4 respectively. The average CO_2 efflux is then lower than the global average of $16\,mmol.m^{-2}.d^{-1}$ for freshwaters lakes (Cole et al. 1994). Efflux of CH_4 is very low in Lake Pavin, regarding the high methane concentration in the monimolimnion, but comparable to other meromictic lakes (Borges et al. 2011). This fact is explained by an efficient bacterial oxidation at the permanent redoxcline in the water column that virtually remove all the dissolved CH_4 (Lopes et al. 2011).

DIC present in the Lake Pavin has two main origins:

- The DIC resulting of the biochemical processes (OM mineralization) occurring in the water column and in the sediment.
- The DIC issuing from volcanic influence, brought to the lake by a mineral water input

A mineral water inflow has been located at $68\pm2\,m$ depth (expressed as the depth of influence on water column at the center of the lake) and the amount of deep originated C is about 1.7 time the amount of biogenic carbon mineralized as DIC in the monimolimnion.

Lake Pavin is geochemically studied for about 40 years. No significant change in its chemical composition has been detected. This confirms the assumption a steady state of the monimolimnion where the estimated residence time of dissolved carbon species is 40 years. The risk of a spontaneous gas outburst in the monimolimnion will be thus considered as negligible for some decades, except if an important landslide occurs on the lake shore (Chapron et al. 2010) or if a large amount of gas rich sediment lying on steep slope within the lake is destabilised. The contribution of gas rich sediment should be precised in the future as a possible natural hazard.

Acknowledgements A part of this work was supported by the Metanox program (ANR 2006–2008, PI. G. Fonty) and EC2CO programs (Methanolac 2008–2009 & Interlac 2009–2010, PI. G. Fonty). Authors thank the Artemis program and facilities for ^{14}C determinations. A big thank you also to Mr and Ms Joubert and their team for their hospitality at Lake Pavin, as well as to Mairie de Besse for their help. Finally, the authors thank Alberto Borges for his comments that helped to improve the manuscript.

References

Abril G, Iversen N (2002) Methane dynamics in a shallow, non-tidal, estuary (Randers Fjord, Denmark). Mar Ecol Prog Ser 230:171–181

Abril G, Guérin F, Richard S, Delmas R, Galy-Lacaux C, Gosse P, Tremblay A, Varfalvy L, Dos Santos MA, Matvienko B (2005) Carbon dioxide and methane emissions and the carbon budget of a 10-year old tropical reservoir (Petit Saut, French Guiana). Global Biogeochem Cycles 19(GB4007). doi:10.1029/2005GB002457

Aeschbach-Hertig W, Hofer M, Kipfer R, Imboden DM, Wieler R (1999) Accumulation of mantle gases in a permanently stratified volcanic lake (Lac Pavin, France). Geochim Cosmochim Acta 63:3357–3372

Aeschbach-Hertig W, Hofer M, Schmid M, Kipfer R, Imboden DM (2002) The physical structure and dynamics of a deep, meromictic crater lake (Lac Pavin, France). Hydrobiologia 487:111–136

Albéric P, Viollier E, Jezequel D, Grosbois C, Michard G (2000) Interactions between trace elements and dissolved organic matter in the stagnant anoxic deep layer of a meromictic lake. Limnol Oceanogr 45:1088–1096

Albéric P, Jézéquel D, Bergonzini L, Chapron E, Viollier E, Massault M, Michard G (2013) Carbon cycling and organic radiocarbon reservoir effect in a meromictic crater lake (Lac Pavin, Puy-de-Dôme, France). Radiocarbon 55(3–4):1–14

Alvinerie J, Degot B, Leveque P, Vigneaux M (1966) Activité en tritium et caractéristiques chimiques du Lac Pavin. C R Acad Sci Paris 262(1220):846–849

Amblard C, Restituito F (1983) Observations complémentaires en faveur de l'origine biogénique de la couche à hydrogène sulfuré d'un Lac de moyenne montagne (Lac Pavin, France). C R Acad Sci Paris 296:1787–1790

Assayag N, Jézéquel D, Ader M, Viollier E, Michard G, Prévot F, Agrinier P (2008) Hydrological budget, carbon sources and biogeochemical processes in Lac Pavin (France): constraints from $\delta^{18}O$ of water and $\delta^{13}C$ of dissolved inorganic carbon. Appl Geochem 23:2800–2816

Bastviken D, Cole JJ, Pace ML, Tranvik L (2004) Methane emissions from lakes: dependence of lake characteristics, two regional assessments, and a global estimate. Glob Biogeochem Cycles 18(GB4009):12. doi:10.1029/2004GB002238

Bastviken D, Cole JJ, Pace ML, Van de Bogert MC (2008) Fates of methane from different lake habitats: connecting whole-lake budgets and CH_4 emissions. J Geophys Res 113(G02024):1–13

Bogard MJ, del Giorgio PA, Boutet L, Garcia Chaves MC, Merante A, Prairie YT, Derry AM (2014) Oxic water column methanogenesis as a major component of aquatic CH4 fluxes. Nat Commun 5:5350

Boivin P, Besson J-C, Briot D, Gourgaud A, Labazuy P, Langlois E, de Larouzière FD, Livet M, Mergoil J, Miallier D, Morel J-M, Vernet G, Vincent P (2009) Volcanologie de la chaîne des Puys – Volcanology of the Chaîne des Puys. Parc Naturel Régional des Volcans d'Auvergne (Ed.)

Bonhomme C (2008) Turbulence et ondes en milieu naturel stratifié. Deux études de cas: – étude du mélange turbulent et des ondes internes du lac Pavin (Auvergne, France); – influence des ondes de Rossby sur la concentration en chlorophylle de surface dans l'upwelling du Pérou. Thesis. Ecole Nationale des Ponts et Chaussées

Bonhomme C, Poulin M, Vinçon-Leite B, Saad M, Groleau A, Jézéquel D, Tassin B (2011) Maintaining meromixis in Lake Pavin (Auvergne, France): the key role of a sublacustrine spring. Compt Rendus Geosci 343(11):749–759

Borges AV, Abril G, Delille B, Descy J-P, Darchambeau F (2011) Diffusive methane emissions to the atmosphere from Lake Kivu (Eastern Africa). J Geophys Res 116:G03032

Borges AV, Morana C, Bouillon S, Servais P, Descy J-P, Darchambeau F (2014) Carbon cycling of Lake Kivu (East Africa): net autotrophy in the epilimnion and emission of CO_2 to the atmosphere sustained by geogenic inputs. PLoS One 9(10):e109500

Brousse R (1969) L'âge de lacs du sud du Mont-Dore. Annales de la Station Biologique de Besse-en-Chandesse 4:201–208

Brousse R, Delibrias G, Labeyrie J, Rudel A (1969) Eléments de chronologie de la Chaine des Puys. Bull Soc Geol Fr 11:770–793

Camus G, De Goër de Hervé A, Kieffer G, Mergoil J, Vincent PM (1973) Mise au point sur le dynamisme et la chronologie des volcans holocènes de la région de Besse-en-Chandesse (Massif Central français). C R Acad Sci Paris 277:629–632

Camus G, Michard G, Olive P, Boivin P, Desgranges P, Jézéquel D, Meybeck M, Peyrus JC, Vinson JM, Viollier E, Kornprobst J (1993) Risques d'éruption gazeuse carbonique en Auvergne. Bull Soc Géol France 164:767–781

Carini P, White AE, Campbell EO, Giovannoni SJ (2014) Methane production by phosphate-starved SAR11 chemoheterotrophic marine bacteria. Nat Commun 5:4346

Chapron E, Albéric P, Jézéquel D, Versteeg W, Bourdier J-L, Sitbon J (2010) Multidisciplinary characterisation of sedimentary processes in a recent Maar lake (Lake Pavin, French Massif Central) and implication for natural hazards. Nat Hazards Earth Syst Sci 10:1815–1827

Cole JJ, Caraco NF (1998) Atmospheric exchange of carbon dioxide in a low-wind oligotrophic lake measured by the addition of SF_6. Limnol Oceanogr 43:647–656

Cole JJ, Caraco NF, Kling GW, Kratz TK (1994) Carbon dioxide supersaturation in the surface waters of lakes. Science 265:1568–1570

Cossa D, Mason RP, Fitzgerald WF (1994) Chemical speciation of mercury in a meromictic lake. In: Watras CJ (ed) Mercury pollution. Lewis Publishers–CRC Press, Boca Raton, pp 57–67

De Goër de Hervé A (1974) Bordure orientale du massif du Mont Dore jusqu'en Limagne. Association Française pour l'Etude du Quaternaire Bulletin 11:45–53

Delebecque A (1898) Les Lacs Français. Chamerot & Renouard, Paris

Delibrias G (1979) Problèmes liés à la datation des sols fossiles par le carbone 14. Bulletin de l'Association française pour l'étude du quaternaire 16(1–2):35–42

Demont JM (1981) Isotopic study of some fossil and actual hydrothermal systems. Thesis, University of Paris 7, 359 p

Devaux J, Lair N, Amblard (1983) Un écosystème lacustre profond: le lac Pavin. In: Problèmes d'Ecologie. Ecosystèmes Limniques. Bourlière, F., Masson, France.

Dussart B (1966) Limnologie: l'Etude des Eaux Continentales. Gauthier-Villars, Paris

Fouillac C (1983) Chemical geothermometry in CO_2-rich thermal waters. Example of the French Massif Central. Geothermics 12:149–160

Freeth SJ, Kay RLF (1987) The lake Nyos gas disaster. Nature 325(6100):104–105

Gal F, Gadalia A, Millot R (2015) Geochemical study of a Crater lake: Lake Pavin, France: a view through Li-OH isotopes. Procedia Earth Planet Sci 13:189–193

Gewelt M, Juvigné E (1988) Téphrochronologie du Tardiglaciaire et de l'Holocène dans le Cantal, le Cézallier et les Monts Dore (Massif Central, France): résultats nouveaux et synthèse. Bulletin de l'Association française pour l'Etude du quaternaire 25(1):25–34

Gran G (1952) Determination of the equivalent point in potentiometric titrations. Analyst 77:661–671

Grossart H-P, Frindte K, Dziallas C, Eckert W, Tang KW (2011) Microbial methane production in oxygenated water column of an oligotrophic lake. Proc Natl Acad Sci U S A 108:19657–19661

Guenet P (1986) Datation par l'analyse pollinique de l'explosion des volcans du groupe du Pavin (Besse-en-Chandesse, Puy de Dome, France). Réunion des Sciences de la Terre

Guenet P, Reille M (1991) Datation par l'analyse pollinique des éruptions des plus jeunes volcans d'Auvergne (Besse-en-Chandesse, Puy-de- Dome, France). Cahiers du Quaternaire 16:175–180

Guimbaud C, Catoire V, Gogo S, Robert C, Chartier V, Laggoun-Défarge F, Grossel A, Albéric P, Pomathiod L, Nicoullaud B, Richard G (2011) A portable infrared laser spectrometer for flux measurements of trace gases at the geosphere-atmosphere interface. Meas Sci Technol 22(7):1–17

Juvigné E, Gewelt M (1987) La narse d'Ampoix comme téphrostratotype dans la chaîne des Puys mériodionale (France). Bull de l'Association Française pour l'étude du Quaternaire 1:37–49

Juvigné E, Gilot E (1986) Ages et zones de dispersion de téphra émises par les volcans du Montcineyre et du Lac Pavin (Massif Central, France). Z Dt Geol Ges 137:313–623

Juvigné E, Lousberg N, Gewelt M (1988) Morphological and sedimentary development of the Narse D'Espinasse Massif Central France. Revue des Sciences Naturelles d'Auvergne 54(1–4):7–14

Lavina P, Rosso-d'Hers T (2006) Les éruptions volcaniques holocènes du système Montcineyre–Montchal–Pavin: les nouvelles découvertes. Réunion des Sciences de la Terre, Dijon

Lefèvre N, Ciabrini JP, Michard G, Briend B, Du Chaffaud M, Merlivat L (1993) A new optical sensor for pCO_2 measurements in sea water. Mar Chem 42:395–409

Lehours A-C, Bardot C, Thenot A, Debroas D, Fonty G (2005) Anaerobic microbial communities in Lake Pavin, a unique Meromictic Lake in France. Appl Environ Microbiol 71(11):7389–7400

Lehours A-C, Evans P, Bardot C, Joblin K, Fonty G (2007) Phylogenetic diversity of archae and bacteria in the anoxic zone of a meromictic lake (Lake Pavin, France). Appl Environ Microbiol 73:2016–2019

Lopes F, Viollier E, Thiam A, Michard G, Abril G, Groleau A, Prévot F, Carrias J-F, Jézéquel D (2011) Biogeochemical modeling of anaerobic *vs.* aerobic methane oxidation in a meromictic crater lake (Lake Pavin, France). Appl Geochem 26(12):1919–1932

Martin J-M (1985) The Pavin crater lake. In: Stumm W (ed) Chemical processes in lakes. Wiley, London, pp 169–188

Martin J-M, Meybeck M, Nijampurka VN, Somayajulu BLK (1992) ^{210}Pb, ^{226}Ra and ^{32}Si in Lac Pavin (Massif Central, France). Chem Geol 94:173–181

Matthews A, Fouillac C, Hill R, O'Nions RK, Oxburgh ER (1987) Mantle-derived volatiles in continental crust: the Massif Central of France. Earth Planet Sci Lett 85(1–3):117–128

Meybeck M, Martin JM, Olive P (1975) Géochimie des eaux et des sédiments de quelques lacs volcaniques du Massif Central français. Verh Int Verein Limnol 19:1150–1164

Michard G, Viollier E, Jézéquel D, Sarazin G (1994) Geochemical study of a crater lake: the lake Pavin, France – Identification, location and quantification of chemical reactions in the lake. Chem Geol 115:103–115

Michard G, Jézéquel D, Viollier E (2003) Vitesses des réactions de dissolution et précipitation au voisinage de l'interface oxydo-réducteur dans un lac méromictique: le Lac Pavin (Puy-de-Dôme, France). Revue des Sciences de l'Eau 16:199–218

Olive P, Boulègue J (2004) Etude biogéochimique d'un lac méromictique: le Lac Pavin France. Géomorphol Relief Process Environ 4:305–316

Olivier L (1952) Sur la présence en été, dans le Lac Pavin, d'une couche dépourvue d'oxygène, vers 70 m de profondeur. C R Acad Sci Paris 234:743–745

Omaly N (1968) Le Lac Pavin: historique et hydrobiologie. Revue des Sciences Naturelles d'Auvergne 34:7–30

Pelletier JP (1968) Un lac méromictique, le Pavin (Auvergne)/a Meromictic Lake Lake Pavin (Auvergne). Annales de la Station Biologique de Besse-en-Chandesse 3:147–170

Perreau A (1948) Le lac – cratère de Pavin. Revue des Sciences Naturelles d'Auvergne Nouvelle Série 14(fasc. 2-3-4):103–106

Podda F, Michard G (1994) Mesure colorimétrique de l'alcalinité. C R Acad Sci Paris 319:651–657

Prévot F, Michard G, Ciabrini JP, Beaumont L, Guillot A, Merlivat L, Brient B, Danguy T (1998) Development of a sensor for carbonate alkalinity in sea water. Marc'h Mor Workshop (Marine Analytical Chemistry for Monitoring and Oceanographic Research) reports, Brest (France), 17–19 November 1998

Reichert P (1994) Aquasim – a tool for simulation and data-analysis of aquatic systems. Water Sci Technol 30:21–30

Restituito F (1984) Study of a layer containing hydrogen sulfide in a meromictic lake Lake Pavin, France. Hydrobiologia 109:235–242

Restituito F (1987) Consequences of redox conditions on the distribution of cations in a meromictic oligotrophic lake. Hydrobiologia 144:63–76

Schettler G, Schwab MJ, Stebich M (2007) A 700 years record of climate change based on geochemical and palynological data from varved sediments (Lac Pavin, French Massif Central). Chem Geol 240:11–35

Schmid M (1997) Physikalische und chemische prozesse im Lac Pavin: ursachen und folgen der satbilen schichtung eines vulkanishen kratersees. Diploma thesis, ETH Zürich

Schmid M, Halbwachs M, Wüest A (2006) Simulation of CO_2 concentrations, temperature, and stratification in Lake Nyos for different degassing scenarios. Geochem Geophys Geosyst 7. doi:10.1029/2005GC001164

Sigurdsson H, Devine JD, Tchua FM, Presser FM, Pringle MKW, Evans WC (1987) Origin of the lethal gas burst from Lake Monoun, Cameroun. J Volcanol Geotherm Res 31(1–2):1–16

Viollier E (1995) Géochimie des éléments traces en milieu lacustre. Ph.D thesis, Université Paris 7

Viollier E, Alberic P, Jézéquel D, Michard G, Pèpe M, Sarazin G (1995) Geochemical study of a crater lake: the lake Pavin, France – trace element behaviour in the monimolimnion. Chem Geol 125:161–172

Viollier E, Michard G, Jézéquel D, Pèpe M, Sarazin G (1997) Geochemical study of a crater lake: lake Pavin, Puy-de-Dôme, France. Constraints afforded by the particulate matter distribution in the element cycling within the lake. Chem Geol 142(3–4):225–241

Whiticar MJ (1999) Carbon and hydrogen isotope systematics of bacterial formation and oxidation of methane. Chem Geol 161:291–314

The Iron Wheel in Lac Pavin: Interaction with Phosphorus Cycle

12

Vincent Busigny, Didier Jézéquel, Julie Cosmidis, Eric Viollier, Karim Benzerara, Noah J. Planavsky, Patrick Albéric, Oanez Lebeau, Gérard Sarazin, and Gil Michard

Abstract

Lac Pavin is a crater lake, characterized by water column stratification, with oxygenated shallow waters lying above anoxic and ferruginous deep waters. In the deep waters, ferrous iron, $Fe(II)_{aq}$, is the main dissolved cation, with concentrations up to 1 mM. Iron is efficiently confined below the oxic-anoxic boundary due to the formation of insoluble ferric iron species, $Fe(III)_s$, by oxidation with O_2 and other oxidants (e.g., NO_3^-, Mn(IV)). The $Fe(III)_s$ particles settle down and are reduced in the anoxic waters and at the lake bottom by reaction with organic matter to soluble $Fe(II)_{aq}$. It then diffuses upward in the water column and finally is re-oxidized to Fe(III) at the redox boundary. This process, known as the "iron wheel", is described in the present paper that reviews available data for dissolved and particulate matter in the water column, settling particles collected by sediment traps and sediment cores. Detailed analyses for some major and trace element concentrations, along with iron speciation and isotope composition, high-resolution microscopy, and geochemical modeling provide a picture of biogeochemical cycling in this Fe-rich aqueous system. At Lac Pavin the P and Fe cycles are tightly coupled. Orthophosphate is sorbed onto Fe oxyhydroxides and/or precipitated as Fe(II)-Fe(III)-phosphates at the redox interface, confining P ions in the deep anoxic waters. Deeper in the water column, particulate Fe concentrations progressively increase due to Fe(II) phosphate (vivianite) formation. In the sediment, Fe is buried as various ferrous minerals, such as vivianite, pyrite and siderite.

Keywords

Anoxic • Ferruginous • Lake • Pavin • Meromictic • Iron • Phosphate • Sulfide • Manganese • Biogeochemical cycle

V. Busigny (✉) • D. Jézéquel • E. Viollier • G. Sarazin
G. Michard
IPGP (Institut de Physique du Globe de Paris), Sorbonne Paris Cité, Univ. Paris Diderot, UMR 7154 CNRS, 1 rue Jussieu, 75005 Paris, France
e-mail: busigny@ipgp.fr

J. Cosmidis
Department of Geological Sciences, University of Colorado, Boulder, CO, USA

K. Benzerara
Institut de Minéralogie, de Physique des Matériaux, et de Cosmochimie (IMPMC), Sorbonne Universités, UPMC Université Paris, 6, CNRS UMR 7590, MNHN, IRD UMR 206, Paris, France

N.J. Planavsky
Department of Geology and Geophysics, Yale University, New Haven, CT, USA

P. Albéric
ISTO (Institut des Sciences de la Terre d'Orléans), Observatoire des Sciences de l'Univers en région Centre, UMR 7327 (CNRS-Université d'Orléans-BRGM), 1A rue de la Férollerie, 45071 Orléans Cedex 2, France

O. Lebeau
Institut de Physique du Globe de Paris, Sorbonne Paris Cité, Université Paris Diderot, UMR 7154 CNRS, Paris, France

T. Sime-Ngando et al. (eds.), *Lake Pavin*, DOI 10.1007/978-3-319-39961-4_12

12.1 Introduction

Lac Pavin is a crater lake in the Massif Central (France, located at 45°29.740 N, 2°53.280 E), with a maximum water depth of 92 m. It is meromictic, *i.e.* permanently stratified with two layers. The upper layer, called mixolimnion, extends from 0 to ~50–60 m depth and contains abundant dissolved oxygen (O_2). According to the season, and in particular to the luminosity, there is a strong peak in oxygen concentrations in the zone of maximum photosynthetic production, between surface water and ~20 m depth. There is diffusional O_2 exchange and O_2 transfer deeper into the lake by mixing during seasonal overturns. The oxygen chemocline ranges from ~50 to ~65 m depth, depending on the efficiency of water mixing. Some year the mixolimnion may be partially mixed, only down to ~30 m depth (Aeschbach-Hertig et al. 2002). Others authors noted that the spring overturn can affect the water column down to 60–70 m depth, depending on the year (Restituito 1987). The autumn overturn is much less intense (Restituito 1987). From ~70 to 92 m depth, a stable layer, called the monimolimnion, does not mix with oxygenated upper water and exchanges solutes by diffusion only. The high organic matter (particulate organic carbon, POC) flux from the surface waters in Lac Pavin results in a complete consumption of oxygen in the monimolimnion. A purported deep ferruginous spring probably feeds this anoxic zone with dissolved ferrous iron (Meybeck et al. 1975; Assayag et al. 2008; Jézéquel et al. 2011). Between the mixolimnion and the monimolimnion, an intermediate layer, called the mesolimnion, is characterized by a sharp gradient of salinity and dissolved chemical species. There are varved, Fe-rich sediments in the bottom of the lake, making Lac Pavin a potential modern analogue to the Earth's oceans and an archive of recent environmental change in the Massif Central (Stebich et al. 2005; Schettler et al. 2007; Busigny et al. 2014). Numerous studies have been published on biogeochemical cycling at Lac Pavin and a wealth of data cover a wide variety of topics such as geochemistry (Michard et al. 1994; 2003; Viollier et al. 1995, 1997; Albéric et al. 2000, 2013), biology and microbiology (Amblard and Bourdier 1990; Biderre-Petit et al. 2011; Lehours et al. 2007, 2009), hydrology and dynamic of the water column (Meybeck et al. 1975; Aeschbach-Hertig et al. 2002; Assayag et al. 2008; Jézéquel et al. 2011; Bonhomme et al. 2011), mineralogy and sedimentology (Schettler et al. 2007; Chapron et al. 2010; Cosmidis et al. 2014).

The Fe cycle at Lac Pavin is driven by redox cycling: there is extensive ferric iron reductive dissolution in the anoxic layer and at the bottom of the lake, upward diffusion of Fe^{2+}, oxidation and precipitation as Fe(III) at the redox interface, and finally Fe particles settling (Michard et al. 2003). This cycling, termed "iron wheel" or "ferrous wheel", is known for other anoxic water systems, where sulfate and sulfide concentrations are low enough so that dissolved Fe can accumulate (Campbell and Torgersen 1980; Davison 1993). The ferrous wheel in anoxic basins has a strong impact on biogeochemical cycles (e.g. Hongve 1997) and plays a key role for bacteria and archaea that derive their chemical energy from redox reactions.

Lac Pavin is a relatively unique analogue to Earth's early oceans—in particular Archean marine systems (Martin 1985; Busigny et al. 2014). At that time, oceans were anoxic, sulfate-poor and ferruginous, and Fe represented a central element driving other biogeochemical cycles such as carbon, sulfur, nitrogen or phosphorus (Konhauser et al. 2011). There is evidence for Archean ferruginous conditions in a wide range of Fe-rich sedimentary deposits, including some shales and iron formations (Poulton and Canfield 2011). Determining the significance of the geochemical signatures recorded in these sedimentary rocks, and extrapolating paleo-environment and -ecosystem conditions, is complicated by the lack of studies on modern analogues (e.g., Crowe et al. 2008; Staubwasser et al. 2013) and laboratory experimental calibration (e.g., Konhauser et al. 2002; 2007). Lac Pavin represents a natural laboratory in which redox sensitive element cycling under ferruginous conditions can be studied.

In the present paper, we present chemical and isotopic data available for dissolved and particulate matter, and recent mineralogical advances on the water column and bottom sediments of Lac Pavin. In particular, we describe the Fe biogeochemical cycle and its interactions with phosphorus (P), manganese (Mn), sulfur (S) and organic matter. A geochemical model consistent with chemical, isotopic and mineralogical data supports our current view of the Fe cycle in Lac Pavin and allows us to quantify some chemical and physical parameters such as chemical reaction rates, the velocity of falling particles, and Fe isotope fractionations associated with various processes in the water column. We also highlight evidence for microbial mediation and outline the links between the water column Fe cycle and sediment Fe-bearing phases.

12.2 Methods

12.2.1 Sampling

Water samples were collected either from a sampling platform anchored near the center of the lake or from a boat, using various types of sampler (syringes, Niskin bottles), equipped with an electronic depth gauge providing a precision of ±0.2 m on the depth position in the water column profile. Waters were filtered to 0.2 µm and acidified with a few drops of ultrapure HNO_3 to pH 1–2, for sample preservation (see Michard et al. 1994; Viollier et al. 1995; Busigny et al. 2014). The filters were either discarded or preserved for analysis of the suspended particulate matter (SPM). Dissolved oxygen, temperature, and specific conductivity were determined in situ using a Seabird SBE 19 Seacat profiler (Sea-Bird Electronics Inc., Washington, DC) and turbidity was measured using a NKE STBD 300 probe.

Settling particles were collected with sediment traps (Uwitec, Austria) disposed at different depths along a vertical profile in the water column from April, 1994, to May, 1995 (Viollier et al. 1997) and June to September, 2011 (Cosmidis et al. 2014).

A sediment core was collected in September 2009 at the bottom of the lake (92 m) using a Uwitec gravity corer (90 mm ext. diameter), allowing to sample a vertical profile at regular intervals below the water-sediment interface (Busigny et al. 2014). Sediment traps and core samples were transferred into glove bags and placed under anoxic conditions (N_2 atmosphere) immediately after collection. Porewaters were separated from solid phase using Rhizons® or centrifugation. During the separation process, oxygen levels were monitored with an Oxi 340i WTW oxygen meter and were always below the detection limit of 0.1 mg/L. Sediments were dehydrated by freeze-drying before mineralogical, chemical and isotopic analyses. Finally, all samples were stored at 4 °C until analyses were conducted.

12.2.2 Chemical and Isotope Analyses

The dissolved Fe concentration and Fe(II)/Fe(III) speciation were measured on board by spectrophotometry (Merck SQ300 spectrophotometer) on filtered water samples following the method by Viollier et al. (2000). Spectrophotometric analyses ($\lambda = 562$ nm) of the iron-ferrozine complexes were performed in a single aliquot before and after a reduction step with hydroxylamine. The procedure was calibrated using Fe(III) standards stable under normal conditions of analysis. Manganese and phosphate concentrations were determined in laboratory on filtered and acidified samples from the water column using inductively coupled plasma atomic emission spectroscopy (ICP-AES; Perkin Elmer Optima 3000) or using inductively coupled plasma mass spectrometry (ICP-MS; Thermo Element2). Sulfate (SO_4^{2-}) concentrations were measured by ion chromatography (Dionex DX-600 IC System) from samples filtered and poisoned with zinc acetate to avoid reoxidation of any sulfide present. Total sulfide concentrations (ΣH_2S) were determined by spectrophotometry using the methylene blue method as described in Bura-Nakic et al. (2009).

Iron isotope compositions were measured on water samples collected from various depths in the lake. The method was described in a previous contribution (Busigny et al. 2014) and can be summarized as follows. Samples were acidified with HNO_3 to ensure that all Fe is in the ferric state. Iron was then separated from matrix elements on anion exchange chromatography in HCl medium (Strelow 1980). Iron concentrations and isotopic compositions were measured using a Neptune ThermoFischer MC-ICP-MS (Multiple Collector Inductively Coupled Plasma Mass Spectrometer). We corrected for instrumental mass discrimination using the conventional sample-standard bracketing (SSB) approach. The $^{56}Fe/^{54}Fe$ ratio is reported in the usual δ notation in per mil (‰) as:

$$\delta^{56}Fe = [(^{56}Fe/^{54}Fe)_{sample} / (^{56}Fe/^{54}Fe)_{standard} - 1] \times 1000$$

where the standard is IRMM-014 (Institute for Reference Materials and Measurements; Taylor et al. 1992). Based on replicate analyses of international rock standards, the external precision and accuracy on $\delta^{56}Fe$ values were always better than 0.06‰ (2SD).

12.2.3 Mineralogical Analyses

The mineralogy of Lac Pavin particles from sediment traps and sediment core samples has been characterized by X-ray diffraction (XRD), Extended X-ray absorption fine structure (EXAFS) spectroscopy, scanning electron microscopy (SEM), scanning transmission X-ray microscopy (STXM) and transmission electron microscopy (TEM) (Viollier et al. 1997; Cosmidis et al. 2014). The coupling between these various methods provides information on the identity of the mineral phases, their chemical composition (including redox state and speciation of iron), size and shape. The detection limit for a specific mineral phase in a complex natural sample is generally considered to be ~1 %.

12.2.4 Geochemical Modeling

A geochemical modeling was developed herein to (i) constrain Fe and P reaction pathways and to (ii) obtain quantitative constraints on physical, chemical and isotopic parameters (such as Fe isotope fractionation associated with various chemical reactions). Our model is a reactive-transport model developed from AQUASIM software (Reichert 1998). It describes spatial and temporal distribution of dissolved and particulate chemical species in the lake water column. The detailed methodology and formalism have been presented in a previous study, focused on the methane biogeochemical cycle in Lac Pavin (Lopes et al. 2011), and applied to ^{14}C profiles modeling of organic and inorganic carbon pools (Albéric et al. 2013) and is briefly summarized below.

In the model, dissolved and particulate species are transported through the water column by vertical mixing, but particles are also transported by sedimentation. Dissolved Fe and P species are supplied to the lake by inflows from sublacustrine input. Dissolved and particulate species are allowed to react in multiple biogeochemical pathways such as oxidation, adsorption, precipitation and dissolution. Two partial differential transport-reaction equations implemented in the AQUASIM code were used to describe these biochemical and physical processes. They were solved for dissolved and particulate matter, respectively, and can be written as follow:

$$\frac{\partial C}{\partial t} = \frac{\partial}{A\partial z}\left(AK_z\frac{\partial C}{\partial z}\right) - \frac{\partial}{A\partial z}(QC) + r_C + \frac{q}{A}C_{in}$$

$$\frac{\partial X}{\partial t} = \frac{\partial}{A\partial z}\left(AK_z\frac{\partial X}{\partial z}\right) - \frac{\partial}{A\partial z}(QX) + r_X + \frac{\partial}{\partial z}(V_{sed}X)$$

where C is the concentration of dissolved species (mol. m^{-3}), X is the concentration of particulate species (mol. m^{-3}), t is the time (s), z is the vertical coordinate pointing downwards (m), A is the cross-sectional area of the lake (m^2), K_z is the vertical mixing coefficient (m^2.s^{-1}), r_C and r_X are the transformation rates of dissolved or particulate species (mol.m^{-3}.s^{-1}), Q is the vertical discharge induced by water inflows in the lake depth (m^3.s^{-1}), C_{in} is the inflow concentration of dissolved species (mol.m^{-3}), V_{sed} is the sedimentation velocity (m.s^{-1}), q is the water discharge at depth into the lake (m^2.s^{-1}). Both equations were solved numerically with a vertical resolution of 1 m (Reichert 1998). Four chemical parameters from the various depth profiles obtained in the water column were considered herein and fitted in our modeling approach: dissolved Fe and P concentrations, particulate Fe concentration and Fe isotope composition of the dissolved Fe(II). In the model, we focused on the processes occurring in the monimolimnion and at the redox transition zone, where the majority of Fe is cycled.

12.3 Results

12.3.1 Chemical and Isotopic Compositions in the Water Column

Dissolved Fraction Figure 12.1 presents chemical and Fe isotope profiles in a depth range from the redox transition zone down to the bottom of the lake (data from July 2007 reported in Table 12.1). Dissolved Fe and P concentrations are low in the oxic zone, respectively <30 nM and ~1 μM, but increase dramatically below the redox transition zone (~60 m) up to values of ~1200 and 370 μM at the lake bottom (Fig. 12.1b). Dissolved Mn concentration also shows a progressive increase with depth but in a lower range, from 0.3 to 26 μM. Manganese concentration increases at a shallower depth level in the redox transition zone, with a distinct peak at ~61 m depth (Fig. 12.1a), suggesting rapid water column dissolution of Mn oxides. Dissolved sulfate (SO_4^{2-}) concentration decreases from ~15 μM in the oxic zone to <1 μM in the anoxic zone, due to sulfate reduction into sulfide (Fig. 12.1a). Total sulfide (ΣH_2S) in the dissolved fraction (water filtered at 0.2 μm) corresponds to a limited amount of "free" dissolved sulfide (H_2S/HS^-) but is dominated by FeS colloids (about 80 % of the total sulfide; Bura-Nakic et al. 2009). Iron isotope composition shows a large increase with depth from−2.14 to +0.31 ‰, with the largest variation near the redox transition zone (Fig. 12.1c).

The relationship between dissolved Fe and P in Lac Pavin water column is illustrated in Fig. 12.2a. It shows a compilation of data from two different periods of time (September 1992 and July 2007). Dissolved Fe and P concentrations display a remarkable linear correlation, with a mean slope of 3.45 passing through the origin of the graph. This suggests a significant connection between Fe and P cycles in the lake, also supported by suspended particulate matter data (see below). Although this correlation suggest that Fe/P ratio is roughly constant in dissolved fraction of water samples at Lac Pavin, a detailed analysis shows some variability along depth profile (Fig. 12.2b). The Fe/P molar ratio is very low (<0.05) in the mixolimnion (0–60 m depth) due to the presence of oxygen, driving Fe oxidation and precipitation. Between 60 and 65 m depth, the Fe/P ratio increases strongly up to values of 5. From 65 m to lake bottom, the Fe/P ratio decreases progressively from about 3.8 to 3.2 (Fig. 12.2b).

Suspended Particulate Matter Suspended particulate matters (SPM) collected in October 2010 were analyzed and compared to turbidity profile. The variation with depth of Fe concentrations of SPM (expressed relative to the volume of water that has been collected and filtered) is consistent with the results of a previous study (Viollier et al. 1995). It shows a good agreement with turbidity variations at the redox interface and in the anoxic zone (Fig. 12.3a). This suggests that (1) the variations of turbidity around the redox boundary are largely related to Fe-bearing particles in SPM at depth > 50 m in Lac Pavin and (2) turbidity measurements can be used as a proxy of these Fe-rich particles. In detail, we observed sometimes that Fe particles collected on SPM extend deeper in the water column than what turbidity measurement would suggest. It may result from the fact that turbidity is a good

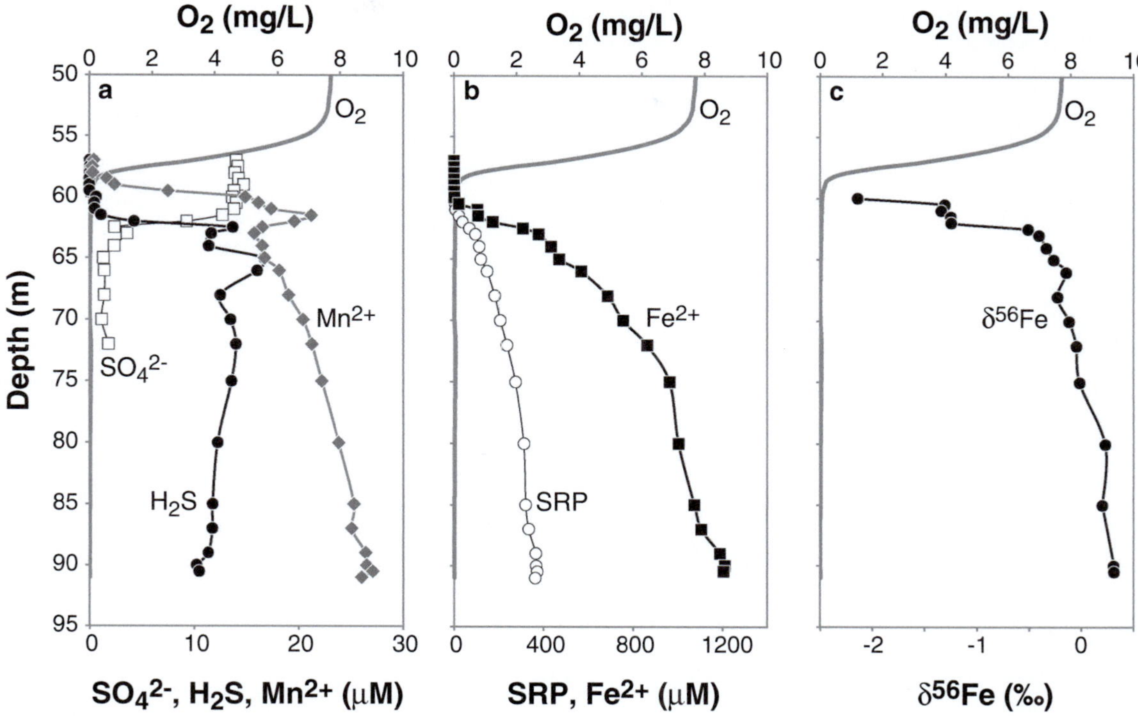

Fig. 12.1 Concentration profiles of selected dissolved chemical species (**a** and **b**) and Fe isotope composition (**c**) in Lac Pavin water samples collected in July 2007. The oxygen (O_2) concentration is reported in all diagrams for highlighting the redox transition zone (between 50 and 60 m) and fully anoxic conditions (below 60 m). The bottom of the lake is at 92 m depth. SRP corresponds to soluble reactive phosphorus

proxy of small-sized Fe particles (formed at the peak of turbidity) but when settling particles collapse and form larger aggregates, the turbidity signal is decreased. Another interesting feature is that Fe/P molar ratio in SPM is roughly constant (around ~3; Fig. 12.3a) although some variations are observed in the redox transition zone. A new dataset for SPM collected in October 2010, November 2011 and June 2013 is presented in Fig. 12.4 (see data in Table 12.2). It shows that, below 50 m depth in the water column, particulate Fe and P concentrations are linearly correlated, with a slope of 3.07 (*i.e.* Fe/P molar ratio), slightly lower than the slope found for the dissolved fraction (3.45 in Fig. 12.2).

Figure 12.3b shows three different profiles of turbidity measured in June 2011, September 2011 and June 2013, selected to represent some of the variability observed in the lake. One or several peaks of turbidity, with various magnitudes, are generally observed from the surface of the lake down to ~25 m. These peaks represent living organisms, such as diatoms, green algae and/or cyanobacteria (Amblard, 1984, 1986, 1988; Stebich et al. 2005). Another peak of turbidity is observed at ~50–60 m depth, and corresponds to a peak of particulate iron (Fig. 12.3). The intensity of this peak is usually between 4 and 6 NTU (Nephelometric Turbidity Unit), but an occasional collapse (*e.g.* down to 1.5 NTU in December 2012) or increase (up to 20 NTU in October 2010) were also observed.

12.3.2 Mineralogy of Particulate Matter in the Water Column and Bottom Sediments

SEM and XRD analyses show that the particles collected in the water column and sediments of Lac Pavin mostly contain diatom frustules (Michard et al. 1994; Viollier et al. 1997; Cosmidis et al. 2014), consistently with early studies of phytoplankton diversity in Lac Pavin (*e.g.*, Devaux 1980). Organic matter also represents a significant portion of the particulate matter, with concentrations ranging from 26 to 34 wt% in sediment traps (Cosmidis et al. 2014). Organic matter content decreases strongly in sediments, down to ~5–10 wt%, due to biological degradation by Fe(III) reduction and methanogenesis metabolisms (Restituito 1984; Lehours et al. 2009; Biderre-Petit et al. 2011; Cosmidis et al. 2014). Detailed mineralogy of Fe-bearing particles collected using sediment traps set up at different depths in the water column of Lac Pavin is presented in Fig. 12.5. It shows that Fe-bearing particles in the shallower oxygenated zone are mostly detrital phyllosilicates and Fe(III)-oxyhydroxides. Poorly crystalline ferric/ferrous phosphates (Cosmidis et al. 2014), possibly with some Fe-oxyhydroxides (Viollier et al. 1997), precipitate at the redox transition zone. In the anoxic zone, a ferrous phosphate (vivianite, $Fe^{2+}_3(PO_4)_2.8(H_2O)$) precipitates increasingly with depth and is the most abundant Fe-bearing phase in the sedi-

Table 12.1 Dissolved iron, phosphate (SRP: soluble reactive phosphorus), manganese, sulfate and sulfide concentrations, together with isotopic composition of the bulk dissolved iron (expressed as $\delta^{56}Fe$) in water samples from Lac Pavin collected in July 2007

Sample #	Depth (m)	SRP (μM)	Fe(II)[a] (μM)	Mn(II)[a] (μM)	SO_4^{2-}[b] (μM)	ΣH_2S[b] (μM)	$\delta^{56}Fe \pm 2SD$[a] (‰)
MX-15_10	57.0	1.0	0.0	0.42	14.0	0.0	
MX-15_11	57.5	1.0	0.0	0.29	14.2	0.1	
MX-15_12	58.0	1.1	0.1	0.31	13.9	0.0	
MX-15_13	58.5	1.2	0.0	1.65	14.3	0.0	
MX-15_14	59.0	1.1	0.0	2.39	14.7	0.0	
MX-15_15	59.5	1.3	0.1	7.49	13.8	0.0	
MX-15_16	60.0	1.6	2.1	14.9	13.7	0.6	−2.14 ± 0.04
MX-15_17	60.5	4.9	21	16.1	14.0	0.4	−1.30 ± 0.08
MX-15_18	61.0	5.9	106	17.4	13.8	0.5	−1.34 ± 0.02
MX-15_19	61.5	20	108	21.2	12.7	1.1	−1.25 ± 0.04
MX-15_20	62.0	37	174	19.6	9.3	4.2	−1.25 ± 0.08
MX-15_21	62.5	70	308	16.5	2.4	13.7	−0.51 ± 0.04
MX-15_22	63.0	97	378	15.7	3.6	11.6	−0.40 ± 0.04
MX-15_23	64.0	113	435	16.5	2.4	11.4	−0.33 ± 0.06
MX-15_24	65.0	120	470	16.7	1.3	16.7	−0.26 ± 0.04
MX-15_25	66.0	149	568	18.1	1.4	16.0	−0.15 ± 0.04
MX-15_26	68.0	183	688	19.0	1.4	12.5	−0.23 ± 0.02
MX-15_27	70.0	208	757	20.4	1.2	13.5	−0.11 ± 0.04
MX-15_28	72.0	237	864	21.2	1.8	14.0	−0.05 ± 0.06
MX-15_29	75.0	275	964	22.2		13.6	−0.01 ± 0.08
MX-15_30	80.0	312	1004	23.8		12.3	0.23 ± 0.04
MX-15_31	85.0	321	1074	25.3		11.8	0.20 ± 0.06
MX-15_32	87.0	335	1105	25.1		11.8	
MX-15_33	89.0	366	1189	26.4		11.4	
MX-15_34	90.0	368	1210	26.5		10.2	0.31 ± 0.08
MX-15_35	90.5	372	1203	27.1		10.5	0.31 ± 0.08

Depth: depth in the water column. The oxic-anoxic interface was at 60 m depth in July 2007
Uncertainties on $\delta^{56}Fe$ correspond to 2SD (standard deviation)
[a]Data from Busigny et al. (2014)
[b]Data from Bura-Nakić et al. (2009)

ment (Fig. 12.5), representing up to 70 wt% of the total solid fraction in some sediment samples (Schettler et al. 2007). Imaging at the micro- and nano-meter scales using SEM, TEM and STXM techniques shows that Fe-phosphates often encrust microbial bodies (Figs. 12.6a, b), suggesting that microorganisms are involved in the formation of these phases (Cosmidis et al. 2014). Pyrite (FeS_2) is a common phase in Lac Pavin sediments (Viollier et al. 1997; Schettler et al. 2007) but is in very low abundance relative to diatoms, iron phosphates and organic matter (<0.2 wt%; Busigny et al. 2014). Fine-grained pyrite occurs in sediments as spherical aggregates (20–30 μm) of bipyramidal crystals (0.5–1 μm) termed framboids (Fig. 12.6c). Siderite ($FeCO_3$) occurrence has also been suggested in lake bottom sediments from bulk chemical analysis, coupling bulk analyses of Fe and CO_3 concentrations (Schettler et al. 2007). Preliminary XRD analysis of the samples from our sediment core collected at the lake bottom support the presence of siderite.

12.3.3 Modeling

The present geochemical model was limited to first order reactions involving iron and phosphates. It was mostly confined to the monimolimnion, with a rather thin (5 m thick) oxic zone replacing the whole mixolimnion (~60 m thick), because chemical reactions between Fe, P and other elements occur only in the 3–4 m in the deepest part of the mixolimnion. The water column was thus represented by two distinct layers: (1) a 5 m-thick oxidizing layer, representing the bottom part of the mixolimnion (redox transition zone), where soluble Fe^{2+} is oxidized to Fe^{3+} which is weakly soluble and precipitates as ferric oxyhydroxides, and phosphate is adsorbed on these ferric particles, and (2) a 30 m-thick reducing layer, representing the monimolimnion, where ferric oxyhydroxides are reduced and dissolved, adsorbed phosphate is released, and vivianite, a ferrous phosphate, is precipitated. Because of the limited amount of Mn and S relative to large Fe and P concentrations,

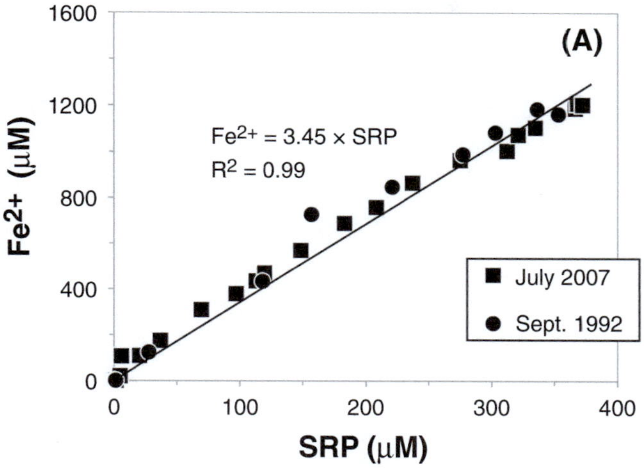

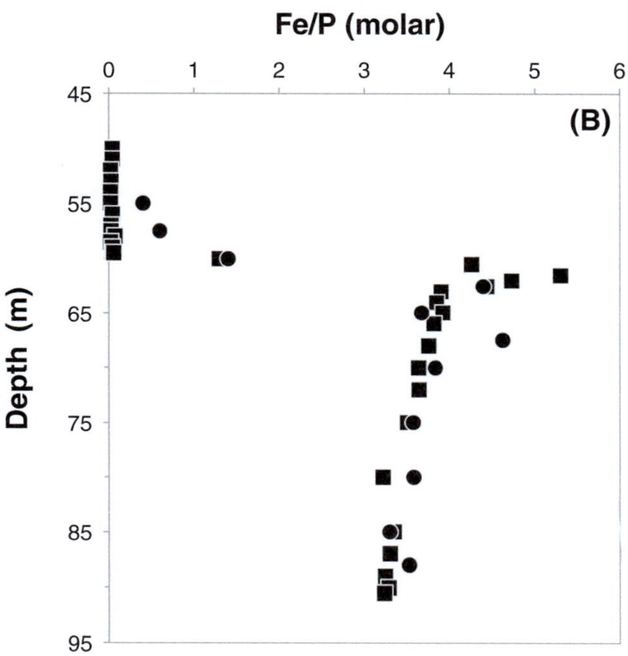

Fig. 12.2 (**a**) Relation between the concentrations of dissolved iron and phosphate (SRP: soluble reactive phosphorus) in water samples from Lac Pavin. *Black squares* and circles represent two different periods of sampling in July 2007 (MX15) and September 1992 (Data from Michard et al. 1994) respectively. (**b**) Fe/P molar ratio in the dissolved fraction of water samples collected at different depths in the water column. The symbols represent two periods of sampling like in (**a**)

we assumed that their effect are negligible for the bulk Fe-P cycles. In the reducing layer, Fe and P are supplied from the sediment to the water column (benthic Fe and P flux from Viollier et al. 1995; Table 12.3), and from a mineralized subsurface spring assumed to be located 8 m below the redox transition zone (water flow estimate from Jézéquel et al. 2011). The Fe isotope composition of this subsurface spring is presently unknown. However, we analyzed two different ferruginous springs in the Pavin area, supposed to be representative of the purported subsurface spring feeding the monimolimnion (*e.g.* Assayag et al. 2008). Both springs, Fontaine Goyon and Source Saint-Anne, have δ^{56}Fe values of -0.80 ± 0.03‰ (n=2) and -1.09 ± 0.02‰ (n=2), respectively. In our modeling, Fe isotope composition of the subsurface ferruginous spring was thus assumed to be -1‰ (Table 12.3). The vertical eddy diffusion coefficients were calculated from temperature and conductivity profiles (Lopes et al. 2011). Other parameters such as chemical reaction rates, isotope fractionations associated with chemical reactions and particle sedimentation velocity were obtained from modeling by adjusting the parameters for a better fit of the experimental data. Table 12.3 provides a list of chemical reactions, kinetic constants and isotope fractionations used in the present model. Figure 12.7 represents the results of the model obtained from values given in Table 12.3. A first observation is that the model reproduces, with good agreement, experimental data collected from the field. Although a test of sensitivity was performed on the model, we cannot rule out the possibility that another set of parameters would also fit the data.

The sensitivity analysis shows that the concentrations of particulate Fe along the water column profile are tightly controlled by particle sedimentation velocity. On the contrary, the concentrations of dissolved species are not significantly affected by the sedimentation velocity. This explains the relative constancy of dissolved Fe^{2+} and SRP (soluble reactive phosphorus) concentrations over time in the water column, while turbidity profiles (tracking particles) in the anoxic zone are variable (Fig. 12.3b). Although the model does not perfectly match with the distribution of particulate Fe, the overall shape is reproduced (Fig. 12.7d) and allows us to estimate an average particle sedimentation velocity of about 1 m.d^{-1}. The variability of particles distribution along the depth profile may reflect time dependent variations related to the type of particles and/or fluxes. For instance, it was demonstrated, from sediment trap monitoring, that the main flux of particles to the sediment is related to diatoms bloom in the spring, which brings about 60 % of the annual particle deposit (Viollier et al. 1997).

Another important result from the model is that the profile of Fe isotope composition (Fig. 12.7c) is mostly –if not only– controlled by the values of Fe isotope fractionation used for reactions of Fe oxidation and reduction, and vivianite precipitation. Thus the model provides an estimate of each of these isotope fractionation values (Table 12.3). The deep ferruginous spring, located at the relative depth of 13 m in the present model (*i.e.* ~65 m depth in Lac Pavin from Jézéquel et al. 2011), is not able to produce any peak in the Fe concentration at this depth, but can modify the whole profile at the scale of the monimolimnion. These modifications are observed only when the dissolved Fe concentration of the ferruginous spring is largely increased (by a factor of 4 or higher), which is not compatible with the depth profile of

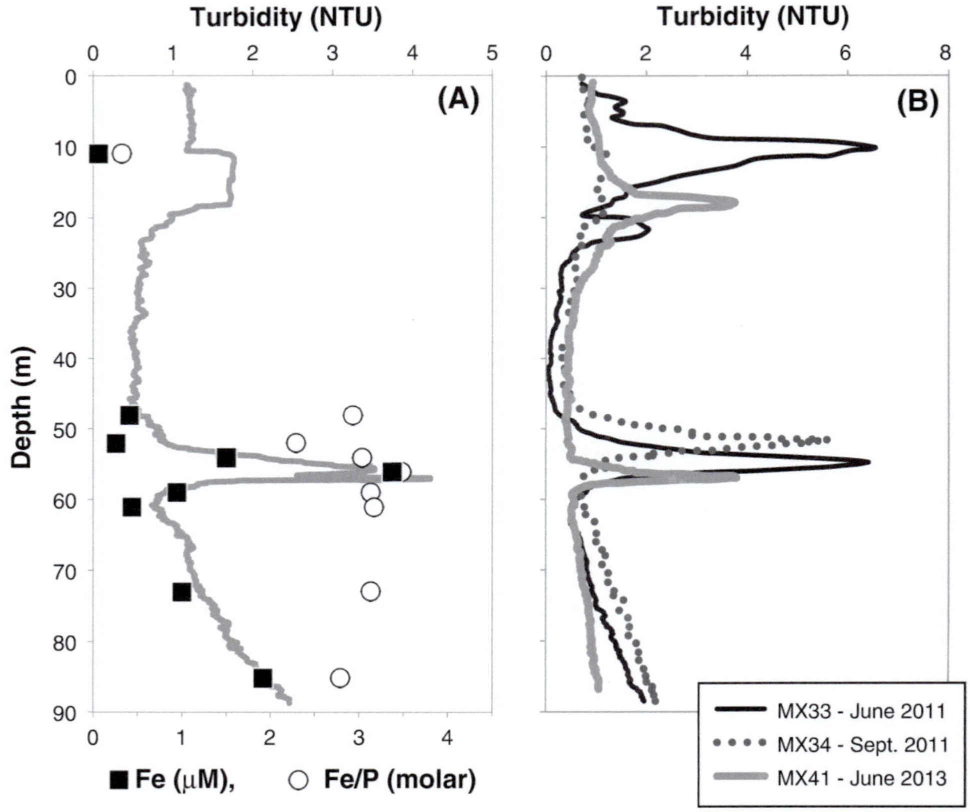

Fig. 12.3 (**a**) Turbidity profiles in Lac Pavin water column, compared to Fe concentrations (*black squares*) and Fe/P ratio (*white circles*) of suspended particulate matter (SPM). Turbidity measurement and SPM collection were both performed in October 2010. (**b**) Three turbidity profiles illustrating some of the variability of particulate matter distribution

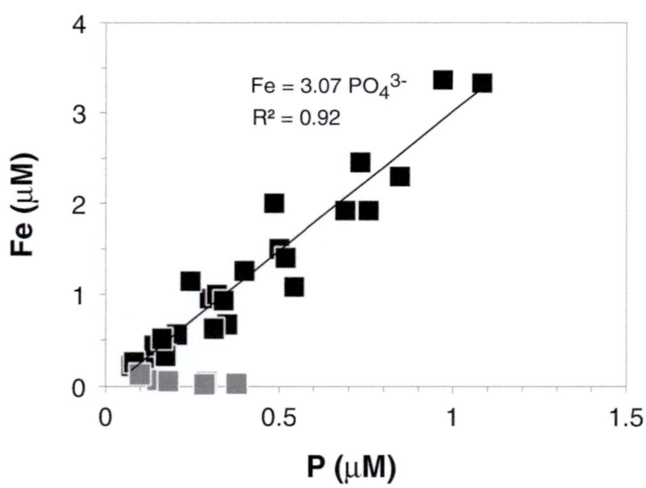

Fig. 12.4 Fe and P concentrations of suspended particulate matter in Lac Pavin water column. The concentrations are reported in μmol.L^{-1}, relative to the volume of water filtered. The *black squares* represent the particles around and below the peak of turbidity (*i.e.* depth > 48 m). The *straight line* is the best fit for these samples. The *grey squares* represent the particles sampled in the mixolimnion, at a depth shallower than 40 m in the water column. The data plotted correspond to a compilation of various samples collected in October 2010 (MX30), November 2011 (MX35) and June 2013 (MX41)

dissolved Fe concentrations measured in the monimolimnion.

In order to explain the variation of turbidity profiles observed in the monimolimnion (Fig. 12.3b), we modified our former steady-state model into a pseudo steady-state model. In this second model, the particle sedimentation flux in the water column was increased during 1 month (from April 11 to May 11) due to a diatom bloom (Fig. 12.8). As expected, the modification of the particle sedimentation flux only affects the particulate matter distribution in the water column, but does not impact the dissolved Fe profile. The disturbance tends to smooth the distribution patterns of particulate matter along the depth profile, with smaller peaks of SPM at the redox boundary and at the lake bottom than during steady state periods. An important result of the model is that the system returns very rapidly to steady state since only 2 months after diatom blooms, the disturbance is not seen anymore (Fig. 12.8).

12.4 Discussion

12.4.1 Global Iron and Phosphorus Biogeochemistry in Lac Pavin

The coupling between chemical, isotopic, mineralogical and modeling studies allows us to propose an integrated view of the Fe and P biogeochemical cycles in Lac Pavin, and their connection to Mn, S and C cycles (Fig. 12.9). In Lac Pavin water column, dissolved Fe(II) and SRP show a strong concentration gradient from high values at the lake bottom to low values at the redox transition zone (Fig. 12.1b). This

Table 12.2 Iron and phosphorus concentrations and Fe/P molar ratio in Lac Pavin suspended particulate matter

Sample #	Depth[a] (m)	Fe (μM)	P (μM)	Fe/P (molar)	Vol[b] (mL)
October 2010 (MX30)					
MX30–SPM1	9	0.07	0.15	0.49	
MX30-SPM2	11	0.06	0.18	0.32	
MX30-SPM3	48	0.41	0.14	2.93	
MX30-SPM4	52	0.26	0.11	2.29	
MX30-SPM5	54	1.50	0.50	3.03	
MX30-SPM6	56	3.37	0.97	3.48	
MX30-SPM7	73	1.00	0.32	3.13	
MX30-SPM8	85	1.92	0.69	2.79	
November 2011 (MX35)					
MX35-SPM1	40.2	0.13	0.10	1.33	500
MX35-SPM2	51.7	1.26	0.40	3.16	500
MX35-SPM3	52.6	2.46	0.73	3.35	500
MX35-SPM4	57.7	0.26	0.08	3.22	500
MX35-SPM5	60.3	0.51	0.16	3.15	500
MX35-SPM6	65.2	1.14	0.24	4.67	500
MX35-SPM7	70.2	2.00	0.49	4.12	500
MX35-SPM8	80.2	3.33	1.08	3.08	500
June 2013 (MX41)					
MX41-SPM1	0	0.05	0.29	0.19	400
MX41-SPM2	10	0.03	0.29	0.09	500
MX41-SPM3	15	0.03	0.38	0.08	420
MX41-SPM4	50	0.22	0.08	2.95	500
MX41-SPM5	54.2	0.93	0.34	2.73	500
MX41-SPM6	55.1	1.40	0.52	2.71	500
MX41-SPM7	56.2	1.92	0.76	2.54	500
MX41-SPM8	57.2	1.08	0.54	2.00	500
MX41-SPM9	58.3	0.67	0.35	1.91	500
MX41-SPM10	59.3	0.62	0.31	1.99	700
MX41-SPM11	60.2	0.33	0.17	1.91	500
July 2013 (MX42)					
MX42-SPM1	57	2.30	0.85	2.72	1000

[a]Depth: depth in the water column. The oxic-anoxic transition zone is at 50–60 m depth
[b]Vol: volume of water filtered for suspended particulate matter collection

indicates that Fe(II) and SRP are released at the bottom of the lake, diffuse through the monimolimnion, and are consumed at the redox boundary (Michard et al. 1994, 2003). The consumption is attributed to Fe(II) oxidation to Fe(III), followed by precipitation of a Fe(III)-bearing solid phase (Fig. 12.9). The oxidation-precipitation process is demonstrated from suspended particulate matter analyses and turbidity profiles, showing a peak of Fe particles at 50–60 m depth (Fig. 12.3). This is also suggested by Fe(II)/Fe_{total} measurements of the solid phase by XANES spectroscopy at different depths (Cosmidis et al. 2014). This is finally supported by the large shift in Fe isotope composition of the dissolved Fe(II), corresponding to a depletion in heavy isotopes after Fe(III) precipitate formation (Busigny et al. 2014). The solid phase formed at 50–60 m depth was previously proposed to be Fe-oxyhydroxides with phosphate adsorption (Michard et al. 1994; Viollier et al. 1995) but a recent mineralogical study also identified the presence of mixed valence amorphous Fe(II)-Fe(III)-phosphate and possibly Fe(III)-phosphate in sediment traps placed at 56 m in the water column (Cosmidis et al. 2014). The linear correlation observed between Fe and P concentrations in SPM provides further support for Fe phosphates and/or Fe-oxyhydroxides with strong phosphate adsorption (Fig. 12.4). The Fe/P molar ratio obtained from bulk SPM analyses (3.07) is significantly higher than the one expected for pure Fe phosphates (between 1.5 and 2) but also lower than the one of hydrous ferric oxides (HFO) with adsorbed P (Fe/P ~ 4; e.g. Michard et al. 1994; Li et al. 2012; Voegelin et al. 2013; Cosmidis et al. 2014). The intermediate Fe/P ratio in SPM may indicate that

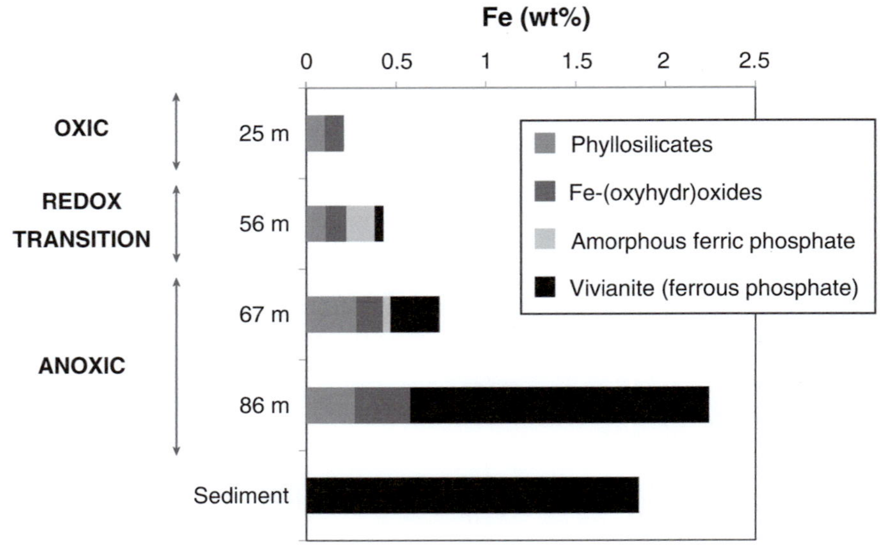

Fig. 12.5 Iron mineralogy in sediment traps from the water column (at 25, 56, 67 and 86 m depth) and sediments from the lake bottom (Modified from Cosmidis et al. 2014). The results were obtained by combining EXAFS spectra treatments with bulk Fe content measurements

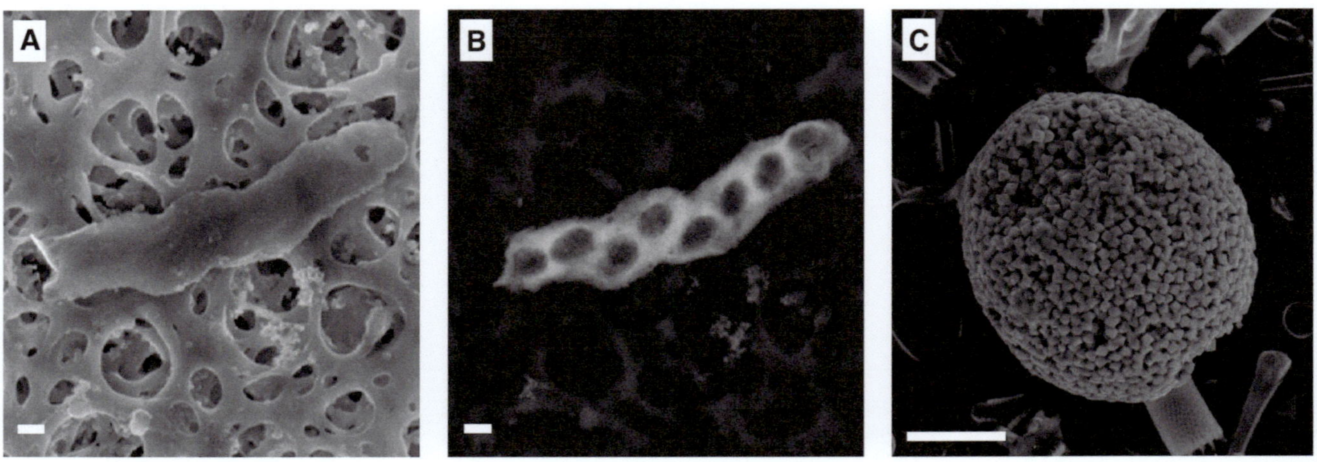

Fig. 12.6 Scanning electron microscopy (SEM) images of Fe particles from Lac Pavin. (**a**) and (**b**): Iron phosphate, collected on a filter, from 67 m depth in the water column. Same view in secondary electron mode (**a**) and back-scattered electron mode (**b**). Iron phosphate particles are made of chain of single cocci-shaped compartments, suggesting microbial control of Fe phosphate precipitation. Scale bar is 1 μm. (**c**) pyrite framboid from lake bottom sediment. *Scale bar* represents 10 μm

the solid phase corresponds to a mixing between P-rich HFO and Fe phosphate. This could be determined in the future by nano-scale mapping of the Fe/P ratio in SPM samples. The Fe oxidation at the redox boundary can be mediated by O_2, particularly during mixing episode of the shallow waters down to the chemocline, but other oxidizing agents such as NO_3^- and Mn(IV) are also likely involved in more quiet periods. For instance, the peak of dissolved Mn(II) at the redox transition zone, interpreted as Mn oxide reductive dissolution provides evidence for Mn mediated oxidation (Fig. 12.1a; Busigny et al. 2014). Moreover, microorganisms may mediate some of these oxidation processes but their identity and the exact involved metabolisms have yet to be identified (Cosmidis et al. 2014).

Solid Fe(III) phases sinking in the anoxic zone experience reductive dissolution in the water column (Fig. 12.9). This finding was supported by mineralogical study of the P- and Fe-rich phases (Cosmidis et al. 2014), although preservation of Fe(III) down to the sediments was observed (Fig. 12.5). This reductive dissolution in the water column is also required from geochemical modeling in order to reproduce the very low turbidity values and the variations of Fe isotope composition in the depth range 60–65 m, just below the peak of Fe(III) particles formation (Fig. 12.3). The progressive increase in turbidity and Fe content of suspended particulate matter below 65 m depth, down to the lake bottom (Fig. 12.3), indicates that an Fe(II) phase precipitates in the monimolimnion and increases progressively the particles flow. Mineralogical analyses have identified well-crystallized vivianite, an Fe(II)-phosphate. Specifically, vivianite was the dominant Fe-bearing phase detected in sediment traps from the anoxic zone and in the lake bottom sediments (Fig. 12.5).

Table 12.3 Description of the parameters and constants used for the biogeochemical model of Lac Pavin

1. Notation of variables	
c_Fe54: concentration of dissolved ^{54}Fe	
c_Fe56: concentration of dissolved ^{56}Fe	
x_Fe54: concentration of particulate ^{54}Fe	
x_Fe56: concentration of particulate ^{56}Fe	
c_P: concentration of dissolved PO_4^{3-}	
2. List of reactions	*(kinetic equations)*
In the mixolimnion	
Iron oxidation	V = v_ox_Fe*c_Fe/(c_Fe-c_M)
Phosphate adsorption	V = v_ads_P*c_P
In the monimolimnion	
Iron reduction	V = v_red_Fe*x_Fe
Vivianite precipitation	V = v_viv_Fe*(c_Fe*c_P-Ks)
3. Kinetic constants and adjusted parameters	
Monod parameter (c_M)	0.01 $mol.m^{-3}$
Fe concentration in the subsurface spring	0.2 $mol.m^{-3}$
PO_4^{3-} concentration in the subsurface spring	0.05 mol/m^3
Benthic Fe flux (from sediment)	9.2×10^{-4} $mol.m^{-2}.day^{-1}$
Benthic PO_4^{3-} flux (from sediment)	4.5×10^{-4} $mol.m^{-2}.day^{-1}$
Vivianite solubility product	0.12 $mol^2.m^{-6}$
Fe oxidation kinetic constant (v_ox_Fe)	1×10^{-3} day^{-1}
PO_4^{3-} adsorption kinetic constant (v_ads_P)	3×10^{-4} $m^3.mol^{-1}.day^{-1}$
Fe reduction kinetic constant (v_red_Fe)	0.2×10^{-4} $m^3.mol^{-1}.day^{-1}$
Vivianite precipitation kinetic constant (v_viv_Fe)	2.7×10^{-4} $m^6.mol^{-2}.day^{-1}$
Fe isotope fractionation factors ($^{56}Fe/^{54}Fe$ ratio)	
oxidation/precipitation	1.0009
reduction/dissolution	0.999
vivianite precipitation	0.9995
Fe isotope compositions ($^{56}Fe/^{54}Fe$ ratio)	
benthic Fe flux	0.2 ‰
subsurface ferruginous spring	−1 ‰

Mineralogical study was not able to detect Fe(III) minerals in the sediments, while they were still present in the deepest sediment trap at 86 m in the water column (Cosmidis et al. 2014). This suggests complete Fe(III) reductive dissolution at the sediment-water interface, probably favored by the high availability of fresh and reactive organic matter. A complete reduction of Fe(III) minerals is also supported by the $\delta^{56}Fe$ values (near 0 ‰) of sediment porewaters (Busigny et al. 2014), because partial Fe(III) reduction would produce dissolved Fe(II) with markedly negative $\delta^{56}Fe$ values, which are not detected (Severmann et al. 2006; Crosby et al. 2005).

An important question to address in future studies is whether Fe(III) is mostly reduced in the water column and/or at the water-sediment interface, and in which proportion? From the profiles obtained on Fe in SPM and turbidity (Fig. 12.3), it appears that a large part of the Fe(III) particles precipitated at the redox transition zone re-dissolves rapidly when particles sink in the anoxic water column. This conclusion is deduced from geochemical modeling (Fig. 12.7d). However, the high concentration of Fe(II) and SRP at the lake bottom requires benthic fluxes from the sediment, which should be derived from the dissolution of an Fe-P-rich phase. This phase could be either (1) a residual Fe(III)-P rich phase formed at the redox transition zone and having escaped water column reductive dissolution, (2) a detrital Fe(III)-phase with adsorbed P (as evidenced in Fig. 12.5), (3) a metastable Fe(II) phosphate, formed in the anoxic Fe-P-rich deep waters, that transforms into vivianite during sediment early diagenesis, or (4) a combination of these three cases. It is worth noting that the flux of Fe(III)-P phase reaching the lake bottom sediment can be increased occasionally for instance if particles located at the redox transition zone are flushed by diatom blooms, resulting in an increase of the particle sedimentation flux. The flux of Fe(III) can also be increased during periods of lake mixing down to the redox boundary, inducing a drastic precipitation of Fe(III)-P phase. In this case, the steady-state regime used for our modeling is no longer valid and these punctual disturbances may explain that a larger fraction of the Fe(III) flux reaches the water-sediment interface. In any case, a benthic flux of Fe(II) and

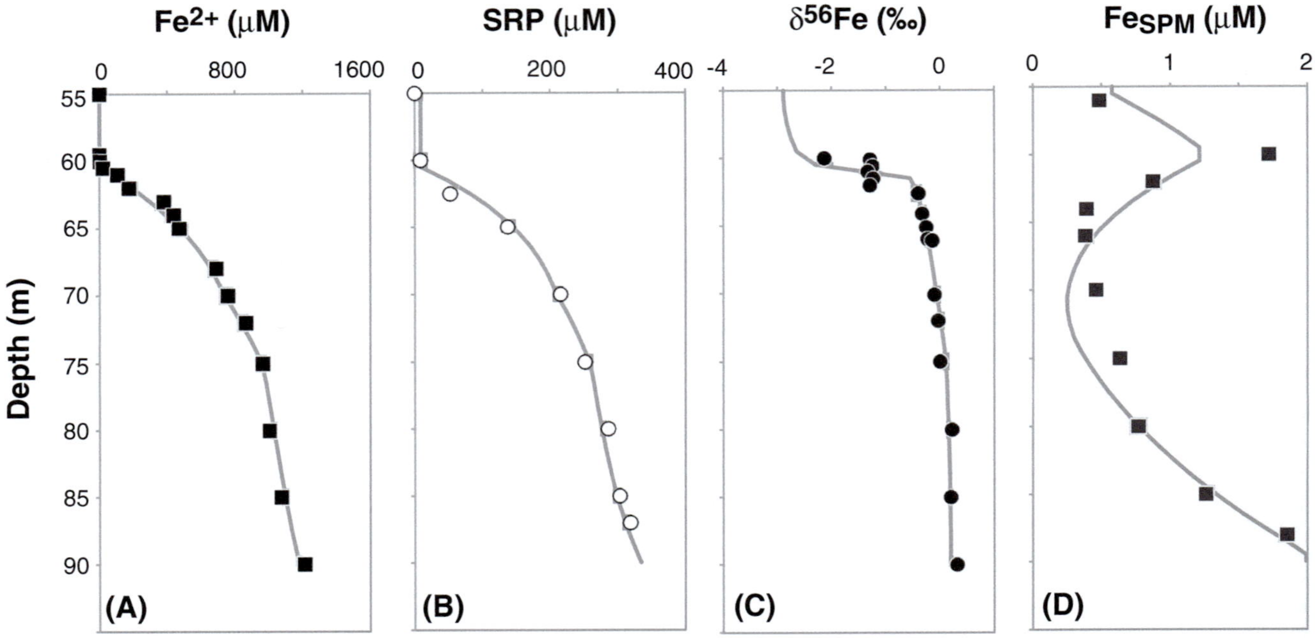

Fig. 12.7 Comparison between the results of the geochemical modeling (*lines*) and data measured (*symbols*) on natural Lac Pavin samples. (**a**) Dissolved Fe(II) concentration, (**b**) dissolved phosphate concentration (SRP: soluble reactive phosphorus), (**c**) isotope composition of the dissolved Fe, (**d**) Fe concentration of suspended particulate matter (SPM)

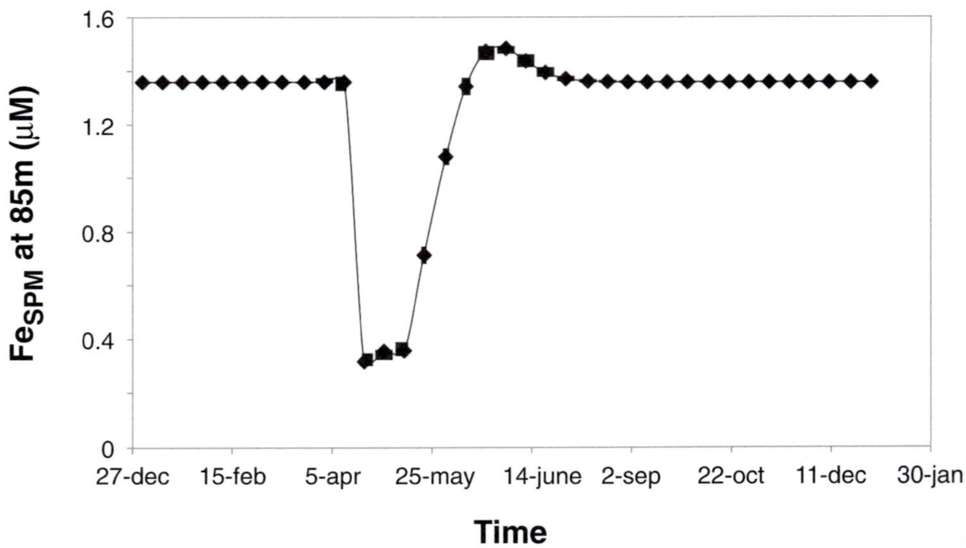

Fig. 12.8 Results of the "pseudo-steady" state geochemical model. Iron concentration in suspended particulate matter (SPM) at 85 m depth in the water column as a function of time. From the 11th of April to 11th of May, diatoms bloom is represented by an increase of the particle sedimentation flux, inducing a sweep of particles in the water column (decrease of Fe_{SPM}). The model shows that the system comes back to equilibrium state in less than 2 months after disturbance

PO_4^{3-} is clearly released from the sediment, and then migrates by diffusion, due to concentration gradients.

The global Fe geochemical cycle described above (release of dissolved Fe(II), diffusion, oxidation-precipitation, Fe(III) particle drop, and reductive dissolution) represents a specific case of the "ferrous wheel" (Campbell and Torgersen 1980), in which phosphate is intimately connected with iron. Most reactions of Fe oxidation, reduction and precipitation in Lac Pavin are likely bacterially-mediated as evidenced from high-resolution images of Fe particles (Fig. 12.6; Cosmidis et al. 2014), involving possibly a large diversity of microorganisms evidenced by molecular diversity analyses (Lehours et al. 2005, 2007, 2009).

In Lac Pavin, dissolved Fe pool of the monimolimnion is actively sequestered in the sediment by vivianite precipitation. Thus, the maintenance of a large dissolved Fe pool requires a ferruginous subsurface spring (Meybeck et al. 1975; Jézéquel et al. 2011). The reaction of vivianite precipitation can be written as:

$$3Fe^{2+}_{(aq)} + 2PO_4^{3-}{}_{(aq)} + 8\,H_2O \rightleftarrows Fe_3(PO_4)_2.8H_2O_{(s)}$$

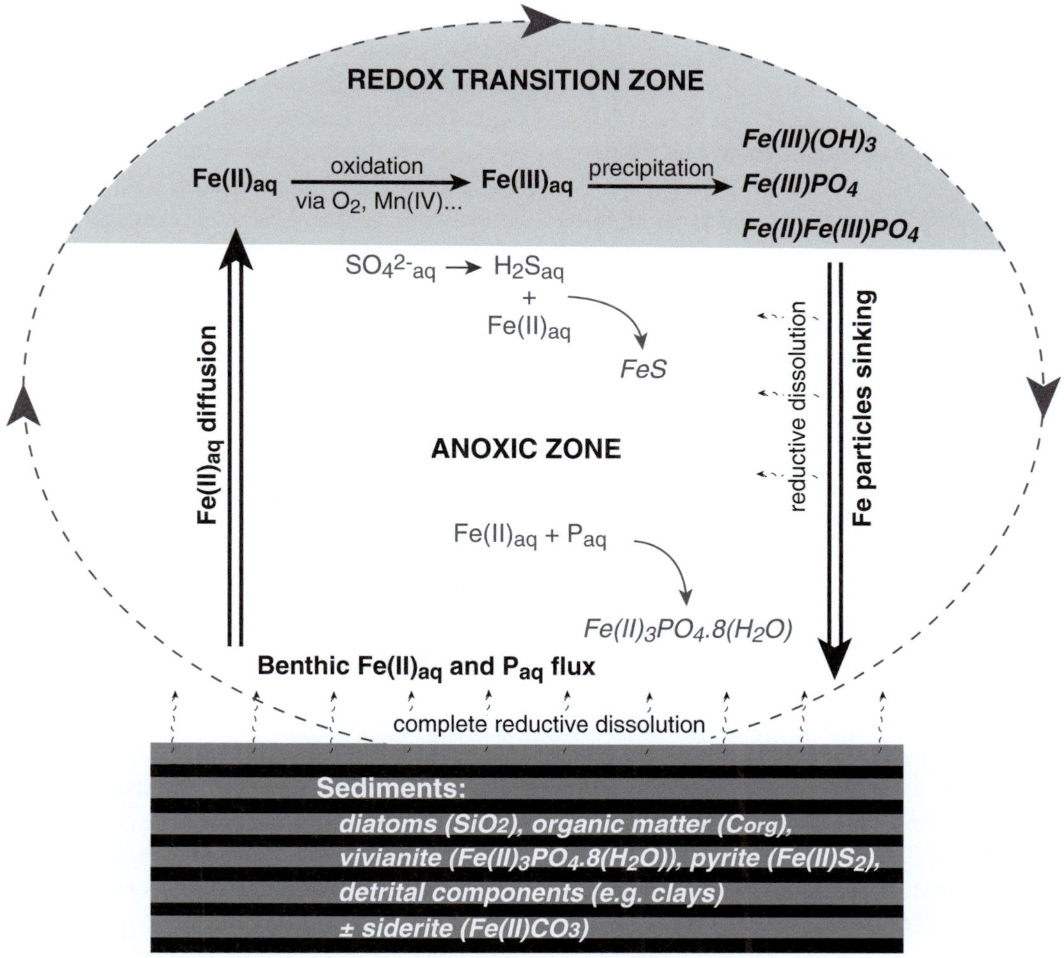

Fig. 12.9 Schematic model illustrating the Fe biogeochemical cycle in Lac Pavin (see main text for description). The so-called "ferrous wheel" is highlighted in *black* and *bold*. Only the processes in the anoxic zone (monimolimnion) and the redox transition zone (identified by the *light grey* area) are represented, while the oxic zone (mixolimnion) is not illustrated here. The chemical reactions in grey correspond to additional interactions with sulfur and phosphate cycles (beside the "iron wheel"). Solid phases are reported in italic. The subscript "aq" stands for aqueous chemical species (*i.e.* dissolved)

with corresponding solubility product of ~10^{-36} at 5 °C (Al-Borno and Tomson 1994), the temperature of Lac Pavin deep waters. Another sink of the dissolved Fe pool to the sediment is driven by the precipitation of pyrite (Fig. 12.6). The mechanism of pyrite formation in Lac Pavin remains unclear but it is likely that pyrite (Fe(II)S(0)S(−II)) or its precursor (e.g. mackinawite, FeS) is formed just below the redox boundary between 60 and 65 m depth, where sulfate is reduced into sulfide, and sulfide reacts with dissolved Fe(II) (Fig. 12.9; Bura-Nakic et al. 2009). This hypothesis was proposed to explain the strong similarity between Fe isotope compositions of sedimentary pyrites and dissolved Fe(II) at the redox boundary (Busigny et al. 2014). The solubility of pyrite in water at ambient temperature is very low. Although various reaction pathways can be considered (e.g. Butler et al. 2005), the equilibrium solubility product is for instance $10^{-7.2}$ for the reaction (Rickard and Luther 2007):

$$Fe^{2+}_{(aq)} + H_2S_{(aq)} + S^0 \rightleftarrows FeS_{2(s)} + 2H^+_{(aq)}$$

The last authigenic Fe-bearing solid phase, which has been identified in Lac Pavin sediments, is siderite ($FeCO_3$; Schettler et al. 2007). Siderite precipitation can be written as:

$$Fe^{2+}_{(aq)} + CO_3^{2-}{}_{(aq)} \rightarrow FeCO_{3(s)}$$

Siderite solubility product (K_{sp}) is strongly dependent on temperature and varies over five orders of magnitude from 25 to 250 °C (Bénézeth et al. 2009). The siderite K_{sp} is not well-known for temperature of ~5 °C but is near 10^{-10}, thus indicating a very low solubility at high Fe(II) concentration such as those measured in Lac Pavin. Although siderite may precipitate during early diagenetic process, at the sediment water interface, when Fe(III) is reduced to Fe(II) and highly reactive organic matter is oxidized to CO_2 by anaerobic microbial respiration, the detailed mechanisms of siderite formation needs to be evaluated in future work.

12.4.2 Iron Isotope Fractionations in Lac Pavin

The geochemical model developed in the present contribution provides an estimate of Fe isotope fractionations for Fe oxidation-precipitation and reduction-dissolution, and vivianite precipitation in Lac Pavin. Contrasting with some geochemical parameters adjusted in our modeling, the tests of sensitivity demonstrate that Fe isotope fractionations are reasonably well constrained, and can be compared to data from the literature.

In Lac Pavin, Fe oxidation-precipitation at the redox transition zone produces an isotope fractionation of ~1‰ (0.9±0.2‰) (Table 12.3), leaving the residual aqueous Fe(II) depleted in heavy isotopes (Fig. 12.7c). This result is compatible with experimental data available for $Fe(II)_{aq}$ oxidation to $Fe(III)_{aq}$, showing that $Fe(III)_{aq}$ is systematically enriched in the heavy isotopes (*e.g.* Johnson et al. 2002; Welch et al. 2003; Anbar et al. 2005; Balci et al. 2006). The enrichment in heavy Fe isotopes during oxidation is generally counterbalanced by a kinetic isotope fractionation during precipitation, enriching Fe(III) precipitate in the light isotopes (Skulan et al. 2002). The net isotope fractionation thus depends on the rate of precipitation but produces in most cases a precipitate enriched in the heavy isotopes of Fe. For instance, in experiments of abiotic precipitation of ferrihydrite from $Fe(II)_{aq}$, the precipitate was enriched in heavy isotopes by ~0.9‰ (Bullen et al. 2001), which is similar to the value deduced from our geochemical modeling. Experimental culture of Fe(II)-oxidizing photoautotrophic bacteria under anaerobic condition showed that Fe oxyhydroxide precipitate was enriched in the heavier isotope relative to $Fe(II)_{aq}$ by ~1.5±0.2‰ (Croal et al. 2004). The close similarity between Fe isotope fractionation in abiotic and biotic experiments prevents any assessment of life involvement in the Fe oxidation process at Lac Pavin. SEM images of particulate matter collected on filters at the redox transition zone shows encrusted microbial shape, suggesting that Fe oxidation may be bacterially-mediated.

Iron reduction also plays a key role in the Fe biogeochemical cycle at Lac Pavin. The geochemical model predicts a Fe isotope fractionation of ~1‰, with an enrichment in light isotopes in the produced $Fe(II)_{aq}$ (Table 12.3). This value is similar in magnitude but reverse in direction to the Fe isotope fractionation determined for oxidation. From various experimental studies, it was shown that Fe(III) substrate reduction and dissolution preferentially release light Fe isotopes into solution, in good agreement with the results of the present study (*e.g.* Crosby et al. 2005). Abiotic mineral dissolution produces either no or limited Fe isotope fractionation, depending on the mechanism (proton-promoted, ligand-controlled or reductive dissolution). In contrast, bacterial dissimilatory iron reduction (DIR) can be associated with large Fe isotope fractionation, up to 3‰, between $Fe(II)_{aq}$ and the ferric substrate (Beard and Johnson 1999; Crosby et al. 2005; Johnson et al. 2008). In Lac Pavin, Fe(III) reduction is probably bacterially mediated, as demonstrated by culture experiments on water samples from the monimolimnion (Lehours et al. 2009). In particular, facultative Fe(III) reducers such as fermentative and sulphate-reducing bacteria were enriched in culture experiments while obligatory Fe(III) reducers, such as *Geobacter*, were not detected. Although not discriminating, the Fe isotope fractionation determined from our geochemical modeling is fully compatible with a bacterial Fe reduction in Lac Pavin.

The last Fe isotope fractionation value determined from modeling corresponds to vivianite precipitation process. To our knowledge, no previous study determined such fractionation, either experimentally or from theoretical calculation. The present results thus provide the first estimate of this fractionation. We find that vivianite should be enriched in the light isotope of Fe by ~0.5‰ relative to dissolved Fe(II) (Table 12.3). This may reflect a kinetic isotope process associated with vivianite precipitation, but needs to be explored further in future studies. An interesting observation is that the spherical shape of vivianite particles formed in the monimolimnion (n) and their close association with organic matter suggest that microorganisms might be involved in the precipitation process (Cosmidis et al. 2014).

12.5 Conclusion and Perspective

Over the last decades, geochemical data and modeling on dissolved and particulate matters coupled with mineralogical analyses have provided a global picture of the biogeochemical machinery in Lac Pavin. Like for other anoxic and ferruginous aquatic systems and seasonally stratified lake, Lac Pavin is characterized by a "ferrous wheel" or "iron wheel", which corresponds to strong Fe cycling through successive reductive dissolution, diffusion and oxidation-precipitation reactions. In Lac Pavin, Fe cycle is intimately associated with P cycle through precipitation of Fe(II)-Fe(III)-phosphates, P-rich HFO and Fe-oxyhydroxides with high amount of adsorbed phosphate, at the redox transition zone, and release of SRP during Fe(III) reductive dissolution under anoxic conditions. This leads to extremely high concentrations of dissolved Fe(II) and SRP at the lake bottom, inducing ferrous phosphate (*i.e.* vivianite) precipitation and sequestration to the sediment.

In addition, phosphorus cycle is also partly controlled by microorganisms in Lac Pavin, which can store high amounts of intracellular polyphosphates and/or catalyze the hydrolysis of these polymers (Miot et al. 2016). How this intracellular polyphosphate/orthophosphate cycle superimposes with Fe-phosphate biomineralization will have to be determined by future studies. Moreover, the identification of the microbial diversity specifically involved in these processes as well as the oxidation and reduction of Fe in Lac Pavin has yet to be assessed precisely.

Iron also interacts with sulfur at the redox transition zone, where sulfate is reduced into sulfide, and forms insoluble FeS particles. The fate of these FeS particles is unclear but they ultimately transform into pyrite (FeS_2), which is found in the lake sediments. An important question to elucidate in the future is the mechanism of pyrite formation, since this would greatly help in understanding the geochemical signature recorded in ancient sediments deposited under ferruginous conditions. Systematic analyses of sulfur isotopes and dedicated mineralogical study may provide strong constraints on pyrite formation pathways in Lac Pavin. Another Fe-bearing mineral of interest for ancient sediment interpretation and paleo-environment reconstruction is siderite, a ferrous carbonate ($FeCO_3$). Contrasting with vivianite, which is ubiquitous in Lac Pavin sediments, siderite has only been identified in restricted layers of the lake bottom sediment (Schettler et al. 2007). Water in the monimolimnion is grossly supersaturated with respect to siderite (Michard et al. 1994) but it remains undetermined whether siderite precipitates in the water column or in the sediment, and what processes can lead to siderite formation. Despite some uncertainties in mineral formation pathways, Lac Pavin provides a well-understood analogue for organic and iron rich sedimentary systems on early Earth. As such Lac Pavin is an ideal spot calibration of novel geochemical and biological signatures.

References

Aeschbach-Hertig W, Hofer M, Schmid M, Kipfer R, Imboden DI (2002) The physical structure and dynamics of a deep, meromictic crater lake (Lac Pavin, France). Hydrobiologia 487:111–136

Albéric P, Viollier E, Jezequel D, Grosbois C, Michard G (2000) Interactions between trace elements and dissolved organic matter in the stagnant anoxic deep layer of a meromictic lake. Limnol Oceanogr 45:1088–1096

Albéric P, Jézéquel D, Bergonzini L, Chapron E, Viollier E, Massault M, Michard G (2013) Carbon cycling and organic radiocarbon reservoir effect in a meromictic crater lake (Lac Pavin, Puy-de-Dôme, France). Radiocarbon 55:1029–1042

Al-Borno A, Tomson MB (1994) The temperature dependence of the solubility product constant of vivianite. Geochim Cosmochim Acta 58:5373–5378

Amblard C (1984) Variations nycthémérales des concentrations en nucléotides adényliques d'un phytoplancton lacustre (Le Pavin, France). Verh Internat Verein Limnol 22:1011–1018

Amblard C (1986) A study of spatial and temporal variability in the adenine nucleotides of lake phytoplankton during a diel cycle (Lac Pavin - France). Hydrobiologia 137:159–173

Amblard C (1988) Seasonal succession and strategies of phytoplankton development in two lakes of different trophic states. J Plankton Res 10(6):1189–1208

Amblard C, Bourdier G (1990) The spring bloom of the diatom Melosira italica subsp. subarctica in Lake Pavin: biochemical, energetic and metabolic aspects during sedimentation. J Plankton Res 12:645–651

Anbar AD, Jarzecki AA, Spiro TG (2005) Theoretical investigation of iron isotope fractionation between $Fe(H_2O)_6^{3+}$ and $Fe(H_2O)_6^{2+}$: implications for iron stable isotope geochemistry. Geochim Cosmochim Acta 69:825–837

Assayag N, Jézéquel D, Ader M, Viollier E, Michard G, Prévot F, Agrinier P (2008) Hydrological budget, carbon sources and biogeochemical processes in Lac Pavin (France): constraints from $\delta^{18}O$ of water and $\delta^{13}C$ of dissolved inotganic carbon. Appl Geochem 23:2800–2816

Balci N, Bullen TD, Witte-Lien K, Shanks WC, Motelica M, Mandernack KW (2006) Iron isotope fractionation during microbially stimulated Fe(II) oxidation and Fe(III) precipitation. Geochim Cosmochim Acta 70:622–639

Beard BL, Johnson CM (1999) High precision iron isotope measurements of terrestrial and lunar materials. Geochim Cosmochim Acta 63:1653–1660

Bénézeth P, Dandurand JL, Harrichoury JC (2009) Solubility product of siderite ($FeCO_3$) as a function of temperature (25–250°C). Chem Geol 265:3–12

Biderre-Petit C, Jézéquel D, Dugat-Bony E, Lopes F, Kuever J, Borrel G, Viollier E, Fonty G, Peyret P (2011) Identification of microbial communities involved in the methane cycle of a freshwater meromictic lake. FEMS Microbiol Ecol 77:533–545

Bonhomme C, Poulin M, Vinçon-Leite B, Saad M, Groleau A, Jézéquel D, Tassin B (2011) Maintaining meromixis in Lake Pavin (Auvergne, France): the key role of a sublacustrine spring. C R Geosci 343:749–759

Bullen TD, White AF, Childs CW, Vivit DV, Schulz MS (2001) Demonstration of significant abiotic iron isotope fractionation in nature. Geology 29:699–702

Bura-Nakic E, Viollier E, Jézéquel D, Thiam A, Ciglenecki I (2009) Reduced sulfur and iron species in anoxic water column of meromictic crater Lake Pavin (Massif Central, France). Chem Geol 266:320–326

Busigny V, Planavsky NJ, Jézéquel D, Crowe S, Louvat P, Moureau J, Viollier E, Lyons TW (2014) Iron isotopes in an Archean ocean analogue. Geochim Cosmochim Acta 133:443–462

Butler IB, Archer C, Vance D, Oldroyd A, Rickard D (2005) Fe isotope fractionation on FeS formation in ambient aqueous solution. Earth Planet Sci Lett 236:430–442

Campbell P, Torgersen T (1980) Maintenance of iron meromixis by iron redeposition in a rapidly flushed monimolimnion. Can J Fish Aquat Sci 37:1303–1313

Chapron E, Albéric P, Jézéquel D, Versteeg W, Bourdier JL, Sitbon J (2010) Multidisciplinary characterisation of sedimentary processes in a recent maar lake (Lake Pavin, French Massif Central) and implication for natural hazards. Nat Hazards Earth Syst Sci 10:1815–1827

Cosmidis J, Benzerara K, Morin G, Busigny V, Lebeau O, Jézéquel D, Noël V, Dublet G, Othmane G (2014) Biomineralization of mixed valence iron-phosphates in the anoxic water column of Lake Pavin (Massif Central, France). Geochim Cosmochim Acta 126:78–96

Croal LR, Johnson CM, Beard BL, Newman DK (2004) Iron isotope fractionation by Fe(II)-oxidizing photoautotrophic barteria. Geochim Cosmochim Acta 68:1227–1242

Crosby HA, Johnson CM, Roden EE, Beard BL (2005) Coupled Fe(II)-Fe(III) electron and atom exchange as a mechanism for Fe isotope fractionation during dissimilatory iron oxide reduction. Environ Sci Technol 39:6698–6704

Crowe SA, Jones CA, Katsev S, Magen C, O'Neill AH, Sturm A, Canfield DE, Haffner GD, Mucci A, Sundby B, Fowled DA (2008) Photoferrotrophs thrive in an Archean ocean analogue. Proc Natl Acad Sci 105:15938–15943

Davison W (1993) Iron and manganese in lakes. Earth Sci Rev 34:119–163

Devaux J (1980) Contribution à l'étude limnologique du Lac Pavin (France): Facteurs abiotiques et phytoplancton. Hydrobiologia 68:167–189

Hongve D (1997) Cycling of iron, manganese, and phosphate in a meromictic lake.

Jézéquel D, Sarazin G, Prévot F, Viollier E, Groleau A, Agrinier P, Albéric P, Binet S, Bergonzini L, Michard G (2011) Bilan hydrique du lac Pavin – water balance of the Lake Pavin. Rev Sci Nat d'Auvergne 74:75–96

Johnson CM, Skulan JL, Beard BL, Sun H, Nealson KH, Braterman PS (2002) Isotopic fractionation between Fe(III) and Fe(II) in aqueous solutions. Earth Planet Sci Lett 195:141–153

Johnson CM, Beard BL, Roden EE (2008) The iron isotope fingerprints of redox and biogeochemical cycling in modern and ancient earth. Annu Rev Earth Planet Sci 36:457–493

Konhauser KO, Hamade T, Raiswell R, Morris RC, Ferris FG, Southam G, Canfield DE (2002) Could bacteria have formed the Precambrian banded iron formations? Geology 30:1079–1082

Konhauser KO, Lalonde SV, Amskold L, Holland HD (2007) Was there really an Archean phosphate crisis? Science 315:1234

Konhauser KO, Kappler A, Roden EE (2011) Iron in microbial metabolisms. Elements 7:89–93

Lehours AC, Bardot C, Thenot A, Debroas D, Fonty G (2005) Anaerobic microbial communities in Lake Pavin, a unique meromictic lake in France. Appl Environ Microbiol 71:7389–7400

Lehours AC, Evans P, Bardot C, Joblin K, Fonty G (2007) Phylogenetic diversity of Archea and bacteria in the anoxic zone of a meromictic lake (Lake Pavin, France). Appl Environ Microbiol 73:2016–2019

Lehours AC, Batisson I, Guedon A, Mailhot G, Fonty G (2009) Diversity of culturable bacteria, from the anaerobic zone of the meromictic Lake Pavin, able to perform dissimilatory-iron reduction in different in vitro conditions. Geomicrobiol J 26:212–223

Li Q, Wang X, Kan D, Bartlett R, Pinay G, Ding Y, Ma W (2012) Enrichment of phosphate on ferrous iron phases during bio-reduction of ferrihydrite. Int J Geosci 3:314–320

Lopes F, Viollier E, Thiam A, Michard G, Abril G, Groleau A, Prévot F, Carrias JF, Jézéquel D (2011) Biogeochemical modeling of anaerobic vs. aerobic methane oxidation in a meromictic crater lake (Lake Pavin, France). Appl Geochem 26:1919–1932

Martin JM (1985) The Pavin Crater lake. In: Stumm W (ed) Chemical processes in lakes, Reviews in mineralogy and geochemistry. Wiley, New York, pp 169–188

Meybeck M, Martin JM, Olive P (1975) Géochimie des eaux et des sédiments de quelques lacs volcaniques du Massif Central français. Verh Internat Verein Limnol 19:1150–1164

Michard G, Viollier E, Jézéquel D, Sarazin G (1994) Geochemical study of a crater lake: Lac Pavin, France – Identification, location and quantification of the chemical reactions in the lake. Chem Geol 115:103–115

Michard G, Jézéquel D, Viollier E (2003) Vitesses des réactions de dissolution et précipitation au voisinage de l'interface oxydo-réducteur dans un lac méromictique : Le lac Pavin (Puy-de-Dôme, France). Revue des Sciences de l'Eau 16:199–218

Miot J, Jézéquel D, Benzerara K, Cordier L, Rivas-Lamelo S, Skouri-Panet F, Férard C, Poinsot M, Duprat E. (2016) Mineral diversity in Lake Pavin: connections with water column chemistry and biomineralization processes. Minerals 6, in press. doi :10.3390/

Poulton SW, Canfield DE (2011) Ferruginous conditions: a dominant feature of the ocean through earth's history. Elements 7:107–112

Reichert P (1998) AQUASIM 2.0 – computer program for the identification and simulation of aquatic systems. Swiss Federal Institute for Environmental Science and Technology (EAWAG), Dubendorf, pp 219

Restituito F (1984) Contribution à l'étude du sédiment d'un lac oligo-mésotrophe d'origine volcanique (Lac Pavin, France). Hydrobiologia 109:229–234

Restituito F (1987) Consequences of redox conditions on the distribution of cations in a meromictic oligotrophic lake. Hydrobiologia 144:63–76

Rickard D, Luther GW (2007) Chemistry of iron sulfides. Chem Rev 107:514–562

Schettler G, Schwab MJ, Stebich M (2007) A 700 years record of climate change based on geochemical and palynological data from varved sediments (Lac Pavin, France). Chem Geol 240:11–35

Severmann S, Johnson CM, Beard BL, McManus J (2006) The effect of early diagenesis on the Fe isotope compositions of porewaters and authigenic minerals in continental margin sediments. Geochim Cosmochim Acta 70:2006–2022

Skulan JL, Beard BL, Johnson CM (2002) Kinetic and equilibrium Fe isotope fractionation between aqueous Fe(III) and hematite. Geochim Cosmochim Acta 66:2995–3015

Staubwasser M, Schoenberg R, von Blanckenburg F, Krüger S, Pohl C (2013) Isotope fractionation between dissolved and suspended particulate Fe in the oxic and anoxic water column of the Baltic Sea. Biogeosciences 10:233–245

Stebich M, Brüchmann C, Kulbe T, Negendank JFW (2005) Vegetation history, human impact and climate change during the last 700 years recorded in annually laminated sediments of Lac Pavin. France Rev Palaeobot Palynol 133:115–133

Strelow FWE (1980) Improved separation of iron from copper and other elements by anion-exchange chromatography on a 4% cross-linked resin with high concentrations of hydrochloric acid. Talanta 27:727–732

Taylor PDP, Maeck R, De Bievre P (1992) Determination of the absolute isotopic composition and atomic weight of a reference sample of natural iron. Int J Mass Spectrom 121:111–125

Viollier E, Jézéquel D, Michard G, Pèpe M, Sarazin G, Albéric P (1995) Geochemical study of a crater lake (Pavin Lake, France): trace-element behaviour in the monimolimnion. Chem Geol 125:61–72

Viollier E, Michard G, Jézéquel D, Pèpe M, Sarazin G (1997) Geochemical study of a crater lake: Lake Pavin, Puy de Dôme France. Constraints afforded by the particulate matter distribution in the element cycling within the lake. Chem Geol 142:225–241

Viollier E, Inglett PW, Hunter K, Roychoudhury AN, Van Cappellen P (2000) The ferrozine method revisited: Fe(II)/Fe(III) determination in natural waters. Appl Geochem 15:785–790

Voegelin A, Senn AC, Kaegi R, Hug SJ, Mangold S (2013) Dynamic Fe-precipitate formation induced by Fe(II) oxidation in aerated phosphate-containing water. Geochim Cosmochim Acta 117:216–231

Welch SA, Beard BL, Johnson CM, Braterman PS (2003) Kinetic and equilibrium Fe isotope fractionation between aqueous Fe(II) and Fe(III). Geochim Cosmochim Acta 67:4231–4250

Part IV

Biology and Microbial Ecology

Télesphore Sime-Ngando

The Significance of Transparent Exopolymeric Particles (TEP) for Microorganisms in Lake Pavin

13

Jean-François Carrias, Christian Amblard, and Télesphore Sime-Ngando

Abstract

Transparent exopolymeric particles (TEP) consist of a matrix of colloidal fibrils and are produced from dissolved carbohydrate polymers exuded by phytoplankton and bacteria. These particles are involved in the formation of lake snow aggregates and are colonized by diverse microorganisms. Abundance, distribution, size spectra and bacterial colonization of TEP in Lake Pavin were studied and compared with data from other freshwater and marine systems. Abundance of TEP ranged from 10^5 to 10^6 particles L^{-1} and the majority of these particles contained attached bacteria. Bacterial density within TEP was related to temperature and decreased with particle size. TEP-associated bacteria were more important in the oligomesotrophic Lake Pavin than in the near-by eutrophic Lake Aydat, indicating that TEP are particularly important in environments with low nutrient loading. The community composition of particle-associated bacteria was different from that of free-living bacteria. In addition, we found that bacteria associated with TEP exhibit higher enzymatic activities than free-living cells in the surrounding water. Finally, the densities of heterotrophic nanoflagellates in Lake Pavin were more significantly related to the densities of TEP than to the densities of bacteria and the bacterial density within TEP. Based on comparisons with other freshwater and marine systems, we conclude that TEP are involved in the phytoplankton – primarily diatom – sedimentation processes and in the dynamics of bacteria and protozoa in the pelagic zone of aquatic ecosystems.

Keywords

Transparent exopolymeric particles • Bacterial colonization • Bacteria • Heterotrophic nanoflagellates • Freshwater • Lake Pavin

13.1 Introduction

Transparent exopolymeric particles (TEP) are colloidal fibrils produced from dissolved carbohydrate polymers secreted by phytoplankton and bacteria (Passow and Alldredge 1994; Passow 2000) and are easily detected by microscopy using an acidic solution of Alcian Blue. These particles are able to attach to each other and to other particles including bacteria, phytoplankton and detritus and then they promote the formation of large organic aggregates or lake snow (Simon et al. 2002). High TEP concentrations in lakes are generally associated with phytoplankton bloom indicating that most of these particles derived from dissolved organic carbon released by phytoplankton. TEP-associated bacteria (Fig. 13.1) generally comprised 1–20 % of the total bacteria. Attached bacteria can account for a large proportion of the bacterial production in lakes. TEP are considered as hotspot microenvironments, and the activity of attached bacteria may stimulate the growth rates of the free-living microorganisms in the vicinity of the particle (Simon et al. 2002). In addition, protozoa and metazooplankton can consume

J.-F. Carrias (✉) • C. Amblard • T. Sime-Ngando
LMGE, Laboratoire Microorganismes: Génome et Environnement, UMR CNRS 6023, Université Clermont-Auvergne, Université Blaise Pascal, BP 80026, 63171 Aubière Cedex, France
e-mail: j-francois.carrias@univ-bpclermont.fr

T. Sime-Ngando et al. (eds.), *Lake Pavin*, DOI 10.1007/978-3-319-39961-4_13

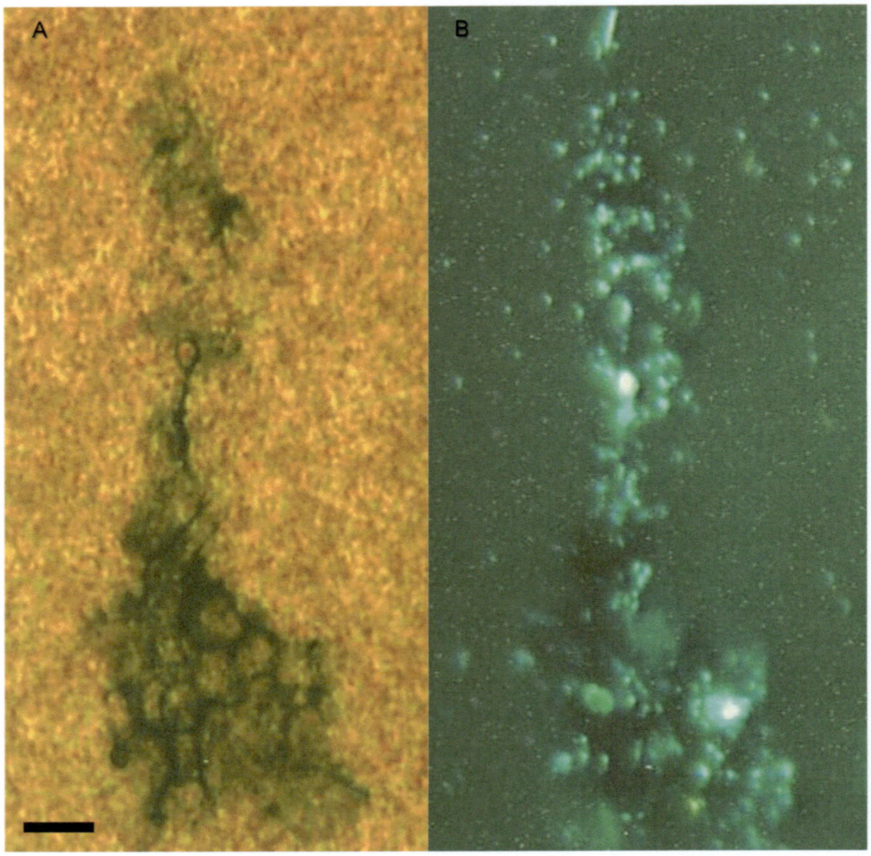

Fig. 13.1 Double-staining procedure showing a particle stained with Alcian Blue (**a**) and the DAPI-stained bacteria associated to this particle (**b**). Scale bar represents around 5 µm. See also http://www.aslo.org/lo/toc/vol_47/issue_4/1202a1.html

TEP and attached bacteria (Passow 2002). Therefore TEP constitute a link to higher trophic levels, influencing the fluxes of matter in lakes.

The importance of TEP in Lake Pavin was assessed during the spring diatom bloom in 2000, 2002, 2005 and 2006. The scopes of these studies were (1) to determine the abundance, the size spectra and the distribution of TEP, (2) to analyse the bacterial density within TEP, and (3) to test the potential impact of TEP on the composition and distribution of both bacteria and protozoa. We reviewed here the main results of these studies and the roles of these particles in Lake Pavin, in the general context of freshwater ecology.

13.2 Abundance and Distribution of TEP

Microscopic counts from Lake Pavin revealed that the abundance of TEP ranged from 0.3×10^5 particles L^{-1} to 13.4×10^5 particles L^{-1}. Annual means ranged from 3.2×10^5 particles L^{-1} to 6.7×10^5 particles L^{-1} (Carrias et al. 2002; Arnous et al. 2010). The abundance of TEP decreased with depth in the mixolimnion (Fig. 13.2) and was related to diatom densities. TEP are however recorded in high numbers at the oxic-anoxic boundary (58–60 m depth), where they are probably produced by the large quantities of dissolved organic carbon from the monimolimnion. Cumulative surface area of TEP averaged 35.5 ± 28.7 mm^2L^{-1} (Range: 4.6–85.9 mm^2L^{-1}). Both the abundance and the surface area of TEP were lower in Lake Pavin than in Lake Aydat (Carrias et al. 2002; Lemarchand et al. 2006) and were related to chlorophyll *a* concentrations and epilimnetic phytoplankton biomass. These features indicate that TEP, which originate from dissolved organic matter, increase with the productivity of the environment. The abundances recorded in Lake Pavin are within the range of those reported from marine environments of similar trophic status (Passow and Alldredge 1994; Schuster and Herndl 1995; Mari and Kiørboe 1996).

13.3 Size Spectra of TEP

The size spectra of TEP were described by the power relationship $dN/dl = kl^{-(\beta+1)}$ where the constant k depends on the concentration of particles and β describes the size spectra; the smaller β is, the smaller the fraction of small particles (Passow and Alldredge 1994; Mari and Kiørboe 1996). In practice, for each size interval, the particle concentration (dN) was normalized to the length of the interval (dl) and

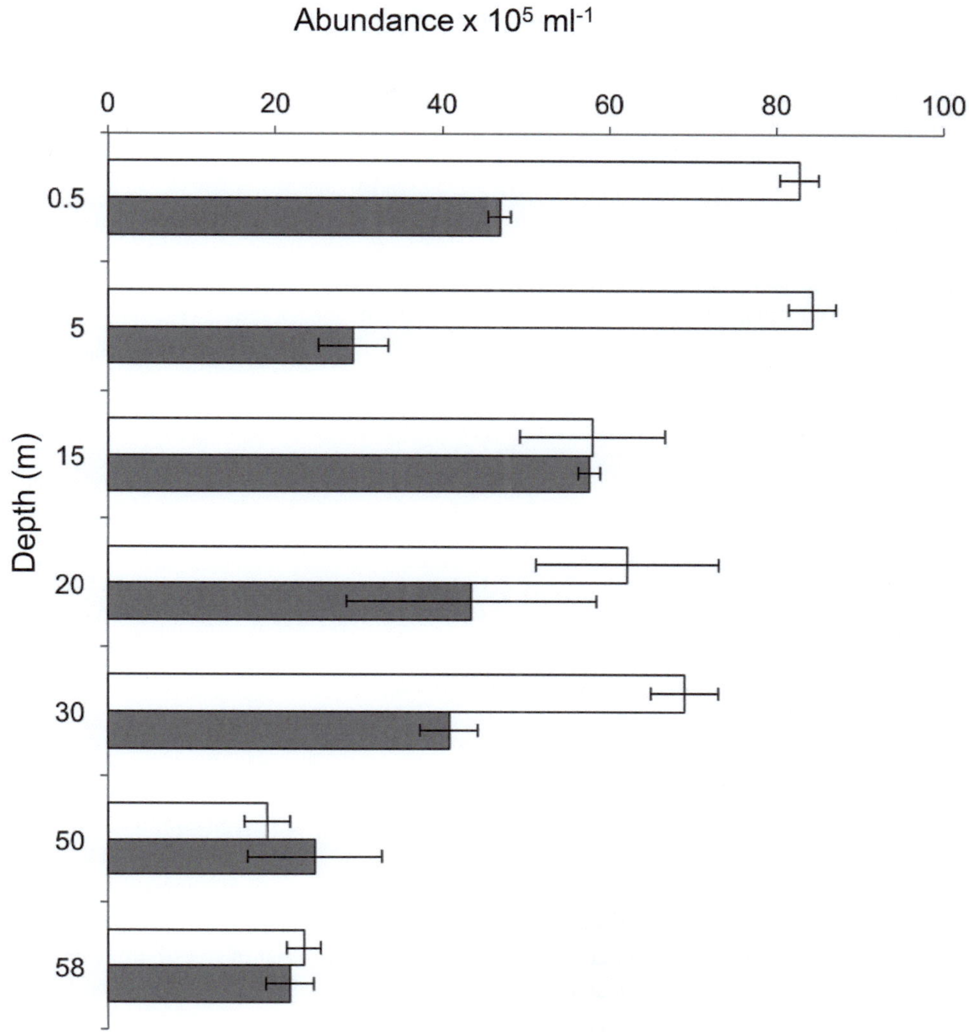

Fig. 13.2 Vertical distribution of the abundance of TEP in April 3 (*white histogram*) and May 4 (*Grey histogram*) during spring 2005 in Lake Pavin

plotted against the arithmetic average length (l) (Carrias et al. 2002). The parameters k and β are estimated by regression after logarithmic transformation. Such analyses of the TEP size spectra revealed that the particles followed a non-steady-state distribution during the spring period. The β values were always lower than 4 and close to the values reported by Mari and Kiørboe (1996) in marine waters.

13.4 Bacterial Within TEP

Double-staining procedure with DAPI and Alcian Blue (Fig. 13.1) indicates that > 97 % of TEP contained attached bacteria. The number of attached bacteria per particle increased with the size of the particle. However, expressed per unit area of particle, the density decreased with the size of the particle (Table 13.1). The bacterial density within TEP can be expressed by the power law $n = ad^b$ (Mari and Kiørboe 1996) where n is the number of attached bacteria per particle, d is the equivalent spherical diameter of the particle, and a and b are constants for a given sample. These constants were obtained from log-log plots of n versus d. The density of TEP-associated bacteria, expressed by the constant a (the slope of the regression lines between the number of bacteria per particle and the equivalent spherical diameter of the particles), was slightly higher in Lake Pavin than in the eutrophic Lake Aydat (Carrias et al. 2002) suggesting that TEP are more suitable substrates for bacteria in the oligomesotrophic environment. We also found that the constant a tends to decrease with depth in relation with decreasing temperature values (Arnous et al. 2010). TEP-associated bacteria represented 0.3–8.9 % of total DAPI-stained bacteria corresponding to about 10^7–10^8 bacteria L^{-1} associated with polysacharidic particles.

Table 13.1 Average values for bacterial colonization of TEP in Lake Pavin during spring 2000

Size interval	Bacteria per particle	Bacteria per unit area (N per μm^2)	Number of attached bacteria (× 10^3 ml^{-1})
2–5 μm	3.05	0.32	0.04
5–10 μm	6.18	0.21	0.81
10–15 μm	10.47	0.17	1.21
15–20 μm	16.28	0.12	0.90

Modified from Carrias et al. (2002)

Table 13.2 Comparison (one-way ANOVA) between free and attached fractions of the various bacterial groups expressed as mean abundances or as percentages of DAPI or EUB338 -stained bacteria in Lake Pavin

	Free	Attached	*P*
Total DAPI cells (nb.ml^{-1})	$3.8 \pm 1.4 \times 10^6$	$9.9 \pm 3.6 \times 10^4$	<0.0001
EUB 338 (cell.ml^{-1})	$1.9 \pm 0.4 \times 10^6$	$7.5 \pm 0.9 \times 10^4$	0.043
EUB338/DAPI (%)	49.4 ± 11.8	76.6 ± 9.3	0.043
ALF 1b (cell.ml^{-1})	$4.2 \pm 1.2 \times 10^5$	$1.1 \pm 0.2 \times 10^4$	<0.0001
ALF1b/EUB338 (%)	21.9 ± 6.4	15.3 ± 2.1	<0.0001
BET42a (cell.ml^{-1})	$5.5 \pm 0.9 \times 10^5$	$3.4 \pm 0.4 \times 10^4$	<0.0001
BET42a/EUB338 (%)	29.1 ± 4.9	45.7 ± 6.1	<0.0001
CF319a (cell.ml^{-1})	$4.7 \pm 0.6 \times 10^5$	$2.4 \pm 0.4 \times 10^4$	0.008
CF319a/EUB338 (%)	24.7 ± 3.1	32.2 ± 5.3	0.008

Mean values were calculated from data of different dates and depths
Modified from Lemarchand et al. (2006)

13.5 Community Composition of Bacteria Associated to TEP

We used combined protocols to observe TEP and the composition of attached and free-living bacterial cells. The bacterial community composition was analysed with fluorescent *in situ* hybridization (FISH) using group-specific rRNA oligonucleotidic probes according to Moter and Göbel (2000). This method was coupled with a coloration using Alcian Blue to visualize TEP. Detailed methods can be found in Lemarchand et al. (2006). The following oligonucleotidic probes were used to analyse bacterial community composition: EUB338 for *Eubacteria*, ALF1b and BET42a for the alpha- and beta-subclasses of *Proteobacteria*, and CF319a for the *Cytophaga-flavobacterium* group (now renamed *Bacteroidetes*, group *Sphingobacteria-Flavobacteria*). Probes were labelled with the indocarbocyanine fluorescent dye Cy3 (MWG-Biotech). Our results indicated that bacteria associated with TEP constituted 2.6 % of the total DAPI-stained bacteria and the proportion of *Eubacteria* is different between the free-living and attached fractions. The contribution of *Eubacteria* to particle-associated and DAPI-stainable bacteria averaged 75 % while their contribution to free-living bacteria accounted for only 50 %. As noted previously in different studies in both marine and freshwater environments (Brachvogel et al. 2001; Bockelmann et al. 2000; Brümmer et al. 2000; Selje and Simon 2003) we also found high proportions of *Betaproteobacteria* and *Cytophaga-Flavobacteria* in particle-associated bacteria compared with free-living cells (Table 13.2). In contrast, the contribution of *Alphaproteobacteria* to particle-associated bacteria was lower than their contribution to free-living cells. Thus, our results clearly indicate that the composition of particle-associated bacteria is different from that of free-living bacteria highlighting the significant role of TEP in driving microbial diversity.

13.6 Activity of Free-Living Bacteria and Particle-Associated Bacteria

Potential ectoenzyme activities by bacterial extracellular enzymes were measured in fractionated water samples incubated at 20 °C in the dark for 2 or 6 h according to the protocol of Hoppe (1983). The 0.2–1.2 μm fraction contained free-floating bacteria while the fractions 1.2–5 μm and > 5 μm contained bacteria associated to detrital organic particles. The α-glucosidase and leucine aminopeptidase activities were estimated on duplicate filters for each size fraction. Most activity was associated with the 1.2–5 μm fraction and glycolytic activities were three times lower than proteolytic activities. Values of *K*m for leucine uptake were on average higher in the 1.2–5 μm fraction than in the 0.2–1.2 μm fraction and the > 5 μm fraction. The *K*m values of α-glucosidase activity were lower than the *K*m of leucine aminopeptidase activity. Glucose uptake affinity had the lowest values in the 0.2–1.2 μm fraction. Cell-specific hydrolysis rates in the attached bacterial fraction (1.2–5 μm and > 5 μm fractions) were significantly higher than those in the free-living bacterial fraction (0.2–1.2 μm fraction). Finally, we found that the concentrations of 2–5 and 5–50 μm TEP were related to the *V*max of glycolytic activities in the 1.2–5 μm and>5 μm fractions. This indicates that TEP-associated bacteria were specialized bacteria of high hydrolytic capacity. It is noteworthy however that the enhanced rates of enzymatic activities associated with particles may be the result of enzymes released by the grazing of bacterivorous protists. *V*max values as well cell-specific activities were on average highest in the 1.2–5 μm fraction, in which bacterivores are often dominant (Carrias et al. 1996).

13.7 Flagellates-TEP Relationships

Freshwater heterotrophic nanoflagellates (HNF) consume bacteria and, due to their high densities, they are considered the main bacterivores in the pelagic food web (Carrias et al. 1996; Simek et al. 2000; Sherr and Sherr 2002). Assuming that the activity of TEP-associated bacteria may stimulate the growth rates of the free-living bacteria in the vicinity of the particle, we hypothesized that HNF abundances could be related to TEP densities and to attached bacteria density. In Lake Pavin, HNF densities (total or taxon-specific) were however uncorrelated to bacterial density within TEP as expressed by coefficient a. However, during 2 years of the study, HNF densities were highly correlated to TEP abundances ($r^2 = 0.63$ in 2005 and 0.61 in 2006, $P < 0.001$ for both years). In assessing factors that can control the dynamics of HNF, the structure of the food web is frequently expected to have a major impact. In particular, the top-down impact of predators (ciliates and metazooplankton) is deemed an important control mechanism of the seasonal dynamics of free-living HNF in lakes (Weisse 1990; Carrias et al. 1998, 2001). To our knowledge, the role of particles has never been evaluated, and there are still too few studies on the interactions between protozoans and TEP. Our results suggest freshwater HNF are at least partially dependent on TEP densities. Particle distribution explained 63 % and 61 % of the variability in HNF abundance in 2005 and 2006, respectively. We thus conclude that TEP is a relevant factor of HNF distribution during the spring diatom bloom in Lake Pavin. Based on studies of organic aggregates in pelagic environments, it has been hypothesized that particles may be involved in the growth of certain HNF species in pelagic environments by grazing on particle-associated bacteria (Caron et al. 1982, 1986). In Lake Pavin, bodonids, *Monas*-like cells and the small-sized dominant taxa are potentially able to feed on attached bacteria. Given that they exhibit the properties of gels (Passow 2002), TEP may constitute a pelagic microhabitat protecting HNF against the ciliate and metazoan predation, resulting in higher HNF densities during spring plankton development. In addition, the smallest TEP may be an alternative food resource for HNF during low bacterial densities and/or periods of strong competition between bacterivorous taxa.

13.8 Conclusions

The inclusion of TEP in food webs will provide us a much better understanding of ecosystem functioning. TEP contain attached bacteria that exhibit pronounced differences compared to free-living bacteria in the surrounding water. Attached bacteria are more active and release inorganic nutrients that can enhance the growth of phytoplankton. Both attached and free bacteria (in the vicinity of the particle) constitute prey for protozoa that in turn are grazed by zooplankton. Finally, phytodretitus (mainly frustules of diatoms) interact with TEP to form large organic aggregates or lake snow that contribute to the sinking flux of matter in deep aphotic layers and the related carbon pump. Thus, based on the case study of Lake Pavin, we conclude that TEP are microbial microhabitats of major importance for biogeochemical cycles in the pelagic zone of aquatic ecosystems.

Acknowledgement We are indebted to Dr Xavier MARI for a critical review of the ms.

References

Arnous M-B, Courcol N, Carrias J-F (2010) The significance of Transparent Exopolymer Particles in the vertical distribution of bacteria and Heterotrophic Nanoflagellates in Lake Pavin. Aquat Sci 72:245–253

Bockelmann U, Manz W, Neu TR, Szewzyk U (2000) Characterization of the microbial community of lotic organic aggregates (river snow) in the Elbe river of German by cultivation and molecular methods. FEMS Microbiol Ecol 33:157–170

Brachvogel T, Schweitzer B, Simon M (2001) Dynamics and bacterial colonization of microaggregates in a large mesotrophic lake. Aquat Microb Ecol 26:23–35

Brümmer IHM, Fehr W, Wagner-Döbler I (2000) Biofilm community structure in polluted rivers: abundance of dominant phylogenetic groups over a complete annual cycle. Appl Environ Microbiol 66:3789–3796

Caron DA, Davis PG, Madin LP, Sieburth JMN (1982) Heterotrophic bacteria and bacterivorous protozoans in oceanic macroaggregates. Science 218:795–797

Caron DA, Davis PG, Madin LP, Sieburth JMN (1986) Enrichment of microbial populations in macroaggregates (marine snow) from surface waters of the North Atlantic. J Mar Res 44:643–565

Carrias J-F, Amblard C, Bourdier G (1996) Protistan bacterivory in an oligomesotrophic lake: importance of attached ciliates and flagellates. Microb Ecol 31:249–268

Carrias J-F, Amblard C, Bourdier G (1998) Seasonal dynamics of free and attached heterotrophic nanoflagellates in an oligomesotrophic lake. Freshw Biol 39:101–111

Carrias J-F, Thouvenot A, Amblard C, Sime-Ngando T (2001) Dynamics and growth estimates of planktonic protists during early spring in Lake Pavin (France). Aquat Microb Ecol 24:163–174

Carrias J-F, Serre J-P, Sime-Ngando T, Amblard C (2002) Distribution, size, and bacterial colonization of pico and nano-detrital organic particles (DOP) in two lakes of different trophic status. Limnol Oceanogr 47:1202–1209

Hoppe HG (1983) Significance of exoenzymatic activities in the ecology of brackish water: measurements by means of methylumbelliferyl-substrates. Mar Ecol Prog Ser 11:299–308

Lemarchand C, Jardillier L, Carrias J-F, Richardot M, Debroas D, SimeNgando T, Amblard C (2006) Community composition and activity of prokaryotes associated to detrital particles in two contrasting lake ecosystems. FEMS Microbiol Ecol 57:442–451

Mari X, Kiørboe T (1996) Abundance, size distribution and bacterial colonization of transparent exopolymeric particles (TEP) during spring in the Kattegat. J Plankton Res 18:969–986

Moter A, Göbel UB (2000) Fluorescence in situ hybridization (FISH) for direct visualization of microorganisms. J Microbiol Met 41:85–112

Passow U (2000) Formation of transparent exopolymer particles, TEP, from dissolved precursor material. Mar Ecol Prog Ser 192:1–11

Passow U (2002) Transparent exopolymer particles (TEP) in the aquatic environments. Prog Oceanogr 55:287–333

Passow U, Alldredge AL (1994) Distribution, size and bacterial colonization of transparent exopolymer particles (TEP) in the ocean. Mar Ecol Prog Ser 113:185–198

Schuster S, Herndl JG (1995) Formation and significance of transparent exopolymer particles in the Northern Adriatic Sea. Mar Ecol Prog Ser 124:227–236

Selje N, Simon M (2003) Composition and dynamics of particle-associated and free-living bacterial communities in the Weser estuary, Germany. Aquat Microb Ecol 30:221–237

Sherr EB, Sherr BF (2002) Significance of predation by protists in aquatic microbial food webs. Anton Leeuw Int J G 81:293–308

Simek K, Jürgens K, Nedoma J, Comerma M, Armengol J (2000) Ecological role and bacterial grazing of *Halteria* sp.: small oligotrichs as dominant pelagic ciliate bacterivores. Aquat Microb Ecol 22:43–56

Simon M, Grossart HP, Schweitzer B, Plough H (2002) Microbial ecology of organic aggregates in aquatic ecosystems. Aquat Microb Ecol 28:175–211

Weisse T (1990) Trophic interactions among heterotrophic microplankton, nanoplankton, and bacteria in Lake Constance (FRG). Hydrobiol 191:111–122

Lake Pavin: A Pioneer Site for Ecological Studies of Freshwater Viruses

14

Télesphore Sime-Ngando, Yvan Bettarel, Jonathan Colombet, Stéphanie Palesse, Angia Sriram Pradeep Ram, Marie Charpin, and Christian Amblard

Abstract

Since the discovery, 2–3 decades ago, that viruses of microbes are abundant in marine ecosystems, aquatic viral ecology has grown increasingly to reach the status of a full scientific discipline in environmental sciences. A dedicated society, the International Society for Viruses of Microorganisms (ISVM) (http://www.isvm.org/), was recently launched. Increasing studies in aquatic viral ecology are a source of novel knowledge related to the biodiversity, the functioning of ecosystems and the evolution of the cellular world. This is because viruses are perhaps the most diverse, abundant, and ubiquitous biological entities in the world's aquatic ecosystems. They exhibit various lifestyles that intimately depend on living cell metabolism, and are ultimately replicated by members of all three domains of life (*Bacteria*, *Eukarya*, *Archaea*). This establishes viruses as microbial killers and manipulators in the hydrosphere. Lake Pavin is one of the pioneer sites where original ecological data were first provided on the qualitative, quantitative and functional significance of both lytic and temperate viruses of prokaryotes in temperate freshwater lakes, taking into account both seasonal and depth-related variability in the water column and in the sediments. These data were acquired by means of original protocols we developed. In addition to these protocols, we herein provide a synthesis of Lake Pavin studies on viral ecology, focusing on: (i) spatio-temporal dynamics of the diversity of viral communities, (ii) the significance of seasonal and depth-related variations of viral abundance and lytic and lysogenic activities, and (iii) the relative importance of lytic viruses and grazers for bacterial mortality and the biogeochemical implications for the food web dynamics. Unexpected and novel putative viruses were discovered in the deep-aged, dark, and permanently anoxic monimolimnic waters and sediments of Lake Pavin, highlighting the possible endemicity of these habitats. Some of these original viruses resembled dsDNA viruses of hyperthermophilic and hyperhalophilic *Archaea*. Unusual types of spherical and cubic virus-like particles (VLPs) were also observed for the first time. Infected prokaryotic cells were detected in deep sedi-

T. Sime-Ngando (✉) • J. Colombet • S. Palesse • A.S.P. Ram
M. Charpin • C. Amblard
LMGE, Laboratoire Microorganismes: Génome et Environnement, UMR CNRS 6023, Université Clermont-Auvergne, Université Blaise Pascal, BP 80026, 63171 Aubière Cedex, France
e-mail: telesphore.sime-ngando@univ-bpclermont.fr

Y. Bettarel
LMGE, Laboratoire Microorganismes: Génome et Environnement, UMR CNRS 6023, Université Clermont-Auvergne, Université Blaise Pascal, BP 80026, 63171 Aubière Cedex, France

Institut de Recherche pour le Développement (IRD), MARBEC Marine Biodiversity, Exploitation & Conservation UMR IRD (n° 248), Ifremer, Univ. Montpellier, CNRS (n° 9190), Montpellier, France

T. Sime-Ngando et al. (eds.), *Lake Pavin*, DOI 10.1007/978-3-319-39961-4_14

ment cores, and their vertical distribution correlated with both viral and prokaryotic abundances. Pleomorphic ellipsoid VLPs were visible in filamentous cells tentatively identified as representatives of the archaeal genus *Methanosaeta*, a major group of methane producers on Earth.

Keywords
Lake Pavin • Viruses • Seasonal and depth related variability • Viral lysis • Lysogeny • Microbial loop

14.1 Introduction

With the discovery, few decades ago, that viruses of microbes are abundant in marine ecosystems (Torrella and Morita 1979; Bergh et al. 1989), aquatic viral ecology has increasingly grown to reach the status of full scientific discipline in environmental sciences, with the recent launch of a dedicated society, i.e. the International Society for Viruses of Microorganisms (ISVM) (http://www.isvm.org/). As infectious agent of potentially all types of living cells, viruses are the most abundant biological entity in the **biosphere**. They are an ubiquitous component of the **microbial food web** dynamics in a great variety of environments, including the most extreme ecosystems (Sime-Ngando and Colombet 2009). Moreover, in spite of the difficulties to routinely observe and describe biological nanoparticles, combined with the absence of conserved evolution tracers such as RNA ribosomal genes, we now consider that viruses represent the greatest reservoir of non-characterized genetic diversity and resources on the earth (Suttle 2007). They contain genes that code for essential biological functions such as photosynthesis (Lindell et al. 2005), making their hosts powerful vehicles for genetic exchanges in the environment. Because lytic viruses kill their hosts, they play fundamental roles in cycling nutrients and organic matter, structuring **microbial food webs**, governing microbial diversity and, to a lesser extent, by being a potential food source for **protists** (Bettarel et al. 2005). Viruses can also form long-lived association with their specific hosts, reducing their fitness, or allowing infected hosts to remain strong competitors through mutualistic symbioses (Roossinck 2011). In addition, the discovery and characterization of the unique group of archaeal viruses are influencing the field of prokaryotic virology, increasing our knowledge on viral diversity and changing perspectives on early stages of evolution (Prangishvili et al. 2006).

The majority of the studies on the ecology of wild viruses in aquatic systems was conducted in marine systems, although some of the original work was done in freshwater systems (Miller et al. 1992). The present chapter sought to review the literature on viruses in freshwater ecosystems, focusing on data collected in a pioneer site, the oligo-mesotrophic Lake Pavin located in the temperate French Massif Central, where studies on freshwater viral ecology had started about two decades ago. These data encompassed different aspects of the whole aquatic viral ecology (Fig. 14.1), including (i) methods for concentrating and counting free-floating and infectious viruses via transmission electron and epifluorescence microscopy (Sime-Ngando et al. 1996; Bettarel et al. 2000; Wommack et al. 2010; Colombet et al 2007; Colombet and Sime-Ngando 2012a), (ii) seasonal and depth-related variability in viral diversity, **metagenomics**, abundance and both lytic and lysogenic activities (Bettarel et al. 2003a; Colombet et al. 2009; Colombet and Sime-Ngando 2012a; Roux et al. 2012), (iii) the role of nutrient inputs on the trade-off between the two major lifestyles (i.e. lysis and **lysogeny**) of planktonic viruses (Pradeep Ram and Sime-Ngando 2010) (iv) the seasonal contribution of viral bacteriolysis vs protistan bacterivory to the fate of prokaryotes and the related biogeochemical cycling (Bettarel et al. 2003b, 2004; Pradeep Ram et al. 2014), (v) the significance of viruses as food source for phagotrophic **protists** (Bettarel et al. 2005), and (vi) the impact of viruses on the metabolism (Pradeep Ram et al. 2013) and community composition (Jardillier et al. 2005) of prokaryotes. Recently, unexpected and novel putative viruses, primarily viruses of Archaea, have also been discovered in the deep-dark permanently anoxic water and sediments of Lake Pavin (Borrel et al. 2012).

14.2 Basic Knowledge on Viruses

Viruses are biological entities consisting each of a single- or a double stranded DNA or RNA surrounded by a protein and, for some of them, a lipid coat. In mesophilic aquatic systems, most virus-like particles (VLPs) are tailed or untailed phages, with capsid diameter typically smaller than 100 nm, based on direct observation with transmission electron microscope (TEM). Viruses have no intrinsic metabolism and for all processes requiring energy, they need the intracellular machinery of a living and sensitive host cell. They have various life cycles, all starting by fixation on specific receptors (often **transporter proteins**) present on the surface of host cells, followed by injection of the viral genome in the cytosol. In the lytic cycle, the viral genome induces the synthesis of

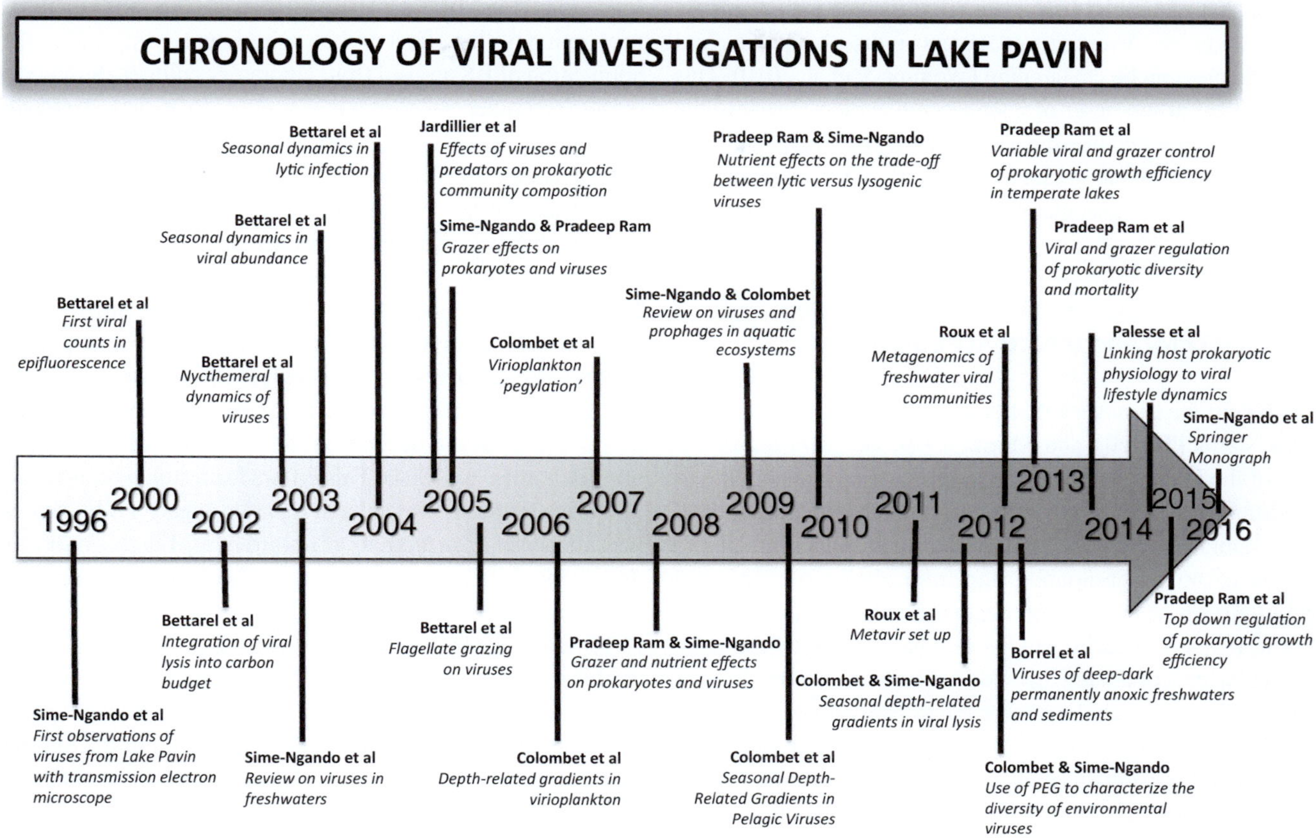

Fig. 14.1 Chronogram of ecological studies of freshwater viruses in Lake Pavin

viral constituents, including the replication of the viral genetic material. A number of progeny viruses are then produced and released in the environment by fatal bursting of the host cell. In the chronic cycle, the progeny viruses are episodically or constantly released from the host cell by budding or extrusion, without immediate lethal events. This cycle is less known in the plankton, but is common in metazoan viruses such as *Herpes* and *Hepatitis* viruses or *Rhabdoviruses*. Chronic viral infection is a dynamic and metastable equilibrium process (Virgin et al. 2009) which ends by the lysis of the host cell after serial budding of lipid membrane coated viruses, as known from the marine *Emiliana huxleyi* host (Mackinder et al. 2009). Recently, chronic infection without host lysis has been reported for the first time in the marine primary producer *Ostreococcus tauri*, where the low rate of viral release through budding (1–3 viruses cell^{-1} day^{-1}) allows cell recovery and stable coexistence of the viruses and their hosts (Clerissi et al. 2012).

In the lysogenic cycle, the viral genome integrates the genome of the host cell to form a 'prophage' which can stay in a dormant/latent state for extended periods until an environmental stress to the immune host cell set off a switch to a lytic cycle. Unlike lytic infection, **lysogeny** provides a mean of persistence for viruses when the abundance of the host cells is very low (Weinbauer et al. 2003). **Prophages** may also affect the metabolic properties of host cells which can acquire immunity to superinfections and new phenotypic characteristics, such as antibiotic resistance, antigenic changes, and virulence factors. A variant to the lysogenic cycle is the so-called carrier state or **pseudolysogenic** cycle where the viral genome is not integrated to the host genome but rather remains in an 'inactive state' within the host cell. There is no replication of the viral DNA which is segregated unequally into progeny cells, likely for few generations. **Pseudolysogenic** viruses likely occur in very poor nutrient conditions where host cells are in starvation and cannot offer the energy necessary to the viral gene expression (Ripp and Miller 1997; Łoś and Węgrzyn 2012).

14.3 Methodological Considerations: Abundance, Basic Infectivity Parameters and Concentration of Viruses for Quantitative and Genomic Studies

Ecological investigations rely on biomass, diversity, and composition of organisms, which, in the case of microbial communities are, primarily, gathered using microscopy and molecular genetic approaches. The diminutive size of viruses means that obtaining viral material sufficient for examining the abundance, diversity and composition of aquatic viral assemblages can be challenging (Wommack et al. 2010). This was critical at the onset of aquatic viral ecology because the need for samples containing a high-density of viruses and viral genomes was critical to challenge the detection thresholds of microscopic and molecular methods for the study of viruses.

14.3.1 Transmission Electron Microscopy (TEM) and Infectivity Parameters

Following the discovery that viruses are abundant in marine ecosystems (Torrella and Morita 1979; Bergh et al. 1989), one of the major challenges for starting the new discipline of environmental virology was to overcome the methodological constraints which in the past have strongly limited the observation of both free-floating and intracellular (i.e. infected cells) viruses in the environment (Bratbak and Heldal 1993; Suttle 1993). The earlier studies that have tackled these challenges were conducted in lake ecosystems of the French Massif Central, including Lake Pavin (Bettarel et al. 2000). In a short communication, we first proposed an optimized method to ultracentrifuge and concentrate planktonic viruses for their examination with transmission electron microscopy (TEM) (Sime-Ngando et al. 1996). Because it is important, for the accuracy of quantitative estimates of viruses and infected cells, to avoid collecting these items in an ultracentrifuge-pellet but on a platform using a swing-out rotor, we propose a platform made from a commercial polymer resin adaptable at the bottom of an **ultracentrifugation** tube, where up to three 400-mesh copper grids (catalog no. A03; Pelanne Instruments®) can be fixed with a double-face tape. This tool allows to directly harvest, viruses and cell-hosts onto the TEM grids after a centrifugation at 4 °C either at 120,000 x g for 2 h to keep free-occurring viruses or at 70,000 x g for 30 min to distinguish between prokaryotic cells with and without intracellular viruses. With our method, we provided one of the first accurate descriptions of freshwater viruses dominated by particles with regular icosaedric capsid with 6 faces and a tail, assimilated to phages that have been previously described in marine environments (Fig. 14.2). These particles were about 10 times more abundant than prokaryotes, corroborating the hypothesis of their role in controlling the community dynamics of bacteria in freshwater systems (Sime-Ngando et al. 1996).

Our method also allowed observation of infected cells, primarily of prokaryotic cells, and the determination of critical parameters in the calculation of the infection prevalence (i.e. the so-called FVIC, **frequency of visibly infected cells**), the related viral production and the **burst size** (BS) or the average number of viruses released when a single host cell lyses. These parameters are generally derived from TEM observation of visibly mature phages within intact or thin-sectioned host cells (Fig. 14.3), a method that has the advantage of accounting for the effect of the environment, compared to the application of BS derived from laboratory cultures. Similar to marine and other freshwater ecosystems, FVIC in Lake Pavin is generally less that 5 % but, at a regional local scale, is higher (Mean 2 %, range 0–4 %) compared to the eutrophic Lake Aydat (Mean 1 %, range 0–3 %). The seasonal variability in both types of lakes is however roughly the same, with maximum in late spring (May to June) and early autumn (September to October). It is thus possible that in oligotrophic lakes, where substrates are in short supply for bacterial production, bacterial lysates might represent an important source of dissolved organic matter or inorganic nutrients, the quantity of which would depend on the frequency of viral infection (Bettarel et al. 2004). This differs from studies conducted in marine systems, which have reported higher FVIC values with increasing productivity (Steward et al. 1992), or no evidence for a relationship between FVIC and productivity (Noble and Fuhrman 1997). In Fig. 14.4, we plotted mean FVIC values for freshwaters in relation to the trophic states of the lakes. Although the gaps between each trophic level on the *x* axis cannot be considered equivalent, there is no apparent trend similar to that which has been found in the marine environment, suggesting that for lakes, productivity alone is a poor predictor of the level of viral infection.

In contrast to FVIC in freshwaters, the number of viruses produced per infected cells (e.g. burst size), as measured in TEM, increased with increasing productivity, ranging from about 5–60 viruses prokaryote^{-1} in oligotrophic marine waters to 5–100 viruses prokaryote^{-1} in productive freshwaters, suggesting that the **burst size** of infected bacteria would be higher in more productive environments (Sime-Ngando et al. 2003). In Lake Pavin, BS averaged about 25 viruses prokaryote^{-1} (range: 15–50 viruses prokaryote^{-1}), with relatively low fluctuations with depth and over the year when excluding the peaks that generally occurred in spring time (Bettarel et al. 2004). The trophy-trend in BS is not unreasonable as both cell size and growth rate are generally greater in eutrophic than in oligotrophic environments, but can be strongly constrained by the variability in capsid size for different occurring viral populations. Based on reports from a variety of different aquatic environments, Parada et al. (2006) calculated a mean BS of 24 and 34 for marine and freshwater environments, respectively.

14.3.2 Epifluorescence Microscopy (EM)

The interest of biologists for the role of viruses in ecological processes has rapidly raised the requirement of reliable methods for estimating the level of these biological entities in aquatic systems. The earlier estimates of viral levels were obtained by concentrating the viruses by **ultracentrifugation** (Børsheim et al. 1990) or by ultrafiltration (Suttle et al.

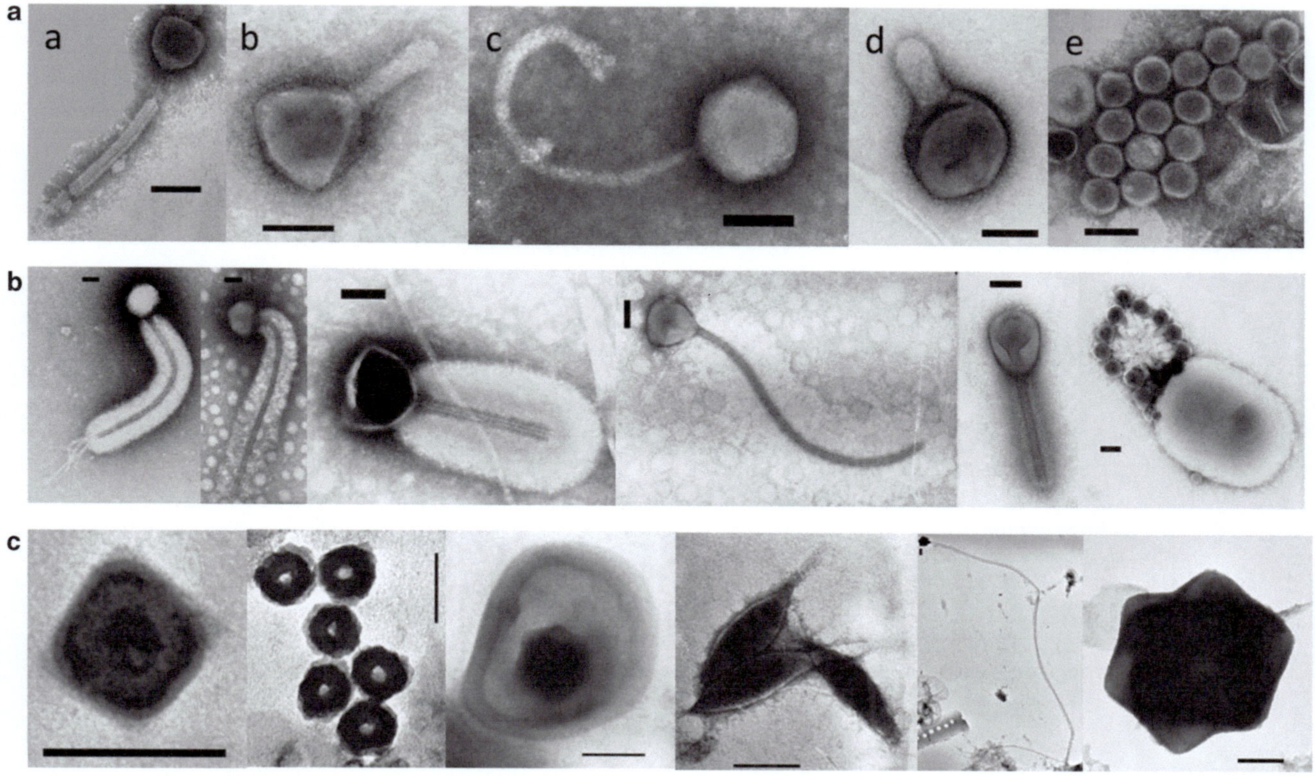

Fig. 14.2 Surface (**a**), deep water (**b**) and sediment (**c**) viruses are apparently different in their morphology in Lake Pavin: overview of the general morphotypes of viruses in the water column, including the characteristic free-occurring pelagic forms [Myoviridae (*a*, *b*), Siphoviridae (*c*), Podoviridae (*d*) and untailed phages (*e*)], which are dominant, primarily in the mixolimnion (**a**), and additionnal and more atypical forms which occurred in the deep waters (**b**) and sediments (**c**). Bar scale 100 nm

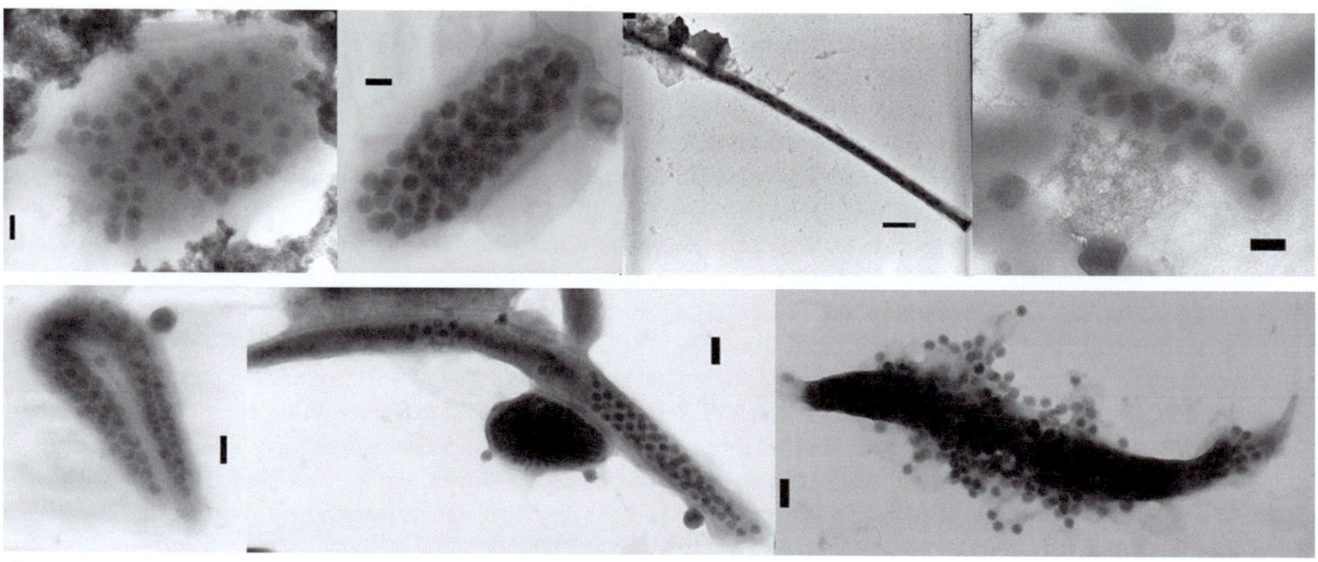

Fig. 14.3 Example of TEM-visibly infected bacterial cells in Lake Pavin. Bar scale 100 nm

1991; Wommack et al. 1995) and then counting them by using TEM, as described above. However, the high costs, the long analysis time required and the difficulty of this method led to the development of alternative, faster and less expensive methods based on the use of epifluorescence microscopy (EM). With these methods, virus-like particles (VLPs) can be observed and counted after they have been labeled with **fluorochromes** that specifically bind to nucleic acids. The most common **fluorochromes** were DAPI (4',6'-diamino-2-phenylindole) (Porter and Feig 1980), YOPRO-1 {4-[3-methyl-2,3-dihydro-(benzo-1,3-oxazole)-2-methylmethyledene]-1-(39-trimethyl ammoniumpropyl)-quinilinium diioide} (Hennes and Suttle 1995), and SYBR Green I (Noble and Fuhrman 1998).

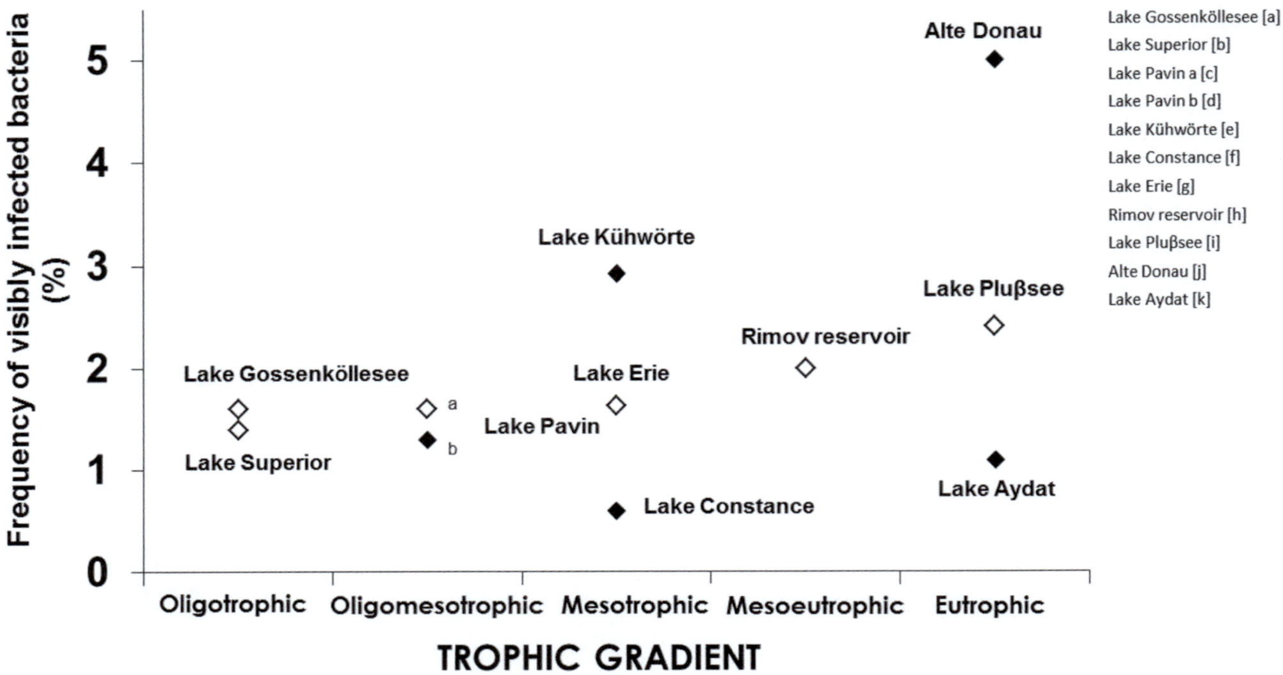

Fig. 14.4 Frequencies of visibly infected bacteria recorded in lakes, along a trophic gradient.◇ correspond to short-term studies (<5 days), ◆ correspond to long-term studies (>3 months) (Sources: [a] Hofer and Sommaruga (2001); [b] Tapper and Hicks (1998); [c] Bettarel et al. (2003a,b); [d, k] Bettarel et al. (2004); [e] Mathias et al. (1995); [f] Hennes and Simon (1995); [g] Wilhelm and Smith (2000); [h] Simek et al. (2001); [i] Weinbauer and Höfle (1998); [j] Fischer and Velimirov (2002))

For the first time, we have provided an accurate comparison of five most used EM and TEM protocols for virus studies based on freshwater communities collected in Lakes Pavin and Aydat. Our results revealed that the quantitative determination of viruses was different according to the type of fluorochrome used (Bettarel et al. 2000). Nowadays, SYBR Green I is the most used fluorochrome, when coupled with a powerful antifading, it allows to improve the observation of viruses in EM (Suttle and Furhman 2010). However, it is important to specify that the latter targets preferentially dsDNA viruses and has very few affinity with ssDNA and RNA viruses for which the use of SYBR Gold would be more suitable. The first protocol for high throughput counting of free-floating viruses using **flow cytometry** and SYBR Green I stain was developed for pelagic oligotrophic marine samples (Marie et al. 1997, 1999), and was later on optimized for both marine (Brussaard 2004) and freshwater samples (Duhamel and Jacquet 2006).

14.3.3 Concentration of Viruses for Microscopic and Molecular Genetic Analyses

The genetic and biological diversity of aquatic viruses remains largely unexplored, mainly because of the methodological difficulties related to the observation of environmental nanoparticles and the weakness of environmental viral gene banks. The study of free-floating viruses in aquatic systems indeed requires a concentration method for estimating their number and their diversity based on both morphologic and genomic features. The first generation of methods used came from the study of human, animal and plant pathogenic viruses. These methods are based on physicochemical approaches such as (i) adsorption of phages onto microporous filters charged positively or negatively, (ii) the use of precipitation agents like calcium phosphate, aluminium, magnesium or acid precipitation, and (iii) the use of organic flocculation (cf. Colombet et al. 2007). The main disadvantage of these techniques for natural samples is their selectivity because viruses may have different adsorptive properties, while electrostatic interactions may affect the viability of concentrated viruses.

As described above, Børsheim et al. (1990) have proposed the first concentration method based on the **ultracentrifugation** of viruses contained in few milliliters of samples directly onto grids, followed by staining before observation under TEM. Disadvantages of this method include the risk of viral disruption due to the high centrifugation speed, the effects of suspended particles and 'virucidal' substances, and the low initial volume of experimental samples that may render TEM observations tedious according to the high magnifications used. The method is thus costly and time-consuming for large volume of samples, and therefore cannot be recommended as a routine procedure (Bettarel et al. 2000). Besides, the **fluorochrome**-based methods described previously used low pressure membrane-concentration. They are well known to provide accurate counts of microbes and viruses, but resulted in a highly poor description of diversity.

An accurate description of the diversity of pelagic viruses usually needs highly concentrated aliquots. The current method available to date is ultrafiltration that allows concentrating viruses contained in several tens of liters of water into small volumes of a few hundreds of milliliters (Suttle et al. 1991; Wommack et al. 1995). However, the application of this procedure requires further concentration by **ultracentrifugation** to obtain viral pellets necessary for TEM observations and for estimating viral diversity. To avoid **ultracentrifugation** and the underlined disadvantages, we conducted, for the first time, a methodological comparative study with samples collected in Lake Pavin and other freshwater sites located in the Massif Central region of France, and have proposed a simple, cheaper and efficient alternative protocol using polyethylene glycol (PEG) to obtain virioplankton precipitates that can be used for different purposes, primarily for TEM observations, electrophoretic plugs, and the downstream molecular genetic analyses (Colombet et al. 2007; Colombet and Sime-Ngando 2012b). The process, named "pegylation", yields the greatest recovery efficiently of free-occurring viruses, allowing the recovery of > two-fold more viruses compared to **ultracentrifugation** and ultrafiltration. In addition, the diversity of virioplankton, based on genomic size profiling using **pulsed field gel electrophoresis**, was higher and better discriminated when we used the PEG method which both concentrates and purifies viral material in water samples. The effects of potential virucidal compounds (e.g. enzymes, antibodies, and other inhibitory substances) are probably minimized or avoided during pegylation. This property of PEG molecules is currently used in medicine to protect drug-carrier viruses, interferons of immunitary system, or antibodies (Harris and Chess 2003). PEG can thus also be used as a conservative for planktonic viruses, thereby avoiding the use of toxic fixatives and the related disadvantages such as losses during storage of samples.

14.4 Diversity and Endemicity of Viruses

14.4.1 Phenotypic Diversity

The first descriptions of the global diversity of viruses were based on their morphological discrimination with transmission electron microscopy (Børsheim et al. 1990; Proctor and Fuhrman 1990; Weinbauer and Höfle 1998). To our knowledge, the phenotypic diversity of viruses in Lake Pavin is apparently highly endemic, with viral morphotypes that typically are different and highly constraint by depth (Fig. 14.2). In the surface mixed waters (i.e. **mixolimnion**), viral phenotypes are limited, mainly including tailed or untailed particles with capsid heads, characteristics of bacteriophages. This is typical in the world aquatic ecosystems: tailed phages belong to the order *Caudovirales*, all of which are double-stranded DNA viruses that generally represent 10 to 40 % of the total abundance of viruses (Sime-Ngando and Colombet 2009). Within *Caudovirales*, three families emerge as quantitatively dominant: *Siphoviridae* with long non-contractile tails (e.g. Phage lambda), *Podoviridae* with a short non-contractile tail (e.g. Phage T7), and *Myoviridae* with contractile tails of variable length (e.g. Phage T4). In Lake Pavin, non-tailed capsids dominate viral abundance in the **mixolimnion** (Colombet and Sime-Ngando 2012a) in accordance with results obtained from metagenomic analysis (Roux et al. 2012). A recent global morphological analysis of marine viruses also suggested that non-tailed viruses, which comprised 50–90 % of the viral particles observed, might represent the most ecologically important component in natural viral communities (Brum et al. 2013). However, we cannot completely exclude a significant effect of mechanic shocks from handling resulting in losses of tails, because 96 % of the 5500 specimens of described bacteriophages are tailed particles (Ackermann 2007).

Our seasonal depth-related studies in Lake Pavin (Colombet et al. 2006, 2009; Colombet and Sime-Ngando 2012a) clearly demonstrated substantial changes in the morphological diversity of pelagic viruses in relation to the depth-related gradients. Indeed, viral communities in the permanently anoxic **monimolimnion** of Lake Pavin comprised a substantial number of atypical bacteriophage morphotypes, larger in capsid size (>60 nm), and with more complex spatial conformation (Fig. 14.2), compared to the mixolimnic viruses, which were more typical of the world aquatic free-occurring bacteriophages dominated by small (capsid size < 60 nm) *Caudovirales* (see above). In addition, an analysis of viral diversity based on the distribution frequency of capsid sizes through the whole water column of Lake Pavin has indicated that viruses were apparently typical and more diversified in the **monimolimnion** than in the surface waters.

This spatial pattern was further enhanced by the recent discovery of the highest diversity and complexity of viral communities in the deep-dark permanently anoxic sediments of Lake Pavin, with the occurrence of unexpected and novel viruses (Borrel et al. 2012). Indeed, the examination of sediment cores encompassing 130 years of sedimentation of the lake has unveiled exceptional morphotypes of viruses, previously never reported in freshwater systems (Fig. 14.2). Some of these resembled dsDNA viruses of **hyperthermophilic** and **hyperhalophilic** archaea. Moreover, unusual types of spherical and cubic virus-like particles (VLPs) were observed. Infected prokaryotic cells were detected in the whole sediment core, and their vertical distribution correlated with both viral and prokaryotic abundances. Pleomorphic ellipsoid VLPs were visible in filamentous cells (Fig. 14.3) tentatively identified as representatives of the archaeal genus *Methanosaeta*, a major group of methane producers on earth. We consider these empirical observations as preliminary spot findings which may pave the way to one of the last frontiers of our knowledge in environmental microbiology: the connection between the viral world and the prokaryotic species that consume or produce greenhouse gases, such as methane. More details on methane cycle and the related prokaryotic actors are provided in Chap. 16.

The above findings which are typical of Lake Pavin corroborate a study in the same lake which has unveiled a complete shift in the composition of the prokaryotic assemblages between the **mixolimnion** and the **monimolimnion**, with a maximal *Archaea*/*Bacteria* ratio that is reached below mixolimnic layers (Lehours et al. 2005). It is thus likely that (i) deep anoxic viruses in Lake Pavin include endemic, typical populations, with different morphology and diversity characteristics compared to surface water viruses, and (ii) the potential viral infection of Archaea which are abundant in the **monimolimnion** and in the sediments of Lake Pavin could explain the depth-related changes in viral morphology.

14.4.2 Molecular Genetic Diversity

Viral species are mostly known from their isolated hosts in laboratory cultures, which, in the case of environmental samples, may not exceed 1 % of the total prokaryotes (Hugenholtz et al. 1998). Molecular approaches are thus critical and offer windows to the largest uncharacterized reservoir of diversity on the earth. Polymerase chain reaction (PCR)-based methods are restricted to chosen viral groups as no gene is universally conserved among viruses, while **metagenomics** gives access to an exhaustive view of uncultured viral diversity, and has so far revealed an important unknown diversity and an unexpected richness of viral communities (Edwards and Rohwer 2005). In aquatic viral ecology, the most used molecular fingerprinting approach is the **pulse-field gel electrophoresis** (PFGE), which separates PCR-generated dsDNA products (Wommack et al. 1999). Application to a series of spring samples collected in the **mixolimnion** of Lake Pavin and concentrated via the "pegylation" approach yielded viral genomes in the size range from about 15 to 290 kbp, with a bulk around 35 kbp in the metalimnion. Visible electrophoretic bands in our agarose gels, calculated both from computer-scanning and densitometry profiling, ranged in number from 7 to 9 (Colombet et al. 2007). Above results are within the ranges known from aquatic ecosystems (Sime-Ngando and Colombet 2009).

In Lake Pavin, we have performed one of the first studies on the composition and diversity of freshwater viral communities through **metagenomics**, based on 454 **pyrosequencing** derived **virome** of 649,290 reads with an average length of 420 bp (Roux et al. 2012). In 0.2 µm filtered samples collected on June and July 2008, the proportion of reads similar to protein sequences of the non-redundant NCBI database was at 14 % (Fig. 14.5), which is among the highest compared to published **viromes** (range 1–28 %, mean 6.3 %). However, when our known fraction was determined using reads randomly reduced to 100 bp as in previous studies, this fraction dramatically dropped to 0.7 %, which is the lowest among aquatic **viromes**. Among the known fraction, the majority of reads (60 %) was most similar to non-viral sequences, whereas our **virome** was not contaminated by bacteria: the absence of 16S rRNA in our sample was checked by PCR amplification and BLAST search for 16S rRNA sequences, and only a paucity of **virome** reads was found to be partly similar to ribosomal protein. This result is consistent with previous studies and is thought to be an indication of both the lack of viral gene annotation and the horizontal gene transfers between viral and host genomes (Edwards and Rohwer 2005).

Phylogenetic analyses performed on the major viral families of our **virome** using different marker genes (Roux et al. 2011) highlight the dominance of unexpected single stranded DNA viruses: *Micro-*, *Nano-* and *Circoviridae*, the later family being the most abundant and represented 60 % of the total sequences (Fig. 14.5). This sheds light on a hitherto undescribed viral diversity for freshwaters. In addition, using 29 previously published **viromes**, the cluster richness in viral communities was shown to vary between different environment types and appeared significantly higher in marine ecosystems than in other biomes. Furthermore, significant genetic similarity between viral communities of related environments was highlighted as freshwater, marine and hypersaline environments were separated from each other despite the vast geographical distances between sample locations within each of these biomes (Roux et al. 2012). Overall, our pioneer viral **metagenomics** in Lake Pavin spotlighted a very broad diversity and previously unknown clades undetectable by PCR and microscopic analyses. The Lake Pavin **virome** appears closely related to other freshwater **viromes**, despite the significant ecological differences between lakes. Furthermore, freshwater viral communities appear genetically distinct from other aquatic ecosystems, demonstrating the specificity of freshwater viruses at a community scale. This was the first time that such results were reported. More details on the metagenetics and metagenomics of prokaryotes and viruses are provided in Chap. 15.

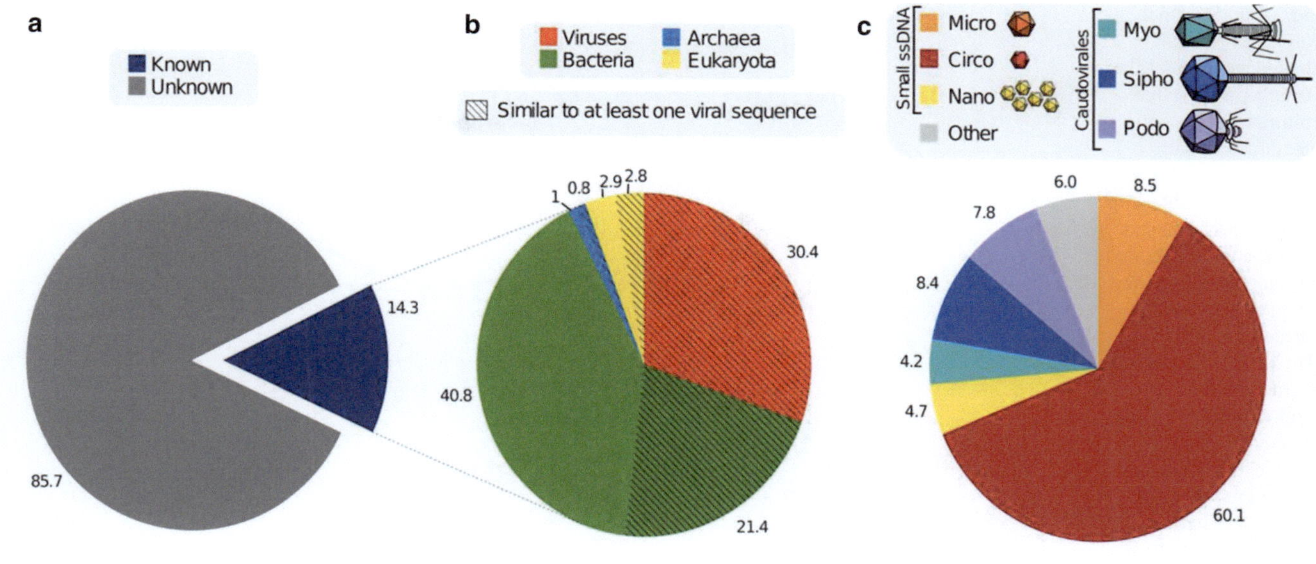

Fig. 14.5 Composition and taxonomic affiliations of Lake Pavin virome reads as determined by similarity to known sequences. (**a**) The percent of "known" virome sequences when compared to the non redundant (NR) protein database. (**b**) Breakdown of the "known" sequences into Viruses, Bacteria, Archaea, or Eukarya using similarity results against NR. Hatched parts were reads having a best BLAST hit against a non-viral sequence, but still presenting significant similarities against a complete virus genome sequence and thus designated as reads "similar to at least one viral sequence". (**c**) Taxonomic composition at the viral family level of these reads "similar to at least one viral sequence". The "Other" category pools families which represented less than 1 % of the full virome sequences. For more details see Roux et al. (2012)

14.5 Standing Stocks and Seasonal Patterns in Viral Dynamics

The first seasonal studies on viruses in Lake Pavin were conducted in 2000 (Bettarel et al. 2003b) following a diel cycle study in 1998 (Bettarel et al. 2002), corresponding to one of the earliest attempts in freshwater ecology that established viruses as ubiquitous, abundant and dynamic component whose role in the functioning of aquatic ecosystems needs to be assessed. At seasonal time scale, the abundances of free-floating viruses in the **mixolimnion** of Lake Pavin typically ranged from 1 to 6 × 10^7 viruses ml^{-1}, which is within the range of those found in other pelagic environments (Wommack and Colwell 2000). Similarly, a small peak was noted in spring followed by a more or less progressive increase during the period of **thermal stratification**, especially between May and October in Lake Pavin (Fig. 14.6A), which also corresponded to the seasonal patterns in other lakes of the region and in other freshwater and marine environments as well, where an autumnal peak is characteristic of the seasonal abundances of aquatic viruses. Wommack and Colwell (2000) interpreted this peak as the consequence of the autumnal phytoplankton bloom. As a consequence, the virus-to-bacteria ratio (VBR) in Lake Pavin (range 3–13, mean 7) also peaked in spring and in autumn, and corresponded to the typical values (range 3–10) reported in pelagic environments (Wommack and Colwell 2000). Typically the VBR is higher; the low VBRs are usually due to bad storage conditions for viruses. The calculated **burst size** from visibly infected cells observed under TEM ranges from 15 to 50, mean 25 viruses $cell^{-1}$. Part of the lytic viral production in the plankton is rapidly removed by complex environmental processes, such as adsorption on particles, enzyme digestion via proteolysis, destruction by UV raditions, grazing etc…

The seasonal abundances of viruses and bacteria in Lake Pavin and other pelagic environments are rather homeostatic as they do not generally vary by more than ten-fold at a seasonal scale. Both variables are generally correlated, suggesting that most viruses are bacteriophages. In our seasonal studies, we found a closer linear relationship between viruses and bacteria in the oligotrophic Lake Pavin ($R^2=0.59$) compared with the eutrophic Lake Aydat ($R^2=0.13$) (Fig. 14.6B), indicating a stronger interdependence between these two communities in less productive lakes (Bettarel et al. 2003b). We expected a stronger coupling of bacterial and viral abundances at high bacterial density in productive ecosystems, given the fact that virus-mediated mortality of cells generally increases with increasing bacterial host density. However, we found an opposite situation which we explain by: (i) an increasing output of dissolved organic matter (DOM) from viral lysis in the more oligotrophic system where DOM from phytoplankton exudates may be a limiting factor for bacterial production, or (ii) an increasing relative abundance of non-bacteriophage viruses, such as **cyanophages**, in productive lakes where planktonic communities are often dominated by **cyanobacteria**.

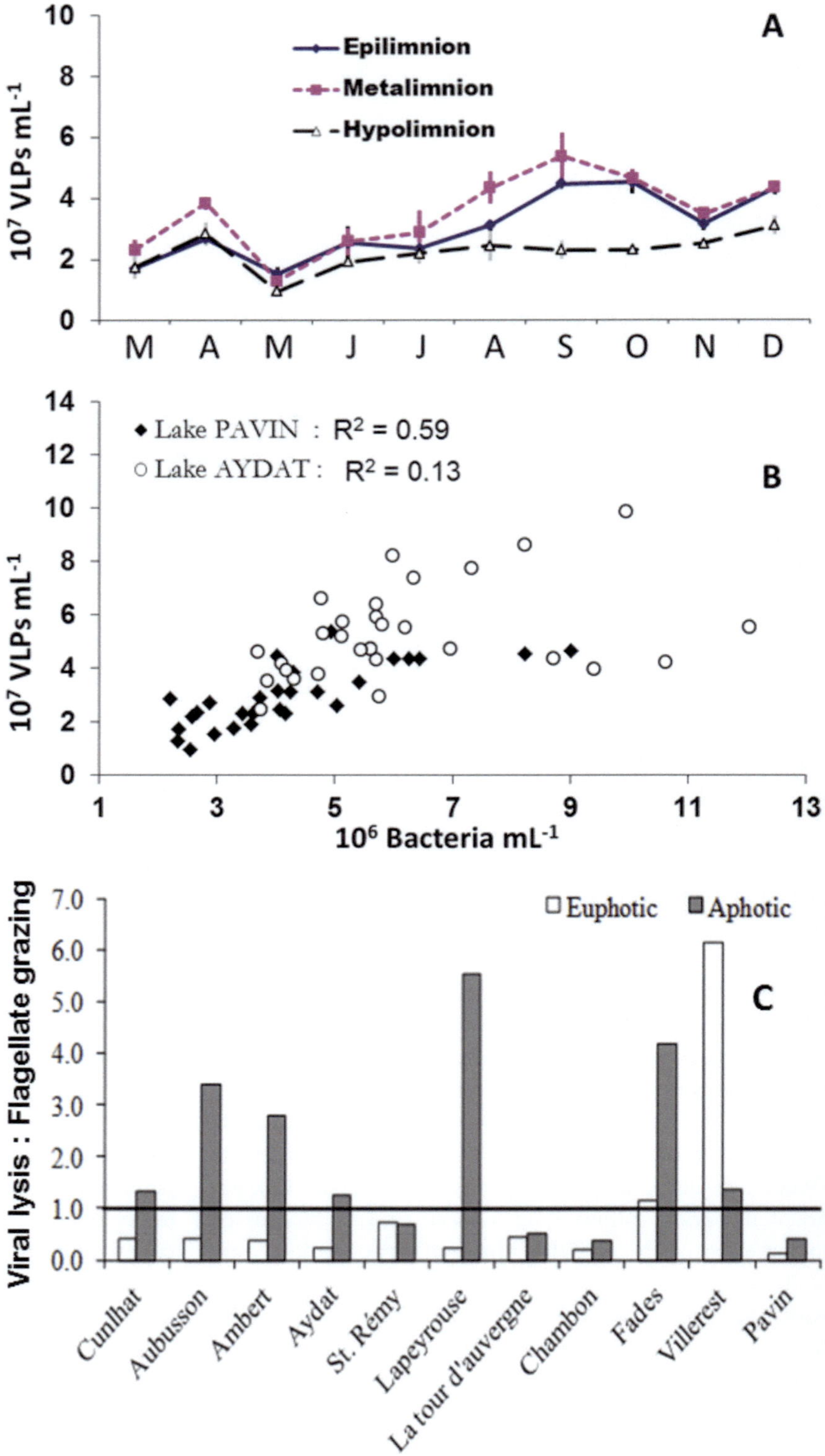

Fig. 14.6 Seasonal variations of viral abundance (Virus-Like-Particles, VLPs) in the epilimnion, metalimnion, and hypolimnion of Lake Pavin (**A**). Relationship between bacterial and viral abundances in the oligomesotrophic Lake Pavin compared to the eutrophic Lake Aydat (**B**). Variation in the average ratio of viral-induced mortality to heterotrophic nanoflagellate grazing-induced mortality of prokaryotes in Lake Pavin, compared to 10 other lakes (**C**) of the same geographical region (the French Massif-Central, refer to Pradeep Ram et al. 2013 for more details on these lakes)

14.6 Activities of Viruses

14.6.1 Lytic Activity and Impact on Bacterial Community Composition and Metabolism

It is now well accepted that lysis is one of the main causes of microbial mortality in aquatic systems. Based on our method (described previously) for direct observation of infected cells, the seasonal study conducted in Lake Pavin in 2000 yielded one of the first estimates of the impact of lytic viruses on bacterial communities (Bettarel et al. 2004). To estimate the impact of viruses on bacterial mortality, as percent of bacterial biomass production loss, a series of conversions must be applied (Binder 1999; Weinbauer et al. 2002).

The levels and the seasonal fluctuations in the basic lytic parameter (i.e. FVIC) in Lake Pavin have been described previously (see section 17.3). The translation into viral production yields an average of 80×10^6 viruses liter^{-1} hour^{-1} produced in the mixolimnion, with the highest values recorded in the metalimnion (Table 14.1). This corresponds to a mean bacteria production loss due to viral lysis of 16 %, which is lower than the grazing impact from flagellates (38 %), but higher than that from bacterivorous ciliates (3 %) (Bettarel et al. 2004). Viral bacteriolysis levels in the mixolimnion of Lake Pavin appear conservative, compared to the world marine and freshwaters where viral-mediated mortality averages 10–50 % of the production of heterotrophic prokaryotes, and is considered approximately equal to the bacterivory from grazers (Fuhrman and Noble 1995; Pradeep Ram et al. 2005). Based on the combined average mortality related to viral lysis and protistan bacterivory, a significant fraction of bacterial production in the mixolimnion of Lake Pavin is controlled by factors other than viral lysis and protistan predation, such as metazoan bacterivory. This is particularly credible during the period of **thermal stratification**, which usually coincides with the maximum development of metazoan zooplankton, which then obtain most of their resources from intense bacterivory (Thouvenot et al. 1999). During autumn, all of the bacterial production in Lake Pavin is balanced by viral lysis and protistan grazing (Bettarel et al. 2004).

The impacts of viral lysis and ciliate grazing in Lake Pavin were highest in the metalimnion, while that from flagellates was highest in the hypolimnion (Table 14.1). From a study conducted in 11 lakes located in the French Massif Central (including Lake Pavin), we found that the magnitude of prokaryotic mortality due to viruses and flagellates varies with sampled depth (Pradeep Ram et al. 2013). The average ratio of viral mortality to flagellate grazing of prokaryotes in the euphotic depth was 0.5, with lowest ratio (0.1) observed in Lake Pavin (Fig. 14.6C). In the aphotic zone, the dominance of viruses as a factor of prokaryotic mortality was largely due to significantly higher viral infection which was accompanied by the significantly lower grazing rates compared to the euphotic zone. Lower levels of dissolved oxygen at the aphotic compared to euphotic depth explained for low flagellate grazing rates. A high-resolution vertical sampling in Lake Pavin (Colombet et al. 2006) has confirmed the pattern of high relative abundance and activity of viruses and bacteria in the **monimolimnion** where protistan flagellates were almost absent (Fig. 14.7). Further seasonal investigations of depth-related gradients in microbial components in Lake Pavin (Colombet et al. 2009; Colombet and Sime-Ngando 2012a) have provided evidence that the ecology of the deep, cold, dark and permanently anoxic monimolimnic waters may be driven by **viral loop** (dissolved organic matter–prokaryotes–viruses) processes, which can sequester organic matter and nutrients for a long-lived turnover time. In general, under anoxic conditions where most of the eukaryotes are absent, the predominance of viruses vs grazers as a factor of bacterial mortality in the aquatic environment is still poorly understood at the community level. Indeed, it has been reported that high viral control of bacterial production occurs either at reduced or high grazing activities (Bettarel et al. 2004), while experimental studies have provided evidence of synergistic interactions between viral bacteriolysis and protistan bacterivory. In these interactions, the presence of grazers appears as a stimulating factor for prokaryotic growth and viral proliferation (Sime-Ngando

Table 14.1 Mean seasonal contributions of viruses, phagotrophic flagellates, and ciliates to bacterial mortality and of flagellates to viral losses in the three main layers of the mixolimnion of Lake Pavin

	% (CV) of bacterial production removed by:			% (CV) of viral production removed by flagellates
Sampled depth	Viral lysis	Flagellate grazing	Ciliate grazing	
Epilimnion	13.6 (63.1)	25.5 (117.4)	1.6 (137.1)	4.8
Metalimnion	18.8 (53.2)	23.4 (112.2)	4.6 (122.0)	3.6
Hypolimnion	16.1 (124.0)	64.2 (88.8)	1.8 (93.6)	3.8
Mean	16.2 (13.1)	37.7 (49.8)	2.7 (50.3)	4.1 (12.3)

CV coefficient of variation

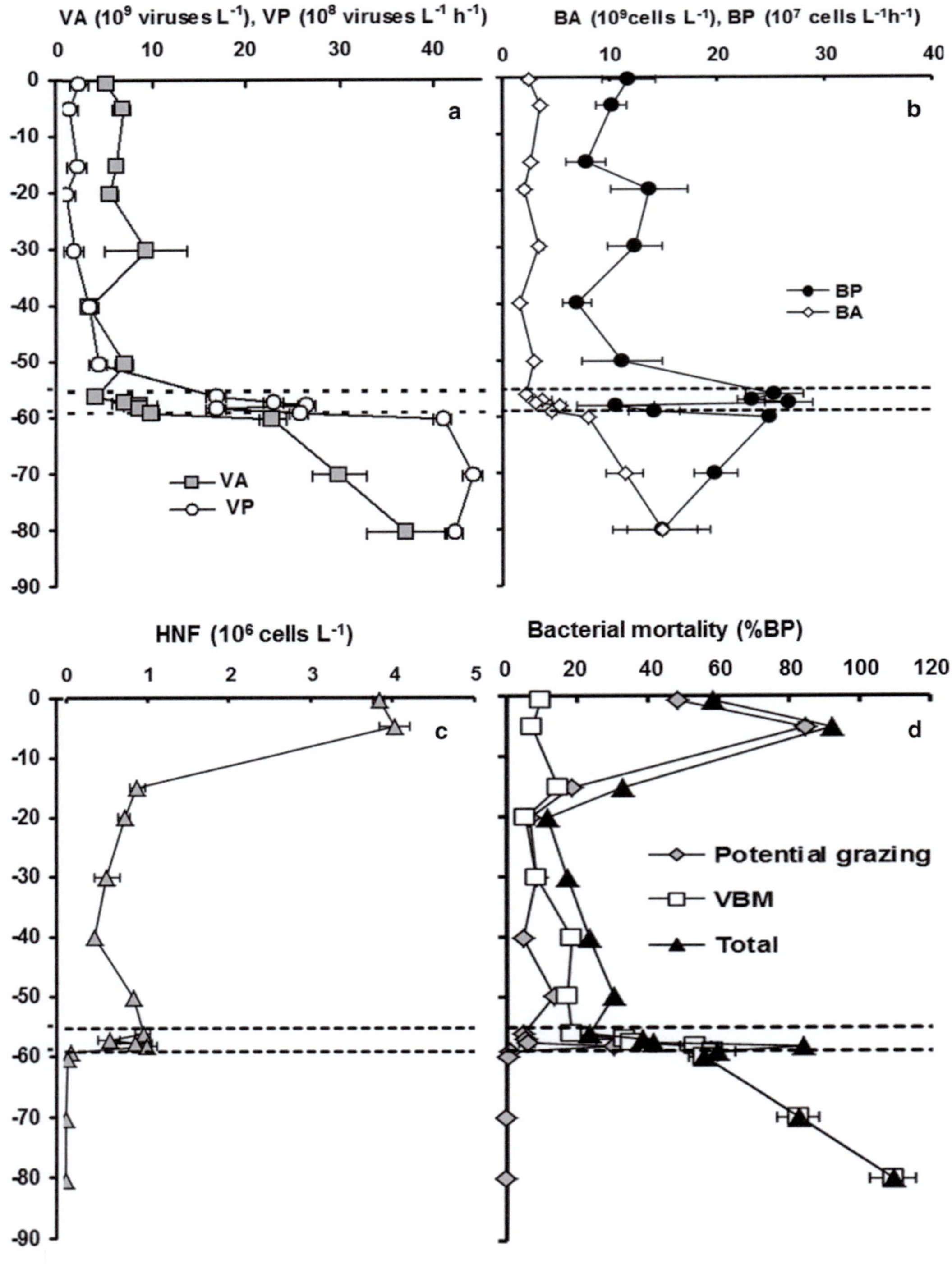

Fig. 14.7 Vertical variations of viral abundance (*VA*) and lytic production (*VP*) (**a**), of bacterial abundance (*BA*) and production (*BP*) (**b**), of the abundance of heterotrophic nanoflagellates (*HNF*) (**c**) and of the contributions of viruses (*VBM*) and potential grazing from HNF to bacterial mortality (*D*) in Lake Pavin (**d**). Error bars represent standard deviations for triplicates

and Pradeep Ram 2005), likely through nutrient regeneration that favors certain bacterial phyla (*alpha-* and *beta-Proteobacteria*) over others (*Archaea* and *Bacteroidetes*) (Pradeep Ram and Sime-Ngando 2008). Such uneven and heterogeneous impact of viral lytic activity for different prokaryotic phenotypes and/or genotypes probably is one of the mechanisms by which viruses shape the host diversity and community structure. We have demonstrated that this can have a strong influence on the processes occurring in the plankton food web dynamics (Pradeep Ram and Sime-Ngando 2010).

A major pathway from which lytic viruses drive prokaryotic community composition (PCC) is through their ultimate dependence on the availability of specific host populations.

This pattern has the strong feedback effect of preventing species dominance and enhanced species cohabitation within microbial communities, i.e. the so-called 'phage kills the winner' hypothesis (Thingstad and Lignell 1997). In early June 2001, we conducted a short-term experiment to examine the impact of viral lysis on bacterial community composition in Lake Pavin using fluorescence *in situ* hybridization (FISH) and terminal restriction fragment length polymorphysm (T-RFLP) (Jardillier et al. 2005). The main findings suggested a control of the PCC (i) by viral lysis of some dominant phylotypes, and (ii) by interspecific competition between resistant strains for the uptake of substrates released by viral lysis. The increase of *Archaea* suggested that these cells benefit such resources.

The dependence of lytic virulent viruses on the availability of specific hosts implies that viruses respond to the growth rate of the most active hosts (Palesse et al. 2014). In our recent study conducted in 11 freshwater lakes (Pradeep Ram et al. 2013), we found that high nucleic acid content cells are the best determinant of viral abundance, and served as the main host target for viral proliferation. Such preferential lysis can have a strong impact on the overall prokaryote community metabolism. We tested the hypothesis of top-down control on prokaryotic growth efficiency (PGE), which we used as an index of physiological and energetic status at the community level. Viruses through lytic infection had a strong antagonistic impact on PGE at the aphotic zone, whereas flagellate grazing was found to enhance PGE in the euphotic zone. These differences may have significant implications on the food web dynamics because the nature of nutrients released through the processes of grazing and viral lysis is different. Grazing activity results in more effective release of inorganic nutrients, whereas viral lysis has been shown to produce elements in organic complexes (Fuhrman 1999; Wilhelm and Suttle 1999).

14.6.2 Lysogeny : A Survival Strategy for Viruses

One of the key explanations for the omnipresence of viruses in natural ecosystems is undoubtedly through the existence of several lifestyles, of which two major pathways, namely **viral lytic** and **lysogenic pathways**, are prevalent in aquatic systems. Examination of natural prokaryotic communities inducible with a mutagenic agent (e.g. **mitomycin C**) has suggested that the fraction of lysogenic bacteria (known as FLC, frequency of lysogenically infected cells) is typically < 50 % (range 0–100 %) of the total abundance in marine environments. In freshwaters, these values fluctuate from 0 to 16 % in temperate and tropical lakes, and from 0 to 73 % in Antarctic lakes (Sime-Ngando and Colombet 2009). In Lake Pavin, analysis of samples collected through the whole water column on 20 April 2004 yielded FLC values between 0.1 and 16 % (Colombet et al. 2006). This suggests that a substantial proportion of bacteria in Lake Pavin contain functional viral genomes. The relative abundance of lysogens appeared negatively correlated with bacterial abundance and production, with the frequency of infected cells, and with viral lytic production, supporting the hypothesis that **lysogeny** is a strategy for survival of phages in environments with low host availability (Colombet et al. 2006). From these findings, we hypothesized that (i) there are environmental characteristics, in relation with potential host densities and availability, favoring one of the two "viral life cycles", and (ii) host availability is prevalent over the physical and chemical environments and favors lytic over lysogenic "viral life cycles."

To test these hypotheses, we designed an experimental study to investigate the effects of added organic and inorganic nutrients on the two viral lifestyles in freshwater microbial (i.e. < 0.8 µm fraction) microcosms collected in Lake Pavin during a nutrient-depleted period, using **mitomycin C** as **prophage** inductor agent (Pradeep Ram and Sime-Ngando 2010). In the absence of **mitomycin C**, viral lytic production increased as a functional response to the strong stimulation of bacterial growth rates (0.7–0.8 day^{-1}) by the added nutrients, primarily the organic nutrients which appeared scarcer than inorganic nutrients. In the presence of **mitomycin C**, temperate phage production (frequency of lysogenically infected bacterial cells, FLC = 17–19 % of total cells) significantly exceeded lytic production (frequency of lytically infected bacterial cells, FIC = 9–11 %) in natural samples (i.e. without nutrient additions) as a result of higher **prophage** induction, which relatively increased with the decreasing contact probability between viruses and their potential hosts. In contrast, addition of nutrients drastically reduced FLC (<4 %) and increased FIC (>22 %). Both variables were antagonistically correlated as was the correlation between FLC and bacterial growth rates. This strongly supports the idea that **lysogeny** may represent a maintenance strategy for viruses in harsh nutrient/host conditions which appeared as major instigators of the trade-off between the two viral lifestyles. Overall, at the community level, we reject the hypothesis that nutrients but **mitomycin C** stimulate temperate phage induction, and retained the hypotheses that nutrients rather (i) stimulate lytic viruses via enhanced host growth and (ii) when limiting, promote lysogenic conversion in natural waters (Pradeep Ram and Sime-Ngando 2010).

14.7 Protistan Grazing of Viruses and Ecological Implications

Despite the fact that bacteria are undoubtedly the principal constituent in the diet of heterotrophic nanoflagellates (HNF) in most aquatic systems, González and Suttle (1993) showed that marine viruses can also contribute significantly to the nutrition of HNF and that being consumed is one possible route for the decay of virioplankton, in addition to degradation by solar radiation, by heat-labile substances, or by adsorption to particulate organic matter. The second and last report on this topic so far was conducted in the oligotrophic Lake Pavin and in the nearby eutrophic Lake Aydat (Bettarel et al. 2005). This study provides the first estimates of viral losses from HNF predation in freshwaters, with removal rates from HNF grazers which were relatively modest, averaging 4.1 and 0.8 % of viral production in Lake Pavin and Lake Aydat, respectively (Table 14.1). Although phagotrophy by nanoflagellates appears of minor importance as a loss process for natural virioplankton communities, our study suggests that viruses are probably of higher nutritional significance for HNF in oligotrophic systems than in productive ones. Based on the amount of carbon (C), nitrogen (N) and phosphorus (P) contained per single virus and bacterium (Gonzàlez and Suttle 1993), we estimated that ingested viruses would constitute only 0.6, 0.7 and 1.4 % of C, N and P, respectively, in the total diet of HNF in Lake Pavin, and only 0.1, 0.2, and 0.4 % in Lake Aydat. This suggests that viruses are not as energetically valuable as bacteria in the HNF diet. Clearly, we conclude that (i) there is a trophic link between HNF and viruses in the functioning of aquatic **microbial food webs**, (ii) where viruses would represent a weak food source for HNF (i.e. compared to bacteria), (iii) but they still may not be inconsequential for the HNF diet, primarily in systems with low productivity (Bettarel et al. 2005). A question that now emerges is to what extent does ingestion of viruses by HNF represent a source of infection for these **protists** ? More details on HNF bacterivory are provided in Chap. 18.

14.8 Conclusions and Perspectives

Aquatic viral ecology is a relatively recent discipline. Pioneer ecological and molecular researches in Lake Pavin have contributed to demonstrate that viruses are omnipresent in aquatic environments, including one of the most extreme biotopes such as the dark permanently anoxic monimolimnic waters and sediments of Lake Pavin. These environments are highly selective for heterotrophic prokaryotes whose seasonal activity offers an optimal and unique resource for thriving viral communities, some of which may be endemic. Because all types of cells in the nature have their specific viruses and offer ecological niches to different viral lifestyles, viruses are considered the greatest reservoir of the uncharacterized biological diversity on the earth, which is being probed and described at an increasingly rapid rate, almost exclusively with molecular sequence data. Studies in Lake Pavin suffer from the lack of such attempts for the exploration of viruses other than viruses of *Bacteria* and the related functions, some of which may be crucial to get better insights into global change processes (e.g. consumption and production of greenhouse gas such as methane etc.), in the seasonal successions of phytoplankton or zooplankton communities, and the sudden mortality events affecting fish, crustacean, mollusks, etc. For example, Hewson et al. (2013) recently used a **metagenomic** approach to identify circular, single-stranded DNA viruses that may be involved in the seasonal dynamics of *Daphnia* spp. in Oneida and Cayuga lakes (upstate New York). Because *Daphnia* plays a critical role in many lake ecosystems, such viruses may have important effects on herbivory and thus carbon flow through the lake ecosystem.

Our conceptual understanding of the function and regulation of aquatic ecosystems, from microbial to global biogeochemical processes, has changed with the study of viruses. Viral-mediated prokaryotic mortality seems lower in Lake Pavin but roughly equals bacterivory from **protists** in the world aquatic ecosystems, which is a significant departure from the traditional view that predation and resource availability are the main factors controlling prokaryotic abundance and production in pelagic systems. Viruses influence both the retention and the export of organic matter in pelagic environments, and thus affect the biogeochemical cycling of the major conserved elements (C, N, P). Besides, given the prevalence of phage-encoded biological functions within host cells as attested by the substantial levels of lysogens in Lake Pavin, and the occurrence of **recombination** between phage and host genes, phage populations serve as gene reservoirs that contribute to niche partitioning of microbial species in aquatic ecosystems. Viral-mediated gene transfers include diverse mechanisms (**transduction**, **transformation**, **conjugation** and **recombination**) that are known to affect gene evolution in the marine environment. It is thus clear that most of the viruses are not pathogens but mutualistic cell partners that provide helper functions, and this must be included in the future agenda of viral studies in freshwater ecosystems.

Acknowledgements This study includes results from PhD and postdoc researches supported by various instances: Région Auvergne, CNRS, Université Blaise Pascal, French Ministère de la Recherche et de la Technologie, and the Ministry of Culture, High School and Research of Grand-Duché de Luxembourg. Part of the results presented, most of which are detailed in different Chapters of this book, were obtained with the collaboration of colleagues: G. Borrel, D. Debroas, F. Enault, L. Jardillier, A. C. Lehours, V. Ravet, S. Roux. The study was also granted from different sources: French National Programme EC2CO

(Project VIRULAC) and ANR Programme Biodiversité (Project AQUAPHAGE). We thank M. G. Weinbauer for critical and proofreading of the manuscript.

References

Ackermann HW (2007) 5500 phages examined in the electron microscope. Arch Virol 152:227–243

Bergh Ø, Børsheim KY, Bratbak G, Heldal M (1989) High abundance of viruses found in aquatic environments. Nature 340:467–468

Bettarel Y, Sime-Ngando T, Amblard C, Laveran H (2000) A comparison of methods for counting viruses in aquatic systems. App Environ Microbiol 66:2283–2289

Bettarel Y, Sime-Ngando T, Amblard C, Carrias J-F, Sargos D, Lavandier P, Garabetian F (2002) The functional importance of bacteriophages in the microbial loop of an oligomesotrophic lake over a diel cycle. Ann Limnol 38:263–269

Bettarel Y, Amblard C, Sime-Ngando T, Carrias JF, Sargos D, Garabetian F, Laveran H (2003a) Viral lysis, flagellate grazing potential and bacterial production in Lake Pavin : a short-term study. Microb Ecol 45:119–127

Bettarel Y, Sime-Ngando T, Amblard C, Carrias JF, Portelli C (2003b) Virioplankton and microbial communities in aquatic systems: a seasonal study in two lakes of differing trophy. Freshw Biol 48:810–822

Bettarel Y, Sime-Ngando T, Amblard C, Dolan J (2004) Viral activity in two contrasting lake ecosystems. Appl Environ Microbiol 70:2941–2951

Bettarel Y, Sime-Ngando T, Bouvy M, Arfi R, Amblard C (2005) Low consumption of virus-sized particles by heterotrophic nanoflagellates in two lakes of the French Massif Central. Aquat Microb Ecol 39:205–209

Binder B (1999) Reconsidering the relationship between viral induced bacterial mortality and frequency of infected cells. Aquat Microb Ecol 18:207–215

Borrel G, Colombet J, Robin A, Lehours A-C, Prangishvili D, Sime-Ngando T (2012) Unexpected and novel putative viruses in the sediments of a deep-dark permanently anoxic freshwater habitat. ISME J 6:2119–2127

Børsheim KY, Bratbak G, Heldal M (1990) Enumeration and biomass estimation of planktonic bacteria and viruses by transmission electron microscopy. Appl Environ Microbiol 56:352–356

Bratbak G, Heldal M (1993) Total count of viruses in aquatic environments. In: Kemp PF, Sherr BF, Sherr EB, Cole JJ (eds) Current methods in aquatic microbial ecology. Lewis Publishers, London, pp 135–138

Brum JR, Schenck RO, Sullivan MB (2013) Global morphological analysis of marine viruses shows minimal regional variation and dominance of non-tailed viruses. ISME J 7:1738–1751

Brussaard CPD (2004) Optimization of procedures for counting viruses by flow cytometry. Appl Environ Microbiol 70:1506–1513

Clerissi C, Desdevises Y, Grimsley N (2012) Prasinoviruses of the marine green alga *Ostreococcus tauri* are mainly species-specific. J Virol 86:4611–4619

Colombet J, Sime-Ngando T (2012a) Seasonal depth-related gradients in virioplankton: lytic activity and comparison with protistan grazing potential in Lake Pavin (France). Microb Ecol 64:67–78

Colombet J, Sime-Ngando T (2012b) Use of PEG, polyethylene glycol, to characterize the diversity of environmental viruses. In: Méndez-Vilas A (ed) Current microscopy contributions to advances in science and technology, vol 1. Formatex Research Center, Badajoz, pp 316–322

Colombet J, Sime-Ngando T, Cauchie HM, Fonty G, Hoffmann L, Demeure G (2006) Depth-related gradients of viral activity in Lake Pavin. Appl Environ Microbiol 72:4440–4445

Colombet J, Robin A, Lavie L, Bettarel Y, Cauchie HM, Sime-Ngando T (2007) Virioplankton 'pegylation': use of PEG (polyethylene glycol) to concentrate and purify viruses in pelagic ecosystems. J Microbiol Methods 71:212–219

Colombet J, Charpin M, Robin A, Portelli C, Amblard C, Cauchie HM, Sime-Ngando T (2009) Seasonal depth-related gradients in virioplankton: standing stock and relationships with microbial communities in Lake Pavin (France). Microb Ecol 58:728–736

Duhamel S, Jacquet S (2006) Flow cytometric analysis of bacteria and virus-like particles in lakes. J Microbiol Methods 64:316–322

Edwards RA, Rohwer F (2005) Viral metagenomics. Nat Rev Microbiol 3:504–510

Fischer UR, Velimirov B (2002) High control of bacterial production by viruses in a eutrophic oxbow lake. Aquat Microb Ecol 27:1–12

Fuhrman JA (1999) Marine viruses and their biogeochemical and ecological effects. Nature 399:541–548

Fuhrman JA, Noble RT (1995) Viruses and protists cause similar mortality in coastal seawater. Limnol Oceanogr 40:1236–1242

González JM, Suttle CA (1993) Grazing by marine nanoflagellates on viruses and virus-sized particles: ingestion and digestion. Mar Ecol Prog Ser 94:1–10

Harris JM, Chess RB (2003) Effect of pegylation on pharmaceuticals. Nat Rev Drug Discov 2:214–221

Hennes KP, Simon M (1995) Significance of bacteriophages for controlling bacterioplankton growth in a mesotrophic lake. Appl Environ Microbiol 61:333–340

Hennes KP, Suttle CA (1995) Direct counts of viruses in natural waters and laboratory cultures by epifluorescence microscopy. Limnol Oceanogr 40:1050–1055

Hewson I, Ng G, Li WF, LaBarre BA, Aguirre I, Barbosa JG, Breitbart M, Greco AW, Kearns CM, Looi A, Schaffner LR, Thompson PD, Hairston NG Jr (2013) Metagenomic identification, seasonal dynamics, and potential transmission mechanisms of a *Daphnia*-associated single-stranded DNA virus in two temperate lakes. Limnol Oceano 58:1605–1620

Hofer JS, Sommaruga R (2001) Seasonal dynamics of viruses in an alpine lake: importance of filamentous forms. Aquat Microb Ecol 26:1–11

Hugenholtz P, Goebel BM, Pace NR (1998) Impact of culture-independent studies on the emerging phylogenetic view of bacterial diversity. J Bacteriol 180:4765–4774

Jardillier L, Bettarel Y, Bardot C, Amblard C, Sime-Ngando T, Debroas D (2005) Effects of viruses and predators on the prokaryotic community composition. Microb Ecol 50:557–569

Lehours A-C, Bardot C, Thenot A, Debroas D, Fonty G (2005) Anaerobic microbial communities in Lake Pavin, a unique meromictic lake in France. Appl Environ Microbiol 71:7389–7400

Lindell D, Jaffe JD, Johnson ZI, Church GM, Chisholm SW (2005) Photosynthesis in marine viruses yield proteins during host infection. Nature 438:86–89

Łoś M, Węgrzyn G (2012) Pseudolysogeny. Adv Virus Res 82:339–349

Mackinder LC, Worthy CA, Biggi G, Hall M, Ryan KP, Varsani A, Harper GM, Wilson WH, Brownlee C, Schroeder DC (2009) A unicellular algal virus *Emiliana huxleyi* virus 86, exploits an animal-like infection strategy. J Gen Virol 90:2306–2316

Marie D, Partensky F, Jacquet S, Vaulot D (1997) Enumeration and cell-cycle analysis of natural populations of marine picoplankton by flow cytometry using the nucleic-acid stain SYBR Green I. Appl Environ Microbiol 63:186–193

Marie D, Brussaard CPD, Thyrhaug R, Bratbak G, Vaulot D (1999) Enumeration of marine viruses in culture and natural samples by flow cytometry. Appl Environ Microbiol 65:45–52

Mathias CB, Kirschner KT, Velimirov B (1995) Seasonal variations of virus abundance and virus control of the bacterial population in backwater system of the Danube River. Appl Environ Microbiol 61:3734–3740

Miller RV, Ripp S, Relicon J, Ogunseitan OA, Kokjohn TA (1992) Virus-mediated gene transfer in freshwater environment. In: Gauthier MJ (ed) Gene transfers and environment. Springer, Berlin, pp 51–62

Noble RT, Fuhrman JA (1997) Virus decay and its causes in coastal waters. Appl Environ Microbiol 63:77–83

Noble RT, Fuhrman JA (1998) Use of SYBR Green I for rapid epifluorescence counts of marine viruses and bacteria. Aquat Microb Ecol 14:113–118

Palesse S, Colombet J, Pradeep Ram AS, Sime-Ngando T (2014) Linking host prokaryotic physiology to viral lifestyle dynamics in a temperate freshwater lake (Lake Pavin, France). Microb Ecol 68:740-750

Parada V, Herndl GJ, Weinbauer MG (2006) Viral burst size of heterotrophic prokaryotes in aquatic systems. J Mar Biol Ass UK 86:613–621

Porter KG, Feig YS (1980) The use of DAPI for identifying and counting aquatic microflora. Limnol Oceanogr 25:943–948

Pradeep Ram AS, Sime-Ngando T (2008) Functional responses of prokaryotes and viruses to grazer effects and nutrient additions in freshwater microcosms. ISME J 2:498–509

Pradeep Ram AS, Sime-Ngando T (2010) Resources drive trade-off between viral lifestyles in the plankton: evidence from freshwater microbial microcosms. Environ Microbiol 12:467–479

Pradeep Ram AS, Boucher D, Sime-Ngando T, Debroas D, Romagoux JC (2005) Phage bacteriolysis, protistan bacterivory potential, and bacterial production in a freshwater reservoir: coupling with temperature. Microb Ecol 50:64–72

Pradeep Ram AS, Colombet J, Perriere F, Thouvenot A, Sime-Ngando T (2015) Viral and grazer regulation of prokaryotic growth efficiency in temperate freshwater pelagic environments. FEMS Microbiol Ecol 91(2):1–12

Pradeep Ram AS, Palesse S, Colombet J, Sabart M, Perriere F, Sime-Ngando T (2013) Variable viral and grazer control of prokaryotic growth efficiency in temperate freshwater lakes (French Massif Central). Microb Ecol 66:906–916

Pradeep Ram AS, Palesse S, Colombet J, Thouvenot A, Sime-Ngando T (2014) The relative importance of viral lysis and nanoflagellate grazing for prokaryote mortality in temperate lakes. Freshw Biol 59:300–311

Prangishvili D, Forterre P, Garrett RA (2006) Viruses of the Archaea: a unifying view. Nat Rev Microbiol 4:837–848

Proctor LM, Fuhrman JA (1990) Viral mortality of marine bacteria and marine bacteria and cyanobacteria. Nature 343:60–62

Ripp S, Miller RV (1997) The role of pseudolysogeny in bacteriophage-host interactions in a natural freshwater. Environment Microbiol 143:2065–2070

Roossinck MJ (2011) The good viruses: viral mutualistic symbiosis. Nat Rev Microbiol 9:99–108

Roux S, Faubladier M, Mahul A, Paulhe N, Bernard A, Debroas D, Enault F (2011) Metavir: a web server dedicated to virome analysis. Bioinformatics 27:3074–3075

Roux S, Enault F, Robin A, Ravet V, Personnic S, Theil S, Colombet T, Sime-Ngando T, Debroas D (2012) Assessing the diversity and specificity of two freshwater viral communities through metagenomics. PLoS One 7:e33641

Simek K, Pernthaler J, Weinbauer MG, Hornak K, Dolan J, Nedoma J, Masin M, Amann R (2001) Changes in bacterial community composition and dynamics and viral mortality rates associated with enhanced flagellate grazing in mesotrophic reservoir. Appl Environ Microbiol 67:2723–2733

Sime-Ngando T, Colombet J (2009) Viruses and prophages in the dynamics of aquatic ecosystems (in French). Can J Microbiol 55:1–15

Sime-Ngando T, Pradeep Ram AS (2005) Grazer effects on prokaryotes and viruses in a freshwater microcosm experiment. Aquat Microb Ecol 41:115–124

Sime-Ngando T, Mignot JP, Amblard C, Bourdier G, Devilettes C, Quilblier-Lloberas C (1996) Characterization of planktonic virus-like particles in a French mountain lake: methodological aspects and preliminary results. Ann Limnol 32:1–5

Sime-Ngando T, Bettarel Y, Chartogne C, Sean K (2003) The imprint of wild viruses on freshwater microbial ecology. Recent Res Dev Microbiol 7:481–497

Steward GF, Wikner J, Smith DC, Cochlan WP, Azam F (1992) Estimation of virus production in the sea: I. Methodology development. Mar Microb Food Webs 6:57–78

Suttle CA (1993) Enumeration and isolation of viruses. In: Kemp PF, Sherr BF, Sherr EB, Cole JJ (eds) Handbook of methods in aquatic microbial ecology. Lewis, Boca Raton, pp 121–134

Suttle CA (2007) Marine viruses – major players in the global ecosystem. Nat Rev Microbiol 5:801–812

Suttle CA, Furhman JA (2010) Enumeration of virus particles in aquatic or sediment samples by epifluorescence microscopy. In: Wilhelm SW, Weinbauer M, Suttle C (eds) Manual of aquatic viral ecology. American Society of Limnology and Oceanography, Waco, pp 145–153, Chapter 15

Suttle CA, Chen AM, Fuhrman JA (1991) Use of ultrafiltration to isolate viruses from seawater which are pathogens to marine phytoplankton. Appl Environ Microbiol 57:721–726

Tapper MA, Hicks RE (1998) Temperate viruses and lysogeny in Lake Superior bacterioplankton. Limnol Oceanogr 43:95–103

Thingstad TF, Lignell R (1997) Theoretical models for the control of bacterial growth rate, abundance, diversity and carbon demand. Aquat Microb Ecol 13:19–27

Thouvenot A, Richardot M, Debroas D, Devaux J (1999) Bacterivory of metazooplankton, ciliates and flagellates in a newly flooded reservoir. J Plankton Res 21:1659–1679

Torrella F, Morita RY (1979) Evidence by electron micrographs for a high incidence of bacteriophage particles in the waters of Yaquina Bay, Oregon: ecological and taxonomical implications. Appl Environ Microbiol 37:774–778

Virgin HW, Wherry EJ, Ahmed R (2009) Redefining chronic viral infection. Cell 138:30–50

Weinbauer MG, Höfle MG (1998) Significance of viral lysis and flagellate grazing as factors controlling bacterioplankton production in a eutrophic lake. Appl Environ Microbiol 64:431–438

Weinbauer MG, Winter C, Höfle MG (2002) Reconsidering transmission electron microscopy based estimates of viral infection of bacterioplankton using conversion factors derived from natural communities. Aquat Microb Ecol 27:103–110

Weinbauer MG, Brettar I, Höfle MG (2003) Lysogeny and virus-induced mortality of bacterioplankton in surface, deep, and anoxic marine waters. Limnol Oceanogr 48:1457–1465

Wilhelm SW, Smith REH (2000) Bacterial carbon production in Lake Erie is influenced by viruses and solar radiation. Can J Fish Aquat Sci 57:317–326

Wilhelm SW, Suttle CA (1999) Viruses and nutrient cycles in the sea. Bioscience 49:781–788

Wommack KE, Colwell RR (2000) Virioplankton: viruses in aquatic ecosystems. Microbiol Mol Biol Rev 64:69–114

Wommack KE, Hill RT, Colwell RR (1995) A simple method for the concentration of viruses from natural water samples. J Microbiol Methods 22:57–67

Wommack KE, Ravel J, Hill RT, Chun J, Colwell RR (1999) Population dynamics of Chesapeake Bay virioplankton: total community analysis by pulsed-field gel electrophoresis. Appl Environ Microbiol 65:231–240

Wommack KE, Sime-Ngando T, Winget DM, Jamindar S, Helton RR (2010) Filtration-based methods for the collection of viral concentrates from large water samples. In: Wilhelm SW, Weinbauer M, Suttle C (eds) Manual of aquatic viral ecology. American Society of Limnology and Oceanography, Waco, pp 110–117, Chapter 12

Study of Prokaryotes and Viruses in Aquatic Ecosystems by Metagenetic and Metagenomic Approaches

15

Gisèle Bronner, Didier Debroas, François Enault, Mylène Hugoni, Isabelle Jouan-Dufournel, Isabelle Mary, Viviane Ravet, Simon Roux, and Najwa Taïb

Abstract

High-throughput sequencing methods have revolutionized the study of microbial communities in ecosytems. In this chapter, we present the use of metagenetics, targeted metagenomics and viral metagenomics to depict the diversity and the role of *Bacteria*, *Archaea* and viruses in the mixolimnion of the lake Pavin. *Actinobacteria* dominated the bacterial community and was represented by two typical clades: acI and acIV. Focusing on this phylum, we describe its genomic properties characterized by a low GC content (42–52 %) and high substitution rates. Phylogeny based on both SSU rRNA gene and concatenated proteins showed that these clades represented different ecotypes. Archaea was a minor but active component among prokaryotes. The Thaumarchaeota Marine Group I, already known to play an important role in ammonia oxidation dominated the archaeal assemblage in surface waters but surprisingly the Miscellaneous Euryarchaeotic Group (MEG) was the most active in particular zones such as the oxycline. Finally, the analysis of viral particles by a metagenomic approach in the lake Pavin was the first study published about freshwater temperate lakes. Among aquatic datasets, genetic richness of marine viromes was higher in average than that of freshwater ones. The viral community was composed of viruses from families of double-stranded (*Caudovirales*) and single-stranded (*Microviridae, Circoviridae, Nanoviridae*) DNA viruses. The phylogeny of the complete *Microviridae* genomes from the Lake Pavin virome led to the description of a new group, the *Pichovirinae*, and new "chimeric" genomes could be assembled, challenging the established view of ancient viral evolutionary history.

Keywords

Bacteria • *Archaea* • Viruses • Metagenetics • Metagenomics • Viromes • High-throughput sequencing

15.1 Introduction

Massive sequencing has provided fundamental insights into the diversity of microbes and their function and roles in ecosystems. Initially, this method was used largely as a way of easily obtaining genomic information about organisms for which culturing technique was unknown (Schloss and Handelsman 2005). However, due to reduction of the cost and decrease of the difficulty of sequencing, « metagenomics » has become a tool for studying any microbial or viral community, regardless of cultivability. Metagenomics can historically be defined as random shotgun sequencing of DNA isolated from environmental samples. From sequence data it is then possible to infer which organisms are present in a sample (*i.e.*, who is there?) as well as their functional potential (*i.e.*, what are they doing?). First metagenomics

G. Bronner • D. Debroas (✉) • F. Enault • M. Hugoni
I. Jouan-Dufournel • I. Mary • V. Ravet • S. Roux • N. Taïb
LMGE, Laboratoire Microorganismes: Génome et Environnement, UMR CNRS 6023, Université Clermont-Auvergne, Université Blaise Pascal, BP 80026, 63177 Aubière Cedex, France
e-mail: didier.debroas@univ-bpclermont.fr

T. Sime-Ngando et al. (eds.), *Lake Pavin*, DOI 10.1007/978-3-319-39961-4_15

experiments were based on sequencing after cloning a DNA fragment (ranging from 30 to 100 kb) in *Escherichia coli*. This method tends to disappear in favour of high-throughput sequencing (pyrosequencing and Illumina methods) without cloning. High throughput sequencing technologies used after PCR amplification of marker gene amplicons (such as 16S/18S rRNA) refered as metabarcoding or metagenetics, has also emerged as a powerful and straightforward means to analyze richness, diversity and composition of microbial communities. In contrast to shotgun metagenomics, metagenetics enable the detection of rare taxa. Metagenomics and metagenetics have been used in lake Pavin to decipher the microbial and viral diversities. These approaches are illustrated here through the analysis of three taxonomic groups of peculiar interest as they correspond to dominant (*Actinobacteria*), minor also referred to as rare (*Archaea*) and uncharacterized (viruses) groups in lacustrine freshwater ecosystems.

The phylum *Actinobacteria* is defined as Gram-positive bacteria with a high mol% GC DNA composition. These bacteria are conventionally found in soils. However, 16S rRNA gene- and FISH-based studies revealed that *Actinobacteria* are also common and often dominant in a variety of freshwater habitats (Newton et al. 2011). Likewise, in a 2 year study of the epilimnion of the Pavin Lake, *Actinobacteria* dominated the main bacterial group, followed by *Betaproteobacteria* (Boucher et al. 2006). The 16S rRNA genes of these freshwater bacteria are distinct from the soil ones and can be considered as typical from lacustrine ecosystem, while the latest review on freshwater bacteria shows that *Actinobacteria* can be divided in nine lineages (Newton et al. 2011). However, the physiology of these bacteria is still poorly understood and the lineages were delineated from a single gene (*i.e.* 16S rRNA). The clades therefore defined may not correspond to distinct ecological populations (*i.e.* ecotypes). However, the isolation and sequencing of DNA fragments with phylogenetic markers among these groups can highlight relations between phylogenetic position and physiological properties.

Another enigmatic lineage in the oxic zone of aquatic ecosystems is the *Archaea* that have originally been isolated from environments considered as extreme with high temperatures, very high salt concentrations or without oxygen (DeLong 1998). First "non extreme" *Archaea* were discovered as free-living mesophilic in oxic zone of ocean by DeLong (1992). Ammonia oxidation by *Archaea* was first shown through metagenomic studies (Treusch et al. 2005) and then through the cultivation of isolates like *Nitrosopumilus maritimus* (Konneke et al. 2005) or *Nitrososphaera gargensis* (Hatzenpichler et al. 2008). Lacustrine freshwater habitats have also emerged as an unsuspected reservoir of archaeal diversity (Galand et al. 2006; Llirós et al. 2010; Hugoni et al. 2013a, b) and abundance, ranging between 1 and 20 % of the total bacterioplankton (Pernthaler et al. 1998). However, the activity of some *Archaea*, like *Euryarchaota*, in the oxic zones is still discussed. Metagenetics can be then used for sequencing transcripts allowing determining the activity of these less abundant taxa.

Finally, viruses, the most abundant biological entities on earth, act in the regulation of microbial community composition by establishing specific interactions with their host. However, if the marine viruses were extensively studied, the freshwater ones were less known until the study of (Roux et al. 2012a) conducted in the lake Pavin. In addition, the presence of specific clades among hosts in a given type of ecosystem (*e.g.* saline or freshwater systems) suggests the existence of specific viral clades. The viral diversity of freshwater systems was thus investigated from several lakes through metagenomics analyses to evaluate this hypothesis.

These three case studies illustrate the power of "omic" approaches in the analysis of microbial and viral communities in lacustrine ecosystems.

15.2 Phylogeny and Ecology of *Actinobacteria* by a Targeted Metagenomic Approach

15.2.1 Phylogeny of the 16S rRNA Gene and Ecology of Actinobacteria

In a recent synthesis (Newton et al. 2011), freshwater *Actinobacteria* diversity was described through 9 clades defined as monophyletic groups of 95 % identity sequences and 32 tribes (97 % identity). Most clades (acI, acIII, acTH1, acTH2, acSTL, Luna1, Luna3) belong to the *Actinomycetales*, while clades acIV and acV were affiliated to the *Acidimicrobiales*. The acI lineage is more abundant in the free-living fraction (Allgaier et al. 2007), and the acIV lineage exhibits more localized dominance (Wu et al. 2007). Clade acI represents about 66 % of the freshwater actinobacterial 16S rRNA gene sequences available in databases, with tribes acI-B1 and acI-A6 representing 28 % and 18 % of this clade, respectively. The occurrence of freshwater actinobacterial clades may depend on some environmental parameters such as pH (acidic for acI-A1, B2 and B3, more basic for clades acI-A2, A6 et B1) (Lindström et al. 2005) or temperature (adaptation to different thermal niches within the Luna2 tribe from calde acIII) (Hahn and Pöckl 2005), while actinobacterial abundance decreases with decreasing oxygen concentration (Garcia et al. 2013). From a metagenomics study on the functional potential of freshwater bacterioplankton (Debroas et al. 2009) *Actinobacteria* were mainly associated with pathways for replication and repair, nucleotide, cofactors, vitamin and energy metabolism. This was confirmed by the genome sequencing of a single cell acI-B1 isolate from lake Mendota (Garcia et al. 2012). These functions seemed indeed prevalent in this latter genome together with protein metabolism, while stress response, motility and chemotaxis were under-represented. Thus, it has been proposed that freshwater *Actinobacteria* (at least clade acI-B) have a « passive » lifestyle, persisting on constant background concentration of resources (Livermore et al. 2014).

Freshwater *Actinobacteria* were suggested to have a limited carbohydrate substrate range of preferentially small simple compounds, being primary polysaccharide degraders or commensalists of polysaccahride degradation products by other heterotrophic microorganisms (Garcia et al. 2012). While amino sugars are a major component of bacteria and algae cells, the amino sugars metabolism was the most represented in the metagenome (Debroas et al. 2009). The implication of members of clade acI in the mineralization of N-acetylglucosamine, a breakdown product of chitin and bacterial cell walls was also reported (Beier and Bertilsson 2011). Genes encoding for transporters of xylose, the most abundant monosaccharide in terrestrial plants, arabinose or ribose, and enzymes for their incorporation into the glycolysis pathway have been detected in the genome of a acI-B1 member (Garcia et al. 2012). The nutrient uptake pattern of acI *Actinobacteria* suggests adaptation to phytoplankton exudates (Salcher et al. 2012), and cyanobacterial blooms could affect the actinobacterial community composition (Parveen et al. 2011). On the other hand, it has been suggested that the comparatively minor fluctuations in the abundance of acI taxa point to a consistent source of energy generated through actinorhodopsin, including the acI and Luna lineages (Sharma et al. 2008). Although, various substrate and incorporation rates observed for the freshwater *Actinobacteria* may reflect preferences of individual tribes for different substrate sources (Buck et al. 2009), it is not known whether genomes from different tribes reflect different adaptation patterns to peculiar conditions.

15.2.2 Phylogenomics of the Main Tribes

Phylogenetic analyses of both 16S rRNA gene and eight proteins coding genes common to eleven actinobacterial genomic fragments sampled from six different lakes (Philosof et al. 2009; Ghylin et al. 2014) confirmed the closeness of clade acIV to *Acidimicrobium*, while the sister group of the AcI clade was less clearcut. In the 16S rRNA phylogeny, clade acI branch as a sister group to all *Actinomycetales* orders except *Frankiale*, which emerges as a paraphyletic group at the root of the *Actinobacteria* class. In the eight protein concatenated phylogeny, the AcI clade is a sister group to *Acidothermus cellulolyticus*, although moderately supported, and shares a common ancestor with all Actinomycetale orders except *Micromonosporales*, *Glycomycetales* (*Stackenbrandtia*) and *Frankiales*, which emerge at the root of the *Actinomycetales* (Fig. 15.1a). Interestingly, freshwater *Actinobacteria* clades previously defined from 16S rRNA sequences analysis were recovered in the protein phylogeny. At the genomic level, clade acI-B has the lowest G+C content (37–43 %) observed in the *Actinobacteria* phylum, while the acI-A and acIV lineages are moderately low G+C (41–50 % and 49–53 %) (Ghai et al. 2012). Whatever the lineage, this low GC is associated with a fast evolutionary rate, an excess of mutations from A/T to G/C in clade acI-B, suggesting a G+C genome enrichment for this clade, and different evolutionary constraints among clades. For all lineages, functional characterization of these genomic fragments was found to be mostly related to nucleic and amino-acid transport and metabolism, translation and ribosomal structure and biogenesis. Among others, an arginine operon upstream the rRNA operon shows a divergent organization depending on freshwater tribes (Fig. 15.1b). While the arginine operon is almost complete except for gene argG in acIV tribes, it is shuffled in the acI clade. Indeed, arginine operons lack genes argG and argF in tribes acI-A1 and acI-A7. ArgJ is also absent from the operon in acI-A6 and the operon in acI-B1 is reduced to genes argHRBC. We can speculate that the arginine genes were scattered along the genome through rearrangements, as argG and argF have been found elsewhere in single-cell genome contigs. However, there is no evidence of the presence of argJ or argD in the acI-B1 or argJ in acI-A6 genomes (but only 97 % genome acI-B1 is assembled – Garcia et al. 2013). It has been postulated that the turnover of operon structure of genes with a close functional relationship can lead to adaptive changes in gene expression (Price et al. 2006). Thus, the regulation of arginine metabolism, which is related to numerous metabolic pathways, might reflect tribes adaptations to the nutrients present in particular niches.

15.2.3 Genomic Features of Actinobacteria

The independent 16S based selection for acI and acIV actinobacterial genomic fragments from the six lakes revealed a similar genomic organization around the rRNA operon, with conserved synteny suggesting that both acI and acIV actinobacterial clades may have a unique rRNA operon. This finding is in agreement with the estimation of a low number of rRNA operon in the genome of freshwater *Actinobacteria* compared to that of their marine counterparts (Roux et al. 2011). Since a relation between rDNA operon number and

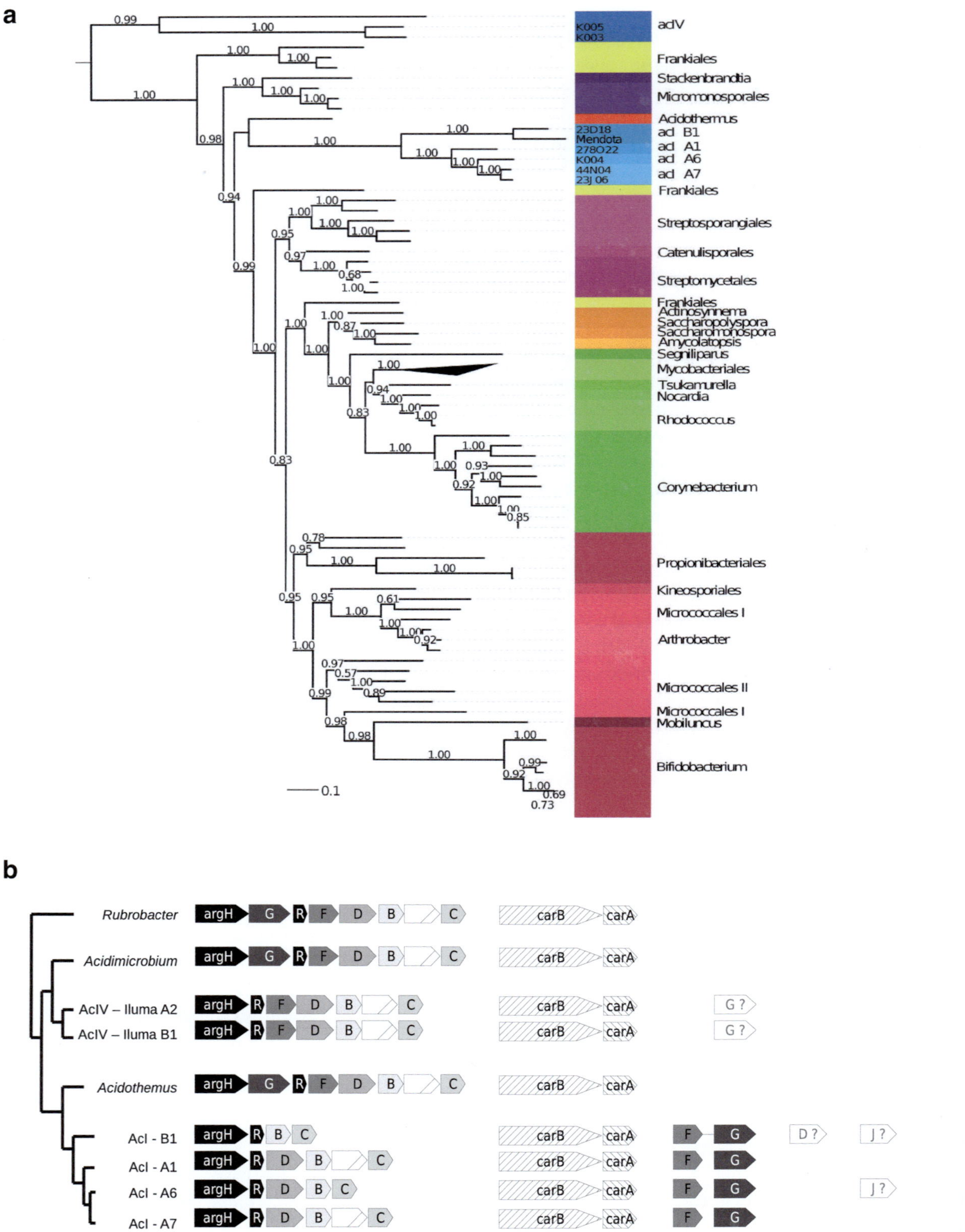

Fig. 15.1 (**a**) Phylogeny of the class *Actinobacteria* from concatenated proteins ArgB, ArgC ArgH, ArgR, InfC, TyrS, PheT and PheS. Freshwater clades are highlighted in *blue*, including representatives from tribes acIV-Iluma-A2 (K003, K005), acI-B1 (Mendota) and acI-A6 (K004). Muscle was used to align each protein separately before alignments concatenation. The resulting alignment was analyzed using PAML to remove sites with substitution rate > 1. Phylogeny was computed using phyML with a WAG + G4 model and 1000 bootstrap. Only bootstrap values > 0.5 (500/1000) are reported here. (**b**) Structural organization of the arginine operon in six freshwater actinobacterial tribes: AcIV-Iluma-A2 (K003, K005), acIV-Iluma-B1 (Mgen15-A23, unpublished results), acI-B1 (23-D18), acI-A1 (278-O22), acI-A6 (K004), acI-A7 (44-N04, 23-J06)

growth rate exists, although not strict, it also corroborates the observation of a moderate to low growth rates (Simek et al. 2006; Livermore et al. 2014). In the same way these results are coherent with the small cell size reported for lacustrine *Actinobacteria* (Hahn et al. 2003) and their presence in the LNA (Low Nucleic Acid) fraction of lacustrine environments. Moreover, freshwater actinobacterial genomes show small genome sizes (1,2 Mb for acI-B1, 3 Mb for acTH2 – Livermore et al. 2014), a low G+C content as compared to known *Actinobacteria*, ranging from 37 to 53 % and small intergenic distances (Ghai et al. 2014). Altogether these observations suggest that some freshwater *Actinobacteria* clades may be subjected to genome streamlining (Garcia et al. 2012), or at least genome reduction. It has been postulated that genome streamlining and small cell size could be driven by selection for living in nutrient poor environments (Giovannoni et al. 2014 and references therein). This hypothesis does not hold for freshwater *Actinobacteria* as they were reported to be prevalent in a large spectrum of trophic status (Newton et al. 2006; Debroas et al. 2009; Humbert et al. 2009). Another hypothesis is that small cell size promotes better escape from bacterivores and phages. Some studies have indeed reported actinobacterial reduced grazing load (see references in Newton et al. 2011). However, the link between genome reduction and selection against predation remains unclear, especially in the designation of which gene could be lost. Loss of non-essential metabolic pathways and regulatory machinery is possible for intracellular bacteria which live in a stable environment (Batut et al. 2014), but it seems difficult to consider lakes as stable environments for free living bacteria. Moreover, for a genome reduction to be adaptive, "selective gene loss" should correlate with ecological niches. Recent literature suggests potential specialization in substrate preference for freshwater *Actinobacteria* (Ghylin et al. 2014). However, work is still to be done to draw the metabolic potential of these organisms and understand the selective cause, if any, of this genome reduction.

15.3 Metagenetics for Assessing the Activity of Archaea

15.3.1 Archaea in Freshwater Ecosystems

Archaea have been the focus of intensive research in the last decades because of their ubiquity and abundance in aquatic ecosystems. Nevertheless, data available for freshwater ecosystems are still scarce compared to marine systems mainly due to the physico-chemical and limnological heterogeneity of the environments studied (rivers, lakes, inland waters). The recovery of archaeal genes encoding the ammonia monooxygenase subunit A (*amoA*), within *Thaumarchaeota* Marine Group I, in many marine and terrestrial environments (Schleper et al. 2005) confirms the global importance of ammonia oxidizing *Archaea* (AOA). AOA are also detected in oxic layers of freshwater ecosystems (*e.g.* Hugoni et al. 2013a, b), however, in lacustrine deeper layers (*i.e.* anoxic zone and above sediments) AOA are rare (Lliròs et al. 2010) and methanogenesis performed by *Euryarchaeota* is often the most important process. Moreover, initially thought to occur only in anoxic environments, methane production has been recently described in a well-oxygenated water column of an oligotrophic lake (Grossart et al. 2011). Thus, *Euryarchaeota* may also be important players in carbon cycling in oxic aquatic environments and water to air methane flux.

15.3.2 Assessing Community Structure of Active Archaeal Populations

Within freshwater ecosystems, meromictic lakes with permanent oxic/anoxic interfaces constitute optimal study sites to monitor archaeal population changes through physico-chemical variations in their habitats. However, most studies investigating lacustrine ecosystems have focused on archaeal diversity based on gene abundance (*i.e.* metagenetics), but only a few studies have examined active archaeal assemblages (La Cono et al. 2013). Recent studies have shown the importance of differentiating the active communities from the total communities (Hugoni et al. 2013a, b). One method to explore an aspect of their activity (*i.e.*, the growth rate for specific taxa) is then to investigate microbial communities at both the 16S rRNA genes (DNA) and 16S rRNA level (RNA). Nevertheless, the use of 16S rRNA should be addressed considering the environmental parameters and specific taxa due to some inconsistent relationships between 16S rRNA and activity (Blazewicz et al. 2013). Thus, the changes in the community structure of active archaeal populations could be evaluated over time by pyrosequencing 16S rRNA genes and 16S rRNA.

This method allowed correlating the presence of some taxonomic groups with their potential activity, as in the epilimnion of Lake Pavin, where archaeal assemblage was clearly dominated by *Thaumarchaeota* Marine Group I (MGI, 83 % of the reads in the 16S rRNA genes dataset) (Fig. 15.2a), which was also the most active group in this zone (99.6 % of the reads in the 16S rRNA dataset). Some other groups, such as the Lake Dagow Sediment (LDS) cluster or the Deep-sea Hydrothermal Vent Eukaryotic Group 6 (DHVEG-6) were present in the 16S rRNA genes fraction in the epilimnion (Fig. 15.2) even if they accounted for a negligible proportion of potentially active archaeal fraction. The LDS cluster is a highly diverse euryarchaeal lineage (Barberán et al. 2011) identified in rivers (Galand et al. 2006) where they accounted for a large proportion of the archaeal

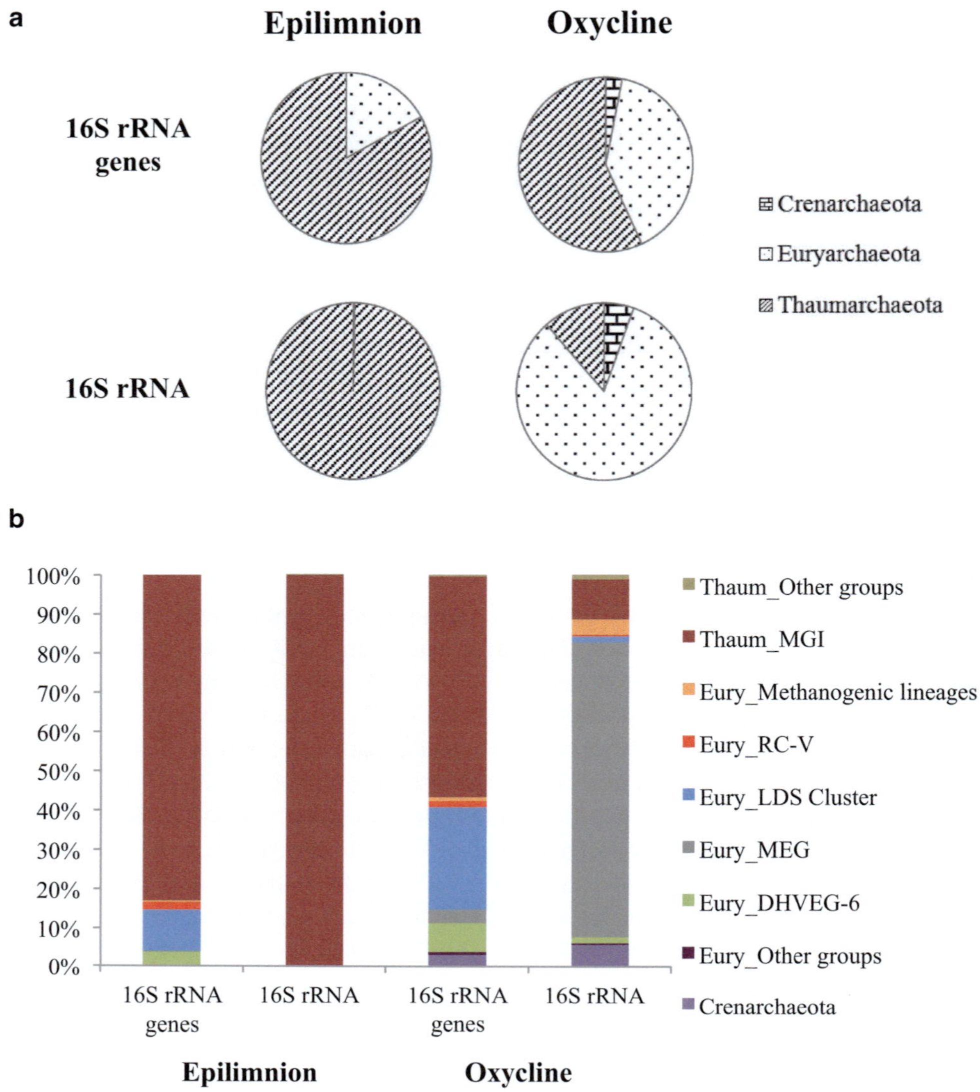

Fig. 15.2 Taxonomic compostion of the archaeal assemblage present in Lake Pavin (**a**) and focus on the different groups retrieved among *Crenarchaeota*, *Euryarchaeota* (Eury_) and *Thaumarchaeota* (Thaum_) (**b**). Thaum_Other groups included SAGMCG-1 and SCG, Eury_Other groups included DSEG, Sm1K20, MHVG, SAGMEG, TMEG, MBGD and DHVEG-1, *Crenarchaeota* included C3, MBGB, MCG, Sulfolobus and Terrestrial Group. **Pyrosequencing amplicons were produced from the V4 to V5 region of the 16S rRNA genes and 16S rRNA using the archaeal primers Arch519F (Herfort et al. 2009) and Arch915R (Casamayor et al. 2002). Pyrosequencing was performed using a Roche 454 GS-FLX system with titanium chemistry**

cell counts (Herfort et al. 2009) and frequently retrieved in lakes (Restrepo-Ortiz et al. 2014). Due to lack of cultivated representatives, it is currently difficult to link the LDS cluster with its ecological role and we do not know its metabolic capacities. In the same way, DHVEG-6 was a largely unknown archaeal group originally described as a hydrothermal vent lineage (Takai and Horikoshi 1999) but that was also detected from relatively anoxic terrestrial soils (Takai et al. 2001) to marine sediments (Teske and Sørensen 2008) and water column (Hugoni et al. 2013a, b).

The approach using both 16S rRNA genes and 16S rRNA also highlighted that groups that are not necessarily the most abundant in freshwater ecosystems may have a potential activity. In the oxycline abundant archaeal major groups were similar from the epilimnion and affiliated with *Thaumarchaeota* Marine Group I (56.4 % of the reads), followed by *Euryarchaeota* LDS cluster, DHVEG-6 (26 and 7.4 % of the reads respectively). Nevertheless, in the oxycline of Lake Pavin, the Miscellaneous Euryarchaeotic Group (MEG) dominated the potentially active archaeal assemblage throughout the entire year, suggesting that this poorly-characterized group could play a key functional role in this aquatic ecosystem (Fig. 15.2b). This group was initially retrieved in terrestrial soil, marine sediments (Takai et al. 2001) and in deep subsurfaces (Hirayama et al. 2007) but recent studies indicated its presence in some Spanish lakes (Auguet et al. 2011) suggesting the existence of a freshwater clade.

15.3.3 *Thaumarchaeota* Marine Group I Versatility

In the oxycline, *Thaumarchaeota* Marine Group I represented 10.2 % of the potentially active archaeal fraction. Overall, thaumarchaeal *amoA* transcripts abundance in both the epilimnion and the oxycline of the Lake Pavin were low suggesting that lacustrine oxic water layers are not an obligatory hotspot for archaeal ammonia oxidation and that MGI could use another metabolic pathway as an energy source. Indeed, it has been proposed that these microorganisms present a large metabolic plasticity, from autotrophy to potential mixotrophic lifestyles (Ingalls et al. 2006). Nevertheless, further works are necessary to better understand the role of *Thaumarchaeota* in ammonia oxidation or other metabolic pathways.

15.4 Metagenomics and Meta-Analysis for Deciphering the Viral Diversity

15.4.1 Viral Richness and Composition

For studying the diversity of viruses in lakes, two viromes from freshwater temperate lakes were generated from samples collected in the epilimnion (at 5 m depth) of lacustrine ecosystems (lakes Pavin and Bourget, Roux et al. 2012a). The genetic diversity and composition of these datasets were assessed and compared to those of published viromes from different ecosystems. Subsequently, two families of single stranded DNA (ssDNA) viruses retrieved in the virome of Lake Pavin were further analyzed : the *Microviridae* (Roux et al. 2012b) and chimeric viruses (Roux et al. 2013).

Species richness (number of different virotypes), gene richness (number of different genes) and the viral gene composition were cross-compared between a large set of viromes (>30) including viromes of Lakes (Roux et al. 2012a). Whereas the species richness remains identical through the different ecosystems, the gene richness varies greatly between the different environments. Among aquatic datasets, genetic richness of marine viromes was higher in average than that of freshwater ones. However, Lake Bourget and Lake Pavin viromes harbor very different genetic richness, as the one from Lake Bourget is comparable to marine viromes and the one from Lake Pavin is much lower (Roux et al. 2012a). These observations are tentatively explained by the general relation between trophic status and microbial richness (Horner-Devine et al. 2003) that impacted the viral richness through the number of hosts.

Despite these differences in terms of genetic richness, the comparison of the 31 viromes reveals potential genetic links between viral communities according to the ecosystem investigated (Roux et al. 2012a). This analysis was here performed using recently published viromes (Hurwitz and Sullivan 2013; Fancello et al. 2013) and the resulting hierarchical clustering tree highlights a clear distinction between seawater and freshwater viral-communities (Fig. 15.3a).

In addition to these differences between datasets, the two lacustrine viromes made it possible to assess the composition and diversity of temperate freshwater viral communities (Roux et al. 2012a). Firstly, the Lake Pavin dataset has a very small "known" viral-gene fraction (<15 %) illustrating that viruses in this ecosystem are a huge source of so-far uncharacterized genetic diversity. Secondly, considering sequences that match known viruses and in accordance with electronic microscope observations, *Caudovirales* (*Myo-*, *Sipho-* and *Podoviridae*) were retrieved in Lake Pavin virome (Fig. 15.3b). Furthermore, a high proportion of ssDNA viruses (*Micro-*, *Circo-* and *Nanoviridae*) virus was recorded. These different families of ssDNA viruses gained more of our attention because they might play an important role in a broad spectrum of environments as they were found in great number among the viral fraction from freshwater to human gut samples (Roux et al. 2012b).

15.4.2 The Pichovirinae, a New Group Among the Microviridae Family

Microviridae are small icosahedral viruses with circular ssDNA genomes that infect bacteria. Based on structural and genomic differences, members of this family were divided into three subgroups: *Microvirinae*, *Gokushovirinae* and *Alpavirinae* (Krupovic and Forterre 2011). Out of the 81 complete *Microviridae* genomes assembled from viromes originating from various ecosystems (Roux et al. 2012b), three were generated out of the Lake Pavin virome (Fig. 15.3c top). The phylogenetic analysis of the major capsid protein revealed that one of these three new genomes was affiliated to the *Gokushovirinae*. Among this diverse and cosmopolitan subfamily (Hopkins et al. 2014), the genome from the Lake Pavin was grouped with other freshwater *Microviridae*, demonstrating that these viruses are specific to this environment. The phylogeny of the other two complete *Microviridae* genomes from the Lake Pavin virome led to the description of a new group, the *Pichovirinae*. Furthermore, all genomes of this new group have an organization of the three core genes different from the rest of the *Microviridae*. Thus, this group that encompasses viruses from various ecosystems (coral, microbialites, marine and freshwater systems) has probably diverged from the common *Microviridae* ancestor a long time ago (Roux et al. 2012b). Therefore, the description of these new genomes expands the existing knowledge on genome evolution, diversity and environmental distribution of this viral family.

15.4.3 Chimeric Viruses: Implication for Viral Evolution?

Small ssDNA viruses encoding replication proteins (Reps) related to known eukaryotic viruses have been repeatedly

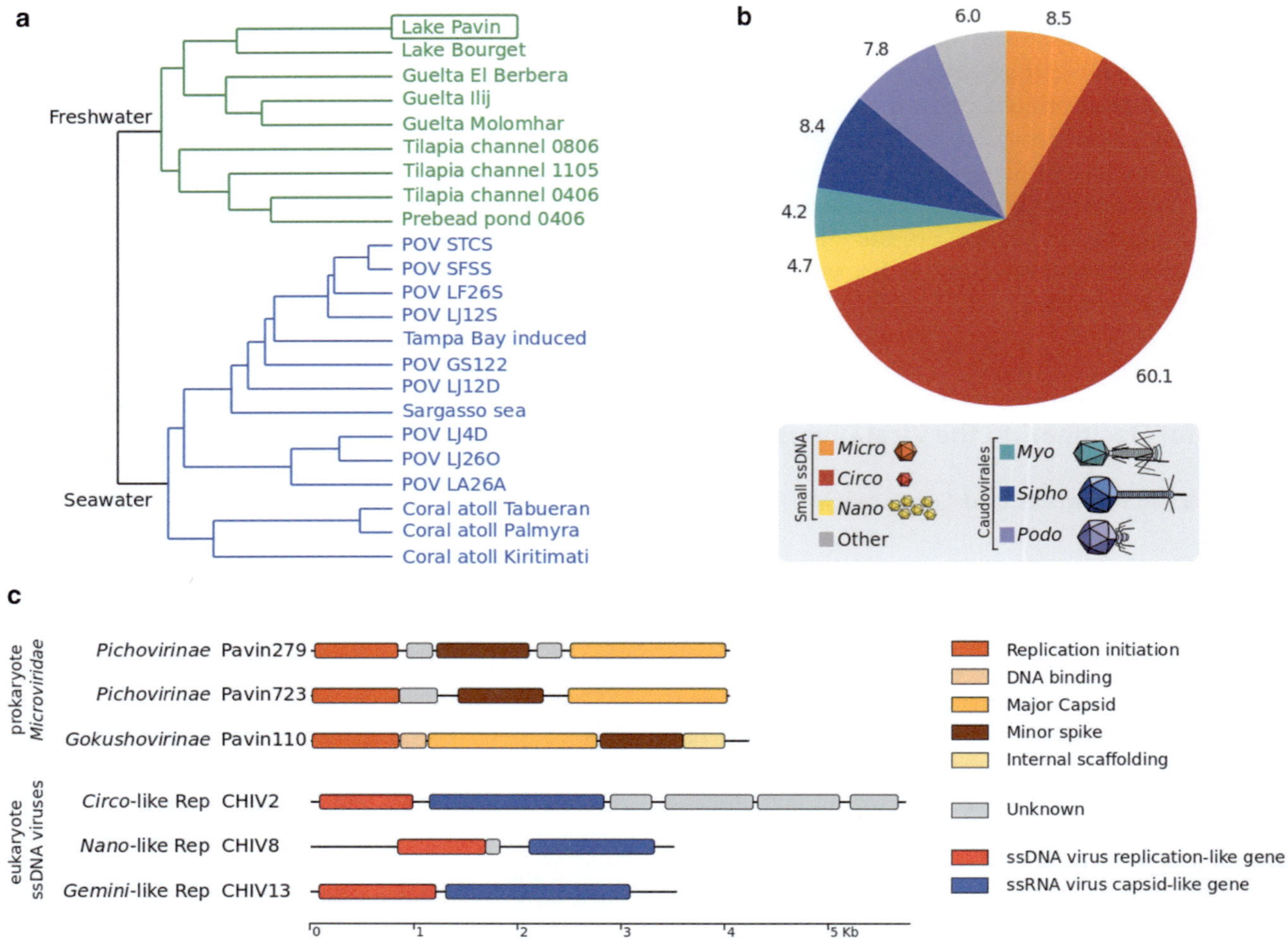

Fig. 15.3 (**a**) Virome hierarchical clustering tree based on sequence similarity. This tree was computed from tBLASTx comparisons of virome subsamples. Marines and freshwater viromes are colored respectively in *blue* and *green*. (**b**) Taxonomic composition of the Lake Pavin virome at the viral family level, determined by similarity to complete virus genomes of the RefseqVirus database (tBLASTx using thresholds of 10^{-3} on e-value and 50 on bit score). The "Other" category pools families which represented less than 1 % of the full virome sequences. (**c**) Genomic maps of the three *Microviridae* and chimeric viruses assembled out the Lake Pavin virome. Conserved genes are colored for each of these two families, the colour key being provided in the figure

isolated from diverse environments, including marine (Dunlap et al. 2013) and freshwater systems (López-Bueno et al. 2009). Recently, one such Rep-encoding genome, recovered in an extreme environment, was suggested to be a recombination between unrelated groups of ssRNA and ssDNA viruses (Diemer and Stedman 2012). The validity of the assembled viral genome and its presence in the lake sediment pore water were confirmed by PCR amplification (Diemer and Stedman 2012). Since this study, 13 chimeric viral genome (CHIV) were assembled from 103 published viromes (Roux et al. 2013) and were grouped according to the type of Rep they encode: *Circo-*, *Gemini-* or *Nanoviridae* like Rep groups. As many as 3 of these 13 CHIV genomes were assembled from the Lake pavin virome (Fig. 15.3c, bottom). These 3 genomes were each affiliated to a different group, highlighting the diversity of these viruses in the Lake Pavin. Analysis of these new genomes indicate a single event of capsid protein gene capture from an RNA virus in the history of this virus group. Such hints on the evolution of this viral family raise some exciting horizons in our understanding of the connections between viruses in the biosphere.

15.5 Summary

Molecular based analyses of microbial communities through metagenomics are a proxy to investigate the metabolic potential and evolutionary trajectory of organisms that are still to be cultured. Here, we have illustrated advances that have been made using these approaches on three compartments of freshwater aquatic microorganisms that are of peculiar interest as they correspond to either dominant, rare or unexplored living bodies.

Phylogenomic profiling and genomic analyses of freshwater Actinobacteria highlighted both reduced, AT-rich genomes and potential adaptation to peculiar nutrient niches. Their capabilities to metabolise simple carbohydrate compounds from degrading organisms, including amino sugar, and potentially produce energy through actinorhodopsin can explain their dominance in freshwater lakes. On the other hand, some driving forces that are usually advanced for genome reduction in free living microorganisms do not hold for freshwater Actinobacteria. Considering Archaea, our work revealed different patterns of active archaeal assemblages in the epilimnion and the oxycline of Lake Pavin and suggested that *Thaumarchaeota* MGI were not obligate ammonia-oxidizers. Moreover, this work highlighted the importance of studying groups considered as "minor groups" that could nevertheless be active. Those taxa, with no cultivated member and present in small proportions can be overrepresented in the active fraction and may thus play key functional role in freshwater ecosystems. Finally, viruses from freshwater viromes belong to families classically observed in aquatic systems such as *Caudovirales* and different families of small ssDNA viruses. Yet, genomic and phylogenetic analyses revealed that viruses are specific to these ecosystems as they form distinct clades for all the cited families. These studies are however hindered by the lack of cultivated references and additional metagenomic datasets will be required to better understand the extent of freshwater virus diversity and their dynamics.

Acknowledgement We are indebted to Dr Steven M. SHORT, University of Toronto (Canada), for a critical review of the manuscript.

References

Allgaier M, Brückner S, Jaspers E, Grossart H-P (2007) Intra- and inter-lake variability of free-living and particle-associated Actinobacteria communities. Environ Microbiol 9:2728–2741

Auguet J-C, Nomokonova N, Camarero L, Casamayor EO (2011) Seasonal changes of freshwater ammonia-oxidizing archaeal assemblages and nitrogen species in oligotrophic alpine lakes. Appl Env Microbiol 77:1937–1945

Barberán A, Fernández-Guerra A, Auguet J-C, Galand PE, Casamayor EO (2011) Phylogenetic ecology of widespread uncultured clades of the Kingdom Euryarchaeota. Mol Ecol 20:1988–1996

Batut B, Knibbe C, Marais G, Daubin V (2014) Reductive genome evolution at both ends of the bacterial population size spectrum. Nat Rev Microbiol Dec 12(12):841–850

Beier S, Bertilsson S (2011) Uncoupling of chitinase activity and uptake of hydrolysis products in freshwater bacterioplankton. Limnol Oceanogr 56:1179–1188

Blazewicz SJ, Barnard RL, Daly RA, Firestone MK (2013) Evaluating rRNA as an indicator of microbial activity in environmental communities: limitations and uses. ISME J 7:2061–2068

Boucher D, Jardillier L, Debroas D (2006) Succession of bacterial community composition over two consecutive years in two aquatic systems: a natural lake and a lake-reservoir. FEMS Microbiol Ecol 55:79–97

Buck U, Grossart H-P, Amann R, Pernthaler J (2009) Substrate incorporation patterns of bacterioplankton populations in stratified and mixed waters of a humic lake. Environ Microbiol 11:1854–1865

Casamayor EO, Massana R, Benlloch S, Ovreas L, Diez B, Goddard VJ et al (2002) Changes in archaeal, bacterial and eukaryal assemblages along a salinity gradient by comparison of genetic fingerprinting methods in a multipond solar saltern. Environ Microbiol 4:338–348

Debroas D, Humbert JF, Enault F, Bronner G, Faubladier M, Cornillot E (2009) Metagenomic approach studying the taxonomic and functional diversity of the bacterial community in a mesotrophic lake (Lac du Bourget–France). Environ Microbiol 11:2412–2424

DeLong EF (1992) Archaea in coastal marine environments. Proc Natl Acad Sci U S A 89:5685–5689

DeLong EF (1998) Everything in moderation: archaea as "non-extremophiles.". Curr Opin Genet Dev 8:649–654

Diemer GS, Stedman KM (2012) A novel virus genome discovered in an extreme environment suggests recombination between unrelated groups of RNA and DNA viruses. Biol Direct 7:13

Dunlap DS, Ng TFF, Rosario K, Barbosa JG, Greco AM, Breitbart M, Hewson I (2013) Molecular and microscopic evidence of viruses in marine copepods. Proc Natl Acad Sci U S A 110:1375–1380

Fancello L, Trape S, Robert C, Boyer M, Popgeorgiev N, Raoult D, Desnues C (2013) Viruses in the desert: a metagenomic survey of viral communities in four perennial ponds of the Mauritanian Sahara. ISME J 7:359–369

Galand PE, Lovejoy C, Vincent WF (2006) Remarkably diverse and contrasting archaeal communities in a large arctic river and the coastal Arctic Ocean. Aquat Microb Ecol 44:115–126

Garcia SL, McMahon KD, Martinez-Garcia M, Srivastava A, Sczyrba A, Stepanauskas R, Grossart H-P, Woyke T, Warnecke F (2012) Metabolic potential of a single cell belonging to one of the most abundant lineages in freshwater bacterioplankton. ISME J 7:137–147

Garcia SL, Salka I, Grossart H-P, Warnecke F (2013) Depth-discrete profiles of bacterial communities reveal pronounced spatio-temporal dynamics related to lake stratification. Environ Microbiol Rep 5:549–555

Ghai R, McMahon KD, Rodriguez-Valera F (2012) Breaking a paradigm: cosmopolitan and abundant freshwater actinobacteria are low GC. Environ Microbiol Rep 4:29–35

Ghai R, Mizuno CM, Picazo A, Camacho A, Rodriguez-Valera F (2014) Key roles for freshwater Actinobacteria revealed by deep metagenomic sequencing. Mol Ecol 23:6073–6090

Ghylin TW, Garcia SL, Moya F, Oyserman BO, Schwientek P, Forest KT, … McMahon KD (2014) Comparative single-cell genomics reveals potential ecological niches for the freshwater acI Actinobacteria lineage. ISME J 8(12):1–14

Giovannoni SJ, Thrash CJ, Temperton B (2014) Implications of streamlining theory for microbial ecology. ISME J 8(8):1553–1565

Grossart H-P, Frindte K, Dziallas C, Eckert W, Tang KW (2011) Microbial methane production in oxygenated water column of an oligotrophic lake. Proc Natl Acad Sci 108:19657–19661

Hahn MW, Pöckl M (2005) Ecotypes of planktonic actinobacteria with identical 16S rRNA genes adapted to thermal niches in temperate, subtropical, and tropical freshwater habitats. Appl Environ Microbiol 71:766–773

Hahn MW, Lünsdorf H, Wu Q, Schauer M, Höfle MG, Boenigk J, Stadler P (2003) Isolation of novel ultramicrobacteria classified as actinobacteria from five freshwater habitats in Europe and Asia. Appl Environ Microbiol 69:1442–1451

Hatzenpichler R, Lebedeva EV, Spieck E, Stoecker K, Richter A, Daims H, Wagner M (2008) A moderately thermophilic ammonia-oxidizing crenarchaeote from a hot spring. Proc Natl Acad Sci U S A 105:2134–2139

Herfort L, Kim J, Coolen MJL, Abbas B, Schouten S, Herndl GJ, Damst JSS (2009) Diversity of archaea and detection of crenarchaeotal amoA genes in the rivers Rhine and Têt. Aquat Microb Ecol 55:189–201

Hirayama H et al (2007) Culture-dependent and -independent characterization of microbial communities associated with a shallow submarine hydrothermal system occurring within a coral reef off Taketomi Island. Japan Appl Environ Microbiol 73:7642–7656

Hopkins M, Kailasan S, Cohen A, Roux S, Tucker KP, Shevenell A, Agbandje-McKenna M, Breitbart M (2014) Diversity of environmental single-stranded DNA phages revealed by PCR amplification of the partial major capsid protein. ISME J 8:2093–2103

Horner-Devine MC, Leibold MA, Smith VH, Bohannan BJM (2003) Bacterial diversity patterns along a gradient of primary productivity. Ecol Lett 6:613–622

Hugoni M, Taib N, Debroas D, Domaizon I, Jouan Dufournel I, Bronner G, Salter I, Agogué H, Mary I, Galand PE (2013a) Structure of the rare archaeal biosphere and seasonal dynamics of active ecotypes in surface coastal waters. Proc Natl Acad Sci 110:6004–6009

Hugoni M, Etien S, Bourges A, Lepère C, Domaizon I, Mallet C, Bronner G, Debroas D, Mary I (2013b) Dynamics of ammonia-oxidizing archaea and bacteria in contrasted freshwater ecosystems. Res Microbiol 164:360–370

Humbert J-F, Dorigo U, Cecchi P, Le Berre B, Debroas D, Bouvy M (2009) Comparison of the structure and composition of bacterial communities from temperate and tropical freshwater ecosystems. Environ Microbiol 11(9):2339–2350

Hurwitz BL, Sullivan MB (2013) The Pacific Ocean virome (POV): a marine viral metagenomic dataset and associated protein clusters for quantitative viral ecology. PLoS One 8:e57355

Ingalls AE, Shah SR, Hansman RL, Aluwihare LI, Santos GM, Druffel ERM, Pearson A (2006) Quantifying archaeal community autotrophy in the mesopelagic ocean using natural radiocarbon. Proc Natl Acad Sci U S A 103:6442–6447

Konneke M, Bernhard AE, de la Torre JR, Walker CB, Waterbury JB, Stahl DA (2005) Isolation of an autotrophic ammonia-oxidizing marine archaeon. Nature 437:543–546

Krupovic M, Forterre P (2011) Microviridae goes temperate: microvirus-related proviruses reside in the genomes of bacteroidetes. PLoS One 6:e19893

La Cono V et al (2013) Partaking of archaea to biogeochemical cycling in oxygen-deficient zones of meromictic saline Lake Faro (Messina, Italy). Environ Microbiol 15:1717–1733

Lindström ES, Agterveld MPK-V, Zwart G (2005) Distribution of typical freshwater bacterial groups is associated with pH, temperature, and lake water retention time. Appl Environ Microbiol 71:8201–8206

Livermore JA, Emrich SJ, Tan J, Jones SE (2014) Freshwater bacterial lifestyles inferred from comparative genomics. Environ Microbiol 16:746–758

Llirós M, Gich F, Plasencia A, Auguet J-C, Darchambeau F, Casamayor EO, Descy J-P, Borrego C (2010) Vertical distribution of ammonia-oxidizing crenarchaeota and methanogens in the epipelagic waters of Lake Kivu (Rwanda-Democratic Republic of the Congo). Appl Environ Microbiol 76:6853–6863

López-Bueno A, Tamames J, Velázquez D, Moya A, Quesada A, Alcamí A (2009) High diversity of the viral community from an Antarctic lake. Science 326:858–861

Newton RJ, Kent AD, Triplett EW, McMahon KD (2006) Microbial community dynamics in a humic lake: differential persistence of common freshwater phylotypes. Environ Microbiol 8:956–970

Newton RJ, Jones SE, Eiler A, McMahon KD, Bertilsson S (2011) A guide to the natural history of freshwater Lake Bacteria. Microbiol Mol Biol Rev 75:14–49

Parveen B, Reveilliez J-P, Mary I, Ravet V, Bronner G, Mangot J-F, Domaizon I, Debroas D (2011) Diversity and dynamics of free-living and particle-associated Betaproteobacteria and Actinobacteria in relation to phytoplankton and zooplankton communities. FEMS Microbiol Ecol 77:461–476

Pernthaler J, Glöckner F-O, Unterholzner S, Alfreider A, Psenner R, Amann R (1998) Seasonal community and population dynamics of pelagic bacteria and archaea in a high mountain lake. Appl Environ Microbiol 64:4299–4306

Philosof A, Sabehi G, Béjà O (2009) Comparative analyses of actinobacterial genomic fragments from Lake Kinneret. Environ Microbiol 11:3189–3200

Price MN, Arkin AP, Alm EJ (2006) The life-cycle of operons. PLoS Genet 2:e96

Restrepo-Ortiz CX, Auguet J-C, Casamayor EO (2014) Targeting spatio-temporal dynamics of planktonic SAGMGC-1 and segregation of ammonia-oxidizing thaumarchaeota ecotypes by newly designed primers and quantitative PCR. Environ Microbiol 16:689–700

Roux S, Enault F, Bronner G, Debroas D (2011) Comparison of 16S rRNA and protein-coding genes as molecular markers for assessing microbial diversity (Bacteria and Archaea) in ecosystems. FEMS Microbiol Ecol 78:617–628

Roux S, Enault F, Robin A, Ravet V, Personnic S, Theil S, Colombet J, Sime-Ngando T, Debroas D (2012a) Assessing the diversity and specificity of two freshwater viral communities through metagenomics. PLoS One 7:e33641

Roux S, Krupovic M, Poulet A, Debroas D, Enault F (2012b) Evolution and diversity of the microviridae viral family through a collection of 81 new complete genomes assembled from virome reads. PLoS One 7:e40418

Roux S, Enault F, Bronner G, Vaulot D, Forterre P, Krupovic M (2013) Chimeric viruses blur the borders between the major groups of eukaryotic single-stranded DNA viruses. Nat Commun 4:2700

Salcher MM, Posch T, Pernthaler J (2012) In situ substrate preferences of abundant bacterioplankton populations in a prealpine freshwater lake. ISME J 7:896–907

Schleper C, Jurgens G, Jonuscheit M (2005) Genomic studies of uncultivated archaea. Nat Rev Micro 3:479–488

Schloss PD, Handelsman J (2005) Metagenomics for studying unculturable microorganisms: cutting the Gordian knot. Genome Biol 6:229

Sharma AK, Zhaxybayeva O, Papke RT, Doolittle WF (2008) Actinorhodopsins: proteorhodopsin-like gene sequences found predominantly in non-marine environments. Environ Microbiol 10:1039–1056

Simek K, Hornák K, Jezbera J, Nedoma J, Vrba J, Straskrábová V, Macek M, Dolan JR, Hahn MW (2006) Maximum growth rates and possible life strategies of different bacterioplankton groups in relation to phosphorus availability in a freshwater reservoir. Environ Microbiol 8(9):1613–1624

Takai K, Horikoshi K (1999) Genetic diversity of archaea in deep-sea hydrothermal vent environments. Genetics 152:1285–1297

Takai K, Moser DP, DeFlaun M, Onstott TC, Fredrickson JK (2001) Archaeal diversity in waters from deep south African gold mines. Appl Env Microbiol 67:5750–5760

Teske A, Sørensen KB (2008) Uncultured archaea in deep marine subsurface sediments: have we caught them all? ISME J 2:3–18

Treusch AH, Leininger S, Kletzin A, Schuster SC, Klenk H-P, Schleper C (2005) Novel genes for nitrite reductase and Amo-related proteins indicate a role of uncultivated mesophilic crenarchaeota in nitrogen cycling. Environ Microbiol 7:1985–1995

Wu QL, Zwart G, Wu J, Kamst-van Agterveld MP, Liu S, Hahn MW (2007) Submersed macrophytes play a key role in structuring bacterioplankton community composition in the large, shallow, subtropical Taihu Lake, China. Environ Microbiol 9:2765–2774

Anaerobic Microbial Communities and Processes Involved in the Methane Cycle in Freshwater Lakes-a Focus on Lake Pavin

16

Anne-Catherine Lehours, Guillaume Borrel, Nicole Morel-Desrosiers, Corinne Bardot, Vincent Grossi, Benoit Keraval, Eléonore Attard, Jean-Pierre Morel, Christian Amblard, and Gerard Fonty

Abstract

The atmospheric concentration of methane (CH_4), a major greenhouse gas, is mainly controlled by the activities of CH_4-producing (methanogens) and CH_4-consuming (methanotrophs) microorganisms. Freshwater lakes are identified as one of the main CH_4 sources, as it is estimated that they contribute to 6–16 % of natural CH_4 emissions. It is therefore critical to better understand the biogeochemical cycling of CH_4 in these ecosystems.

In this vein, the Lake Pavin provides a useful microbial ecosystem to investigate CH_4 cycle in freshwater systems. Despite a significant production of CH_4 in the deep anoxic water column and sediment, the amounts of CH_4 emitted by Lake Pavin to the atmosphere are several orders of magnitude lower than those of temperate lakes suggesting intense consumption activities of this gas.

This chapter focuses on CH_4 cycle, but as methanogenesis and anaerobic methanotrophy build competitive and cooperative relationships with a number of bacterial metabolic groups, we also address bacterial processes that are tightly coupled with CH_4 cycle (*e.g.*, ferric iron reduction). Three main sections constitute this chapter:

- A presentation of CH_4 cycle, including methanogenesis and methanotrophy, in freshwater systems and particularly in Lake Pavin,
- The relationships between CH_4 cycle and some other biogeochemical processes in Lake Pavin (ferric iron reduction, sulfate reduction and fermentation), including a brief overview of anaerobic microbial metabolisms,
- Sections on methodologies enabling to access informations on the anaerobic metabolisms (*e.g.*, biomarkers, isotopes, microcalorimetry, nucleic acid molecular markers, magnetoFISH).

A.-C. Lehours (✉) • G. Borrel • N. Morel-Desrosiers • C. Bardot
B. Keraval • J.-P. Morel • G. Fonty
LMGE, Laboratoire « Microorganismes: Génome & Environnement », UMR CNRS 6023, Université Clermont-Auvergne, Université Blaise Pascal,
24 avenue des Landais, BP 80026, 63171 Aubière, France

V. Grossi
LGLTPE, Laboratoire de « Géologie de Lyon, Terre, Planètes, Environnement », UMR CNRS 5276, Université Claude Bernard, Lyon1, Campus de la Doua, bât. GEODE, 2, rue Raphaël Dubois, 69622 Villeurbanne, France

E. Attard
LMGE, Laboratoire « Microorganismes: Génome & Environnement », UMR CNRS 6023, Université Blaise Pascal,
24, avenue des Landais, 63171 Aubière Cedex Clermont-Ferrand, France

Equipe Environnement et Microbiologie, UMR CNRS-IPREM 5254, Université de Pau et des Pays de l'Adour,
BP 1155, 64013 Pau Cedex Pau, France

C. Amblard
LMGE, Laboratoire Microorganismes: Génome et Environnement, UMR CNRS 6023, Université Clermont-Auvergne, Université Blaise Pascal, BP 80026, 63171 Aubière Cedex, France

T. Sime-Ngando et al. (eds.), *Lake Pavin*, DOI 10.1007/978-3-319-39961-4_16

Keywords
Methane cycle • Methanogens • Methanotrophs • Ferric iron reduction • Fermentation • Anaerobic food chain • Biogeochemistry • Anoxic freshwater layers

List of Abbreviations

ANME	Anaerobic Methanotrophs
AOM	Anaerobic Oxidation of Methane
CARD	Catalyzed Reporter Deposition
CLSM	Confocal Laser Scanning Microscopy
DAPI	Diamidino-2-phenylindole
FISH	Fluorescent *in situ* hybridization
FRB	Ferric Iron Reducing Bacteria
HRP	Horseradish Peroxidase
MBGD	Marine Benthic Group D
MMO	Methane Mono-Oxygenase
MPR	Methane Production Rate
NRB	Nitrate Reducing Bacteria
pMMO	Particulate Methane Mono-Oxygenase
p.p.m.	Parts Per Millions
PLFA	Phospholipids Fatty Acids
rRNA	Ribosomal RNA
sMMO	Soluble Methane Mono-Oxygenase
SRB	Sulfate Reducing Bacteria
TEM	Transmission Electron Microscopy

16.1 Introduction

"*A bacterium, an amoeba…what destiny can they dream of other than forming two bacteria, two amoebae…?*" This rhetorical question of François Jacob illustrates the "*leitmotiv*" of microorganisms: optimizing their growth and, achieving this objective implies a remarkable efficiency and adaptability to environmental conditions and fluctuations. Their adaptive strategies determine the effectiveness and the nature of their regulatory processes. The optimization of the prokaryotic cellular machinery is visible on the scale of ecosystems, in which microorganisms are involved in the biogeochemical cycles of many elements. The influence of the prokaryotic component on the biotope is probably exacerbated in anoxic ecosystems in which microorganisms are almost hegemonic. In these environments, the use of a variety of electron acceptors confers to the anaerobic prokaryotes a key role in the distribution of a large number of minerals (*e.g.*, iron, sulfur, manganese) and implies the establishment by these communities of metabolic processes and functional specializations as well as numerous cooperative and competitive interactions.

The microbial ecological studies in marine or freshwater environments have focused primarily on the aerobic layers, in which the roles of heterotrophic microorganisms in the aquatic food web were particularly analyzed. In these ecosystems, the processes occurring in the anoxic zones, often reduced to sediment and, in some cases, to deep pelagic layers, may appear "decoupled" from processes of the microbial loop. However, anaerobic prokaryotes are likely to play important roles in aquatic systems allowing the regeneration of bioactive elements, deriving some of the carbon and energy, and providing access through the sedimentation process to a complete mineralization of the organic material. These anaerobic microbial processes are therefore critical to maintain the homeostasis of aquatic systems and are important drivers of biogeochemical cycles.

The present article focuses on the microorganisms and processes involved in the biogeochemical cycle of methane (CH_4) in freshwater systems with a focus on Lake Pavin. CH_4 is a key metabolite in freshwater lakes where methanogenesis (*i.e.*, biogenic CH_4 production) accounts for 10 to 50 % of organic matter mineralization (Bastviken et al. 2008). The diversity and the physiology of the microorganisms involved in the biogeochemical cycle of CH_4 and the environmental factors that regulate their activities are detailed in the following sections. As all anaerobic microbial metabolic groups build competitive and cooperative relationships, some bacterial processes tightly linked to CH_4 cycle are also addressed. These elements are presented in a conceptual context, that of freshwater lakes and in particular that of Lake Pavin for which past and current researches aspire to (1) identify microbial agents involved in CH_4 biogeochemical cycle, (2) identify novel microbial processes responsible of CH_4 production and/or consumption, (3) link *in situ* observations with the ecophysiology of the targeted microorganisms, (4) develop complementary cognitive and methodological approaches to access to these informations.

16.2 Brief Overview of Methane Cycle

16.2.1 Methane

16.2.1.1 Ecological Importance of CH_4 in the Context of Climate Change

CH_4 has played an important role since the beginning of life on Earth, helping the planet to remain warm and habitable. It still plays a key role in ecosystems, mainly in the biogeochemical carbon cycle but also in the cycles of many inorganic elements. CH_4 is therefore at the center of scientific and environmental concerns but also of societal ones. In the context of climate change, CH_4 is one of the several greenhouse gases responsible for increased radiative forcing on the climate system. Atmospheric CH_4 concentration has increased by about 1000 p.p.m. (parts per million) since the beginning of the industrial era representing the fastest changes in CH_4 atmospheric concentrations over the last 80,000 years (IPCC 2007). Despite a short lifetime in the atmosphere of approximately 8 years, CH_4 is 25 more potent than carbon dioxide (CO_2) as a greenhouse gas over a 100-year horizon due to its higher efficiency in trapping radiation (Shindell et al. 2009; Nazaries et al. 2013). Whereas atmospheric CH_4 has stabilized at a value of 1.77–1.78 p.p.m since 2005, an increase in the atmospheric concentration of CH_4 to 2.55 p.p.m. is predicted by 2050 (Lelieved et al. 1998; Nazaries et al. 2013).

16.2.1.2 Budget and Sources of Methane[1]

The global budget of atmospheric CH_4 is on the order of 500–600 $Tg.yr^{-1}$ (Lelieveld et al. 1998; Wang et al. 2004; Conrad 2009). CH_4 is emitted from a range of natural and anthropogenic (relating to human activity) sources as a result of the anaerobic decomposition of organic matter, land use changes and fossil fuel emissions. Whereas much attention is currently focused on the anthropogenic sources of the greenhouse gases, there is ample evidence that emissions of these gases from natural sources have also changed over time (EPA 2010). Natural sources of CH4 are estimated to produce 37 % of the total CH4 flux into the atmosphere every year. The largest source of natural CH4 emissions is natural wetlands (170 Tg $CH4.yr^{-1}$). Several other sources contribute substantially as well, including geologic emissions (42 to 64 Tg $CH4.yr^{-1}$), lakes (10–50 Tg $CH4.yr^{-1}$), and vegetation (20 to 60 Tg $CH4.yr^{-1}$) (EPA 2010). About 70 % of all CH_4 formation is the result of microbial processes showing that most of the atmospheric CH_4 originates from microbial metabolism.

16.2.1.3 Methane Sinks

The global CH_4 sources are balanced by sinks of similar magnitude. The largest sink (>80 % of the total) is the photochemical oxidation of CH_4 initiated by the reaction with •OH radicals (Cicerone and Oremland 1988). The loss of CH_4 in the stratosphere accounts for ~ 7 % of the sink and finally CH_4 is also eliminated from the atmosphere uptake in upland soils due to microbial oxidation (~5 % of the sink, Conrad 2009). Noteworthy that rates of CH_4 production are much larger than rates of CH_4 emission since a large fraction (50–90 %) of the initially produced CH_4 is oxidized by microorganisms and does not reach the atmosphere (Frenzel 2000; Reeburgh 2003; Kvenvolden and Rogers 2005).

16.2.2 Freshwater Lakes as Methane Sources

Lakes and ponds (excluding impoundments and reservoirs) are natural sources of CH_4. They contribute substantially to CH_4 emissions, although analysis of this source has been limited to date. Current emissions to the atmosphere are estimated to 10–50 Tg $CH_4.yr^{-1}$ but one key uncertainty involves the total surface area of lakes and ponds (for comparison the oceans emit 9 Tg $CH_4.yr^{-1}$, EPA 2010). The total lake surface area is estimated to 0.9 % of the Earth surface; although the number and total area of large lakes is well known, the number and total area of small lakes and ponds is not. It is estimated that there are about 300×10^6 natural lakes and ponds worldwide, 90 % of which are smaller than 1 ha. Because small lakes and ponds generally emit more CH_4 per unit area than large lakes, a major uncertainty in the global estimation of CH_4 emission relies on the precise estimation of the total lake surface area (EPA 2010). Contribution of freshwater lakes to CH_4 emissions is thus probably underestimated.

Climate warming impacts on permafrost and the development of thermokarst lakes could substantially affect future CH_4 emissions from lakes. A long-term decline of emissions from lakes north of 45°N should be observed due to lake area loss and permafrost thaw. But, a period of increase of CH_4 emissions associated with thermokarst lake development in the zone of continuous permafrost will precede this decline. CH_4 emission rates from northern lakes should rise to 50 to 100 Tg $CH_4.yr^{-1}$ during this transitional period, which would last hundreds of years (EPA 2010).

16.2.3 Methane Cycle in Freshwater Lakes[2]

CH_4 cycling in lakes contributes to 6–16 % of the non-anthropogenic emissions of this gas. It is therefore potentially important for the present global CH_4 budget (Bastviken et al. 2004). CH_4 emissions are dependent on the relative rates of CH_4 production and consumption, but also on the dominant pathways for its transport to the water-atmosphere interface.

16.2.3.1 Biological Production of Methane (Methanogenesis)

CH_4 is a major product of carbon metabolism in lakes and is produced by the activity of CH_4-generating microbes (*i.e.,*

[1] For detailed informations see Conrad 2009; EPA 2010.

[2] For more details, see Bastviken et al. 2004, 2008, 2009.

methanogens). These anaerobic microorganisms are encountered in the profound and epilimnetic sediments as well as in anoxic water columns. As discussed later in the paper, their activities and the related methane production rates (MPRs) depend on temperature, organic matter availability and isolation from oxygen. These factors are influenced by climate, lake size and depth, and ecosystem productivity (EPA 2010). For example, CH_4 fluxes measured in tropical lakes (2023 mmol.m^{-2}.yr^{-1}) are higher than those of boreal (331 mmol.m^{-2}.yr^{-1}) and temperate (1110 mmol.m^{-2}.yr^{-1}) lakes (Bastviken 2009). Moreover, CH_4 production is not uniform over the entire lake surface. For example, substantially higher CH_4 production can occur in littoral relative to profound sediments (Michmerhuizen et al. 1996; Murase et al. 2005).

Considering that anaerobic carbon mineralization processes account for as much as 20–60 % of the overall carbon mineralization in freshwater environments, and that methanogenesis is responsible for 30–80 % of anaerobic carbon mineralization in waters and sediments, methanogenesis likely accounts for 10–50 % of the overall carbon mineralization (Bastviken et al. 2008). Several studies also indicate that 20–59 % of the sestonic carbon imputs to lake sediments is converted into CH_4 (Wetzel 2001).

16.2.3.2 Methane Transport

There are four pathways for CH4 transport from lakes: bubbling, diffusion, seasonal overturning and plant-mediated transport. Wind speed is an important control on gas exchange between a lake and the atmosphere. Flux rates by all pathways generally increase with increasing wind speed.

- **The bubbling process** has been determined to be the dominant pathway for the release of CH4 accounting for more than 50 % of CH4 emissions from lakes (Bastviken et al. 2004). Bubbling occurs when CH_4 concentrations exceed the limit of solubility which relies on the hydrostatic pressure and temperature (at 20 °C and atmospheric pressure, CH_4 solubility is 0.023 g.kg^{-1} of water compared to 1.7 g.kg^{-1} of water for CO_2). Sediments situated under a shallow water column are the major contributors of this process, due to lower hydrostatic pressure and wave-induced perturbations which favor the formation and release of the CH_4 bubbles (Bastviken et al. 2004).
- Most of the CH_4 that reaches the upper mixed layer of the water column is transported by **diffusive flux.** Diffusion rates are lower in areas of physical and chemical gradients (*e.g.,* thermocline, chemocline) and at the water-sediment interface. In epilimnetic areas, CH_4 transport can be accelerated by the turbulence associated with wind. Several studies suggest that epilimnetic dissolved CH_4 may be derived from epilimnetic sediments rather than from hypolimnetic waters (*e.g.,* Murase et al. 2005). Diffusion and turbulent transport may be responsible for 26–48 % of CH_4 emissions from lakes.
- CH_4 accumulating in anoxic hypolimnia during a period of stratification can be rapidly released to the atmosphere upon **lake overturning**. For example, temperate dimictic lakes mix from top to bottom during spring and fall, resulting in significant emissions of CH_4 to the atmosphere (accounting for 24–86 % of total CH_4 emissions from lakes).
- CH_4 produced in sediment along lake margins can also be transported by the aerenchyma of emergent vegetation (**plant-mediated transport**). This flux component depends on CH_4 production and oxidation in the sediments, and on vegetation characteristics (Bastviken et al. 2004).

16.2.3.3 Methane Oxidation

As soon as CH_4 reaches oxic sediment or water, a large proportion (30–99 %) is likely oxidized by CH_4-oxidizing bacteria (Bastviken et al. 2002). It is noteworthy that, in the last decades, consortia of anaerobic microbes have been described that convert CH_4 to CO_2 while reducing sulfate, nitrate, manganese or iron. These processes are discussed further in following sections.

16.2.4 Methane Profile in Lake Pavin

16.2.4.1 Origin of Methane

The average carbon ($\delta^{13}C$) and hydrogen (δD) isotopic composition of CH_4 in the monimolimnion of Lake Pavin are −60 ‰ and −276 ‰, respectively (**Agrinier P.** Personal communication). These values are indicative of a biogenic origin of CH_4 rather than of a geothermal or magmatic origin (Whiticar 1999).

16.2.4.2 Methane in the Profundal Sediment (Fig. 16.1a)

CH_4 concentrations in the first 40 cm deep sediment, collected at 92 m water depth, range from 4 to 9 mM (Borrel et al. 2012a). Modelisation of CH_4 production in Lake Pavin suggests that 90 % of CH_4 is formed in lake sediments (Lopes et al. 2011).

16.2.4.3 Methane in the Water Column (Fig. 16.1a)

CH_4 concentrations extend from few micromolars at 60 m to 4 mM at 92 m water depths. It has been estimated that the monimolimnion of Lake Pavin contains 260.10^3 kg of CH_4 (Jezequel et al. 2010). CH_4 concentrations in the oxic part of the water column are near the boundary of the detection threshold of 3 nM.

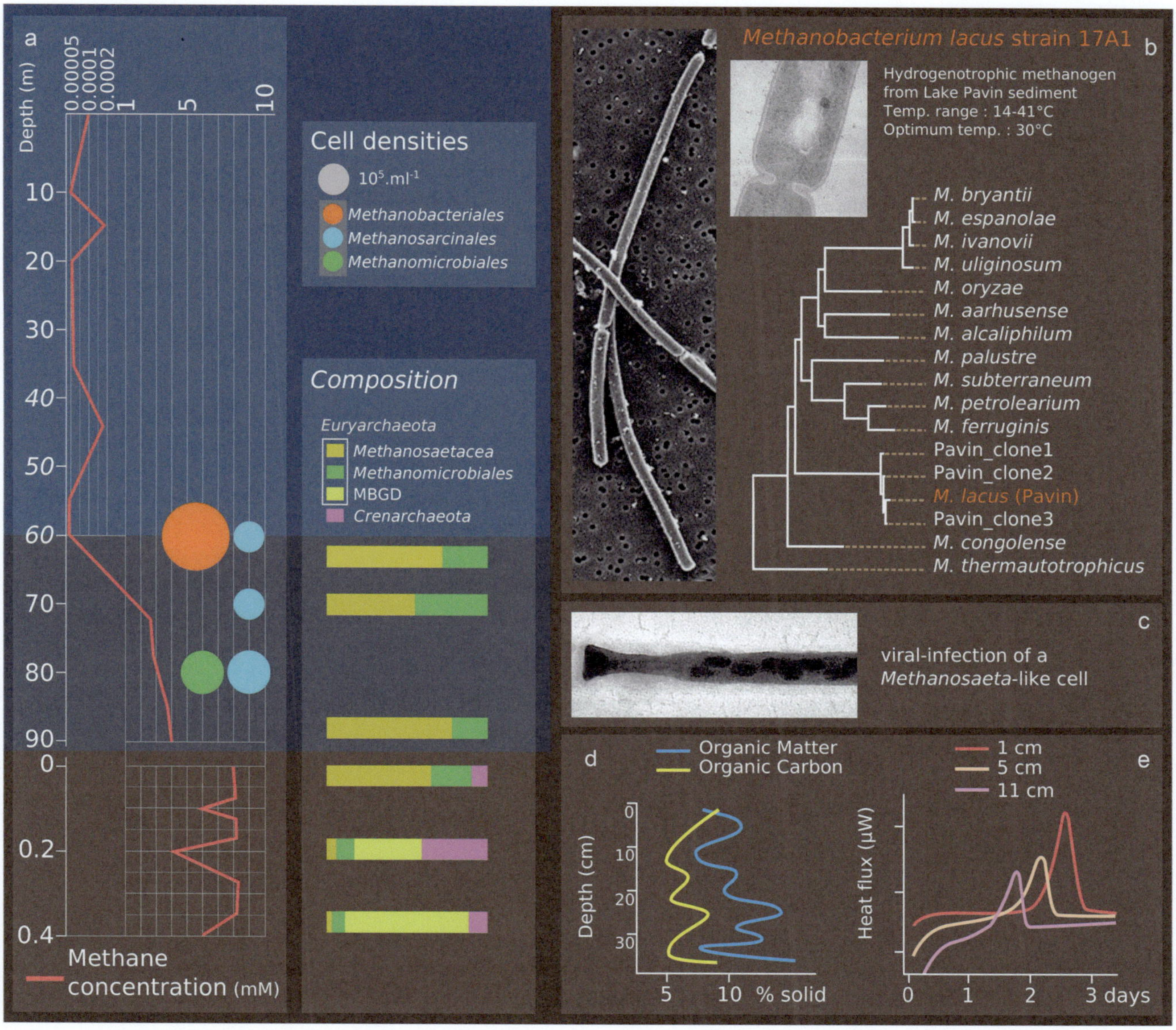

Fig. 16.1 Methane profiles and methanogens in Lake Pavin. (**a**) Schematic representation of CH_4 profiles in the oxic water column (0–60 m), in the anoxic water column (60–92 m) and in the sediment (0–40 cm) (Note that the scales for CH_4 concentrations are different between the three layers). The abundances of the *Methanobacteriales*, *Methanosarcinales* and *Methanomicrobiales* are indicated according to Lehours et al. (2005). The composition of archaeal clone libraries [calculated as follows: (number of sequences of the targeted group/number of total archaeal sequences) x 100] is also indicated for the anoxic water column and the sediment according to Lehours et al. (2007) and Borrel et al. (2012a), respectively. (**b**) Transmission electron micrographs of the strain *Methanobacterium lacus* isolated from the sediments of Lake Pavin (adapted from Borrel et al. 2011). *M. lacus* is representative of the population of *Methanobacteriales* in these sediments as the *mcrA* sequences of *M. lacus* exhibited > 98 % similarity with those of *Methanobacteriales* inhabiting Pavin sediments. (**c**) Transmission electron micrographs of pleiomorphic ellipsoid infective viruses in a visibly infected filamentous sheathed prokaryote, identified as a *Methanosaeta*-like cell, observed in the sediment of Lake Pavin (adapted from Borrel et al. 2012c). (**d**) Schematic representation of organic matter (OM) and organic carbon (OC) content in the fortieth first centimeters of Lake Pavin sediments (adapted from Borrel et al. 2012a). (**e**) Thermograms representing the heat production in three sediment layers (1, 5 and 11 cm), measured using microcalorimetry (Box 16.2). The total heat produced is 460 mJ, 365 mJ and 255 mJ for the sediment samples collected at 1, 5 and 11 cm depth, respectively (unpublished results). This decrease of heat production according to depth is the result of a decrease in the microbial activity. As the organic matter content does not significantly change with depth, the decrease in microbial activity could be attributed to a change in organic matter quality

16.2.4.4 Methane Emissions to the Atmosphere

The CH_4 concentrations at the water surface being higher than the concentrations in the atmosphere, CH_4 is emitted from the lake. These emissions, estimated at ~ 22 mmol.m^{-2}.yr^{-1} (Guimbaud et al. 2011), are several orders of magnitude below the average emissions of temperate lakes (see Sect. 16.2.3.1).

16.3 Biogenic Methane Production

"*The details of microbial physiology can have large impacts on global biogeochemical cycles and the planet's climate system*" (Schimel 2004).

16.3.1 The Anaerobic Foodweb

Anaerobic microorganisms perform a variety of fermentation processes, and also have the ability to couple the oxidation of organic substrates with the reduction of diverse inorganic compounds other than O_2 (*e.g.*, nitrate, manganese, iron, sulfate, CO_2). However, their bioenergetic capacities compel them to establish a tissue of interactions including cooperation, syntrophy and competition.

16.3.1.1 Carbon and Electron Fluxes Between Metabolic Types

In anoxic environments, the degradation of organic matter is performed in several steps (Fig. 16.2) involving different metabolic groups of anaerobic *Bacteria* and *Archaea*.

(1) The polymers (polysaccharides, proteins, nucleic acids, lipids) are converted to oligo- and monomers (sugars, amino acids, fatty acids, glycerol, purines, pyrimidines) through extracellular hydrolytic enzymes produced by the hydrolytic bacteria.
(2) The fermentative bacteria ferment these monomers in volatile fatty acids with short carbon chains (acetate, propionate, butyrate, valerate, isovalerate, etc..), alcohols, aromatic compounds, *etc*. The simplest metabolites (acetate, H_2, CO_2, or other one-carbon compounds) are

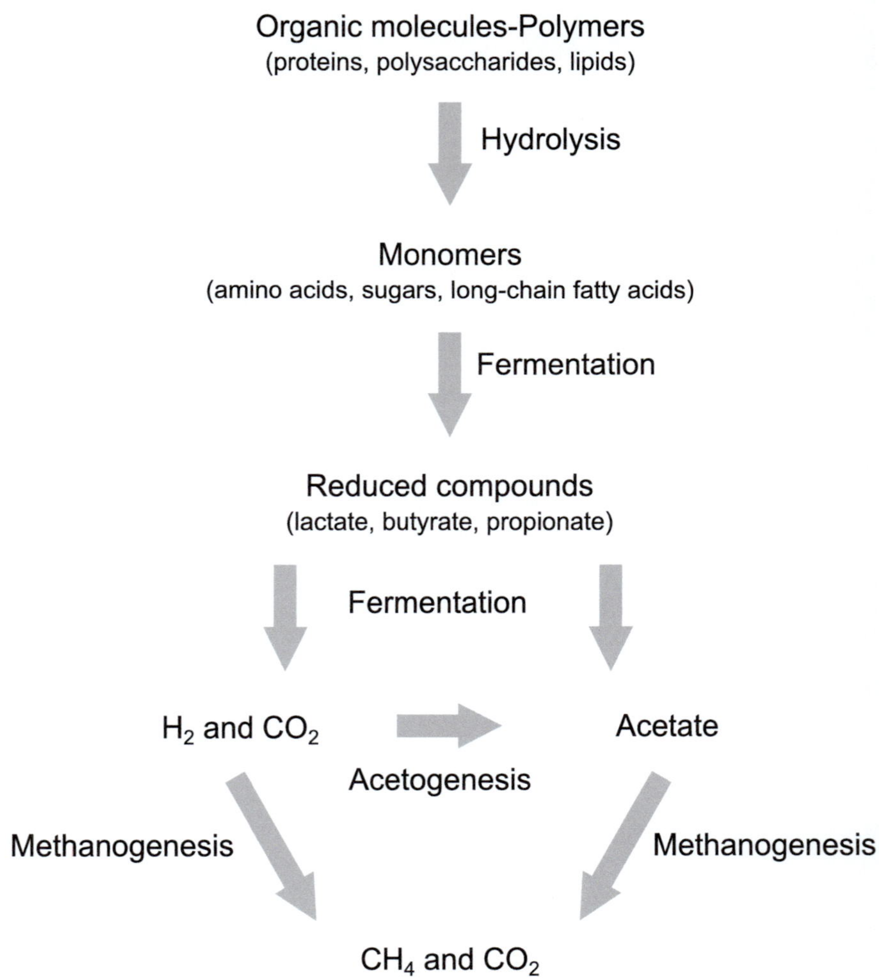

Fig. 16.2 Sequential pattern of microbial degradation of complex organic matter in anoxic environments. Macromolecules are hydrolysed by hydrolytic bacteria. The monomers are fermented by fermentative bacteria into a range of fermentation products. The acetate is produced directly by fermentation or indirectly by acetogenesis. Fermentation products are consumed by methanogens or others microorganisms performing terminal electron accepting processes such as sulfate reduction, ferric-iron reduction, denitrification

directly assimilated by different microbial respiratory processes (sulfate-reduction, methanogenesis, denitrification, dissimilatory metal reduction).

(3) The more complex metabolites (e.g., alcohols and fatty acids with more than two carbon atoms, aromatic and branched fatty acids) are secondarily converted by oxidative bacteria into substrates consumed through different anaerobic metabolic pathways.

(4) The terminal step of the anaerobic degradation of organic matter leads to the production of CO_2, CH_4 and other reduced compounds (H_2S, Fe(II), *etc.*). This step depends mainly on the availability of various electron acceptors (see following sections).

In sulfate-poor freshwater environments where methanogenesis is a dominant process, biomass is converted mainly to CH_4 and CO_2 by a complex community of fermentative bacteria in cooperation with methanogens and homoacetogens which keep the hydrogen partial pressure in the range of 10^{-4}–10^{-5} atm. (Schink 1997; Schink and Stams 2001). Under such conditions, the fermentation of sugars by fermentative bacteria is shifted from the production of reduced sideproducts such as butyrate or ethanol to nearly exclusive formation of acetate, CO_2 and H_2 (Ianotti et al. 1973; Zeikus 1977, 1983; Tewes and Thauer 1980). Alternatively, formate could be formed instead of H_2 with a similar energy yield (−202 kJ mol^{-1}, Müller et al. 2008).

16.3.1.2 Energetic Considerations

In anoxic environments, the use of a variety of electron acceptors by microorganisms affects the biogeochemical cycles of numerous inorganic elements (*e.g.*, hydrogen, nitrogen, sulfur, iron, manganese). Equilibrium constants and free energy diagram can be used to predict which reactions are thermodynamically possible. These thermodynamic considerations allow suggesting that the most energetically favorable process is realized first and condition the subsequent redox sequences. For example, a reducer such as organic carbon (*e.g.*, CH_2O) can be first degraded through oxidation with O_2, followed by the successive use of NO_3^-, Mn^{4+}, Fe^{3+}, SO_4^{2-} and CO_2 as final electron acceptors. Methanogenesis, by reducing CO_2, is therefore the least energetically favorable process among anaerobic metabolisms. These sequences are, within certain limits, a realistic description of what happens in stable ecosystems (*e.g.*, sediment) and add a spatial dimension to thermodynamic considerations. However, many other conditions such as kinetic factors and syntrophic couplings are also required for the occurrence of some reactions (see following sections).

16.3.2 Methanogens and Methanogenesis

As aforementioned, methanogens act in the terminal step of the anaerobic food chain, converting methanogenic substrates to CH_4 through methanogenesis. To date, none methanogen has been identified that can grow without producing CH_4. These *Archaea* are all obligate CH_4-producers that are uniquely specialized for this lifestyle (Hedderich and Whitman 2006).

16.3.2.1 Evolutionary Aspects of Methanogenesis

Methanogens constitute the largest described group within the domain *Archaea* and are among the most ancient of extant forms of life. Growing evidences suggest that methanogenic microbes evolved close to the time of the origin of life providing a window on the early evolution of Earth's biosphere (Battistuzzi et al. 2004). Studies on the origin of methanogenesis [estimated between −4.11 Ga and −3.78 Ga, Battistuzzi et al. 2004] suggest that methanogens were present on Earth during the Archean period which is consistent with the CH_4 greenhouse theory (Pavlov et al. 2000).

16.3.2.2 Taxonomic Diversity of Methanogens[3]

Methanogens are exclusive to the *Euryarchaeota* kingdom of the *Archaea* domain. They are currently classified into 5 classes (*Methanopyri*, *Methanococci*, *Methanobacteria*, *Methanomicrobia*, and *Thermoplasmata*) and 7 orders (*Methanobacteriales*, *Methanococcales*, *Methanomicrobiales*, *Methanosarcinales*, *Methanopyrales*, *Methanocellales*, *Methanomassiliicoccales* (Paul et al. 2012; Iino et al. 2013).

16.3.2.3 Metabolic Diversity of Methanogens[4]

Methanogens derive their metabolic energy from the conversion of a restricted number of substrates to CH_4. The three main methanogenic substrates are H_2, acetate and methylated compounds (such as methanol, methylated amines and methylated sulfides). Therefore, several distinct pathways for CH_4 production exist and none of these methanogenic metabolisms have been currently found in *Bacteria* or eukaryotes. The different pathways for CH_4 production are:

[3]For detailed informations, see Nazaries et al. 2013.

[4]For more informations, see Hedderich and Whitman 2006; Borrel et al. 2011; Nazaries et al. 2013.

- ***Hydrogenotrophic methanogenesis***: reduction of CO_2 with hydrogen (hydrogenotrophic methanogens) or formate (formatotrophic methanogens) as electron donors. The CO_2 reduction into CH_4 involves 4 H_2 and proceeds *via* carrier-bound one carbon intermediates along the C1-reductive pathway (Hedderich and Whitman 2006). Hydrogenotrophic methanogens belong to the orders *Methanopyrales*, *Methanococcales*, and *Methanobacteriales* (Lang et al. 2015). The hydrogenotrophic pathway is found also in most of the derived lineages of methanogens (*Methanomicrobiales* and *Methanocellales*) and was most probably present already in the common ancestor of the *Euryarchaeota* (Bapteste et al. 2005). Hydrogenotrophic methanogenesis constitute the main sink of H_2 produced by fermentative bacteria (see Sect. 16.3.3.3). Methanogens and fermentative bacteria often grow in syntrophic consortia allowing methanogens to maintain low H_2 concentrations, a necessary condition for the fermentation process (Stams and Plugge 2009). Hydrogenotrophic methanogenesis is a commonly used pathway and contributes to a huge part of CH_4 production in many environments including freshwater lakes. Theoretical values predict that hydrogenotrophic methanogenesis accounts for 30 % of overall methanogenesis in freshwater lakes, and the measured rates of this process range from 0 to 100 % (Conrad 1999).
- ***Acetotrophic methanogenesis***: catabolization of acetate by cleavage (*i.e.*, disproportionation), with the carboxyl group oxidized into CO_2 and the methyl group reduced into CH_4. This pathway is only performed by members of two genera, *Methanosarcina* and *Methanosaeta* (both belonging to the order *Methanosarcinales,* Thauer et al. 2008a). This pathway of CH_4 production often represents the most important CH_4 source in cold and temperate freshwater lakes (Conrad 1999).
- ***Methylotrophic methanogenesis***: disproportionation of methyl compounds. The methyl group of C1-compounds such as methanol, methylamines, dimethylsulfide, or methanethiol is converted to CH_4 and CO_2. Methylotrophic methanogens are a phylogenetically and biochemically heterogeneous group comprising members of the *Methanosarcinaceae*, *Methermicoccaceae* (both belonging to the order *Methanosarcinales,* Sprenger et al. 2000; 2005), the genus *Methanosphaera* and some species of the genus *Methanobacterium* (order *Methanobacteriales,* Miller and Wolin 1985; Fricke et al. 2006), and members of the recently discovered seventh order of methanogens, the *Methanomassiliicoccales* (Paul et al. 2012; Borrel et al. 2013, 2014). Unlike hydrogenotrophic and acetoclastic methanogens, methylotrophic methanogens do not compete with sulfate reducing bacteria for substrate uptake (so-called "noncompetitive substrate", Oremland and Polcin 1982) and may thus grow in zones where alternative electron acceptors are not depleted. This process is expected to be low in freshwater lakes since precursors of methyl compounds (*e.g.,* pectin, cholin, osmoregulators such as glycine betaine) are not abundant (Lomans et al. 1997, 2001; Lovley and Klug 1983; Zinder and Brock 1978). However, methylotrophic methanogenesis was observed in numerous freshwater lake sediments (Lomans et al. 2001).

16.3.2.4 Biochemical Aspects of Methanogenesis[5]

The complexity and uniqueness of methanogenesis as a form of anaerobic respiration reside in the requirement of six unusual coenzymes [ferredoxin (Fd), methanofuran (MFR), tetrahydromethanopterin (H4MPT), coenzyme *F*420 (*F*420), coenzyme M (CoM) and coenzyme B (CoB)]; a multistep pathway and several unique membrane-bound enzyme complexes coupled to the generation of a proton gradient driving ATP synthesis (Ferry 2010; Nazaries et al. 2013). Although the intermediates and enzymatic reactions of the hydrogenotrophic, acetotrophic and methylotrophic pathways are different, they share common features in the final steps of CH_4 production. The final enzymatic step is catalyzed by the methyl-coenzyme M reductase (MCR) which is unique to methanogens and, as developed later in this paper, to archaeal anaerobic methanotrophs. Hence, MCR constitutes a functional marker of microorganisms involved in these metabolisms and is used to investigate the diversity, structure, distribution and ecology of methanogenic communities in freshwater lakes (Box 16.1).

16.3.2.5 Environmental Distribution of Methanogens

Methanogens are the only currently cultivated *Archaea* that are truly cosmopolitan. The distribution of methanogens in natural environments is highly dependent on their adaptation to various temperature, pH and salinity ranges (Garcia 1990). Methanogens are found in mesophilic as well as extreme environments (Ollivier et al. 1994). Interestingly, methanogenic archaea have also been found in oxic environments such as various aerated soils (Angel et al. 2012) and the oxygenated water column of an oligotrophic lake (Grossart et al. 2011). These observations do not question the anaerobic character of methanogens. In aerated soils, methanogens become active under wet anoxic conditions whereas in oxygenated lake waters methanogens can be attached to photoautotrophs (*e.g., Cyanobacteria, Chlorella*), which may provide methanogenic substrates and anaerobic niches.

[5]For detailed informations, see Ferry 2010.

Box 16.1: Which Nucleic-Acids Biomarkers May Be Used to Investigate the Diversity of Methanogens and Methanotrophs?

Investigation of prokaryotic composition in environmental samples by culturing techniques is laborious and prone to strong bias since the growth requirements of many bacteria are still unknown. Therefore, nucleic-acid biomarker sequencing has become the gold standard in environmental microbiology to explore the diversity of microorganisms. One obvious marker frequently used in molecular ecology particularly for the taxonomic identification and the phylogenetic affiliation of prokaryotes is the 16S rRNA gene. A complementary option is to use functional genes encoding proteins (i.e., specific enzymes). This option allows (in some cases) establishing a relationship between the taxonomic identity and the functional role of microorganisms in ecosystems.

The 16S rRNA gene as a phylogenetic gene

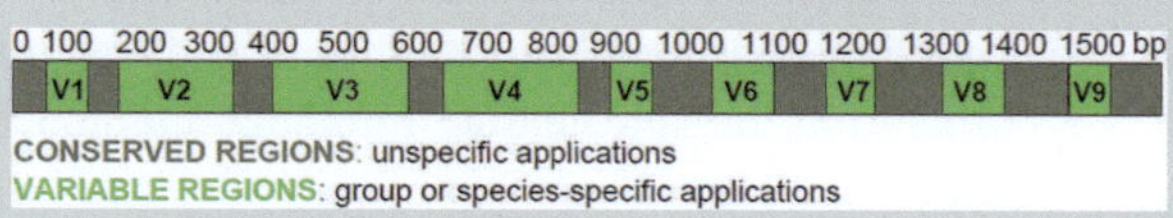

16S rRNA gene illustrating the conserved (grey) and variable (green) regions

The 16S rRNA gene emerged, since early, as a suitable phylogenetic marker. This gene is universally present in prokaryotes, is not submitted to lateral gene transfert, and has a relatively short sequence (~1.5 kb) making its sequencing fast and (relatively) cheap.

Moreover, the degree of conservation differs considerably within 16S rRNA gene. Conserved regions of the gions contain specific sites unique to a species. The uniqueness enables designing primers targeting different taxonomic ranks (i.e., *Archaea* → *Euryarcha eota* → *Methanosarcinales* → *Methanosaeta* → *Methanosaeta concilii*).

rpoB gene as an alternative to 16S rRNA

Various artifacts are reported with the used of 16S rRNA (*e.g.,* multiple copies within a genome, heterogeneity of rates of change between the sequences, Brochier-Armanet et al. 2008). The increase in protein sequence data, through the sequencing of complete genomes enables to make phylogenies based on protein markers. For example, alternative core-house keeping genes, such as the RNA polymerase β subunit gene (*rpoB*), have emerged as an alternative choice

(continued)

Box 16.1 (continued)

(Case et al. 2007). The *rpoB* inferred phylogenies are congruent with those based on 16S rRNA, show a higher resolution of some nodes and are less sensitive to artifact reconstruction.

The mcrA gene as a diagnostic indicator of methanogenesis (and of anaerobic methanotrophy)

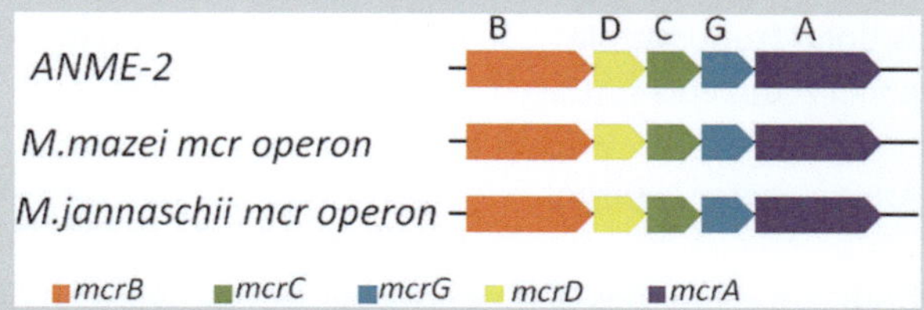

Schematic representation of the *mcr* operon structure (Adapted from Hallam et al. 2003)

The methyl-coenzyme M reductase is an enzymatic complex encoding by a gene cluster. The genomes of all known methanogens contain at least one copy of the *mcrA* which is a highly conserved gene. This property makes *mcrA* a suitable candidate to study the diversity of methanogens and of anaerobic methanotrophs. Furthermore, *mcrA* and 16S rRNA phylogenies are congruent (Hallam et al. 2003; Luton et al. 2002).

The pmoA gene constitutes a functional marker of aerobic methanotrophs

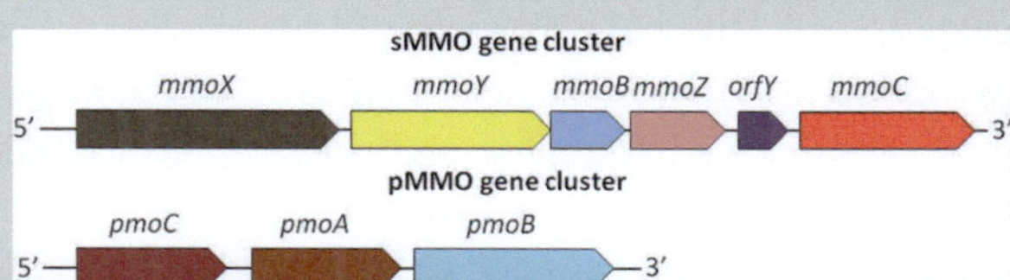

sMMO and pMMO gene clusters from *Methylocystis sp.* (Adapted from K. Iwasaki (http://www.nies.go.jp/kenko/biotech/topi_1-3/iwakiku.html))

The methane monooxygenase (MMO) is an enzymatic complex unique to aerobic methanotrophs. Two distinct forms of the MMO complex have been identified at different cellular locations, a cytoplasmic soluble form (sMMO) and a particulate form generally bound to the intracytoplasmic membrane (pMMO) (Lipscomb 1994). Those enzymes are encoded by different gene clusters.

A large dataset of *pmoA* sequences is available in public database and *pmoA* and 16S rRNA phylogenies are congruent; therefore *pmoA* is a suitable candidate to study the diversity of aerobic methanotrophs.

16.3.2.6 Diversity, Distribution and Abundance of Methanogens in Lake Pavin

In freshwater lakes, analyses of clone libraries and/or quantification of methanogenic communities by fluorescent *in situ* hybridization (FISH) revealed that the *Methanomicrobiales* and the *Methanosarcinales* dominate the methanogenic communities (*e.g.,* Auguet et al. 2010; Briee et al. 2007; Chan et al. 2005) and that the *Methanobacteriales* occur scarcely (*e.g.,* Conrad et al. 2010; Ye et al. 2009). These patterns are consistent with observations done from samples collected in the anoxic water column and sediments of Lake Pavin.

– ***Anoxic water column***

Both clone libraries and fluorescent *in situ* hybridization (FISH) approaches based on 16S rDNA sequences revealed similar distribution patterns of archaeal communities in the anoxic water column of Lake Pavin (– 60 to –92 m) (Lehours et al. 2005, 2007). Archaeal communities exhibit a homogeneous distribution along the water column and are dominated by sequences related to the *Methanosarcinales* and to the *Methanomicrobiales* with a ratio of 2:3 and 1:3, respectively (Fig. 16.1a). The abundance of *Methanosarcinales* (including *Methanosaeta* and *Methanosarcina*) increased with depth in the monimolimnion and the *Methanomicrobiales* peaked below the chemocline (Fig. 16.1a). *Methanobacteriales* were only quantifiable at the interface of the oxic and anoxic zones of the water column (Fig. 16.1a).

CH_4 concentrations in the anoxic water layer of Lake Pavin are significantly correlated ($r=0.8$, $P<0.01$) with the abundance of *Methanosarcinales*, which are mainly represented (71 %) by sequences related to *Methanosaeta concilii*, a strictly acetotrophic methanogen. Acetotrophic methanogenesis is therefore probably a major process in the anoxic zone of the water column of Lake Pavin. The carbon and hydrogen stable isotopic compositions of CH_4 [$\delta^{13}C\text{-}CH_4=-60$ ‰, δD $CH_4=-276$ ‰, Agrinier P. Personal communication] support this hypothesis because such isotopic signatures are similar to that of the methyl group of acetate.

– ***Sediment***

The composition of archaeal communities within the anoxic sediments of Lake Pavin was investigated along a 40-cm sediment core collected at the vertical of the 92 m isobath (Borrel et al. 2012a). The composition of the methanogenic communities within the sediments was similar to that detected in the anoxic water column (Fig. 16.1a). However, differences in distribution of the methanogenic communities are observed within both compartments. Unlike the water column the distribution of methanogens is not homogeneous down core. Methanogenic lineages are the dominant *Archaea* in the uppermost sediment layer (66 % of acetoclastic *Methanosaetaceae* and 25 % of hydrogenotrophic *Methanomicrobiales*) while below 5 cm sediment depth, clone libraries were mainly composed of uncultured archaeal lineages (Marine Benthic group B (MBG-D) and Miscelleanous Group G (MCG), Borrel et al. 2012a).

The main archaeal function in the superficial layer of the sediment of Lake Pavin monimolimnion is CH_4 production. Methanogens occurrence in the surface sediment layer is consistent with Lake Pavin meromicticity, as the reduction of the main inorganic electron acceptors (*e.g.,* sulfate, ferric iron, nitrate) occurs in the anoxic water column (Lopes et al. 2011). These observations, based on the relative proportions of sequences in clone libraries, were confirmed by quantification of these archaeal groups by quantitative PCR performed on genomic DNA and complementary DNA (Borrel et al. 2012a).

16.3.3 Factors Influencing Methanogens and Methanogenesis (Fig. 16.3)

Many factors impact the CH_4 production rates (MPRs) by affecting methanogenic populations either directly or indirectly by structuring the surrounding microbial community which contains suppliers of methanogenic substrates and competitors for these substrates. Environmental factors can also act on the dominant methanogenic pathways in freshwater lakes (Schulz and Conrad 1996; Nüsslein et al. 2003; Glissmann et al. 2004; Chan et al. 2005; Nozhevnikova et al. 2007).

16.3.3.1 Brief Overview of Physiological Factors- Bottom-Up Regulations

The intrinsic characteristics of each microorganism determine its survival, its growth and its abundance but also the interactions that it establishes with other life forms and with the abiotic environment. This section on the factors related to the physiology of microorganisms is not exhaustive; a simple enumeration of potentially important factors is proposed with some examples related to methanogenic populations. These bottom-up regulations impact methanogens directly or indirectly by structuring the surrounding microbial community.

– ***The growth rate*** is highly variable from a microbial species to another and depends besides intrinsic characteristics, on various parameters such as the energy source, the pH, the temperature, the concentration of nitrogen source, the transfer through membranes, *etc.* For example, oxygen concentrations above 10 p.p.m. completely inhibit most methanogens because several cofactors and enzymes involved in their metabolism (*e.g.,* CO dehydrogenase acetyl-CoA synthase, F420) are oxygen-sensitive (Schonheit *et al.* 1981; Ragsdale and Kumar 1996). Temperature may affect composition patterns of

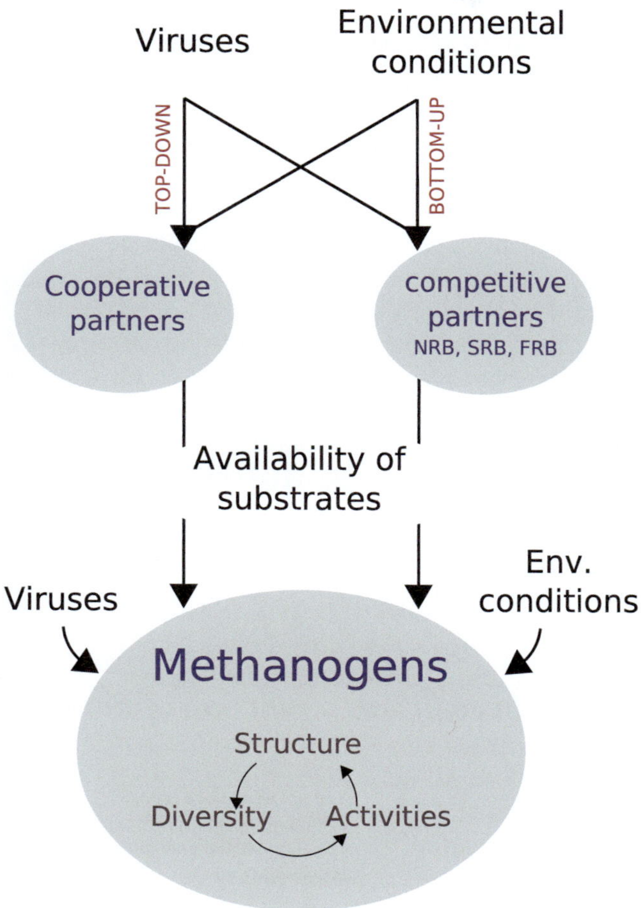

Fig. 16.3 Schematic representation of some factors (abiotic and biotic) influencing methanogens and consequently methane production rates (MPRs). To summarize, environmental conditions such as temperature and oxygen levels affect the methanogenic communities according to their physiology. But these factors also affect the syntrophic (*e.g.*, fermentative bacteria) and competitive communities of methanogens and consequently the availability of substrates. The affinity for the available substrates of different methanogenic communities determines the structure and the diversity of methanogenic assemblages. Thus methanogens are subjected to bottom-up regulation, and depend on cooperative and competitive interactions. More works remain to be done to get the methane cycle right, notably a big missing in methanogens ecology is our ignorance of top-down regulation (impact of viruses)

methanogenic communities as well as MPRs (Borrel et al. 2011). Accordingly, cold adapted methanogens dominate in the hypolimnion and the profound sediment of mid-latitude lakes and are different from methanogens identified in tropical and equatorial lakes or in shallow sediments of temperate lakes (Conrad et al. 2010). Moreover, the growth rate of methanogens may be directly limited by *in situ* temperature (Yvon-Durocher et al. 2014), which is generally below the growth optimum of methanogens (Borrel et al. 2011) and/or indirectly by limiting the production rate of methanogenic substrates (Schulz and Conrad 1996).

- ***The preferential substrate utilization*** by microbial species allows them to select the most efficient growth substrate while limiting the enzymatic synthesis (catabolic repression). This phenomenon allows understanding the diversity of ecosystems; it explains why several species occupying apparently similar ecological niches can coexist in an ecosystem.
- ***Affinity for the substrate***: when nutrient availability is limited, microorganisms cannot grow at their maximum growth rate and the affinity for the substrate becomes a major determinant of growth and competition between species. For example, the concentrations of H_2 and acetate in the methanogenic zone of freshwater lakes, close to the minimal threshold for substrate uptake by *Methanosaeta* (Jetten et al. 1992) and hydrogenotrophic methanogens (Zinder 1993) reflect the limitation of MPRs by substrate availability.
- ***Energy requirements***: microorganisms need energy for growth and to ensure their cellular integrity (maintenance energy) but a decoupling of energy (*i.e.*, a fermentation without growth) may be observed. When the growth rate is high, the proportion of the energy required for maintenance is low, but when the growth rate decreases (*e.g.*, nutritional deficiency), all the energy may be used for cell maintenance. The maintenance energy varies greatly between microbial species. When substrate is limiting, species with low maintenance energy needs are favored and can become predominant in the ecosystem.
- ***Others factors***: in complex ecosystems, the growth, development and eventually the survival of microorganisms depend on many other factors such as: energy yields (the effectiveness of ATP use varies from species to species), resistance to various environmental factors (*e.g.*, pH, heavy metals, antibiotics, xenobiotics), adherence to substratum and ability to form biofilms, mobility, response to concentration gradients of physico-chemical parameters, distance between cells, resistance to fast regeneration systems or storage energy, mutations in the stationary phase and quorum sensing.

16.3.3.2 Example of Bottom-Up Factors Affecting the Composition of Methanogenic Communities in the Sediment of Lake Pavin

- ***Substrate availability***: in Lake Pavin sediment, the abundance of *Methanosaetaceae* significantly decreased with depth along the sediment core (Fig. 16.1a) whereas the content in organic matter and organic carbon did not decrease significantly with depth (Fig. 16.1d). This trend is attributed to the decrease in content and lability of organic matter which becomes more and more depleted and recalcitrant with depth due to its longer exposure to degradation processes (Borrel et al. 2012a). This hypothesis of a gradient of microbial activities along the sediment core is sustained by the decrease of the 16S rRNA to 16S rDNA ratio and by thermograms of sediment samples (Fig. 16.1e, Box 16.2).

Box 16.2: Microcalorimetry: A Powerful Tool for Monitoring Metabolic Activities

Since all living organisms produce heat, calorimetry appeared very early as a useful instrument for the observation of metabolic activities. The first operative calorimeter, which was devised by Lavoisier and Laplace (1780), revealed that the energetic output of a Guinea pig amounts to about 3 W per kg of body weight, which is 2.5 times larger than the value found later for man using a respiration chamber (Ravussin et al. 1982). Microorganisms are even more energetic (resting yeast cultures, for instance, produce about 5 W per kg of living mass or 25 W per kg of dry mass (Birou and von Stockar 1989) but, due to their size, their global heat-output is extremely small and difficult to quantify at bench-scale. The gigantic experiment carried out by Dubrunfaut (1856) was probably the first quantitative measurement of microbial heat production (the heat of alcoholic fermentation was determined by adding enough yeast inoculum to a culture medium consisting of 21,400 L of a molasses solution containing 2559 kg of crystallizable sugar, all that being put in a huge vat of 3 m deep and 3 m diameter) but this type of experiment could not be carried out routinely! Quantitative microbiological calorimetry thus remained almost inaccessible for a very long time and it is only after the Second World War that were devised heat fluxmeters sufficiently sensitive to enable miniaturization (Belaich 1980; Lamprecht 1980; Gustafsson 1991).

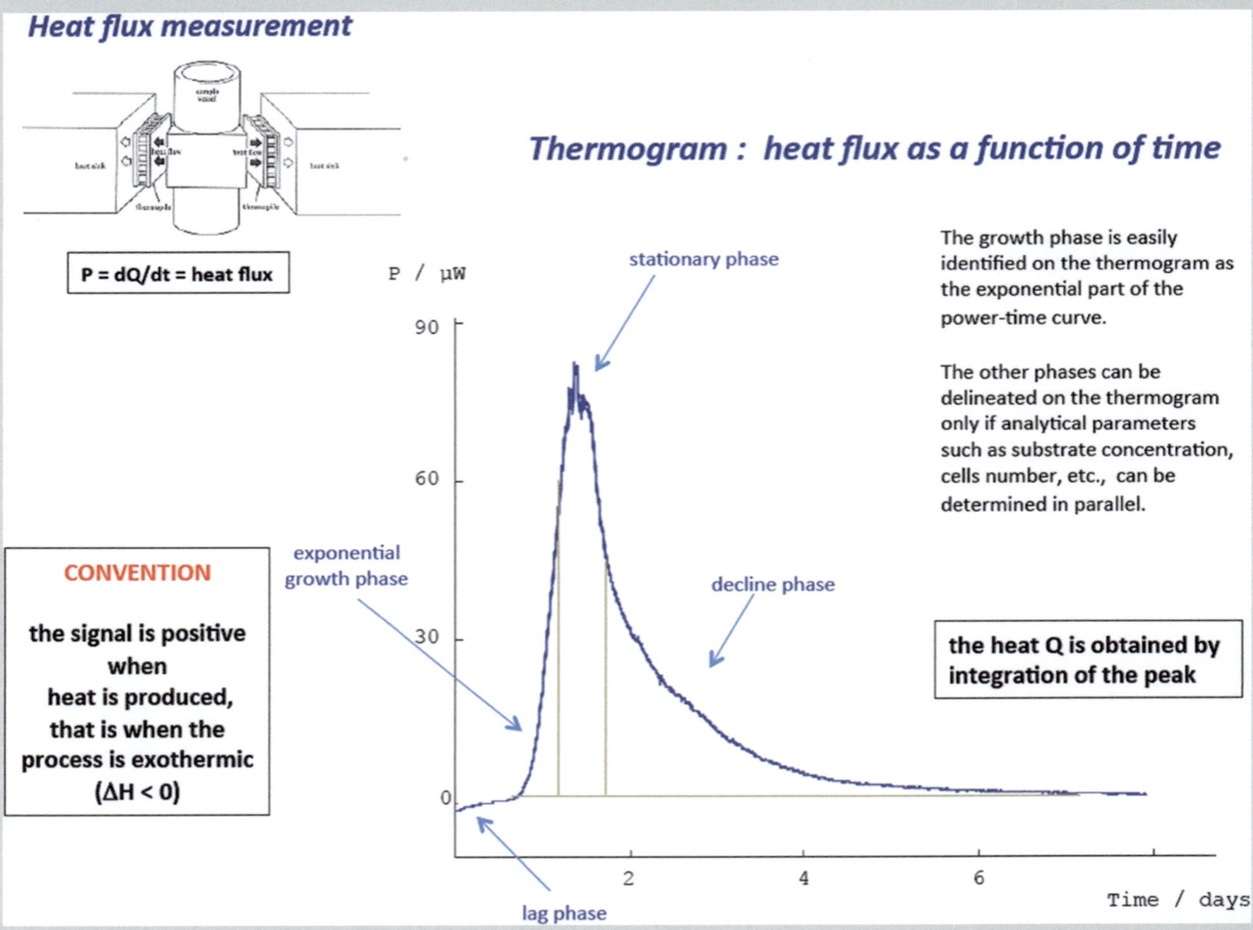

With nowadays instruments, it is possible to work on volumes of the order of the milliliter and detect heat effects at the nanowatt-level. Different types of calorimeters are available, each of them having its own advantages and drawbacks (Wadsö 1985; von Stockar and Marison 1989). Isothermal microcalorimeters of the heat-conduction type, which enable the study at constant temperature of slow processes, allow non-stop and real-time monitoring of the heat flux (also called thermal power output, P=dQ/dt) produced by as few as 10^4–10^5 active bacterial cells.

Multicalorimeters (holding, for instance, six independent minicalorimeters) are particularly interesting

(continued)

Box 16.2 (continued)

for the simultaneous investigation of several samples of a given living system (Wadsö 1985). It is with this type of instrument (TAM III, Thermometric – TA Instruments) that the samples coming from lake Pavin were examined (in closed ampoules and under unstirred conditions). Since the thermal power output corresponds to the sum of all the heat fluxes arising from the different chemical and physical processes that occur during incubation, the shape of the power-time curve is very dependent on the experimental conditions (Lamprecht 1980; Braissant et al. 2013; Bricheux et al. 2013). During the growth phase, the power-time curve shows the same exponential shape as the biomass curve but beyond that phase it is modulated by transition periods. Using complementary analyses (cells density, for instance), the main metabolic periods can be identified and the heat Q produced during any of them or during the whole process can be obtained by integration of the corresponding portion of the power-time curve. In batch incubation, the thermal power output curve finally returns to its baseline when the metabolism of the exogeneous substrates ceases; in most cases, a very small but non-zero final baseline is observed indicating that the cells are still alive but that they have switched over to time-limiting endogenous metabolism (Lamprecht 1980). Recently, this type of microcalorimetry showed to be not only a very effective tool for the determination of growth rate constants (Bricheux et al. 2013) but also a valuable probe for a rapid detection of the metabolic perturbations induced by xenobiotics (Lehours et al. 2010 ; Bricheux et al. 2013; Lescure et al. 2013).

– ***Temperature***: the cultural approaches performed from Pavin sediment samples led to the enrichment and isolation of several methanogens including *Methanobacteriales* representatives (Borrel et al. 2012b). However, this order was not detected in clone libraries from the same samples (Fig. 16.1a) suggesting that the relative species are minoritary among the methanogenic community. This hypothesis was verified by quantification of 16SrDNA and 16S rRNA transcripts. Nevertheless, a *Methanobacteriales* strain, representative of the population inhabiting Pavin sediments, was isolated and described as *Methanobacterium lacus* (Fig. 16.1b, Borrel et al. 2012b). The physiological preferenda of *M. lacus* are compatible with the environmental conditions prevailing in the sediment of Lake Pavin in terms of pH and salinity. However, the *in situ* temperature (~5 °C) of Pavin sediments is largely below the optimum temperature (30 °C) of this strain might partially explain why *Methanobacteriales* are poorly represented in this environment.

16.3.3.3 Cooperative Interactions with Other Anaerobic Microbes (Fig. 16.3)

This section mainly focus on the "classical" fermentative process and, because, of the central role of acetate in the anaerobic food chain, the acetogenesis process is also discussed.

- ***The particular case of syntrophy***[6]

Amongst heterotrophs it is as anaerobes that bacteria specially excel.... It is in the use of hydrogen acceptors that bacteria are specially developed as compared with animals and plants." (Stephenson 1947).

[6]For detailed information's, see McInerney et al. 2008, 2009, 2011; Sieber et al. 2012.

The methanogenic degradation of complex polymeric materials involves a number of diverse, interacting microbial species (see Sect. 16.3.1.1). The mutual dependence between interacting species can be so extreme that species cannot function without the activity of its partner, and the partners perform together functions that species cannot do alone. Syntrophy is a specialized case of tightly coupled mutualistic interactions between two metabolic types, which depend on each other to degrade certain substrates by transferring one or more metabolic intermediates (such as H_2) between partners (Fig. 16.2). For example, methanogens can interact with obligate hydrogen producing bacteria. The latter phenomenon is termed "interspecies hydrogen transfer" (Ianotti et al. 1973). Different genera of syntrophic bacteria have been described as partnerships of methanogens, contributing to the oxidation of fatty acids, benzoic acid, or fructose. In the absence of a hydrogen scavenger, these reactions are endergonic and cannot develop. When H_2, is consumed, the reaction becomes exergonic and the syntroph (*i.e.,* methanogens) can grow and oxidize the substrate. Syntrophic interactions enable methanogenesis when methanogenic substrates are limiting, sometimes leading to increased MPRs. The global biogeochemical impact of syntrophic interactions involving methanogenic *Euryarchaeota* is considerable as they enable the complete degradation of complex organic molecules to CO_2 and CH_4 in methanogenic habitats (Sieber et al. 2012).

- ***Fermentation***

Fermentation is a metabolism in which energy is derived from the partial oxidation of an organic compound using organic intermediates as electron donors and electron acceptors. No external electron acceptor is involved; no membrane or electron transport system is required; all ATP is produced by substrate level phosphorylation.

Syntrophic associations between fermentative bacteria and methanogenic archaea involve the exchange of hydrogen, formate or acetate between the partners. These metabolites are produced by the fermentative syntrophic metabolizer from its growth substrates (*e.g.,*, propionate, butyrate, and benzoate) and are consumed by the methanogenic partner keeping them at concentrations low enough to permit the metabolism of the substrate by the syntrophic metabolizer to be sufficiently exergonic to support adenosine triphosphate (ATP) synthesis, anabolism, and growth.

For example, the syntrophic transformation of butyrate to acetate and H_2 [at pH 7, 1 atm H_2 (101 kPa), and 1 M of acetate and butyrate] is thermodynamically unfavorable with a Gibbs free energy change of +48.6 kJ per mole of butyrate metabolized. When hydrogen and formate concentrations are kept low by methanogens or other hydrogen/formate users, butyrate degradation becomes thermodynamically favorable with a Gibbs free energy change of −39.2 kJ per mole of butyrate.

- ***Acetogenesis***[7]

Acetogenesis has only been described in organisms belonging to the *Bacteria* and, whereas most acetogens are affiliated to the *Firmicutes*, they constitute paraphyletic groups including *Spirochaetes*, δ-*Proteobacteria* like *Desulfotignum phosphitoxidans*, and *Acidobacteria* like *Holophaga foetida*. Encountered in most anaerobic environments, including extreme ones, acetogens reduce one-carbon compounds using the reductive acetyl-CoA (or Wood-Ljungdahl) pathway as their main mechanism for energy conservation and for the synthesis of acetyl-CoA, a metabolic precursor of acetate and biomass.

If acetogenesis is often perceived as a fermentative process, the use of CO_2 as final electron acceptor implies a strong dissimilarity with conventional fermentation process. Acetogens are sometimes called "homoacetogens" (meaning that they produce only acetate as fermentation product) or "CO_2-reducing acetogens". Important for the biology of ecosystems, particularly of freshwater lakes, acetogens contribute globally to 10 % of the output of acetate on Earth (Ragdsdale and Pierce 2008).

16.3.3.4 Example of Interactions Between Abiotic Factor (Iron)-Fermentation and Methanogenesis in Lake Pavin

A wide variety of fermentative microorganisms are able to reduce Fe(III) during anaerobic growth, but, Fe(III) reduction appears to be a minor pathway for electron flows in fermentative microorganisms that are not considered to be conservers of the process energy (Lovley 1987). Cultures of these microorganisms can accumulate significant amounts of Fe(II) but less than 5 % of the reducing equivalents are transferred to Fe(III) (Lovley 1987, 2006; Lovley and Phillips 1988). Although thermodynamic calculations demonstrated that fermentation is more energetically favorable with than without Fe(III) reduction (Lovley and Phillips 1988), it has been postulated that the minor electron transfer from reducing equivalents to Fe(III) during fermentation does not cause any increase the cell yield (Lovley 2006).

A comparative study has been performed on the growth, fermentative profile and heat production of the facultative iron-reducing bacterial strain BS2 (Fig. 16.4e), isolated from the anoxic zone of the iron-rich Lake Pavin and cultured on glucose in presence and absence of Fe(III), was performed (Fig. 16.4a–d, Lehours et al. 2010). The possible ecological benefit for this microorganism to use Fe(III) as an electron acceptor was clearly demonstrated, and modifications of the fermentative balance between glucose and glucose + Fe(III) growth conditions were recorded (*e.g.,* a higher gas production (H_2 and CO_2) and differences in acetate and lactate amounts, Fig. 16.4f). Given that fermentation products are used as electron donors for terminal electron-accepting processes like methanogenesis, sulfate-reduction or dissimilatory Fe(III) reduction, their availability may have a major impact on terminal electron-accepting processes. Previous studies on the anoxic zone of Lake Pavin have highlighted that acetate might be the main precursor for methanogenesis (Lehours et al. 2007, 2009) and that lactate and H_2 favor the Fe(III) reduction process (Lehours et al. 2009), suggesting that Fe (III) reduction by fermentative bacteria may not only affect their physiology but also the competitive relationships among terminal electron-accepting microorganisms of the anaerobic microbial food-web.

16.3.3.5 Competitive Interactions with Other Anaerobic Microbes (Fig. 16.3)

This section only presents the main terminal-electron-accepting processes that may be competitive of methanogenesis.

- ***Denitrification***[8]: Nitrogen is available in nature at different degrees of oxidation. Denitrification in the strict sense of the term corresponds to the reduction of nitrate to nitrite which is often used as respiratory acceptor and is reduced, in turn, into gaseous compounds such as NO, N_2O, N_2.
- ***Sulfate-reduction***[9]: Sulfate-reducing bacteria (SRB) use sulfate (SO_4^{2-}) as a terminal electron acceptor for the deg-

[7] For a detailed review, see Drake et al. 1994, 2006; Müller et al. 2004; Ragsdale and Pierce 2008.

[8] For detailed informations see Zumft 1992, 1997; Burgin and Hamilton 2007.

[9] For detailed informations, see Widdel 1988; Muyzer and Stams 2008.

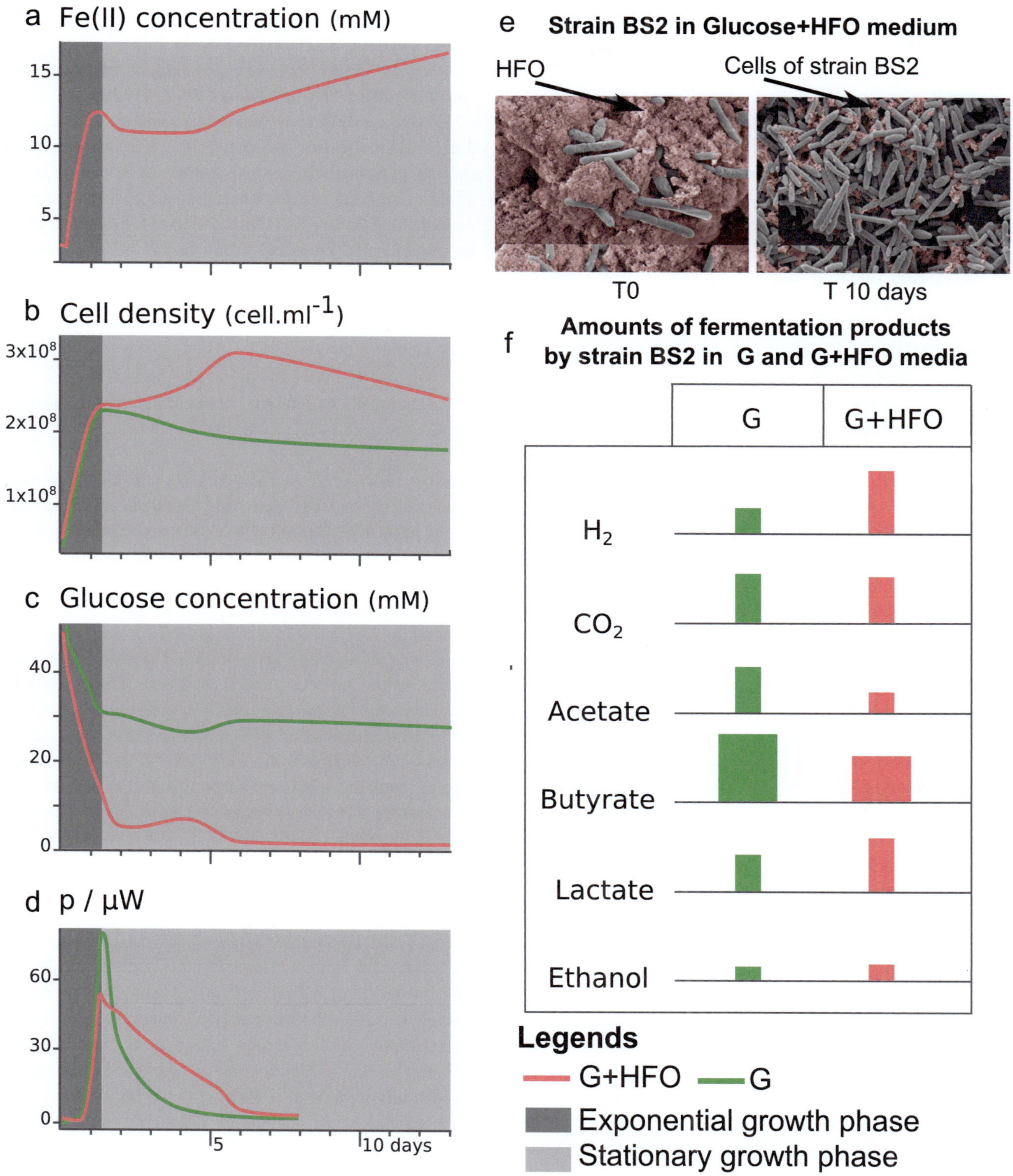

Fig. 16.4 Schematic representation of some culture parameters of the fermentative strain BS2 isolated from Lake Pavin. (**a**) Evolution of Fe(II) concentrations (mM) in glucose+iron (III) oxide-hydroxide (G+HFO) medium. (**b**) Evolution of cell densities (cell.mL^{-1}) in glucose (G) and G+HFO media. (**c**) Evolution of glucose concentrations (mM) in G and G+HFO media. (**d**) Thermograms (thermal power-time curves) of BS2 grown anaerobically in G or in G+HFO media measured using microcalorimetry (Box 16.2). (**e**) Transmission electron micrographs of BS2 strain in G+HFO medium at the beginning of the growth phase (T0) showing the attachment of BS2 cells to HFO particles and after 10 days of growth (T10) showing HFO particles almost completely reduced to soluble Fe(II). (**f**) Organic acid, H_2 and CO_2 amounts by strain BS2 (in mol.h^{-1}) in glucose and glucose+HFO media during the exponential and stationary/death phases (Adapted from Lehours et al. 2010)

radation of organic compounds, resulting in the production of sulfide (H_2S).

- ***Ferric iron reduction***[10]: Ferric iron (Fe^{3+}) in soluble or insoluble (*e.g.,* FeOOH, Fe_2O_3-H_20) forms may be microbiologically reduced to ferrous iron (Fe^{2+}). Dissimilatory iron reduction may be a form of respiration in which ferric iron serves as a dominant or exclusive terminal electron acceptor; or it may also accompany fermentation in which ferric iron is used as a supplementary electron acceptor (see above the example of strain BS2, Sect. 16.3.3.4) as opposed to dominant or exclusive terminal electron acceptor.

16.3.3.6 Do Sulfate-Reducing and Ferric-Iron Reducing Bacteria Compete with Methanogens in the Water Column of Lake Pavin?

- ***Sulfate-reducing bacteria (SRB) in Lake Pavin*** **:** using fluorescent in situ hybridization (FISH) method, SRB were demonstrated to account for 44 % of the *Bacteria* in the chemocline of Lake Pavin (Lehours et al. 2005). These high densities support a biogenic origin for the H_2S found in the anoxic zone of Lake Pavin. The genera *Desulfobulbus*, *Desulfobacter*, and *Desulfovibrio*, targeted with specific probes, represented only a small fraction of the SRB quantified in the chemocline meaning that other groups of SRB are involved in the sulfate-reduction activity in this zone. The genera *Desulfobulbus*, *Desulfobacter*, and *Desulfovibrio* presented density peaks at the aerobic/anaerobic interface (>10^5 cells ml^{-1}). The great abundance of the genus *Desulfobulbus* suggests a major role of this genus in sulfate-reduction activity.
- ***Ferric-Iron-reducing bacteria (FRB) in Lake Pavin*** **:** Despite high concentrations of Fe (II) in the anoxic zone of lake Pavin, both cloning-sequencing analyses and culture-dependent approaches did not allow the detection of members of known obligatory FRB, such as *Geobacteriaceae* (Lehours et al. 2007, 2009). This does not mean that there is no known or unknown obligatory FRB in the water column of lake Pavin, but suggests that facultative FRB may substantially contribute to Fe(III) reduction in this environment. In Fe(III)-enrichment cultures, bacteria affiliated to the *Planctomycetes*, the *Firmicutes*, the *Actinobacteria,* the *Spirochaetes*, and to the five classes within the *Proteobacteria* were retrieved and probably used ferric-iron reduction as an alternative dissimilatory pathway. For example, several *Pseudomonas*, *Aeromonas* and *Serratia* species, previously shown to reduce iron under anaerobic conditions (Lovley 2006), were enriched. Moreover, *Desulfovibrio* members, largely detected in the chemocline of Lake Pavin (Lehours et al. 2005), were also enriched suggesting that some SRB may be involved in ferric-iron reduction in Lake Pavin anoxic zone.
- ***Hypotheses explaining why acetoclastic methanogens are the main CH4 producers in Lake Pavin*** **:** Data acquired from both the water column and the sediment of Lake Pavin revealed that acetotrophic methanogenesis is an important mineralization pathway and the main source of CH_4 in this ecosystem. It is rather surprising when considering that acetotrophic methanogenesis is the less thermodynamically favorable pathway for methanogenesis (Garcia et al. 2000). We formulate two hypotheses, in relation to competitive interactions, to explain this observation:

(1) FRB and methanogens are not in competition for acetate uptake

In most freshwater environments, Fe (III) reduction and methanogenesis are the dominant processes but Fe (III) reducers have a higher affinity for acetate providing them some competitive advantages over methanogenesis (van Bodegom et al. 2004). However, in the steady state environment of Lake Pavin anoxic zone, acetate may be preferentially used for methanogenesis while ferric-iron reducers use preferentially H_2 and alternative intermediate products such as lactate or fumarate (Lehours et al. 2009). This hypothesis is consistent with the dominance of *Methanosaeta concilii*, an acetoclastic methanogenic species, within the methanogenic communities (Lehours et al. 2005, 2007). This hypothesis also agrees with the results of van Bodegom et al. (2004) who demonstrated that hydrogenotrophic methanogens are more directly inhibited by Fe (III) than *Methanosaeta concilii.*

(2) SRB are outcompeted by methanogens for acetate uptake

Acetate-utilizing sulfate reducers may also compete with acetoclastic methanogens. However, this competition is not straightfoward as for hydrogen. In experiments in which sulfate was added to a fully methanogenic anaerobic bioreactor, the acetotrophic *Methanosaeta* species were out-competed by sulfate reducers only several years after the beginning of the incubation (Omil et al. 1998).

Is Lake Pavin an exception? Probably not. Indeed, Methanosaeta species are probably the predominant CH_4-producers on Earth and up to two-thirds of biologically produced CH_4 released in the atmosphere each year is derived from the methyl group of acetate (Smith and Ingram-Smith 2007). To understand why acetate is the main precursor of methanogenesis, we also needed to come back to the physiology of microorganisms which is important for modeling the variation in MPRs according to environmental constraints. For example, pH and temperature conditions prevailing in Pavin sediments favor acetogenesis over hydrogenotrophic methanogenesis (Nozhevnikova et al. 2007).

[10]For detailed informations, see Ehrlich and Newman 2008.

16.3.3.7 Top-Down Regulations: Impact of Virus

The impact of viruses on methanogens and then on MPRs is potentially of importance but remains completely overlooked. Whereas viruses infecting some mesophilic and thermophilic methanogens have been isolated since 20 years, their involvement in the regulation of this key functional group has not yet been established. The available data (mainly on viruses of thermophilic methanogens), suggest that these viruses belong to *Caudovirales*. Unlike other archaeal viruses, the viruses of methanogens seem predominantly lytic and therefore could have a significant impact in the regulation of methanogenic communities.

The first study, conducted a few years ago with Lake Pavin sediments, revealed that viruses are abundant and exhibit unexpected phenotypes. Among those phenotypes, we detected viruses that exhibit shapes similar to known archaeal viruses (Borrel et al. 2012c). A filamentous prokaryote infected with ellipsoid pleomorphic viruses was also observed (Fig. 16.1c). The infected hosts exhibited sheath and flat ends, similar to those observed with the cultured archaeal methanogenic species *Methanosaeta concilii* (Patel and Sprott 1990). Although there is no certitude about the identity of the infected cell from Lake Pavin sediments, its tentative affiliation to *Methanosaeta* was based on morphological similarities and on the predominance of members of this archaeal group in Lake Pavin (Lehours et al. 2007: Borrel et al. 2012a). To further investigate the role of viruses in the regulation of methanogenic communities (*e.g.,* by determining the number of infected cells), we currently develop an approach coupling magneto-FISH and electron microscopy (Box 16.3).

Box 16.3: The Magneto-Fish Method[11]: Does It Allow the Study of Viral Infection of Targeted Groups of Micro-Organisms (*e.g.,* Methanogens)?

Magneto-FISH allows physical separation of microorganisms of interest from environmental samples. This method is based on 16S rRNA catalyzed reporter deposition (CARD)–FISH (identity) (Pernthaler et al. 2002) and immunomagnetic bead capture (separation) (Pernthaler et al. 2008; Pernthaler and Orphan 2010). Magneto-FISH was applied to study interspecies partnerships between anaerobic methane-oxidizing (ANME) *Archaea* and sulfate-reducing *Deltaproteobacteria* (SRB) in anoxic marine sediments (Pernthaler et al. 2008) illustrating that this method is adapted for complex environmental samples.

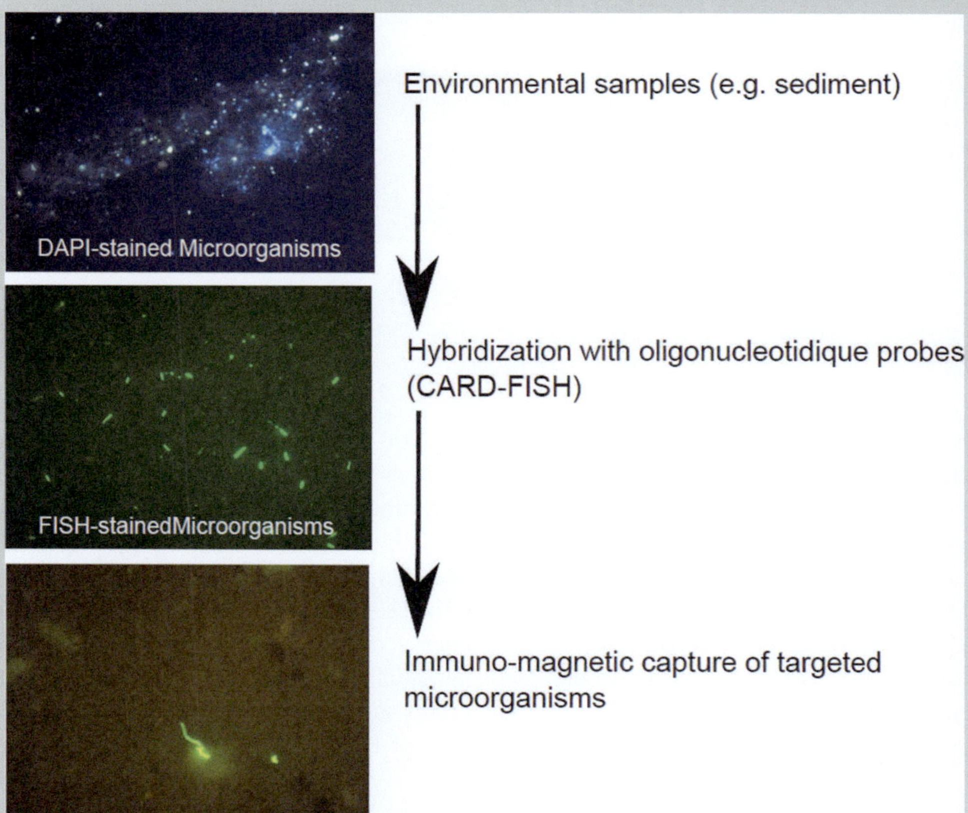

[11] For a detailed protocol see Trembath-Reichert et al. (2013)

*TEM picture J. Colombet

** CLSM picture O. Bardot.

(continued)

Box 16.3 (continued)

Step 1: CARD-FISH

FISH is a printing method targeting classically the cellular ribosomal RNA (rRNA) by hybridization of a probe complementary to the rRNA sequence of a targeted cell. The probe is labeled with a fluorochrome allowing the observation by epifluorescence microscopy. CARD-FISH combines CARD (catalysed reported deposition) of fluorescently labeled tyramides with single-cell identification by FISH. The hybridization involves a single oligonucleotide that is covalently crosslinked to a horseradish peroxidase (HRP) label. Amplification of the signal relative to that achieved with probes that are labeled with a single fluorochrome is based on the radicalization of multiple tyramide molecules by a single horseradish peroxidase (Amann and Fuchs 2008).

Step 2: Magnetic capture

To induce the cellular capture after cell hybridization, monoclonal mouse anti-fluorochrome-antibody labeled to paramagnetic beads are added. A complex labeled cell-paramagnetic bead is created and those complexes are "captured" by the application of a magnetic field (with a magnet) allowing to enrich a cellular fraction of targeted cells in environmental samples.

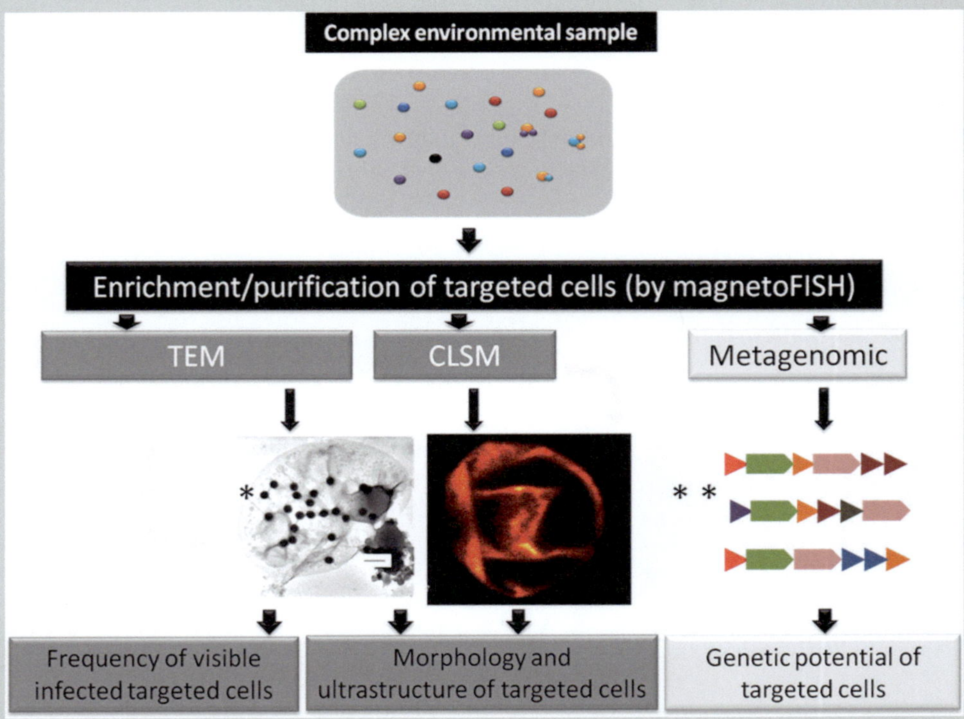

Step 3: Observation of "captured"-cells by transmission electron microscopy (TEM)

TEM is a useful method to appreciate the cell ultrastructure and the viral infection of cells. A key problem to determine the cellular infection of a targeted microbial community (*e.g., Euryarchaeota*) is that all the prokaryotic community is observed in an environmental sample by TEM without any information on their identity. We managed to perform magnetoFISH (using oligonucleotide probes designed for targeted group such as *Euryarchaeota*) prior to TEM observations. First results show that both methods are compatible.

Further work will be done to determine the frequency of visible infected cells of different groups of methanogens but also to investigate the morphological characteristics (by TEM and confocal laser scanning microscopy (CLSM)) of targeted cells and the genetic potential (by metagenomic) of some uncultivated groups of microorganisms (*e.g.,* MBGD) encountered in Lake Pavin.

16.4 Biological Methane Consumption

Biological CH_4 oxidation is carried out by methanotrophs which are estimated to consume 60 % of the biogenic CH_4 produced in the environment. In freshwater lakes, methanotrophs oxidizes 30 to 99 % of the CH_4 produced (Bastviken et al. 2008), and plays a fundamental role in regulation of CH_4 emissions. Two biological pathways may be involved in CH_4 oxidation[12]:

- The ***aerobic methane oxidation*** which was first described at the beginning of the twentieth century (Kaserer 1905; Sohngen 1906). The activity of aerobic CH_4-oxidizing bacteria depends on the availability and concentrations of both CH_4 and O_2. The highest CH_4 consumption rates are located at oxic/anoxic interfaces where opposite fluxes of CH_4 and O_2 occur (Fig. 16.5).
- The ***anaerobic methane oxidation*** pathway which, until the 1970s, was thought to be unrealistic. The microorganisms involved in this process have only been identified during the last decade, but have not been isolated in pure culture so far. They are anaerobic microorganisms, such as methanogens, and their activities are located in the anoxic compartments (Fig. 16.5).

16.4.1 Aerobic Methanotrophy

16.4.1.1 Aerobic Methanotrophs

All known aerobic methanotrophs belong to the domain *Bacteria* and constitute a polyphyletic group belonging to 4 families affiliated to the *γ-Proteobacteria* (*Methylococcaceae*, *Methylocystaceae*), the *α-Proteobacteria* (*Beijerinckiaceae*) and to the *Verrucomicrobia* (*Methylacidiphilaceae*).

- ***Methylococcaceae*** (known as Type I methanotrophs): the intracytoplasmic membrane (ICM) forms bundles of vesicular disks distributed throughout the cell. These methanotrophs use the ribulose monophosphate pathway

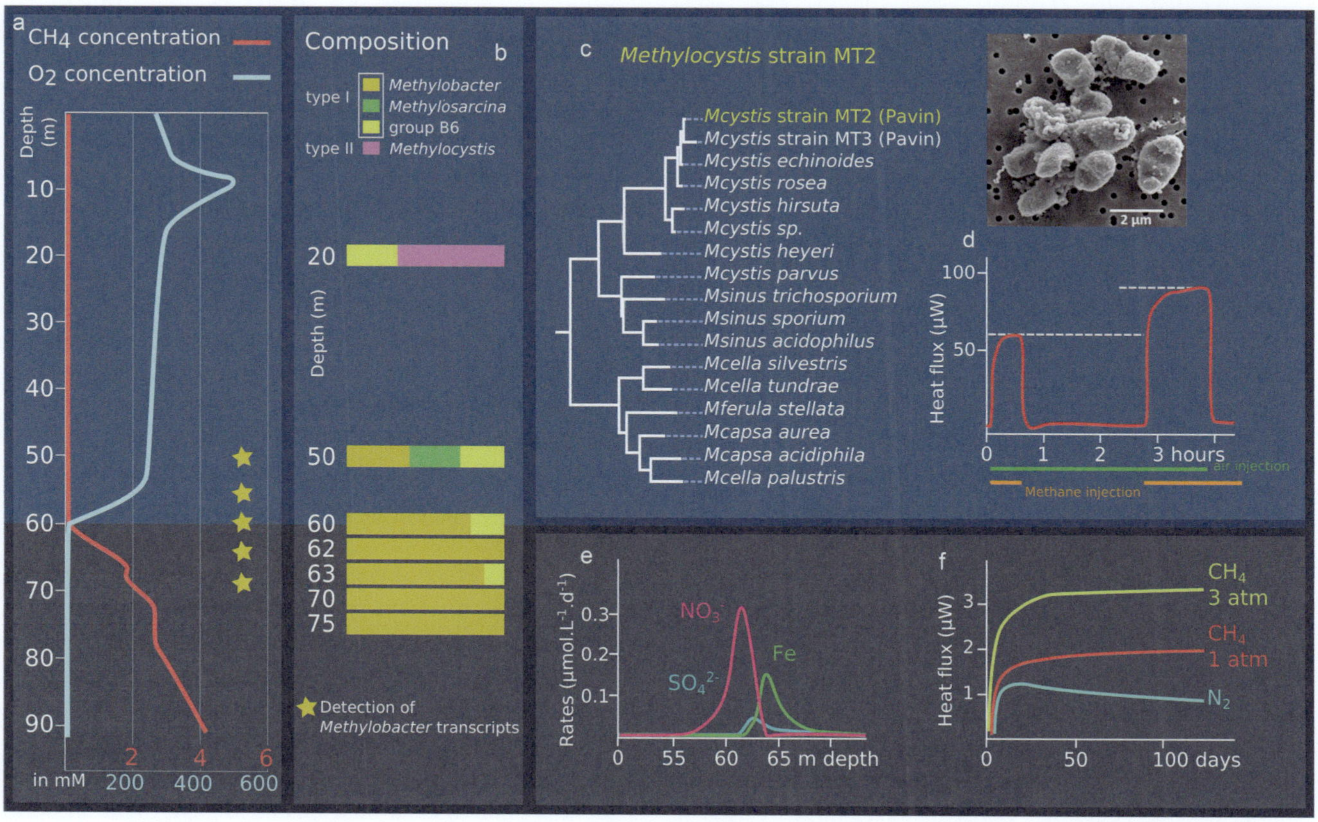

Fig. 16.5 Methanotrophs in Lake Pavin. (**a**) Schematic representation of CH_4 and O_2 profiles along the water column of Lake Pavin. (**b**) Composition of methanotrophic communities, determined from *pmoA* gene clones libraries. Depths for which *pmoA* transcripts of *Methylobacter* were detected are indicated by a star (adapted from Biderre-Petit et al. 2011). (**c**) Transmission electron micrograph and phylogenetic affiliation (based on 16S rRNA) of methanotrophic strains MT2 and MT3 (Mcystis = *Methylocystis)* isolated from the water column of Lake Pavin. (**d**) Thermograms representing the heat production by the strain MT2 (unpublished results), measured using microcalorimetry (Box 16.2). (**e**) Schematic representation of importance of the anaerobic oxidation of methane processes in the water column of Lake Pavin (adapted from Lopes et al. 2011). (**f**) Thermograms representing the heat production of water samples collected at 70 m depth in the water column of Lake Pavin (unpublished results), measured using microcalorimetry (Box 16.2). Incubations were performed under N_2 and CH_4 (1 atm and 3 atm pressure)

[12]For more details, several reviews can be found on aerobic (Hanson and Hanson 1996; Trotsenko and Murrell 2008) and anaerobic oxidation of methane (Knittel and Boetius 2009; Thauer and Shima 2008.

for carbon fixation and their phospholipid fatty acid (PLFA) composition is generally dominated by saturated and monosaturated C_{16} fatty acids.

- ***Methylocystaceae*** (known as Type II methanotrophs): the ICM is parallel to the cell membrane. They use the serine pathway for carbon fixation and PLFAs are dominated by C_{18} isomers, except for *Methylocystis heyeri* which also contains large amounts of C_{16} fatty acids (Dedysh et al. 2007).
- ***Beijerinckiaceae***: this family is metabolically diversified, with obligate methanotrophs, facultative methanotrophs and non-methanothrophs (Dunfield et al. 2010).
- ***Methylacidiphilaceae***: only three acidophilic and thermotolerant species are presently affiliated with this family (Dunfield et al. 2007; Islam et al. 2008; Op den Camp et al. 2009; Pol et al. 2007).

16.4.1.2 Metabolic Pathways and Enzymes Involved

While several facultative methanotrophs (belonging to the genera *Methylocella* (Dedysh et al. 2005), *Crenothrix* (Stoecker et al. 2006), *Methylocystis* (Belova et al. 2011; Im et al. 2011), and *Methylocapsa* (Dunfield et al. 2010) were recently isolated, most of the aerobic methanotrophs areobligatory C1-users growing only on CH_4 or methanol. The aerobic CH_4 oxidation pathway contains four sequential steps with methanol, formaldehyde and formate as metabolic intermediates. Formaldehyde plays a central role in the metabolism of aerobic methanotrophs, since it also constitutes the main carbon source for the anabolic pathway. The first step of the aerobic oxidation of CH_4 (*i.e*., conversion to methanol) is catalyzed by the methane monooxygenase (MMO) enzymatic complex (See Box 16.1).

16.4.1.3 Examples of Environmental Factors Affecting Aerobic Methanotrophy in Freshwater Lakes

- ***Availability of* CH4/O2:** The activity of aerobic CH_4-oxidizing bacteria depends on the availability and concentrations of both CH_4 and O_2. The highest CH_4 consumption rates are located at oxic/anoxic interfaces where opposite fluxes of CH_4 and O_2 occur. In freshwater lakes, such interfaces are found in the upper layers of sediment overlaid by oxic water or in the oxycline of stratified water columns. Consequently, a large fraction of CH_4 diffusing from the lacustrine anoxic sediments is oxidized at the sediment interface (66–95 %, Liikanen et al. 2002) and at the oxic/anoxic interface of the water column (45–100 %, Bastviken et al. 2002).
- ***Nitrogen***: The effect of nitrogen input on methanotrophs and CH_4 oxidation in freshwater lakes has been investigated in the context of anthropogenic-accelerated eutrophication (Liikanen and Martikainen 2003; Nold et al. 1999). Addition of ammonium generally inhibits CH_4 oxidation (Murase and Sugimoto 2005) and incorporation of CH_4-derived carbon into lipids of methanotrophs (Nold et al. 1999). This inhibition is attributed to the competition between ammonium and CH_4 for binding sites on MMO, because of their similar chemical structure (Bedard and Knowles 1989).
- ***Top-down regulation***: Depending on local and seasonal conditions, methanotrophic bacteria may represent a substantial fraction of planktonic and/or benthic biomasses in freshwater lakes (Bastviken et al. 2003; Costello et al. 2002). A number of observations led to the hypothesis that a significant fraction of the diet of aquatic consumers may consist of CH_4-derived carbon. Accordingly, several studies based on isotopic labeling (*e.g.*, ^{13}C-labeled-CH_4) found evidence of CH_4-derived carbon transfer through zooplanktons and *Chironomidae* (Deines et al. 2007; Kankaala et al. 2006) or detected lipid biomarkers specific of methylotrophs (*i.e.,* specific PLFA) in *Chironomidae* tissues (Kiyashko et al. 2004).

16.4.1.4 Composition of Aerobic Methanotrophic Communities in Lake Pavin

The family *Methylococcaceae* was found to dominate methanotrophic communities inhabiting temperate and boreal lakes (average annual temperature < 12 °C). Borrel et al. (2011) demonstrated that the genus *Methylobacter* is retrieved in all mid-latitude lakes and that methanotrophs affiliated to *Methylocystaceae* dominated the water column of a tropical dam lake characterized by temperatures comprised between 25 and 30 °C.

The methanotrophic communities present in the water column of Lake Pavin was investigated using the particulate methane monooxygenase gene (*pmoA*) (Biderre-Petit et al. 2011). The analysis of *pmoA* transcripts revealed that the majority of aerobic CH_4 oxidation is carried out by a small number of phylotypes affiliated to *Methylobacter*, and particularly to *M. psychrophilus* (Fig. 16.5). The temperature of the water column, ranging between 4 and 5 °C, may explain the predominance of phylotypes close to psychrophilic species in the water column of Lake Pavin. Few sequences of *Methylosarcina sp.* and *Methylocystis sp.* were retrieved but, surprisingly, *Methylocystis sp.* was recovered at 20 m depth, a O_2-rich zone but almost entirely CH_4-depleted (Fig. 16.5).

The *in-situ* analysis of aerobic methanotrophs was completed by a cultural approach in order to isolate representatives of this metabolic group. Samples for enrichments were collected in strata where an optimal activity of methanotrophy was observed. A wide range of growing conditions was achieved by combining changes in temperature, concentration of nutrients and pH. Due to the dominance of *Methylobacter* species in the water column of Lake Pavin, we therefore optimized the culture conditions for the isolation of Type I methanotrophs. Unfortunately, all strains isolated were affiliated to Type II methanotrophs, and more especially to the genus *Methylocystis*. We are currently investigating an isolation strategy based on culture with opposite fluxes of CH_4 and oxygen, in order to favor the physiology of the strains dominating the methanotrophic community of Lake Pavin.

Microcalorimetry studies performed to characterize the physiological response of isolated Methylocystis strains to varying concentrations of CH_4 and O_2. As illustrated in Fig. 16.5, the response of the methanotrophic strains to CH_4 input is immediate, even after a period of deprivation, suggesting that:

(i) The enzymes involved in the methanotrophic pathway are continuously synthesized, even in the absence of substrate. This hypothesis might explain the detection of transcripts of pmoA in strata with very low CH_4 concentrations (Fig. 16.5),
(ii) the strain investigated has the capacity to lower its maintenance energy and therefore to increase its stationary growth phase.

16.4.2 Anaerobic Methanotrophy

16.4.2.1 Brief History

From a biochemical perspective, CH_4 is relatively difficult to activate. Its aerobic activation is catalyzed by monooxygenases. These enzymes that make use of high-potential oxygen radical chemistry are unavailable to anaerobic life. On another hand, the enrichment or detection of organisms capable of anaerobic growth on CH_4 has remained unsuccessful for a long time. This led to the idea that CH_4 was inert under anoxic conditions, a belief that persisted among most microbiologists throughout most of the twentieth century. This assumption was challenged during the 1970s. Several independent studies showed that, in marine sediments, CH_4 concentrations decreased upward while sulfate concentrations concomitantly decreased downward. This indicated that sulfate-reduction might be coupled to the anaerobic oxidation of methane (AOM) with sulfate acting as the electron acceptor according to the equation 1 (Eq. 16.1). But it took more than 30 years before the AOM in marine sediments could be unambiguously demonstrated thanks to the analysis of the stable carbon isotopic composition ($\delta^{13}C$) of lipid biomarkers specific to *Archaea* (*e.g.*, isoprenoid glycerol ether lipids) and *Bacteria* (particular fatty acids, Box 16.4). In the following years, a number of publications reported the occurrence of AOM in different sedimentary settings, highlighting the importance of this process in the biosphere (for review, see Knittel and Boetius 2009).

16.4.2.2 Pathways of Anaerobic Oxidation of Methane (AOM)

Microbial communities and metabolisms involved in AOM are diversified since at time three different processes are described in various environments. To our knowledge, nitrate-dependent methane oxidation (NDMO) was exclusively observed in freshwater environments, whereas sulfate dependent methane oxidation (SDMO) and iron/manganese-dependent methane oxidation were reported from both marine and freshwater environments.

– **AOM coupled to sulfate-reduction**

Visual identification of cells hybridized with fluorochrome-labeled specific oligonucleotide probes in marine sediments revealed conspicuous aggregates of *Archaea* and SRB (AOM consortia), representing > 90 % of the total microbial community (Boetius et al. 2000). This finding supported the hypothesis that CH_4 could be used as a substrate source *via* the cooperation of *Archaea* able to activate CH_4 and SRB able to provide an electron sink (see Eq. 16.1). These consortia consist of archaea, performing a reverse methanogenesis, which form the phylogenetically distinct cluster ANME-2 (ANaerobic MEthanotrophs) within the methanogenic order *Methanosarcinales*, and of sulfate-reducing bacteria from the *Desulfosarcina and Desulfococcus* branches of *Delta-Proteobacteria.* Subsequently, similar consortia were found with *Archaea* of the clusters ANME-1, which is distantly related to *Methanosarcinales* and *Methanomicrobiales* (Orphan et al. 2001; Knittel et al. 2005), and ANME-3, which is related to the genera *Methanococoides* and *Methanolobus* within the *Methanosarcinales* (Lösekann et al. 2007).

$$CH_4 + SO_4^{2-} \rightarrow HCO_3^- + HS^- + H_2O \quad (\Delta_r G^\circ = -17\,kJ\,mol^{-1} CH_4) \tag{16.1}$$

Box 16.4: Prokaryotic Lipids as Indicators of Methanotrophic Communities

The combined study of the distribution and stable carbon isotopic composition ($\delta^{13}C$ values) of diagnostic lipids derived from bacteria and archaea in environmental samples is a powerful mean for demonstrating the implication of prokaryotes into the past and present methane cycle. Due to the strongly depleted $\delta\ ^{13}C$ values of biogenic (but also thermogenic) methane, the biomass and cellular constituents of prokaryotes using methane as a carbon source consequently appear ^{13}C-depleted. For example, $\delta\ ^{13}C$ values as low as −100 ‰ of different hopanoid compounds derived from aerobic bacteria (see an example of structure), classically serve as indicators of aerobic oxi-

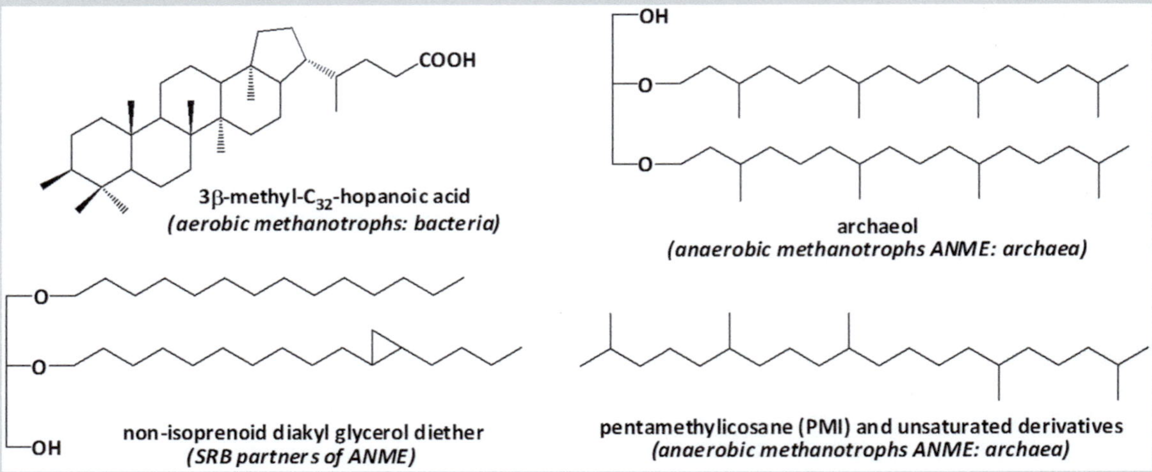

dation of methane (*i.e.,* by aerobic or microaerophilic methanotrophs).

Examples of lipid structures produced by aerobic and anaerobic methanotrophs

The strongly depleted carbon isotopic composition ($-130‰ < \delta^{13}C < -50‰$) of lipid biomarkers specifically synthesized by *Archaea* (see examples of structures) has provided the first irrefutable evidence of the involvement of these organisms in the anaerobic oxidation of methane (AOM) in habitats where the process was previously inferred. The stable carbon isotope composition of other lipid biomarkers (see an example of structure) suggested to derive from the bacterial partners of ANME involved in sulfate-dependent AOM further supported a syntrophic association between ANME and SRB. These bacterial lipids are usually less depleted ($-90‰ < \delta^{13}C < -40‰$) than the archaeal lipids. Differences in biomarker profiles from one AOM site to the other typify distinct anaerobic microbial consortia involved in AOM. In such settings, the diversity of ^{13}C-depleted biomarkers of prokaryotic origin are indicative of methane carbon transferring via ANME to several SRB populations and then, eventually, to other surrounding prokaryotic and/or eukaryotic populations. Our understanding of AOM, as well as demonstrations of its occurrence in different ecosystems and of its direct or indirect involvement in biogeochemical processes (*e.g.,* precipitation of carbonates or iron sulfide nodules), are steadily increasing, and partly rely on the analysis of lipid biomarker ^{13}C composition.

– **AOM coupled to denitrification**

CH_4 oxidation coupled to denitrification was first demonstrated in an anaerobic enrichment culture from a freshwater contaminated aquifer containing high nitrate concentrations (Smith et al. 1991). Evidence of this process was later supported by experimental observations (Islas-Lima et al. 2004). Moreover Raghoebarsing et al. (2006) subsequently demonstrated AOM coupled to nitrite reduction (Eq. 16.2) in enrichment cultures obtained from sediments of two freshwater ecosystems with high nitrate concentrations (up to 1 mM).

$$3CH_4 + 8NO_2^- + 8H^+ \rightarrow 3CO_2 + 4N_2 + 10H_2O \quad (\Delta_r G^\circ = -928\,\text{kJ mol}^{-1}CH_4) \tag{16.2}$$

Enrichment cultures consisted of 10 % archaea of the order *Methanosarcinales* distantly related to ANME-2, named ANME-2d, the rest being bacteria affiliated to the new bacterial division NC10 (Raghoebarsing et al. 2006). Hence, it was plausible to assume that anoxic CH_4 oxidation with nitrate would operate similarly as with sulfate, *i.e.,* the archaea oxidize CH_4 to CO_2 and transfer the electrons to the denitrifiers reducing nitrate or nitrite to N_2. This hypothesis was also supported by microscopic observations of consortia of *Bacteria* and *Archaea* (Raghoebarsing et al. 2006). However, it turned out that the oxidation of CH_4 is catalyzed exclusively by the denitrifiers without participation of the archaeal members (Ettwig et al. 2008). Indeed, ANME disappeared after prolonged incubation (Ettwig et al. 2009), while AOM continued suggesting that ANME were not obligatory involved in this process. This process, driven by members of the NC10 bacterial division, appears to be completely different from SDMO, since it does not seem to involve reverse methanogenesis. Complete genome analysis of *Candidatus Methylomirabilis oxyfera*, the dominant bacterium affiliated with NC10, and labeling experiments pointed to a new pathway of 'intra-aerobic denitrification' (Ettwig et al. 2010). This pathway leads to the intracellular production of O_2 by disproportionation of 2 NO coming from the nitrite. The enzyme and genes involved in NO disproportionation are not yet clearly identified. Most of the genes involved in the classical pathway of aerobic methanotrophy were retrieved in the genome of *Candidatus M. oxyfera*. Therefore, the O_2 produced by NO disproportionation is likely to oxidize CH_4 intracellularly. Subsequently, nitrate-dependent AOM was also demonstrated (Eq. 16.3.) from an enrichment culture largely dominated by a representative of the ANME-2d lineage, *Candidatus Methanoperedens nitroreducens* Haroon et al. 2013). To oxidize CH_4, this *Archaea* only reduce nitrate to nitrite which in turn is used by *Methylomirabilis oxyfera* or *Candidatus Kuenenia stuttgartiensis*, an anaerobic ammonium oxidizer using nitrite as electron acceptor (Haroon et al. 2013).

$$3CH_4 + 4NO_3^- \rightarrow CO_2 + 4NO^-_2 + 2H_2O \quad (\Delta G^\circ = -503\,\text{kJ mol}^{-1}CH_4) \tag{16.3}$$

– **AOM coupled to iron and manganese reduction**

AOM was shown to be coupled with iron and manganese reduction (Eqs. 16.4 and 16.5) in meCthane-seep sediments through an enrichment culture (Beal et al. 2009). The communities involved were dominated by an uncultured archaeal group named Marine Benthic Group-D (MBG-D). ANME-1 and 2 were also identified in these sediments. AOM coupled to iron and manganese reduction was suggested to largely contribute to the global oceanic AOM. However, the process and the identity of microorganisms involved are still unclear.

$$CH_4 + 8Fe(OH)_3 + 15H^+ \rightarrow HCO_3 \; + 8Fe^{2+} + 21H_2O(\Delta_r G^\circ = \; 572\,\text{kJ mol}\;{}^{1}CH_4) \tag{16.4}$$

$$CH_4 + 4MnO_2 + 7H^+ \rightarrow HCO_3^- + 4Mn^{2+} + 5H_2O(\Delta_r G^\circ = -790\,\text{kJ mol}^{-1}CH_4) \tag{16.5}$$

16.4.2.3 Mechanisms of Anaerobic Oxidation of Methane

In addition to the close phylogenetic affiliation between ANMEs and methanogens, evidence for metabolic similarities comes from different studies. The *mcrA* gene (coding for the a-subunit of MCR, see previous Sect. 16.3.2.4) was shown to be associated with the ANME community (see Box 16.1, Hallam et al. 2003) and metagenomic analyses of ANME-rich sediments revealed that nearly all genes involved in the methanogenesis pathways are present in ANME-1 (Hallam et al. 2004). Proteomic approaches led to isolation of an abundant enzymatic complex with a high degree of similarity with MCR (Kruger et al. 2003). This complex, named Ni-protein I, is associated with a modified F430 cofactor (172-methyl-thio-F430, Mayr et al. 2008) and may be involved in AOM. More recently, Scheller et al. (2010) showed that an MCR extract from a methanogen was able to cleave the strong C-H bond of CH_4 to form methyl-coenzyme M. All these data support the hypothesis that anaerobic CH_4 oxidation is performed by a reverse methanogenesis process.

16.4.2.4 Preliminary Microbiological Studies Supporting AOM in Lake Pavin

Geochemical data and reactive transport modeling of CH_4 in Lake Pavin reveal that CH_4 produced in the monimolimnion is partly oxidized under anaerobic conditions (Lopes et al. 2011). The water layer between 65 and 70 m, in which environmental conditions are not favorable to the activity of aerobic methanotrophs, may provide favorable conditions for the growth of microorganisms performing AOM: anoxia, availability of CH_4 and of alternative electron acceptors other than O_2.

Microcalorimetric measurements (Box 16.2) performed with water samples collected in the chemocline of Lake Pavin revealed a greater metabolic activity in samples incubated under 1 and 3 atm of CH_4 compared to those incubated under N_2 (Fig. 16.5). These results, indicating a CH_4-induced metabolic activity under anoxic conditions, were completed by enrichment cultures. Different culture conditions were performed using anoxic water and sediment samples as inocula and labeled CH_4. ^{13}C-CH_4 was shown to be consumed and partly transformed into ^{13}C-CO2 in both the sediment and the water column. We also observed that CH_4 consumption was greater when NO_3^- was provided as an electron acceptor (Attard E. Personal communication), fitting well with the model of CH_4 consumption provided by Lopes and coworkers (2011) (Fig. 16.5).

16.4.2.5 Hypothetical Identity of Microbial Actors Involved in AOM Process in Lake Pavin

The microorganisms involved in AOM processes in Lake Pavin are not presently identified. The model developed by Lopes et al. (2011) shows that, probably, all currently known AOM mechanisms occur (AOM coupled with sulfate-, nitrate- and ferric-iron reduction), suggesting that different microorganisms are able to perform AOM in this ecosystem (Fig. 16.5).

– ***Observations and results favoring the involvement of Archaea members***

None sequence of 16S rDNA and of *mcrA* affiliated to the ANME lineages was retrieved in the studies performed from samples collected in the anoxic water column and sediment of Lake Pavin (Lehours et al. 2007; Biderre-Petit et al. 2011). However the AOM process in lacustrine environments remains poorly known and may involve other archaeal populations than ANME.

• *Are Marine Benthic group D (MBG-D) members involved in AOM in the sediment?*

Sequences belonging to the euryarchaeotal MBG-D, a *Thermoplasmatales*-related group, were retrieved from the sediments of Lake Pavin (Fig. 16.1a, Borrel et al. 2012a). Members of the MBG-D represented the unique archaeal members in clone library from the bottom of the AOM zone of the Lake Cadagno (Schubert et al. 2011) and were detected in several other marine environments and cultures where AOM occurs (Inagaki et al. 2006; Harrison et al. 2009; Webster et al. 2011; Zhang et al. 2011). These observations coupled with that of Beal et al. (2009) who noted that MBG-D clones represented up to 40 % of the archaeal community in clone libraries of their AOM enrichments from methane-seep sediment might imply the involvement of MBG-D members in AOM. This hypothesis is consistent with the high activities of this group detected in the sediment layer where CH_4 concentrations decreased (Fig. 16.1a). However, the recent genomic analysis of three incomplete genomes of MBG-D representatives suggests that representatives of this lineage are more likely involved in the degradation of detrital protein (Lloyd et al. 2013). Future complete genomes and activities determined through cultural approaches will be necessary to have confirmation of their metabolism.

• *Are Methanomassiliicoccales members involved in AOM in the water column?*

Sequence of *mcrA* and 16S rRNA affiliated to the *Methanomassiliicoccales* were detected in the anoxic water column of the Lake Pavin (Biderre-Petit et al. 2011) and belong to a separate cluster than the sequences of cultured representatives of this order (Borrel et al. 2013). These representatives are all obligate methylotrophic methanogens requiring an external H_2 source to reduce methyl-compound (*e.g.,* methanol, methylated amines) to CH_4 (Dridi et al. 2012; Poulsen et al. 2013; Brugère et al. 2013). Due to the low number of data on this order and to the affiliation of the sequences from Lake Pavin to a large sub-cluster of uncultured representatives, it is not excluded that the member of the *Methanomassiliicoccales* occurring in the Lake Pavin might be involved in anaerobic methanotrophy.

– ***Observations and results which do not favor the involvement of archaeal members***

We recently started to investigate the distribution and stable carbon isotopic composition ($\delta^{13}C$ values) of diagnostic prokaryotic lipids in the anoxic sediment and the different redox zones of the water column (oxic, dysoxic, fully anoxic) of Lake Pavin, in order to decipher which prokaryotic communities are involved in the CH_4 cycle in this singular meromictic Lake (see Box 16.4 for a rationale of this approach). Preliminary results show a low and homogenous diversity of specific archaeal lipid biomarkers in both the superficial sediments and the anoxic waters of the lake, and the absence of most biomarkers diagnostic of ANME (*e.g.,* crocetane, pentamethylicosane or hydroxyarchaeol). The stable carbon isotopic composition of isoprenoid glycerol ether lipids present in the sediment (i.e., $\delta^{13}C$ archaeol = −18‰ and $\delta^{13}C$ caldarchaeol = −31‰) argue against CH_4 consumption by archaeal members. The only ^{13}C-depleted (−70‰ < $\delta^{13}C$ < −55‰) lipid biomarkers detected so far in the superficial sediments and in the dysoxic and anoxic waters of Lake Pavin are bacterial C_{16} fatty acids and hopanoids (diploptene, diplopterol, C_{32} homohopanols and homohopanoic acid). Such ^{13}C-depleted lipids are generally thought to be specific of (micro-) aerophilic methanotrophic bacteria (*e.g.*, Birgel and Peckmann 2008). However, the more depleted $\delta^{13}C$ values of some hopanoids in the anoxic sediment of Lake Pavin compared to those of hopanoids from the anoxic water column may indicate the presence of a bacterial population capable of oxidizing CH_4 under strict anaerobiosis. This intriguing hypothesis clearly needs to be further investigated. Large volumes of water sampled at high depth resolution in the oxic, dysoxic and anoxic parts of the water column are currently being studied and should help assessing the unconventional CH_4 consumers of Lake Pavin.

16.5 Conclusion and Perspectives

The Lake Pavin is an eccentric and original environment and the study of the anaerobic biota of this biotope is a particularly exciting prospect. In the point of view of the CH_4 cycle, due to its specific characteristics, Lake Pavin is an exceptional site of high interest for scientists. In contrast to most of freshwater lakes, Lake Pavin emits very low amounts of CH_4 in the atmosphere although high concentrations of this gas are formed in the sediment and at the bottom of the water column. The characteristics of the water column of Lake Pavin provide, undoubtedly, particularly favorable conditions to the oxidation of this metabolite. In particular, the strong physicochemical gradients present at the oxic/anoxic interface probably slow down the flow of CH_4 from the monimolimnion, enabling a greater efficiency in its consumption. Indeed, diffusion is the pathway of CH_4 which gives the most accessibility to its consumption by methanotrophic microorganisms (Bastviken 2009). Moreover, the geomorphological characteristics of Lake Pavin, conditioning its meromixis, may limit the contribution of other pathways of CH_4 transport. As the overturning of the water column is incomplete in Lake Pavin, only a small portion of the CH_4 accumulated in the anoxic zone is transported during seasonal mixing events. In addition, the low epilimnetic sediment surface limits the bubbling process. Finally, considering the small area covered by emergent macrophytes, transport by plants should be negligible. In addition to the relative importance of the different pathways of CH_4 transport, we showed that the combination of the two CH_4 oxidation processes (aerobic and anaerobic) contributes to limit dramatically emissions of this gas from the Lake Pavin.

Surprisingly, a small number of species are involved in the biogeochemical CH_4 cycle in Lake Pavin, at least both in the production and in the aerobic consumption of this metabolite; those involved in the anaerobic consumption being not yet identified. This is probably the result of a long term evolution of anaerobic microbial communities since the formation of the Lake. We can postulate that the steady state of the monimolimnion illustrates that microbial communities reach the climax in this anoxic ecosystem. Moreover, bottom-up regulations are also probably crucial in determining the species composition. Particularly, the temperature seems to play an important role in the community composition of methanogens and aerobic methanotrophs in Lake Pavin illustrating, if it was necessary, that physiological factors are crucial to understand the structure, the diversity and the activities of microbes in the environment.

A strong work remains to be done to identify the microorganisms involved in the AOM in this Lake. Actors involved in AOM in different other ecosystems (marine and freshwater systems) are not detected in the water column and in the sediment of Lake Pavin. It seems therefore plausible that the investigation of AOM in Lake Pavin will extend the knowledge of the microorganisms and of the mechanisms acting to consume CH_4 anaerobically in the biosphere. At time, our hypothesis favor a coupling between AOM and denitrification as results from geochemical modeling (Lopes et al. 2011) and enrichments cultures encourage such hypothesis.

Due to the greenhouse effect of CH_4 and in the context of global warming, the understanding of microbial processes involved in CH_4 biogeochemical cycle is of crucial importance both in terms of fundamental science and applications.

Acknowledgements This study includes results from Master, PhD and postdoc researches supported by various instances: Région Auvergne, CNRS, Université Blaise Pascal, Ministère de la Recherche et de la Technologie. Part of the results presented was obtained with the collaboration of colleagues: Didier Jézéquel, Gilles Mailhot, Jonathan Colombet, Keith Joblin, Paul Evans, Annie Guedon, Corinne Biderre-Petit, Marion Rabiet, Karl Rockne, Christophe Guimbaud, Frédéric Savoie, Stéphane Chevrier, Pierre Agrinier, Nelly Assayag, François Prévot, Jean-Claude Romagoux, Guy Demeure, Aurélie Thénot, Télesphore Sime-Ngando. The studies were granted from different sources: French National Program EC2CO (Projects INTERLAC, METHANOLAC) and ANR Program (Project METANOX). We thank the reviewer, Bernard Ollivier, for his constructive comments, which helped us to improve the manuscript.

References

Amann R, Fuchs BM (2008) Single-cell identification in microbial communities by improved fluorescence *in situ* hybridization techniques. Nat Rev Microbiol 6:339–348

Angel R, Claus P, Conrad R (2012) Methanogenic archaea are globally ubiquitous in aerated soils and become active under wet anoxic conditions. ISME J 6:847–862

Auguet JC, Barberan A, Casamayor EO (2010) Global ecological patterns in uncultured *Archaea*. ISME J 4:182–190

Bapteste E, Brochier C, Boucher Y (2005) Higher-level classification of the *Archaea*: evolution of methanogenesis and methanogens. Archaea 1:353–363

Bastviken D (2009) Methane. In: Likens G (ed) Encyclopedia of inland waters. Elsevier, Oxford, pp 783–805

Bastviken D, Ejlertsson J, Tranvik L (2002) Measurement of methane oxidation in lakes- a comparison of methods. Environ Sci Technol 36:3354–3361

Bastviken D, Ejlertsson J, Sundh I, Tranvik L (2003) Methane as a source of carbon and energy for lake pelagic food webs. Ecology 84:969–981

Bastviken D, Cole J, Pace M, Tranvik L (2004) Methane emissions from lakes: dependence of lake characteristics, two regional assessments, and a global estimate. Global Biogeochem Cycles 18:GB4009

Bastviken D, Cole JJ, Pace ML, Van de Bogert MC (2008) Fates of methane from different lake habitats: connecting whole-lake budgets and CH_4 emissions. J Geophys Res 113:G02024

Battistuzzi FU, Feijao A, Hedges SB (2004) A genomic timescale of prokaryote evolution: insights into the origin of methanogenesis, phototrophy, and the colonization of land. BMC Evol Biol 4:44

Beal EJ, House CH, Orphan VJ (2009) Manganese- and iron-dependent marine methane oxidation. Science 325:184–187

Bedard C, Knowles R (1989) Physiology, biochemistry, and specific inhibitors of CH_4, NH_4^+, and CO oxidation by methanotrophs and nitrifiers. Microbiol Rev 53:68–84

Belaich JP (1980) Growth and metabolism in bacteria. In: Beezer AE (ed) Biological microcalorimetry. Academic, London, pp 1–42

Belova SE, Baani M, Suzina NE, Bodelier PLE, Liesack W, Dedysh SN (2011) Acetate utilization as a survival strategy of peat-inhabiting *Methylocystis spp*. Environ Microbiol Rep 3:36–46

Biderre-Petit C, Jézéquel D, Dugat-Bony E, Lopes F, Kuever J, Borrel G, Viollier E, Fonty G, Peyret P (2011) Identification of microbial communities involved in the methane cycle in a freshwater meromictic lake. FEMS Microbiol Ecol 77:533–545

Birgel D, Peckmann J (2008) Aerobic methanotrophy at ancient marine methane seeps: a synthesis. Org Geochem 39:1659–1667

Birou B, von Stockar U (1989) Application of bench-scale calorimetry to chemostat cultures. Enzyme Microb Technol 11:12–16

Boetius A, Ravenschlag K, Schubert CJ, Rickert D, Widdel F, Gieseke A, Amann R, Jorgensen BB, Witte U et al (2000) A marine microbial consortium apparently mediating anaerobic oxidation of methane. Nature 407:623–626

Borrel G, Jézéquel D, Biderre-Petit C, Morel-Desrosiers N, Morel JP, Peyret P, Fonty G, Lehours AC (2011) Production and consumption of methane in freshwater lake ecosystems. Res Microbiol 162:832–847

Borrel G, Lehours AC, Crouzet O, Jézéquel D, Rockne D, Kulczak A, Duffaud E, Joblin K, Fonty G (2012a) Stratification of *Archaea* in the deep sediments of a freshwater meromictic lake: vertical shift from methanogenic to uncultured archaeal lineages. PlosOne 7:e43346

Borrel G, Joblin K, Guedon A, Colombet J, Tardy V, Lehours AC, Fonty G (2012b) *Methanobacterium lacus* sp. nov., isolated from the profundal sediment of a freshwater meromictic lake. Int J Syst Evol Microbiol 62:1625–1629

Borrel G, Colombet J, Robin A, Lehours AC, Prangishvili D, Sime-Ngando T (2012c) Unexpected and novel putative viruses in the sediments of a deep-dark permanently anoxic freshwater habitat. ISME J 6:2119–2127

Borrel G, O'Toole PW, Harris HM, Peyret P, Brugère JF, Gribaldo S (2013) Phylogenomic data support a seventh order of methylotrophic methanogens and provide insights into the evolution of methanogenesis. Genome Biol Evol 5:1769–1780

Borrel G, Parisot N, Harris HMB, Peyrtaillade E, Gaci N, Tottey W, Bardot O, Raymann K, Gribaldo S, Peyret P, O'Toole PW, Brugère JF (2014) Comparative genomics highlights the unique biology of *Methanomassiliicoccales*, a *Thermoplasmatales*-related seventh order of methanogenic archaea that encodes pyrrolysine. BMC Genomics 15:679

Braissant O, Bonkat G, Wirz D, Bachmann A (2013) Microbial growth and isothermal microcalorimetry: growth models and their application to microcalorimetric data. Thermochimica Acta 555:64–71

Bricheux G, Bonnet JL, Bohatier J, Morel JP, Morel-Desrosiers N (2013) Microcalorimetry: a powerful and original tool for tracking the toxicity of a xenobiotic on *Tetrahymena pyriformis*. Ecotoxicol Environ Saf 98:88–94

Briee C, Moreira D, Lopez-Garcia P (2007) Archaeal and bacterial community composition of sediment and plankton from a suboxic freshwater pond. Res Microbiol 158:213–227

Brochier-Armanet C, Boussau B, Gribaldo S, Forterre P (2008) Mesophilic *Crenarchaeota*: proposal for a third archaeal phylum, the *Thaumarchaeota*. Nat Rev Microbiol 6:245–252

Brugère JF, Borrel G, Gaci N, Tottey W, O'Toole PW, Malpuech-Brugère C (2013) Archaebiotics: proposed therapeutic use of archaea to prevent trimethylaminuria and cardiovascular disease. Gut Microbes 5:5–10

Burgin AJ, Hamilton SK (2007) Have we overemphasized the role of denitrification in aquatic ecosystems? A review of nitrate removal pathways. Front Ecol Environ 5:89–96

Case RJ, Boucher Y, Dahllof I, Holmstrom C, Doolittle WF, Kelleberg S (2007) Use of 16S rRNA and rpoB genes as molecular markers for microbial ecology studies. Appl Environ Microbiol 73:278–288

Chan OC, Claus P, Casper P, Ulrich A, Lueders T, Conrad R (2005) Vertical distribution of structure and function of the methanogenic archaeal community in Lake Dagow sediment. Environ Microbiol 7:1139–1149

Cicerone RJ, Oremland RS (1988) Biogeochemical aspects of atmospheric methane. Global Biogeochem Cycles 2:299–327

Conrad R (1999) Contribution of hydrogen to methane production and control of hydrogen concentrations in methanogenic soils and sediments. FEMS Microbiol Ecol 28:193–202

Conrad R (2009) The global methane cycle: recent advances in understanding the microbial processes involved. Environ Microbiol Rep 1:285–292

Conrad R, Klose M, Claus P, Enrich-Prast A (2010) Methanogenic pathway, C-13 isotope fractionation, and archaeal community composition in the sediment of two clear-water lakes of Amazonia. Limnol Oceanogr 55:689–702

Costello AM, Auman AJ, Macalady JL, Scow KM, Lidstrom ME (2002) Estimation of methanotroph abundance in a freshwater lake sediment. Environ Microbiol 4:443–450

Dedysh SN, Knief C, Dunfield PF (2005) *Methylocella* species are facultatively methanotrophic. J Bacteriol 187:4665–4670

Dedysh SN, Belova SE, Bodelier PLE, Smirnova KV, Khmelenina VN, Chidthaisong A, Trotsenko YA, Liesack W, Dunfield PF (2007) *Methylocystis heyeri* sp. nov., a novel type II methanotrophic bacterium possessing 'signature' fatty acids of type I methanotrophs. Int J Syst Evol Microbiol 57:472–479

Deines P, Grey J, Richnow HH, Eller G (2007) Linking larval chironomids to methane: seasonal variation of the microbial methane cycle and chironomid delta C-13. Aquat Microb Ecol 46:273–282

Drake HL, Daniel SL, Matthies C, Küsel K (1994) Acetogenesis, acetogenic bacteria, and the acethyl-CoA pathway: past and current perspectives. In: Drake HL (ed) Acetogenesis. Chapman and Hall, New York, pp 3–60

Drake HL, Küsel K, Matthies C (2006) Acetogenic prokaryotes. In: Dworkin M, Falkow S, Rosenberg E, Schleifer KH (eds) The prokaryotes, vol 2, Ecophysiology and biochemistry. Springer, New York, pp 354–420

Dridi B, Fardeau ML, Ollivier B, Raoult D, Drancourt M (2012) *Methanomassiliicoccus luminyensis* gen. nov., sp. nov., a methanogenic archaeon isolated from human faeces. Int J Syst Evol Mirobiol 62:1902–1907

Dubrunfault M (1856) Note sur la chaleur et le travail mécanique produits par la fermentation vineuse. Compt Rend 42:945–948

Dunfield PF, Yuryev A, Senin P, Smirnova AV, Stott MB, Hou SB, Ly B, Saw JH, Zhou ZM, Ren Y, Wang JM, Mountain BW, Crowe MA, Weatherby TM, Bodelier PLE, Liesack W, Feng L, Wang L, Alam M (2007) Methane oxidation by an extremely acidophilic bacterium of the phylum *Verrucomicrobia*. Nature 450:879–882

Dunfield PF, Belova SE, Vorob'ev AV, Cornish SL, Dedysh SN (2010) *Methylocapsa aurea* sp. nov., a facultative methanotroph possessing a particulate methane monooxygenase, and emended description of the genus *Methylocapsa*. Int J Syst Evol Microbiol 60:2659–2664

Ehrlich HL, Newman DK (2008) Geomicrobiology of iron. In: Ehrilich HL, Newman DK (eds) Geomicrobiology, vol 5. CRC Press, Boca Raton, pp 279–329

EPA (United States Environmental Protection Agency) (2010) Methane and nitrous oxide emissions from natural sources. Office of atmospheric programs, Washington. Available at http://www.epa.gov/outreach/pdfs/Methane-and-Nitrous-Oxide-Emissions-From-Natural-Sources.pdf

Ettwig KF, Shima S, van de Pas-Schoonen KT, Kahnt J, Medema MH, Op den Camp HJM, Jetten MSM, Strous M (2008) Denitrifying bacteria anaerobically oxidize methane in the absence of *Archaea*. Environ Microbiol 10:3164–3173

Ettwig KF, van Alen T, van de Pas-Schoonen KT, Jetten MSM, Strous M (2009) Enrichment and molecular detection of denitrifying methanotrophic bacteria of the NC10 phylum. Appl Environ Microbiol 75:3656–3662

Ettwig KF, Butler MK, Le Paslier D, Pelletier E, Mangenot S, Kuypers MMM, Schreiber F, Dutilh BE, Zedelius J et al (2010) Nitrite-driven anaerobic methane oxidation by oxygenic bacteria. Nature 464:543–548

Ferry JG (2010) The chemical biology of methanogenesis. Planet Space Sci 58:1775–1783

Frenzel P (2000) Plant-associated methane oxidation in rice fields and wetlands. Adv Microb Ecol 16:85–114

Fricke WF, Seedorf H, Henne A, Kruer M, Liesegang H, Hedderich R, Gottschalk G, Thauer RK (2006) The genome sequence of *Methanosphaera stadtmanae* reveals why this human intestinal archaeon is restricted to methanol and H_2 for methane formation and ATP synthesis. J Bacteriol 188:642–658

Garcia JL (1990) Taxonomy and ecology of methanogens. FEMS Microbiol Rev 87:297–308

Garcia JL, Patel BKC, Ollivier B (2000) Taxonomic, phylogenetic and ecological diversity of methanogenic archaea. Anaerobe 6:205–226

Glissmann K, Chin KJ, Casper P, Conrad R (2004) Methanogenic pathway and archaeal community structure in the sediment of eutrophic Lake Dagow: effect of temperature. Microb Ecol 48:389–399

Grossart HP, Frindte K, Dziallas C, Eckert W, Tang KW (2011) Microbial methane production in oxygenated water column of an oligotrophic lake. Proc Natl Acad Sci U S A 108:19657–19661

Guimbaud C, Catoire V, Gogo S, Robert C, Chartier M, Laggoun-Défarge F, Grossel A, Albéric P, Pomathiod L, Nicoullaud B, Richard G (2011) A portable infrared laser spectrometer for flux measurements of trace gases at the geosphere-atmosphere interface. Meas Sci Technol 22:1–17

Gustafsson L (1991) Microbiological calorimetry. Thermochimica Acta 193:145–171

Hallam SJ, Girguis PR, Preston CM, Richardson PM, DeLong EF (2003) Identification of methyl coenzyme M reductase A (*mcrA*) genes associated with methane-oxidizing archaea. Appl Environ Microbiol 69:5483–5491

Hallam SJ, Putnam N, Preston CM, Detter JC, Rokhsar D, Richardson PM, DeLong EF (2004) Reverse methanogenesis: testing the hypothesis with environmental genomics. Science 305:1457–1462

Hanson RS, Hanson TE (1996) Methanotrophic bacteria. Microbiol Rev 60:439–471

Haroon MF, Hu S, Shi Y, Imelfort M, Keller J, Hugenholz P, Yuan Z, Tyson GW (2013) Anaerobic oxidation of methane coupled to nitrate reduction in a novel archaeal lineage. Nature 500:567–570

Harrison BK, Zhang H, Berelson W, Orphan VJ (2009) Variations in archaeal and bacterial diversity associated with the sulfate-methane transition zone in continental margin sediments (Santa Barbara Basin, California). Appl Environ Microbiol 75:1487–1499

Hedderich R, Whitman W (2006) Physiology and biochemistry of the methane-producing *Archaea*. In: Dworkin M, Falkow S, Rosenberg E, Schleifer KH, Stackebrandt E (eds) The prokaryotes. Springer, New York, pp 1050–1079

Ianotti EL, Kafkewitz D, Wolin MJ, Bryant MP (1973) Glucose fermentation products of *Ruminococcus albus* grown in continuous culture with *Vibrio succinogenes*: changes caused by interspecies transfer of H_2. J Bacteriol 114:1231–1240

Iino T, Tamaki H, Tamazawa S, Ueno Y, Ohkuma M, Suzuki K, Igarashi Y, Haruta S (2013) Candidatus *Methanogranum caenicola*: a novel methanogen from the anaerobic digested sludge, and proposal of *Methanomassiliicoccaceae* fam. nov. and *Methanomassiliicoccales* ord. nov., for a methanogenic lineage of the class *Thermoplasmata*. Microbes Environ 28:244–250

Im J, Lee SW, Yoon S, DiSpirito AA, Semrau JD (2011) Characterization of a novel facultative *Methylocystis* species capable of growth on methane, acetate and ethanol. Environ Microbiol Rep 3:174–181

Inagaki F, Nunoura T, Nakagawa S, Teske A, Lever M, Lauer A, Suzuki M, Takai K, Delwiche M et al (2006) Biogeographical distribution and diversity of microbes in methane hydrate-bearing deep marine sediments on the Pacific Ocean Margin. Proc Natl Acad Sci U S A 103:2815–2820

IPCC (2007) Climate change 2007: the physical science basis. In: Solomon S, Qin D, Manning M, Chen Z, Marquis M, Averyt KB et al (eds) contribution of working group I to the fourth assessment report of the intergovermmental panel on climate change (IPCC). Cambridge University Press, Cambridge, UK/New York

Islam T, Jensen S, Reigstad LJ, Larsen O, Birkeland NK (2008) Methane oxidation at 55 degrees C and pH 2 by a thermoacidophilic bacterium belonging to the *Verrucomicrobia* phylum. Proc Natl Acad Sci U S A 105:300–304

Islas-Lima S, Thalasso F, Gómez-Hernandez J (2004) Evidence of anoxic methane oxidation coupled to denitrification. Water Res 38:13–16

Jetten MSM, Stams AJM, Zehnder AJB (1992) Methanogenesis from acetate - a comparison of the acetate metabolism in *Methanothrix soehngenii* and *Methanosarcina spp*. FEMS Microbiol Rev 88:181–197

Jézéquel D, Michard G, Viollier E, Prévot F, Groleau A, Sarazin G, Lopes F (2010) Le cycle du carbone et les risques d'éruption gazeuse au Pavin. Rev Sci Nat Auver 74:67–86

Kankaala P, Taipale S, Grey J, Sonninen E, Arvola L, Jones RI (2006) Experimental delta C-13 evidence for a contribution of methane to pelagic food webs in lakes. Limnol Oceanogr 51:2821–2827

Kaserer H (1905) Ueber die oxydation des wasserstofes und des methane durch mikroorganismen (Sur l'oxydation de l'hydrogène et du méthane par les microorganismes). Z landw Versuchsw in Osterreich 8:789–792

Kiyashko SI, Imbs AB, Narita T, Svetashev VI, Wada E (2004) Fatty acid composition of aquatic insect larvae Stictochironomus pictulus (Diptera: Chironomidae): evidence of feeding upon methanotrophic bacteria. Comp Biochem Physiol B Biochem Mol Biol 139:705–711

Knittel K, Boetius A (2009) The anaerobic oxidation of methane-progress with an unknown process. Annu Rev Microbiol 63:311–334

Knittel K, Losekann T, Boetius A, Kort R, Amann R (2005) Diversity and distribution of methanotrophic archaea at cold seeps. Appl Environ Microbiol 71:467–479

Kruger M, Meyerdierks A, Glockner FO, Amann R, Widdel F, Kube M, Reinhardt R, Kahnt R, Bocher R, Thauer RK, Shima S (2003) A conspicuous nickel protein in microbial mats that oxidize methane anaerobically. Nature 426:878–881

Kvenvolden KA, Rogers BW (2005) Gaia's breath-global methane exhalations. Mar Pet Geol 22:579–590

Lamprecht I (1980) Growth and metabolism in yeasts. In: Beezer AE (ed) Biological microcalorimetry. Academic, London, pp 43–112

Lang K, Schuldes J, Kligl A, Poehlein A, Daniel R, Brune A (2015) New mode of energy metabolism in the seventh order of methanogens as revealed by comparative genome analysis of *"Candidatus Methanoplasma termitum"*. Appl Environ Microbiol 81:1338–1352

Lavoisier AL, de Laplace PS (1780) Mémoire sur la chaleur. Mémoires de l'Académie des Sciences, Paris. Available online by CRHST/CNRS

Lehours AC, Bardot C, Thenot A, Debroas D, Fonty G (2005) Anaerobic microbial communities in Lake Pavin, a unique meromictic lake in France. Appl Environ Microbiol 71:7389–7400

Lehours AC, Evans P, Bardot C, Joblin K, Gerard F (2007) Phylogenetic diversity of archaea and bacteria in the anoxic zone of a meromictic lake (Lake Pavin, France). Appl Environ Microbiol 73:2016–2019

Lehours AC, Batisson I, Guedon A, Mailhot G, Fonty G (2009) Diversity of culturable bacteria, from the anaerobic zone of the meromictic lake pavin, able to perform dissimilatory-iron reduction in different in vitro conditions. Geomicrobiol J 26:212–223

Lehours AC, Rabiet M, Morel-Desrosiers N, Morel JP, Jouve L, Arbeille B, Mailhot G, Fonty G (2010) Ferric iron reduction by fermentative strain BS2 isolated from an iron-rich anoxic environment (Lake Pavin, France). Geomicrobiol J 27:714–722

Lelieved J, Crutzen PJ, Dentener FJ (1998) Changing concentration, lifetime and climate forcing of atmospheric methane. Tellus B Chem Phys Meteorol 50:128–150

Lescure T, Carpentier A, Battaglia-Brunet F, Morel-Desrosiers N (2013) Oxidation of As(III) by the bacterial community of a marine sediment monitored by microcalorimetry. Geomicrobiol J 30:540–548

Liikanen A, Martikainen PJ (2003) Effect of ammonium and oxygen on methane and nitrous oxide fluxes across sediment-water interface in a eutrophic lake. Chemosphere 52:1287–1293

Liikanen A, Huttunen JT, Valli K, Martikainen PJ (2002) Methane cycling in the sediment and water column of mid-boreal hyper-eutrophic Lake Kevätön, Finland. Arch Hydrobiol 154:585–603

Lipscomb JD (1994) Biochemistry of the soluble methane monooxygenase. Annu Rev Microbiol 48:371–399

Lloyd KG, Schreiber L, Petersen DG, Kjeldsen KU, Lever MA, Steen AD, Stephanauskas R, Ritcher M, Kleindienst S, Lenk S, Schramm A, Jørgensen BB (2013) Predominant *Archaea* in marine sediments degrade detrital proteins. Nature 496:1–6

Lomans BP, Smolders AJP, Intven LM, Pol A, denCamp H, vanderDrift C (1997) Formation of dimethyl sulfide and methanethiol in anoxic freshwater sediments. Appl Environ Microbiol 63:4741–4747

Lomans BP, Luderer R, Steenbakkers P, Pol A, van der Drift C, Vogels GD, Op den Camp HJM (2001) Microbial populations involved in cycling of dimethyl sulfide and methanethiol in freshwater sediments. Appl Environ Microbiol 67:1044–1051

Lopes F, Viollier E, Thiam A, Michard G, Abril G, Groleau A, Prévot F, Carrias JF, Albéric P, Jézéquel D (2011) Biogeochemical modeling of anaerobic *vs.* aerobic methane oxidation in a meromictic crater lake (Lake Pavin, France). Appl Geochem 26:1919–1932

Lösekann T, Knittel K, Nadalig T, Fuchs B, Niemann H, Boetius A, Amann R (2007) Diversity and abundance of aerobic and anaerobic methane oxidizers at the Haakon Mosby Mud Volcano, Barents Sea. Appl Environ Microbiol 73:3348–3362

Lovley DR (1987) Organic matter mineralization with the reduction of ferric iron: a review. Geomicrobiol J 5:375–399

Lovley DR (2006) The dissimilatory Fe(III)- and Mn(IV)-reducing prokaryotes. The Prokaryotes, vol. 2 (Springerlink eds.), Springer, New York, pp 635–658

Lovley DR, Klug MJ (1983) Methanogenesis from methanol and methylamines and acetogenesis from hydrogen and carbon-dioxide in the sediments of a eutrophic lake. Appl Environ Microbiol 45:1310–1315

Lovley DR, Phillips EJP (1988) Novel mode of microbial energy metabolism: organic carbon oxidation coupled to dissimilatory reduction of iron and manganese. Appl Environ Microbiol 54:1472–1480

Luton PE, Wayne JM, Sharp RJ, Riley PW (2002) The *mcrA* gene as an alternative to 16S rRNA in the phylogenetic analysis of methanogen populations in landfill. Microbiol 148:3521–3530

Mayr S, Latkoczy C, Kruger M, Gunther D, Shima S, Thauer RK, Widdel F, Jaun B (2008) Structure of an F430 variant from archaea associated with anaerobic oxidation of methane. J Am Chem Soc 130:10758–10767

Mc Inerney MJ, Struchtemeyer CG, Sieber J, Mouttaki H, Stams AJM, Schink B, Rohlin L, Gunsalus RP (2008) Physiology, ecology, phylogeny, and genomics of microorganisms capable of syntrophic metabolism. Ann NY Acad Sci 1125:58–72

McInerney MJ, Sieber JR, Gunsalus RP (2009) Syntrophy in anaerobic global carbon cycles. Curr Opin Biotechnol 20:623–632

McInerney MJ, Sieber JR, Gunsalus R (2011) Microbial syntrophy: ecosystem-level biochemical cooperation. Microbe 6:479–485

Michmerhuizen CM, Striegl RG, McDonald ME (1996) Potential methane emission from north-temperate lakes following ice melt. Limnol Oceanogr 41:985–991

Miller TL, Wolin MJ (1985) *Methanosphaera stadtmaniae* gen. nov., sp. nov.: a species that forms methane by reducing methanol with hydrogen. Arch Microbiol 141:116–122

Müller V, Imkamp F, Rauwolf A, Küsel K, Drake HL (2004) Molecular and cellular biology of acetogenic bacteria. In: Nakano MM, Zuber P (eds) Strict and facultative anaerobes: medical and environmental aspects. Horizon Bioscience, Wymondham, pp 251–281

Müller N, Griffin BM, Stingl U, Schink B (2008) Dominant sugar utilizers in sediment of Lake Constance depend on syntrophic cooperation with methanogenic partner organisms. Environ Microbiol 10:1501–1511

Murase J, Sugimoto A (2005) Inhibitory effect of light on methane oxidation in the pelagic water column of a mesotrophic lake (Lake Biwa, Japan). Limnol Oceanogr 50:1339–1343

Murase J, Sakai Y, Kametani A, Sugimoto A (2005) Dynamics of methane in mesotrophic Lake Biwa, Japan. Ecol Res 20:377–385

Muyzer G, Stams AJM (2008) The ecology and biotechnology of sulphate-reducing bacteria. Nat Rev 6:441–454

Nazaries L, Murrell JC, Millard P, Baggs L, Singh BK (2013) Methane, microbes and models: fundamental understanding of the soil methane cycle for future predictions. Environ Microbiol 15:2395–2417

Nold SC, Boschker HTS, Pel R, Laanbroek HJ (1999) Ammonium addition inhibits C-13-methane incorporation into methanotroph membrane lipids in a freshwater sediment. FEMS Microbiol Ecol 29:81–89

Nozhevnikova AN, Nekrasova V, Ammann A, Zehnder AJB, Wehrli B, Holliger C (2007) Influence of temperature and high acetate concentrations on methanogenensis in lake sediment slurries. FEMS Microbiol Ecol 62:336–344

Nüsslein B, Eckert W, Conrad R (2003) Stable isotope biogeochemistry of methane formation in profundal sediments of Lake Kinneret (Israel). Limnol Oceanogr 48:1439–1446

Ollivier B, Caumette P, Garcia JL, Mah RA (1994) Anaerobic bacteria from hypersaline environments. Microbiol Rev 58:27–38

Omil F, Lens P, Visser A, Hulshoff Pol LW, Lettinga G (1998) Long-term competition between sulfate reducing and methanogenic bacteria in UASB reactors treating volatile fatty acids. Biotechnol Bioeng 57:676–685

Op den Camp HJM, Islam T, Stott MB, Harhangi HR, Hynes A, Schouten S, Jetten MSM, Birkeland N-K, Pol A, Dunfield PF (2009) Environmental, genomic and taxonomic perspectives on methanotrophic *Verrucomicrobia*. Environ Microbiol Rep 1:293–306

Oremland RS, Polcin S (1982) Methanogenesis and sulfate reduction: competitive and noncompetitive substrates in estuarine sediments. Appl Environ Microbiol 44:1270–1276

Orphan VJ, House CH, Hinrichs KU, McKeegan KD, DeLong EF (2001) Methane-consuming archaea revealed by directly coupled isotopic and phylogenetic analysis. Science 293:484–487

Patel GB, Sprott GD (1990) *Methanosaeta concilii* gen. nov., sp. nov. ("*Methanothrix concilii*") and *Methanosaeta thermoacetophila* nom. rev., comb. nov. Int J Syst Bacteriol 40:79–82

Paul K, Nonoh JO, Mikulski L, Brune A (2012) "Methanoplasmatales", Thermoplasmatales-related archaea in termite guts and other environments, are the seventh order of methanogens. Appl Environ Microbiol 78:8245–8253

Pavlov AA, Kasting JF, Brown LL, Rages KA, Freedman R (2000) Greenhouse warming by CH_4 in the atmosphere of early Earth. J Geophys Res 105:11981–11990

Pernthaler A, Orphan VJ (2010) U.S. Patent No. 7736855. Process for separating microorganisms. U.S. Patent and Trademark Office, Washington, DC

Pernthaler A, Pernthaler J, Amann R (2002) Fluorescence *in situ* hybridization and catalyzed reporter deposition for the identification of marine bacteria. Appl Environ Microbiol 68:3094–3101

Pernthaler A, Dekas AE, Brown CT, Goffredi SK, Embaye T, Orphan VJ (2008) Diverse syntrophic partnerships from deep-sea methane vents revealed by direct cell capture and metagenomics. Proc Natl Acad Sci U S A 105:7052–7057

Pol A, Heijmans K, Harhangi HR, Tedesco D, Jetten MSM, den Camp H (2007) Methanotrophy below pH1 by a new *Verrucomicrobia* species. Nature 450:817–874

Poulsen M, Schwab C, Jensen BB, Engberg RM, Spang A, Canibe N, Højberg O, Milinovich G, Fragner L, Schleper C, Weckwerth W, Lund P, Schramm A, Urich T (2013) Methylotrophic methanogenic *Thermoplasmata* implicated in reduced methane emissions from bovine rumen. Nat Commun 4:1428

Raghoebarsing AA, Pol A, van de Pas-Schoonen KT, Smolders AJP, Ettwig KF, Rijpstra WIC, Schouten S, Damste JSS, Op den Camp HJM et al (2006) A microbial consortium couples anaerobic methane oxidation to denitrification. Nature 440:918–921

Ragsdale SW, Kumar M (1996) Nickel-containing carbon monoxide dehydrogenase/acetyl-CoA synthase. Chem Rev 96:2515–2539

Ragsdale SW, Pierce E (2008) Acetogenesis and the Wood-Ljundahl pathway of CO_2 fixation. Biochim Biophys Acta 1784:1873–1898

Ravussin E, Burnand B, Schutz Y, Jéquier E (1982) Twenty-four-hour energy expenditure and resting metabolic rate in obese, moderately obeses and control subjects. Am J Clin Nutr 35:566–573

Reeburgh WS (2003) Global methane biogeochemistry. In: Keeling RF, Holland HD, Turekian KK (eds) Treatise on geochemistry, vol 4, The atmosphere. Elsevier-Pergamon, Oxford, pp 65–89

Scheller S, Goenrich M, Boecher R, Thauer RK, Jaun B (2010) The key nickel enzyme of methanogenesis catalyses the anaerobic oxidation of methane. Nature 465:606–608

Schimel J (2004) Playing scales in the methane cycle: from microbial ecology to the globe. Proc Natl Acad Sci USA 101:12400–12401

Schink B (1997) Energetics of syntrophic cooperations in methanogenic degradation. Microbiol Mol Biol Rev 61:262–280

Schink B, Stams AJM (2001) Syntrophism among prokaryotes. In: Dworkin M, Falkow S, Rosenberg E, Schleifer K-H, Stackebrandt E (eds) The prokaryotes: an evolving electronic resource for the microbiological community, 3rd edn. Springer, New York, p 25

Schonheit P, Keweloh H, Thauer RK (1981) Factor F420 degradation in *Methanobacterium thermoautotrophicum* during exposure to oxygen. FEMS Microbiol Lett 12:347–349

Schubert CJ, Vazquez F, Lösekann-Behrens T, Knittel K, Tonolla M, Boetius A (2011) Evidence for anaerobic oxidation of methane in sediments of a freshwater system (Lago di Cadagno). FEMS Microbiol Ecol 76:26–38

Schulz S, Conrad R (1996) Influence of temperature on pathways to methane production in the permanently cold profundal sediment of Lake Constance. FEMS Microbiol Ecol 20:1–14

Shindell DT, Faluvegi G, Koch DM, Schmidt GA, Unger N, Bauer SE (2009) Improved attribution of climate forcing to emissions. Science 326:716–718

Sieber JR, McInerney MJ, Gunsalus RP (2012) Genomic insights into syntrophy: the paradigm for anaerobic metabolic cooperation. Annu Rev Microbiol 66:429–452

Smith KS, Ingram-Smith C (2007) *Methanosaeta*, the forgotten methanogen? Trends Microbiol 15:150–155

Smith RL, Howes BL, Garabedian SP (1991) *In situ* measurement of methane oxidation in groundwater by using natural-gradient tracer tests. Appl Environ Microbiol 57:1997–2004

Sohngen NL (1906) Uber bakterien welche methan ab kohlenstoffnahrung und energiequelle gerbrauchen (Les bactéries qui utilisent le méthane comme source d'énergie et de carbone). Z Bakteriol Parazitenk (Infektionster) 15:513–517

Sprenger WW, van Belzen MC, Rosenberg J, Hackstein JHP, Keltjens JT (2000) *Methanomicrococcus blatticola* gen. nov., sp. nov., a methanol- and methylamine-reducing methanogen from the hindgut of the cockroach *Periplaneta americana*. Int J Syst Evol Microbiol 50:1989–1999

Sprenger WW, Hackstein JHP, Keltjens JT (2005) The energy metabolism of *Methanomicrococcus blatticola*: physiological and biochemical aspects. Antonie Van Leeuwenhoek 87:289–299

Stams AJM, Plugge CM (2009) Electron transfer in syntrophic communities of anaerobic bacteria and archaea. Nat Rev Microbiol 7:568e577

Stephenson M (1947) Some aspects of hydrogen transfer. Ant v Leeuwenhoek 12:33–48

Stoecker K, Bendinger B, Schoning B, Nielsen PH, Nielsen JL, Baranyi C, Toenshoff ER, Daims H, Wagner M (2006) Cohn's *Crenothrix* is a filamentous methane oxidizer with an unusual methane monooxygenase. Proc Natl Acad Sci U S A 103:2363–2367

Tewes FJ, Thauer RK (1980) Regulation of ATP synthesis in glucose fermenting bacteria involved in interspecies hydrogen transfer. In: Gottschalk G, Pfennig N, Werner H (eds) Anaerobes and anaerobic infections. G. Fischer, Stuttgart, pp 97–104

Thauer RK, Shima S (2008) Methane as fuel for anaerobic microorganisms. Ann NY Acad Sci 1125:158–170

Thauer RK, Kaster AK, Seedorf H, Buckel W, Hedderich R (2008) Methanogenic archaea: ecologically relevant differences in energy conservation. Nat Rev Microbiol 6:579–591

Trembath-Reichert E, Green-Saxena A, Orphan VJ (2013) Whole cell immunomagnetic enrichment of environmental microbial consortia using rRNA-targeted Magneto-FISH. Methods Enzymol 531:21–44

Trotsenko YA, Murrell JC (2008) Metabolic aspects of aerobic obligate methanotrophy. Adv Appl Microbiol 63:183–229

Van Bodegom PM, Scholten JCM, Roden EE, Stams AJM (2004) Direct inhibition of methanogenesis by ferric iron. FEMS Microb Ecol 49:261–268

Von Stockar U, Marison IW (1989) The use of calorimetry in biotechnology. In: Fiechter A (ed) Advances in biochemical engineering/biotechnology, vol 40. Springer, Berlin, pp 93–134

Wadsö I (1985) Isothermal biocalorimetry. A status report. Thermochimica Acta 88:35–48

Wang JS, Logan JA, McElroy MB, Duncan BN, Megretskaia IA, Yantosca RM (2004) A 3-D model analysis of the slowdown and interannual variability in the methane growth rate from 1988 to 1997. Global Biogeochem Cycles 18:B3011. doi:10.1029/2003GB002180

Webster G, Sass H, Cragg BA, Gorra R, Knab NJ, Green CJ, Mathes F, Fry JC, Weightman AJ et al (2011) Enrichment and cultivation of prokaryotes associated with the sulphate methane transition zone of diffusion controlled sediments of Aarhus Bay, Denmark under heterotrophic conditions. FEMS Microbiol Ecol 77:248–263

Wetzel RG (2001) Limnology – Lake and river ecosystems, 3rd edn. Academic, San Diego, p 1006

Whiticar MJ (1999) Carbon and hydrogen isotope systematics of bacterial formation and oxidation of methane. Chem Geol 16:1291–1314

Widdel F (1988) Microbiology and ecology of sulfate- and sulfur-reducing bacteria. In: Zehnder AJB (ed) Biology of anaerobic microorganisms. Wiley, New York, pp 469–585

Ye WJ, Liu XL, Lin SQ, Tan J, Pan JL, Li DT, Yang H (2009) The vertical distribution of bacterial and archaeal communities in the water and sediment of Lake Taihu. FEMS Microbiol Ecol 70:263–276

Yvon-Durocher G, Allen AP, Batsviken D, Conrad R, Gudasz C, St-Pierr A, Thanh-Duc N, del Giorgio P (2014) Methane fluxes show consistent temperature dependence across microbial to ecosystem scales. Nature 507:488–491

Zeikus JG (1977) The biology of methanogenic bacteria. Bacteriol Rev 41:514–541

Zeikus, JG (1983) Metabolic communication between biodegradative populations in nature. In: Slater JH, Whittenbury R, Wimpenny, JWT (eds) Microbes in their natural environments. Soc Gen Microbiol Symp 34:423–462. Cambridge, UK: Cambridge University Press

Zhang Y, Maignien L, Zhao X, Wang F, Boon N (2011) Enrichment of a microbial community performing anaerobic oxidation of methane in a continuous high-pressure bioreactor. BMC Microbiol 11:137–145

Zinder SH (1993) Physiological ecology of methanogens. In: Ferry JG (ed) Methanogenesis. Chapman & Hall, New York, pp 128–206

Zinder SH, Brock TD (1978) Production of methane and carbon dioxide from methane thiol and dimethyl sulfide by anaerobic lake-sediments. Nature 273:226–228

Zumft WG (1992) The denitrifying prokaryotes. In: Balows A, Trüper HG, Dworkin M, Harder W, Schleifer KH (eds) The prokaryotes. Springer, New York, pp 554–582

Zumft WG (1997) Cell biology and molecular basis of denitrification. Microbiol Mol Biol Rev 61:533–616

New Insights into the Microbial Contribution to the Chlorine Cycle in Aquatic Ecosystems

17

Eric Dugat-Bony, Pierre Peyret, and Corinne Biderre-Petit

Abstract

Microorganisms hold key positions in ecosystem functioning, and thus in **biogeochemical cycles**. Among these cycles, some, such as chlorine (Cl), are still poorly understood. Recent works have revealed that natural chlorination and dechlorination of organic matter (OM) in most of the ecosystems were much more extensive and ubiquitous than previously suggested. Currently, there are clear evidences that natural chlorination is tightly linked to different defence mechanisms and antagonistic reactions among microorganisms. Likewise, it has been clearly demonstrated that organochlorine (Cl_{org}) formation is also linked to OM degradation, possibly affecting carbon cycle. The chlorination rate of OM depends on several parameters including OM content and quality, microbial activity, chloride (Cl^-) input and pH. Once produced, Cl_{org} undergoes oxidative or reductive degradation in the environment depending on the surrounding physico-chemical conditions. Among all enzyme-mediated processes described, the organohalide respiration (an anaerobic bacterial respiratory process) is the only known mechanism leading to the removal of halogens from highly chlorinated compounds, transforming them into biodegradable metabolites. However, despite a significant growth in the literature since the early 1990s, the biogeochemistry of Cl in natural environment is still poorly documented. For instance, the Cl cycling in aquatic environments including Cl^- and Cl_{org} pools in sediment and water, are largely missing. The present chapter seeks to review the literature on the natural Cl cycling in environment, with a focus on a freshwater ecosystem, the Lake Pavin.

Keywords

Freshwater • Chlorine • Biogeochemical cycle • Chlorination • Dechlorination • Organohalide respiration • Abiotic • Biotic

E. Dugat-Bony
INRA-AgroParisTech, UMR782 Génie et Microbiologie des Procédés Alimentaires, route Lucien Brétignière, Thiverval-Grignon 78850, France

P. Peyret
Clermont-Université, Université d'Auvergne, 28 place Henri Dunant, EA CIDAM 4678, BP 10448, Clermont-Ferrand 63001, France

C. Biderre-Petit (✉)
LMGE, Laboratoire Microorganismes: Génome et Environnement, UMR CNRS 6023, Université Clermont-Auvergne, Université Blaise Pascal, Campus Universitaire des Cézeaux, 1 Impasse Amélie Murat – TSA 60026, Aubière Cedex CS 60026, 63178, France
e-mail: corinne.petit@univ-bpclermont.fr

17.1 Introduction

Chlorine (Cl) is one of the twenty most abundant elements on earth. Together with fluorine (F), bromine (Br), and iodine (I), Cl belongs to the **halogen group** in the periodic table. Cl is ubiquitous and naturally present in our environment as well as carbon (C), hydrogen (H), oxygen (O), nitrogen (N) phosphorus (P), and sulphur (S). Cl is essential to most forms of life for diverse reasons (Winterton 2000). Chloride (Cl^-) is the only stable ionic form of Cl. Cl^- is the main anion in blood, and is present at ~100 mmol . L^{-1} in plasma and interstitial fluid (Yunos et al. 2010). Cl^- takes part to the

T. Sime-Ngando et al. (eds.), *Lake Pavin*, DOI 10.1007/978-3-319-39961-4_17

osmoregulation of cells (White and Broadley 2001), and is an important electrolyte for regulation of muscle function and synaptic transmission in the neural system of animals. Cl^- also functions as an essential co-factor in enzymes involved in photosynthesis related to the oxidation of water by the PSII photosystem (Dismukes 1986; Winterton 2000). Thereby, Cl is a critical nutrient and a suggested minimum requirement of Cl for crops is 1 g . kg^{-1} dry mass (d.m.) (White and Broadley 2001).

Some decades ago, the general opinion was that chlorinated organic compounds (Cl_{org}) detected in ecosystems should have **anthropogenic** sources and toxic effects. Indeed, many of the most debated organic pollutants are chlorinated (Godduhn and Duffy 2003). But since then, several surveys showed that organic-bound chlorine is more widespread than previously identified and that not all Cl_{org} can be explained by pollution. This assumption was confirmed by a lot of studies. Currently nearly 5000 naturally produced Cl_{org} have been identified and chemically characterized. The production of Cl_{org} has been associated with bacteria, fungi, lichen, plants, marine organisms of all types, insects, and higher animals including humans (Öberg 2002; Gribble 2003; Wagner et al. 2009; Gribble 2010). Some of these have well known physiological functions, including several important antibiotics such as vancomycin and chloramphenicol. Others, such as short-chain volatile Cl_{org} (VOCls), have important effects in the environments because they can, for instance, enhance atmospheric ozone destruction (Winterton 2000). In addition, Cl represents the sixth or seventh most common element of soil OM (SOM) (0.01–0.5 %), at the same level as for P (0.03–0.2 %) and only slightly lower than N (1–5 %) and S (0.1–1.5 %) (Hjelm et al. 1995; Öberg 2002). However, the ecological functions of most Cl_{org} in nature, and the reasons for its production, are largely unknown.

On the earth's surface, the largest Cl reservoirs are the crust and the ocean (Graedel and Keene 1996). Inorganic Cl by far dominates these reservoirs. Soil (pedosphere), freshwater (rivers, lakes and groundwater), oceans, cryosphere and atmosphere (troposphere and stratosphere) (Graedel and Keene 1996; Öberg 2003; Svensson et al. 2007) are the other important reservoirs. Estimates for these reservoirs are also largely based on Cl^- concentration measurements. This assumption of a general dominance of Cl^- is problematic for the pedosphere because Cl_{org} levels have been shown to range from 11 to near 100 % of the total Cl pool in a large range of soil types (Johansson et al. 2003; Redon et al. 2011, 2013; Gustavsson et al. 2012).

All these compartments are linked by the **hydrological cycle**. Although the greatest quantity and variety of naturally-produced chlorinated compounds are found in the marine environment, it is now well accepted that terrestrial and non-marine aquatic ecosystems receive significant fluxes of Cl from the deposition of sea-salt aerosol, through wet and dry precipitations, from **leaching** processes or from biological processes linked to the formation or the degradation of the OM. However, the terrestrial biogeochemistry of Cl and its cycle between soil, water, biota and dead OM is still ill-understood. Most of previous studies have primarily dealt with Cl^- and Cl_{org} separately because they were considered not connected to each other. However, currently, there is irrefutable evidence that Cl undergoes a more complex **biogeochemical cycling** than expected in terrestrial environments where Cl^- can be biotically or abiotically transformed into Cl_{org} in nature, and vice versa.

17.2 Chlorine Compounds Origin and Formation in Aquatic Ecosystems

17.2.1 Chloride Ions

17.2.1.1 Chloride Ions in the Environment

Cl^- has long been believed to take part in geochemical processes only, *i.e.* carried from oceans *via* soil back to the oceans again, being only negligibly affected by biological processes or interactions with OM. Cl^- in the environment can originate from some common chloride salts such as sodium chloride (NaCl), potassium chloride (KCl) and magnesium chloride ($MgCl_2$) (used for de-icing of roads and walkways), calcium chloride ($CaCl_2$) (used as a dust suppressant on roads), aluminium chloride ($AlCl_2$) (used in municipal drinking water and wastewater treatment facilities) as well as ferric chloride ($FeCl_3$) (used at municipal wastewater treatment plants). Cl^- containing compounds are highly soluble in water and hence they easily dissociate and tend to remain in their ionic forms once dissolved in water (*e.g.* Na^+ and Cl^-). Cl^- is the dominating chlorine pool globally and has a high enrichment factor when comparing oceanic and riverine concentrations (*i.e.* sea water concentrations are in the order of 2500 times larger than freshwater concentrations; Winterton 2000). At the first glance this indicates that Cl^- is unreactive in ecosystems and this has been a prevailing view for a long time (*e.g.* White and Broadley 2001). Accordingly, Cl^- has been seen as an inexpensive and suitable tracer of soil and ground water movements (Herczeg and Leaney 2011; Hruška et al. 2012) and studies using Cl^- as a water tracer has been a foundation for contaminant transport models (*e.g.* Kirshner et al. 2000). However, this view has gradually been revised during the last few decades and there is now clear evidence that Cl^- is highly reactive in some ecosystems (Öberg 2002; Lovett et al. 2005). For instance, in soil, some type of 'sorption' process is present and the retention of Cl^- in this compartment has been reported (Bastviken et al. 2007). Hence, the concentrations of Cl_{org} in surface soil layers are in most cases higher than Cl^- levels (Johansson et al. 2003; Redon et al. 2011, 2013;

Gustavsson et al. 2012). Thus, a large proportion of Cl^- deposited in terrestrial ecosystems can be transformed into Cl_{org} in soil, but also certainly in other compartments such as aquatic ones, although the underlying processes are not yet completely elucidated (Öberg 2002; Öberg et al. 2005).

In contrast to soils, Cl^- concentrations generally exceed Cl_{org} concentrations in water. For instance, the Cl^- concentration in diverse waters is measured in mg . L^{-1}, while Cl_{org} is typically measured in µg . L^{-1} and VOCls are in the range of ng . L^{-1} (Eriksson 1960; Asplund and Grimvall 1991; Enell and Wennberg 1991; McCulloch 2003; Svensson et al. 2007). This means that the atmospheric deposition of Cl_{org} is in the order of 1000-fold lower than deposition of Cl^- and thereby often assumed to be negligible from a bulk chlorine perspective. While ground water has the highest Cl^- concentrations in comparison with rain water and surface waters, Cl_{org} and VOCl concentrations can be highest in surface waters. The environmental quality criteria with regard to Cl^- levels in groundwater published by Swedish food agency use a Cl^- threshold level of 100 mg. L^{-1}. Regarding the lakes, the ambient Cl^- concentrations in the Atlantic region of Canada are normally <10 mg . L^{-1} in inland lakes, with concentrations as high as 20–40 mg . L^{-1} in lakes located closer to coastal areas (Mayer et al. 1999). Unimpacted lakes on Canadian shield of Canada's central region have measured Cl^- concentrations of <1–7 mg . L^{-1}, with higher concentrations (20–40 mg . L^{-1}) measured in the lower Great Lakes and St Lawrence River. However, Cl^- concentrations above background are commonly detected in densely populated areas, and result usually from human activities. Indeed, anthropogenic Cl^- input from irrigation and fertilization can represent substantial inputs to terrestrial ecosystems. Other anthropogenic sources include application of chloride brine solutions for dust in summer, water softeners, industrial effluent, domestic sewage, or yet landfill leachate. Moreover, since the start of de-icing of roads in mid-twentieth century, studies have shown increased Cl^- concentrations in both surface water and groundwater in the vicinity of roads. In Canada, elevated concentrations of Cl^- associated with de-icing have been documented in groundwater, wetlands, streams and ponds adjacent to snow dumps and salt-storage areas, and also those draining major roadways and urban areas. In the Laxemar-Simpevarp area in South East Sweden, 35–56 % of the total Cl^- input was estimated to come from road salt (Tröjbom et al. 2008).

17.2.1.2 Chloride ions in Lake Pavin

Still to date, the water balance of Lake Pavin is not really well constrained except for the water inputs by direct precipitation (Q_p) onto the lake surface which are estimated to be 18 L . s^{-1} (Fig. 17.1). Those from the main surface streams feeding the lake (Q_r) are estimated to be about 20 L . s^{-1} (Aeschbach-Hertig et al. 2002). The water outflow (Q_{out}) *via* the surface outlet is estimated to a mean value of 50 L . s^{-1} and the water outputs by evaporation (Q_{ev}), to be 8 L . s^{-1}. The difference between water inputs and outputs leads to a deficit of about 20 L . s^{-1}, which is assumed to be balanced by two sub-surface springs, one located in the mixilimnion (Q_{45}) and the second in the monimolimnion (Aeschbach-Hertig et al. 2002; Assayag et al. 2008). Concerning Cl content in Lake Pavin, only Cl^- concentrations were measured regularly in water column over the past decades because Cl^- was used as a tracer of water circulation. Its concentrations vary from 1.7 to 2.1 mg . L^{-1} along the water column. Cl^- content is almost 30 % higher in the deep anoxic waters than in surface waters (Assayag et al. 2008; Jézéquel et al. 2012). The main streams and springs feeding the lake display similar Cl^- contents (~1–4 mg . L^{-1}). No significant increase of Cl^- concentrations was detected between 1992 (~1.6 mg . L^{-1} in the mixolimnion and ~2.1 mg . L^{-1} in the monimolimnion) (Michard et al. 1994) and 2006 (~1.7 mg . L^{-1} in the mixolimnion and ~2.1 mg . L^{-1} in the monimolimnion) (Jézéquel et al. 2012) which may suggest a relatively good preservation of this ecosystem from human activities.

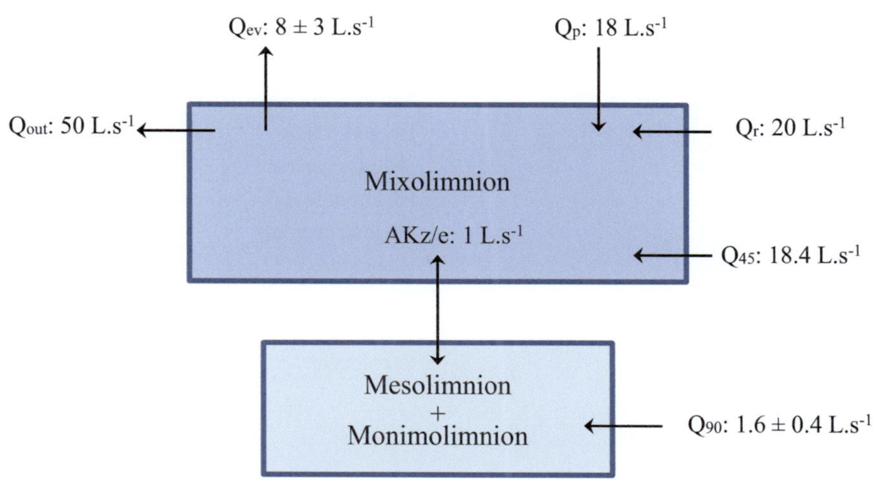

Fig. 17.1 Summary figure recapitulating the hydrological budget of lake Pavin, with the water inputs: precipitation (Qp), surface streams (Qr), sub-surface springs located in the mixolimnion (Q45) and in the monimolimnion (Q90) and water outputs: evaporation (Qev), output (Qout), and their respective water flow rates. AKz/e is the turbulent water exchange term between the monimolimnion and the mixolimnion (from Assayag et al. 2008)

17.2.2 Processes of OM Chlorination

Oceans, with a water volume of about 1.36×10^8 km^3 and a Cl content of approximately 26×10^{15} tons (t), are the primary reservoir for this element in Earth's hydrosphere. In comparison, surface waters (including lakes and rivers) have a volume estimated at about 1×10^5 km^3 and contain approximately 58×10^7 t of Cl (Graedel and Keene 1996; Bastviken et al. 2013). As ecosystems are open systems, Cl can be supplied and be taken away from the system by various ways. In surface waters, Cl_{org}, like Cl^-, originate from atmospheric precipitations but it is not certainly the main source. Other inter-reservoir transfer fluxes have been identified (Table 17.1, Graedel and Keene 1996; Winterton 2000). Rock-water interactions (dissolution and desorption), thermal and mineral springs in volcanic areas but also, to a lesser extent, **watershed runoff** or leaching can further increase Cl_{org} concentrations in surface waters (Svensson et al. 2007). In addition, higher molecular weight OM found in soil, groundwater and sediment (from rivers, reservoirs and lakes) has a significant Cl_{org} content believed to be of natural and sometimes ancient origin. These Cl_{org} range from peptides, polyketides, indoles, terpenes, acetogenins and phenols to volatile VOCls (for example chloroform) that are produced on a very large scale (Edwards et al. 2004; Gribble 2010). The majority, not biologically active, can be easily transformed into smaller chlorinated by-products or completely degraded by various organisms (Öberg 2002). Finally, a multitude of complex biotic and abiotic transformation processes takes place directly in surface waters and can also lead to the production of a wide variety of Cl_{org}.

17.2.2.1 Abiotic Chlorination

Although most of the natural chlorinated compounds identified so far are without doubt the results of reactions mediated by enzymes, there is also some evidence of the existence of abiotic formation processes leading to a large number of Cl_{org} (Keppler et al. 2000; Hamilton et al. 2003). These compounds include the majority of chlorinated hydrocarbons and their derivatives formed as a result of various geothermal (*e.g.* volcano, burning, lightning, erosion) and geochemical (*e.g.* Fenton reaction) processes.

Sources from Geothermal Processes

Several geothermal processes can lead to the production of Cl_{org}. Among these ones, volcanism may be a significant natural source of an extraordinarily large array of Cl_{org}. Indeed, few hundreds of organic substances were detected in fumarolic and lava gases of several volcanoes, of which at least 100 were chlorinated (Jordan et al. 2000). The chloromethanes, chloroethenes, and chlorobenzenes represent the most concentrated molecule families. For instance, the chloroform concentrations ($CHCl_3$) detected in volcanic gases (up to 40 nmol . L^{-1}) are between 1.5 and 2 orders of magnitude higher than those observed above the oceans (Isidorov 1990; Laturnus et al. 2002). Chlorofluorocarbons (CFCs) were also identified but in negligible concentrations as compared to the anthropogenic CFC burden (Jordan et al. 2000).

Another source of Cl_{org} is rocks where they are present in gas pockets or are components of some minerals (apatite, biotite, hornblende among others). Thus when rocks are crushed or as a result of **wheathering** processes, small quantities of Cl_{org} such as methyl chloride, dichloromethane

Table 17.1 Estimated flux (10^6 t per year) of chlorine between reservoirs (Graedel and Keene 1996; Winterton 2000)

Inter-reservoir transfer		Flux
Mantle to troposphere		2
Pedosphere to troposphere	Mineral aerosol	15
	Biomass burning	3
	Bioproduced	0.5
Crust to freshwater		175
Pedosphere to freshwater	Precipitation passtrough	34
	Evaporite beds	11
Ocean to troposphere	Seasalt injection	6000
	HCl from seasalt (a proportion of seasalt injection flux)	25
	Magma intrusion	4
	Bioproduced	2
Troposphere to surface	Oceans	5990
	Pedosphere	34
	Cryosphere	6
Troposphere to stratosphere		0.03
Stratosphere to troposphere		0.03
Oceans to crust		17

(DCA) or also carbon tetrachloride can be released (Gribble 2004). Some human activities can also accelerate these natural processes. For instance, we estimate that potassium-salt mining alone liberates thousands of tons of $CHCl_3$ per year.

Finally, significant amount of abiotic Cl_{org} originates from biomass burning, though a minor fraction of these events is considered to be entirely natural, caused for example by volcanic eruptions or lightning. Forest fires are known to produce chloromethane, polychlorinated dibenzodioxins (PCDDs) and polychlorinated dibenzofurans (PCDFs) (Kim et al. 2003).

Regarding Lake Pavin, water inputs by direct precipitation onto the lake surface are currently reasonably well constrained (Aeschbach-Hertig et al. 2002) and could supply the ecosystem in a wide variety of Cl_{org} of abiotic origin emanating both from natural sources and anthropogenic activities (Fig. 17.2). Furthermore, because the lake is set in a maar crater and thus has volcanic structures and because it is fed by several sub-surface mineral springs in the monimolimnion and in the mixolimnion (Aeschbach-Hertig et al. 2002; Assayag et al. 2008), it is reasonable to consider erosion or weathering processes as possible sources of Cl_{org}, but certainly in negligible amount.

OM Chlorination Through Chemical Processes

In forest soil, spontaneous **chlorination** of OM can occur during degradation by an oxidant (*e.g* FeIII) in presence of Cl^- but without microbial mediation (Keppler et al. 2000). The thermodynamically labile OM is oxidized and the redox partner (for example iron) is reduced. During this process, halides such as Cl^- are methylated and the resulting methyl chlorides represent degradation products of oxidized OM. One other way of chlorination can be explained by the Fenton reaction. In the presence of hydrogen peroxide (H_2O_2) and iron, Cl^- can react with OM to form chloroacetic acids. There is evidence for the formation of four classes of Cl_{org} through these processes: volatyl alkyl chlorides, chloroacetates, PCDDs and chlorinated humic substances (Fahimi et al. 2003).

Lake Pavin being continuously fed by OM, essentially plant material (leaves and plant debris) coming from the watershed densely covered by mixed deciduous-coniferous forest, several processes may support the presence of Cl_{org} in the water column (Fig. 17.2). First, OM decomposition in the water column could lead to the release of Cl_{org} in this zone. Furthermore, chemical processes described above, *i.e.* spontaneous chlorination during oxidative degradation of OM and Fenton reaction, could also take place, though this has never been demonstrated. Indeed, labile OM might also react with Cl^- and iron hydroxides, both present in the water column (Viollier et al. 1995; Bura-Nakic et al. 2009), to produce Cl_{org}.

17.2.2.2 Biochlorination

To date, evidence suggests that the chlorination rate is largely controlled by biosynthetic processes which are mainly conducted by prokaryotes and single cell eukaryotes (Bastviken et al. 2007). They use this strategy to increase the biological activity of secondary metabolites, compounds that are often effective like drugs (Bengtson et al. 2009, 2013). The Cl_{org} could thus be used by the organism (i) to depolimerize the

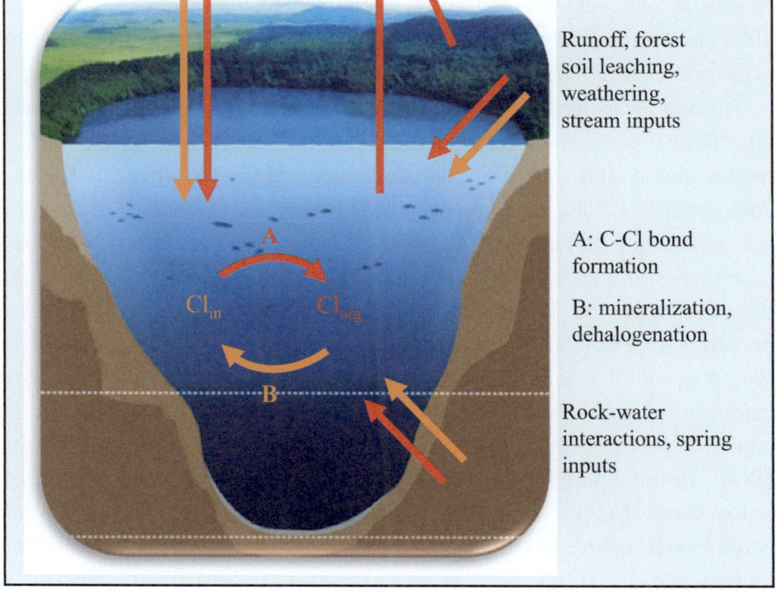

Fig. 17.2 Chlorine cycle in lake Pavin area with transformation processes of chloride (Cl^-) depicted in orange and organically bound chlorine (Cl_{org}) in red

lignocellulose in order to access carbon, a limited resource in some environments, (ii) to dissolve cell wall for cell penetration, (iii) as a chemical defence (*e.g.* antibiotics), (iv) to compete with other organisms for limited resources by means of antagonistic interactions or (v) yet as a defence against oxygen stress.

Although literature is most impressive on marine algae and terrestrial fungi, bacteria and lichens, reports on halogenating organisms living in fresh water also exist. Aquatic organisms can produce Cl_{org} by enzymatic-mediated processes associated with the transformation and degradation of OM, particularly fulvic and humic acids (Asplund and Grimvall 1991; Öberg 2002). However, Cl^- is not particularly reactive unless it is activated, typically by oxidation. Chlorinating enzymes have been discovered from a broad range of organisms and mainly can be grouped into two classes, *i.e.* the less specific chloroperoxidases (CPOs) utilizing hydrogen peroxide (H_2O_2) and the highly substrate-specific halogenases requiring O_2 for enzymatic activity. In O_2-dependent halogenases, either flavin (FADH2-dependent halogenases) or α-ketoglutarate (αKG) (non-heme iron (Fe_{NH})/αKG/O_2-dependent halogenases) (Vaillancourt et al. 2005a, b) are found to function as co-substrates. Furthermore, other enzymes requiring S-adenosyl-L-methionine (SAM) as catalyst have been identified to be involved in chlorination (Eustaquio et al. 2008).

OM Chlorination Through Chloroperoxidases (CPOs)

Haloperoxidases have traditionally been classified on the basis of the most electrophilic halide that is readily oxidized. Thus, CPOs oxidize chloride, bromide and iodide by H_2O_2, whereas iodoperoxidases oxidize only iodide in this way. Hydrogen peroxide lacks the thermodynamic potential to oxidize fluoride; thus, enzymes catalysing fluorination are not peroxidases. The overall stoichiometry of the haloperoxidase reaction is consumption of one equivalent of H_2O_2 per halogenated compound produced. The reaction is:

$$X^- + H_2O_2 + R-H + H^+ \rightarrow R-X + 2H_2O\left(X:\text{halogen atom}, R:\text{alkyl group}\right).$$

However, CPO can also catalyse a number of oxidation reactions in the absence of Cl^-. These oxidation reactions are generally carried out at neutral pH, whereas chlorination occurs only at low pH. Indeed, the structural features shared with both heme peroxidases and cytochromes P450 make CPOs the most versatile of the known heme enzymes. In addition to catalyzing chlorination, CPOs also catalyse characteristic reactions of heme peroxidases (dehydrogenation), catalases (H_2O_2 dismutation) and cytochromes P450 (monooxygenation) (Grisham 1991; Rai et al. 2001). Most importantly, CPOs are especially adept in catalyzing the stereoselective epoxidation of alkenes (Allain et al. 1993), benzylic hydroxylation (Zaks and Dodds 1995), propargylic oxidation of 2-alkyenes to chiral alcohol (Hu and Hager 1999) and oxidation of organic sulfides to chiral sulfoxides (Trevisan et al. 2004).

Among the current known CPOs, the heme-depedent and non-heme vanadium-dependent ones require the transition metals heme and vanadium as cofactors, while the few cofactor-free CPOs, also named perhydrolase enzymes, do not require any metal cofactor.

Heme-iron CPOs The first halogenating enzyme to be discovered, in the 1960s, was the heme (iron-containing porphyrin) CPO from the terrestrial fungus *Caldariomyces fumago*, which produces the chlorinated natural product caldariomycin (Shaw and Hager 1959). The optimal pH for chlorination turnover by heme CPO is pH 2.7 (Hager et al. 1966). In this reaction, the heme iron centre functions as a redox catalyst (Fig. 17.3). Hydrogen peroxide oxidizes the heme Fe(III) centre to compound I, the Fe(IV)-oxo π-cation radical species (O=Fe(IV)-heme+•), *via* the short-lived compound-0 state, characterized as a peroxo-anion complex, HOO–Fe(III)-heme (Wagenknecht and Woggon 1997). Glutamate residue 183 is proposed to assist in formation of both compound 0 and compound I. Compound I oxidizes Cl^- by two electrons, reforming the heme Fe(III) centre and generating an oxidized chlorine intermediate that is formally at the oxidation level of the hypochlorite anion (OCl^-). The oxidized chlorine intermediate can then chlorinate the organic substrate or react with a second equivalent of H_2O_2, producing O_2 (in the singlet excited state).

Heme CPOs are produced by organisms generally associated with dead plant material (wood and litter decomposers) such as bacteria belonging to *Actinomycetes* and various fungi such as ascomycetes and most of decomposing basidiomycetes (Verhagen et al. 1996). This process is also largely used by the bacteria to synthesize antibiotics.

Vanadium chloroperoxidases The second sub-class of CPOs, called vanadium chloroperoxidases (V-CPOs), requires the transition metal vanadium as the necessary cofactor, instead of a heme-iron group in its reactive center. V-CPOs have been isolated from fungi (van Schijndel et al. 1993) and are predicted to be present in marine bacteria (Winter et al. 2007). Although chlorinated natural products have not been isolated from the fungi containing V-CPO, these enzymes may be important in fungus-mediated chlorination and degradation of lignin (Ortiz-Bermúdez et al. 2007). In the terrestrial environment, V-CPOs are widespread among plant-associated microorganisms (living plants or decomposing plant material) such as *Streptomyces* and *Pseudomonas*, and symbiotic or parasitic *Bradyrhizobium, Burkholderia, Cupriavidus, Frankia* and *Rhizobium* spp.

They may catalyze the synthesis of chlorinated antibiotics in bacteria (Bernhardt et al. 2011).

V-CPOs seem to be exclusively exo-enzymes. In contrast to the heme CPOs that function as redox catalysts, V-CPOs function as Lewis acid catalysts of Cl^- oxidation by H_2O_2. The catalytic reaction is initiated by coordination of one equivalent of H_2O_2 to the resting V(v) state of the enzyme (Fig. 17.4) (Butler and Sandy 2009). The X-ray structure of the peroxo adduct of V-CPO reveals that a Lysine side chain is hydrogen bonded to the coordinated peroxide. This is probably an essential feature of the catalytic reaction because it would increase the potential of the oxoperoxo-V(v) centre for Cl^- oxidation. The oxo-peroxo-V(v) species can then oxidize Cl^- by two electrons, forming an oxidized halogen that is formally at the OCl^- oxidation state. Electrophilic halogenation results from reaction of OCl^- either with the organic substrate or, in the absence of a good organic substrate, with a second equivalent of H_2O_2, forming O_2 and Cl^-.

Cofactor-free CPOs (or perhydrolases) The perhydrolases had been previously known as prosthetic group-free bacterial haloperoxidases, particularly non-metal and non-heme haloperoxidases (Song et al. 2006). They have the same catalytic triad as serine hydrolases (Ser-Asp-His) and belong to the large and diverse α/β hydrolase fold enzyme class. These enzymes catalyze the reversible formation of peroxyacids from carboxylic acids and H_2O_2 in aqueous

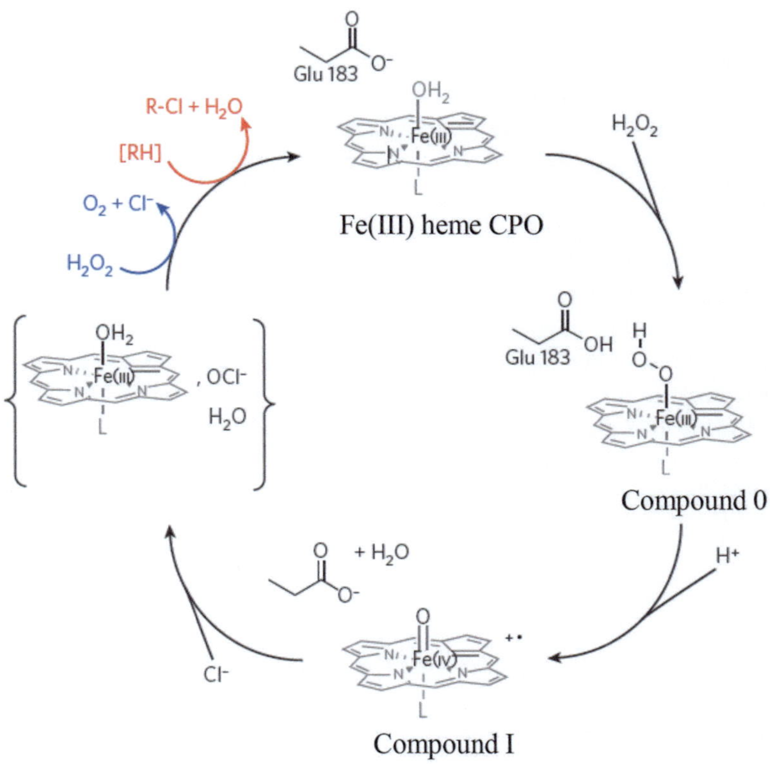

Fig. 17.3 Proposal catalytic cycle for heme CPOs (from Butler and Sandy 2009). The Fe(III)-heme resting state is oxidized by H_2O_2, forming compound I. Compound I oxidizes Cl^- by two electrons, reforming the heme Fe(III) centre and generating an oxidized chlorine intermediate that is formally at the oxidation level of OCl^-. This oxidized chlorine intermediate could chlorinate the organic substrate (shown in red) or oxidize a second equivalent of H_2O_2 (shown in blue), depending on the reaction conditions. L, cysteine

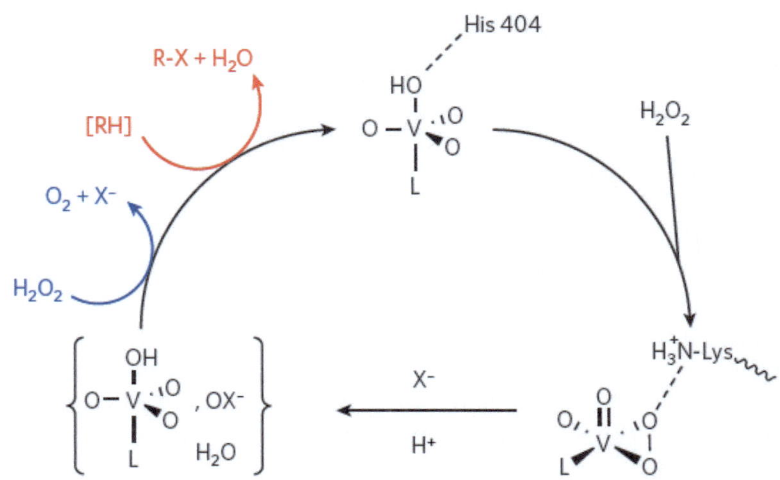

Fig. 17.4 Proposal catalytic cycle for Vanadium-CPOs (from Butler and Sandy 2009). The catalytic reaction is initiated by coordination of one equivalent of H_2O_2 to the resting V(v) state of the enzyme. In red: chlorination of an organic substrate (RH); in blue: oxidation of a second equivalent of H_2O_2, depending on the reaction conditions. L, ligand

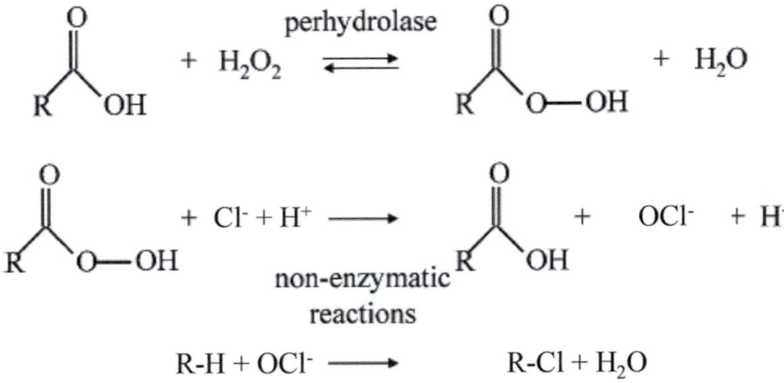

Fig. 17.5 Scheme of the halogenation reaction catalyzed by perhydrolases (Song et al. 2006)**.** The enzymatic formation of peroxyacids in the presence of H_2O_2 is followed by the non-enzymatic oxidation of substrates, for example that of halide ions (Cl^-) to hypohalite ions (OCl^-). The subsequent chlorination reaction is finished by the incorporation of OCl^- into organic compounds (R-H). This perhydrolase reaction that can provide the peroxyacid for the oxidation is not regio-, chemo- or stereoselective

media (Fig. 17.5) (Kirk and Conrad 1999). The peroxyacids oxidize Cl^- to OCl^-, which then halogenates organic substrates non-specifically. Only few bacterial perhydrolase genes have been sequenced and characterized from *Pseudomonas* (Wiesner et al. 1988; Kirner et al. 1996), *Serratia* (Burd et al. 1995), *Burkholderia* (Song et al 2006) and *Streptomyces* genera (Bantleon et al. 1994; Pelletier et al. 1994).

OM Chlorination Through $FADH_2$-Dependent Halogenases (FDHs)

The second class of enzymes generating group of halogenating enzyme using oxidative mechanisms is FDHs. These enzymes used O_2 as the oxidant, flavin as the redox active co-factor and the presence of a nicotinamide adenine dinucleotide (NADH)-dependent reductase to reduce flavin adenine dinucleotide (FAD) (Keller et al. 2000). The reaction is (X: halogen atom, R, R': alkyl group):

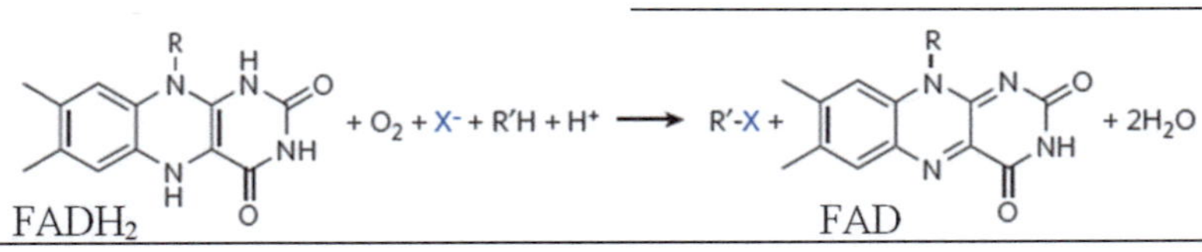

In contrast to CPOs, these enzymes have a high degree of substrate specificity and are able of regioselective halogen incorporation. The first FDH found was the tryptophan 7-halogenase PrnA which catalyzes the chlorination of tryptophan to 7-chlorotryptophan, the first step in pyrrolnitrin biosynthesis (Keller et al. 2000). In this reaction, the flavin reductase produces $FADH_2$ from FAD and reduced NADH. $FADH_2$ is bound by PrnA where it reacts with O_2 to form a flavin hydroperoxide. A single Cl^- is bound close to the isoalloxazine ring of the FAD and attacks the flavin hydroperoxide leading to the formation of hypochlorous acid (HOCl) (Dong et al. 2005; Lang et al. 2011). However, since the substrate tryptophan is bound about 10 **Å** away from the isoalloxazine ring, HOCl is guided through a "tunnel" towards the substrate. The lysine residue, located close to the substrate and conserved in all flavin-dependent halogenase, is suggested to react with HOCl to form a chloramine as the halogenating intermediate (Yeh et al. 2007; Lang et al. 2011).

This class of enzymes is responsible for the halogenation of many bacterial secondary metabolites including many antibiotics, growth hormones as well as antitumor and antifungal compounds. Nearly all known FDHs are involved in the halogenation of aromatic or **heteroaromatic ring** molecules. Two distinct subgroups of enzymes exist: one uses phenol or pyrol as substrates and the other tryptophan (Murphy 2006). Homologues of FDHs have been found in various bacterial phyla including the *Actinobacteria*, *Cyanobacteria*, *Planctomycetes* and *Proteobacteria* (Murphy 2006; Bayer et al. 2013).

Additional Chlorination Processes

Dramatic advances in deciphering the logic of halogenation enzymes have occurred in the recent past through bacterial genomic and bioinformatics analyses which allow identification of new classes of enzymes, *i.e.* halogenases of the mononuclear non-heme iron family and chlorinases.

The mononuclear non-heme iron-containing enzymes are a new class of αKG-dependent Fe_{NH} halogenase enzymes. The first example was the enzyme SyrB2, which generates a 4-chloro-L-threonine residue incorporated into the framework of the nonribosomal lipopeptidolactone syringomycin produced by *Pseudomonas syringae* (Vaillancourt et al. 2005a). Detailed evaluation of cofactor requirement established that Fe^{2+}, O_2, αKG, and Cl^- are required for enzymatic activity. The requirement for Fe^{2+}, O_2 and αKG is the hallmark of the two-His, one-carboxylate non-Fe_{NH} αKG-dependent oxygenase reactions, and SyrB2 falls into that protein superfamily. However, this enzyme effects chlorination rather than hydroxylation of unactivated methyl groups in substrates (Fujimori and Walsh 2007; Neumann et al. 2008; Butler and Sandy 2009). The reaction mechanism involves the binding of Cl^- directly to iron before decarboxylation of a αKG to produce succinate, CO_2 and a high-energy ferryl-oxo intermediate that is very reactive and acts as a hydrogen-abstracting species (Blasiak et al. 2006). The reaction is:

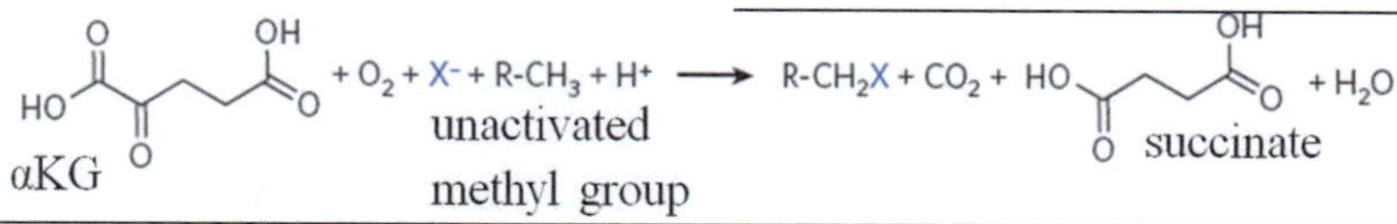

Another class of halogenases is that of non-metallo-halogenating enzymes which use SAM as a co-substrate. The first of the SAM-dependent halogenases to be discovered was the methyl chloride transferase enzyme in seaweeds and other plants, which is responsible for the production of profuse levels of methyl chloride (Ni and Hager 1998), raising questions about its biological function. Other SAM-dependent halogenases, including chlorinase (Eustaquio et al. 2008), were recently discovered. The reaction is:

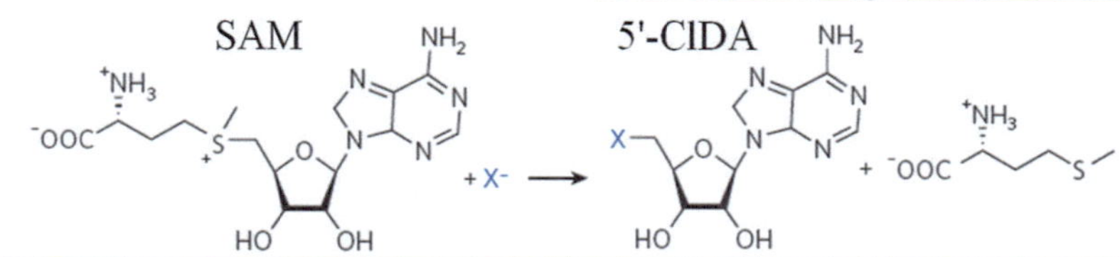

These enzymes catalyse nucleophilic halogenation reactions, rather than electrophilic or radical halogenation reactions. The SAM-dependent chlorinase SalL, from the marine bacterium *Salinospora tropica*, chlorinates SAM by nucleophilic Cl^- addition generating 5′-chloro-5′-deoxyadenosine (5′-ClDA) as a precursor to chloroethylmalonyl-CoA, which is ultimately incorporated into salinosporamide A (Dong et al. 2004).

Although not yet described, the presence of all necessary precursors in the water column of Lake Pavin assumes that enzymatic chlorination is plausible in this ecosystem (Fig. 17.6). Over the past decade, numerous studies exploring microbial diversity in pelagic zone of this ecosystem revealed the presence of parasitic and saprophytic fungi belonging to the three major divisions, *i.e., Chytridiomycota, Basidiomycota* and *Ascomycota* (Monchy et al. 2011; Jobard et al. 2012), well known to synthetize various CPOs. Regarding bacteria, the main phyla involved in OM chlorination through enzyme-mediated processes (*Actinobacteria, Proteobacteria, Cyanobacteria, Planctomycetes*) were also identified (Biderre-Petit et al. 2011a, b; Lehours et al. 2007).

17.3 Chlorinated OM Degradation

As stated above, several thousands of chlorinated compounds are naturally produced in the environment, including non-marine aquatic ecosystems whereas tens of thousands are synthetized by human. Some of them are aromatics, *i.e.*, containing one or more aromatic rings with one or more chlorine atoms while others are aliphatics *i.e.*, formed by carbon chains without benzene ring but with one or more hydrogen atoms replaced by a chlorine atom. Because the freshwaters (lakes, rivers, groundwaters among others) display different and complex geochemical and biological patterns, they may exhibit different potential to degrade these chlorinated compounds which can be from both natural and human origin. Like chlorination, dechlorination can occur through both biotic and abiotic processes with abiotic dechlorination usually slower than microbial one. Because bio-mineralization of chlorinated compounds occurs frequently through multiple-step processes producing various intermediate molecules, it usually requires the involvement of microbial **consortia** (Men et al. 2014).

Most of the processes involved in the biodegradation of chlorinated compounds are catalyzed by enzymes carried by microorganisms. Indeed, microorganisms, due to the immense reservoir of their metabolic capacities, can adapt to almost every environmental conditions found on Earth (Guerrero and Berlanga 2006). Many of them can use Cl_{org} as a main carbon and/or energy source through enzymes with specific dehalogenase activities. Over the past decades, investigation of the microbial degradation capacities allowed the identification of a broad range of this type of enzymes. All are capable to cleave carbon-halogen bonds through

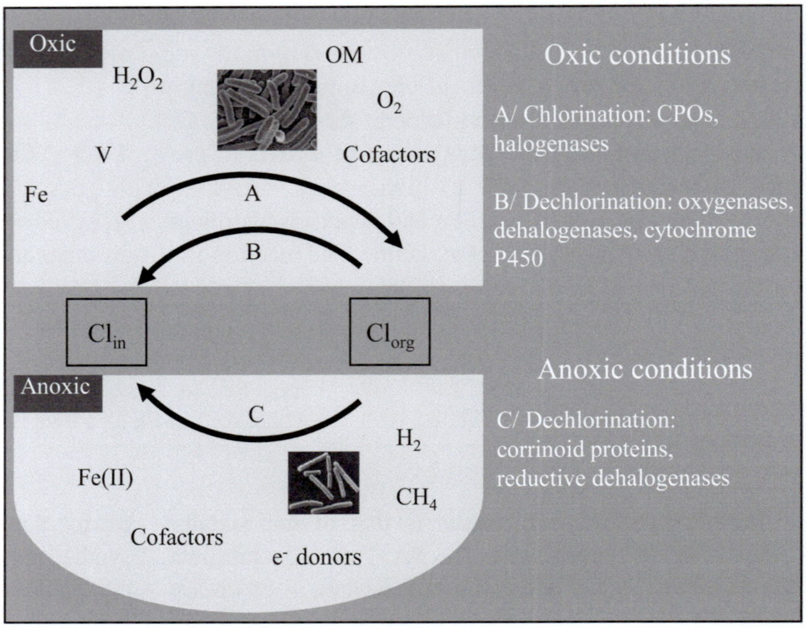

Fig. 17.6 Schematic representation of main processes of chlorination and dechlorination mediated by microorganisms potentially occurring in water column of lake Pavin (*OM* organic matter; *V* vanadium, *e−* electrons, *CPOs* chloroperoxidases, Cl_{in} inorganic chlorine, *Clorg* organic chlorine)

totally different reaction mechanisms, under aerobic or anaerobic conditions (Janssen et al. 2001; de Jong and Dijkstra 2003; van Pée and Unversucht 2003). Furthermore, some microorganisms are also able to degrade chlorinated compounds through fortuitous reactions mediated by non-specific enzymes, also called co-metabolism. While relatively lightly chlorinated compounds can generally be degraded by aerobes, heavily chlorinated relatives are often recalcitrant to biodegradation under aerobic conditions. However, overall, the aerobic processes are low for most chlorinated compounds.

17.3.1 Thermodynamic Feasibility of Dehalogenation Mechanisms

The thermodynamic properties of a large number of biochemical compounds and reactions have been calculated experimentally under different conditions (Thauer et al. 1977; Alberty 2003; Goldberg et al. 2004; Finley et al. 2009) but for most of the known biochemical compounds such information is lacking. Dolfing and Janssen (1994) used group contribution to classify conversion steps in a biodegradation pathway of halogenated compounds as endergonic or exergonic, and determine whether the reactions yield adequate energy to sustain microbial growth. Thermodynamics has also been used to compare biodegradation routes involving reductive dechlorination to those involving oxidative and fermentative degradation reactions (Dolfing 2000, 2003; Smidt and de Vos 2004).

Estimation of Gibbs free energies (ΔG0') and redox potentials ($E0'_0$) for a wide range of aliphatic and aromatic Cl_{org} have indicated that chlorinated compounds should be excellent electrons acceptors, yielding ΔG0' ranging between −130 and −180 kJ . mol^{-1} of Cl removed by hydrogenolytic reductive dehalogenation (Dolfing 1990; Dolfing and Harrisson 1992; Dolfing and Janssen 1994). Corresponding E0' range between +300 and +550 mV which is considerably higher than that of the reduction of sulfate (SO_4^{2-}/H_2S, E0'=−220 mV) and comparable to the potential of NO_3^-/NO_2^- (E0'=+443 mV) (Table 17.2) (Löffler et al. 1999). From these thermodynamic considerations it has been predicted that reductive dehalogenation should also be occurring, though rarely, under aerobic conditions (Dolfing 2003).

While hydrogenolysis is the main mechanism involved in reductive dehalogenation reactions, a second type has also been observed for nonaromatic Cl_{org}, *i.e.* the dichloroelimination. In this reaction, two rather than one chlorine groups are simultaneously removed and the aliphatic bond C-C is converted into a C=C bond (Fig. 17.7). Because this reaction requires only one mole of reducing equivalent (H_2), as opposed to two moles of H_2 in hydrogenolysis, its energy balance is more favorable. However, while dichloroelimination should prevail over hydrogenolysis under hydrogen-limiting conditions, addition of easily available reducing equivalents may decrease this advantage by making compe-

Table 17.2 H_2 threshold concentrations, ΔG0′ values, and redox potentials (E0′) of different redox couples with H_2 as the electron donor (Löffler et al. 1999)

TEAP	H_2 threshold concentration		ΔG0′ (kJ/mol of H_2)	E0′ (mV)
	ppmv	nM		
Acetogenesis	430–4660	336–3640	−26.1	−280
Methanogenesis	6–120	5–95	−33.9	−240
Sulfate reduction ($SO_4^{2-} \rightarrow HS^-$)	1.3–19	1–15	−38.0	−220
Ammonification ($NO_3^- \rightarrow NH_4^+$)	0.02–0.03	0.015–0.025	−149.9	+36
Nitrate reduction ($NO_3^- \rightarrow N_2O, N_2$)	<0.06	<0.05	−240	+790
Fe(III) reduction	0.13–1	0.1–0.8	−228.3	+770
Chlororespiration	<0.4	<0.3	−130 to − 187	+300 to + 550

TEAP terminal electron-accepting process, ΔG0′ Gibbs free energy, E0′ redox potentials

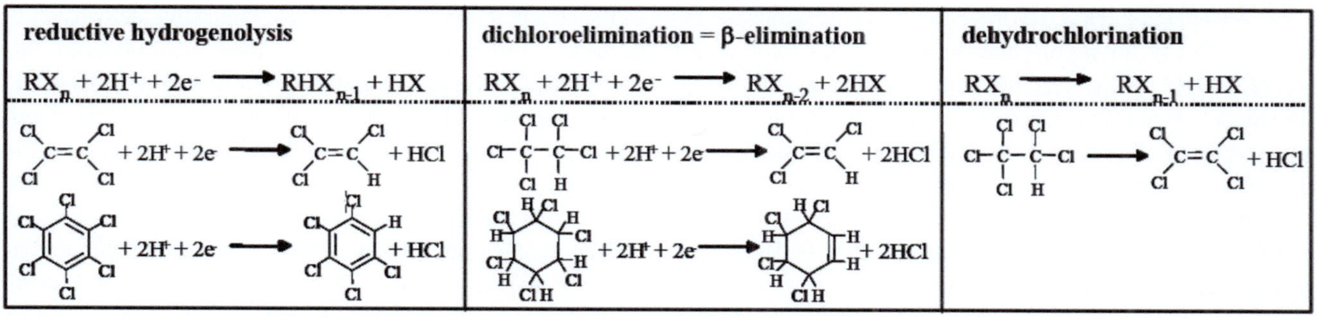

Fig. 17.7 Examples of (reductive) redox dehalogenation reactions occurring under anaerobic conditions

tition for reducing equivalents less important (Dolfing 1999).

From a thermodynamic point of view, dehalogenation *via* routes other than reductive dehalogenation is also feasible. Anaerobic microorganisms should degrade Cl_{org} by oxidative or fermentative pathways, but so far, evidence for their functionality is rather scarce when compared with the knowledge that has been accumulated on reductive dehalogenation. Let us take chlorinated ethylenes as an example. Organisms can grow on their fermentative degradation in a type of reaction in which the chloro substituent serves as an internal electron acceptor (Bradley and Chapelle 2000; Dolfing 1999). Chlorinated ethylenes can, for instance, be fermented to ethane and CO_2 or acetate (Dolfing 1988). The exergonicity of such fermentation implies the potential existence of new types of dechlorinating bacteria that would not depend on the presence of external sources of reducing equivalents. Likewise, oxidative degradation of chlorinated ethylenes to H_2, CO_2 and HCl is an exergonic reaction (−62 to −545 kJ . mol^{-1} of Cl) which is energetically independent of the activity of H_2-consuming organisms.

17.3.2 Abiotic Degradation of Chlorinated Compounds

Although abiotic degradation does not require the mediation of microorganisms, the biotic processes (which change pH and redox potential, for instance) are frequently required to stimulate abiotic reactions. Abiotic dechlorination is usually a slow process but plays an important role in the degradation of some of the aliphatic Cl_{org} (mainly chloroalkanes and chloroalkenes) and of a few aromatic Cl_{org}. The main mechanisms are **hydrogenolysis**, **dichloroelimination**, **hydrolysis** and **dehydrochlorination** (Fig. 17.7) (for review see Tobiszewski and Namiesnik 2012). These mechanisms and the degradation rate strongly depend on the redox conditions found in the surrounding environment.

17.3.2.1 Reductive Pathways of Degradation

Hydrogenolysis and dichloroelimination are both reductive pathways and occur under anaerobic conditions, though dichloroelimination was also demonstrated under partially aerobic conditions (Chen et al. 1996). Hydrogenolysis reactions can be represented by the formula: $RCl_n + 2H^+ +$

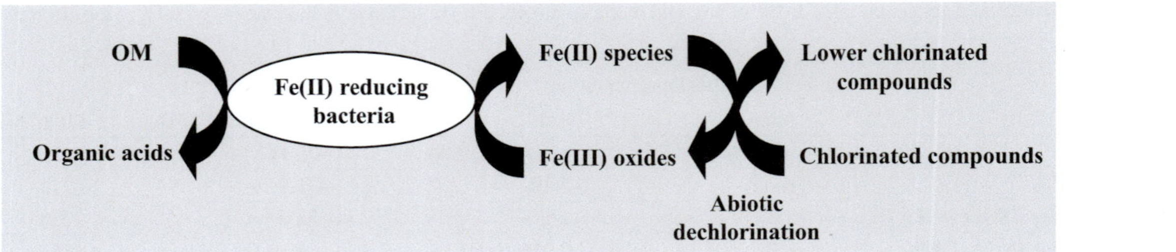

Fig. 17.8 Main mechanisms of anaerobic chlorinated compounds transformation through abiotic or coupled biotic-abiotic processes involving iron species (Modified from Xu et al. 2014). *OM* organic matter

$2e^- \rightarrow RHCl_{n-1} + HCl$ and dichloroelimination by: $RCl_n + 2H^+ + 2e^- \rightarrow RHCl_{n-2} + 2HCl$. Hydrogenolysis is the major degradation pathway for highly chlorinated alkenes (Nobre and Nobre 2004). In this reaction, the chlorinated compound, considered as an electron acceptor, is reduced by an electron donor. Among all potential electron donors involved in chlorinated compounds hydrogenolysis, the more efficient are zero valent metals (*e.g.*, Fe^0, Zn^0, Sn^0) (Scherer et al. 1998). However, recently, reductive dechlorination of PCBs, pentachloroethane (PCA), PCE, TCE, and carbon tetrachloride by nano-sized zero valent iron (nFe^0) has shown promising application due to its higher reactivity than that of Fe^0 (Amir and Lee 2011). Likewise, biogenic or chemogenic ferrous iron (FeII) species have been the focus of many studies due to their enhanced reactivity that favors the reductive transformation of a number of Cl_{org} such as carbon tetrachloride, hexachloroethane, 1,1,1-trichloroethane, pentachloronitrobenzene, pentachlorophenol, dichlorodiphenyltrichloroethane and 2,4-dichlorophenoxyacetic acid through coupled biotic-abiotic or abiotic processes (Fig. 17.8). For example, biogenic magnetite (Fe_3O_4), created by the iron reducing bacterium *Geobacter metallireducens*, can cause the abiotic reductive dechlorination of carbon tetrachloride, and carbonate green rust formed by the iron reducing bacteria *Shewanella putrefaciens* CN32 can cause that of cis-DCE. Because both reactive minerals and microorganisms are usually present in aquatic ecosystems, both abiotic and biotic reductive dechlorinations have the potential to occur simultaneously (Liu et al. 2013; Xu et al. 2014). Therefore, abiotic transformation, though less rapid than microbial, may be crucial if the concentration of reactive minerals is high and/or the activity of dechlorinating bacteria is low (Tobiszewski and Namiesnik 2012).

17.3.2.2 Hydrolysis and Dehydrochlorination Pathways

Beside reductive pathways, hydrolysis and dehydrochlorination are two abiotic processes that may degrade chlorinated compounds under either aerobic or anaerobic conditions. Hydrolysis in natural waters is an extremely slow process. The reaction is summarized by the formula: $RCl + H_2O \rightarrow ROH + HCl$. Generally, this reaction happens when organic molecule reacts with water, resulting in the formation of a new covalent bond with OH^- and the cleavage of the covalent bond with chlorines. Regarding dehydrochlorination, chlorinated compounds may also undergo this mechanism in certain conditions. The formula is: RHCCl–CRH2 → RHC = CHR + HCl. In this reaction, HCl is eliminated from the solvent molecule, which results in the formation of double or triple carbon bonds and less saturated and less chlorinated compounds.

The axonic zone of Lake Pavin provides suitable conditions for abiotic reductive dechlorination processes. Indeed, this ecosystem is permanently redox-stratified, with anoxic and ferruginous deep waters (from 60 to 92 m depth) topped by oxic shallow waters (from 0 to 60 m). Lake Pavin contains dissolved Fe(II) up to 1.2 mM below the chemocline at 60 m (Viollier et al. 1995). More details on the iron cycle are provided in Chap. 14. The Fe(II) form is released from the sediment and either diffuses towards the mixolimnion to precipitate as ferric iron (FeIII) at the redox interface (Michard et al. 2003) or reacts with sulfides or phosphates to form FeS colloids and secondary minerals such as pyrite, vivianite or siderite (Bura-Nakic et al. 2009). These chemogenic Fe(II) species may therefore favor the abiotic reductive transformation of a number of Cl_{org} in the water column of Lake Pavin. Moreover, it has been demonstrated that the large concentration gradient associated with negative Fe isotope composition observed below the oxic-anoxic interface could be interpreted as the signature of intense bacterial dissimilatory Fe reduction due to, for instance, the presence of the well-known obligatory Fe(III) reducers *Geobacter* which are present in the anoxic compartment of this ecosystem (personal data). Therefore, reductive dechlorination of the chlorinated compounds in the anoxic zone of Lake Pavin might also occur through a coupled biotic-abiotic process (Fig. 17.8).

17.3.3 Biotic Degradation of Chlorinated Compounds

17.3.3.1 Chlorinated Compounds Degradation Under Aerobic Conditions

Aerobic dehalogenation is one form of biodegradation which enable the microorganisms in soil and water to utilize chlorinated substances as carbon and electron source. Over the past decades, many investigations focused on microbial degradation of aromatic and aliphatic Cl_{org}, leading to the identification of an important diversity of dehalogenation mechanisms and dehalogenase enzymes (Janssen et al. 1994; Fetzner 1998). In these enzymatic processes, chlorine substituents are removed to form non-halogenated intermediates that can then enter into general metabolic pathways, such as the tricarboxylic acid cycle (TCA). The main processes of aerobic dehalogenation involve hydrolytic, reductive and oxidative mechanisms (van Pée and Unversucht 2003). Furthermore, a wide range of Cl_{org} can also be microbially degraded under aerobic conditions by means of co-metabolic transformation reactions (mainly oxidations), yielding no carbon or energy benefits to the transforming cells (Horvath 1972).

Enzymatic-Mediated Hydrolytic Dehalogenation

Hydrolytic dehalogenation involves the replacement of a chlorine atom by an OH group derived from water and results in the formation of a primary alcohol (Janssen et al. 1994). There are several hydrolytic dehalogenase families which belong to different protein superfamilies of which the members catalyze diverse reactions, mostly with non-chlorinated compounds.

Haloalkane Dehalogenases

These enzymes are the best studied hydrolytic dehalogenases. They belong to the α/β hydrolase superfamily and contain a catalytic triad consisting of one histine and two aspartate residues which are involved in the nucleophilic substitution of Cl^- by water (Ollis et al. 1992). Although these haloalkane dehalogenases can convert a broad range of chlorinated alkanes and alkenes to their corresponding alcohols, they can also dehalogenate aromatic compounds, such as the LinB enzyme of *Sphingomonas paucimobilis* UT26 which catalyzes chlorinated cyclic dienes degradation (Marek et al. 2000).

Haloacid Dehalogenases

These enzymes are the second most common group of aliphatic dehalogenases. These enzymes are dimers and define the so-called HAD (haloacid dehalogenase) superfamily of hydrolases. They are responsible of the hydrolysis of chlorinated carboxylic acids. Like haloalkane dehalogenases, they use an aspartate-based catalytic mechanism but there is no histidine residue in the catalytic site to activate a nucleophilic water molecule.

Representatives of these two enzyme families have been isolated from various bacteria including *Moraxella, Mycobacterium, Pseudomonas, Xanthobacter, Sphingomonas, Sphingobium* and *Rhodococcus* species (van Pée and Unversucht 2003).

Hydrolytic Dehalogenases Specific to Chlorinated Aromatics

At present, only two hydrolytic dehalogenases targeting specifically chlorinated aromatics have been reported. The first is the 4-chlorobenzoyl-CoA dehalogenase (CbzA), belonging to the enoyl hydratase superfamily. It is a component of the 4-chlorobenzoate degradation system. This system comprises three separate enzymes and requires CoA and ATP as cofactors (Benning et al. 1998; Pieper et al. 2010). In the reaction, the compound is first conjugated to CoA and further dechlorinated by CbzA which catalyzes the displacement of chlorine by a nucleophilic addition-elimination mechanism. This process was demonstrated in few species belonging to *Pseudomonas, Rhodococcus* and *Acinetobacter* genera (Janssen et al. 2005). The second enzyme specific to chlorinated aromatics is the chlorothalonil hydrolytic dehalogenase (Chd) of *Pseudomonas* sp. CTN-3 strain. This enzyme contains a conserved domain of metallo-β-lactamase superfamily and catalyzes the dechlorination of chlorothalonil through a not yet described mechanism independent of CoA and ATP (Wang et al. 2010).

Reductive Dehalogenation

Reductive dehalogenation of chlorinated compounds can occur under aerobic conditions. This process can be mediated by two major enzyme families, *i.e.* glutathion-S-transferase and cytochrome P450 type enzymes.

Glutathion-S-transferase (GST) Some aliphatic and aromatic chlorinated compounds can be reductively dehalogenated by thiolytic substitution in the presence of glutathione. During this reaction, the chlorine atoms are displaced by the nucleophilic attack of the thiolate anion of glutathione through the GST activity (Wilce and Parker 1994). The most studied enzyme of this group involved in chlorinated aliphatics degradation is the dichloromethane dehalogenase (DcmA). It catalyses the conversion of DCA to HCl and formaldehyde, a central intermediate of methylotrophic growth used for biomass and energy production (Kayser et al. 2002). This enzyme can be found in a large variety of methylotrophic bacteria belonging to the *Proteobacteria* phylum such as *Methylophilus, Methylorhabdus, Methylobacterium, Hyphomicrobium, Methylopila, Albibacter, Paracoccus, Ancylobacter* and *Pseudomonas* species (Fetzner and Lingens 1994; Torgonskaya et al. 2011). A glutathione-dependent biotransformation was also reported for various chlorinated aromatics including the chlorothalonil and the tri-/tetrachloro-*p*-hydroquinones, through the action of the bacterial GST of species belonging to the *Ochrobactrum, Flavobacterium* and *Spingbium* genera (Kim et al. 2004).

C-type cytochromes Respiratory c-type cytochromes such as cytochrome P-450_{CAM} may also mediate reductive dechlorination of chlorinated aliphatics but only under very low O_2

concentrations. These cytochromes have been proposed to mediate hexachloroethane, pentachloroethane and tetrachloromethane reduction (Walsh et al. 2000). This mechanism can be found in *Pseudomonas* sp. and in the purple nonsulfur bacteria *Rhodopseudomonas* and *Rhodospirillum*.

Oxidative Dehalogenation

Microbial oxygenases are key enzymes for aerobic biodegradation of chlorinated compounds, which are thus used as carbon and energy source (Pérez-Pantoja et al. 2012). Oxidative dehalogenation involves the replacement of a chlorine atom by an OH group whose oxygen atom is derived from O_2. Two classes of oxygenases type dehalogenases have been identified: (i) the monooxygenases which incorporate one oxygen atom to chlorinated compounds to remove chlorine atom whereas the second oxygen atom is reduced to water and (ii) the dioxygenases which add two atoms of O_2 into the substrate to remove the chlorine atom (for review see Arora et al. 2010). Though the majority of dehalogenating oxygenases have been demonstrated for aromatic substrates (Fetzner and Lingens 1994), some have also been shown for aliphatic relatives such as vinyl chloride (VC) and dichloroethene (DCE) (Bradley and Chapelle 2000).

The Monooxygenases

These enzymes are classified into two subclasses depending on the cofactor they used. Flavin-dependent monooxygenases contain flavin as prostethic group and require NADP or NADPH as coenzyme (Arora et al. 2010). Examples of well-studied enzymes of this type are the pentachlorophenol-4-monooxygenase (PcpB), the chlorophenol 4-monooxygenase and the 2,4-dichlorophenol monooxygenase (Fetzner 1998; Arora et al. 2010). These enzymes, mainly involved in polychlorinated compound dehalogenation, have been characterized in various microorganisms including *Pseudomonas, Sphingobium, Sphingomonas, Burkholderia, Ralstonia* and *Azotobacter* genera. The second subclass is represented by the P450 monooxygenases whose best example is the P-450_{CAM} monooxygenase system, already mentioned in the reductive dehalogenation but which can also catalyze oxidative dehalogenation reactions under O_2-rich conditions. From Nishino et al. (2013), this monooxygenase might also be involved in the DCE dechlorination in *Polaromonas chloroethenica* JS666. Finally, a third mechanism results in the VC and ethene assimilation through epoxidation. It is catalyzed by an alkene monooxygenase (AkMO), followed by conversion of the (chloro)epoxide to a 2-hydroxyalkyl coenzyme M by an epoxyalkane coenzyme M transferase (EaCoMT) with the release of the chlorine atom (Coleman and Spain 2003). The by-product is then metabolized through the TCA cycle. Bacteria involved in this mechanism belong to *Mycobacterium*, *Nocardioides*, *Pseudomonas*, *Ochrobactrum* and *Ralstonia* genera.

The Dioxygenases

The aerobic degradation of aromatic compounds is frequently initiated by Rieske non-heme iron oxygenases. These enzymes are multicomponent complexes composed of a terminal oxygenase component (iron-sulfur protein) and of different electron transport proteins. They catalyze the incorporation of two oxygen atoms into the aromatic ring to form arene *cis*-diols which spontaneously rearomatize along with chloride elimination, yielding a (chloro)catechol product. This compound then undergoes the *ortho*-cleavage pathway to form β-ketoadipate (β-KAP), a central metabolite in the degradation of aromatic compounds. This molecule, presumed to have lost Cl^- during one of the above listed-reactions, is then subjected to further transformations before entering the TCA cycle (Ogawa et al. 2003). Nowadays, two-component dioxygenase enzymes that dehalogenate 4-chlorophenylacetate and 2-halobenzoate but also a three-component dioxygenase system that dehalogenates both Z-chlorobenzoate and 2,4-dichlorobenzoate have been described in different strains of *Pseudomonas* (Copley 1997). Likewise, biphenyl 2,3-dioxygenases (BphAs) are of crucial importance for the successful metabolism of PCBs in environment. These enzymes are able to oxidize a wide range of PCB congeners, from mono-chlorobiphenyls to 2,3,4,5,2′,5′-hexachlorobiphenyl, with an increased preference for dioxygenation linked to a decrease of chlorine atom number (Pieper and Seeger 2008; Pieper et al. 2010). This reaction has been observed in a variety of microorganisms belonging to *Alcaligenes, Acinetobacter, Rhodococcus* and *Burkholderia* genera (Fetzner 1998; Pérez-Pantoja et al. 2012).

Biodegradation by Co-metabolism

Besides specific dehalogenation pathways, aerobic co-metabolism represents a significant process responsible for aliphatic and aromatic chlorinated compound biodegradation in the environment (Field and Sierra-Alvarez 2008; Jechorek et al. 2003; Little et al. 1988). It involves microbial oxygenase enzymes, *e.g.* methane monooxygenases (MMOs), toluene mono- and dioxygenases, ammonia monooxygenases and biphenyl monooxygenases among others and does not provide any benefit to the microorganisms (Furukawa 2000; Hazen 2010). In general, these enzymes have low specificity to chlorinated compounds but this limitation is overcome by the increase in enzyme expression levels in presence of the growth substrate for the microorganism. MMOs have been the most widely studied and are present under two forms, the soluble form (sMMO) found in a few selected methanotrophs and the particulate form (pMMO) found in most methanotrophs. These enzymes can oxidize more than 300 different compounds, most of them chemically distinct from methane, the natural substrate of the enzyme (Hazen 2010).

Co-metabolism of Chlorinated Aliphatics

Various chlorinated aliphatic compounds including VC, DCE and trichloroethene (TCE) can be metabolized through a co-metabolism process in the presence of growth-substrates such as methane, butane, propane, ammonium, ethylene, ethane, phenol, benzene, isopropylene or toluene (Hazen 2010; Nzila 2013). For instance, TCE can be converted by the bias of the MMOs in a non-stable epoxide which is spontaneously degraded into stable water-soluble products (acetic, formic, oxalic, glyoxylic, and mono-, di-, and trichloroacetic acids) and CO_2 (Little et al. 1988). In this reaction, chlorine atoms are liberated under the Cl^- form. Some bacteria can also use VC as primary substrate to co-metabolize DCE and to a lesser extent, TCE (Broholm et al. 2005). Likewise, other aliphatic molecules such as chloroform and DCA have been reported to be co-metabolized by bacteria using alkane (methane or butane) or ammonia as growth substrates (Hamamura et al. 1997). Biodegradation of chlorinated aliphatic compounds through co-metabolic reactions have been shown for many bacteria belonging to the *Xanthobacter, Rhodococcus, Nitrosomonas, Pseudomonas, Mycobacterium, Methylobacterium, Methylocystis, Ralstonia* genera.

Co-metabolism of Chlorinated Aromatics

Like aliphatic compounds, various chlorinated aromatic compounds can be biodegraded in the context of co-metabolism by a wide variety of microorganisms (Hazen 2010). Thus, co-metabolism of chlorophenols has been reported in bacteria using phenol (Aktas and Cecen 2009) but also glucose or dextrose (Ziagova et al. 2009) as growth substrates. This mechanism has also been described for the oxidation of chlorobenzenes, chlorobenzoates and PCBs in presence of acetate, glucose, fluorobenzenes, but also of low chlorinated chlorobenzoates and PCBs as growth substrates. Bacteria involved in chlorinated aromatics degradation include members of the *Burkholderia, Alcaligenes, Arthrobacter, Nocardia, Acinetobacter, Pseudomonas* and *Staphylococcus* genera among others. In general, aerobic biodegradation pathways of aromatic compounds are catalyzed, by dioxygenases, though involvement of MMOs has been reported for some chlorobenzenes (Jechorek et al. 2003; Hazen 2010).

At present, although there is no evidence of chlorinated compounds biodegradation in the oxic zone of Lake Pavin, most of microorganisms involved in the different mechanisms responsible for dechlorination have been detected in the water column of this ecosystem (Biderre-Petit et al. 2011a; Lehours et al. 2007). It is particularly the case of methylotrophic bacteria (Borrel et al. 2011), a common group characterized in most freshwaters. We can thus hypothesize that some of these methylotrophs might also carry enzymes related, for instance, to the DcmA dehalogenase and thus be involved in chlorinated aliphatics biodegradation through the aerobic reductive dehalogenation pathway. Moreover, a recent study focusing on methane cycling in this ecosystem has revealed the presence of a wide diversity of methanotrophs harboring genes which encode various pMMOs (Biderre-Petit et al. 2011b). Consequently, these bacteria could represent potential key-players in the halogen cycle in this environment by performing co-metabolism oxidation.

17.3.3.2 Chlorinated Compounds Degradation Under Anaerobic Conditions

Although abiotic processes might be involved in reductive transformations of chlorinated compounds under anoxic conditions, the majority of these reactions are biologically catalyzed. Reductive dehalogenation has been reported for many aliphatic (*e.g.* chloromethanes, chloroethanes, chloroethenes, chlorinated acetic acids) and aromatic (*e.g.* chlorobenzoates, chlorophenols, chlorobenzenes, dioxins, PCBs) chlorinated compounds. These molecules can serve in three metabolic functions in different anaerobic bacteria: (i) as carbon or energy source or both, (iii) as terminal electron acceptor in anaerobic respiration process and (ii) as substrate for co-metabolic activity. The respiration process is the most widespread in environment but also the best documented because its wide use in bioremediation strategies (Holliger et al. 1998b; Hiraishi 2008).

Chlorinated Compounds as Carbon or/and Energy Source

Some microorganisms have ability to grow under anaerobic conditions by using the chlorinated compounds as an electron donor and/or carbon source (Kuntze et al. 2011). These reactions may result in the complete mineralization of these molecules to CO_2 or in their fermentation in products such as acetate and formate. Only very little is known about the degradation pathways and enzymes involved. Anaerobic growth on chlorinated compounds as electron donors or carbon source was clearly the least common form of biodegradability. Indeed, evidence for this type of metabolism was limited to five aliphatic compounds, *i.e.* chloromethane, DCA, VC, *cis*-DCE and chloroacetic acids and to only few aromatic compounds in chlorobenzoate and chlorophenol categories. For both chloromethane and DCA degradation, the reaction involves corrinoid proteins (vitamin B12 containing proteins) and the transfer of the methyl or methylene group, depending on the substrate, onto tetrahydrofolate which is an important coenzyme of chlorine-metabolizing organisms (Magli et al. 1998; Harper 2000). This mechanism has been shown in few isolated microorganisms (*e.g. Acetobacterium dehalogenans* and *Dehalobacterium formicoaceticum*), but also in mixed cultures (methanogenic/acetogenic). DCA degradation can also occur under denitrifying conditions by *Acinetobacter* and *Hyphomicrobium* species. Finally, chloroacetic acids can be used as sole substrate for growth by the phototrophic anaerobe, *Rhodospirillum photometricum*. For chloroethenes and chlorinated aromatics, their mineralization into CO_2 and Cl^- was essentially observed in mixed cultures under methanogenic, iron-reducing, humus-reducing and manganese-reducing conditions (for review see Field and Sierra-Alvarez 2004). Moreover, in previous studies *Thauera chlorobenzoica* strain 3CB-1 was reported to utilize 3-chlorobenzoate as sole sources of cell carbon and energy growth substrate under denitrifying conditions (Song et al. 2001; Kuntze et al. 2011).

Reductive Dehalogenation Through Organohalide Respiration

Organohalide respiration (OHR) is the more efficient and the well-studied biodegradation pathway of chlorinated compounds under anaerobic conditions. In this respiratory process, bacteria use chlorinated species as the terminal electron acceptors during electron-based energy conservation (Holliger et al. 1998b). Highly reduced anaerobic environments (indicated by a low redox potential), typical for methanogenesis and sulfate-reduction, have been found to be a requisite for the reductive dechlorination of halogenated compounds (*i.e.*, the substitution of halogen atoms by hydrogen atoms) (Stuart et al. 1999; Olivas et al. 2002). Almost all chlorinated compounds, whatever the molecular structure and the number of substitutions, can be degraded through this microbial process.

OHR is performed by organohalide-respiring bacteria (OHRBs) that have been identified in various taxa including the δ-*Proteobacteria*, ε-*Proteobacteria*, *Firmicutes* and *Chloroflexi*. These bacteria are divided into two categories, *i.e.* facultative and obligate OHRBs. Besides Cl_{org}, facultative OHRBs are able to use a wide diversity of electron acceptors (*e.g.* fumarate, nitrate, sulfate, thiosulfate, Fe(III), Mn(IV), U(VI), S^0, selenate, arsenate) but also several electron donors (H_2, pyruvate, acetate, lactate, butyrate, succinate, formate, ethanol, glycerol, crotonate). They include members of the *Desulfomonile*, *Geobacter*, *Sulfurospirillum* and *Desulfitobacterium* genera (Bradley and Chapelle 2000; Smidt and de Vos 2004; Hiraishi 2008). The second category of ORHBs, and possibly the most intriguing, is composed of extremely specialized bacteria requiring an organohalide molecule as terminal electron acceptor, H_2 or acetate as electron donor and vitamin B12 as cofactor to perform OHR. Currently, obligate OHRBs are present in the *Chloroflexi* phylum with representatives of the "*Dehalococcoidia*" class such as *Dehalococcoides* and *Dehalogenimonas* species (Löffler et al. 2013) and in the *Firmicutes*, within the *Dehalobacter* genus (Holliger et al. 1998a).

OHR reactions are catalyzed by the reductive dehalogenases (RDase), an iron-sulfur and corrinoid containing family of enzymes, which is very diverse and whose number is continually growing, suggesting that the diversity of this protein family is much deeper than is currently accounted for (Hug and Edwards 2013). Reductive dehalogenase genes typically comprise an operon containing *rdhA* (gene for catalytically active enzyme), *rdhB* (gene for a putative membrane-anchoring protein) and sometimes one or more associated genes (*rdhTKZECD*) (Smidt et al. 2000). The *rdhA* genes have been identified in a wide variety of strictly anaerobic microorganisms, in which a unique archaeal *Ferroglobus* species (Hug and Edwards 2013; Hug et al. 2013). Genomic analysis and genome comparisons between obligate and facultative OHRBs recently revealed specific features such as the presence of multiple putative RDase genes (up to 39) in the genome of obligate OHRBs. These genes are generally localized in high plasticity regions suggesting an important role of gene transfer and/or genomic rearrangement in the adaptation of these microorganisms (Kube et al. 2005; McMurdie et al. 2009; Kruse et al. 2013; Richardson 2013).

Dehalogenation by Anaerobic Co-metabolism

Anaerobic co-metabolism results in the partial or complete reductive dehalogenation of chlorinated compounds. This kind of metabolism has been shown for a wide variety of lightly and heavily chlorinated aliphatics while it is more rarely shown for chlorinated aromatics (for review see Field and Sierra-Alvarez 2004). Regarding aliphatics, the majority of studies reveal that they can be slowly co-metabolized by pure or mixed cultures involving mainly methanogens but also acetogenic, fermentative, sulfate-reducing and iron-reducing bacteria. Rapid anaerobic co-metabolism was observed in a few cases (*e.g.* pentachloroethane, chloroethenes) due to an enzymatic reduction by a reductive dehalogenase expressed for another chlorinated compound. However, the more common, slow anaerobic co-metabolism results from the direct reaction of the chlorinated compound with commonly occurring reduced enzyme cofactors (*e.g.* vitamin B12, a common cofactor of strict anaerobes, especially those involved in chlorine metabolism, or yet the nickel containing coenzyme F430 of methanogens). Regarding aromatic compounds, anaerobic co-metabolism degradation has essentially been observed for chlorobenzenes and PCBs. For the formers, best examples are tetrachlorobenzene and hexachlorobenzene dehalogenation by *Staphylococcus epidermidis* (Tsuchiya and Yamaha 1984) but also in anaerobic sewage sludge respectively (Yuan et al. 1999). About PCBs, only a few congeners were demonstrated to be subject to co-metabolic reductive dechlorination as long as more highly chlorinated parent compounds were present in the non-methanogenic mixed culture (May et al. 2006).

Methanogenesis is a very active process in Lake Pavin. Although CH_4 production has been demonstrated in the anoxic water layers, CH_4 concentration is mainly due to CH_4 flux from the sediment. More details on methane cycle and

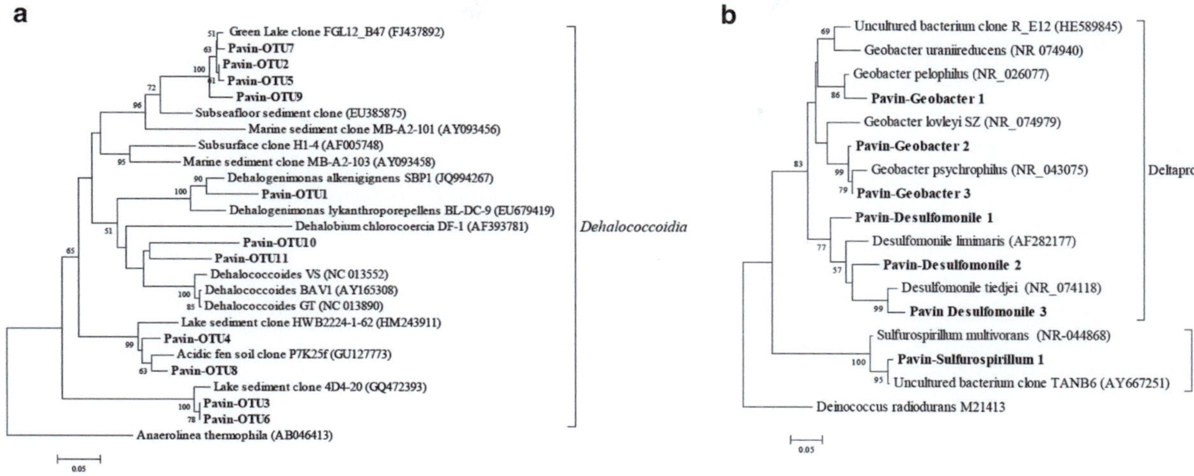

Fig. 17.9 **Phylogenetic trees showing the position of obligate (a) and facultative OHRBs (b) 16S rRNA genes sequences** recovered from water column of Lake Pavin (unpublished data). The trees were based on the neighbour-joining algorithm within the MEGA v6 package (Tamura et al. 2013). Nodes with bootstraps values ≥50 % are indicated. Scale bar represents 5 % sequence divergence

the related prokaryotic actors are provided in Chap. 19. As described in the above paragraph, OHR generally occurs under methanogenic conditions in most environments (Vogel and McCarty 1985); hence Lake Pavin provides ideal conditions for this process. Likewise, OM degradation in the water column can provide a wide variety of electron donors such as short-chain fatty acids (*e.g.* acetate, butyrate, propionate) and H_2 which are essential to anaerobic microbial respiratory processes such as OHR. Nowadays, various microorganisms able to synthesize these molecules have been characterized in the anoxic zone of Lake Pavin (Biderre-Petit et al. 2011a) *e.g. Syntrophus* species which are hydrogen-producing partners. These species are known to live in syntrophic association with hydrogen-using microorganisms, such as OHRBs of the class *Dehalococcoidia* (Bunge et al. 2007). Moreover, in Lake Pavin, *Syntrophus* species co-occur with several phylotypes closely affiliated with representing obligate OHRBs of this class (Biderre-Petit et al. 2011a) (Fig. 17.9). All together, these data strongly support the possible occurrence of both abiotic and biotic reductive dehalogenation of chlorinated compounds in this lake although this remains to be confirmed by further studies.

17.4 Conclusions and Perspectives

Research works conducted over the last decade on most ecosystems revealed that natural chlorination and dechlorination of OM were much more extensive and ubiquitous than previously suggested. As reviewed in this Chapter, a vast array of biotic and abiotic processes can lead to the transformation of Cl^- into Cl_{org}, resulting in the production of thousands of natural chlorinated compounds, and vice versa. Moreover, the high abundance of a natural organic Cl pool in the environment implies that the Cl transformation and cycling are of fundamental and general importance to most living-organisms and to maintain ecosystem properties. However, though the existence of Cl cycle is now well accepted, its regulation remains poorly understood. Likewise, while knowledge is now available for a few terrestrial ecosystems, data from freshwater systems are still scarce.

Lake Pavin, with (i) its volcanic rocks that may contain high Cl contents, (ii) its dense surrounding vegetation suggesting a high chlorinated OM load and degradation within the water column, (iii) its division into three distinct vertical zones (oxic, suboxic and permanently anoxic zones) favourable both to oxic and anoxic processes of chlorination and dechlorination, (iv) its redox conditions, Fe(II) species concentrations and CH_4 content in the bottom layers, all favorable to reductive dechlorination processes, and finally (v) its important microbial diversity (fungi and prokaryotes), constitutes an ideal natural field laboratory to study this biogeochemical cycle.

Thus far, the Cl cycle in Lake Pavin is largely unknown due to lack of data. But this is also true for most, if not all, aquatic environments for which the most important aspects regarding this cycling – including Cl^- and Cl_{org} pools in sediment and water, are largely missing. In Lake Pavin, the only data available is about the measurement of Cl^- concentrations along the water column (Viollier et al. 1995). Another study, aiming at describing microbial community composition along the water column, provides evidence that putative dehalogenating phylotypes are present in the monimolimnion (Fig. 17.9). Among other, members of the *Dehalococcoidia* class, methanogenic *Archaea* and methylotrophic bacteria are the most promising candidates to take part directly or indirectly at the transformation of Cl_{org}.

Future works aiming at identifying Cl_{org} present in Lake Pavin as well as characterizing microbial populations and enzymatic mechanisms responsible of Cl transformation will help to fill these gaps. In addition to provide a better understanding of the Cl cycle in freshwater ecosystems, it will certainly results in the discovery of new reaction mechanisms and enzymes with many potential applications, for instance, in bioremediation.

Acknowledgements This study includes results from PhD researches supported by various instances: Auvergne Region, CNRS, DGA, Blaise Pascal University and French Ministry of Research and Technology. The study was also granted from different sources: French National Programme INSU-EC2CO (Project EChlore) and ANR Programme PRECODD (Project EVASOL). Finally, we are indebted to Dr Lorenz Adrian, Helmholtz Centre for Environmental Research at Leipzig, for a critical review of the ms.

References

Aeschbach-Hertig W, Hofer M, Schmid M, Kipfer R, Imboden D (2002) The physical structure and dynamics of a deep, meromictic crater lake (Lac Pavin, France). Hydrobiologia 487:111–136

Aktas O, Cecen F (2009) Cometabolic bioregeneration of activated carbons loaded with 2-chlorophenol. Bioresour Technol 100:4604–4610

Alberty RA (2003) Thermodynamics of biochemical reactions. Wiley, Hoboken

Allain EJ, Hager LP, Deng L, Jacobsen EN (1993) Highly enantioselective epoxidation of disubstituted alkenes with hydrogen peroxide catalyzed by chloroperoxidase. J Am Chem Soc 115:4415–4416

Amir A, Lee W (2011) Enhanced reductive dechlorination of tetrachloroethene by nano-sized zero valent iron with vitamin B_{12}. Chem Eng J 170:492–497

Arora PK, Srivastava A, Singh VP (2010) Application of monooxygenases in dehalogenation, desulphurization, denitrification and hydroxylation of aromatic compounds. J Biorem Biodegrad 1:1–8

Asplund G, Grimvall A (1991) Organohalogens in nature. Environ Sci Technol 25:1346–1350

Assayag N, Jézéquel D, Ader M, Viollier E, Michard G, Prévot F, Agrinier P (2008) Hydrological budget, carbon sources and biogeochemical processes in Lac Pavin (France): constraints from $\delta^{18}O$ of water and $\delta^{13}C$ of dissolved inorganic carbon. Appl Geochem 23:2800–2816

Bantleon R, Altenbuchner J, van Pee KH (1994) Chloroperoxidase from *Streptomyces lividans*: isolation and characterization of the enzyme and the corresponding gene. J Bacteriol 176:2339–2347

Bastviken D, Thomsen F, Svensson T, Karlsson S, Sandén P, Shaw G, Matucha M, Öberg G (2007) Chloride retention in forest soil by microbial uptake and by natural chlorination of organic matter. Geochim Cosmochim Acta 71:3182–3192

Bastviken D, Svensson T, Sandén P, Kylin H (2013) Chlorine cycling and fates of ^{36}Cl in terrestrial environments. In Svensk Kärnbränslehantering AB Swedish nuclear fuel and waste management co technical report TR-13-26

Bayer K, Scheuermayer M, Fieseler L, Hentschel U (2013) Genomic mining for novel FADH2-dependent halogenases in marine sponge-associated microbial consortia. Mar Biotechnol 15:63–72

Bengtson P, Bastviken D, De Boer W, Öberg G (2009) Possible role of reactive chlorine in microbial antagonism and organic matter chlorination in terrestrial environments. Environ Microbiol 11:1330–1339

Bengtson P, Bastviken D, Oberg G (2013) Possible roles of reactive chlorine II: assessing biotic chlorination as a way for organisms to handle oxygen stress. Environ Microbiol 15:991–1000

Benning MM, Wesenberg G, Liu RQ, Taylor KL, Dunaway-Mariano D, Holden HM (1998) The three-dimensional structure of 4-hydroxybenzoyl-CoA thioesterase from *Pseudomonas* sp. strain CBS-3. J Biol Chem 273:33572–33579

Bernhardt P, Okino T, Winter JM, Miyanaga A, Moore BS (2011) A stereoselective vanadium-dependent chloroperoxidase in bacterial antibiotic biosynthesis. J Am Chem Soc 133:4268–4270

Biderre-Petit C, Boucher D, Kuever J, Alberic P, Jézéquel D, Chebance B, Borrel G, Fonty G, Peyret P (2011a) Identification of sulfur-cycle prokaryotes in a low-sulfate lake (Lake Pavin) using aprA and 16S rRNA gene markers. Microb Ecol 61:313–327

Biderre-Petit C, Jézéquel D, Dugat-Bony E, Lopes F, Kuever J, Borrel G, Viollier E, Fonty G, Peyret P (2011b) Identification of microbial communities involved in the methane cycle of a freshwater meromictic lake. FEMS Microbiol Ecol 77:533–545

Blasiak LC, Vaillancourt FH, Walsh CT, Drennan CL (2006) Crystal structure of the non-haem iron halogenase SyrB2 in syringomycin biosynthesis. Nature 440:368–371

Borrel G, Jezequel D, Biderre-Petit C, Morel-Desrosiers N, Morel JP, Peyret P, Fonty G, Lehours AC (2011) Production and consumption of methane in freshwater lake ecosystems. Res Microbiol 162:832–847

Bradley PM, Chapelle FH (2000) Aerobic microbial mineralization of dichloroethene as sole carbon substrate. Environ Sci Technol 34:221–223

Broholm K, Ludvigsen L, Jensen TF, Ostergaard H (2005) Aerobic biodegradation of vinyl chloride and cis-1,2-dichloroethylene in aquifer sediments. Chemosphere 60:1555–1564

Bunge M, Kleikemper J, Miniaci C, Duc L, Muusse MG, Hause G, Zeyer J (2007) Benzoate-driven dehalogenation of chlorinated ethenes in microbial cultures from a contaminated aquifer. Appl Microbiol Biotechnol 76:1447–1456

Bura-Nakic E, Viollier E, Jézéquel D, Thiam A, Ciglenecki I (2009) Reduced sulfur and iron species in anoxic water column of meromictic crater Lake Pavin (Massif Central, France). Chem Geol 266:311–317

Burd W, Yourkevich O, Voskoboev AJ, van Pee KH (1995) Purification and properties of a non-haem chloroperoxidase from Serratia marcescens. FEMS Microbiol Lett 129:255–260

Butler A, Sandy M (2009) Mechanistic considerations of halogenating enzymes. Nature 460:848–854

Chen C, Puhakka JA, Ferguson JF (1996) Transformations of 1,1,2,2-tetrachloroethane under methanogenic conditions. Environ Sci Technol 30:542–547

Coleman NV, Spain JC (2003) Epoxyalkane: coenzyme M transferase in the ethene and vinyl chloride biodegradation pathways of mycobacterium strain JS60. J Bacteriol 185:5536–5545

Copley SD (1997) Diverse mechanistic approaches to difficult chemical transformations: microbial dehalogenation of chlorinated aromatic compounds. Chem Biol 4:169–174

de Jong RM, Dijkstra BW (2003) Structure and mechanism of bacterial dehalogenases: different ways to cleave a carbon-halogen bond. Curr Opin Struct Biol 13:722–730

Dismukes GC (1986) The metal centers of the photosynthetic oxygen-evolving complex. Photochem Photobiol 43:99–115

Dolfing J (1988) Acetogenesis. In: Zehnder AJB (ed) Biology of anaerobic microorganisms. Wiley Interscience, New York

Dolfing J (1990) Reductive dechlorination of 3-chlorobenzoate is coupled to ATP production and growth in an anaerobic bacterium, strain DCB-1. Arch Microbiol 153:264–266

Dolfing J (1999) Comment on "Methane as a product of chloroethene biodegradation under methanogenic conditions". Environ Sci Technol 33:2302–2303

Dolfing J (2000) Energetics of anaerobic degradation pathways of chlorinate aliphatic compounds. Microb Ecol 40:2–7

Dolfing J (2003) Thermodynamic consideration for dehalogenation. In: Haggblom MM (ed) Dehalogenation: microbial processes and environmental applications. Kluwer academic publishers, Secaucus, pp 89–114

Dolfing J, Harrisson BK (2000) Gibbs free energy of formation of halogenated aromatic compounds and their potential role as electron acceptors in anaerobic environments. Environ Sci Technol 26:2213–2218

Dolfing J, Janssen DB (1994) Estimates of Gibbs free energies of formation of chlorinated aliphatic compounds. Biodegradation 5:21–28

Dong C, Huang F, Deng H, Schaffrath C, Spencer JB, O'Hagan D, Naismith JH (2004) Crystal structure and mechanism of a bacterial fluorinating enzyme. Nature 427:561–565

Dong C, Flecks S, Unversucht S, Haupt C, van Pèe KH, Naismith JH (2005) Tryptophan 7-halogenase (PrnA) structure suggests a mechanism for regioselective chlorination. Science 309:2216–2219

Edwards DJ, Marquez BL, Nogle LM, McPhail K, Goeger DE, Roberts MA, Gerwick WH (2004) Structure and biosynthesis of the jamaicamides, new mixed polyketide-peptide neurotoxins from the marine cyanobacterium *Lyngbya majuscula*. Chem Biol 11:817–833

Enell M, Wennberg L (1991) Distribution of halogenated organic-compounds (AOX) – Swedish transport to surrounding sea areas and mass balance studies in 5 drainage systems. Water Sci Technol 24:385–395

Eriksson E (1960) The yearly circulation of chloride and sulfur in nature; metoerological, geochemical and pedological implications. Part II. Tellus 12:63–109

Eustaquio AS, Pojer F, Noel JP, Moore BS (2008) Discovery and characterization of a marine bacterial SAM-dependent chlorinase. Nat Chem Biol 4:69–74

Fahimi IJ, Keppler F, Schöler HF (2003) Formation of chloroacetic acids from soil, humic acid and phenolic moieties. Chemosphere 52:513–520

Fetzner S (1998) Bacterial dehalogenation. Appl Microbiol Biotechnol 50:633–657

Fetzner S, Lingens F (1994) Bacterial dehalogenases: biochemistry, genetics, and biotechnological applications. Microbiol Rev 58:641–685

Field JA, Sierra-Alvarez R (2004) Biodegradability of chlorinated solvents and related chlorinated aliphatic compounds. Rev Environ Sci Biotechnol 3:185–254

Field JA, Sierra-Alvarez R (2008) Microbial transformation and degradation of polychlorinated biphenyls. Environ Pollut 155:1–12

Finley SD, Broadbelt LJ, Hatzimanikatis V (2009) Thermodynamic analysis of biodegradation pathways. Biotechnol Bioeng 103:532–541

Fujimori DG, Walsh CT (2007) What's new in enzymatic halogenations. Curr Opin Chem Biol 11:553–560

Furukawa K (2000) Engineering dioxygenases for efficient degradation of environmental pollutants. Curr Opin Biotechnol 11:244–249

Godduhn A, Duffy LK (2003) Multi-generational health risks of persistent organic pollution in the Far North. Environ Sci Policy 6:341–353

Goldberg RN, Tewari YB, Bhat TN (2004) Thermodynamics of enzyme-catalyzed reactions – a database for quantitative biochemistry. Bioinformatics 20:2874–2877

Graedel TE, Keene WC (1996) The budget and cycle of earth's natural chlorine. Pure Appl Chem 68:1689–1697

Gribble GW (2003) The diversity of naturally produced organohalogens. Chemosphere 52:289–297

Gribble GW (2004) Amazing organohalogens. Am Sci 92:342–349

Gribble GW (2010) Naturally occurring organohalogen compounds-a comprehensive update. In: Progress in the chemistry of organic natural products. Springer, Wien, pp 256–327

Grisham MB (1991) Chloroperoxidase: a review. In: Everse J, Everse KE, Grisham MB (eds) Peroxidase in chemistry and biology. CRC Press, Boca Raton, pp 85–137

Guerrero R, Berlanga M (2006) Life's unity and flexibility: the ecological link. Int Microbiol 9:225–235

Gustavsson M, Karlsson S, Öberg G, Sandén P, Svensson T, Valinia S, Thiry Y, Bastviken D (2012) Organic matter chlorination rates in different boreal soils: the role of soil organic matter content. Environ Sci Technol 46:1504–1510

Hager LP, Morris DR, Brown FS, Eberwein H (1966) Chloroperoxidase. II. Utilization of halogen anions. J Biol Chem 241:1769–1777

Hamamura N, Page C, Long T, Semprini L, Arp DJ (1997) Chloroform cometabolism by butane-grown CF8, *Pseudomonas butanovora*, and *Mycobacterium vaccae* JOB5 and methane-grown *Methylosinus trichosporium* OB3b. Appl Environ Microbiol 63:3607–3613

Hamilton JTG, McRoberts WC, Keppler F, Kalin RM, Harper DB (2003) Chloride methylation by plant pectin: an efficient environmentally significant process. Science 301:206–209

Harper DB (2000) The global chloromethane cycle: biosynthesis, biodegradation and metabolic role. Nat Prod Rep 17:337–348

Hazen TC (2010) Cometabolic bioremediation. In: Handbook of hydrocarbon and lipid microbiology. Springer, Berlin Heidelberg, pp 2505–2514

Herczeg AL, Leaney FW (2011) Review: environmental tracers in arid-zone hydrology. Hydrogeol J 19:17–29

Hiraishi A (2008) Biodiversity of dehalorespiring bacteria with special emphasis on polychlorinated biphenyl/dioxin dechlorinators. Microbes Environ 23:1–12

Hjelm O, Johansson MB, Öberg-Asplund G (1995) Organically bound halogens in coniferous forest soil – distribution pattern and evidence of *in situ* production. Chemosphere 30:2353–2364

Holliger C, Hahn D, Harmsen H, Ludwig W, Schumacher W, Tindall B, Vazquez F, Weiss N, Zehnder AJ (1998a) *Dehalobacter restrictus* gen. nov. and sp. nov., a strictly anaerobic bacterium that reductively dechlorinates tetra- and trichloroethene in an anaerobic respiration. Arch Microbiol 169:313–321

Holliger C, Wohlfarth G, Diekert G (1998b) Reductive dechlorination in the energy metabolism of anaerobic bacteria. FEMS Microbiol Rev 22:383–398

Horvath RS (1972) Microbial co-metabolism and the degradation of organic compounds in nature. Bacteriol Rev 36:146–155

Hruška J, Oulehle F, Šamonil P, Šebesta J, Tahovská K, Hleb R, Houška J, Šikl J (2012) Longterm forest soil acidification, nutrient leaching and vegetation development: linking modelling and surveys of a primeval spruce forest in the Ukrainian Transcarpathian Mts. Ecol Model 244:2–37

Hu S, Hager LP (1999) Highly enantioselective propargylic hydroxylations catalyzed by chloroperoxidase. J Am Chem Soc 121:872–873

Hug LA, Edwards EA (2013) Diversity of reductive dehalogenase genes from environmental samples and enrichment cultures identified with degenerate primer PCR screens. Front Microbiol 4:1–16

Hug LA, Maphosa F, Leys D, Löffler FE, Smidt H, Edwards EA, Adrian L (2013) Overview of organohalide-respiring bacteria and a proposal for a classification system for reductive dehalogenases. Philos Trans R Soc Lond B Biol Sci 368:20120322

Isidorov VA (1990) Chapter 3: Organic chemisty of the earth's atmosphere. In: Natural source of organic components of the atmosphere. Springer, Berlin

Janssen DB, Pries F, van der Ploeg JR (1994) Genetics and biochemistry of dehalogenating enzymes. Annu Rev Microbiol 48:163–191

Janssen DB, Oppentocht JE, Poelarends GJ (2001) Microbial dehalogenation. Curr Opin Biotechnol 12:254–258

Janssen DB, Dinkla IJ, Poelarends GJ, Terpstra P (2005) Bacterial degradation of xenobiotic compounds: evolution and distribution of novel enzyme activities. Environ Microbiol 7:1868–1882

Jechorek M, Wendlandt KD, Beck M (2003) Cometabolic degradation of chlorinated aromatic compounds. J Biotechnol 102:93–98

Jézéquel D, Sarazin G, Prévot F, Viollier E, Groleau A, Michard G (2012) Bilan hydrique du lac Pavin. In Le lac Pavin. Revue des Sciences Naturelles d'Auvergne 74–75: 73–96

Jobard M, Rasconi S, Solinhac L, Cauchie HM, Sime-Ngando T (2012) Molecular and morphological diversity of fungi and the associated functions in three European nearby lakes. Environ Microbiol 14:2480–2494

Johansson E, Sandén P, Öberg G (2003) Organic chlorine in deciduous and coniferous forest soils in southern Sweden. Soil Sci 168:347–355

Jordan A, Harnisch J, Borchers R, Le Guern F, Shinohara H (2000) Volcanogenic halocarbons. Environ Sci Technol 34:1122–1124

Kayser MF, Ucurum Z, Vuilleumier S (2002) Dichloromethane metabolism and C1 utilization genes in *Methylobacterium* strains. Microbiology 148:1915–1922

Keller S, Wage T, Hohaus K, Holzer M, Eichhorn E, van Pèe KH (2000) Purification and partial characterization of tryptophan 7-halogenase (PrnA) from *Pseudomonas fluorescens*. Angew Chem Int Ed 39:2300–2302

Keppler F, Eiden R, Niedan V, Pracht J, Scholer HF (2000) Halocarbons produced by natural oxidation processes during degradation of organic matter. Nature 403:298–301

Kim EJ, Oh JE, Chang YS (2003) Effects of forest fire on the level and distribution of PCDD/Fs and PAHs in soil. Sci Total Environ 311:177–189

Kim Y-M, Park K, Joo G-J, Jeong E-M, Kim J-E, Rhee I-K (2004) Glutathione-dependent biotransformation of the fungicide chlorothalonil. J Agric Food Chem 52:4192–4196

Kirchner JW, Feng X, Neal C (2000) Fractal stream chemistry and its implications for contaminant transport in catchments. Nature 403:524–527

Kirk O, Conrad LS (1999) Metal-free haloperoxidases: fact or artifact? Angew Chem Int Ed Engl 38:977–979

Kirner S, Krauss S, Sury G, Lam ST, Ligon JM, van Pee KH (1996) The non-haem chloroperoxidase from *Pseudomonas fluorescens* and its relationship to pyrrolnitrin biosynthesis. Microbiology 142:2129–2135

Kruse T, Maillard J, Goodwin L, Woyke T, Teshima H, Bruce D, Detter C, Tapia R, Han C, Huntemann M, Wei CL, Han J, Chen A, Kyrpides N, Szeto E, Markowitz V, Ivanova N, Pagani I, Pati A, Pitluck S, Nolan M, Holliger C, Smidt H (2013) Complete genome sequence of *Dehalobacter restrictus* PER-K23(T.). Stand Genomic Sci 8:375–388

Kube M, Beck A, Zinder S, Kuhl H, Reinhardt R, Adrian L (2005) Genome sequence of the chlorinated compound-respiring bacterium *Dehalococcoides* species strain CBDB1. Nat Biotechnol 23:1269–1273

Kuntze K, Kiefer P, Baumann S, Seifert J, von Bergen M, Vorholt JA, Boll M (2011) Enzymes involved in the anaerobic degradation of meta-substituted halobenzoates. Mol Microbiol 82:b758–b769

Lang A, Polnick S, Nicke T, William P, Patallo EP, Naismith JH, van Pèe KH (2011) Changing the regioselectivity of the tryptophan 7-halogenase PrnA by site-directed mutagenesis. Angew Chem Int Ed Engl 50:2951–2953

Laturnus F, Taselmann KFH, Borch T, Groen C (2002) Terrestrial natural sources of trichloromethane (chloroform, CHCl3) – An overview. Biogeochemistry 60:121–139

Lehours AC, Evans P, Bardot C, Joblin K, Fonty G (2007) Phylogenetic diversity of archaea and bacteria in the anoxic zone of a meromictic lake (Lake Pavin, France). Appl Environ Microbiol 73:2016–2019

Little CD, Palumbo AV, Herbes SE, Lidstrom ME, Tyndall RL, Gilmer PJ (1988) Trichloroethylene biodegradation by a methane-oxidizing bacterium. Appl Environ Microbiol 54:951–956

Liu Y, Li F-B, Xia W, Xu J-M, Yu X-S (2013) Association between ferrous iron accumulation and pentachlorophenol degradation at the paddy soil-water interface in the presence of exogenous low-molecular-weight dissolved organic carbon. Chemosphere 91:1547–1555

Löffler FE, Tiedje JM, Sanford RA (1999) Fraction of electrons consumed in electron acceptor reduction and hydrogen thresholds as indicators of halorespiratory physiology. Appl Environ Microbiol 65:4049–4056

Löffler FE, Yan J, Ritalahti KM, Adrian L, Edwards EA, Konstantinidis KT, Muller JA, Fullerton H, Zinder SH, Spormann AM (2013) *Dehalococcoides mccartyi* gen. nov., sp. nov., obligately organohalide-respiring anaerobic bacteria relevant to halogen cycling and bioremediation, belong to a novel bacterial class, *Dehalococcoidia* classis nov., order *Dehalococcoidales* ord. nov. and family *Dehalococcoidaceae* fam. nov., within the phylum *Chloroflexi*. Int J Syst Evol Microbiol 63:625–635

Lovett GM, Likens GE, Buso DC, Driscoll CT, Bailey SW (2005) The biogeochemistry of chlorine at Hubbard Brook, New Hampshire, USA. Biogeochemistry 72:191–232

Magli A, Messmer M, Leisinger T (1998) Metabolism of dichloromethane by the strict anaerobe *Dehalobacterium formicoaceticum*. Appl Environ Microbiol 64:646–650

Marek J, Vévodova J, Smatanova IK, Nagata Y, Svensson LA, Newman J, Takagi M, Damborsky J (2000) Crystal structure of the haloalkane dehalogenase from *Sphingomonas paucimobilis* UT26. Biochemistry 39:14082–14086

May HD, Cutter LA, Miller GS, Milliken CE, Watts JE, Sowers KR (2006) Stimulatory and inhibitory effects of organohalides on the dehalogenating activities of PCB-dechlorinating bacterium o-17. Environ Sci Technol 40:5704–5709

Mayer T, Snodgrass WJ, Morin D (1999) Spatial characterization of the occurrence of road salts and their environmental concentrations as chlorides in Canadian surface waters and benthic sediments. Water Qual Res J Can 34:545–574

McCulloch A (2003) Chloroform in the environment: occurrence, sources, sinks and effects. Chemosphere 50:1291–1308

McMurdie PJ, Behrens SF, Muller JA, Goke J, Ritalahti KM, Wagner R, Goltsman E, Lapidus A, Holmes S, Löffler FE, Spormann AM (2009) Localized plasticity in the streamlined genomes of vinyl chloride respiring *Dehalococcoides*. PLoS Genet 5, e1000714

Men Y, Seth EC, Yi S, Allen RH, Taga ME, Alvarez-Cohen L (2014) Sustainable growth of *Dehalococcoides mccartyi* 195 by corrinoid salvaging and remodeling in defined lactate-fermenting consortia. Appl Environ Microbiol 80:2133–2141

Michard G, Viollier E, Jézéquel D, Sarazin G (1994) Geochemical study of a crater lake: Pavin Lake, France — identification, location and quantification of the chemical reactions in the lake. Chem Geol 115:103–115

Michard G, Jézéquel D, Viollier E (2003) Vitesses de réaction de dissolution et précipitation au voisinage de l'interface oxydo-réducteur dans un lac méromictique : le lac Pavin (Puy de Dôme, France). Revue des sciences de l'eau 16:199–218

Monchy S, Sanciu G, Jobard M, Rasconi S, Gerphagnon M, Chabé M, Cian A, Meloni D, Niquil N, Christaki U, Viscogliosi E, Sime-Ngando T (2011) Exploring and quantifying fungal diversity in freshwater lake ecosystems using rDNA cloning/sequencing and SSU tag pyrosequencing. Environ Microbiol 13:1433–1453

Murphy CD (2006) Recent developments in enzymatic chlorination. Nat Prod Rep 23:147–152

Neumann CS, Fujimori DG, Walsh CT (2008) Halogenation strategies in natural product biosynthesis. Chem Biol 15:99–109
Ni X, Hager LP (1998) cDNA cloning of *Batis maritima* methyl chloride transferase and purification of the enzyme. Proc Natl Acad Sci U S A 95:12866–12871
Nishino SF, Shin KA, Gossett JM, Spain JC (2013) Cytochrome P450 initiates degradation of cis-dichloroethene by *Polaromonas* sp. dtrain JS666. Appl Environ Microbiol 79:2263–2272
Nobre RC, Nobre MM (2004) Natural attenuation of chlorinated organics in a shallow sand aquifer. J Hazard Mater 110:129–137
Nzila A (2013) Update on the cometabolism of organic pollutants by bacteria. Environ Pollut 178:474–482
Öberg G (2002) The natural chlorine cycle-fitting the scattered pieces. Appl Microbiol Biotechnol 58:565–581
Öberg G (2003) The biogeochemistry of chlorine in soil. In: Gribble G (ed) The handbook of environmental chemistry. The natural production of organohalogen compounds, vol 3, part P. Springer, Berlin, pp 43–62
Öberg G, Holm M, Sandén P, Svensson T, Parikka M (2005) The role of organic-matter-bound chlorine in the chlorine cycle: a case study of the Stubbetorp catchment, Sweden. Biogeochemistry 75:241–269
Ogawa N, Miyashita K, Chakrabarty AM (2003) Microbial genes and enzymes in the degradation of chlorinated compounds. Chem Rec 3:158–171
Olivas Y, Dolfing J, Smith GB (2002) The influence of redox potential on the degradation of halogenated methanes. Environ. Toxicol Chem 21:493–499
Ollis DL, Cheah E, Cygler M, Dijkstra B, Frolow F, Franken SM, Harel M, Remington SJ, Silman I, Schrag J, Sussman JL, Verschueren KHG, Goldman A (1992) The α/β hydrolase fold. Protein Eng 5:197–211
Ortiz-Bermúdez P, Hirth KC, Srebotnik E, Hammel KE (2007) Chlorination of lignin by ubiquitous fungi has a likely role in global organochlorine production. Proc Natl Acad Sci U S A 104:3895–3900
Pelletier I, Pfeifer O, Altenbuchner J, van Pee KH (1994) Cloning of a second non-haem bromoperoxidase gene from *Streptomyces aureofaciens* ATCC 10762: sequence analysis, expression in Streptomyces lividans and enzyme purification. Microbiology 140:509–516
Pérez-Pantoja D, Donoso R, Agulló L, Córdova M, Seeger M, Pieper DH, González B (2012) Genomic analysis of the potential for aromatic compounds biodegradation in *Burkholderiales*. Environ Microbiol 14:1091–1117
Pieper DH, Seeger M (2008) Bacterial metabolism of polychlorinated biphenyls. J Mol Microbiol Biotechnol 15:121–138
Pieper DH, González B, Cámaral B, Pérez-Pantoja D, Reineke W (2010) Aerobic degradation of chloroaromatics. In: Timmis KN (ed) Handbook of hydrocarbon and lipid microbiology. Springer, Berlin/Heidelberg, pp 839–864
Rai GP, Sakai S, Florez AM, Mogollon L, Hager LP (2001) Directed evolution of chloroperoxidase for improved epoxidation and chlorination catalysis. Adv Synth Catal 343:638–645
Redon P-O, Abdesselam A, Bastviken D, Cecchini S, Nicolas M, Thiry Y (2011) Chloride and organic chlorine in forest soils: storage, residence times, and influence of ecological conditions. Environ Sci Technol 45:7202–7208
Redon P-O, Jolivet C, Saby NPA, Abdelouas A, Thiry Y (2013) Occurrence of natural organic chlorine in soils for different land uses. Biogeochemistry 114:413–419
Richardson RE (2013) Genomic insights into organohalide respiration. Curr Opin Biotechnol 24:498–505
Scherer MM, Blako BA, Gallagher DA, Tratnyek PG (1998) Correlation analysis of rate constants for dechlorination by zero valent ion. Environ Sci Technol 32:3026–3033
Shaw PD, Hager LP (1959) Biological chlorination IV. Peroxidative nature of enzymatic chlorination. J Am Chem Soc 81:6527–6528
Smidt H, de Vos WM (2004) Anaerobic microbial dehalogenation. Annu Rev Microbiol 58:43–73
Smidt H, Akkermans AD, van der Oost J, de Vos WM (2000) Halorespiring bacteria-molecular characterization and detection. Enzyme Microb Technol 27:812–820
Song B, Palleroni NJ, Kerkhof LJ, Häggblom MM (2001) Characterization of halobenzoate-degrading, denitrifying *Azoarcus* and *Thauera* isolates and description of *Thauera chlorobenzoica* sp. nov. Int J Syst Evol Microbiol 51:589–602
Song J, Ahn H, Kim H, Song B (2006) Molecular cloning and expression of perhydrolase genes from *Pseudomonas aeruginosa* and *Burkholderia cepacia* in *Escherichia coli*. Biotechnol Lett 28:849–856
Stuart SL, Woods SL, Lemmon TL, Ingle JD (1999) The effect of redox potential changes on reductive dechlorination of pentachlorophenol and the degradation of acetate by a mixed, methanogenic culture. Biotechnol Bioeng 63:69–78
Svensson T, Sandén P, Bastviken D, Öberg G (2007) Chlorine transport in a small catchment in southeast Sweden during two years. Biogeochemistry 82:181–199
Tamura K, Stecher G, Peterson D, Filipski A, Kumar S (2013) MEGA6: molecular evolutionary genetics analysis version 6.0. Mol Biol Evol 30:2725–2729
Thauer RK, Jungermann K, Decker K (1977) Energy conservation in chemotrophic anaerobic bacteria. Bacteriol Rev 41:100–180
Tobiszewski M, Namiesnik J (2012) Abiotic degradation of chlorinated ethanes and ethenes in water. Environ Sci Pollut Res Int 19:1994–2006
Torgonskaya ML, Doronina NV, Hourcade E, Trotsenko YA, Vuilleumier S (2011) Chloride-associated adaptive response in aerobic methylotrophic dichloromethane-utilising bacteria. J Basic Microbiol 51:296–303
Trevisan V, Signoretto M, Colonna S, Pironti V, Strukul G (2004) Microencapsulated chloroperoxidase as a recyclable catalyst for the enantioselective oxidation of sulfides with hydrogen peroxide. Angew Chem Int Ed 43:4097–4099
Tröjbom M, Söderbäck B, Kalinowski B (2008) Hydrochemistry of surface water and shallow groundwater. Site descriptive modeling, SDM-Site Laxemar. SKB R-08-46, Svensk Kärnbränslehantering AB
Tsuchiya T, Yamaha T (1984) Reductive dechlorination of 1, 2, 4-trichlorobenzene by *Staphylococcus epidermidis* isolated from intestinal contents of rats. Agric Biol Chem 48:1545–1550
Vaillancourt FH, Yeh E, Vosburg DA, O'Connor SE, Walsh CT (2005a) Cryptic chlorination by a non-haem iron enzyme during cyclopropyl amino acid biosynthesis. Nature 436:1191–1194
Vaillancourt FH, Yin J, Walsh CT (2005b) SyrB2 in syringomycin E biosynthesis is a nonheme FeII α-ketoglutarate- and O_2-dependent halogenase. Proc Natl Acad Sci U S A 102:10111–10116
van Pée KH, Unversucht S (2003) Biological dehalogenation and halogenation reactions. Chemosphere 52:299–312
van Schijndel JW, Vollenbroek EG, Wever R (1993) The chloroperoxidase from the fungus *Curvularia inaequalis*; a novel vanadium enzyme. Biochim Biophys Acta 1161:249–256
Verhagen FJM, Swarts HJ, Kuyper TW, Wijnberg JBPA, Field JA (1996) The ubiquity of natural adsorbable organic halogen production among basidiomycetes. Appl Microbiol Biotechnol 45:710–718
Viollier E, Jézéquel D, Michard G, Pèpe M, Sarazin G, Alberic P (1995) Geochernical study of a crater lake (Pavin Lake, France): Trace-element behaviour in the monimolimnion. Chem Geol 125:61–72
Vogel TM, McCarty PL (1985) Biotransformation of tetrachloroethylene to trichloroethylene, dichloroethylene, vinyl chloride, and carbon dioxide under methanogenic conditions. Appl Environ Microbiol 49:1080–1083

Wagenknecht HA, Woggon WD (1997) Identification of intermediates in the catalytic cycle of chlororperoxidase. Chem Biol 4:367–372

Wagner C, El Omari M, Konig GM (2009) Biohalogenation: nature's way to synthesize halogenated metabolites. J Nat Prod 72:540–553

Walsh ME, Kyritsis P, Eady NAJ, Hill HAO, Wong L-L (2000) Catalytic reductive dehalogenation of hexachloroethane by molecular variants of cytochrome P450cam (CYP101). Eur J Biochem 267:5815–5820

Wang G, Li R, Li S, Jiang J (2010) A novel hydrolytic dehalogenase for the chlorinated aromatic compound chlorothalonil. J Bacteriol 192:2737–2745

White PJ, Broadley MR (2001) Chloride in soils and its uptake and movement within the plant: a review. Ann Bot 88:967–988

Wiesner W, van Pee KH, Lingens F (1988) Purification and characterization of a novel bacterial non-heme chloroperoxidase from *Pseudomonas pyrrocinia*. J Biol Chem 263:13725–13732

Wilce MCJ, Parker MW (1994) Structure and function of glutathione S-transferases. Biochim Biophys Acta 1205:1–18

Winter JM, Moffitt MC, Zazopoulos E, McAlpine JB, Dorrestein PC, Moore BS (2007) Molecular basis for chloronium-mediated meroterpene cyclization: cloning, sequencing, and heterologous expression of the napyradiomycin biosynthetic gene cluster. J Biol Chem 282:16362–16368

Winterton N (2000) Chlorine: the only green element – towards a wider acceptance of its role in natural cycles. Green Chem 2:173–225

Xu Y, He Y, Feng X, Liang L, Xu J, Brookes PC, Wu J (2014) Enhanced abiotic and biotic contributions to dechlorination of pentachlorophenol during Fe(III) reduction by an iron-reducing bacterium *Clostridium beijerinckii* Z. Sci Total Environ 473–474:215–223

Yeh E, Blasiak LC, Koglin A, Drennan CL, Walsh CT (2007) Chlorination by a long-lived intermediate in the mechanism of flavindependent halogenases. Biochemistry 46:1284–1292

Yuan SY, Su CJ, Chang BV (1999) Microbial dechlorination of hexachlorobenzene in anaerobic sewage sludge. Chemosphere 38:1015–1023

Yunos NM, Bellomo R, Story D, Kellum J (2010) Bench-to-bedside review: chloride in criticalillness. Crit Care 14:226

Zaks A, Dodds DR (1995) Chloroperoxidase-catalyzed asymmetric oxidations: substrate specificity and mechanistic study. J Am Chem Soc 117:10419–10424

Ziagova M, Kyriakou G, Liakopoulou-Kyriakides M (2009) Co-metabolism of 2,4- dichlorophenol and 4-Cl-m-cresol in the presence of glucose as an easily assimilated carbon source by *Staphylococcus xylosus*. J Hazard Mater 163:383–390

18 The Importance of Phagotrophic Protists in Lake Pavin

Jean-François Carrias, Christian Amblard, Gilles Bourdier, and Télesphore Sime-Ngando

Abstract

Phagotrophic protists include unicellular taxa from different lineages of eukaryotes. Most of these freshwater microorganisms are free-living, heterotrophic forms, consuming bacteria in the water column of lakes. In Lake Pavin, we focused on the seasonal and spatial dynamic and the diversity of phagotrophic protists and estimated *in situ* specific-grazing rates of both ciliates and flagellates. We shown that a large part of bacterivory was the result of the phagotrophy of attached (i.e. epiphytic) ciliates and flagellates. Pigmented flagellates have also an important grazing impact on bacteria, with clearance rates equivalent to those estimated for heterotrophic taxa of similar size. Total impact of phagotrophic protists on bacteria indicates that bacterial productivity might be totally consumed by unicellular eukaryotes. In a short-term study, we estimated that flagellate bacterivory dominated the bacterial mortality in Lake Pavin while viral lysis accounted for less than 10 % of the bacterial mortality. Finally, the use of microcosms suggested that large-size ciliates ingest heterotrophic nanoflagellates and may in turn be a key resource for metazoa during spring. We conclude that phagotrophic protists are involved in the conversion of bacterial biomass and then constitute a relevant source of carbon for metazoans in the planktonic food web of Lake Pavin.

Keywords

Protists • Flagellates • Ciliates • Bacterivory • Plankton • Lake • Freshwater

18.1 Introduction

Freshwater phagotrophic protists comprise predominantly heterotrophic nanoflagellates, phagotrophic phytoflagellates, and ciliates. These unicellular eukaryotes are recognized as the primary consumers of bacterioplankton and picophytoplankton in freshwater pelagic ecosystems (Sanders et al. 1989; Pace et al. 1990; Carrias et al. 1996). Because of the lack of accurate methods to count these communities, most of these cells were usually ignored until 1980–1990s. Improved interest and new methods have shown the large diversity of theses planktonic taxa and their key role in the food web. Firstly, they form a link between picoplankton and higher trophic levels. Secondly, due to their high metabolic activity, they play an important role in the recycling of nutrients that limit algal growth (Caron et al. 1988). During the last two decades freshwater protistan ecology has received increasing attention (Amblard et al. 1993; Bennett et al. 1990; Carrias et al. 1996, 2001; Riemann and Christoffersen 1993). In Lake Pavin, we assessed the distribution of heterotrophic nanoflagellates (Carrias et al. 1998a), the main bacterivores in lakes, and ciliates (Carrias et al. 1998b). We estimated the impact of phagotrophic protists on bacteria using latex beads as tracers with special attention to attached protists (Carrias et al. 1996). The relative importance of viral lysis and protist bacterivory in bacterial losses was also

J.-F. Carrias (✉) • C. Amblard • G. Bourdier • T. Sime-Ngando
LMGE, Laboratoire Microorganismes: Génome et Environnement, UMR CNRS 6023, Université Clermont-Auvergne, Université Blaise Pascal, BP 80026, 63171 Aubière Cedex, France
e-mail: j-francois.carrias@univ-bpclermont.fr

T. Sime-Ngando et al. (eds.), *Lake Pavin*, DOI 10.1007/978-3-319-39961-4_18

estimated during a short-term study (Bettarel et al. 2003). Finally, we used small microcosms to access growth rates of protists and their potential trophic relationships with metazoa (Carrias et al. 2001).

18.2 Distribution and Composition of Heterotrophic Nanoflagellates (HNF)

The seasonal distribution of free and attached HNF was assessed in the epilimnion of Lake Pavin from April to November 1991. Total abundance varied from 0.1×10^3 to 2.5×10^3 cells ml^{-1} and averaged 0.8×10^3 cells ml^{-1} (Fig. 18.1). Free-living HNF largely dominated the community, accounting for 80 % of the cells. *Monas*-like cells and small-unidentified cells were the dominant taxa during spring (Fig. 18.2). Bicoecids and choanoflagellates were attached to large diatoms (*Aulacoseira italica*, *Asterionella formosa*) and to the colonial cyanobacteria *Anabaena flos-aquae*, and accounted for 9 and 11 % of HNF densities, respectively. The dynamics of the free-living HNF were different from that of epiphytic HNF. Free-living HNF reached peak abundance in spring at the onset of thermal stratification and the sedimentation of the spring diatom bloom. The development of epiphytic HNF was recorded at the end of spring during the decline of free-living HNF and in autumn (Fig. 18.1). During these periods the free-living forms were probably subjected to high predation pressure from metazooplankton while large size diatoms and colonial cyanobacteria provided attached forms a refuge against metazooplankton predation. Because attached flagellates had higher ingestion rates than free-living bacterivorous flagellates (see below), their bacterivory may be of great significance in Lake Pavin.

18.3 Spatio-Temporal Distribution and Community Structure of Ciliates

The distribution of ciliated protozoa was investigated during the period of April to December 1992 in the epi- (1 m depth), meta- (15 m depth), and hypolimnion (40 m depth) of Lake Pavin. The densities of ciliates ranged from 200 to 24,000 ciliates L^{-1} (Mean = 4000 ciliates L^{-1}) and are typical of unproductive lakes. Values were comparable at depths of 1 and 15 m, with highest densities during the period of water overturn. Values remain low during the thermal stratification suggesting that there was intense predation pressure by the macrozooplankton (*Cyclops abyssorum prealpinus*, *Acanthodiaptomus denticornis*, *Daphnia longispina*, *Ceriodaphnia quadrangula*) during this period. A total of 28 genera or species of ciliates were identified during this study. Oligotrichs (*Strobilidium* spp., *Strombidium viride*, *Pelagohalteria viridis*) largely dominated the community at 1 and 15 m depth (Fig. 18.3) accounting for 40 % of total abundance. Small-sized algivorous prostomatids (*Balanion planctonicum* and *Urotricha* spp.) were relatively more abundant at 1 m (33 % of total abundance) than at 15 m (22 %). These ciliates are probably the main predators of nanoplanktonic algae. Bacterivorous scuticociliates (mainly *Cyclidium* spp.) accounted for 18 % of total abundance at 1 m and 19 % at 15 m. In addition to these three dominant groups, various taxa (*Vorticella* spp., *Colpoda* sp., *Askenasia volvox*, *Chilodonella* sp., *Lembadion magnum*) developed for short periods and occurred especially at 15 m depth. In the deep hypolimnion (40 m depth), the highest densities occurred during the major spring development of the scuticociliate *Uronema nigricans*. Scuticociliates largely dominated the hypolimnetic ciliate community, accounting for 64 % of the total abundance. In term of biomass, the ciliate community represented from 0.2 to 16.4 µgC L^{-1}.

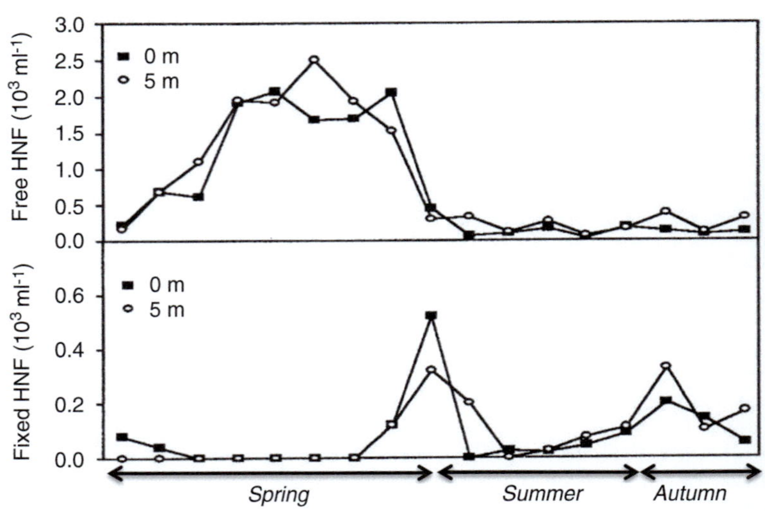

Fig. 18.1 Seasonal distribution of free and attached (fixed) HNF at 0 and 5 m depth in the epilimnion of Lake Pavin (Modified from Carrias et al. (1998a))

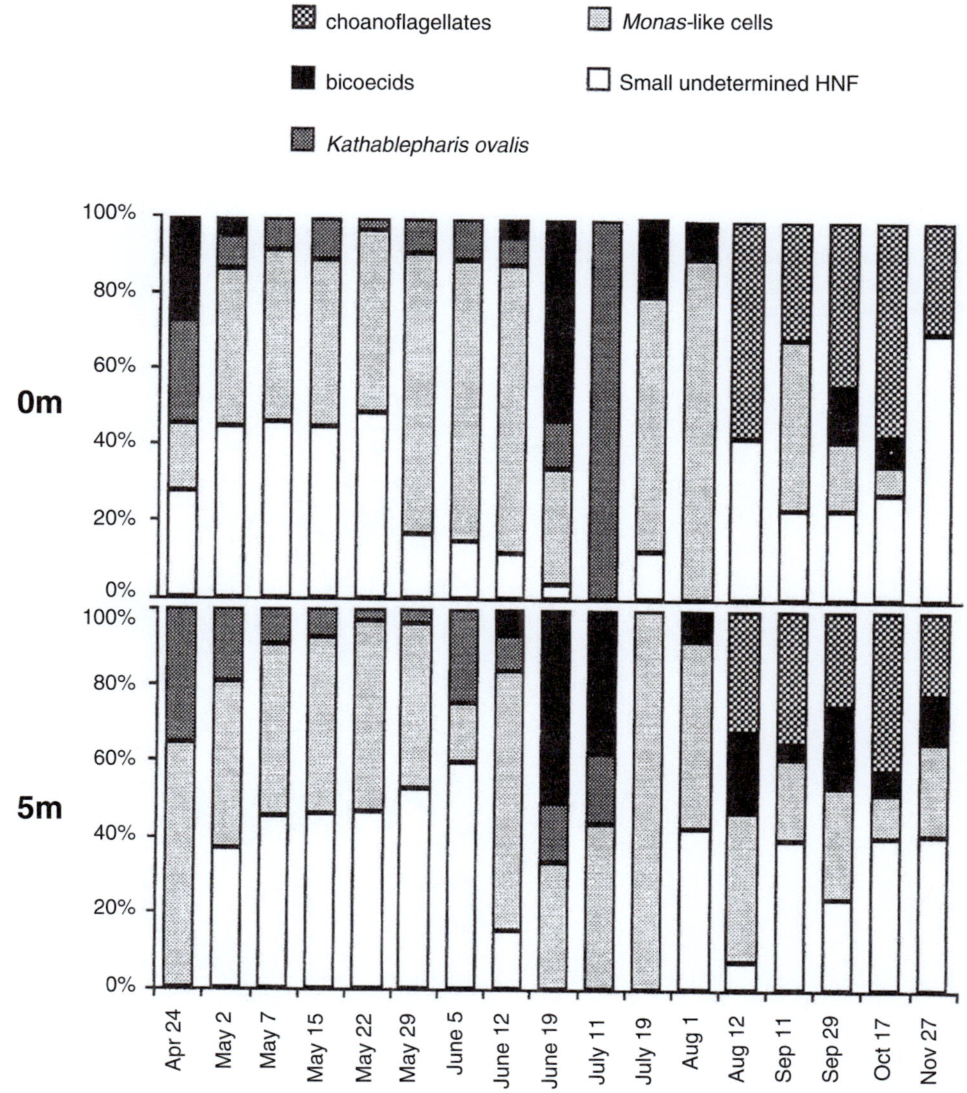

Fig. 18.2 Relative contribution of different HNF taxa to total HNF abundance. From Carrias et al. (1998a). This material is reproduced with permission of John Wiley & Sons, Inc

Mixotrophic oligotrichs (*Strombidium viride*, *Pelagohalteria viridis, Strobilidium caudatum, Strombidium* sp1*)* largely dominated the ciliate biomass at 1 m and 15 m depth during summer. These ciliates contain pigmented endosymbionts and are dependent on light for their growth. In spring and autumn, large-sized ciliates (*Colpoda* sp., *Lembadion magnum*, *Cyclotrichium* sp., *Paradileptus elephantinus*) and the prostomatid *Urotricha* sp. formed most of the ciliate biomass at 1 m and 15 m depth. In the deep hypolimnion, the bacterivorous scuticociliate *Uronema nigricans* dominated the biomass, but only in May and June (Fig. 18.3).

18.4 Bacterivory by Phagotrophic Protists

We assessed bacterivory by phagotrophic protists using fluorescent microspheres as tracers. We evaluated cell-specific grazing rates of the different taxa and estimated the importance of different groups of protists to total protistan bacterivory. These measures were done on seven occasions from June to November 1993 and at three depths representative of the epilimnion (5 m), metalimnion (15 m) and hypolimnion (40 m). Protistan bacterivory was dominated by phagotrophic nonpigmented flagellates (Table 18.1), accounting for 84 % of the total protistan bacterivory. About 62 % of nonpigmented flagellates taxa were found to ingest particles. The highest grazing impact was 18.9×10^6 bacteria L^{-1} h^{-1} during a peak of heterotrophic flagellates in June. At this time, bacterial productivity might be totally consumed by heterotrophic flagellates. Grazing impact was higher in the metalimnion than in the epi- and hypolimnion. A clear seasonal pattern was observed with highest values of grazing impact during June and in autumn, at the three depths. *Monas*-like cells were the dominant grazers, followed by small attached flagellates (bicoecids and choanoflagellates), fixed mainly

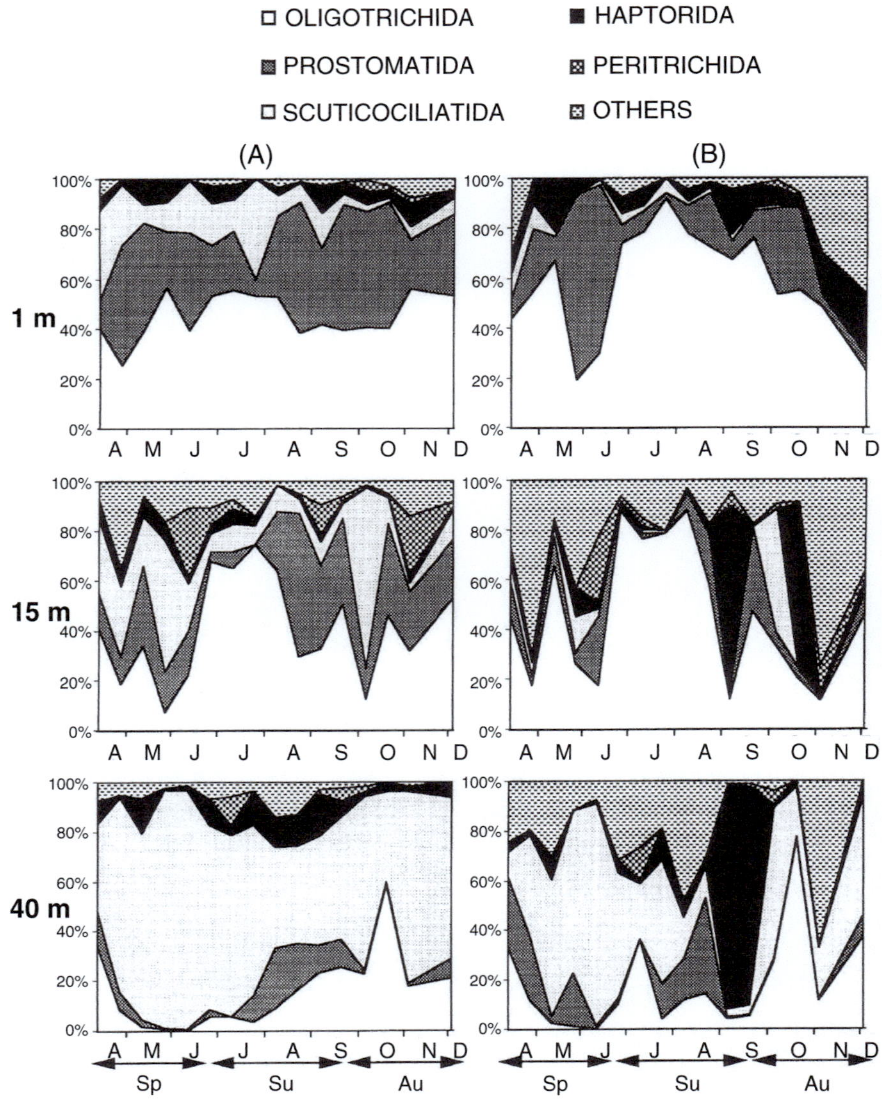

Fig. 18.3 Relative importance of different groups of ciliates, as a percentage of abundance (**a**) and biomass (**b**). Sp: Spring, Su: Summer, Au: Autumn. From Carrias et al. (1998b). This material is reproduced with permission of Schweizerbart'sche Verlagsbuchhandlung (www.schweizerbart.de)

on the diatom *Asterionella formosa*. *Katablepharis ovalis*, bodonids, and small uniflagellated cells were not found to ingest fluorescent beads during the study. *Katablepharis ovalis* may be specialized in the uptake of relatively larger size particles (e.g., nanophytoplankton). Bodonids were present only at 40 m depth and reached low densities. They are specialized as surface grazers. Attached bacteria were abundant at 40 m depth where decaying algae and detritus were usually present (see Carrias et al. chapter 13). Other heterotrophic flagellates that were not found to ingest particles consisted of small (2–5 μm in size) undetermined nonpigmented cells (Table 18.1). DAPI-stained bacteria and ingested picocyanobacteria were never seen inside these cells. Osmotrophy (see Sherr 1988) might be a potentially important source of organic carbon for these small-sized heterotrophic flagellates. Another explanation is that these cells belong to spores of fungi, especially spores of chytrids (see Sime Ngando et al. chapter 20).

Pigmented flagellates that were found to ingest particles included *Dinobryon cylindricum*, *Ochromonas* sp., and *Uroglena*. These phagotrophic algae appeared as common taxa in freshwater plankton (Bird and Kalff 1986; Bennett et al. 1990). Maximum grazing impact of pigmented flagellates was 4.5×10^6 bacteria L^{-1} h^{-1} during a bloom of *Dinobryon cylindricum*. Pigmented flagellates may have accounted for 25 % of total protistan bacterivory at some depth, but for all samples they contributed to only 3 % of the total protistan bacterivory. Bacterivorous ciliates included species of oligotrichs, scuticociliates, and peritrichs (Table 18.2). This community dominated total protistan bacterivory at some dates but accounted for only 13 % of total bacterivory for all dates sampled. The maximum grazing impact of

Table 18.1 Major flagellates taxa of the plankton of Lake Pavin, mean length, biovolume, and the range of ingestion and clearance rates

	Mean length (μm)	Biovolume (μm^3)	Ingestion (bacteria ind^{-1} h^{-1})	Clearance (nl ind^{-1} h^{-1})
Flagellates				
Pigmented				
Cryptomonas sp.	25.5	2800		
Mallomonas reginae	20.4	1100		
Dinobryon cylindricum	14.4	300	2.4–35.3	0.7–26.9
Ochromonas sp.	12.2	950	6.4–87	2.4–13.1
Rhodomonas minuta	10.4	110		
Chrysosphaerella conradii	8.3	300		
Uroglena sp.	8.0	150	1.3–3.9	0.5–1.2
Mallomonas sp.	7.4	210		
Mallomonas akrokomos	7.3	22		
Chrysidalis peritaphrena	3.9	24		
Chlamydomonas sp.	3.0	14		
Undetermined	2.5	8		
Nonpigmented				
Katablepharis ovalis	7.5	63		
Bodonids	6.0	37		
Choanoflagellates	5.0	38	1.7–33.6	1.1–13.0
Bicoecids	4.8	34	6.0–92.4	3.9–24.6
Monas-cells like	3.3	19	1.6–27	0.7–11.5
Undetermined 1	3.5	16	2.0–11.5	2.1–3.0
Undetermined spp.	3.0	14		

Modified from Carrias et al. (1996)

ciliates was 4×10^6 bacteria L^{-1} h^{-1}. Epilimnetic bacterivorous ciliates were dominated largely by small heterotrophic (*Strobilidium* spp. and *Halteria* sp.) and mixotrophic (*Pelagohalteria viridis*) oligotrichs. Due to the vertical zonation of mixotrophic and small algivorous forms the relative contribution of bacterivorous ciliates to the ciliate community increased with depth. In the meta- and hypolimnion attached vorticellids represented up to 66% of the total ciliate bacterivory. Grazing impact of scuticociliates was negligible due to their relative low rates of bacterivory.

An unusual result of this study concerns the importance of attached protozoa as bacterivores. Small heterotrophic flagellates (bicoecids and choanoflagellates) and vorticellids were found almost exclusively on *Asterionella formosa*, and their maximal densities occurred during the sedimenting period of this diatom. These attached protozoa had higher ingestion rates than free bacterivorous protozoa (Tables 18.1 and 18.2) suggesting that attachment to surfaces is the most favorable situation for suspension-feeding protozoa (Fenchel 1986). Attached protozoa were not numerically dominant but were of great significance in the fate of bacteria (Table 18.3). Therefore, even in low numbers, epiphytic protozoa may have a major grazing impact on free bacteria. Due to their attachment to large colonial algae, and the tests of some taxa (bicoecids), epibiotic protozoa are probably difficult for zooplankton to ingest. Consequently, attached protozoa would constitute a sink for carbon rather than a significant link to larger consumers.

18.5 Flagellates Grazing Potential and Viral Lysis in the Fate of Bacteria

Bacterial heterotrophic production, viral lysis, and potential flagellate grazing impacts on bacteria were estimated during a short-term study in the epilimnion and metalimnion of Lake Pavin. The main objective of this study was to estimate the relative bacterial mortality rates inferred from estimated viral infected cells and assumed flagellate grazing rates. On average, viral lysis represented 6.4% (range=3.5–10.3%) of bacterial production in the epilimnion and 15.6% (range=6.0–33.7%) in the metalimnion. Bacterial mortality due to flagellates grazing represented 38.7% (range=0.5–115.4%) of bacterial production in the epilimnion, and 66.7% (range=0.7–97.5%) in the metalimnion. Therefore, flagellates consumed a larger proportion of bacterial production than was lost to viral lysis. The summed bacterial mortality due to viral lysis and flagellate grazing ranged from 4.8 to 125.7% in the epilimnion, and from 10.3 to 103.5% in the metalimnion, suggesting that these two factors may at times control most of the bacterial production, beside other potential bacterivores such as ciliates and cladocerans.

Table 18.2 Major ciliates taxa of the plankton of Lake Pavin, mean length, biovolume, and the range of ingestion and clearance rates

	Mean length (μm)	Biovolume (μm^3)	Ingestion (bacteria $ind^{-1} h^{-1}$)	Clearance (nl $ind^{-1} h^{-1}$)
Ciliates				
Oligotrichida				
Strombidium sp.	84.0	91,700		
Strobilidium gyrans	65.9	65,000		
Strombidium viride	40.1	9500		
Pelagohalteria viridis	29.0	6400	110–1320	29–225
Strobilidium spp.	24.2	3500	6–130	5.4–60
Halteria sp.	20.0	1800	12–440	10–72
Prostomatida				
Urotricha sp.1	59.4	82,400		
Urotricha spp.	19.7	2100		
Pseudobalanion planctonicum	13.7	850		
Prorodon sp.	60.0	13,000		
Scuticociliatida				
Uronema sp.	22.8	1600	5.8–320	4.8–42
Cyclidium spp.	13.1	830	14–64	4.8–54
Haptorida				
Cyclotrichium sp.	197.0	1,200,000		
Trachellophyllum sp.	42.0	3200		
Mesodinium pulex	35.0	10,000		
Askenasia volvox	32.0	8700		
Mesodinium sp.1	24.7	1400		
Peritrichida				
Vorticella sp.1	24.5	4900	30–1610	25–211
Vorticella sp.2	48.0	38,100	460–5910	430–790
Others				
Colpoda sp.	34.6	10,800		
Undetermined	150.0	470,000		

Modified from Carrias et al. (1996)

Table 18.3 Relative abundance (Ab) and grazing impact (Gr) of attached flagellates and attached ciliates as a percentage of total abundance and grazing impact of total bacterivorous nonpigmented flagellates and total bacterivorous ciliates (Lake Pavin 1993)

	Attached flagellates		Attached ciliates	
Depths	Ab	Gr	Ab	Gr
5 m	14.0	22.9	0.9	8.1
15 m	11.5	23.2	20.9	56.8
40 m	15.1	26.9	22.7	41.4

Modified from Carrias et al. (1996)

18.6 Phagotrophic Ciliates as a Trophic Link Between Nanoplankton and Metazoa

The *in situ* growth rates of the dominant planktonic protists were estimated in unfiltered (i.e. raw, with ciliates and zooplankton) and fractionated (<50 μm: without metazooplankton, and < 10 μm: without large-size ciliates and metazooplankton) water samples incubated in diffusion chambers. We assumed that differences in the growth rates between treatments reflected the mortality caused by metazooplankton grazing. Experiments were conducted on five dates during spring to estimate both flagellate and ciliate mortality due to naturally occurring densities of potential consumers. We found that the growth rates of heterotrophic nanoflagellates in the < 10 μm fraction were significantly higher than in the < 50 μm fraction whereas no significant difference was recorded for autotrophic nanoflagellates. We concluded that heterotrophic nanoflagellates were subjected to high predation by large-sized ciliates (*Urotricha pelagica*, *Strombidium viride*, tintinnids, *Colpoda* sp.), unlike autotrophic nanoflagellates. We recorded a strong negative correlation between the growth

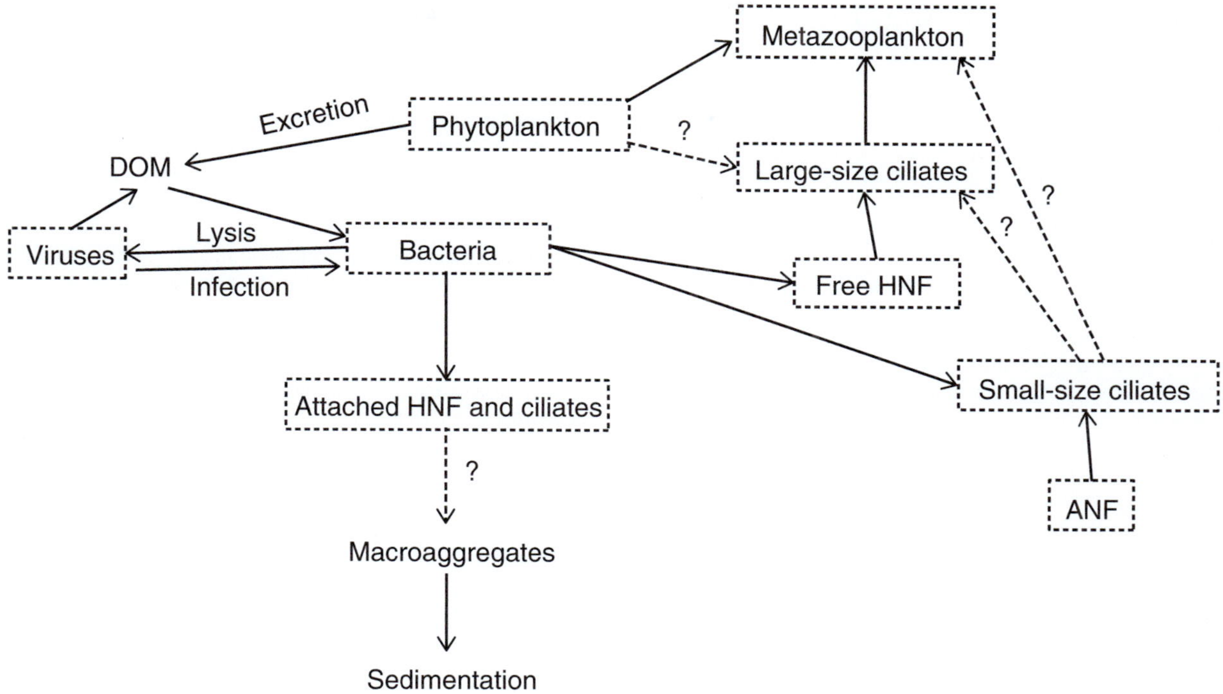

Fig. 18.4 Relationships among lake Pavin planktonic communities including viruses and protists. *DOM* dissolved organic matter, *HNF* heterotrophic nanoflagellates, *ANF* autotrophic nanoflagellates

rates of small algivorous *Urotricha* spp. and autotrophic nanoflagellates in all of the fractions studied, supporting the hypothesis that small *Urotricha* is a very important nanoalgivore in lacustrine pelagic environments (Weisse et al. 1990; Sommaruga and Psenner 1993). The growth rates of large-sized ciliates were higher in the < 50 μm fraction than in the unfiltered water. By assuming that this difference reflected the mortality caused by metazooplankton, we concluded that *Cyclops abyssorum prealpinus*, the dominant metazooplankton species, had a preponderant impact on large-sized ciliates during spring. Therefore, our results suggest that large-size ciliates may be able to efficiently transfer energy between heterotrophic nanoflagellates and metazooplankton.

18.7 Conclusions

The roles of phagotrophic protists in pelagic ecosystems are summarized in Fig. 18.4. The precise position of different groups of phagotrophic protists in the planktonic food web will provide us a much better understanding of microbial ecology and ecosystem functioning. Heterotrophic nanoflagellates are the main consumers of bacteria. Free heterotrophic nanoflagellates are particular important during spring while attached heterotrophic nanoflagellates appeared in summer and autumn. Because of their attachment to large-size colonial algae, attached flagellates and ciliates are probably difficult to ingest by ciliates and metazooplankton. In contrast, free heterotrophic nanoflagellates and autotrophic nanoflagellates are grazed by large-size ciliates and small-size algivorous ciliates, respectively. Large-size ciliates are in turn consumed by copepods (*Cyclops abyssorum prealpinus)* in spring and probably by cladocerans (*Daphnia longispina* and *Ceriodaphnia quadrangula*) during summer and autumn. Although we have no data, we assume that small-size ciliates are potential prey for metazooplankton (Sanders et al. 1996). Thus, phagotrophic protists play a pivotal role in the planktonic food webs by transforming small-size prey (bacteria, small flagellates) into larger preys (flagellates and large-size ciliates), which can then be consumed by metazooplankton.

Acknowledgements We are grateful to Robert W. Sanders for comments and valuable suggestions to improve the manuscript.

References

Amblard C, Sime-Ngando T, Rachiq S, Bourdier G (1993) Importance of ciliated protozoa in relation to the bacterial and phytoplanktonic biomass in an oligo-mesotrophic lake during the spring diatom bloom. Aquat Sci 55(1):1–9

Bennett S-J, Sanders RW, Porter KG (1990) Heterotrophic, autotrophic, and mixotrophic nanoflagellates: seasonal abundances and bacterivory in a eutrophic lake. Limnol Oceanogr 35:1821–1832

Bettarel Y, Amblard C, Sime-Ngando T, Carrias J-F, Sargos D, Garabetian F, Lavandier P (2003) Viral lysis, flagellate grazing

potential and bacterial production in Lake Pavin: a short-term study. Microb Ecol 45:119–127
Bird DF, Kalff J (1986) Bacterial grazing by planktonic algae. Science 231:493–495
Caron DA, Goldman JC, Dennett MR (1988) Experimental demonstration of the roles of bacteria and bacterivorous protozoa in plankton nutrient cycles. Hydrobiologia 159:27–40
Carrias J-F, Amblard C, Bourdier G (1996) Protistan bacterivory in an oligomesotrophic lake: importance of attached ciliates and flagellates. Microb Ecol 31:249–268
Carrias J-F, Amblard C, Quiblier-Lloberas C, Bourdier G (1998a) Seasonal dynamics of free and attached heterotrophic nanoflagellates in an oligomesotrophic lake. Freshw Biol 39:101–111
Carrias J-F, Amblard C, Bourdier G (1998b) Seasonal dynamics and vertical distribution of planktonic ciliates and their relationship to microbial food resources in the oligomesotrophic Lake Pavin. Arch Hydrobiol 143:227–255
Carrias J-F, Thouvenot A, Amblard C, Sime-NGando T (2001) Dynamics and growth estimates of planktonic protists during early spring in Lake Pavin (France). Aquat Microb Ecol 24:163–174
Fenchel T (1986) The ecology of heterotrophic microflagellates. Adv Microb Ecol 9:57–97
Pace ML, McManus GB, Findlay SEG (1990) Planktonic community structure determines the fate of bacterial production in a temperate lake. Limnol Oceanogr 35:795–808
Riemann B, Christoffersen K (1993) Microbial trophodynamics in temperate lakes. Mar Microb Food Webs 7:69–100
Sanders RW, Porter KG, Bennett SJ, Debiase AE (1989) Seasonal patterns of bacterivory by flagellates, ciliates, rotifers, and cladocerans in a freshwater planktonic community. Limnol Oceanogr 34:673–687
Sanders RW, Williamson CE, Stutzman PL, Moeller RE, Goulden CE, Aoki-Goldsmith R (1996) Reproductive success of "herbivorous" zooplankton fed algal and nonalgal food resources. Limnol Oceanogr 41:1295–1305
Sherr EB (1988) Direct use of high molecular weight polysaccharide by heterotrophic flagellates. Nature 335:1225–1227
Sommaruga R, Psenner R (1993) Nanociliates of the order Prostomatida: their relevance in the microbial food web of a mesotrophic lake. Aquat Sci 55:179–187
Weisse T, Müller H, Pinto-Coelho RM, Schweizer A, Springmann D, Baldringer G (1990) Response of the microbial loop to the phytoplankton spring bloom in a large prealpine lake. Limnol Oceanogr 35:781–794

Diversity and Biogeography of Picoeukaryotes: New Insights into the Rare Biosphere

19

Cécile Lepère, Isabelle Domaizon, and Didier Debroas

Abstract

Picoeukaryotes (unicellular eukaryotes smaller than 5 μm) are recognized as important members of microbial assemblages in aquatic ecosystems in terms of both biomass and activity. During the past decade, molecular techniques have begun to provide insights into the structure of picoeukaryote communities in various lacustrine environments. In particular, cloning and sequencing of nuclear small subunit rRNA genes have revealed a very high diversity within this assemblage and the presence of many novel eukaryotes. Understanding the spatial distribution of aquatic microbial diversity and the underlying mechanisms that underpin differences in community composition is a challenging and central goal for ecologists. Recent insights into picoeukaryotes diversity and ecology are increasing the debate on their spatial distribution. The application of novel high-throughput sequencing technologies have provided new data and new perspectives on this debate, allowing improved sampling of phylotype diversity and demonstrating a community structure composed of a complex rare biosphere. Understanding the activities of the rare biosphere may (or may not) be the key to deriving predictive models of picoeukaryotes community structure and function.

Keywords

Picoeukaryotes • Lakes • Diversity • Biogeography • Rare biosphere • 18S rRNA • Metagenetics • Pyrosequencing

19.1 Introduction

Planktonic microorganisms are categorized into classes based on their size for operational purposes. Initially, only prokaryotes were included in the smallest class (picoplankton: cells 0.2 to 5 μm) and microbial eukaryotes were included in the nanoplankton (5 to 20 μm) or microplankton (20 to 200 μm). This changed radically with the advent of molecular tools, particularly culture-independent sampling of ribosomal RNA encoding gene sequences. The first studies on the *in situ* diversity of picoeukaryotes by cloning and sequencing environmental 18S rRNA genes were published in 2001 and were focused on marine ecosystems (Lopez-garcia et al. 2001; Moon-van der Staay et al. 2001). These studies showed that aquatic picoeukaryotes include a large phylogenetic diversity and many novel lineages. In fact, environmental picoeukaryote SSU rDNA sequences sampled from freshwater lakes are distributed throughout the eukaryotic tree (Fig. 19.1).

Picoeukaryotes have a typical eukaryotic cell structure in a miniaturized cell. The smallest known eukaryote, *Ostreococcus tauri*, has a diameter of 0.8 μm. This includes

C. Lepère (✉)
LMGE, Laboratoire Microorganismes: Génome et Environnement, 6023, Université Clermont-Auvergne, Université Blaise Pascal, BP 80026, 63177 Aubière Cedex, France

CNRS, UMR 6023, LMGE, F-63171 Aubière, France
e-mail: cecile.lepere@univ-bpclermont.fr

I. Domaizon
INRA, UMR 42 CARRTEL, F-74200 Thonon les Bains, France

D. Debroas
LMGE, Laboratoire Microorganismes: Génome et Environnement, 6023, Université Clermont-Auvergne, Université Blaise Pascal, BP 80026, 63177 Aubière Cedex, France

T. Sime-Ngando et al. (eds.), *Lake Pavin*, DOI 10.1007/978-3-319-39961-4_19

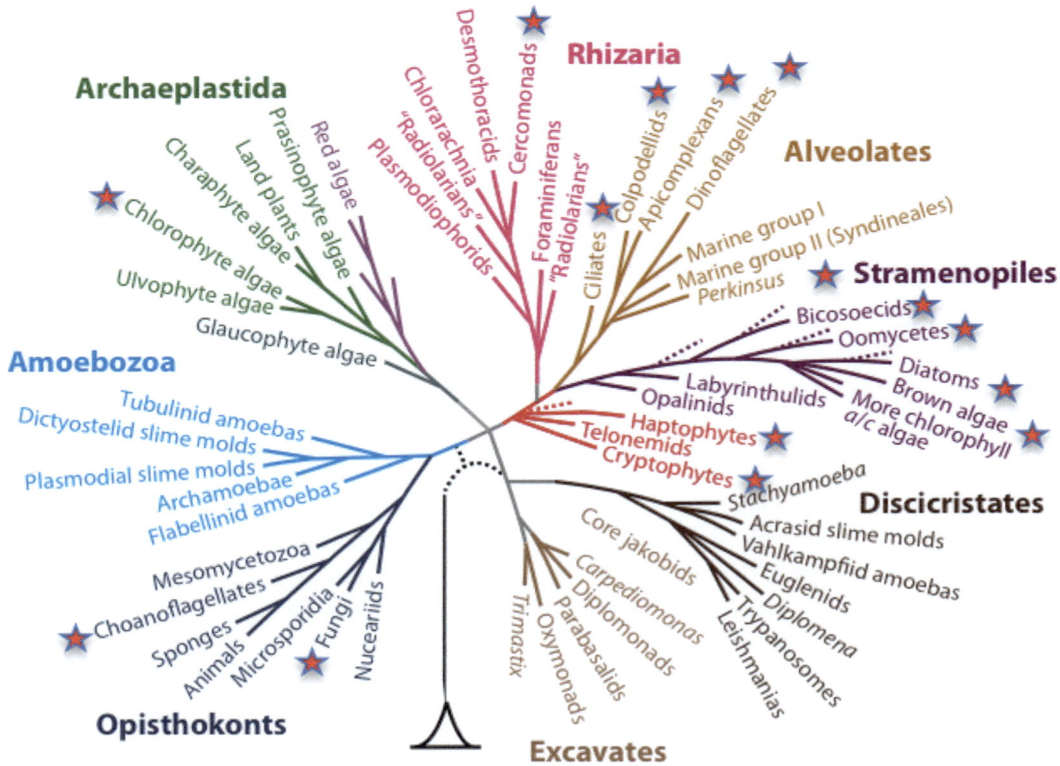

Fig. 19.1 Phylogenetic breadth among microbial eukaryotes adapted from Baldauf (2008). Groups where environmental picoeukaryote SSU rDNA sequences sampled from freshwater lakes have been found are highlighted by a star

the presence of a nucleus, an endomembrane system (endoplasmic reticulum, Golgi, and vesicles), mitochondria, and in the case of photosynthetic picoeukaryotes, a chloroplast. Due to this small size, picoeukaryotes are largely indistinguishable by light microscopy and their study was restricted to the use of electronic microscopy until the revolution of molecular techniques. Consequently, very few picoeukaryotes have been characterized. The picoeukaryote assemblage is formed by picoalgae, which participate in primary production (Stockner and Antia 1986), by colorless heterotrophic cells, mostly flagellates, which are considered to be important grazers of prokaryotic and eukaryotic cells (Caron et al. 1999) and also play a significant role in the mineralization of organic matter, and finally by some eukaryotes which can be mixotrophs (Hartman et al. 2013) and parasites (Guillou et al. 2008; Lepère et al. 2008) (Fig. 19.2).

Among picoplankton the relative abundance of eukaryotic cells fluctuates according to the systems (open ocean, coastal, lakes), the trophic states, and the spatio-temporal dynamics of these communities. Epifluorescence microscopy as well as flow cytometry allowed the determination of the abundance and the distribution of picoeukaryotes from a wide range of regions (Andersen et al. 1996). The densities of picoeukaryotes reach typically 10^2 to 10^4 cells. ml^{-1} in the euphotic zone of lakes and oceans (Caron et al. 1999). In lacustrine environments, picoeukaryotes abundances determined by TSA-FISH (Tyramide Signal Amplification-Fluorescent *in situ* hybridisation) varied on average from 5285 cells ml^{-1}(1414–9428 cells ml^{-1} in the oligomesotrophic lake Pavin) to 31593 cells ml^{-1} (16452–42832 cellsml^{-1} in the eutrophic lake Aydat) (Lepère et al. 2010). When the lake is stratified (summer period) a clear gradient of densities is generally observed from the top to the bottom of the water column (Lepère et al. 2010; Comte et al. 2006).

19.2 Lacustrine Picoeukaryotes Diversity

A main goal in microbial ecology is to assess how individuals are organized in taxonomic units and how they contribute to ecological processes. Picoeukaryotes are integral members of aquatic ecosystems in terms of cell abundance, activity, and diversity and therefore play crucial roles in trophic food webs and biogeochemical cycles (Fig. 19.2). Diversity and physiology of microorganisms in lakes, especially picoeukaryotes, are poorly studied compared with other ecosystems, such as the marine environment.

Lakes provide unique sentinels and integrators of events in their catchments and airsheds and in the total landscapes in which they are embedded (Schindler 2009). Furthermore, these ecosystems, which can be viewed as interconnected islands, are *de facto* excellent models of choice to study

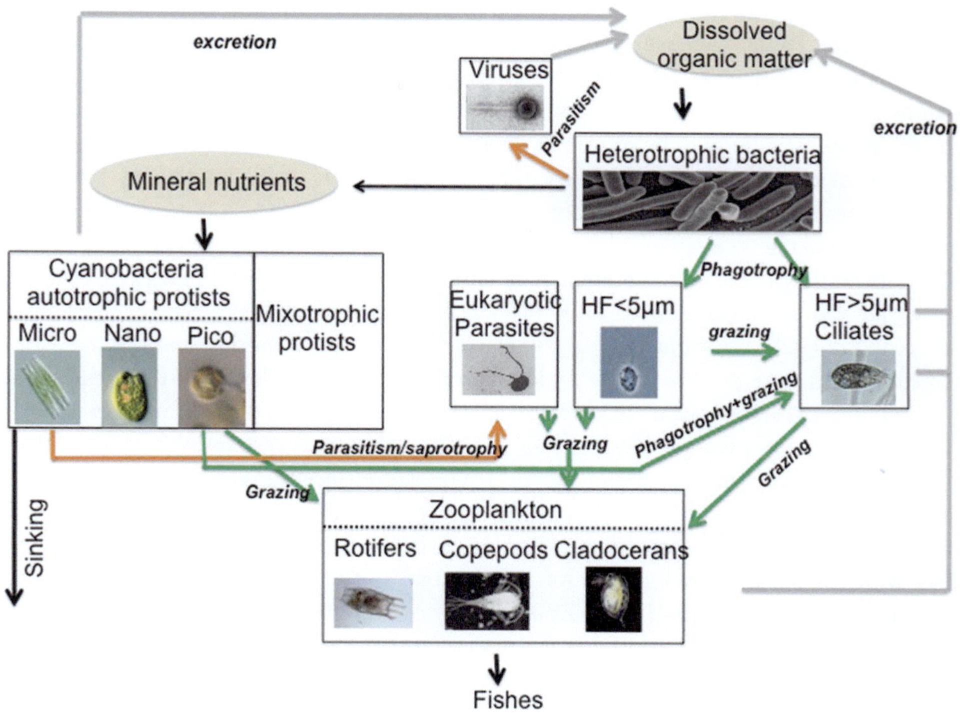

Fig. 19.2 Model of carbon flows between primary producers, grazers zooplankton, and microorganisms in lacustrine pelagic ecosystems. HF:heterotrophic flagellates; DOM: dissolve organic matter

microbial diversity and spatial distribution. One of the first model lake used to study picoeukaryotes was the lake Pavin (Carrias et al. 1996). The lake Pavin is a typical crater mountain lake with a maximum depth of 92 m. It is characterized by the presence of two permanent stratified layers. The upper layer (mixolimnion) and the deepest layer (monimolimnion), which is characterized by anoxic conditions (for more description of the lake, see Chap. 1). Picoeukaryotes are present throughout the water column (including the anoxic zone).

19.2.1 Black Box of Non-pigmented Picoeukaryotes: Unexpected Importance of Parasites

Since the emergence of the 'microbial loop' concept (Azam et al. 1983; Amblard et al. 1995) heterotrophic flagellates have received particular attention as grazers in aquatic ecosystems (Boenigk and Arndt 2002). These microorganisms have historically been considered incorrectly as a homogeneous group of bacterivorous protists (e.g Carrias et al. 1996; Massana et al. 2009). More recently, environmental rDNA surveys in the pelagic zone of lake ecosystems showed that dominant phyla found by molecular studies differed from morphological studies and a substantial part of the sequences cannot be assigned to bacterivorous but to parasitic and saprophytic organisms, such as zoosporic true fungi (chytrids) and alveolate parasites (Perkinsozoa) (Lepere et al. 2008; Lefèvre et al. 2008). These findings changed our view of the trophic web and highlight the potential importance of the parasitism usually left out of aquatic trophic network functioning.

In freshwater lakes, fungi were discovered a long time ago (Canter and Lund, 1948), and culture collections have been steadily supplemented by freshwater fungi in the last 50 years however not with many chytrids. The advent of molecular methods highlighted a high diversity of fungal sequences in lakes and most of them were affiliated to chytrids (Lefranc et al. 2005; Lefèvre et al. 2008; Lepère et al. 2008; Monchy et al. 2011). These fungi are inferred based on phylogenetic position as zoosporic organisms. These fungi were obviously overlooked in previous conventional microscopic studies because of their small size and the lack of conspicuous ultrastructural features (Lefèvre et al. 2008; Lepere et al. 2008). Jobard et al. (2010) showed by using FISH method that these fungi contributed up to 60 % of the total abundance of heterotrophic flagellates in lake Pavin. Multiple roles for zoosporic fungi in lake ecosystems have been suggested, including parasitism, especially with chytrids, which can parasite a wide range of phytoplanktonic hosts (For more details on Fungi role, please see the Chap. 20). Fungi 18S rDNA sequences were identified in lakes with a large range of trophic level (from oligotrophic to eutrophic lakes). In lake Pavin and in other lakes, Fungi OTUs (Operational Taxonomic Units) were for the majority affiliated by phylogenetic analyses with the lineage of chytrids, although Basidiomycota and Ascomycota were also found. On average 25 % of the picoeukaryotes sequences (all sequences including pigmented and non pigmented picoeukaryotes) in lakes are affiliated to fungi. This percentage is around 20 % for the lake Pavin whatever the methods used (Sanger sequencing (Lefèvre et al. 2008) and pyrosequencing (Monchy et al. 2011). In this lake, statistic analysis showed the distribution of fungi to be related to depth (Lepère et al. 2010). In lake Pavin, chytrids infected diverse phytoplankton host communities, primarily diatoms, chlorophytes, and colonial and filamentous cyanobacteria (Rasconi et al. 2009). The data on the prevalence and intensity of chytrid infection show that the prevalence increased with increasing trophic status and ranged from <1 to 16 % in lake Pavin (oligomesotrophic) and from 1 to 24 % in lake Aydat (eutrophic).

Cryptomycota (Jones et al. 2011) were detected in several lakes, with an average of 195 cells ml^{-1} (3.3 % of total picoeukaryotes (i.e eukaryotic cell <5 μm)) (Lepère et al. 2010). Globally, the highest abundance along vertical profiles was found in the meta- or hypolimnion, even though they were not specifically found in the anoxic zone as suggested by Lara et al. (2010). In lake Bourget, they could represent up to 51 % of the small heterotrophs targeted in the deeper zone of the column water (110 m) (Lepère et al. 2010). Members of the new phylum Cryptomycota were proposed to represent intermediate fungal forms, lacking a chitinous cell wall during feeding and known almost exclusively from ubiquitous environmental ribosomal RNA sequences that cluster at the base of the fungal tree (Jones et al. 2011; Lara et al. 2010). By using a culture of the only described genus assigned to Cryptomycota, Rozella, James et al. (2014) sequenced the first Cryptomycotan genome (the endoparasite Rozella allomycis) and unite the Cryptomycota with another group of endoparasites, the microsporidia, based on phylogenomics and shared genomic traits. Tyramide signal amplification coupled with group-specific fluorescence *in situ* hybridization reveals that they are picoeukaryotes of 3–5 μm in length, capable of forming a microtubule-based flagellum. Co-staining with cell wall markers demonstrates that representatives from this clade do not produce a chitin-rich cell wall during any of the life cycle stages observed and therefore do not conform to the standard fungal body plan (Jones et al. 2011).Although their functional role is still unknown, these organisms seem to be associated with the decomposition of phytoplanktonic organisms (microalgae and cyanobacteria), and could therefore contribute to the decomposition of organic compounds in oligotrophic and oligo-mesotrophic systems (Van Hannen et al. 1999). Recently, Lara et al. (2010) reported that the Cryptomycota encompassing Rozella form the deepest branching clade in the fungi and highlighted the hypothesis that the two groups might be composed to a large extent (if not entirely) of parasites. Only one study (Jones et al. 2011) showed Cryptomycota associated with diatoms (Phytoplankton). Even if it would be premature to draw any conclusion on the lifestyle and ecology of these organisms on the basis of only environmental sequences, a number of open questions arose about Cryptomycota ecological role and phylogenetic position.

Among putative parasitic groups, the Perkinsozoa (for a review see Mangot et al. 2011), already known to play a significant role as parasite in marine systems (Moore et al. 2008; Norén et al. 1999) is of special interest, since it has only recently been detected in lakes by constructing 18S clone libraries (Lefranc et al. 2005; Lepère et al. 2008). Using Sanger technique, sequences affiliated to the Perkinsozoa can account for up to 15.2 % of the OTUs. In lake Pavin the greatest abundance was found by TSA-FISH in the epilimnion with an average of 178 cells ml^{-1}. In marine systems, this group is known for parasiting molluscs or phytoplanktonic species, but their functional importance in freshwater environments is still largely unknown. Only one study showed an infection of a cryptophyte by a Perkinsozoa (*Rastrimonas subtilis*) in a river environment (Brugerolle 2002, 2003). This group of putative parasites detected in pelagic environment is characterized by a zoospore stage (Mangot et al. 2011). Recent studies show the quantitative importance of these zoospores in euphotic zones of several lakes (Lepere et al. 2010; Mangot et al. 2013) and their activity inferred by the sequencing of transcripts (RNA) and the high RNA/DNA ratios associated to Perkinsozoan sequences (unpublished data). In lakes environment, the phytoplankton is certainly the most likely host for Perkinsozoans (Mangot et al. 2013), however, despite of certain indications (eg phylogenetic position, punctual microscopic observations), the identification of the host remains incomplete and so their putative role as pathogens remains unknown.

Another lineage relatively well represented in 18S SSU rDNA databanks in lakes is the Cercozoa, a group defined by Cavalier-Smith (1998). Environmental sequences affiliated to these organisms were found in all studied lakes with relatively important proportions both in the euphotic and deeper zones (Lefèvre et al. 2007). These microorganisms can represent more than half of characterized OTUs in the epilimnion of the lake Pavin in summer (Lepère et al. 2006) and more than a third of the sequences in the Autumnal oxycline of the same lake (Lefèvre et al. 2007). Three different clades were determined among these Cercozoa, two are strictly environmental, the third one contain sequences affiliated to the order of Cercomonadida and more exactly to the genus *Cercomonas* and *Heteromita* (Lepère 2007). Broadly bacterivores with thecamebians and the order of Cercomonadida, Cercozoa can also be parasites of phytoplankton. This is the case in particular of the genus *Cryothecomonas,* which are nanoflagellates, found associated with diatoms (Schnepf and Kühn 2000).

Of course the diversity mentioned in this section is far from being exhaustive and some other groups are also retrieved such as Chrysophyceae and ciliates. Chrysophyceae are a mixotrophic class characterized by plastidic and colorless cells. On average, in lake environment, the class of Chrysophyceae represents about 18 % of the diversity described in the epilimnion by molecular techniques. Sequences of Chrysophyceae obtained in lakes allowed phylogenetic analysis of these environmental clades (Richards et al. 2005; Lepère et al. 2008; Tarbe et al. 2011). The majority of these Chrysophyceae sequences are close to non-pigmented species such as *Spumella*, *Oïkomonas*, *Paraphysomonas* or *Poterioochromonas* (Lefranc et al. 2005; Richard et al. 2005). However, some sequences phylogenetically close to photosynthetic cells (eg *Mallomonas*) are also found (Tarbe et al. 2011). Although ciliates are generally known to have a size superior to 5 µm (Bourrelly 1970), their presence in this size fraction of freshwater lakes could represent the detection of previously unisolated groups or the results of cell damages during filtration steps. On average, in lake environment, 8 % of the OTUs characterized in the epilimnion are affiliated to ciliates whatever the lake and the trophic level (Lepère et al. 2008).

Results obtained with molecular and microscopic approaches in lakes showed differences with oceanic ecosystems. Indeed, parasites also seem to be important in marine environments, but most of them are affiliated with Syndiniales (Alveolates) (Guillou et al. 2008). The exclusively marine provenance of this group is confirmed by the fact that their 18S rRNA gene sequences have not been retrieved from lake PCR surveys using eukaryotic general primers.

19.2.2 Pigmented Picoeukaryotes

Studies conducted in lakes using the cloning-sequencing method with generalist primers targeting gene coding for 18S SSU rRNA encoding gene highlighted that putatively pigmented picoeukaryotes sequences were always present at a very low proportion in clone libraries (Richards et al. 2005; Chen et al. 2009; Lefèvre et al. 2008; Lepere et al. 2008). Most of the sequences affiliated to pigmented eukaryotes were retrieved from epilimnic samples and were represented by Cryptophyta, Chlorophyta, and Haptophyta. However, using FISH method, Mangot et al. (2009) showed during temporal study (1-year) that Chlorophyta were well represented, accounting for 17.9 % of small eukaryotes abundance in lake Bourget. Surprisingly, Chlorophyta are present all along the water column, even in the deepest points out of the euphotic zone; up to 950 cells ml^{-1} were counted at 50 m in lake Pavin (Lepère et al. 2010). Similarly, Chen et al. (2009) observed, by morphological observations, that some pigmented taxa, such as *Tetraedron* sp., were abundant in the Meiliang Bay, but were not detected by molecular analysis.

FISH data also showed the presence of members of Prymnesiophyceae (Haptophyta) in lakes (Lepère et al. 2010) although they were present at very low proportions in 18S sequences bank (Lefranc et al. 2005; Lepère et al. 2008; lefèvre et al. 2008; Richards et al. 2005) which is interesting, since, described species of Prymnesiophyceae are generally considered to be larger than 5 µm (Vaulot et al. 2008). Recently, this group was recognized as a major component of the eukaryotic picoplankton in marine water and particularly in oligotrophic waters (Liu et al. 2009; Lepère et al. 2009; Cuvelier et al. 2010). Haptophyta are well known in marine systems (Moon-van der Staay et al. 2000; Iglesias-Rodriguez et al. 2002), especially because of their ability to form toxic blooms (Gjøsæter et al. 2000; Baker et al. 2007), only a dozen Haptophyta species have been previously described from freshwater or terrestrial habitats (John et al. 2002). However, some haptophyte blooms have been previously recorded in different lacustrine systems (Nicholls et al. 1982; Hansen et al. 1994), suggesting an importance of this group in other freshwater ecosystems. Statistic analysis revealed an opposite distribution between Chlorophyta and Haptophyta groups (Lepère et al. 2010). Haptophyta seems to be confined to surface waters, 0–20 m, but they can also be detected below the photic zone in lake Pavin and other lakes (Lepère et al. 2010, unpublished data). On average, Haptophyta contribution to the total diversity seems less important than chlorophytes in all the lakes studied, and sequence contribution seems to decrease with the trophic level.

According to these methods (microscopy, SSU rDNA PCR and FISH), Cryptophyta is an important group among the pigmented lacustrine picoeukaryotes. For example, Mangot et al. (2009) showed by the TSA–FISH method that cryptophytes accounted on average for 9.5 % of the total picoeukaryotes (<5 µm) and they could represent up to 30 % of the sequences in 18S libraries (Lepère et al. 2008; Lefèvre et al. 2008). These sequences revealed four clades where some of them seem to be restricted to oligo- and mesotrophic systems (Lepère et al. 2008).

Although the use of molecular approaches, especially sequencing of the 18S rRNA gene, has greatly improved our understanding of the diversity and distribution of aquatic picoeukaryotes, all these data suggest that 18S rRNA clone libraries construction have underestimated pigmented cells, and because of their limitations, molecular approaches based on PCR may not reflect the real diversity. Quantitative approach by the TSA–FISH method confirmed these potential PCR biases, revealing the relative importance of Chlorophyta and Haptophyta in lakes. Indeed, nuclear rDNA PCR-based studies of eukaryotic communities are subject to selective amplification biases due to GC content (Liu et al. 2009). Gene copy number is another factor that must be taken into account when considering the bias of clone libraries (Zhu et al. 2005). In order to focus on phototrophs, several strategies have been developed, including studies targeting plastid genes (Fuller et al. 2006; Lepere et al. 2009), the use of specific primers for photo- synthetic taxa (Viprey et al. 2008) and the construction of clone libraries from flow cytometry-sorted populations (Shi et al. 2009; Marie et al. 2010). Finally, the use of 454 pyrosequencing allowed to obtain a very efficient sequencing, and phylogenetic analyses applied to NGS data allowed to shed light on clades of freshwater picoeukaryotes rarely or never detected (including pigmented taxa) with classical molecular ecology approaches (Debroas et al. 2015). For instance Taib et al. (2013) showed that less than 5 % of the OTUs identified from massive sequencing data had previously been detected in lakes.

Anyway, all approaches have led to the common conclusion that photosynthetic picoeukaryotes are more diverse than previously thought and have highlighted their importance in aquatic environments. As already highlighted for non-pigmented picoeukaryotes, community composition revealed in lakes highlights a difference in the composition of pigmented picoeukaryotes between these environments and oceans. Indeed, the main pigmented class identified in the ocean, the Mamiellophyceae (Marin and Melkonian, 2010) with the three genera *Ostreoccocus, Bathycoccus*, and *Micromonas,* have not been detected in lakes in the picoeukaryotic fraction so far (Lepère et al. 2008; Debroas et al. 2015).

The presence of pigmented cells, traditionally considered as photoautotrophs, in the deepest zone of lakes (Lepere et al. 2010, 2016) clearly suggests mixotrophic behaviour of these taxa. Chlorophyta sequences found in lakes were mostly affiliated to Chlamydomonadales (Lepere et al. 2008). Recent experiments by Tittel et al. (2009) showed that heterotrophy occurred in Chlamydomonas to exploit dissolved organic carbon by osmotrophy. Moreover, Ukeles and Rose (1976) showed the mixotrophy of various strains of Chlorophyta. In ocean environments, Liu et al. (2009) showed that the phylogenetic position of pico-Haptophyta implies that they are photophagotrophic, in agreement with the recent discovery of dominant bacterivory by small eukaryotic phytoplanktons in oceans (Zubkov and Tarran 2008; Hartman et al. 2012, 2013). According to Zubkov and Tarran (2008), small picoalgae carry out 40–95 % of the bacterivory in the euphotic layer of the temperate North Atlantic Ocean in summer, suggesting the global significance of mixotrophy. This finding reveals that even the smallest algae have less dependence on dissolved inorganic nutrients than previously thought, obtaining a quarter of their biomass from bacterivory. Moreover, phagotrophy in photosynthetic Haptophyta was well described in genus Chrysochromulina (Legrand et al. 2001), a genus observed in various lakes (Temponeras et al. 2000). Mixotrophy may then provide a competitive advantage over both purely phototrophic microalgae and non-pigmented protists (Stickney et al. 2000; Domaizon et al. 2003; Troost et al. 2005; Kamjunke et al. 2007) in oligotrophic systems and/or in situation of phosphorous depletion (epilimnion in summer stratification) or light limitation.

Even more unexpected is the presence of pigmented cells in the anoxic zone of lake Pavin. Indeed, 454 pyrosequencing, quantitative PCR as well as microscopic observations showed the presence of pigmented picoeukaryotes at 80 m in lake Pavin (Lepère et al. 2016). The groups identified are mainly Haptophyta and Chlorophyta. Anaerobic metabolic pathways allow unicellular organisms to tolerate or colonize anoxic environments. Over the past 10 years, genome sequencing projects have brought a new light on the extent of anaerobic metabolism in eukaryotes. A surprising development has been that free-living unicellular algae capable of photoautotrophic lifestyle are, in terms of their enzymatic repertoire, among the best equipped eukaryotes known when it comes to anaerobic energy metabolism. Among phytoplankton, the green algae *Chlamydomonas reinhardtii* and Chlorella have the most extended set of fermentative enzymes reported so far (Atteia et al. 2013).

19.3 Picoeukaryotic Rare Biosphere

Even though the notion of rare taxa is known for a long time in ecology, this notion is fairly new in the microbes world. The recent application of NGS technologies has aided in the discovery of a tremendous diversity of undescribed microbes and gave us a way of sequencing the rare microbial biosphere (Pedros alio 2012). Sogin et al. (2006) provided the first evidence of a myriad of bacterial OTUs found to exist in low abundance, which has resulted in the ecological significance of rare microorganisms becoming one of the hottest topics in microbial ecology in which many questions remain unanswered. The studies about the rare microbial biosphere are mainly focused on bacteria and Archaea and fail to include the microbial eukaryotes (ie. picoeukaryotes), which are involved in the main biogeochemical cycles of the earth and cover all of the functional roles (discussed in the previous section: phototrophs, parasites, saprotrophs, phagotrophs). The first attempt to study the dynamics of the rare microbial eukaryotes was based on the analyses of DNA sequences obtained by Sanger sequencing method (Caron and Countway 2009), demonstrating that it is likely that the rare microbial biosphere was underestimated. Recently, the NGS approach demonstrated a stable predominance of a few highly abundant taxa at different temporal scales in lakes (Nolte et al. 2010; Mangot et al. 2013), while rare microorganisms make up the majority of the diversity observed within eukaryotic assemblages. Although the ecological roles of the rare microorganisms remain unclear; it is now quite clear that we have to distinguish different categories of 'rareness'. Some rare microorganisms are likely on their way to local extinction or are transient taxa in an environment; others may be active at low abundance, providing important functions in the systems; others may be dormant or inactive, awaiting favorable environmental conditions to grow (Shade et al 2014). These rare inactive microbial eukaryotes (ie. dormant taxa) have indeed been observed to become dominant with changing environmental conditions according to the model of Jones and Lennon (2010). Therefore, this study shows that the variation of total phosphorus had no effect on the proportion of dormant microbial eukaryotes that were detected by T-RFLP in lakes, and the transition between activity and dormancy plays a more important role in shaping bacterial communities than eukaryotic communities. The contrast between both domains could be due to the ability to form resting stages or to the differential sinking rates. The switch between active and dormant stages could be insignificant for protists, and this fact could have an effect on the activity of rare eukaryote taxa that have not been studied with the exception of a recent work in marine environments (Logares et al. 2014). Picoeukaryotes were characterized by numerous new lineages as underlined in the previous section and a biogeography of the rarest taxa discussed thereafter. Thus, the richness would be higher than expected and additional lineages could be likely found when described from supplemental molecular inventories.

To display a general view of the picoeukaryote rare biosphere in lakes, the pyrosequencing data obtained from temporal dynamics (1-year period) of the lake Pavin were analysed with those from the lake Bourget (lake located in Alps 45°44'N; 5°51'E). After discarding the singletons and OTUs included only rRNA reads, 6446 OTUs were defined representing 229723 reads (rRNA and rDNA). The taxonomic annotation was conducted according two ways: the nearest neighbour and the last common ancestror (methods. The closest OTUs defined monophyletic groups also named phylogenetic units. These OTUs delineated 686 monophyletic groups in term of OTUs number, the main taxonomic groups were the Alveolta (1923 OTUs), Fungi (1564), Stramenopiles (1165) and Viridiplantae (776). The rDNA

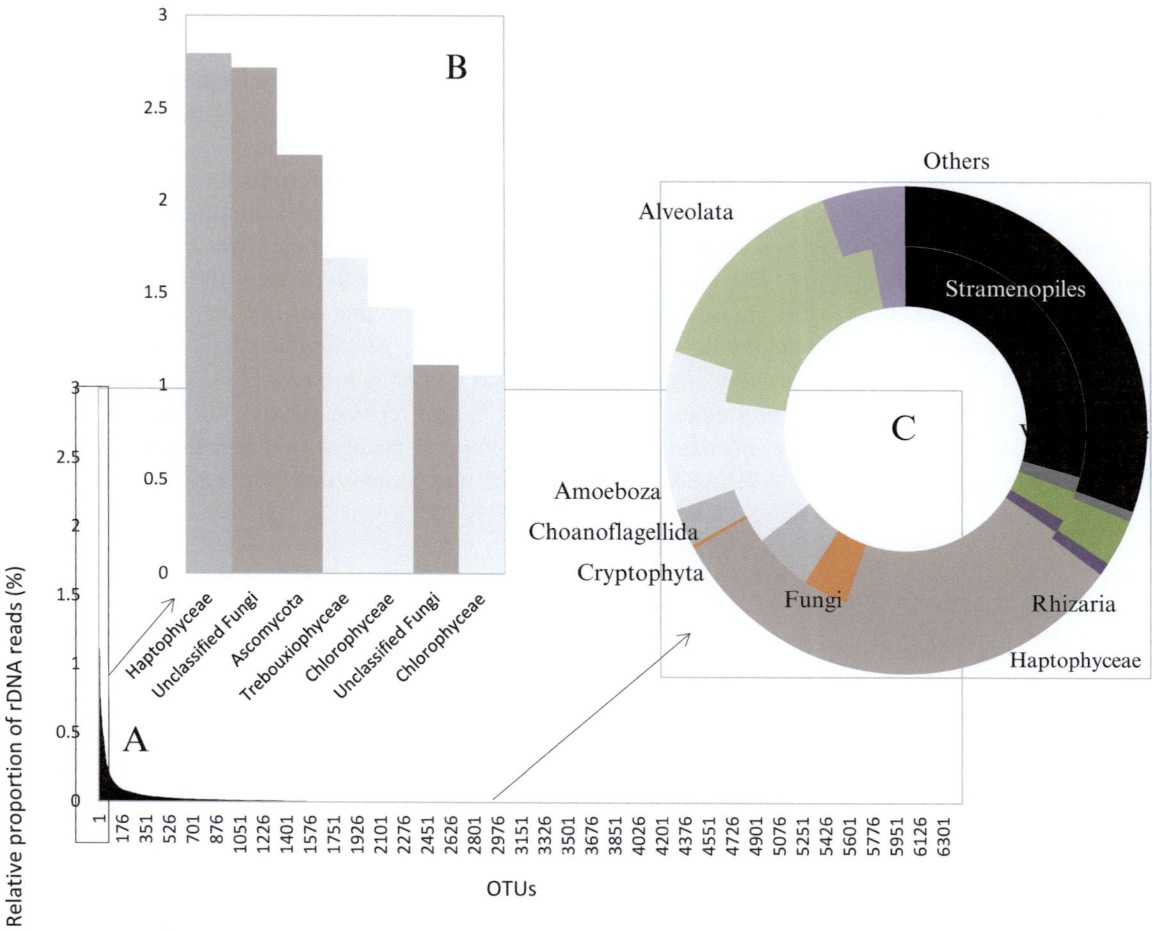

Fig. 19.3 Rank-abundance curves based on the rDNA reads (A and B) and main active and inactive taxa (C) from the two lakes studied. The panel B displays the OTU taxonomy for the dominant ones (>1 %). In the panel C, the outside of the circle corresponds to the active OTUs (OTUs with rRNA reads) and inside the inactive OTUs (including only rDNA reads). The data were obtained from pyrosequencing of 18S rDNA and rRNAamplicons. Primers used amplification conditions and-bioinformatic tool were described in detail in Taib et al. (2013)

reads allowed to build a rank-abundance curve (Fig. 19.3) showing that only 7 OTUs had an abundance greater than 1 % belonging to Haptophyceae, Chlorophyta and Fungi. We defined as abundant the OTUs characterized by a proportion of DNA reads greater than 0.1 %. The other OTUs belonging to the rare biosphere were divided in two classes: the OTUs that are always rare in all samples rare and the others (cycling fraction) being sometimes rare (≤0.1 %) but becoming other times not rare (>0.1 %). The rare and the cycling fraction represented 77.2 % and 20.4 % respectively of the 6446 OTUs. When the data were available, the proportion of rare taxa was lower for marine bacteria (44.7 %) (Campbell et al. 2011) and Archaea (40.1 %) (Hugoni et al. 2013). The contribution of the rare fraction of picoeukaryotes sequences in the lacustrine ecosystems is congruent with the results obtained in marine system, which showed that the rare taxa ranged between 66.2 % and 76.6 % (Logares et al. 2014).

19.3.1 Is There New Taxa in the Rare Biosphere?

Many investigations using NGS have uncovered a large number of unclassified sequences with low abundances (Bartram et al. 2011). However very few studies explore the link between rarity of an OTU and its phylogenetic distance to the nearest neighbor reference (sequences from cultured or uncultured taxa available in public database, ie Silva, NCBI, …) whereas this diversity could represent new lineages with new physiological properties. The always rare Archaea in the coastal surface waters were characterised by a group of OTU with low identity with the public database sequences (Hugoni et al. 2013). For picoeukaryotes, in the always rare category, 13.9 % of OTUs have an identity < 92 % whereas this proportion is only 6.4 % for the abundant OTUs. In addition, the lowest similarities found by BLAST search were observed for MGs including only always rare OTUs. For the phylogenetic indices (MNNDs and patristic distances), the greatest values corresponded to the MGs including only always rare OTUs. Therefore, these results underlined that these MGs presented the most divergent indices from the nearest neighbour and could represent newly detected lineages of which the SSU rDNA phylogeny is indicative that they represent previously undescribed taxonomic groups at the level of Genus. Interestingly, these clades were active and distributed throughout all phylogenetic groups, and thus belonged to various functional groups (parasites, autotrophs, saprotrophs, etc.). Therefore, they are present in groups particularly studied by the cultural approach (Fungi, phytoplankton), which highlights that an important fraction of the eukaryotes' biodiversity likely remains unknown. Finally, the presence of the unknown taxa among the rare biosphere and the existence of a biogeography among the rarest picoeukaryotes (Lepère et al. 2013) suggest that their richness is certainly higher than expected.

19.3.2 An Active Biosphere

There are numerous debates about the fact that the rare biosphere is active or is composed by inactive taxa (dormant cells, dead cells, pseudogenes etc…). From DNA analysis it is not possible to confirm that the rare and highly divergent 18S rRNA gene sequences amplified represent active microorganisms. By studying the presence or not of rRNA in each OTU, it is possible to define active and inactive OTUs. The abundant, the cycling fraction and always rare OTUs represented 40.2 %, 32.3 % and 27.5 % of the active biosphere depicted by rRNA respectively, whereas the most abundant OTUs (>1 %) represented only 13.3 % of the total activity. The temporal changes of the active OTUs on lake Pavin showed that composition of the always rare OTUs differed

Fig. 19.4 Dynamics of the active monophylogenetic groups (MGs) including only always rares OTUs (AllwR) considered as highly frequent (observed at least in 70 % of the samples) in Pavin. Each MG is expressed as normalized rRNA reads according to the following formula: (x–μ)/∫ where x is the number of rRNA reads at a given date, μ and ∫ are the mean and the standard deviation computed for all the dates for the MG considered. A MG is defined by its taxonomy and an number in the brackets

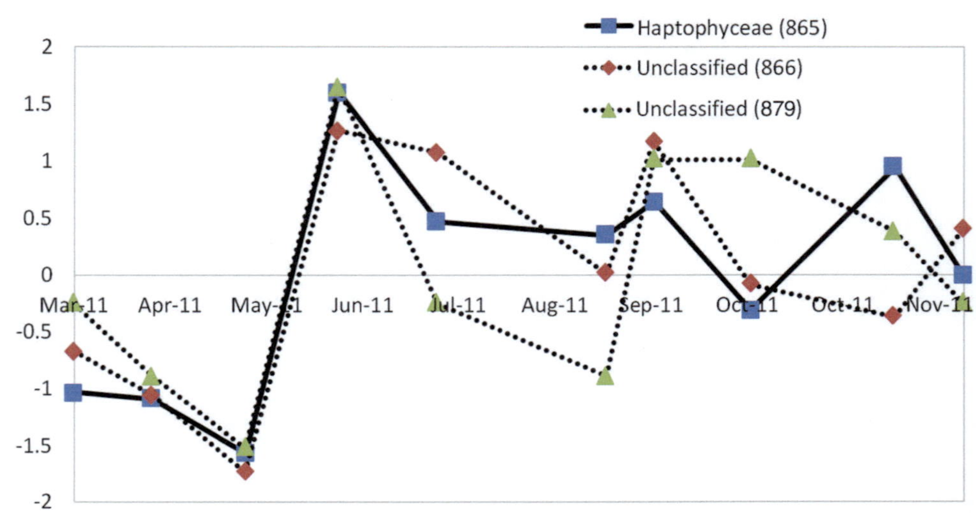

from the composition observed in the others categories. Interestingly, some active PUs constituted exclusively by always rare OTUs and highly frequent in time were discriminated (Fig. 19.4). The dynamics of these clades at 2 m depth were negatively correlated to the Cyanobacteria dynamics ($P<0.05$) and, for the PU named "Unclassified (879)" a positive correlation ($P<0.05$) with ammonium concentration was detected.

The rarest fraction, and more particularly, the always rare OTUs, represented a significant part of the active picoeukaryotes in lacustrine ecosystems. In comparison, the same fraction among the Archaea represented only 1.6% of the total activity measured by the total rRNA reads (Hugoni et al. 2013). Interestingly, all the taxonomic groups defined from the always rare OTUs were active and the activity, described by the ratio rRNA:rDNA, increased with the rarity: from 0.74 for the abundant OTUs to 2.2 for the always rare OTUs. The variation in this ratio, generally a decrease, is expected for species able to produce endospores, cysts, conidia, etc. Nevertheless, some microbes can reduce metabolism without producing specialised cellular structures. This ratio increased with the rarity in all taxonomic groups defined and particularly in groups known for the presence of resting stages, such as Fungi, for example, which are known to disseminate in the form of conidia/spores and could have a low activity (Jones and Lennon 2010). The high ratio measured shows that rare picoeukaryotes were active and likely able to grow; this was verified over the whole year in both lakes (Pavin and Bourget) and for both the epilimnion and hypolimnion zones. Therefore, these data confirm the hypothesis that, in contrast to bacteria, dormancy does not play an important role in planktonic eukaryote communities (Jones and Lennon 2010).

In conclusion, the rare picoeukaryotes represents an active fraction in the lacustrine ecosystems, which is more important than that described for the bacteria and Archaea. Indeed, the rare picoeukaryotes do not act only as a seed bank of dormant cells waiting for condition changes (Debroas et al. 2015). Using methodologies that specifically target this rare biosphere should unveil new lineages with new physiological properties.

19.4 Biogeography of Picoeukaryotes: Not All Is Everywhere

Probably the most important overriding features of microbes are their exceptional diversity and ability to occupy every possible habitat for life. Indeed, what we consider possible is challenged constantly by the discovery of new microbial communities in habitats previously thought inhospitable, or carrying out processes that we had no idea. There is considerable debate about the extent of diversity and biogeographical distribution of microorganisms. It has been argued that given the huge population sizes and the potential for distant dispersal, most microbial species must be cosmopolitan and the total number of species must be relatively low (Massana 2009). However, the existence of biogeographic patterns has been highlighted in microbial ecology mainly by distance–decay or taxa–area relationships (TAR) (Green and Bohannan 2006).

Spatial distribution of microbial eukaryotes has received much less attention than bacteria, and the studies have concentrated mainly on a single taxonomic group. However, their total diversity and distribution in nature are currently the focus of active debates (Finlay and Fenchel 2004; Foissner 2006; Pinel-Alloul and Ghadouani 2007; Caron 2009; Nolte et al. 2010). As well as for bacteria, picoeukaryotes are characterized by a tiny size and are likely to disperse easily (ex: capacity of survival during transport for dormant cells) and could have a cosmopolitan distribution, leading to the classical dictum 'everything is everywhere, but the environment selects' (Baas-Becking 1934). However, they constitute complex assemblages, and one can wonder if the concept of biogeographic diversity is applicable to small-sized heterogeneous group, which is diverse in terms of physiologies, life cycles, phylogenetic positions, with the ability to reproduce sexually and with capacities of dispersal–colonization, which are likely not the same (ie cyst, endospores easily transported in atmosphere/aerosols).

In order to test the different hypotheses, Lepère et al. (2013) analyzed *alpha-* (richness) and *beta* (composition)-diversity of lacustrine picoeukaryotes distribution patterns in view of geographic distances, lake areas, and habitat variables, using two molecular methods including high-throughput sequencing as well as T-RFLP. According to Martiny et al. (2006), one can expect that environmental factors (physical, chemical, and biological local parameters) tested have significant effects in shaping microbial composition at regional scale, whereas distance (dispersal limitation) should be the main structuring factor across continents. In Lepère et al. (2013) study, lacustrine picoeukaryotes composition and richness were first assessed at a regional scale, six French lakes were considered including lakes Pavin. All lakes were characterized by stable summer stratification and were sampled in the epilimnetic zone in the same way. Moreover, the results are based on an equal sampling effort and/or a time series, which allowed to take into account temporal variations in the community to analyze both spatial distribution and environment effects. The study was secondly extended to 11 worldwide lakes.

19.4.1 Picoeukaryotes Spatial Patterns Are Not Linked to Environmental Factors at a Regional Scale

According to the cosmopolitan view of the microbial world, one might expect to find similar microbial community structure (richness, diversity, and composition) in similar habitats and differentiated microbial communities along an environmental gradient (Green and Bohannan 2006). The use of statistical analysis allows to estimate the impact of geographic distance *vs.* environmental conditions on assemblage composition (Martiny et al. 2006), factors rarely taken into account simultaneously. Partial correlations and Mantel test results (Lepère et al. 2013) clearly showed that variations of the *alpha-* and *beta*-diversity were significantly influenced by the geographic distance between lakes or areas rather than the environmental factors analyzed, such as bottom-up factors (ie nutrients) or potential predators (ie heterotrophic nano-flagellates and metazooplankton). However, even though the parameters explored are those commonly considered when attempting to explain the spatial partitioning of aquatic microorganisms, the analysis did not include all potential controlling factors. For example, Schiaffino et al. (2011) showed that light penetration and dissolved organic carbon had a structuring effect on microorganism populations in 45 lakes. Overall, our results contradict the hypothesis of a general microbial cosmopolitanism. These patterns have already been observed for bacterial communities (Reche et al. 2005; Martiny et al. 2011). In addition, analysis performed on monthly samples showed that temporal variations in the composition of the picoeukaryotes community do not affect the importance of geographic distances.

19.4.2 Distribution of Dominant Picoeukaryotes Is Linked to Lake Area and Distance Between Lakes

The TAR is a fundamental pattern in ecology and an essential tool for biogeography studies. Most of what we know of taxa–area curves is derived from analyses of terrestrial systems, but lakes and ponds can be considered as discrete habitats with definable borders that are comparable in some ways to oceanic islands (Dodson 1992). In Lepère et al. (2013) study, the linear relationships observed between lake area and picoeukaryotes richness (determined by the fingerprinting method (T-RFLP)) were significant for total, dominant, and rare T-RFs, whereas the same relationships determined from pyrosequencing results involved mainly the dominant species (ie OTUs). The TAR varied with the molecular method used, as already highlighted (eg Zhou et al. 2008), and therefore with the taxonomic resolution. However, all methods suggested that the spatial distribution of dominant picoeukaryotes follows a significant TAR. Significant TAR was recently found for the richness of phytoplankton (Smith et al. 2005) as well as bacteria (Reche et al. 2005). However, as emphasized above, this theory seems restricted to the dominant taxa (ie OTUs) and to a specific range of lake area. Finally, the results contradict for dominant populations the advocates of microorganism cosmopolitan distribution, which suggests that microorganisms should be characterized by a flat TAR. Moreover, if the richness (*alpha*-diversity) seems to be constant for total OTUs (dominants and rares), picoeukaryotes composition shows important changes. Therefore, even if the pyrosequencing data highlighted that the *alpha*-diversity did not vary with the habitat size, *beta*-diversity was strongly associated with distance for total, dominant, and rare OTUs. Hillebrand et al. (2001) showed also a distance decay relationship for diatoms and ciliates, but these authors did not consider the putative effects of environmental parameters.

19.4.3 Biogeographic Patterns of the Rare Picoeukaryotes

While the TAR does not work for rare OTUs (pyrosequencing data), the composition (*beta*-diversity) was linked to the geographic distances for the rarest T-RFs and OTUs. Thus, rare community composition varies between ecosystems but not the richness. Similarly, Galand et al. (2009) reported that the rare biosphere of the archaeal community followed patterns similar to those of the most abundant members of the community and has a biogeographic pattern. Based on the assumption that rare OTUs are taxa with low abundance, we assume that rare taxa could be more impacted by dispersal limitation because the probability of immigrating and growing in a new ecosystem is limited compared to abundant taxa (Green et al. 2004; Weisse 2008). The 'rare biosphere' has traditionally been thought to indicate the presence of a seed bank of potential new colonizers, according to the 'everything is everywhere' hypothesis. However, statistical analysis showed no correlation between environmental variables and the rare T-RFs/OTUs. Galand et al. (2009) showed that regardless of spatial or temporal scales, most of the rare phylotypes are always rare within an ecosystem and the few rare phylotypes that are sometimes detected as abundant represent traces of phylotypes that are highly abundant in some habitats.

19.4.4 What's Happening at a Larger Geographic Scale?

To extend the discussion to a more global scale (across continent), we used sequences available on the same planktonic

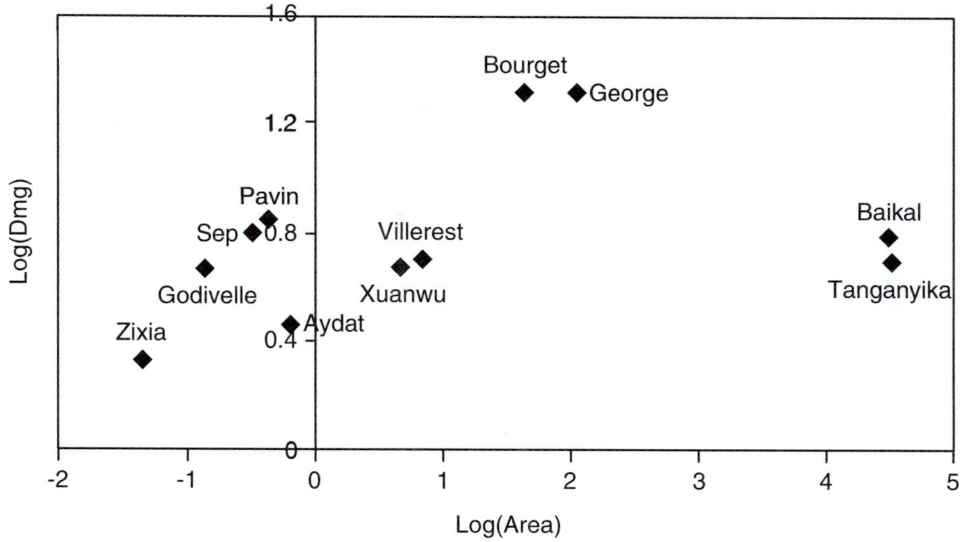

Fig. 19.5 Relationship between lake areas and richness expressed by a diversity index (Margalef:Dmg) calculated from cloning–sequencing method (six French lakes + five worldwide lakes = Baikal (sequence accesion numbers: JN547261-JN547327), George (AY919677-AY919829), Tanganyika (GU290066-GU290116), Zixia (FJ939033-FJ939124), Xuanwu (FJ939033-FJ939124))

size fraction from public database. These sequences were obtained by the traditional cloning-sequencing method from 11 worldwide lakes (Fig. 19.5). Although with this approach the sampling is far from exhaustive and many more taxa (especially rare) might be present at the sampling sites, statistical analysis (UniFrac analysis; Lozupone and Knight 2005) showed 18S rRNA gene OTUs compositions significantly different among most of lakes, and this difference does not seem to be related to the trophic status at any spatial scales. Such results were expected because the effect of dispersal limitation would occur at the largest geographic scales (across continents) rather than at regional scales where environment selection could theoretically be predominant (Martiny et al. 2006). At the global scale, a significant linear relationship with lake areas is obtained but only when using the smaller lake areas (lower than 114 km^2 corresponding the lakes Bourget and George; Fig. 19.5) (Lepère et al. 2013).

19.5 Concluding Remarks

19.5.1 Including Picoeukaryotes in the Debates on Biodiversity and Ecosystem Functioning: Issues of Scale and Trophic Complexity

A central goal in ecology is to understand the patterns and processes of biodiversity. As aquatic microbial ecologists, we still lack a solid understanding of the characteristic scales of variation in aquatic eukaryotic community composition, and this hampers our ability to develop theories about how the stability of the functions (eukaryotes-mediated functions) are maintained across space and time in lakes.The effects of the main drivers (local and global pressures) affecting biodiversity involve changes in species at different trophic level. Evaluating how ecosystem processes are likely to change following species rearrangements (or loss) at multiple trophic levels is important to advance knowledge on biodiversity–ecosystem functioning links.The eukaryotic microbes have certainly to be included in such research work, due to the variety of eukaryotic functional groups involved in trophic network functioning.

Research on the relationship between biodiversity and ecosystem functioning is entering a new phase with the application of DNA/RNA tools used to reveal the huge diversity of microbial communities. The taxa recently revealed within the picoeukaryotic fraction have undoubtedly to be included in the studies aiming at identifying the effects of the main drivers on aquatic biodiversity and the resulting changes in ecosystemic functioning.

Acknowledgments The authors thank, very sincerely, Tom Richards for his review and the quality of its evaluation on this chapter.

References

Amblard C, Carrias J-F, Bourdier G, Maurin N (1995) The microbial loop in a humic lake: seasonal and vertical variations in the structure of the different communities. Hydrobiologia 300–301:71–84

Andersen RA, Bidigare RR, Keller MD, Latasa M (1996) A comparison of HPLC pigment signatures and electron microscopic observations for oligotrophic waters on the north Atlantic and Pacific Oceans. Deep-Sea Res 43:517–537

Atteia A, van Lis R, Tielens AG, Martin WF (2013) Anaerobic energy metabolism in unicellular photosynthetic eukaryotes. Biochim Biophys Acta 1827:210–223

Azam F, Fenchel T, Field JG, Gray JS, Meyer LA, Thingstad F (1983) The ecological role of water column microbes in the sea. Mar Ecol Prog Ser 10:257–263

Baas-Becking LGM (1934) Geobiologie of Inleiding Tot de Milieukunde. W.P. Van Stockum& Zoon, The Hague, The Netherlands (in Dutch)

Baker JW, Grover JP, Brooks BW, Urena-Boeck F, Roelke DL, Errera R (2007) Growth and toxicity of prymnesiumparvum (haptophyta) as a function of salinity, light, and temperature. J Phycol 43:219–227

Baldauf SL (2008) An overview of the phylogeny and diversity of eukaryotes. J Syst Evol 46:263–273

Bartram AK, Lynch MDJ, Stearns JC, Moreno-Hagelsieb G, Neufeld JD (2011) Generation of multimillion-sequence 16S rRNA gene libraries from complex microbial communities by assembling paired-end illumina reads. Appl Environ Microbiol 77:3846–3852

Boenigk J, Arndt H (2002) Bacterivory by heterotrophic flagellates: community structure and feeding strategies. Antonie Van Leeuwenhoek 81:465–480

Bourelly P (1970) Les alguesd'eaudouce, Initiation à la systématique. Tome III: les alguesbleueset rouges: Paris-VIe, p.291pp

Brugerolle G (2002) Cryptophagus subtilis: a new parasite of cryptophytes affiliated with the Perkinsozoa lineage. Eur J Protistol 37:379–390

Brugerolle G (2003) Apicomplexan parasite Cryptophagus renamed Rastrimonas gen. nov. Eur J Protistol 39:101–101

Campbell BJ, Yu L, Heidelberg JF, Kirchman DL (2011) Activity of abundant and rare bacteria in a coastal ocean. Proc Natl Acad Sci 108:12776–12781

Canter HM, Lund JWG (1948) Studies on plankton parasites. I. Fluctuations in the numbers of Asterionella formosa Hass. in relation to fungal epidemics. New Phytol 47:238–261

Caron D (2009) Past president's address: protistan biogeography: why all the fuss? J Eukaryot Microbiol 56:105–112

Caron DA, Countway PD (2009) Hypotheses on the role of the protistan rare biosphere in a changing world. Aquat Microb Ecol 57:227–238

Caron DA, Peele ER, Lim EL, Dennett MR (1999) Picoplankton and nanoplankton and their trophic coupling in the surface waters of the Sargasso Sea south of Bermuda. Limnol Oceanogr 44:259–272

Carrias JF, Amblard C, Bourdier G (1996) Protistanbacterivory in an oligomesotrophic lake: importance of attached ciliates and flagellates. Microb Ecol 31:249–268

Cavalier-Smith T (1998) A revised six-kingdom system of life. Biol Rev 73:203–266

Chen M, Chen F, Zhao B, Wu QL, Kong F (2009) Seasonal variation of microbial eukaryotic community composition in the large, shallow, subtropical Taihu Lake, China. Aquat Ecol 44:1–12

Comte J, Jacquet S, Viboud S, Fontvieille D, Millery A, Paolini G, Domaizon I (2006) Microbial community structure and dynamics in the largest natural French lake (Lake Bourget). Microbial Ecol 52:72–89

Cuvelier ML, Allen AE, Monier A, McCrow JP, Messie´ M, Tringe SG et al (2010) Targeted metagenomics and ecology of globally important uncultured eukaryotic phytoplankton. Proc Natl Acad Sci U S A 107:14679–14684

Debroas D, Hugoni M, Domaizon I (2015) Evidence for an active rare biosphere within freshwater protists community. Mol Ecol 24:1236–1247

Dodson S (1992) Predicting crustacean zooplankton species richness. Limnol Oceanogr 37:848–856

Domaizon I, Viboud S, Fontvieille D (2003) Taxon- specific and seasonal variations in flagellates grazing on heterotrophic bacteria in the oligotrophic Lake Annecy-importance of mixotrophy. FEMS Microb Ecol 46:317–322

Finlay BL, Fenchel T (2004) Cosmopolitan metapopulations of free-living microbial eukaryotes. Protist 155:237–244

Foissner W (2006) Biogeography and dispersal of micro- organisms: a review emphasizing protists. Acta Protozool 45:111–136

Fuller NJ, Campbell C, Allen DJ, Pitt FD, Zwirglmaierl K, Le Gall F et al (2006) Analysis of photosynthetic picoeukaryote diversity at open ocean sites in the Arabian Sea using a PCR biased towards marine algal plastids. Aquat Microb Ecol 43:79–93

Galand PE, Casamayor EO, Kirchman DL, Lovejoy C (2009) Ecology of the rare microbial biosphere of the Arctic Ocean. Proc Natl Acad Sci U S A 106:22427–22432

Gjosaeter J, Lekve K, Stenseth NC, Leinaas HP, Christie H, Dahl E et al (2000) A long- term perspective on the Chrysochromulina bloom on the Norwegian Skagerrak coast 1988: a catastrophe or an innocent incident? Mar Ecol Prog Ser 207:201–218

Green J, Bohannan BJM (2006) Spatial scaling of microbial biodiversity. Trends Ecol Evol 21:501–507

Green JL, Holmes AJ, Westoby M, Oliver I, Briscoe D, Dangerfield M, Gillings M, Beattie AJ (2004) Spatial scaling of microbial eukaryote diversity. Nature 432:747–750

Guillou L, Viprey M, Chambouvet A, Welsh RM, Kirkham AR, Massana R (2008) Widespread occurrence and genetic diversity of marine parasitoids belonging to Syndiniales (Alveolata). Environ Microbiol 10:3349–3365

Hansen LR, Kristiansen J, Rasmussen JV (1994) Potential toxicity of the freshwater Chrysochromulina species C. parva (Prymnesiophyceae). Hydrobiologia 287:157–159

Hartmann M, Grob C, Tarran GA, Martin AP, Burkill PH, Scanlan DJ et al (2012) Mixotrophic basis of Atlantic oligotrophic ecosystems. Proc Natl Acad Sci U S A 109:5756–5760

Hartmann M, Zubkov MV, Scanlan DJ, Lepère C (2013) *In situ* interactions between photosynthetic picoeukaryotes and bacterioplankton in the Atlantic Ocean: evidence for mixotrophy. Environ Microbiol Rep 5:835–840

Hillebrand H, Watermann F, Karez R, Berninger UG (2001) Differences in species richness patterns between unicellular and multicellular organisms. Oecologia 126:114–124

Hugoni M, Taib N, Debroas D, Domaizon I, Dufournel IJ, Bronner G et al (2013) Structure of the rare archaeal biosphere and seasonal dynamics of active ecotypes in surface coastal waters. Proc Natl Acad Sci. doi:10.1073/pnas.1216863110

Iglesias-Rodriguez MD, Garcia S, Groben R (2002) Polymorphic microsatellite loci in global populations of the marine coccolithophorid Emilianiahuxleyi. Mol Ecol Notes 2:495–497

James TY, Pelin A, Bonen L, Ahrendt S, Sain D, Corradi N, Stajich JE (2014) Shared signatures of parasitism and phylogenomics unite cryptomycota and microsporidia. Curr Biol 23:1548–1553

Jobard M, Rasconi S, Sime-Ngando T (2010) Diversity and functions of microscopic fungi: a missing component in pelagic food webs. Aquat Sci 72:255–268

John DM, Whitton BA, Brook AJ (2002) The freshwater algal Flora of the British Isles—An identification guide to freshwater and terrestrial algae, vol 1. International Society for Microbial Ecology, Cambridge

Jones SE, Lennon JT (2010) Dormancy contributes to the maintenance of microbial diversity. Proc Natl Acad Sci 107:5881–5886

Jones MDM, Forn I, Gadelha C, Egan MJ, Bass D, Massana R et al (2011) Discovery of novel intermediate forms redefines the fungal tree of life. Nat Adv Online Publ 474:200–203

Kamjunke N, Henrichs T, Gaedke U (2007) Phosphorus gain by bacterivory promotes the Mixotrophic flagellates Dinobryon spp. during the re-oligotophication. J Plank Res 1:39–46

Lara E, Moreira D, Lopez-Garcia P (2010) The environmental Clade LKM11 and Rozella form the deepest branching clade of fungi. Protist 161:116–121

Lefèvre E, Roussel B, Amblard C, Sime-Ngando T (2008) The molecular diversity of freshwater picoeukaryotes reveals high occurrence of putative parasitoids in the plankton. PLoS One 3:e2324

Lefèvre E, Bardot C, Noel C, Carrias JF, Viscogliosi E, Amblard C et al (2007) Unveiling fungal zooflagellates as members of freshwater picoeukaryotes: evidence from a molecular diversity study in a deep meromictic lake. Environ Microbiol 9:61–71

Lefranc M, Thenot A, Lepere C, Debroas D (2005) Genetic diversity of small eukaryotes in lakes differing by their trophic status. Appl Environ Microbiol 71:5935–5942

Legrand C, Johansson N, Johnsen G, Borsheim KY, Grane´li E (2001) Phagotrophy and toxicity variation in the mixotrophicPrymnesium patelliferum (Hapto- phyceae). Limnol Oceanogr 46:1208–1214

Lepère C (2007) Vol Doctoratd'université en Biologie des Populations. Université de Savoie: Chambéry, France, pp 189

Lepère C, Boucher D, Jardillier L, Domaizon I, Debroas D (2006) Succession and regulation factors of small eukaryote community composition in a lacustrine ecosystem (Lake Pavin). Appl Environ Microbiol 72:2971–2981

Lepère C, Domaizon I, Debroas D (2008) Unexpected importance of potential parasites in the composition of the freshwater small-eukaryote community. Appl Environ Microbiol 74:2940–2949

Lepère C, Domaizon I, Hugoni M, Vellet A, Debroas D (2016) Diversity and dynamics of active microbial eukaryotes in the anoxic zone of a freshwater meromictic lake (Pavin, France). Front Microbiol. doi:10.3389/fmicb.2016.00130

Lepère C, Masquelier S, Mangot J-F, Debroas D, Domaizon I (2010) Vertical structure of small eukaryotes in three lakes that differ by their trophic status: a quantitative approach. ISME J 4:1509–1519

Lepère C, Vaulot D, Scanlan DJ (2009) Photosynthetic picoeukaryote community structure in the South East Pacific Ocean encompassing the most oligotrophic waters on Earth. Environ Microbiol 11:3105–3117

Lepère C, Domaizon I, Taib N, Bronner G, Boucher D, Debroas D (2013) Geographic distance and ecosystem size determine the distribution of smallest protists in lacustrine ecosystems. FEMS Microb Ecol 85:85–94

Liu H, Probert I, Uitz J, Claustre H, Aris-Brosou S, Frada M et al (2009) Extreme diversity in noncalcifyinghaptophytes explains a major pigment paradox in open oceans. Proc Natl Acad Sci U S A 106:12803–12808

Logares R, Audic S, Bass D, Bittner L, Boutte C, Christen R et al (2014) Patterns of rare and abundant marine microbial Eukaryotes. Curr Biol. doi:10.1016/j.cub.2014.02.050

Lopez-Garcia P, Rodriguez-Valera F, Pedros-Alio C, Moreira D (2001) Unexpected diversity of small eukaryotes in deep-sea Antarctic plankton. Nature 409:603–607

Lozupone C, Knight R (2005) UniFrac: a new phylogenetic method for comparing microbial communities. Appl Environ Microbiol 71:8228–8235

Mangot JF, Lepere C, Bouvier C, Debroas D, Domaizon I (2009) Community structure and dynamics of small eukaryotes targeted by new oligonucleotide probes: new insight into the lacustrine microbial food web. Appl Environ Microbiol 75:6373–6381

Mangot J-F, Debroas D, Domaizon I (2011) Perkinsozoa, a well-known marine protozoan flagellate parasite group, newly identified in lacustrine systems: a review. Hydrobiologia 659:37–48

Mangot JF, Domaizon I, Taib N, Marouni N, Duffaud E, Bronner G, Debroas D (2013) Short-term dynamics of diversity patterns: evidence of continual reassembly within lacustrine small eukaryotes. Environ Microbiol 2013 Jun;15(6):1745–1758. doi:10.1111/1462-2920.12065

Marie D, Shi XL, Rigaut-Jalabert F, Vaulot D (2010) Use of flow cytometric sorting to better assess the diversity of small photosynthetic eukaryotes in the English Channel. FEMS Microbiol Ecol 72:165–178

Marin B, Melkonian M (2010) Molecular phylogeny and classification of the Mamiellophyceae class. nov. (Chlorophyta) based on sequence comparisons of the nuclear- and plastid-encoded rRNA operons. Protist 161:304–336

Martiny JBH, Bohannan BJM, Brown JH, Colwell RK, Fuhrman JA, Green JL et al (2006) Microbial biogeography: putting microorganisms on the map. Nat Rev Microbiol 4:102–112

Martiny JBH, Eisenb JA, Pennc K, Allisona SD, Horner- Devine MC (2011) Drivers of bacterial b-diversity depend on spatial scale. Proc Natl Acad Sci USA 108:850–7854

Massana R (2009) Picoeukaryotes. In: Schmidt TM, Schaechter M (eds) Topics in ecological and environmental microbiology. Amsterdam, Academic Press, pp 76

Massana R, Unrein F, Rodriguez-Martinez R, Forn I, Lefort T, Pinhassi J (2009) Grazing rates and functional diversity of uncultured heterotrophic flagellates. ISME J 3:588–596

Monchy S, Sanciu G, Jobard M, Rasconi S, Gerphagnon M, Chabé M et al (2011) Exploring and quantifying fungal diversity in freshwater lake ecosystems using rDNA cloning/sequencing and SSU tag pyrosequencing. Environ Microbiol 13:1433–1455

Moon-van der Staay SY, Van der Staay GWM, Guillou L, Vaulot D (2000) Abundance and diversity of prymnesiophytes in the picoplankton community from the equatorial Pacific Ocean inferred from 18S rDNA sequences. Limnol Oceanogr 45:98–109

Moon-van der Staay SY, De Wachter R, Vaulot D (2001) Oceanic 18S rDNA sequences from picoplankton reveal unsuspected eukaryotic diversity. Nature 409:607–610

Moore RB, Obornik M, Janouskovec J, Chrudimsky T, Vancova M, Green DH et al (2008) A photosynthetic alveolate closely related to apicomplexan parasites. Nature 451:959–963

Nicholls KH, Beaver JL, Estabrook RH (1982) Lakewideodours in Ontario and New Hampshire caused by ChrysochromulinabreviturritaNich. (Prymnesiophyceae). Hydrobiologia 96:91–95

Nolte V, Pandey RV, Jost S, Medinger R, Ottenwalder B, Boenigk J, Schlo#tterer C (2010) Contrasting seasonal niche separation between rare and abundant taxa conceals the extent of protist diversity. Mol Ecol 19:2908–2915

Norén F, Moestrup O, Rehnstam-Holm AS (1999) Parvilucifera infectans noren et moestrup gen. Et sp. Nov. (perkinsozoa phylum nov.): a parasitic flagellate capable of killing toxic microalgae. Eur J Protistol 35:233–254

Pedros-Alio C (2012) The rare bacterial biosphere. Ann Rev Mar Sci 4:449–466

Pinel-Alloul B, Ghadouani A (2007) Spatial heterogeneity of planktonic microoraganisms in aquatic systems: mutiscale patterns and processes. In: Franklin RB, Mills AL (eds) The spatial distribution of microbes in the environment. Springer, The Netherlands, pp 203–310

Rasconi S, Jobard M, Jouve L, Sime-Ngando T (2009) Use of calcofluor white for detection, identification, and quantification of phytoplanktonic fungal parasites. Appl Environ Microbiol 75:2545–2553

Reche I, Pulido-Villena E, Morales-Baquero R, Casamayor EO (2005) Does ecosystem size determine aquatic bacterial richness? Ecology 86:1715–1722

Richards TA, Vepritskiy AA, Gouliamova DE, Nierzwicki-Bauer SA (2005) The molecular diversity of freshwater picoeukaryotes from an oligotrophic lake reveals diverse, distinctive and globally dispersed lineages. Environ Microbiol 7:1413–1425

Schiaffino MR, Unrein F, Gasol JM, Massana R, Balagu_e V, Izaguirre I (2011) Bacterial community structure in a latitudinal gradient of lakes: the roles of spatial versus environmental factors. Freshw Biol. doi:10.1111/j.1365-2427.2011.02628.x

Schindler DW (2009) Lakes as sentinels and integrators for the effects of climate change on watersheds, airsheds, and landscapes. Limnol Oceanogr 54:2349–2358

Schnepf E, Kühn SF (2000) Food uptake and fine structure of Cryothecomonaslongipes sp. nov., a marine nanoflagellateincertaesedis feeding phagotrophically on large diatoms. Helgol Mar Res 54:18–32

Shade A, Jones SE, Caporaso JG, Handelsman J, Knight R, Fierer N, Gilbert JA (2014) Conditionally rare taxa disproportionately contribute to temporal changes in microbial diversity. mbio 5(4):e01371–0e1414

Shi X, Marie D, Jardillier L, Scanlan DJ, Vaulot D (2009) Groups without cultured representatives dominate eukaryotic picophytoplankton in the oligotrophic South East Pacific Ocean. PLoS One 4:7657

Smith VH, Foster BL, Grover JP, Holt RD, Leibold MA, De Noyelles F (2005) Phytoplankton species richness scales consistently from laboratory microcosms to the world's oceans. Proc Natl Acad Sci U S A 102:4393–4396

Sogin ML, Morisson HG, Huber JA et al (2006) Microbial diversity in the deep sea and the underexplored "rare biosphere". Proc Natl Acad Sci U S A 103:12115–12120

Stickney HL, Hood RR, Stoecker DK (2000) The impact of mixotrophy on planktonic marine ecosystems. Ecol Model 125:203–230

Stockner JG, Antia NJ (1986) Algal picoplankton from marine and freshwater ecosystems: a multidisciplinary perspective. Can J Fish Aquat Sci 43:2472–2503

Taib N, Mangot J-F, Domaizon I, Bronner G, Debroas D (2013) Phylogenetic affiliation of SSU rRNA genes generated by massively parallel sequencing: new insights into the freshwa- ter protist diversity. PLoS ONE 8, e58950

Tarbe AL, Stenuite S, Balagu V, Sinyinza D, Descy JP et al (2011) Molecular characterisation of the small-eukaryote community in a tropical Great Lake (Lake Tanganyika, East Africa). Aquat Microb Ecol 62:177–190

Temponeras M, Kristiansen J, Moustaka-Gouni M (2000) Seasonal variation in phytoplankton composition and physical-chemical features of the shallow Lake Doı ̈rani, Macedonia, Greece. Hydrobiologia 424:109–122

Tittel J, Wiehle I, Wannicke N, Kampe H, Poerschmann J, Meier J et al (2009) Utilisation of terrestrial carbon by osmotrophic algae. Aquat Sci 71:46–54

Troost TA, Kooi BW, Kooijman SA (2005) Ecological specialization of mixotrophic plankton in a mixed water column. Am Nat 166:45–61

Ukeles R, Rose WE (1976) Observations on organic carbon utilization by photosynthetic marine micro- algae. Mar Biol 37:11–28

Van Hannen EJ, Mooij W, van Agterveld MP, Gons HJ, Laanbroek HJ (1999) Detritus- dependent development of the microbial community in an experimental system: qualitative analysis by denaturing gradient gel electrophoresis. Appl Environ Microbiol 65:2478–2484

Vaulot D, Eikrem W, Viprey M, Moreau H (2008) The diversity of small eukaryotic phytoplankton (<3 μm) in marine ecosystems. FEMS Microbiol Rev 32:795–820

Viprey M, Guillou L, Ferreol M, Vaulot D (2008) Wide genetic diversity of picoplanktonic green algae (Chloroplastidia) in the Mediterranean Sea uncovered by a phylum-biased PCR approach. Environ Microbiol 10:1804–1822

Weisse T (2008) Distribution and diversity of aquatic protists: an evolutionary and ecological perspective. Biodivers Conserv 17:243–259

Zhou J, Kang S, Schadt CW, Garten CT (2008) Spatial scaling of functional gene diversity across various microbial taxa. PNAS 105:7768–7773

Zhu F, Massana R, Not F, Marie D, Vaulot D (2005) Mapping of picoeucaryotes in marine ecosystems with quantitative PCR of the 18S rRNA gene. Fems Microbiol Ecol 52:79–92

Zubkov MV, Tarran GA (2008) High bacterivory by the smallest phytoplankton in the North Atlantic Ocean. Nature 455:224–226

Molecular Diversity Studies in Lake Pavin Reveal the Ecological Importance of Parasitic True Fungi in the Plankton

20

Télesphore Sime-Ngando, Mélanie Gerphagnon, Jonathan Colombet, Marlène Jobard, Emilie Lefevre, Sébastien Monchy, Serena Rasconi, Delphine Latour, Jean-François Carrias, and Christian Amblard

Abstract

Parasitism is one of the most common symbiotic interactions, occurring in almost all environments. Microbial parasites are typically characterized by their small size, a short generation time, high rates of reproduction, and a simple life cycle usually completed within a single host. Microbial parasites are phylogenetically diverse and ubiquitous in aquatic ecosystems, comprising viruses, prokaryotes and eukaryotes. Pioneering environmental 18S-rDNA surveys of microbial eukaryotes in Lake Pavin, France, revealed the presence of a high diversity of undescribed eukaryotes, primarily affiliated to the fungal phylum Chytridiomycota (chytrids), and more than likely playing major roles as infecting agents in the system. These early diverging fungi produce free-swimming dispersal propagules characterized by a small size (2–8 μm) and a single, posterior flagellum. These characteristics make chytrids part of the so-called 'zoosporic fungi'. Chytrids are particularly adapted to a planktonic lifestyle and have been shown to infect a wide variety of hosts, including fish, eggs, zooplankton, other aquatic fungi, and primarily phytoplankton in aquatic environments. Related ecological implications are important for pelagic food webs. Released organic particles resulting from the death of infected hosts can be used as substrates for other microbial processes, and zoospores themselves can provide nutrient-rich particles for planktonic grazers. The application of the plankton ecology group (PEG) model on Lake Pavin indicated that chytrid epidemics could represent an important driving factor for phytoplanktonic seasonal successions. Besides, the observation that phytoplankton chytridiomycosis preferentially impacts larger size species (e.g., filamentous cyanobacteria) suggests that bloom of such species may not represent a trophic bottleneck for the system as previously thought. In this chapter, based on studies we conducted in Lake Pavin and other lakes in the vicinity, we summarize knowledge on diversity, community structure, quantitative

T. Sime-Ngando (✉) • M. Gerphagnon • J. Colombet • M. Jobard
D. Latour • J.-F. Carrias • C. Amblard
LMGE, Laboratoire Microorganismes: Génome et Environnement, UMR CNRS 6023, Université Clermont-Auvergne, Université Blaise Pascal, BP 80026, 63171 Aubière Cedex, France
e-mail: telesphore.sime-ngando@univ-bpclermont.fr

E. Lefevre
LMGE, Laboratoire Microorganismes: Génome et Environnement, UMR CNRS 6023, Université Clermont-Auvergne, Université Blaise Pascal, BP 80026, 63171 Aubière Cedex, France

Present address: Department of Civil and Environmental Engineering, Duke University, 121 Hudson Hall Box, 90287, 27708-0287 Durham, NC, USA

S. Monchy
LMGE, Laboratoire Microorganismes: Génome et Environnement, UMR CNRS 6023, Université Clermont-Auvergne, Université Blaise Pascal, BP 80026, 63171 Aubière Cedex, France

Present address: INSU-CNRS, UMR 8187 LOG, Laboratoire d'Océanologie et de Géosciences, Université du Littoral Côte d'Opale, ULCO, 32 avenue Foch, F-62930 Wimereux, France

S. Rasconi
LMGE, Laboratoire Microorganismes: Génome et Environnement, UMR CNRS 6023, Université Clermont-Auvergne, Université Blaise Pascal, BP 80026, 63171 Aubière Cedex, France

Present address: WasserCluster – Biological Station Lunz, Inter-university Centre for Aquatic Ecosystem Research, A-3293 Lunz am See, Austria

T. Sime-Ngando et al. (eds.), *Lake Pavin*, DOI 10.1007/978-3-319-39961-4_20

importance, and functional roles of planktonic chytrids. We primarily focus on parasitic chytrids of phytoplankton, the potential ecological implications for food web dynamics, as well as the methodological challenges related to their study. We conclude that phytoplankton chytridiomycosis is an important but still overlooked ecological driving force in aquatic food web dynamics.

Keywords
Lake Pavin • Zoosporic fungi • Chytrids • Microbial parasites • Phytoplankton hosts • Food webs • Microbial ecology • Aquatic ecosystems

20.1 Introduction

Since the emergence of the '**microbial loop**' concept, heterotrophic nanoflagellates (HNF) have received particular attention as grazers in aquatic ecosystems (Amblard et al. 1998). These protists have historically been wrongly regarded as a homogeneous group of **bacterivorous protists**, most of which (>90 % of the total abundance) includes motile, non-photosynthetic and pigmented microbes with a diameter of approximately ≤ 5 µm (Strom 2000). However, in terms of trophic modes, HNF were considered a black box until the early 2000s. The introduction of molecular methods to the study of prokaryotes in the mid-1980s (Pace et al. 1986), and their application to small eukaryotes in the early 2000s (Lopez-Garcia et al. 2001; Moon-van der Staay et al. 2001), allowed the detection of uncultured microbes in diverse natural environments and greatly improved our knowledge on microbial diversity and putative functions. More recently, progresses made in sequencing technologies and the rapid development of **high throughput sequencing** platforms have revealed an even more impressive taxonomic, genetic and functional diversity hidden in natural aquatic ecosystems (Sogin et al. 2006; Huse et al. 2008). The resulting methodological, conceptual and empirical progresses made in aquatic microbial ecology are arguably the greatest advances in environmental sciences.

The first molecular surveys of small eukaryotes, mostly conducted in marine environments and based on the cloning and sequencing of **18S rDNA**, revealed an unexpected high diversity of eukaryotic microbes in the plankton (Moreira and Lopez-Garcia 2002). In freshwater environments, similar surveys, including the ones we conducted in Lake Pavin and other lakes in the vicinity, also revealed a high diversity of novel small eukaryotes (Lefranc et al. 2005; Lefèvre et al. 2007) and provided new insights in the understanding of microbial food web functioning. Our most remarkable results are that: (i) the dominant phyla in these studies differed significantly from those found using traditional microscopy, (ii) the majority of the retrieved **phylotypes** affiliated generally to well-established eukaryotic clades, but represented a diverse and novel diversity within these clades and (iii) a substantial part of the retrieved sequences more than likely belonged to parasitic and saprophytic organisms, such as **zoosporic true fungi** (chytrids), other **fungi**, fungus-like organisms (stramenopiles), and parasitic alveolates (*Perkinsozoa* and *Amoebophrya* spp.). All these microorganisms are known to produce small flagellated **zoospores** during the dispersive phase of their life cycle. Owing to their small size and the lack of distinct morphological features, we hypothesized that zoosporic species had probably often been misidentified as bacterivorous protists in previous studies (Lefèvre et al. 2007, 2008). Indeed, many zoospores are approximately 5 µm in diameter and the **thallus** of zoosporic fungi (i.e., sporangia) is hardly distinguishable from the **lorica** of sessile flagellates such as choanoflagellates and bicosoecids (Rasconi et al. 2009). These observations provided evidence that HNF thriving in pelagic systems are not restricted to protists or **bacterivores**, and suggest that **parasitism** and **saprophytism** represent important functions in freshwater ecosystems (Sime-Ngando et al. 2011).

Both 18S rDNA environmental cloning surveys (Lefèvre et al. 2007, 2008; Jobard et al. 2012) and the use of SSU rRNA hypervariable tag sequencing (Monchy et al. 2011) for the study of microbial eukaryotes unveiled major infectious agents in Lake Pavin, which consisted primarily of zoosporic fungal order of Chytridiales (chytrids). Chytrids are external eucarpic parasites that infect diverse eukaryotic and prokaryotic species, primarily diatoms and filamentous phytoplankton (Rasconi et al. 2011; Sime-Ngando 2012). They produce specialized **rhizoidal systems** through host cells, i.e. the nutrient conveying system for the formation of fruit bodies (sporangia) from which propagules (motile uniflagellated zoospores) are released into the environment. These findings have stimulated our efforts to study microbial parasites of phytoplankton, and we primarily directed our focus towards their community structure, and their quantitative importance and functional roles in the pelagic system. This also included methodological developments, which have been crucial for the study of these inconspicuous eukaryotic parasites. In this chapter, we primarily focus on parasitic chytrids of phytoplankton, discussing the potential ecological implications for food web dynamics, as well as the methodological challenges we had to face for their study. We conclude that phytoplankton chytridiomycosis represents an important but still overlooked ecological driving force in aquatic food web dynamics.

20.2 Molecular Diversity Studies of Small Eukaryotes in Lake Pavin Revealed a High Occurrence of Putative Parasitoids in the Plankton

Our molecular surveys of small eukaryotes in Lake Pavin were stimulated by the fact that most studies that used traditional microscopy to characterize HNF community composition cluster HNF smaller than 5 μm together as 'unidentified HNF' (Carrias et al. 1996). In addition, grazing experiments showed that the bulk of the 'unidentified HNF', which were considered as main bacterivores in aquatic systems until early in this century (Strom 2000), were unable to ingest bacterial-size particles (Fig. 20.1). Our surveys were also motivated by a comparative study conducted by Gasol and Vaqué (1993) demonstrating a lack of trophodynamic coupling between HNF and their presumed bacterial preys in a range of marine and freshwater habitats. Major hypotheses proposed to explain this lack of correlation were that HNF (i) were not the only organisms that feed on bacteria, (ii) they could also ingest viruses (see Chap. 14) and colloidal organic matter (Sherr 1988), and (iii) were not a phylogenetically coherent group since they included organisms from many branches of the eukaryotic tree of life (Patterson 1993). From the above observations, we hypothesized that the so-called 'unidentified HNF' group could be hiding an unexplored diversity and as yet unknown trophic modes. Therefore, in order to characterize the diversity of these small unidentified HNF, pelagic samples from Lake Pavin were repeatedly collected during autumn 2004 and spring/summer 2005, and the eukaryotic community within the size range of 0.6 and 5 μm characterized using the cloning and sequencing of their 18S rDNA (Lefèvre et al. 2007, 2008).

Unexpectedly, most of the retrieved sequences affiliated to 8 major eukaryotic phyla, among which three taxonomic groups represented 66 % of the detected diversity: Alveolates (33 %), Fungi (19 %), and Stramenopiles (14 %) (Lefèvre et al. 2007, 2008). Each of the five other groups (i.e., Cryptomonads, Chlorophyceae, Cercozoa, Telonemia, and Choanozoa) represented less than 10 % of the total diversity (i.e. non redundant **OTU**s or **Operational Taxonomic Units**), and affiliated to HNF taxa previously detected using more traditional microscopic methods. Among Alveolates, 75 % of our sequences belonged to lineages known to contain members of small colorless flagellates with two unequal flagella during the motile stage of their life cycle (i.e., *Perkinsea* and *Dinophyceae*). Among the Fungi kingdom, 75 % of the sequences were related to the parasitic or saprotrophic Chytridiomycota phylum, which contains organisms that produce small uniflagellated cells (zoospores) during their life cycle. Among Stramenopiles, 75 % of our sequences were affiliated to the small heterotrophic biflagellated Bicosoecids and uniflagellated Hyphochytrids. Overall, between 50 and 70 % of the sequences retrieved in Lake Pavin fell into ecologically important taxonomic groups known to contain small heterotrophic flagellated forms: Alveolates, Fungi, Stramenopiles, Cercozoa and Telonemia. Unlike with previous studies conducted in other aquatic environment, our sequences affiliated to known lineages (Lefèvre et al. 2007, 2008). However, 83 % of our sequences were not highly similar to any previously published sequences, and included 23 sequences that belonged to novel clades (i.e. currently containing no cultured representatives), suggesting that eukaryotic microbial diversity was still poorly described in freshwater environments.

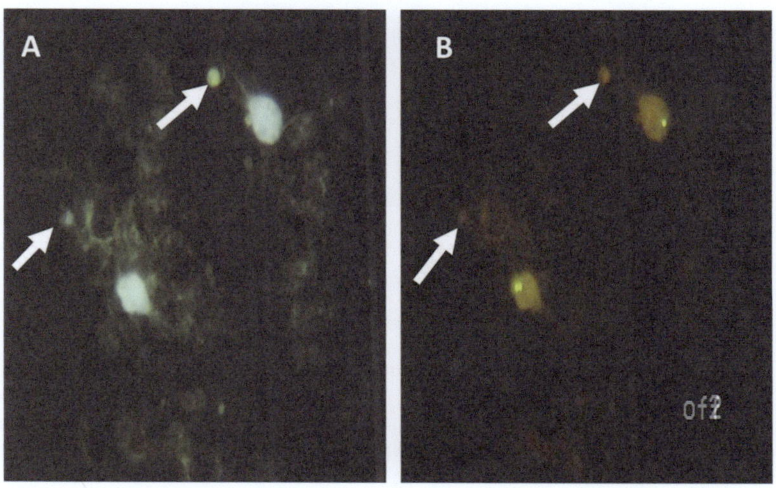

Fig. 20.1 Typical micrographs of heterotrophic nanoflagellates (HNF) observed under UV light for the whole phenotypic visualisation (**A**, forms and flagella) and under *blue light* for visualization of ingested bacterial particles within vacuoles (**B**, the *yellow spots* in two large cells). *White arrows* show that the smallest undetermined forms cannot ingest bacteria

Because our **cloning-sequencing** studies in Lake Pavin were designed to characterize the taxonomic diversity of 'small unidentified' HNF and their related functions, their potential trophic strategies were inferred from the trophic behaviors displayed by their phylogenetically closest described representatives (Lefèvre et al. 2007, 2008). This exercise yielded the following relative trophic mode affiliations for our genetic sequences: 19–31 % for parasites, 15–20 % for saprophytes and (or) parasites, and 4–26 % for **bacterivores**. In addition, there were 6–12 % novel clades of cercozoan-affiliated for which we were not able to assign any trophic role.

Because Fungi had rarely been taken into account in microbial food web processes in pelagic systems, the presence of a high percentage of Fungi in our retrieved sequences was our most unexpected result. Therefore we decided to direct our efforts towards the characterization of the pelagic fungal community of the Lake Pavin using two approaches: (i) the cloning/sequencing of the 18S, ITS1, 5.8S, ITS2 and 28S regions using **primers** specifically targeting Fungi, and (ii) the pyrosequencing of the 18S rDNA hypervariable V2, V3 and V5 regions using both universal eukaryotic **primers** and Fungi-specific **primer** (Monchy et al. 2011). The cloning-sequencing approach yielded 146 complete **SSU rDNA** gene sequences that clustered into 46 OTUs at 99 % similarity level. Half of these OTUs belonged to Fungi with 15, 7 and 1 representatives for *Chytridiomycota*, *Ascomycota* and *Basidiomycota*, respectively. The other half of our sequences affiliated to Viridiplantae (11), Cryptophyta (5), Alveolata (4), Telonemida (1), Ichtyosporea (1) and Stramenopile (1). Sample **rarefaction curves**, however, did not reach saturation, indicating that the biodiversity was underestimated. Our pyrosequencing approach greatly improved this issue and a better diversity coverage was obtained in our samples. When using eukaryotic and fungal primers, a total of 23 519 and 18 545 reads were obtained, respectively. Using eukaryotic primers, the most represented groups were Stramenopile (27 % of reads), Chlorophyta (19 %), Metazoa (15 %), Fungi (12 %), Alveolata (10 %), Telonemida (8 %), and Cercozoa (7 %). Although the use of fungal primers resulted in an enrichment of fungal sequences (15 % of the pyrosequence reads) and a decrease in non-targeted reads such as metazoan and Viridiplantae, untargeted groups such as Katablepharidophyta and Cryptophyta, still represented 60 % and 10 % of the total reads, respectively (Fig. 20.2). Primer specificity is one of the biggest issue in molecular diversity surveys and although the use of a fungal primer set resulted in a substantial enrichment of fungal reads, including 4 new chytrid OTUs (and 3 other fungi), the issue related to primer specificity towards pelagic

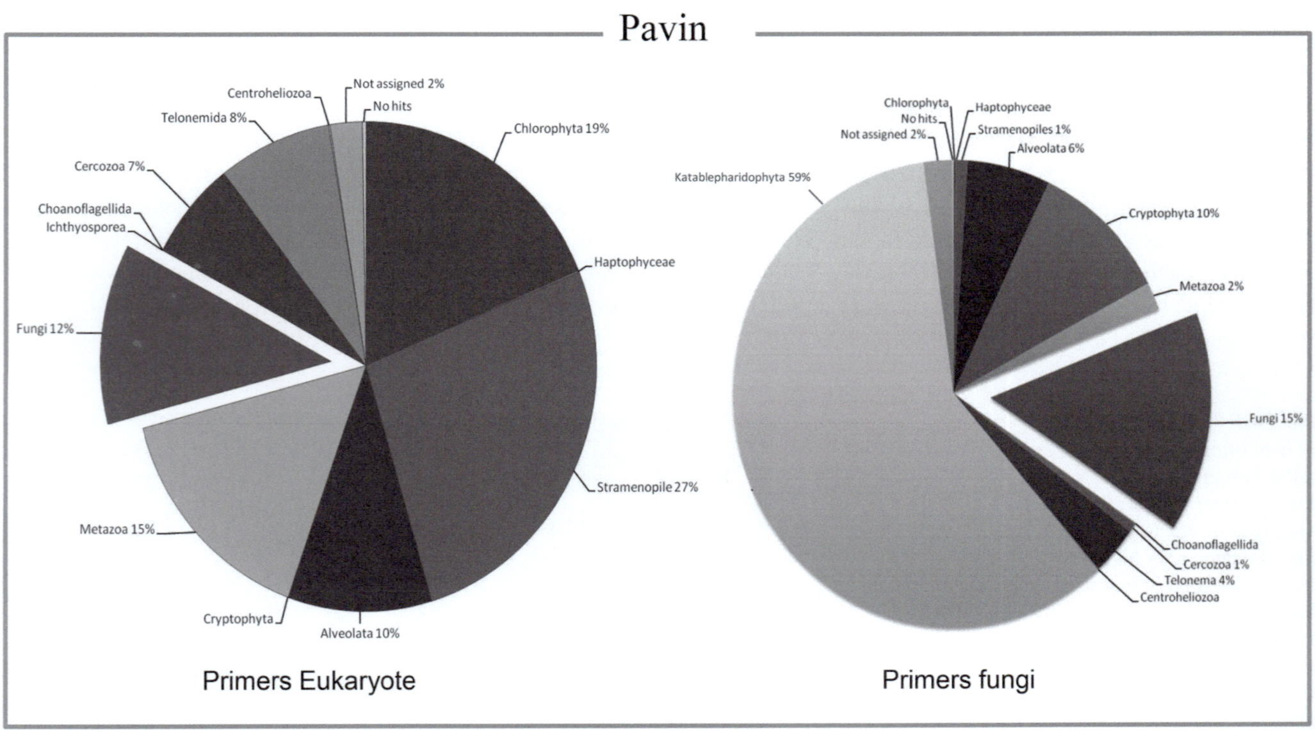

Fig. 20.2 Proportion of taxonomic groups identified in Lake Pavin using pyrosequencing of 18S rRNA gene hypervariable regions amplified by primer sets targeting two different taxonomic levels (Eukaryotes and Fungi). The pie diagram displayed the proportion of reads, obtained from two sampling stations and using the two primers sets (eukaryote and fungus), belonging to a particular phylum. 'No hit' corresponds to reads having no homologous sequence in database. 'Not assigned' corresponds to reads having a match in database but without a precise taxonomic phylum assignment. For more details, see the main text and Monchy et al. (2011)

fungus rRNA genes needed to be addressed before future studies could be conducted (Monchy et al. 2011). Therefore, we used longer and more phylogenetically informative sequences, such as ITS1, and ITS2 regions, generated using our cloning-sequencing approach, to design more specific primer sets that will be used to follow the dynamics of particular fungal species in natural waters.

In a follow-up study we conducted in 2007 in Lake Pavin and other lakes in the vicinity (Jobard et al. 2012), we used the fungal-specific NS1 and ITS4 PCR primers for the amplification of the ribosomal ITS 1 and 2 regions (White et al. 1990). Our study yielded 17 OTUs that belonged to three fungal phyla: *Chytridiomycota* (50 %), *Ascomycota* (40 %), and *Basidiomycota*. Similar to our previous pyrosequencing study (Monchy et al. 2011), no sequence belonging to the division *Zygomycota* was retrieved. This observation is consistent with the current knowledge that in aquatic ecosystems, these fungi are restricted to the group of *Trichomycetes* known as obligate symbionts in the gut of arthropods (Lichtwardt et al. 2003). In Lake Pavin, 60 % of the fungal sequences retrieved were absent from the DNA sequence databases, and mainly affiliated to zoosporic true fungi (i.e. *Chytridiomycota*), primarily to the genera *Rhizophydium* and *Zygorhizidium* known to contain host-specific parasites of phytoplankton in freshwater lakes (Jobard et al. 2010a; Rasconi et al. 2011). This clearly confirms the omnipresence of these **parasites** in aquatic systems, and the need to develop analytical tools for their quantitative and ecological study.

20.3 Methodological Challenges in the Quantitative Assessment of the Ecology of Parasitic Chytrids on Phytoplankton

Until recently, the identification of zoosporic fungi was mainly based on direct observations and culturing techniques (i.e., baiting techniques; recently reviewed in Marano et al. 2012). The main characteristics used for the identification of parasitic chytrids were based on the observation of their sporangia using light or phase-contrast microscopy (Ingold 1940; Canter 1949, 1950, 1951) of living samples or samples preserved with Lugol's iodine (Rasconi et al. 2011; Sen 1988a, b). Subsequently, transmission electron microscopy was used to describe zoospore ultrastructure and spore differentiation (Rasconi et al. 2011; Beakes et al. 1992a, b). The chytrid *Blastocladiella* sp. was the first fungal model for detailed structural studies on **sporogenesis** (Lovett 1963). The conformation of the flagellar rootlets and the spatial distribution of organelles in zoospores provided the basis for chytrid taxonomy (review in Gleason and Lilje 2009).

Ecological investigations of chytrid population dynamics in natural environments greatly improved with the use of epifluorescence microscopy. Based on samples regularly collected in Lake Pavin in 2006 and 2007, we developed a routine protocol based on size fractionation of pelagic samples and the use of the fluorochrome calcofluor white (CFW; $C_{40}H_{44}N_{12}O_{10}S_2$), a chemofluorescent agent which binds to β-1,3 and β-1,4 polysaccharides (Rasconi et al. 2009). CFW is commonly used as a whitening agent in the paper industry and selectively binds to cellulose and chitin. The dye fluoresces when exposed to UV light and offers a very sensitive method for direct microscopic examination of fungal elements on skin scrapings, hairs, nails, and other clinical specimen for fungal elements. Our CFW-based protocol for diagnosing, identifying, and counting chitinaceous fungal parasites (i.e., the sporangia of chytrids) of phytoplankton is provided in Sime-Ngando et al. (2013b). Results from the proposed protocol indicated that CFW penetrates infected host cells remarkably well and is more efficient than other stains such as the protein stain fluorescein isothiocyanate (FITC) for the observation and photomicrography of the complete **rhizoidal system** of chytrid parasites (Rasconi et al. 2009), which is another important criterion for their identification. These preliminary results also highlighted a higher diversity of infected host and parasite communities in Lake Pavin compared to that described in previous studies. Moreover, we showed that chytrid parasites were omnipresent, infecting diverse phytoplankton host communities, primarily diatoms, chlorophytes, and colonial and filamentous cyanobacteria. The diversity and numerical abundance of sporangia and infected hosts, as well as the prevalence of infection (range, <1–24 % of total host cells), increased from the oligotrophic Lake Pavin to the eutrophic Lake Aydat. In addition, we showed that the temporal changes in the abundance and activity of the parasites were mainly influenced by the host community composition (Rasconi et al. 2009).

More recently, we developed an improved CFW protocol using the combination of two fluorochromes, CFW (chitin stain) and SYTOX green (nucleic acid stain) coupled with epifluorescence microscopy for counting, identifying, and investigating the fecundity (i.e. the number of zoopores per sporangium) of parasitic fungi of phytoplankton (Gerphagnon et al. 2013a). This double-staining protocol targeting chitin and nucleic acids allows the visualization of both sporangium and zoospore content and enables the determination of fungal fecundity (Fig. 20.3). The method was applied to freshwater samples collected over two successive years during the end of autumnal cyanobacterial blooms in a eutrophic lake. The study focused on the uncultured host-parasite pair *Anabaena macrospora* (cyanobacterium) and *Rhizosiphon akinetum* (Chytridiomycota). Our results showed that up to 37 % of cyanobacterial akinetes were infected by fungi. Simultaneously, we determined the number of zoospores per sporangium and found out that both host size and intensity of infection conditioned the final size and hence the fecundity

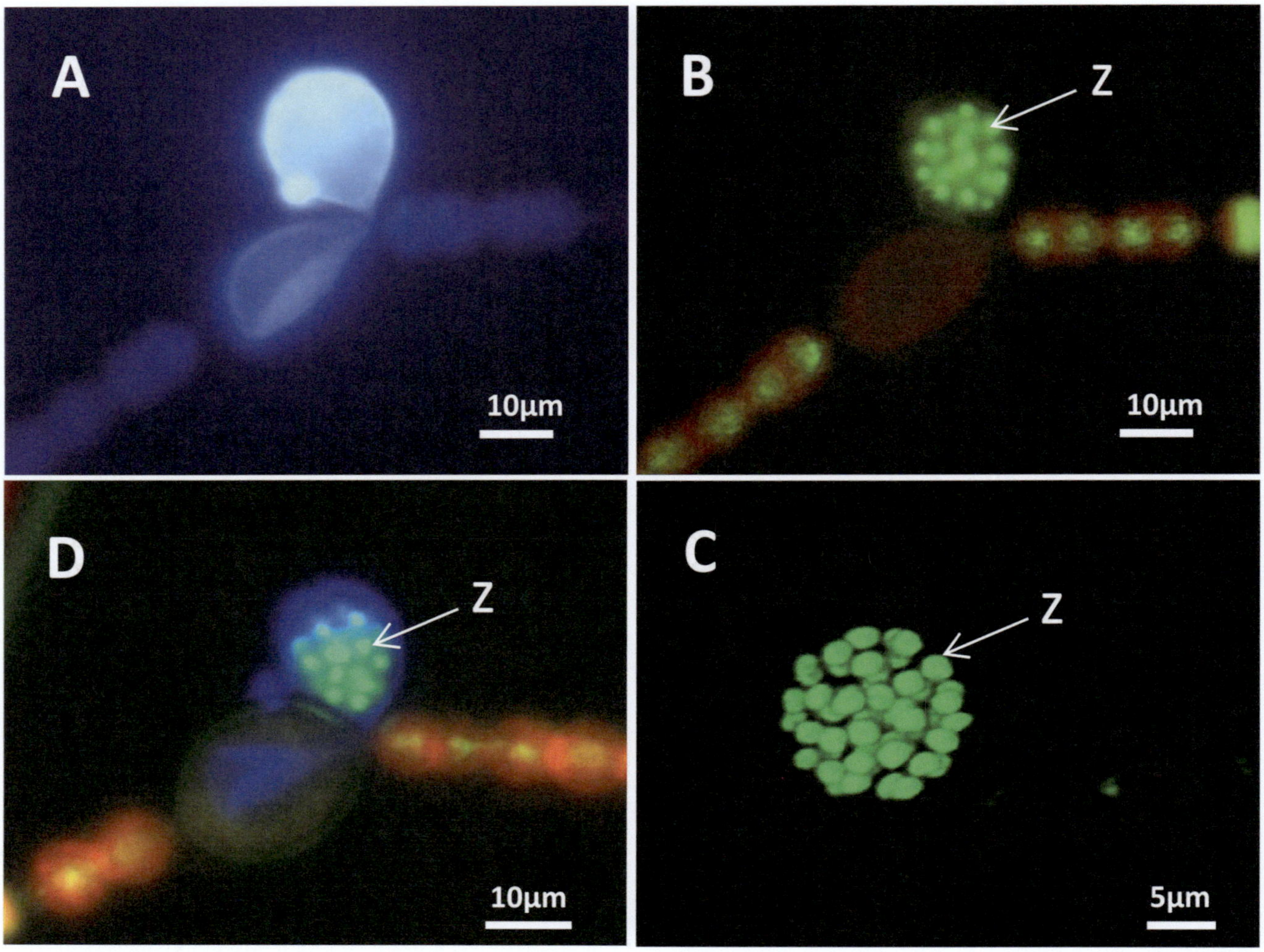

Fig. 20.3 Mature sporangium of *Rhizosiphon akinetum* containing zoospores (**A**, **B**, **C**, **D**), stained by the double staining method (CFW and SYTOX *green*) and excited by UV light (**A**), *blue light* (**B**, **C**), or both UV and *blue light* (**D**) as observed by optical microscopy. Z, zoospores

of chytrids. We also established that the relationships linking host size, final parasite size, and chytrid fecundity were conserved from year to year and seemed to be host-chytrid pairing specific (Gerphagnon et al. 2013a).

In contrast to sporangia, zoospores lack a chitinaceous cell wall, are small, and possess very few phenotypic characteristics that can be used for their identification. Therefore, molecular biology approaches offer an alternative for the quantitative assessment of both chytrid sporangia and zoospores in nature. Molecular approaches are commonly used for the reconstruction of chytrid phylogenies (James et al. 2006), but can also be applied (i.e., **oligonucleotide probes** and primers) to the quantitative ecological study of aquatic parasites. **Fluorescent *in situ* hybridization (FISH)** with horseradish peroxidase activation by fluorescent tyramide (also known as catalyzed reporter deposition, CARD FISH or TSA FISH) is a reliable approach to describe the dynamics of chytrid parasites, especially the zoosporic stage. The optimization of the FISH method requires the use of positive controls, ideally pure cultures that are challenging to obtain especially for chytrid parasites. To circumvent this issue, we used the clone-FISH approach, which uses transformed clones of *Escherichia coli* containing a gene or DNA region of interest (e.g., 18S rDNA sequence of a parasitic chytrid obtained from some of our previous environmental cloning surveys; Jobard et al. 2010b). Using molecular sequences generated from our previous 18 rDNA environmental surveys in Lake Pavin and other lakes in the vicinity, we designed oligonucleotidic probes targeting members of the Chytridiales (i.e., of the groups of Chytridiomycota containing the highest number of described species). We then used the clone-FISH approach to optimize the hybridization conditions of our designed probes before application to natural samples using the CARD-FISH approach (Jobard et al. 2010b). Our protocols are presented in a methodological paper by Sime-Ngando et al. (2013a). Using molecular sequences derived from 18S rDNA cloning/sequencing surveys in Lake Pavin and other lakes in the vicinity, we designed oligonucleotidic probes specific for Chytridiales (i.e. the largest group of the true-fungal division of *Chytridiomycota*). We adapted a clone-FISH approach known from prokaryotes to optimize the hybridization conditions of our designed probes before application to natural samples using the CARD-FISH approach (Jobard et al. 2010b). When applied to natural samples, our CARD-FISH assay demonstrated high specificity and sensitivity, and we showed that 60 % of the total abundance of heterotrophic flagellates were actually fungal zoospores. Although the results from the CARD-FISH approach were preliminary, our findings were consistent with ecological considerations known from pelagic habitats, and recurrent ecological patterns in two contrasting lake ecosystems were observed.

We further developed a **quantitative real-time PCR (qPCR)** assay for the quantification of fungal zoopsores (Lefèvre et al. 2010; also provided in Sime-Ngando and Jobard 2013). QPCR is a very sensitive tool for the detection of less abundant or rare species. When optimal DNA extraction protocol and qPCR conditions were applied, our qPCR assay demonstrated high specificity and sensitivity, with a detection limit as low as 100 18S rDNA copies per reaction (Lefèvre et al. 2010). Application of the method to samples collected in Lake Pavin and other natural samples gave us important clues as to what roles these organisms could have in the pelagic system.

More recently, the application of next generation sequencing technologies has not only advanced our understanding of zoosporic fungal diversity, but also provided insights into their ecological functions through the analysis of their genomes and gene expression (Monchy et al. 2011). Nevertheless, it is still necessary to complement these approaches with culturing methods in order to gain a deeper understanding of the ecological and physiological roles of zoosporic fungi (Marano et al. 2012).

Overall, the methodological difficulties are increasingly being overcome, and it is becoming evident that molecular biology approaches are crucial for a better understanding of zoosporic fungi ecology. The challenge of matching molecular sequence to microscopic phenotypes is increasigly being met (Jobard et al. 2012; Monchy et al. 2012). Furthermore, to assess the functional impact of chytrid parasites on host populations, we applied population dynamic models commonly used by parasitologists such as the ones predicting prevalence and the intensity of infection for the ecological studies of chytridimycosis (Rasconi et al. 2012). Input parameters for these models are derived from direct microscopy and are essential for assessing life cycle, community structure of parasites, their interactions with hosts, as well as the potential impact of fungal parasites on the food web dynamics. Understanding the environmental factors including epidemics can also be inferred through empirical correlations or by using epidemiological approaches such as the changes in incidence rates (i.e. the number of new cases of infections occurring during a given time) or the occurrence of epidemics (i.e. a widespread outbreak of an infection) within host populations (Fox 2003).

20.4 Chytrid Life Cycle and Adaptation to Planktonic Lifestyle

Chytrid species have an atypical life cycle in the context of the pelagic realm where the two main stages (i.e. sporangium and zoospore) have different effects on the food web dynamics. Most members reproduce asexually by releasing swimming zoospores with a single posteriorly-directed whiplash flagellum (Sparrow 1960; Barr 2001). A typical chytrid parasite life cycle begins with the encystment of a free-swimming zoospore on a host followed by the development of a rhizoidal system inside the host cell. The encysted zoospore develops into a mature sporangium drawing its energy from the host, and new zoospores are eventually released into the environment. Because water is necessary for chytrids to complete their life cycle, these water-borne fungi are ubiquitous in bodies of waters, primarily in freshwater environments, and in wet soils.

In a recent study conducted in the eutrophic Lake Aydat located in the vicinity of Lake Pavin (Gerphagnon et al. 2013b), we were able to reconstruct phenotypic stages of host-chytrid systems using the double staining method mentioned above. We sampled at a relatively high frequency (every 3 days) the filamentous cyanobacterium *Anabaena macrospora* during a bloom. Based on the morphology of the sporangium and on the type of host cells, we were able to identify two species of chytrids parasites of this cyanobacterium, both belonging to the same genus, *Rhizosiphon*. While *R. crassum* infected both vegetative cells and resistant forms (i.e. akinetes), and was responsible for the death of cells within host filaments, *R. akinetum*, which only infected akinetes, may have affected the survival of cyanobacterial hosts and their proliferations from year to year. We divided the asexual life cycles of the two parasite species into six stages grouped in 3 phases: young, maturation, and empty phases (Fig. 20.4). These phases correspond to the growth phases (i.e. encystement, germination, growth and maturation) known from the general life cycle of the Chytridiomycota.

Chytridiomycosis epidemics are known to produce massive numbers of zoospores, now known as valuable food source for zooplankton (Kagami et al. 2007a, b). The consideration of the two main development stages (i.e., attached sporangia and free-swimming zoospores) of chytrids in the context of the pelagic system thus highlights two overlooked potential trophic paths in food web dynamics: (i) parasitic predation of host populations, most of which are inedible (i.e. unexploited by grazers), and (ii) the subsequent trophic link via the release of energy-rich zoospores that can serve as food for zooplankton. In addition, we recently shown that chytrid parasitism within the filaments of cyanobacteria during bloom events can result in a mechanical fragmentation of the inedible filaments into shorter-size edible filaments

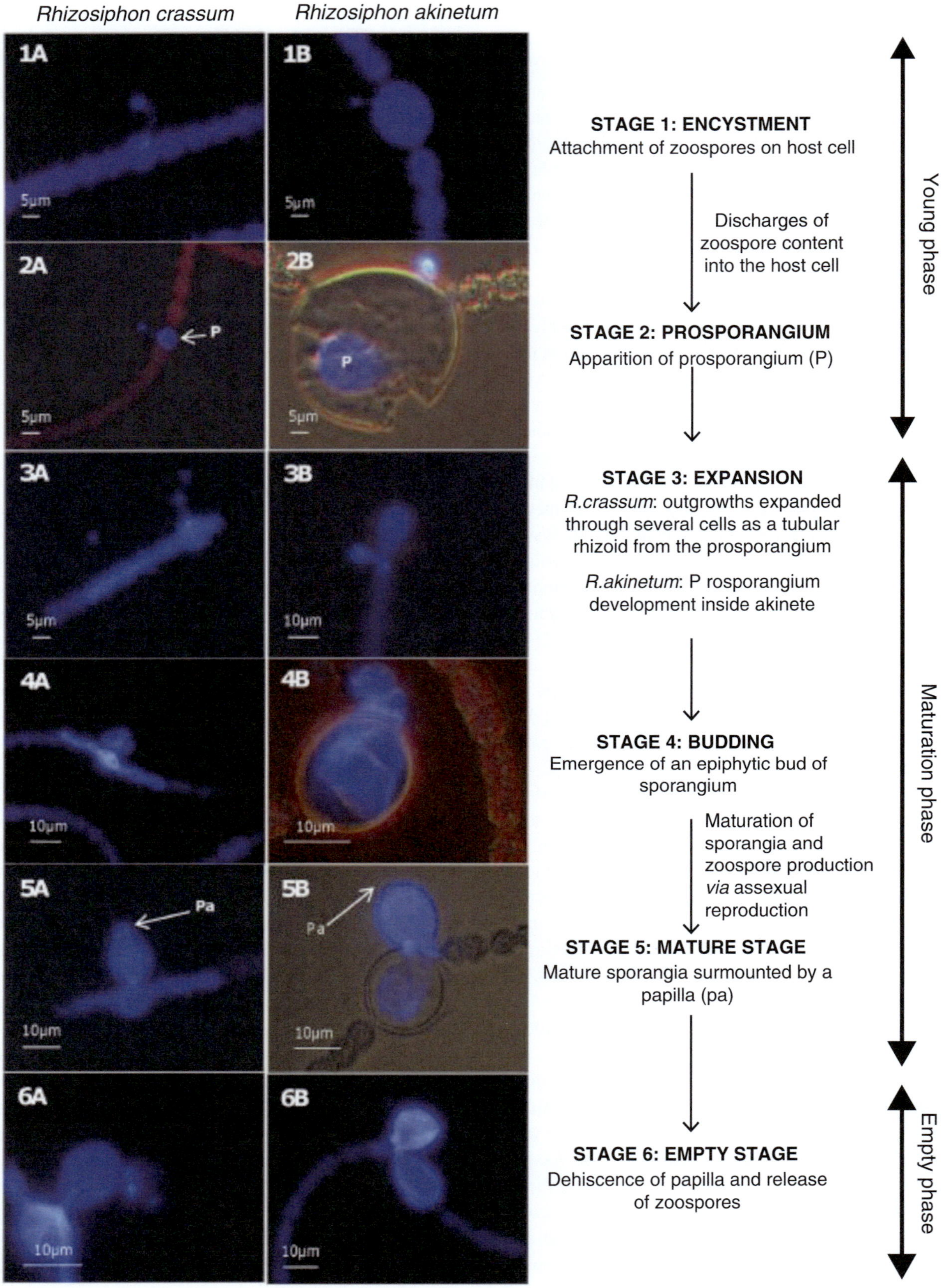

Fig. 20.4 The six different life stages of the two chytrid species, *Rhizosiphon crassum* (**A**) and *Rhizosiphon akinetum* (**B**) parasitizing the cyanobacterium *Anabaena macrospora* from the productive Lake Aydat: Stage1: Encystment; Stage 2: Prosporangium; Stage 3: Expansion stage; Stage 4: Budding; Stage 5: Mature stage; Stage 6: Empty stage. The six life stages were grouped into three different phases: Young phase (Stages 1 and 2), Maturation phase (Stages 3, 4 and 5) and Empty phase (Stage 6). Prosporangium (P) and Papilla (Pa). For more details, see the main text and Gerphagnon et al. (2013b)

(Gerphagnon et al. 2013b), thereby enhancing the contribution of fungal parasites to the bloom decline. This so-called "mechanistic fragmentation" of cyanobacterial filaments (Gerphagnon et al. 2013b) can decrease the resistance of hosts to grazing, which could further accelerate the decline of cyanobacterial blooms. Thus, these results extend the role of chytrid zoospores which are known to upgrade the biochemical diet of zooplankton, establishing zoosporic fungi as potential key players in the food web dynamics, as inferred from intensive fungal studies in the Lakes of the French Massif Central during the past 10 years (Rasconi et al. 2009; Jobard et al. 2010a; Sime-Ngando 2012).

20.5 Trophic Dynamics and PEG-Model Application to Chytrid-Phytoplankton Pairing

We recently presented an extensive seasonal study of chytrid-phytoplankton **trophodynamics** in the oligotrophic Lake Pavin compared to the eutrophic Lake Aydat (Rasconi et al. 2012), based on our CFW staining protocol of infective sporangia and phenotypic identification (Rasconi et al. 2009). We were able to identify up to 15 different chytrid species on diverse host populations, with specific biovolume ranging from 7 to 72 μm^3 $sporangium^{-1}$. Seasonal abundances increased from 0.0005 – 0.4 × 10^6 sporangia l^{-1} in Lake Pavin to 0.0 – 32 × 10^6 sporangia l^{-1} in Lake Aydat. In both lake systems, sporangium abundances peaked with the development of preferred diatom hosts in spring (Pavin) and cyanobacteria in autumn (Aydat), the autumn peak being higher in eutro- compared to oligotrophic conditions, when a monospecific bloom of the filamentous cyanobacterium *Anaebaena* occurred.

All 15 identified species were monocentric (i.e. with one center of growth and development) and eucarpic (using part of the thallus for the fruit-body formation and with a specialized rhizoidal system), typical of the class *Chytridiomycetes*, and belonged to two orders: the *Rhizophidiales* which contained one genus (*Rhizophidium*), and the *Chytridiales* which contained three genera (*Chytridium*, *Zygorhizidium*, and *Rhizosiphon*). The species of *Rhizophidium* spp. infected a wide diversity of hosts, including both large size (e.g. the Chlorophyta *Staurastrum* spp. and the diatoms *Asterionella formosa*, *Synedra* spp. and *Fragilaria crotonensis*) and small size algae (e.g. the diatom *Cyclotella* spp. and the Chlorophyta *Chodatella ciliata* and *Ankistrodesmus convolutus*). The species *Chytridium* spp. infected the chlorophyte *Oocystis* sp, the diatom *F. crotonensis* and the cyanobacterium *Microcystis* sp, while the species of *Zygorhizidium* infected the diatoms *Melosira* spp. The genus *Rhizosiphon* comprised two species that are specific parasites of vegetative cells and akinetes (*R. crassum*) and of akinetes alone (*R. akinetum*). These two species correspond to different niches offered by the filamentous cyanobacterium host *Anabaena macrospora* in the productive Lake Aydat (Gerphagnon et al. 2013b). Although almost all chytrid species were observed in all lakes ranging from oligo- to eutrophic conditions, the seasonal fungal community composition was largely dominated by species of the genus *Rhizophidium* (90 % of total sporangium abundance) in oligotrohic conditions (Pavin), and of the genera *Rhizophidium* (56 %), *Zygorizidium* (22 %), *Chytridium* (19 %), and *Rhizosiphon* (14 %) in eutrophic conditions (Aydat) (Rasconi et al. 2012).

The community structure of natural chytrids is intimately linked to the availability of hosts (Ibelings et al. 2004). However, there is still no study assessing the fungal species successions in natural environments. Based on the seasonal study in Lakes Pavin and Aydat (Rasconi et al. 2012), we proposed a general empirical model on chytrid seasonality and trophodynamics (i.e. with their hosts) based on the theoretical **PEG (i.e. Plankton Ecology Group) model** of seasonal succession of planktonic events in freshwaters (Sommer et al. 1986). During winter, the development and activities of both chytrid parasites and their phytoplanktonic hosts were at their lowest levels, because of low temperature, freezing or ice-cover. Between late winter and spring, the environmental conditions, primarily the increase in water temperature and in mixing-derived nutrient availability, favor the development of host communities, with the dominance of **k-strategists** (e.g. large diatoms) towards spring. As a consequence, the probability of host – parasite contact increases, and led to higher chytrid infectivity, and the production of large amount of zoospores. Enhanced infection prevalence subsequently limits the development of large diatoms and induces their decline, liberating niches for a diversified phytoplankton community of small-size **r-strategists**. The abundance of chytrid sporangia reaches their lowest level, while the availability of food (i.e. small phytoplankton and fungal zoospores) benefits the development of grazers and the establishment of a typical clear-water phase at the end of spring. During the summer months, favorable environmental conditions, together with a high grazing pressure, allow the development of a diversified and complex plankton community. Small edible hosts are inhibited by the grazing pressure, while the availability of large size hosts favors the proliferation of different chytrid species towards the end of the summer. From there, oligotrophic lakes (Pavin) significantly diverge from productive waters (Aydat). In oligotrophic situations, autumnal overturn promotes species coexistence and phytoplankton diversity leads to the association of different chytrid species that parasitize chlorophytes and diatoms, but with general low infection prevalence due to a balanced host – parasite growth. In eutrophic lakes, nutrient conditions and persistent stratification favor the bloom of filamentous cyanobacteria from the end of the summer period towards early autumn, with the development of a monospecific community of chytrids (i.e. *Rhizosiphon* spp.). During that period, the highest infection prevalence is reached, followed by the decline of cyanobacteria – chytrid system towards the late seasonal phase (for more details, see Fig. 7 in Rasconi et al. 2012).

Our analysis of chytrid-phytoplankton pairing in Lakes Pavin and Aydat provided evidence that host species composition and their size are critical factors for chytrid infectivity, the larger hosts being more vulnerable, including pennate diatoms, desmids, and filamentous cyanobacteria. Such host species are more vulnerable because their size naturally increases the host – parasite contact rates. In addition, larger host cell are expected to excrete more attractants substances known to enhance recognition of suitable hosts by zoospores (Canter and Jaworski 1981). Larger algae also contain more resources for parasites, which may explain the heavy infection of larger cell size algal species, even at lower population density (Lund 1957; Holfeld 1998). The prevalence of infection was also related to the productivity of the lake that offers good substrate conditions for the growth of parasite-host systems (Rasconi et al. 2012). Thus, both the abundances and cell volumes of hosts seem important features in determining the amplitude of chytrid epidemics within natural phytoplankton. These parameters also appeared to be related to the tolerance threshold of infection, i.e. the critical prevalence or the level of prevalence from which the standing stock of phytoplankton starts to decline (Bruning et al. 1992). At low host abundance, the critical infection prevalence is generally below 20 %, but increases with increasing host abundance. This is one of the potential mechanisms through which parasites regulate host populations.

20.6 Phytoplankton Chytridiomycosis in Temperate Freshwater Lakes and Potential Implications for Food Web Energy Flow

Based on our seasonal study in Lakes Pavin and Aydat (Rasconi et al. 2012), the prevalences of infection (% of infected host cells) typically average around 20 %, with no significant variation with the trophic status of lakes. These values increased to reach about 100 % when monospecific blooms of infected hosts occurred in natural conditions. Chytrid infection commonly leads to the death of their host cells (Canter and Lund 1951; Sen 1988a, b; Kudoh and Takahashi 1992; Bruning et al. 1992; Holfeld 1998, 2000; Ibelings et al. 2004, 2011). This is enhanced by the intensity of infection (number of parasites per host cell), which can go well above 1. Empty sporangia are commonly found attached on dead phytoplankton cells (Rasconi et al. 2012), which is suggestive of the lethal issue of chytrid infection. There are several lines of evidence showing that parasitism inhibits the development of sensitive species, and particular attention has been paid to the occurrence of fungi on diatoms, and to the effects of parasitism on their seasonal distributions (Canter and Lund 1948; Van Donk and Ringelberg 1983). For example, in Lake Pavin, the spring development of the diatoms *Asterionella* and *Synedra* was inhibited by the chytrid *Rhizophidium planktonicum*. In Lake Aydat, another diatom, *Fragilaria*, became abundant but the proliferation of their parasites, *Rhizophidium fragilariae*, interrupted their development (Rasconi et al. 2012).

In natural phytoplankton communities, the parasitized populations are often replaced by others with similar ecologically requirements, which can lead to an unchanged standing stocks of phytoplankton hosts in the ecosystem, with no visible obvious damage to the total community (Reynolds 1940). However, chytrids seem to preferentially infect large and less edible phytoplankton species, as discussed previously. Some examples provided in the literature suggest that large infected diatoms such as *Asterionella* sp. and *Fragilaria* sp. (with a mean length of 50 and 70 μm, respectively) can be replaced by small centric diatoms such as *Stephanodiscus* spp. (10 μm) (Van Donk and Ringelberg 1983; Sommer 1987). This implies that the development of large species is inhibited by infection, while smaller algae proliferate. Active parasitism may thus act on the host standing stock in a continuum ranging from minor to significant changes which, in all cases, will affect the phytoplankton community structure. In the context of phytoplankton seasonal dynamics and species successions, this can have profound ecological implications (Van Donk 1989). For example, due to chytrid parasites, the phytoplankton community can shift from a mature development stage typically dominated by large, **k-strategist species** towards a pioneer stage of succession that favours the development of small, **r-strategist species**. In addition, by controlling phytoplankton dynamics, chytrids can significantly affect the primary production of aquatic systems, as suggested by a negative correlation between the primary production and the per sporangium biovolume of chytrids in Lakes Pavin and Aydat (Rasconi et al. 2012).

More importantly, phytoplankton chytridiomycosis produce very high number of propagules (i.e. zoospores). Fungal zoospores are valuable food sources for zooplankton because their cytoplasm stores carbohydrates such as glycogen, proteins, and a wide range of fatty acids, phospholipids, sterols and other lipids (Gleason et al. 2009; Sime-Ngando 2013). When chytrids reproduce, most of the cytoplasm is converted into zoospores that swim away to colonize new substrates or infect new hosts. Lipids are high energy compounds, some of which are important for energy storage. Indeed, lipids are present mainly in the form of endogenous reserves, often as membrane bound vesicles called lipoid globules that can easily be seen in the cytoplasm of fungal zoospores using both light and electron microscopy (Gleason and Lilje 2009; Sime-Ngando 2013). The size and number of lipoid globules within zoospores vary, and their ultrastructure is complex. The chemical composition of lipids, including both fatty acids and sterols, has been characterized in a number of genera of zoosporic fungi (Southall et al. 1977; Wheete et al. 1989). These endogenous reserves are consumed during the motile phase. They presumably provide energy for the movement of flagella which can last for up to several hours, as well as for the attachment and germination of zoospores on the appropriate substrates or hosts. Besides, many zoosporic fungi can grow in the laboratory on minimal synthetic media containing one carbon source such as cellulose, xylan, starch or chitin along with salts containing nitrate, sulfate, and phosphate (Gleason et al. 2008; Sime-Ngando 2013). This establishes fungi as potential competitors of bacteria and primary producers for essential minerals.

There are other significant functions for the high energy compounds found in fungal zoospores, especially as food resources for zooplankton and probably for many other consumers in aquatic ecosystems (Rasconi et al. 2014). Fungal spores and hyphae in general are known to be eaten by a large number of different consumers in both aquatic and soil ecosystems, including a variety of mycophagous protozoa such as amoebae and flagellates, detritivores, grazers such as filter feeding zooplankton, and benthic suspension feeders (Gleason et al. 2009; Sime-Ngando 2013). Since most of these consumers do not discriminate between food resources except by size, we would expect zoospores as well as hyphae and non-motile spores to be eaten by many of these consumers, although published records are lacking. Fungal zoospores are well within the suitable size range of particles (2–3 μm in diameter) for zooplankton feeding behavior and consequently, when fed upon, transfer matter to higher trophic levels in the food chain. For example, zoospores are efficiently grazed by crustacean zooplankton such as *Daphnia* spp. (Kagami et al. 2007a, b, 2011; Sime-Ngando 2013), before they grow into a mature thallus (i.e. body). Thus zoospores may provide organic compounds containing nitrogen, phosphorus and sulfur, mineral ions and vitamins to grazing zooplankton.

More interestingly, given their particularly high nutritional qualities, zoospores constitute a good source of food for many consumers, which must obtain at least some essential nutrients that cannot be produced *de novo*. One example is found in the cladoceran *Daphnia*. Recent research has shown that zoospores of the parasitic chytrid, *Zygorhizidium*, are quite rich in polyunsaturated fatty acids (PUFAs) and cholesterols, which are essential for the growth of *Daphnia* (Kagami et al. 2007a, b). These zoospores facilitate the trophic transfer from the inedible large diatom hosts, *Asterionella* sp., and the growth of *Daphnia*. PUFAs and cholesterol are known to promote growth and reproduction in crustacea. This phenomenon, known as the 'trophic upgrading concept', is of significant importance in aquatic food webs because it highlights not only the quantity but also the quality of the matter being transferred *via* fungal zoospores (Sime-Ngando et al. 2011; Sime-Ngando 2012, 2013; Rasconi et al. 2014).

20.7 Phytoplankton Chytridiomycosis and the Impact on Food Web Stability in Lake Pavin: A Linear Inverse Modeling Analysis

Given that food webs are central to ecological concepts (Pascual and Dunne 2005), it is important to establish the role of parasites in the structure and function of food webs. In theory, parasites can have a variety of effects. Lafferty et al. (2006, 2008) suggested that parasites affect food web properties and topology since they double the connectance (defined as the number of observed links divided by the number of possible links) and quadruple the number of links. Others have postulated that parasites drive an increase in species richness, trophic levels, and trophic chain length of the food web (Huxham et al. 1995). These properties may stabilize community structure (Hudson et al. 2006). However, the potential effects of parasites on food web stability are a complex and unresolved issue since the concept of stability is at the center of a perhaps infinite debate in community ecology (May 1972; Pimm 1984; McCann 2000; O'Gorman and Emmerson 2009; Hosack et al. 2009). Based on the ideas of May (1973), parasites should lead to a destabilized trophic network because they increase species diversity and connectance. In addition, adding parasites to food webs extends the length of trophic chains, which can decrease food-web stability (Williams and Martinez 2004). However, the addition of long loops of weak interactions, which may be a characteristic of parasites with complex life cycles, might offset the destabilizing effects of increased connectance (Neutel et al. 2002).

To investigate ecosystem properties and ecological theories, the application of mathematical models, is useful and allows **trophic network** representation through carbon flows. For the first time, we included parasitic chytrids of phytoplankton as an individualized compartment in the pelagic **food web** of Lake Pavin, and quantified their impact on matter flow through a trophic network (Grami et al. 2011). To describe the food web, models representative of carbon flows were built, including chytrid parasitism and the amount of primary production channeled in food web *via* chytrid infection. Carbon flows between the complete food web including parasitic chytrids (MWC, Model with Chytrids) were compared to the same model that did not consider the presence of chytrids and the resulting flows (MWOC, Model without Chytrid), as traditionally done in previous plankton food-web analysis (e.g. Niquil et al. 2006). MWC and MWOC models were constructed on the basis of the same data set corresponding to the spring bloom period in Lake Pavin (i.e. March to June 2007). These models were built using the Linear Inverse Modeling procedure (LIM, (Vézina and Platt 1988)) recently modified into the LIM-Monte Carlo Markov Chain (LIM-MCMC, (Van den Meersche et al. 2009)). This method allows reconstruction of missing flow values and alleviates the problem of under-sampling using the principle of conservation of mass, i.e. the quantity of carbon coming into each compartment considered as equal to the amount leaving it (Vézina and Platt 1988). Thanks to recent development of the inverse analysis into LIM – MCMC, a probability density function covering the range of possible values was generated for each flow. The results of this exercise are summarized in Fig. 20.5 where the inclusion of the two life stages of chytrid parasites of microphytoplankton (>20 μm) increases the number of compartments and flows. These parasites were able to short-circuit about 20 % of the gross primary production, of which 15 % is transferred to grazers with high throughput.

In addition, for each calculated set of flows generated by the Linear Inverse Modeling procedure, there is a set of calculated indices which allows application of statistical tests. The flows obtained from the models were used for calculations of Ecological Network Analysis indices that characterize the structure of the food web, and help reveal emergent properties (Ulanowicz 1986, 2003; Ulanowicz et al. 2009). The use of ecological indices moreover, allows an indirect evaluation of the effects of network properties on the stability of the ecosystem, as several authors have proposed theoretical links between structural properties and local stability (cf. Ulanowicz 2003). On this basis, the model results support recent theories on the probable impact of parasites on food web function. In the lake, during spring, when 'inedible' algae (unexploited by planktonic herbivores) were the dominant primary producers, the epidemic growth of chytrid parasites significantly reduced the sedimentation loss of algal carbon from 21 to 10 % of gross primary production (Fig. 20.5). Furthermore, from the review of some theories about the potential influence of parasites on ecological network properties, we argue that parasitism contributes to longer carbon path lengths, higher levels of activity and specialization, and lower recycling. We conclude that considering the "structural asymmetry" hypothesis as a stabilizing pattern, chytrids should contribute to the stability of aquatic food webs (Grami et al. 2011).

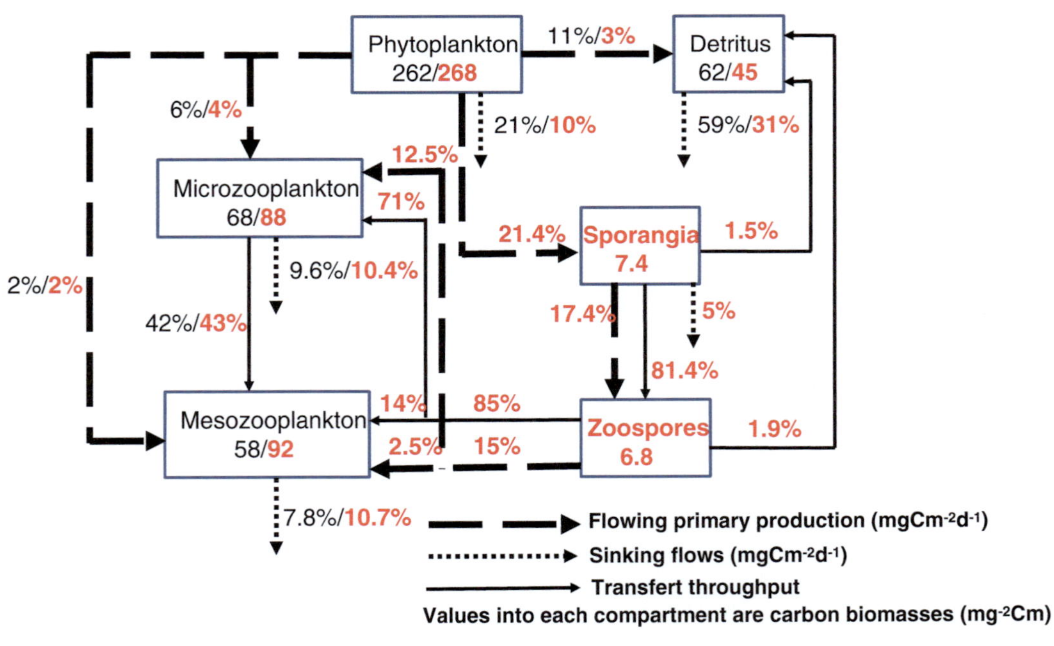

Fig. 20.5 Impact of parasitic chytrids on the microbial loop: flowing and sinking carbon from the gross primary production of phytoplankton (>20 μm) during the spring diatom bloom in Lake Pavin, France. The effects of infective fungal sporangia and their propagules (zoospores) are highlighted in *red color* (i.e. the number in *black* correspond to values without fungi). The diagram corresponds to steady state models of the euphotic zone of the lake generated from a linear inverse modeling analysis. For more details, see the main text and Grami et al. (2011)

20.8 Conclusions and Perspectives

Chytrid and their trophic modes, primarily parasitism, are ubiquitous in aquatic ecosystems, including marine habitats (Gleason et al. 2011). Pioneering ecological and molecular research in Lakes Pavin and Aydat have helped demonstrating that chytrid parasites of phytoplankton are extremely diversified in freshwater habitats, with different taxa characterized by different biological characteristics and requirements that determine their distributions in response to environmental parameters, but primarily to the seasonal dynamics of their hosts. Host abundance, size and biomass establish the threshold for the critical prevalence of infection and the related decline in host communities. Associated ecological implications are considerable, because chytrid parasites can kill their hosts, release substrates for microbial processes, and provide nutrient-rich particles as zoospores and short fragments of filamentous cyanobacterial hosts for the grazer food chain. This implies that cyanobacterial blooms, and other large-size inedible phytoplankton blooms as well, may not necessarily represent trophic bottlenecks. Based on the observation that phytoplankton fungal parasitism preferentially impacts larger species (i.e. characteristics of climax populations), chytrid epidemics represent an important driving factor in phytoplankton successions and maturation, in addition to seasonal forcing. The activity of chytrid parasites of phytoplankton thus represents an important but yet overlooked ecological driving force in aquatic food web dynamics. In addition to their capabilities to resist adverse conditions and use different sources of carbon and nutrients, chytrid parasites can affect the plankton food web functions, ecosystem properties and topology, such as stability and trophic transfer efficiency. We are perhaps approaching paradigm-shift in the development of aquatic microbial ecology.

However, studies on phytoplankton chytridiomycosis remain restricted to a few temperate lakes, and extensive studies in the world's aquatic ecosystems, at wider geographical and time scales, are needed. Besides, the identification of chytrid species based on phenotypic features requires time and experience, and the chytrid diversity currently described is probably underestimated. In this context, the increasing development of molecular tools is instrumental and has already been improving linkage between cell identity and function, which is critical for the consideration of microbial parasites in the pelagic food webs in terms of carbon flows, and the impact on biogeochemical cycling in aquatic ecosystems. Furthermore, the parasitic lifestyle is generally highly subtle and can, for example, control competition by dominant species for resources, thereby promoting species coexistence and diversity. Parasites can also form long-lived associations with hosts, reducing their survival, or allowing infected hosts to remain strong competitors, although few models exist for microbial fungus-host pairings.

Acknowledgements This study includes results from PhD and postdoc researches supported by various instances: Région Auvergne, CNRS, Université Blaise Pascal, French Ministère de la Recherche et de la Technologie, and the Ministry of Culture, High School and Research of Grand-Duché de Luxembourg. Thanks are owed to my colleagues, particularly G Boutheina, FH Gleason, and M Kagami for fruitful discussions and inspiring collaborative papers. We thank Prof. Felix Bärlocher, Mt Allison University, Sackville NB, Canada, for critical and proofreading of the manuscript. This is a contribution to the French ANR Programme Blanc ROME – *Rare and overlooked microbial diversity* - Coordinator TSN, PIs: U Christaki, C Gachon, S Monchy, N Niquil, and E Viscogliosi.

References

Amblard C, Boisson JC, Bourdier G, Fontvielle D, Gayte X, Sime-Ngando T (1998) Ecologie microbienne en milieu aquatique: des virus aux protozoaires. In: Erb F, Villeneuve JP (eds) – Les sciences de l'eau: bilan et perspectives. J Water Sci 11:145–162

Barr DJS (2001) Chytridiomycota. In: McLaughlin DJ, McLaughlin EG, Lemke PA (eds) The Mycota, vol. VII, Part A. Springer, New York, pp 93–112

Beakes GW, Canter HM, Jaworski GHM (1992a) Comparative ultrastructural ontogeny of zoosporangia of *Zygorhizidium affluens* and *Z. planktonicum*, chytrid parasites of the diatom *Asterionella formosa*. Mycol Res 96:1047–1059

Beakes GW, Canter HM, Jaworski GHM (1992b) Ultrastructural study of operculation (discharge apparatus) and zoospore discharge in zoosporangia of *Zygorhizidium* affluens and *Z. planktonicum*, chytrid parasites of the diatom *Asterionella formosa*. Mycol Res 96:1060–1067

Bruning K, Lingeman R, Ringelberg J (1992) Estimating the impact of fungal parasites on phytoplankton populations. Limnol Oceanogr 37:252–260

Canter HM (1949) The importance of fungal parasitism in limnology. Verh Int Verein Limnol 10:107–108

Canter HM (1950) Fungal parasites of the phytoplankton. I: studies on British chytrids, X. Ann Bot 14:263–289

Canter HM (1951) Fungal parasites of the phytoplankton. II (Studies on British Chytrids, XII). Ann Bot 15:129–156

Canter HM, Jaworski GHM (1981) The effect of light and darkness upon infection of *Asterionella formosa* Hassall by the chytrid *Rhizophydium planktonicum* Canter emend. Ann Bot 47:13–30

Canter HM, Lund JWG (1948) Studies on plankton parasites I. Fluctuations in the numbers of *Asterionella Formosa* Hass in relation to fungal epidemics. New Phytol 47:238–261

Canter HM, Lund JWG (1951) Studies on plankton parasites: III. Examples of the interaction between parasitism and other factors determining the growth of diatoms. Ann Bot 15:359–371

Carrias JF, Amblard C, Bourdier G (1996) Protistan bacterivory in an oligomesotrophic lake: importance of attaced ciliates and flagellates. Microb Ecol 31:249–268

der Staay Sy M-v, De Wachter R, Vaulot D (2001) Oceanic 18S rDNA sequences from picoplankton reveal unsuspected eukaryotic diversity. Nature 409:607–610

Fox A (2003) Glossary of epidemiological terms. Internet J Pediatr Neonatol 3. www.ispub.com

Gasol JM, Vaqué D (1993) Lack of coupling between heterotrophic nanoflagellates and bacteria: a general phenomenon across aquatic systems? Limnol Oceanogr 38:657–665

Gerphagnon M, Latour D, Colombet J, Sime-Ngando T (2013a) A double staining method using SYTOX-green and calcofluor white for studying fungal parasites of phytoplankton. Appl Environ Microbiol 79:3943–3951

Gerphagnon M, Latour D, Colombet J, Sime-Ngando T (2013b) Fungal parasitism: life cycle, dynamics and impact on cyanobacterial blooms. PloS One 8:e60894

Gleason FH, Lilje O (2009) Structure and function of fungal zoospores: ecological implications. Fungal Ecol 2:53–59

Gleason FH, Kagami M, Lefevre E, Sime-Ngando T (2008) The ecology of chytrids in aquatic ecosystems: roles in food web dynamics. Fungal Biol Rev 22:17–25

Gleason FH, Kagami M, Marano AV, Sime-Ngando T (2009) Fungal zoospores are valuable food resources in aquatic ecosystems. Feature article, newsletter of the Mycological Society of America, Inoculum, Suppl Mycolog 60:1–3

Gleason FH, Kupper F, Amon J, Picard K, Gachon CMM, Marano A, Sime-Ngando T, Lilje O (2011) Zoosporic true fungi in marine environments, a review. Mar Freshwat Res 62:383–393

Grami B, Rasconi S, Niquil N, Jobard M, Saint Beat B, Sime-Ngando T (2011) Functional effects of parasites on food web properties during the spring diatom bloom in Lake Pavin: a linear inverse modelling analysis. PloS One 6:e23273

Holfeld H (1998) Fungal infections of the phytoplankton: seasonality, minimal host density, and specificity in a mesotrophic lake. New Phytol 138:507–517

Holfeld H (2000) Relative abundance, rate of increase, and fungal infections of freshwater phytoplankton. J Plankton Res 22:987–995

Hosack GR, Li HW, Rossignol PA (2009) Sensitivity of system stability to model structure. Ecol Model 220:1054–1062

Hudson PJ, Dobson AP, Lafferty KD (2006) Is a healthy ecosystem one that is rich in parasites? Trends Ecol Evol 21:381–385

Huse SM, Dethlefsen L, Huber JA, Mark Welch D, Relman DA, Sogin ML (2008) Exploring microbial diversity and taxonomy using SSU rRNA hypervariable tag sequencing. PLoS Genet 4:e1000255

Huxham M, Raffaelli D, Pike A (1995) Parasites and food web patterns. J Anim Ecol 64:168–176

Ibelings BW, De Bruin A, Kagami M, Rijkeboer M, Brehm M, Donk EV (2004) Host parasite interactions between freshwater phytoplankton and chytrid fungi (chytridiomycota). J Phycol 40:437–453

Ibelings BW, Gsell AS, Mooij WM, Van Donk E, Van Den Wyngaert S, De Senerpont DL (2011) Chytrid infections and diatom spring blooms: paradoxical effects of climate warming on fungal epidemics in lakes. Freshw Biol 56:754–766

Ingold CT (1940) *Endocoenobium eudorinae* gen. et sp. nov., a chytridiaceous fungus parasitizing *Eudorina elegans* Ehrenb. New Phytol 39:97–103

James TY, Letcher PM, Longcore JE, Mozley-Standridge Porter D, Powell MJ, Griffith GW, Vilgalys R (2006) A molecular phylogeny of the flagellated fungi (Chytridiomycota) and description of a new phylum (Blastocladiomycota). Mycologia 98:860–871

Jobard M, Rasconi S, Sime-Ngando T (2010a) Diversity and functions of microscopic fungi: a missing component in pelagic food webs. Aquat Sci 72:255–268

Jobard M, Rasconi S, Sime-Ngando T (2010b) Fluorescence *in situ* hybridization of uncultured zoosporic fungi: testing with clone-FISH and application to freshwater samples using CARD-FISH. J Microbiol Methods 83:236–243

Jobard M, Rasconi S, Solinhac L, Cauchie M-H, Sime-Ngando T (2012) Molecular and morphological diversity of fungi and the associated functions in three European nearby lakes. Environ Microbiol 14:2480–2494

Kagami M, de Bruin A, Ibelings B, Van Donk E (2007a) Parasitic chytrids: their effects on phytoplankton communities and food-web dynamics. Hydrobiologia 578:113–129

Kagami M, Von Elert E, Ibelings BW, De Bruin A, Van Donk E (2007b) The parasitic chytrid, *Zygorhizidium*, facilitates the growth of the cladoceran zooplankter, Daphnia, in cultures of the inedible alga, *Asterionella*. Proc R Soc Biol Sci Ser B 274:1561–1566

Kagami M, Helmsing NR, Van Donk E (2011) Parasitic chytrids could promote copepod survival by mediating material transfer from inedible diatoms. Hydrobiologia 659:49–54

Kudoh S, Takahashi M (1992) An experimental test of host population size control by fungal parasitism in the planktonic diatom *Asterionella formosa* using mesocosms in a natural lake. Arch Hydrobiol 124:293–307

Lafferty KD, Dobson AP, Kuris AM (2006) Parasites dominate food web links. Proc Natl Acad Sci U S A 103:11211–11216

Lafferty KD, Allesina S, Arim M, Briggs CJ, Leo G et al (2008) Parasites in food webs: the ultimate missing links. Ecol Lett 11:533–546

Lefèvre E, Bardot C, Noël C, Carrias JF, Viscogliosi E, Amblard C, Sime-Ngando T (2007) Unveiling fungal zooflagellates as members of freshwater picoeukaryotes: evidence from a molecular diversity study in a deep meromictic lake. Environ Microbiol 9:61–71

Lefèvre E, Roussel B, Amblard C, Sime-Ngando T (2008) The molecular diversity of freshwater picoeukaryotes reveals high occurrence of putative parasitoids in the plankton. PloS One 3:e2324

Lefèvre E, Jobard M, Venisse JS, Bec A, Kagami M, Amblard C, Sime-Ngando T (2010) Development of a real-time PCR essay for quantitative assessment of uncultured freshwater zoosporic fungi. J Microbiol Methods 81:69–76

Lefranc M, Thenot A, Lepere C, Debroas D (2005) Genetic diversity of small eukaryotes in lakes differing by their trophic status. Appl Environ Microbiol 71:5935–5942

Lichtwardt RW, White MM, Cafaro MJ (2003) Freshwater trichomycetes and their arthropod hosts. In: Tsui CKM, Hyde KD (eds) Freshwater Mycology. Fungal Diversity Press, Hong Kong, pp 81–100

López-García P, Rodriguez-Valera F, Pedros-Alio C, Moreira D (2001) Unexpected diversity of small eukaryotes in deep-sea Antarctic plankton. Nature 409:603–607

Lovett JS (1963) Chemical and physical characterization of "nuclear caps" isolated from *Blastocladiella* zoospores. J Bacteriol 85:1235–1246

Lund JWG (1957) Fungal diseases of plankton algae. In: Horton-Smith C (ed) Biological aspects of the transmission of diseases. Oliver and Boyd, Edinburgh, pp 19–23

Marano AV, Edwards JE, Gleason FH, Bärlocher F, Pires-Zottarelli CL, Lilje O, Schmidt SK, Rasconi S, Kagami M, Barrera MD, Sime-Ngando T, Boussiba S (2012) Quantitative methods for the analysis of zoosporic fungi. J Microbiol Methods 89:22–32

May RM (1972) Will a large complex system be stable? Nature 238:413–414

May RM (1973) Stability and complexity in model ecosystems. Princeton University Press, Princeton. 265 p

McCann KS (2000) The diversity-stability debate. Nature 405:228–233

Monchy S, Jobard M, Sanciu G, Rasconi S, Gerphagnon M, Chabe M, Cian A, Meloni D, Niquil N, Christaki U, Viscogliosi E, Sime-Ngando T (2011) Exploring and quantifying fungal diversity in freshwater lake ecosystems using rDNA cloning/sequencing and SSU tag pyrosequencing. Environ Microbiol 13:1433–1453

Monchy S, Grattepanche JD, Breton E, Meloni D, Sanciu G, Chabe M, Delhaes L, Viscogliosi E, Sime-Ngando T, Christaki U (2012) Microplanktonic community structure in a coastal system relative to a *Phaeocystis* bloom inferred from morphological and tag pyrosequencing methods. PloS One 7:e39924

Moreira D, Lopez-Garcia P (2002) The molecular ecology of microbial eukaryotes unveils a hidden world. Trends Microbiol 10:31–38

Neutel AM, Heesterbeek JAP, de Ruiter P (2002) Stability in real food webs: weak links in long loops. Science 296:1120–1123

Niquil N, Bartoli G, Urabe J, Jackson GA, Legendre L, Dupuy C, Kumagai M (2006) Carbon steady state model of the planktonic food web of Lake Biwa, Japan. Freshw Biol 8:1570–1585

O'Gorman EJ, Emmerson MC (2009) Perturbations to trophic interactions and the stability of complex food webs. Proc Natl Acad Sci USA 106:13393–13398

Pace NR, Stahl DA, Lane DJ, Olsen GJ (1986) The analysis of natural microbial populations by ribosomal RNA sequences. Adv Microb Ecol 9:1–55

Pascual M, Dunne J (eds) (2005) Ecological networks: linking structure to dynamics in food webs. Oxford University Press, Oxford/New York, p 416

Patterson DJ (1993) The current status of the free-living heterotrophic flagellates. J Eukaryot Micribiol 40:606–609

Pimm SL (1984) The complexity and stability of ecosystems. Nature 307:321–326

Rasconi S, Jobard M, Jouve L, Sime-Ngando T (2009) Use of calcofluor white for detection, identification and quantification of phytoplanktonic fungal parasites. Appl Environ Microbiol 75:2545–2553

Rasconi S, Jobard M, Sime-Ngando T (2011) Parasitic fungi of phytoplankton: ecological roles and implications for microbial food webs. Aquat Microb Ecol 62:123–137

Rasconi S, Niquil N, Sime-Ngando T (2012) Phytoplankton chytridiomycosis: community structure and infectivity of fungal parasites in aquatic systems. Environ Microbiol 14:2151–2170

Rasconi S, Boutheina G, Niquil N, Jobard M, Sime-Ngando T (2014) Parasitic chytrids sustain zooplankton growth during inedible algal bloom. Front Microbiol 5:229. doi:10.3389/fmicb.2014.00229

Reynolds N (1940) Seasonal variations in *Staurastrum paradoxum* eyen. New Phytol 39:86–89

Sen B (1988a) Fungal parasitism of planktonic algae in Shearwater. V. Fungal parasites of the green algae. Arch Hydrobiol Suppl 79:185–205

Sen B (1988b) Fungal parasitism of planktonic algae in Shearwater. IV. Parasitic occurrence of a new chytrid species on the blue-green alga *Microcystis aeruginosa* Kuetz. emend. Elenkin. Arch Hydrobiol Suppl 79:177–184

Sherr EB (1988) Direct use of high molecular weight polysaccharide by heterotrophic flagellates. Nature 335:348–351

Sime-Ngando T (2012) Phytoplankton chytridiomycosis: fungal parasites of phytoplankton and their imprints on the food web dynamics. Front Microbiol 3:361. doi:10.3389/fmicb.2012.00361

Sime-Ngando T (2013) Fungal zoospores in aquatic ecosystems. In: McGraw-Hill Yearbook of science & Technology 2013 and the online edition. http://www.accessscience.com. The McGraw-Hill Companies, New York, pp 174–176

Sime-Ngando T, Jobard M (2013) Development of a real-time quantitative PCR assay for the assessment of uncultured zoosporic fungi. In: Gupta VK, Tuohy M, Manimaran A, Turner KM, O'Donovan A (eds) Laboratory protocols in fungal biology: current methods in fungal biology. Springer, New York, pp 421–426

Sime-Ngando T, Lefèvre E, Gleason FH (2011) Hidden diversity among aquatic heterotrophic flagellates: ecological potentials of zoosporic fungi. Hydrobiologia 659:5–22

Sime-Ngando T, Jobard M, Rasconi S (2013a) Fluorescence in situ hybridization of uncultured zoosporic fungi. In: Gupta V, Tuohy M, Manimaran A, Turner K, O'donovan A (eds) Laboratory protocols in fungal biology: current methods in fungal biology. Springer, New York, pp 231–233

Sime-Ngando T, Rasconi S, Gerphagnon M (2013b) Diagnose of parasitic fungi in the plankton: technique for identifying and counting infective chytrids using epifluorescence microscopy. In: Gupta VK, Tuohy M, Manimaran A, Turner KM, O'donovan A (eds) Laboratory protocols in fungal biology: current methods in fungal biology. Springer, New York, pp 169–174

Sogin ML, Morrison HG, Huber JA, Mark Welch D, Huse SM, Neal PR, Arrieta JM, Herndl GJ (2006) Microbial diversity in the deep sea and the underexplored 'rare biosphere'. Proc Natl Acad Sci USA 103:12115–12120

Sommer U (1987) Factors controlling the seasonal variation in phytoplankton species composition. A case study for a deep, nutrient rich lake. Prog Phycol Res 5:123–178

Sommer U, Maciej Gliwicz Z, Lampert W, Duncan A (1986) The PEG-model of seasonal succession of planktonic events in fresh waters. Arch Hydrobiol 106:433–471

Southall G, Robarts R (1977) Nutrient limitation of phytoplankton growth in seven tropical man-made lakes, with special reference to Lake McIlwaine, Rhodesia. Arch Hydrobiol 64:246–252

Sparrow FK (1960) Aquatic phycomycetes, 2nd edn. University of MichiganPress, Ann Arbor

Strom SL (2000) Bacterivory: interactions between bacteria and their grazers. In: Kirchman DL (ed) Microbial ecology of the oceans. Wiley-Liss, New York, pp 351–386

Ulanowicz RE (1986) Growth and development: ecosystems phenomenology. Springer, New York, 203 p

Ulanowicz RE (2003) Some steps toward a central theory of ecosystem dynamics. Comput Biol Chem 27:523–530

Ulanowicz RE, Goerner SJ, Lietaer B, Gomez R (2009) Quantifying sustainability: resilience, efficiency and the return to of information theory. Ecol Complex 6:27–36

Van den Meersche K, Soetaert K, Van Oevelen D (2009) An R function for sampling linear inverse problems. J Stat Softw 30:1–15

Van Donk E (1989) The role of fungal parasites in phytoplankton succession. In: Plankton ecology. Springer, Berlin, pp 171–194

Van Donk E, Ringelberg J (1983) The effect of fungal parasitism on the succession of diatoms in Lake Maarsseveen I(The Netherlands). Freshw Biol 13:241–251

Vézina AF, Platt T (1988) Food web dynamics in the ocean. I. Best-estimates of flow networks using inverse methods. Mar Ecol Prog Ser 42:269–287

Weete J, Fuller M, Huang M, Gandhi S (1989) Fatty acids and sterols of selected Hyphochytriomycetes and Chytridiomycetes. Exp Mycol 13:183–195

White TJ, Bruns T, Lee S, Taylor J (1990) Amplification and direct sequencing of fungal ribosomal RNA genes for phylogenetics. In: Innis MA, Gelfand DH, Sninsky JJ, White TJ (eds) PCR protocols: a guide to methods and application. Academic, San Diego, pp 315–322

Williams R, Martinez ND (2004) Simple rules yield complex food webs. Nature 404:180–183

21 History of the Fish Fauna of Lake Pavin: A Population Heavily Influenced by Man?

François Desmolles

Abstract

Over the last 20 years, changes in agricultural practices on the surrounding plateaus have driven rapid change in the trophic status of the lake, with knock-on effects on the evolution of its fish population. The original fish stock, badly known, was strongly influenced by numbers of different species introduction in the nineteenth century by Rico and Lecoq. The introduction of Arctic Char (*Salvelinus alpinus*) was particularly successful and this species became the iconic fish of the lake. In contrast to major scientific studies at Lake Pavin for more than 100 years, fish populations have been the subject of very few studies, most of which have focused on specific segments of the population. However, the fish population and stock have been tracked under numerous programs targeted on the Arctic char population, which holds strong economic and symbolic value. Since 2003, the lake and its groundwater inputs have been regularly monitored for their physico-chemical and biological indicators in implementation of the European Water Framework Directive. These surveys show a strong and pervasive anthropic influences which should foster active lake conservation measures and more in-depth studies of fish populations in a lake with such a rich, and at times mysterious piscicultural history?

Keywords

Pavin Lake • History of fish fauna • Fish introduction • Arctic char • Fish management • Eutrophication

21.1 Lake Pavin Harbors an Original Fish Fauna?

The search for the original fish fauna of the lake runs into many contradictions. The first written records tend to describe the lake devoid of life. Eusébio and Reynouard (1925) cite Elie Jaloustre (1884) in his translation of The Godivel Manuscript (late 17th) mentioning the lake, which on its fish fauna claims *"…as for the rest, no fish large or small has ever been seen or caught here, not even the smallest tuft of grass grows at its extremities…"*. In his summary of the history of Auvergne by an Auvergnat (Desbois 1826), the author, on the subject of Lake Pavin, says *"This lake was, we are told, the mouth of one of the most terrible volcanoes of Auvergne, and sustains no fish like Lake Chambon, one of the biggest in the country"*. These claims were largely linked to legends and beliefs that the lake was bottomless and that a huge whirlpool triggered thunder and storms, rendering it unable to support life. The abbot Delarbre (1795), director of the Botanical Garden of Clermont, denounced these legends and confirmed having seen fish: *"It has been advanced that the lake contains no fish; but is this true? I would not guarantee it, but I have seen fish break the surface of the water"*. Chevalier, Inspector of Bridges and Causeways, made one of the first sounding surveys of the lake in order to debunk the belief that it was bottomless and explain the dull thuds heard on the lake, which he attributed to the cracking of ice during heavy freezes.

F. Desmolles (✉)
Departmental Federation for Fisheries and the Protection of Aquatic Environments, 14 Allée des Eaux et Forêts, Site de Marmilhat, 63450 Lempdes, France
e-mail: f.desmolles@gmail.com

T. Sime-Ngando et al. (eds.), *Lake Pavin*, DOI 10.1007/978-3-319-39961-4_21

Despite this work, the beliefs lingered on, which explains the lack of descriptions of its fish fauna which, it was thought, simply could not exist. Fears tied to the lake, together with the absence of catches made with fishing gear, allowed only dubious knowledge of its littoral fish fauna.

Any discussion of original fish fauna needs to bear in mind that man has always taken a certain pleasure in transporting species from site to site, and Lake Pavin is no exception— despite popular belief that the lake was haunted, a millstream and an antique millstone have been found on its banks, and the lake may well have had fish from nearby rivers introduced into it.

For Eusébio and Reynouard (1925), the natural fauna of the lake included minnows, stickleback, and gudgeon, although it is unclear whether these descriptions are from Jaloustre (1884) or Chevalier who may have spotted spiny fish with shimmering colors. They claim the gudgeon disappeared in 1925 after a period of strong growth, whereas minnows and stickleback persisted into the early twentieth century. These writings contradict Meybeck (2011) who states that the original fish fauna of Lake Pavin was composed of bleak and gudgeon. However, the description of the fish fauna given by Eusebio & Reynouard (1925) appears closer to reality, as minnows are very common in the lakes and streams around Lake Pavin, as are gudgeon, even though currently located at much lower altitude and farther downstream in our waterways. The presence of stickleback seems surprising but fits with Chevalier's description. He may have confused stickleback with male minnows in breeding colors with their bright red and green markings and white tubercles on the head. However, Olivier (1939), in his Materials for the Limnological Knowledge of the Lakes of Mont Dore, claims that in most lakes the primitive fauna included tench and perch, but for Lake Pavin he cites the presence of stickleback (*Gasterosteus leiurus*) and gudgeon (*Gobio fluviatilis*). Stickleback is no longer listed among the species present in Lake Pavin, but in the early twentieth century, many fish catches (char and trout) had stickleback in their digestive tract (Eusébio and Reynouard 1925), which would confirm that it was indeed present. It may seem surprising to find stickleback present in such apical lakes and streams, but keep in mind that this species is currently associated with downstream typological levels. Note that Emile Blanchard (1866) cites the abundance of this species in all small streams of low gradient and vegetation-rich water as well as in bodies of water possessing identical habitat characteristics. Emile Blanchard (1866) makes particular reference to sampling done in the Sioule at Pontgibaud, a river situated at the foot of the Sancy.

Refocusing on brown trout (*Salmo trutta*), it may very well be possible that it frequented the lake. As mentioned above, the early descriptions of lake fauna concerned the littoral zone, so it is not surprising that trout were not mentioned as being present. However, the spillway of Lake Pavin converges on the Couze Pavin river after running a straight line for 600 m, and this small emissary is a particularly popular breeding site for brown trout. Though the transfer of the species is not currently possible, it was only in the second half of the nineteenth century that the dike was created and grilles were installed. Old prints (Fig. 21.1) show, at certain

Fig. 82. — Lac Pavin (Auvergne).

Fig. 21.1 Ancient representation of Lake Pavin Tissandier (1873) times, a lower natural weir or spillway which may have been passable with suitable water flow speeds (Tissandier 1873).

This natural weir has been constantly modified over time. There would seem to have originally been a much higher weir, as during periods of drought, the millers of Besse would dig into this weir to increase the flow of water, thus lowering the level of the lake and thereby reducing the difference in water level between the lake and neighboring waterways (Eusébio and Reynouard 1925). Bruyant (1909) cites excavation work by Reynouard who updated an irrigation canal linking the borders of the lake to Olpilière from which the canal could join the Rif stream. Nothing, therefore, would have prevented the species from reaching the lake over time. Remember abbot Delarbre's (1795) words: "*I would not guarantee it, but I have seen fish break the surface of the water*"—this type of behavior hardly resembles that of minnows and stickleback. While the presence of brown trout cannot be definitively excluded, the site nevertheless possesses no favorable breeding grounds for brown trout (lack of tributaries), and it is currently impossible to establish a lifecycle in the lake for this species.

21.2 The Nineteenth Century, A Turning Point for the Fish Population

This period is very well documented, with Berthoule (1890), Eusébio and Reynouard (1925) and Olivier (1939) playing important roles in stocking the lake with fish. Rico, a lake fish farmer, joined Lecoq, who was director of Clermont's botanical garden and the museum of natural history, dean of the city's College of Science, vice-chairman of the Puy-de-Dôme Central Agricultural Society, and head of the Clermont-Ferrand piscicultural laboratory. In 1859, Rico, Lecoq and Delmas introduced:

- 92000 brown trout (*Salmo trutta*)
- 20000 salmon (*Salmo salar*)
- 18 huchen (*Hucho hucho*)
- 8000 Arctic char (*Salvelinus alpinus*)
- 130 tench or common roach (*Rutilus rutilus* or *Tinca tinca*?)
- 200 adult crayfish

Other species were also introduced, whether in 1859 or in subsequent years remains unknown. Olivier (1939) reports the introduction of eel (*Anguilla anguilla*).

According to Machino (1991), the initial batch of Arctic char came from the Imperial Pisciculture of Huningue (68), as did a great number of fish at the time given it status as a pioneer in salmon farming and an important source of dissemination throughout France. The *Pisciculture Départementale de Clermont Ferrand* was at that time engaged in the production of fry for the department's lakes and rivers. Bruyant (then deputy director of the Besse biological station) further developed the laboratory created by Rico by combining a production laboratory with the piscicultural research facility at the Besse biological station. The fish released into the department's lakes and streams had been raised in the basement of Besse biological station with the help of Jean Baptiste Eusébio (General Secretary of the *Société de Pêche et de Pisciculture du Puy-de-Dôme*). Later on (between 1900 and 1906), some of the char used to supply the Besse fish farm came from the *Pisciculture Domaniale de Rives* at Thonon-les-Bains on Lake Geneva.

> *Thus it was that Ch. Bruyant was first to imagine creating at Besse a station both practical and scientific: scientific from the point of view of studying the fauna and flora of the lakes, their geological formation, the physical phenomena which shaped their contours; practical from a fish farming perspective, these magnificent natural water pools to be used:*
> *First, to harvest the breeders which will spawn thousands of fry for the laboratory to use to repopulate the streams of the department, and*
> *Second, to make rational and intensive use of the best species. This means, from a piscicultural perspective, resuming Rico's wonderful program. The Conseil Général du Puy-de-Dôme quickly understood the interest presented in this test of scientific decentralization, unprecedented in France at that time; it sponsored the work, subsidized it in great part without becoming discouraged by the slowness of the process of trial and error which inevitably marks the beginnings of such a project, and never lost confidence in Ch. Bruyant until he was at last able to find the form and definitive installation that best suited the station and would allow him to begin seriously achieving results.*

Today, Mr. Bruyant, departmental director of the fish farm service, has created, again thanks to the Conseil Général du Puy-de-Dôme, a model facility that he baptized The Limnological Station of Besse and placed under the patronage of the French Association for the Advancement of Science at the closing of the Congress held in Clermont-Ferrand in August, 1908, noting that the previous Congress of 1876 had similarly inaugurated another original, scientific work, the meteorological observatory at the summit of the Puy-de-Dôme, which has since grown in stature and gained the notoriety it enjoys today. The similarity is striking, and is a good omen (Reynouard 1909*).*

Following a flood of the lake in 1861, work was begun to prevent fish from the lake swimming into the Couze Pavin. Lake farmer Rico installed grilles and a spillway. It is interesting to note that this work was not undertaken to protect the river from the introduction of non-native species, which is the current rule.

In 1874, a 14.5 kg huchen was caught in a fishing net (the specimen had minnows, stickleback, and crayfish in its stomach)—its head is still on display at Besse biological station. Many outstanding catches have been reported, with lake farmers having captured 2984 trout, salmon or Arctic char for a total weight of 1600 kg (Eusébio and Reynouard 1925). Catches of other species varied, and catches of Atlantic salmon and huchen, like brown trout, started to dwindle. However, these different species introduced were already being challenged.

In his book The Freshwater Fish of France, Emile Blanchard (1866) writes "*Mr. Lecoq, of Clermont-Ferrand, received from the Conseil Général du Puy-de-Dôme the same assistance as Mr. Gervais (in charge of propagating salmonids in the Hérault department ndl) and also reported on his fish farming trials. In the same department, Mr. Rico took charge of propagating salmon in still waters, and Mr. Gillet de Grandmont, reporting on his experiences, confirms that two specimens of this species were caught in Lake Pavin, one having attained a weight of 500 g, the other weighing 700 g*". Blanchard questions the value of this introduction and expresses regret that those responsible had not read the history of salmon in order to avoid a predictable failure. He denounced the profusion with which the Pisciculture of Huningue released salmon, whitefish, and other Arctic char, claiming that "*reading the history of salmon would have reduced expectations which were based upon the lake of Bois de Boulogne, Saint-Cucufa pond near Paris, Lake Pavin in Auvergne, etc.…*".

Over time, different lake farmers released many other species, including Scottish trout (Loch Leven), brook trout (*Salvelinus fontinalis*), rainbow trout (*Oncorhynchus mykiis*), whitefish (*Coregonus fera* and *maroena*), and eel (*Anguilla anguilla*).

Charles Bruyant (1908), in his "Rectification of the Ichthyologic Fauna of Auvergne" assesses the situation concerning whitefish and Arctic char in these words: "*On the affirmation of our predecessors, we had indicated in a previous work the presence of Coregonus Fera (Jur) in some of our lakes, such as Chauvet. Mr. Berthoule, in his work on the lakes of Auvergne, also mentions 25,000 Coregonus Fera fry from Huningue released into the lake Landie. As noted by Professor Forel, these Feras from Huningen certainly couldn't be Coregonus Fera (Jur.) from Lake Geneva, but rather* Blaufelchen, *Coregonus Whartmanni (Bloch) from Lake Constance, which Huninge has been disseminating for a long time now; it is only recently that the Thonon Pisciculture Station has been successful in the very difficult incubation of Fera eggs. We are therefore bound to apply to Coregonus Whartmanni (Bloch) that said previously of the Fera imported into the Auvergne under these conditions. The introduction into our lakes of these Coregones does not seem to have produced great results. The same cannot be said for the Arctic char which is thriving in Lake Pavin. It was in 1860 that the first fry were released into the lake by Lecoq and Rico. Currently, we catch this species far more often than trout. Hybrids of trout and "Ombre" obtained by Rico in 1872, and reported in Raveret-Wattel's classic treatise (T. II, p. 224), are actually hybrids of trout and* Salvelinus, *not* Thymallus *which has never existed in the Pavin. The error is once again due to the confusion between the two names, "Ombre" and "Omble" (Rico wrote "Ombre"), which continues today. C. Bruyant* 1908.

By the early twentieth century, roach, tench, eel, salmon, and rainbow trout seem to have disappeared, which given what is known of the life cycles of some of these species, is not really surprising. Evidence from several sources (Eusébio and Reynouard 1925; Olivier 1939) suggests that whitefish disappeared. Crayfish thrived, but the author states out that they did not go on to populate the Couze Pavin.

Luc Olivier (1939) drew up a table reviewing the introductions made into many lakes around the Mont-Dore based on previous data. He concluded that:

- the natural species are gudgeon, minnows, and stickleback
- the introduced species that since disappeared are eel, whitefish (*Coregonus fera* and *Coregonus maraena*), roach, huchen, salmon, brook trout, rainbow trout, and trout from Loch Leven (*Lochleveni*)
- tests on lake trout (*fario lacustris*) and Arctic char were conclusive.

21.3 Evolution of the Population Between Introduction and Trophic Changes

The species described by Olivier (1939) remained present in the lake until the 1950s. Few studies were conducted before the 1990s, when many lake compartments were studied. The fish fauna, considered artificial, was neglected; with lake farmers only setting nets to catch char to eat—catches that included a few brown trout and white-clawed crayfish. The lake is the property of the inhabitants of Besse, and its management is delegated to the municipality, which allows the farmer to catch char for his restaurant.

The biggest changes to have taken place since the writings of Olivier (1939) are:

- The appearance of a very large population of European perch (*Perca fluviatilis*), which may have been introduced into the lake in the 1950s by a lake farmer. The total disappearance of stickleback may be linked to the presence of a predator such as the common perch.
- The more recent appearance of species whose presence is linked to increased recreational fishing on the lake.

21.3.1 State-of-Knowledge Review of the Current Population

For many years, the only people sampling the fish fauna of Lake Pavin were lake farmers and recreational fishermen. These samplings were quite selective, being done in order to capture specifically targeted species, whether by net or on a line. The species targeted were first Arctic char, then trout (brown trout and rainbow trout from releases), followed by perch. No studies following a scientific protocol were conducted on the fish fauna of Lake Pavin before 1995.

Between 1988 and 1991, we were able to follow numerous fish nettings performed by the lake farmer as well as by the salmon farmers of the Besse federal fish farm in order to capture spawning adults of the Arctic char species. Artificial spawning, fertilization and hatching was carried out on the premises below where the restaurant standards today; the pools are fed by the waters of the lake through an intake located at approximately 10 m. In all that time, we only ever caught Arctic char, European perch, brown trout, rainbow trout, or crayfish. Minnows, loach, and white-clawed crayfish were visually abundant in littoral areas. Fish were netted using a 27 mm mesh size, which produced strong selectivity for the species caught and does not allow an objective picture of the species present.

One of the first scientific studies of the lake's fish fauna was conducted by Jamet (1995), who set gill nets (20-24-30 mm) from April to October 1992 to study the condition coefficients, reproduction, and diet of the Arctic char. Unfortunately, Jamet did not report what other species were captured during this time.

The years 1996 and 1997 also witnessed the near disappearance of the white-clawed crayfish (*Austropotamobius pallipes*) population. At the time, there was no attempt to identify the cause of this decline, but the rapidity of mortality and the sheer size of the event (dead crayfish clotted the outlet grilles) pointed to *Aphanomyces*. With hindsight from current studies on trophic change in the lake and the coincidence between this change and the mass mortality of crayfish, we can now entertain other hypotheses, but unfortunately the species has all but disappeared. In 2006, divers observed a crayfish in the benthic zone; after examining a photo, it seems likely that this individual belongs to *Austropotamobius pallipes*.

Grandjean's thesis on the mitochondrial DNA typing of crayfish varieties (1997) indicates two subspecies of *Austropotamobius pallipes*, the subspecies *pallipes* and *italicus*. The Italian (Alpine) origin of this subspecies could be explained by the importation of fish from the Thonon-station salmon farm or by the successive introduction of individuals from the Alps, but local strains had also been present in many rivers in the Besse region decades ago.

21.3.2 Piscicultural Tracking from 1995 to 2012

21.3.2.1 Tracking Performed by the FDPPMA 63

The FDPPMA (Departmental Federation of Fisheries and Protection of the Aquatic Environment) of the Puy-de-Dôme carried out these catches between 1995 and 2005 in order to count number of fish taken by species and implement a methodological protocol for tracking. However, the protocol was not a scientific protocol:

- catches were taken from March to December (depending on ice cover), which is not year-long monitoring
- sampling was dependent on the demand for char in the restaurant
- sampling sites corresponded to those known to be the most productive
- and, in contrast to scientific sampling in which several mesh sizes are used to capture all age groups of a species that may be present, the manager used meshes allowing him to capture fish large enough to serve to customers (approximately 30 cm). The mesh sizes used were 26 mm, 27 mm, and 32 mm, which led to strong catch selectivity.

The idea was to track the catches and age groups concerned and compare their evolution. We decided to track abundance indexed by Catch Per Unit Effort (*CPUE*; Ricker 1980; Rivier 1996, Gerdeaux 2001b), where the surface of the nets used is multiplied by how long they are set in the water, which determines a certain number of Units of Fishing Effort (*UFE*), a fishing effort corresponding to 1 m^2 of net set for 1 h (Fig. 21.2). Catches are counted and reported in *CPUE*, which serves to track evolution of catches at constant effort (Rivier 1996).

Despite giving fishing managers entry sheets and methodological instructions, fishing times were often variable from catch to catch, mesh sizes were inappropriate, and so on.

21.3.2.2 Tracking According to Directive NF EN 14757 WFD

This standardized protocol was implemented when the Water Framework Directive was adopted in order to qualify the status of European bodies of water both above and below ground. Tracking of fish in water bodies is performed according to the standard *WFD* protocol NF EN 14757. Two samplings were performed following this protocol—one in 2003 (Asconit 2006) and one in 2012 by the ONEMA (the French National Agency for Water and Aquatic Environments) (Olivier 2012). This method provides a fairly comprehensive view of the fish population and its spatial distribution using horizontal multi-mesh nets at different depths.

Fig. 21.2 Annual evolution of Units of Fishing Effort

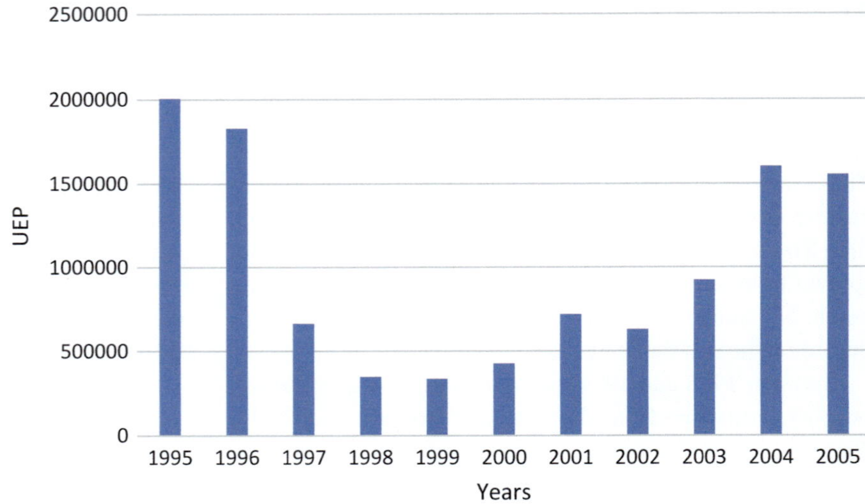

21.3.2.3 Evolution of the List of Species Present Over Time

Review of the literature emerges four separate periods in which a list of species present in the lake was established more or less objectively (Table 21.1):

In 1939, L. Olivier does not specify what methods were used for sampling.

For tracking carried out by the FDPPMA 63, (i) the high selectivity in the nets used in conjunction with the specific targeting of Arctic char had the effect of inducing a sampling bias, and (ii) littoral areas were not sampled, with only visual observations used to complete the list (italics).

For WFD NF EN 14757 (AFNOR 2005) sampling, the protocol is standardized but does not cater to sampling littoral areas of the lake; visual observations are not referenced.

In comparison to the list compiled by Olivier (1939), the big charges are the disappearance of stickleback, reported as an original species, and the disappearance of crayfish, reported as having been introduced in 1859. Over the past 20 years, allowing for the fact that the samplings are not comparable, the species list has essentially changed little. Note the absence of catches of brown trout and minnows with the WFD protocol whereas anglers report regularly catching or observing these species.

The standout feature is the appearance of many introduced species:

- voluntarily by managers. Perch were introduced in the 1950s by a lake farmer. Rainbow trout (*Oncorhynchus mykiss*) were introduced 1997, with individuals sourced from Lake Bouillouses (Pyrénées-Orientales). This species is largely maintained by yearly releases. For 10 years, every individual released was tagged, but some untagged fish were caught between 1997 and 2005, which in conjunction with the observation of reproductive activity on the bank leads us to believe that a small part of the population is the product of reproduction *in situ*.
- by anglers, tourists, children, etc. using the lake for recreational purposes. Crucian carp, rudd, chub, and roach were introduced in this way. Many of these cyprinids were probably used as live bait for fishing and dumped in the lake when no longer needed. Regarding roach, note that Olivier (1939) reports its introduction in 1859 as a failure. Do the individuals caught come from recent releases or have they managed to persist in the lake's waters since that date? Note too that anglers recently reported taking ides (*Leusciscus idus*). These fish are naturally present in the northern watershed of the Loire and are often sold as livebait by fishing tackle shops. These catches would appear to confirm the hypothesis related to the practice of dumping livebait (despite regulations in place prohibiting its use).

The presence of tench and gudgeon also raises questions. If gudgeon have visibly persisted since the inventories taken by Olivier (1939) and perhaps even since the original fish fauna, tench were reported in 1939 as a failed attempt at introduction, then is the capture of two individuals in the space of 30 years to be ascribed to recent introductions or to the persistence of individuals released in 1859?

21.3.2.4 Evolution of Numerical Abundance

We compiled all the data obtained by the FDPPMA 63 over an 11-year period and compared the numbers against the relative proportions of these species in studies performed for WFD monitoring in 2005 and again in 2012 (Fig.21.3).

Despite the different sampling protocols, just two species dominate the fish population, i.e. Arctic char and perch, although their order of abundance is reversed between sampling methods. Species such as rainbow trout have remained stable in terms of representation while chub, rudd and roach

Table 21.1 Evolution of the fish population by study period

Olivier (1939)	FDPPMA63 (1995–2005)	ASCONIT (2006)	ONEMA 2012
Austropotamobius pallipes	*Austropotamobius pallipes*		
	Carrassius carrassius		
	Cyprinus carpio		
Gasterosteus aculeatus			
Gobio gobio	*Gobio gobio*	*Gobio gobio*	*Gobio gobio*
	Squalius cephalus	*Squalius cephalus*	*Squalius cephalus*
	Neimachulus barbatulus		
	Oncorhynchus mykiss	*Oncorhynchus mykiss*	*Oncorhynchus mykiss*
Phoxinus phoxinus	*Phoxinus phoxinus*		
	Perca fluviatilis	*Perca fluviatilis*	*Perca fluviatilis*
	Rutilus rutilus	*Rutilus rutilus*	*Rutilus rutilus*
Salmo trutta fario	*Salmo trutta fario*		
Salvelinus alpinus	*Salvelinus alpinus*	*Salvelinus alpinus*	*Salvelinus alpinus*
	Scardinius erythrophtalmus		*Scardinius erythrophtalmus*
	Tinca tinca		*Tinca tinca*

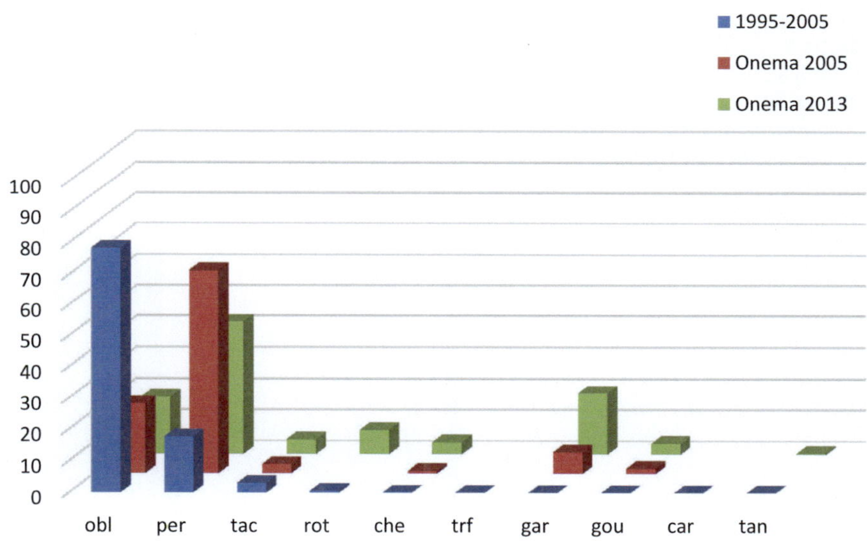

Fig. 21.3 Evolution of numerical abundance in percent (Obl: *Salvelinus alpinus*; Per: *Perca fluviatilis*; Tac: *Oncorhynchus mikiss*; Rot: *Scardinius erythrophtalmus*; Che: *Squalius cephalus*; Trf: *Salmo trutta*; Gar: *Rutilus rutilus*; Car: *Carassius auratus*; Tan: *Tinca tinca)*

appear to be surging upward. In terms of capture rates, there are currently more number being caught than Arctic char.

If we look at the catches taken by the FDPPMA from 1995 to 2005 in terms of UFE (×1000), char and perch are also dominant species (Fig. 21.4). However, these results are skewed by the fact that they relate to species most likely to be caught by the selective range of mesh sizes used, and most importantly by the fact that they correspond to targeting at spots and depths specific to char, a species much sought after by the lake farmer.

This selectivity is particularly evident when comparing these results against comprehensive catches made following the WFD protocol NF EN 14757 (AFNOR 2005), which showed a reverse order of dominant species and a surge in numbers of certain species that had been rarely caught between 1995 and 2005 but have risen fast between 2005 and 2013 (Fig. 21.5).

Although the selectivity in catch methods used between 1995 and 2005 has a huge influence on species representation rates, it does also show a relative stability in Arctic char catches. The value given for catches taken by the FDPPMA is an 11-year cumulative CPUE compared against CPUEs obtained by the standardized method.

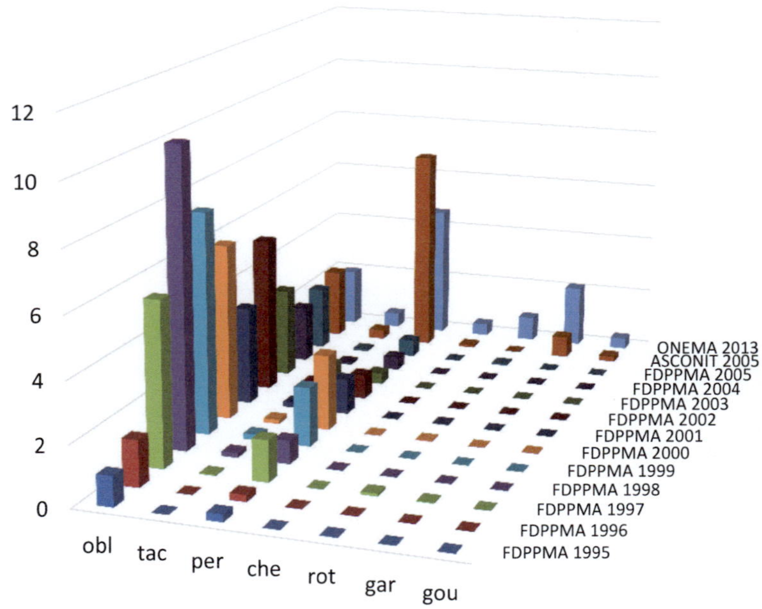

Fig. 21.4 Annual evolution of CPUE (Catch Per Unit Effort * 1000) by species and year(Obl: *Salvelinus alpinus*; Tac: *Oncorhynchus mikiss*; Per: *Perca fluviatilis;* Rot: *Scardinius erythrophtalmus*; Che: *Squalius cephalus*; Gar: *Rutilus rutilus; Gou: Gobio sp)*

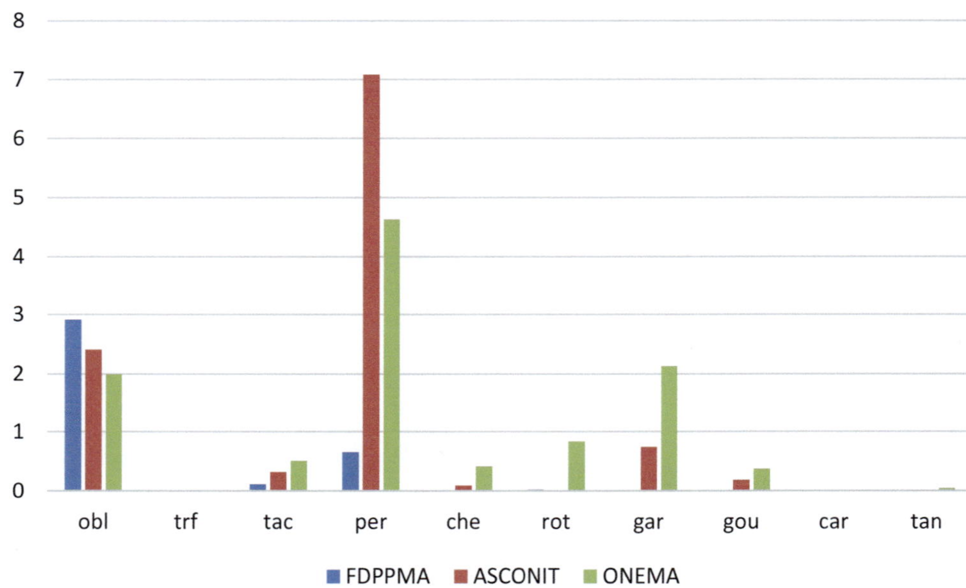

Fig. 21.5 Comparison on a CPUE (*1000) basis(Obl: *Salvelinus alpinus*; Trf: *Salmo trutta*; Tac: *Oncorhynchus mikiss; Per: Perca fluviatilis*; Che: *Squalius cephalus*; Rot: *Scardinius erythrophtalmus* Gar: *Rutilus rutilus*; Gou: *Gobio sp*; Car: *Carassius auratus*; Tan: *Tinca tinca)*

21.4 Study of Arctic Char

Most of the fish fauna monitoring done on the lake has focused on its most emblematic species—the Arctic char. Indeed, Arctic char is as highly prized by anglers as it is by the manager of the lake restaurant. Relatively large stocks have been taken for years, but changes in the trophic status of the lake are now a cause for major concern over the future of this species.

21.4.1 Biology of the Species

Its biological requirements have been well documented in a great number of lakes, but not in Lake Pavin where few studies have focused on fish. In 1995, Jamet tracked the reproduction, feeding, and body weight patterns of the Arctic char in Lake Pavin. In 1997, Léveillé led a study on the use of fatty acids as organic markers of transfer in the phytoplankton, zooplankton, and Arctic char food chains.

21.4.1.1 Feeding

After studying the stomach contents of 200 individuals caught from April 1992 to December 1992, Jamet (1995) concluded that feed is numerically dominated by the year-round consumption of *Daphnia longispina*. The biggest contributors to weight gain are seasonally available food opportunities, essentially *Asellus aquaticus* from April to September and, char eggs from October to December, but chironomid larvae, larval stoneflies, and *Cyclops abyssorum* are also consumed. May and August marks the period of peak feeding activity.

21.4.1.2 Reproduction

Tracking shows that both males and females increase Gonado Somatic Index (*GSI*) in August (Jamet 1995) and again in September. In October, *GSI* decreases in males but increases again in females. Female reproduction begins in October–November, when *GSI* starts gradually decreasing until December. Males appear to be present on spawning grounds in Lake Pavin earlier than the females—a phenomenon frequently noted in salmonids with the youngest individuals first to show when spawners start to gather (Gillet 2001). The presence of males and females in differing proportions following the reproductive period has already been reported in scientific literature (Dussart 1955).

Arctic char as a species displays very broad diversity in its breeding habits. Rubin and Buttiker (1992), Dussart (1955) and Gillet (2001) report this diversity both in the breeding season, in substrata ranging from blocks to fine sediment, and in the depth of reproduction in the littoral zone, from 0.5 m down to bottom depths of nearly 100 m. One criterion for selection is the presence of non-silted substrates (Gillet 2001).

Lake Pavin has two well-known reproduction zones. For years, successive lake farmers have fished these two zones to increase the trout production and produce fry by artificial fertilization in the basement of the hotel restaurant. Many diving forays over the decades have confirmed the presence of these two main char breeding grounds, and reproductive activity has been filmed a number of times (D Chassain CAP 1993–2003).

These egg-laying sites are located at the following coordinates: X 0643.232, Y 2055.553, and X 0643.369, Y 2055.384; (Extended Lambert II). These two sites correspond to talus zones consisting of blocks ranging in size from a few dozen centimeters to a meter and situated at a depth of between 10 and 18 m. Eggs are laid in rock crevices that have been cleared of all fine sediment. While the physical characteristics of the char breeding grounds are different, all authors agree that the reproductive sites remain well-frequented regardless of changing lake conditions (rising and subsiding water levels).

Echo sounding surveys were carried out in December 1995 (Fig. 21.6). The screenshot below shows detections performed while passing at a 90° angle over a supposed char breeding ground. The equipment used was a SIMRAD EY500 (70 kHz, 14°) echo sounder. Analysis of recorded echo readings shows a high density of fish that the echoes were able to distinguish as individuals. The focal concentration of adult fish is situated at a depth of between 8 and 20 m and confirms the observations reported from dives. However, the survey was unable to find any other major sites than those already known and so does not exclude the presence of smaller-sized char.

Spawning and its success are largely dependent on the quality of the substrates, especially the oxygenation conditions (Guillard et al 1992). A minimum concentration of 8 mg l^{-1} is essential for the entire duration from incubation to resorption of the yolk sac, or a period of 450°*days*.

21.4.1.3 Growth

Growth has been analyzed by several methods.

The study of otoliths conducted by Audinot and Jamet (1998) cannot serve to plot a reliable growth curve. The small number of individuals observed and uncertainties in measurements in conjunction with inadequate equipment preclude any attempt to propose a growth model here.

Similarly, a study of growth using a statistical method was conducted as an initial approach (Bhatthacharya 1967), but the results remain questionable because despite a relatively large number of individuals captured, the nets out-selected the youngest and oldest individuals, which interfered with their distribution by size class. A growth model established using this method would lead to over- or under-estimating the K and Lα parameters of the Von Bertalanffy model.

Analysis of scales allows us to obtain a coherent growth curve. In 2002, during monitoring on catches conducted by the *FDPPMA 63*, 232 individuals taken from the catches were analyzed. The growth curve obtained from their real and back-calculated ages is globally comparable to that obtained from statistical analysis. Compared to other lakes, Arctic char growth in Pavin is quite slow, especially from the second year (Fig. 21.7) (Rubin 1990; Gerdeaux 2001a; Zanella 2003), but on a par with that of Lake Chauvet (63) located a few miles from Lake Pavin, for which we carried out a study in 2009.

21.4.2 Arctic Char Population Tracking

Given its economic importance and also its sensitivity to the quality of its living environment, the numerical evolution of the Arctic char population was tracked from 1995 to 2005. In the 1990s, shrinking catch rates for Arctic char led managers to scale back commercial fishing and stock the lake with fingerlings to supplement natural reproduction. These fingerlings are produced from broodstock caught in the lake and raised in captivity until release.

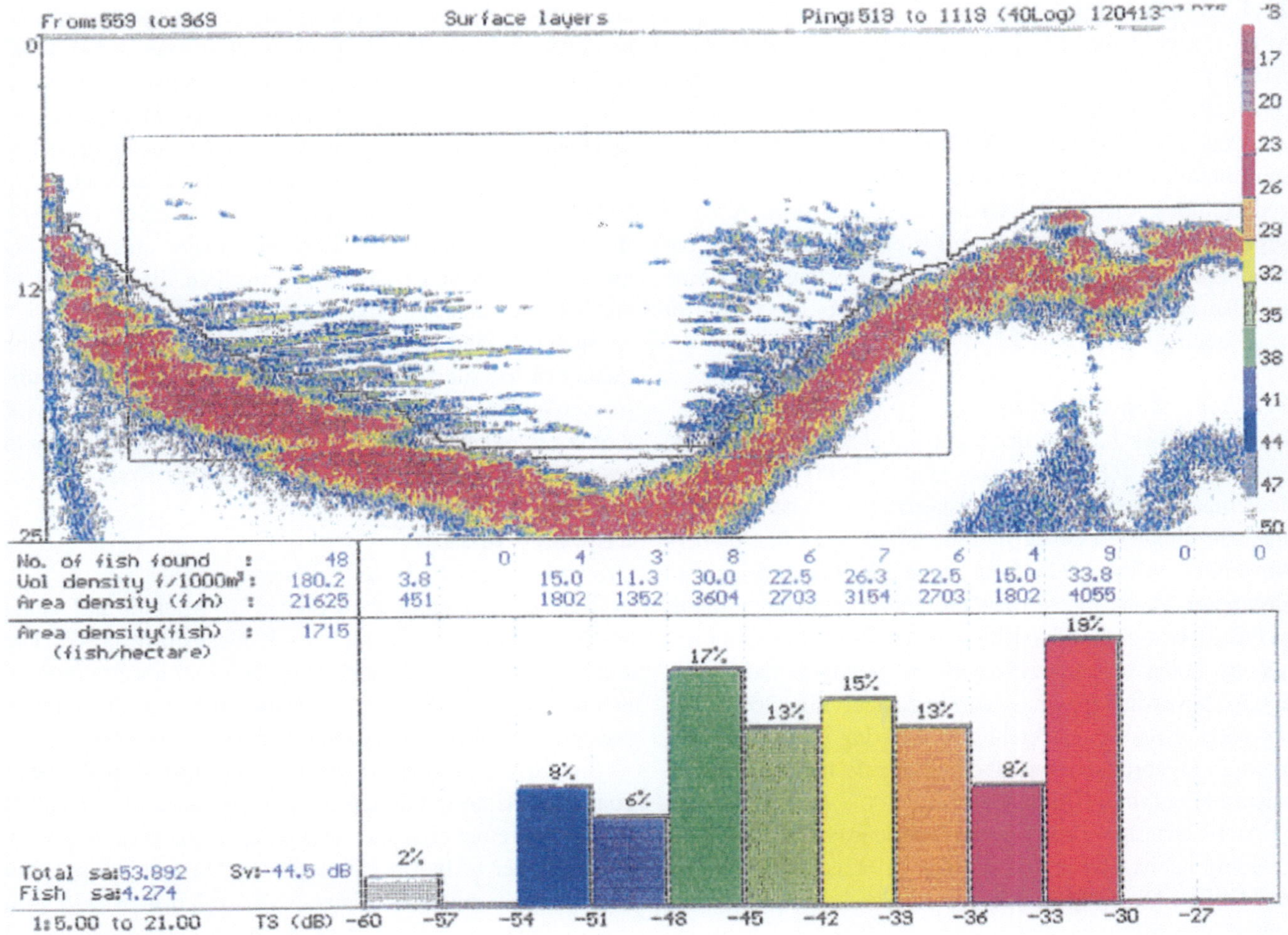

Fig. 21.6 Screenshot of an echogram of an arctic char spawning area in Lake Pavin(TS: Target Strenght in dB. Higher is the TS, bigger is the echo (the fish), 48 fishes were isolate in the selected area, the diagram represent the percent of each echo class)

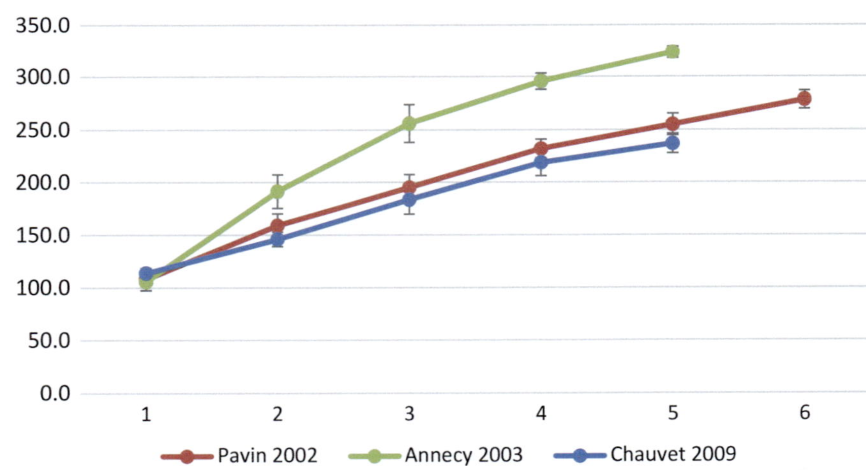

Fig. 21.7 Growth comparison (in mm) of *S. alpinus*

21.4.2.1 Tracking of Catches by Anglers

Anglers on the lake were required to keep a catch log, which they systematically filled out during the first few years of the initiative. This catch log initiative enabled us to:

- Estimate the total number of fish caught in the lake, all species included
- Estimate of size of structures
- Track tagging efforts

However, soon enough, with a lack of verification and the reluctance of the anglers, the catch books became less informative, and now 20 years after they were first introduced, logs and records are no longer kept. However, during the early years of the catch log initiative, monitoring on these records allowed us to estimate the number of char caught by anglers. Knowing the total number of fishermen and the number of fishermen who filled out their catch log, we considered that the fishing pressure and catch efficiency of those having filled out their logs was the same for all fishers. This allowed us to calculate that between 1997 and 2000, when at least 10 % of fishermen reported their catches, anglers caught between 3000 and 3500 Arctic char per fishing season. Catches were concentrated in May, June, and October. During the hottest months of the year, when the thermocline forms, the char stay deeper and more difficult to catch.

We have had no way of monitoring catches during the last few years, but given that fishermen are limited by daily catch quotas and the number of fishermen continues to decline, it is very likely that the catch pressure exerted by fishermen has followed the same trend.

21.4.2.2 Tracking of Catches Made Using Fishing Nets

The lake manager was also required to declare his catches by filling out a capture file. The manager caught between 2500 and 3500 char in the period between 1995 and 2005, for an annual biomass of between 400 and 500 kg (Fig. 21.8). Despite the negative evolution of fish stock in recent years, the manager maintained his catch levels by significantly increasing his fishing efforts (Fig. 21.2).

Variations between the number of fish caught and their weight are linked to modifications in the mesh size of the fishnets used (increasing or decreasing the catch size), and to the accuracy of the fish farmer's declarations.

21.4.2.3 Evolution of Catch Per Unit Effort (CPUE)

As stated above, CPUE was tracked from 1995 to 2005 (Fig. 21.9). After this date, the farmer and the owner of the lake were no longer willing to continue the work following the necessary methodology, and the data produced became unreliable and so has not been included here.

Starting in 1995, commercial fishing restrictions and stocking efforts quickly bore fruit, with a significant increase in the number of captures, peaking in 1998. After 1996, for economic reasons, the owner wanted to reduce the number of juveniles released, and catch numbers plummeted while the farmer increased his fishing efforts to meet his allowed quota of char. Thereafter, despite constant efforts to release fry into Lake Pavin and the establishment of catch quotas, the CPUE values have continued to fall, which prompts us to believe that the proportion of Arctic char produced by the Lake continues to decline.

We consequently implemented tracking on the proportion of char released into the lake compared to the number produced by the natural reproduction cycle of Lake Pavin (Rubin and Buttiker 1993: Champigneulle et al 2001; Zanella 2003). This could only be accomplish if all fry released into the lake were at a similar growth stage. Despite our requests, and for various reasons, the fry were released at different stages from May to September each year. This constraint means we can-

Fig. 21.8 Annual evolution of catches in biomass and numbers

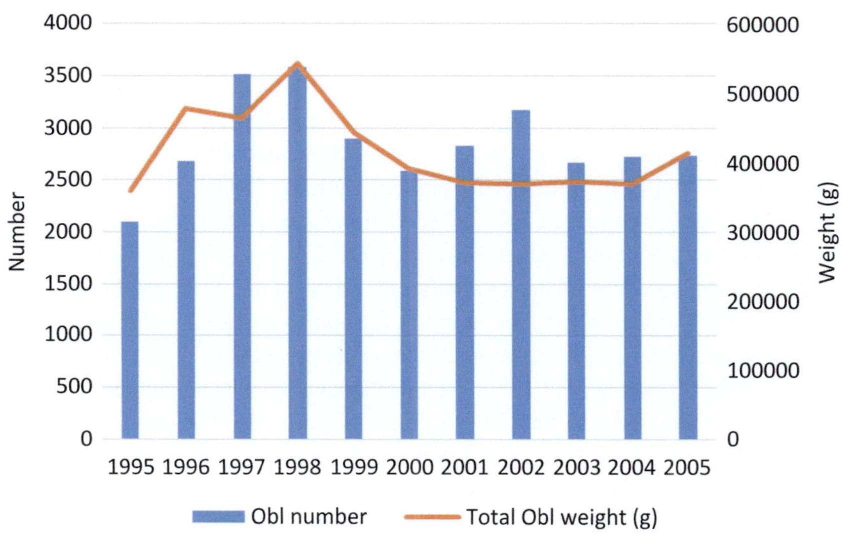

Fig. 21.9 Evolution of *S. alpinus* CPUE (*1000) from 1995 to 2005

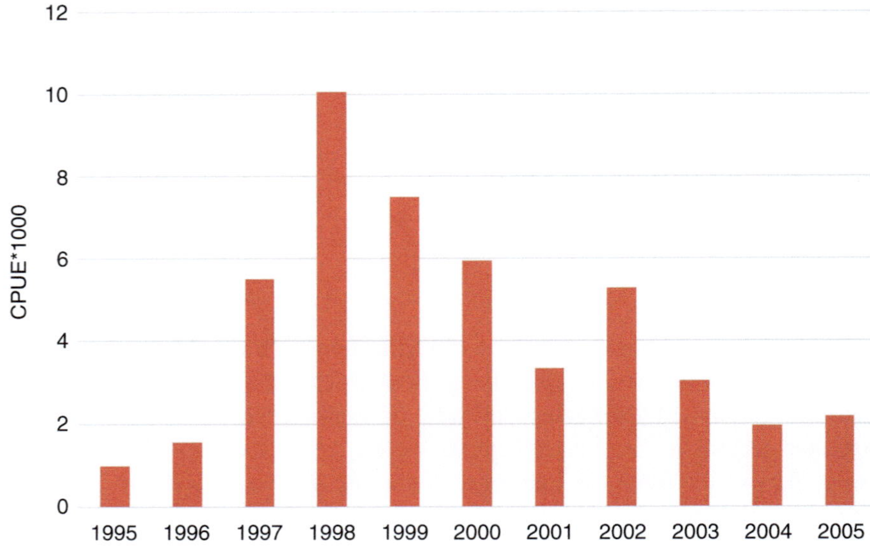

Fig. 21.10 Comparisons of upper (*blue*) and lower (*red*) estimates in 1⁺ equivalents

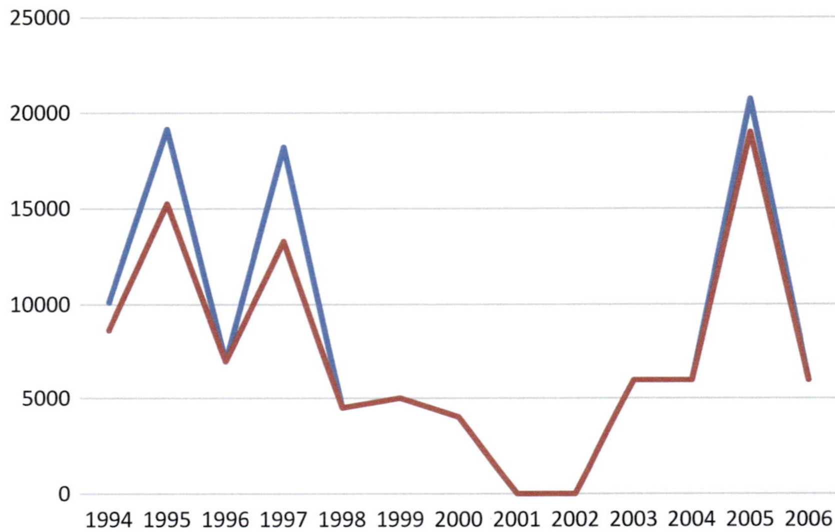

not compare the relationship between number of fish released and number caught without estimating the theoretical survival rate of released individuals. We estimated the survival rate of individuals at various stages of development following the Geiger mortality reference curves (1964) framed according to two mortality hypotheses (Rubin and Buttiker 1993). As 1^+ fry were also released, we elected to transform all the released fry into 1^+ equivalents in May (Fig. 21.10).

As these values were relatively similar, we chose to use an average value for the remainder of the study. Comparison of the average value of the 1^+ equivalent number against the number of fish caught 2 years later (3^+) showed a relationship between fish stocking efforts and number of fish caught (Fig. 21.11).

Applying a year-on-year mortality rate of 50 %, in line with much of the literature (Maisse and Baglinière 1991; Richard 1998), we created a matrix of annual fish catchable. There was a very good match between the CPUE and estimated adult stock ($r=0.84$ $n=8$) (Figs. 21.12, and 21.13) which reflects a strong correlation between CPUE and number of fish released, thus confirming our hypothesis of a sharp decline in char reproduction *in situ*.

21.4.2.4 Tagging

These values suggest that if the influence of stocking is so strong, it is because the number of char produced from natural reproduction is so low. In order to estimate the number of fish from the lake and the number of fish from stocking, we tagged the char released into the lake. This allowed us to model the total number of these cohorts at the time tagged individuals were released into the water, via the Lincoln Index method according to the Peterson formula (Ricker 1971; Rubin and Buttiker 1993; Zanella 2003). Due to a

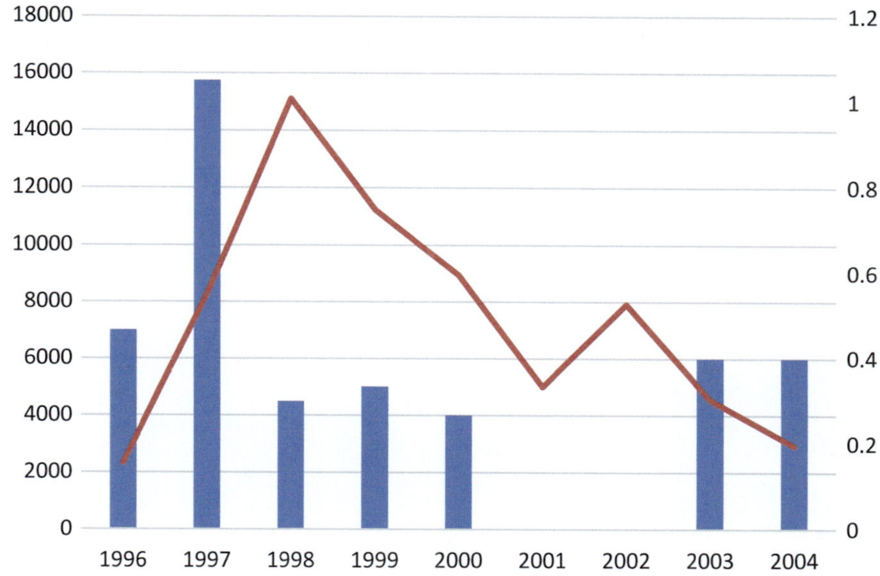

Fig. 21.11 Ratio of 1^+ juveniles number (*blue*) to CPUE (×100) 2 years later (red)

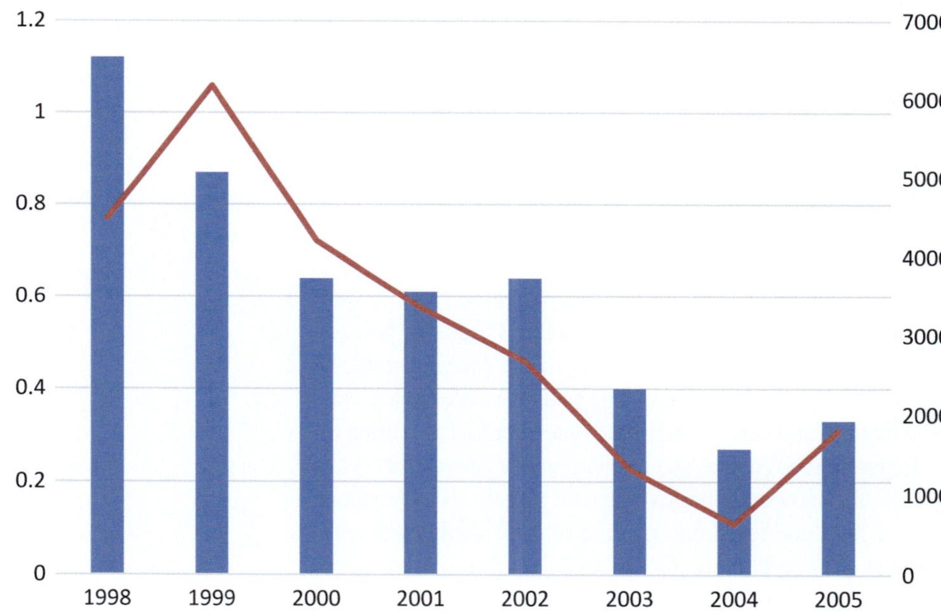

Fig. 21.12 Comparison of changes in CPUE (*100, *blue*) and estimated number of catchable fish from stocking (*red*)

misunderstanding with the people responsible for carrying out the work, individuals got tagged several consecutive years with the same type of tag (adipose ablation). The first years of tagging are not exploitable from a statistical point of view, and we can draw no information from this type of tag.

In the second type of tagging realized, either the adipose fin is clipped or color marking is used, thus making it possible to differentiate separate cohorts. While this tagging methodology makes it possible to specify the year of release, the people in charge of emptying the nets forgot to record the type of tag as well as the size of the fish caught, thus rendering any attempt to follow the stock by tag futile.

21.4.2.5 Tracking According to WFD Protocol NF EN 14757

This standardized protocol was implemented when the Water Framework Directive was adopted in order to qualify the status of European bodies of water both above and below ground. Monitoring on the fish populations in water bodies is performed according to standard WFD protocol NF EN 14757 (AFNOR 2005). The engineering consulting firm *ASCONIT* (ASCONIT 2006) and the ONEMA (Olivier 2012) performed two samplings on Lake Pavin in 2005 and 2012.

This method provides a fairly comprehensive picture of the fish population and its spatial distribution.

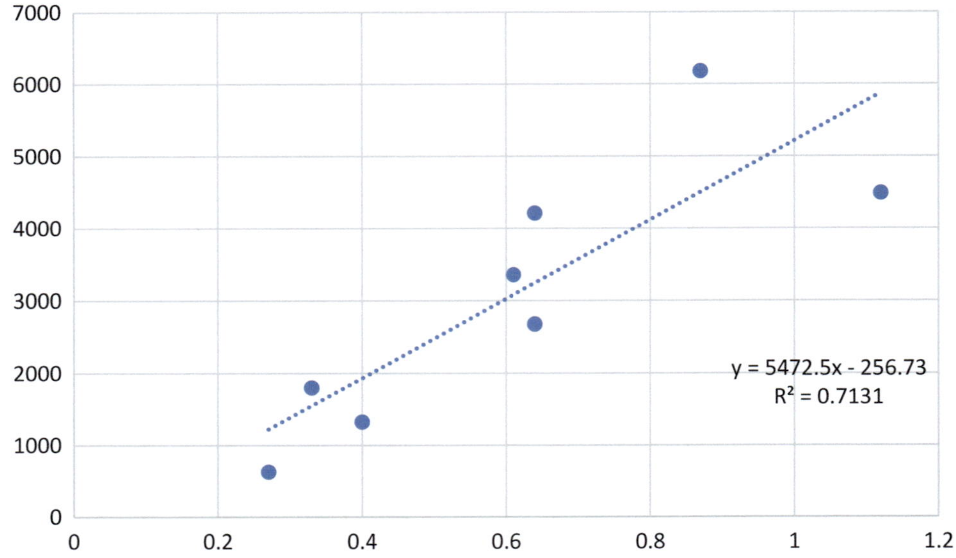

Fig. 21.13 Relationship between number of catchable fish from stocking and CPUE values

The 2012 study by the ONEMA (Olivier 2012) shows:

- a numerical distribution dominated by three species: European perch (42 %), roach (19.5 %), and Arctic char (18.2 %)
- a more even weight distribution, with roach accounting for the largest share of biomass (25 %), then perch (19.2 %), Arctic char (16.4 %) and chub (15 %).

While the share of perch and char was as expected and in line with numbers found in many deep lakes at high altitude, the share of cyprinids is cause for concern. These fish find current thermal and trophic conditions in the lake favorable to their development. Note too that these species are ecologically very flexible and can adapt to many *mesological* conditions.

Regarding the comparison of distributions between 2005 and 2012, while number of species increased slightly with the appearance of rudd and tench, the number of individuals captured is identical (234 in 2005 *vs*. 236 in 2012., Total weight, however, increased significantly (24975 g in 2005 *vs*. 33153 g in 2012).

The evolution in number and weight of different species clearly shows cyprinids thriving at the expense of perch. The situation of the Arctic char remains stable, but remember that the char population is propped up by regular stocking. Note too that no juvenile char were caught, which confirms our doubts over the natural reproductive capacity of this species.

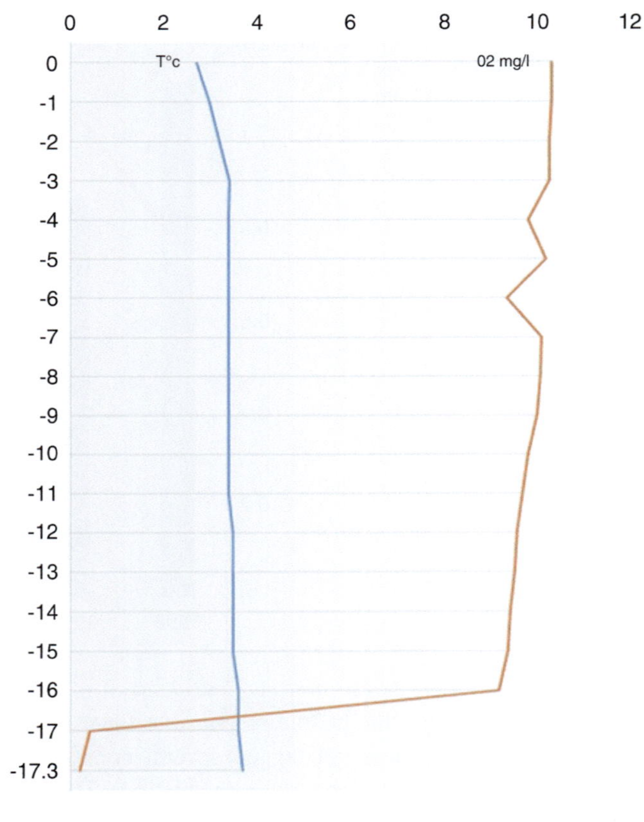

Fig. 21.14 Oxygen/temperature profile of an *S. alpinus* breeding site (East cliff) on January 21st, 2014

21.4.3 Modification of Trophic and Physico–Chemical Conditions

Recent studies show a rapid evolution in the trophic level of the lake in connection with agricultural practices (Ingé Conseil 2006). At the same time, there has a been a concomitant evolution in the fish population and a fairly significant decline in Arctic char sparking heavy doubt over the success of natural reproduction. During winter, the weather conditions make it difficult to follow the evolution of oxygenation in the breeding zones, which is an essential component of breeding success for Arctic char (Guillard 1992). During the breeding season, oxygen measurements at spawning sites show significant deoxygenation of the water at greater depths (at the bottom and 1 m above) (Fig. 21.14). Given the mode of reproduction of Arctic char (eggs deposited on the lake bottom), the chances of success seem low.

Fig. 21.15 Sub-aquatic view of a spawning area of *S. alpinus*

Underwater photos taken by the Arvernes Diving Club (D Chassain) show a change in bottom conditions in the breeding grounds. In the early 1990s, breeding ground bottoms were free of vegetation and the small amounts of deposits were easy for spawners to clean. In 2003, a similar dive showed a lake bottom clogged with stringy algae, making cleaning, incubation, and hatching uncertain (Fig. 21.15). The degradation of this excess organic matter is probably the reason of the lack of dissolved oxygen concentration and breeding success.

21.5 Ongoing Studies, Incubators, Tracking, Monitoring

Several studies and surveys have been implemented and/or are currently underway as part of the Territorial Contract framework governing lakes fed by the Couze Pavin river. Lake Pavin is monitored under Water Framework Directive following a standardized protocol performed by the ONEMA (Olivier 2012). The most important fish-related issue facing Lake Pavin is the tracking of its Arctic char population and its evolution. It is for this reason that the protocol we propose is based primarily on the study of this fish.

21.5.1 Tracking the Population by Fishing Expedition

Three annual fishing expeditions will be organized—one in the spring, one in summer, and one in autumn, for 3 years. Two exploration methods will be implemented:

- Fishing with two batteries of two-meter-wide vertical gill nets, one 50 m deep, the other 25 m deep. This method is designed to explore the entire column of oxygenated water (50 m) and the most productive zone (25 m). Each battery will use a range of 7 mesh sizes (10 mm, 15 mm, 20 mm, 30 mm, 40 mm, 50 mm, 60 mm), giving a total of 14 vertical nets. These nets will always be set in the same place from expedition to expedition. The advantages of this method are that it limits catches and situates fish at specific depths, as the side ropes are depth-graduated.
- Fishing by horizontal nets, called "spiders". For each excursion, we will set four horizontal nets extending from 25 to 6 m (spider type) with mesh sizes of 16 mm, 22 mm, and 32 mm in order to target different age-groups and obtain individuals in a range of sizes from 12 cm to 40 cm. The 'spiders' will be placed specifically to target char. The nets will be positioned at depths of between 8 and 25 m in the zones most frequented by char. Their positioning will remain identical from expedition to expedition.

The following protocol will be implemented for all nets and fish:

- The location of each net will be determined using GPS and depth of laying area determined by echo sounding.
- Each fish will be recorded (including, for vertical nets, depth at which it was caught), individually measured and weighed, and scale samples will be taken from the standardized area for each species.
- Tagged fish will be identified (by type of tag), weighed, and measured.
- Live fish will be released back into the lake. Dead fish will be made available to the owner of the lake.

The first year of fishing is not designed to capture tagged fish and will serve as calibration in order to optimize catches (Rivier 1996).

21.5.2 Marking Char

In December, flash (2-h) catches of char will be performed by the FDPPMA 63. The captured spawners will be stocked at the local salmon farm at Besse in quarantine. Reproduction will be carried out at the salmon farm. The fry produced will be raised until August or September, then released into Pavin. For this protocol, we propose to tag fish before release with a distinctive year-specific tag for the years 2014–2016.

- Every individual fish will be tagged before being released into the lake.
- When catches are made for population tracking, number of tagged individuals and type of tag will be recorded.
- This tagging and tracking by gill net catch will allow us to determine, via statistical analysis, the number of Arctic char resulting from releases *versus* the number from natural reproduction *in situ*.

- This tagging program will also allow us to assess the annual growth of released individuals.

21.5.3 Implementation of Incubators

The goal is to determine whether environmental conditions permit the normal egg development in char breeding grounds.

Concentration of dissolved oxygen and the amount of silt covering the eggs are key parameters for survival, as is length of the development period. The longer the eggs take to develop, the greater the chances that they will be exposed to conditions adverse to their survival. The biggest difficulty is to continuously monitor the physico-chemical conditions of the Arctic char breeding sites throughout the winter to determine whether conditions remain favorable for hatching. We have therefore chosen to monitor egg development via an immersion method. Again, iceover of the lake makes it very difficult to monitor physico–chemical conditions.

When catches of spawners are completed, reproduction will be hosted at the nearby national salmon farm. There are two windows in which when eggs can be manipulated for immersion—two hours after laying, or at embryonic stage. The embryonic stage is later (early February), so if the lake freezes over, it will be impossible to immerse the eggs. Immersion will therefore be effectuated immediately after fertilization.

The experimental incubators are PVC boxes used in salmon farming (Vibert boxes). The boxes have the openings that help keep the eggs inside the incubator while maintaining an exchange with the external environment, allowing the fry to develop to the swimming stage. Side vents allow the fry to swim out of the box. Each incubation line is located on a known char breeding ground. There will be three incubation lines, and each line will have three levels of incubators (bottom, 1 m above bottom, and 8 m above bottom) with three incubators per level. Each incubator will be seeded with 30 eggs. The lines will be plotted by GPS and equipped with submerged buoys. The incubation line will be removed as soon as the lake becomes accessible. This experimental protocol will enable us to assess whether the char eggs can still develop naturally, and at what depth. This experiment must not be considered has a substitution of natural reproduction. During all this incubation period, six oxygen recorders will be immerged at 12,14,16,18 m depth on the bottom and two others at 1 m(15 m) and 5 m (11 m) over the bottom on the 16 m incubation line. One value of oxygen is record every half hour for each recorder between December to May. This monitoring will allow us to demonstrate the relation between dissolved oxygen and success of incubation.

21.6 Conclusion

The piscicultural history of Pavin Lake is highly influenced by man, first by the initial introduction after 1859 of several key species,trouts (*Salmo trutta*), common salmons (*Salmo salar*), Heusch salmons (*Hucho hucho*), chars (*Salvelinus Umbla*) and some coregons (*Coregonus Fera*) (See Sect. 1.4.2).The Arctic Char introduction has been the most successful on the long term and this species became emblematic throughout the twentieth century.

Over the last 20 years another marked change occurred in the lake watershed where its groundwaters are coming from. It is a shift of land use and agricultural practices, from extentive pasture to intensive harvested pasture, requiring numerous inputs of mineral and organic nutrients. In the mid-1990s the increased occurrence of blooms of *Anabaena spiroïdes* raised the concern of some limnologists, but the quantification and origin of the nutrient inputs to Pavin were only established in 2005 (IngéConseil 2006). The resulting modification of the fish population,which was studied at the same period, revealed mass mortality of the crayfish population and a decrease of the Arctic Char landings. This disruption in the lake ecology can be multifactorial, inappropriate fish management, excessive fishing pressure, impact of additional fish species introduction, climate warming…, however we believe that trophic state degradation by nutrient excess is the prominent factor.

In order to restore the original oligotrophic of Pavin Lake which made it chosen as a typical oligotrophic lake in several international projects in the 1960s (See Sect. 1.5.1), a reduction of nutrients inputs from agriculture practices is needed. All local authorities together with the farmers agreed to establish an operational land management within Pavin watershed, named Territorial Contract including two steps:

- Realize an initial survey of the situation with all the stakeholders (*LMGE*, *ONEMA*, *IRSTEA*, FDPPMA63, and *ATHOS* ecological engineers), under the supervision of the Auvergne Regional Park (PNRVA) and the financial support of the Loire-Bretagne Water Authority (AELB), the Puy de Dôme department and the Auvergne region. During 3 years the physico-chemical quality, the chlorophyll, the algal counts, the dissolved oxygen and temperature continuous monitoring by in situ probes should be surveyed and complemented by the study of the Arctic Char population.
- Modify the present land use to gradually return to extensive pasture without external nutrients inputs, by exchange or purchase of the concerned agricultural plots.

The first plot exchange have started but the land property rights complexity for the 120 ha concerned make this second point more slow and difficult than expected.

It is hoped that the trend monitoring does not just show a slow and inflexible degradation of the lake quality and trophic state and that the expected restoration can be quoted as a success story in few decades.

Acknowledgements Many thanks to Prof. Michel Meybeck, Pierre & Marie Curie University Paris VI, France, for critical and proofreading of the manuscript.

References

AFNOR (2005) NF EN 14757. Echantillonnage des poissons à l'aide des filets maillants, 26 p

ASCONIT Consultants (2006) Programme 2005 de suivi et de caractérisation des masses d'eau de plan d'eau. Lot 1: Plans d'eau d'origine naturelle. AELB, pp 140–152

Audinot G, Jamet JL (1998) Etude d'un écosystème limnique oligotrophe. Rapport d'université, 15 p

Berthoule A (1890) Les lacs d'Auvergne. Orographie, faune naturelle, faune introduite. Rev. Soc. Nationale d'Acclimatation, Paris, 136 p

Bhatthacharya CG (1967) A simple method of resolution of a distribution into gaussian components. Biometrics 23(1):115–135

Blanchard E (1866) Les poissons des eaux douces de la France. J.B.Baillière &fils, Paris, 656 p

Bruyant Ch (1908) Rectification à la faune ichtyologique de l'auvergne. Annales de la station limnologique de Besse (III, IV), p 381

Bruyant Ch (1909) Annales de la station limnologique de Besse. (II), p 199–206

Champigneulle A, Michoud M, Brun JC (2001) Pacage lacustre de salmonidés (omble chevalier, corégone et truite) dans le lac Léman et le lac du Bourget. In: Gestion piscicole des grands plans d'eau. INRA Editions, Paris, pp 349–421

Delarbre A (Abbé) (1795) Flore d'Auvergne ou recueil des plantes de cette ci devant province. Suivi de la description du lac Pavin. B. Beauvert et L. Deschamp, Clermont-Ferrand, 337 p

Desbois (1826) Résumé de l'histoire d'Auvergne par un auvergnat. Lecointre & Durey, Paris, 554 p

Dussart B (1955) Un poisson trop peu connu en France: l'omble chevalier. La terre et la vie 3:129–158

Eusébio A, et Reynouard J (1925) Le lac Pavin et le creux de Soucy. Guide du touriste et du naturaliste. Station limnologique de Besse, 55 p

Geiger W (1964) Besatz und fang in forellengewâssern. *Schweiz, Fisherei-Zeitung* (72), p 115–119

Gerdeaux D (2001a) Détermination de l'âge et de la croissance des poissons. In: Gestion piscicole des grands plans d'eau. INRA Editions, Paris, pp 283–292

Gerdeaux D (2001b) Les statistiques de pêche. In: Gestion piscicole des grands plans d'eau. INRA Editions, Paris, pp 423–433

Gillet C (2001) Le déroulement de la fraie des principaux poissons lacustres. In: Gestion piscicole des grands plans d'eau. INRA, Paris, pp 241–281

Grandjean F (1997) Variabilité morphométrique et génétique des populations de l'écrevisse à pattes blanches, *Austropotamobius pallipes*: implications biogréographiques. Thése de Biologie animale. Université de Poitiers

Guillard J, Gillet C, Champigneulle A (1992) Principales caractéristiques de l'élevage de l'omble chevalier (*Salvelinus alpinus* L.). Bull Fr Peche Piscic 325:47–68

Ingé Conseil (2006) Etude hydrogéologique et hydrologique du bassin versant du lac Pavin, 23 p

Jaloustre E (1884) Le Beffroi de Besse. Mém. Acad. Sci. Belles-Lettres et Arts, Clermont-Ferrand (BMIU, Clermont A34 425), p 113–146

Jamet JL (1995) Reproduction, condition and food of adult arctic charr (Salvelinus alpinus, L.) in lake Pavin (Massif Central, France). Hydrobiologia 300–301:279–288

Léveillé JC (1997) Flux de matière en milieu pélagique lacustre: utilisation des acides gras comme marqueurs organiques naturels du transfert dans la chaîne trophique phytoplancton – zooplancton – alevins d'omble chevalier (*Salvelinus alpinus*). Thése Biologie des populations et Ecologie. Université de Clermont-Ferrand

Machino Y (1991) Répartition géographique de l'omble chevalier (Poisson, Salmonidae, *Salvelinus alpinus*) en France. Diplôme Supérieur de Recherches. Lab. Zool. –Hydrobiol. Université Joseph Fourrier de Grenoble, 438 p

Maisse G, Baglinière JL (1991) Biologie de la truite commune (*Salmo trutta* L.) dans les rivières françaises. In: La truite biologie et écologie. INRA-ENSA, Paris, pp 25–45

Meybeck M (2011) Le Pavin et ses voisins, histoire d'un patrimoine européen unique. Revue des Sciences Naturelles d'Auvergne 74–75:46

Olivier L (1939) Matériaux pour la connaissance limnologique des lacs Mont Doriens. Imp. De Bussac, Clermont-Ferrand, 79 p

Olivier G (2012) Rapport d'étude du suivi piscicole du lac Pavin (63). Mise en œuvre du protocole NF EN 14 757 RCS. ONEMA DIR, Massif Central. 20 p

Reynouard J (1909) Besse aujourd'hui autrefois. Annales de la station limnologique de Besse (I):6–11. Internal publication of Besse limnological and biological station, Clermont-Ferrand, France

Richard A (1998) Gestion piscicole. Interventions sur les populations de poissons. Repeuplement des cours d'eau salmonicoles. CSP, Collection mise au Point, 256 p

Ricker WE (1971) Methods of assessment of fish production in freshwater. IBP Handbook No. 3, Blackwell Scientific Publications, Oxford and Edinburgh, 348 p

Ricker WE (1980) Calcul et interprétation des statistiques biologiques des populations de poissons. Bull Fish Res Board Can 191 F: 409 p

Rivier B (1996) Lacs de haute altitude: méthodes d'échantillonnage ichtyologique, gestion piscicole. Collection Etudes du Cemagref, série gestion des milieux aquatiques, n°11. Le Tholonet, Cemagref, Irstea

Rubin JF (1990) Biologie de l'omble chevalier (*Salvelinus alpinus* L.) dans le Léman (Suisse). Thèse Docteur ès Sciences en Biologie, Univ. De Lausanne, 169 p

Rubin JF, Buttiker B (1992) Les sites de fraye de l'omble chevalier, *Salvelinus alpinus* (L.) dans le Léman. Bull Fr Peche Piscic 325:69–82

Rubin JF, Buttiker B (1993) Quelle est la proportion d'ombles chevaliers, *Salvelinus alpinus* (L.), issus de reproduction naturelle ou de repeuplement, dans le Léman ? Bull Fr Peche Piscic 329: 221–229

Tissandier G (1873) L'Eau. 3ème edition. Hachette, 356 p

Zanella D (2003) Bilan de la situation actuelle de l'omble chevalier sur le lac d'Annecy et sa gestion halieutique. D.E.S.S. R.E.M.A.C. Université Blaise Pascal Clermont II, 40 p

Part V

Sedimentology and Paleolimnology

Emmanuel Chapron

Lake Pavin Sedimentary Environments

22

Emmanuel Chapron, Léo Chassiot, Patrick Lajeunesse,
Grégoire Ledoux, and Patrick Albéric

Abstract

In the recent years, Lake Pavin sedimentary basin has been intensively studied by several acoustic surveys (high resolution seismic reflection profiling, multibeam bathymetry) and gravity coring campaigns. This new data set combining acoustic images and multidisciplinary study of sediment cores allows characterizing contrasted subaquatic sedimentary environments along the littoral slopes, a subaquatic plateau (close to the lake **outlet**), steep slopes and its deep central basin. Two main types of lacustrine sediments are identified (i) between the lake shore and 26 m water depth (massive organic rich sandy silts), and (ii) below 26 m water depth when the lake floor slopes are less than 15° steep (organic rich laminated **diatomites**). A large and recent **slide scar** is in particular clearly identified at the edge of the plateau just above the deep central basin. Evidences of former gravity reworking phenomena within the **crater ring** draining into Lake Pavin also include a large subaquatic **slump deposit** accumulated on the subaquatic plateau and several small scale **rock fall** deposits originating from outcropping **lavas** within the **crater ring**. The identification of two recent outstanding **erosive** sandy layers in Pavin littoral environment also suggests that some of this gravity reworking phenomena have been associated with unusually violent waves and/or abrupt lake level drop. Lake Pavin geomorphology and sedimentary environments are in addition compared to the ones of the nearby Lake Chauvet based on a similar acoustic and sedimentary data base in order to highlight the influence of **maar** age and geomorphology on the development of sedimentary environments and Natural Hazards in this volcanic region of the French Massif Central. Lake Chauvet is comparatively to Lake Pavin characterized by a shallower central basin, less steep slopes and no subaquatic plateaus. A recent and relatively large **mass wasting deposit** is, however, clearly identified along the slopes of a small delta facing the only tributary of this maar lake. This work

E. Chapron (✉)
Present Address: ISTO Institut des Sciences de la Terre d'Orléans, Observatoire des Sciences de l'Univers en région Centre, UMR 7327 (CNRS-Université d'Orléans-BRGM), Campus Géosciences, 1A rue de la Férolerie, 45071 Orléans Cedex 2, France

Present Affiliation: GEODE UMR 5602 CNRS, Université de Toulouse 2 Jean Jaurès, 5 Allée A. Machado, 31058 Toulouse Cedex 9, France
e-mail: emmanuel.chapron@univ-tlse2.fr

L. Chassiot • P. Albéric
ISTO Institut des Sciences de la Terre d'Orléans, Observatoire des Sciences de l'Univers en région Centre, UMR 7327 (CNRS-Université d'Orléans-BRGM), Campus Géosciences, 1A rue de la Férolerie, 45071 Orléans Cedex 2, France

P. Lajeunesse • G. Ledoux
CEN Centre d'études nordiques, Département de Géographie, Université Laval, Québec G1VOA6, Canada

T. Sime-Ngando et al. (eds.), *Lake Pavin*, DOI 10.1007/978-3-319-39961-4_22

suggest that maar lakes from the study area are concerned by subaquatic slope stabilities, especially in Lake Pavin were slope failures may in addition impact the development or the stability of its **meromicticity**.

Keywords
Lake Pavin • Lake Chauvet • Sediment • Slope stabilities • Bathymetry • Seismic stratigraphy

22.1 Introduction

Lake Pavin have been intensively investigated in terms of limnology and volcanology (Parts I, II and III, this volume) but little is still known about its subaquatic sedimentary environments and their dynamics (Chapron et al. 2010). Such a characterization of sedimentary environments in this young crater lake (or **maar** lake, see Chap. 5, this volume) can be achieved through a **limnogeological** approach of its sedimentary archives. This chapter aims thus at addressing an up-to date presentation of available knowledge on Lake Pavin sedimentary environments based on the integration of two complementary acoustic mapping techniques (multibeam bathymetry and seismic reflection mapping) and a multidisciplinary characterization of short sediment cores (sediment color, magnetic susceptibility, total organic carbon content and organic geochemistry) retrieved at key locations.

In this chapter, first results from a similar approach recently performed in nearby **maar** lake, Lake Chauvet, (Fig. 22.1) are also presented in order to highlight the main specificities of Lake Pavin's sedimentary environments and to further discuss their influence on the development of natural hazards in this touristic part of the French Massif Central.

Pavin is the only **meromictic maar** lake in France (Chaps. 1, 6 and 10, this volume) and is surrounded by several contrasting older small lakes of volcanic origin as shown in Fig. 22.1. The other **maar lake,** Lake Chauvet, for example, formed during the last glacial period is today 63 m deep and contains several regional **tephra** layers from the Late Glacial and Early Holocene periods (Juvigné 1992). Lake Montcineyre formed in the Early Holocene is nowadays

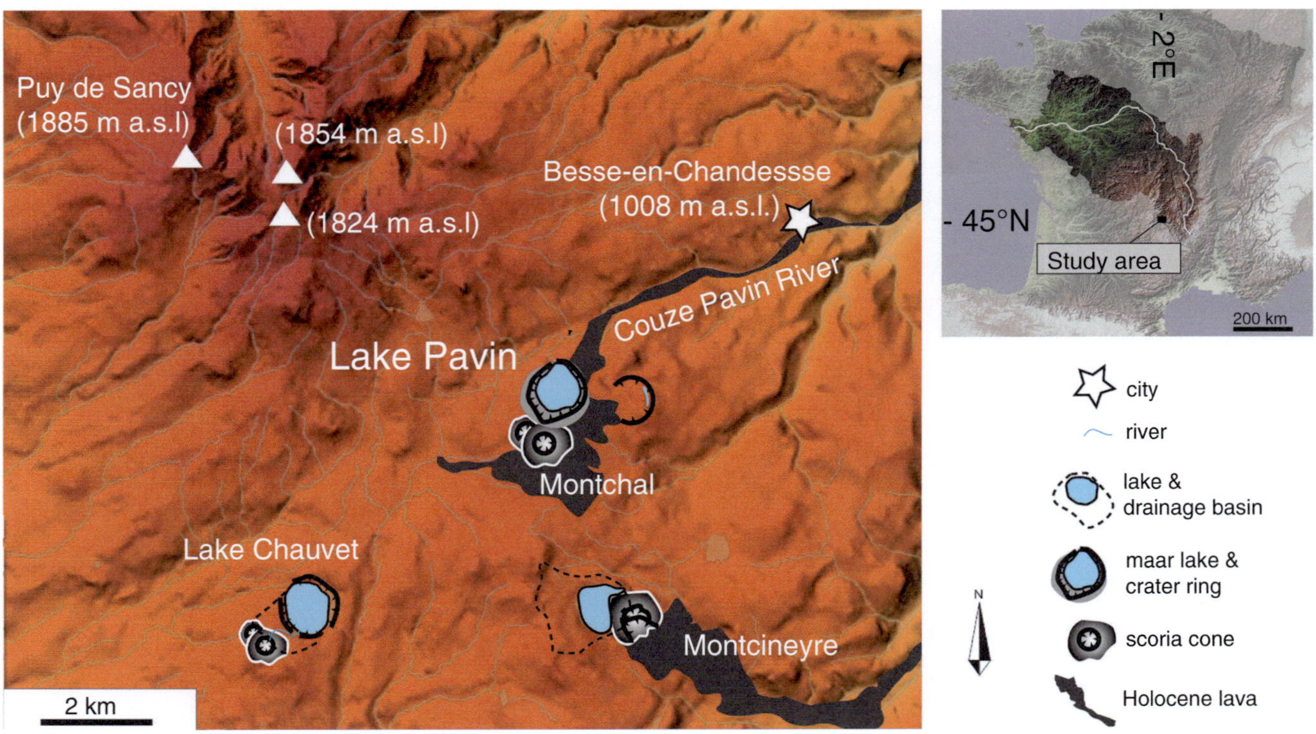

Fig. 22.1 General location of maar lakes Lake Pavin and Lake Chauvet in the French Massif Central and in the Loire river drainage basin (*right*) and simplified Holocene volcanic context south of the Puy de Sancy strato volcano (*left*) (Note that lakes Pavin and Montcineyre are part of the Allier River watershed draining into the Loire River watershed towards the north east, while Lake Chauvet drains into the Dordogne River towards the west)

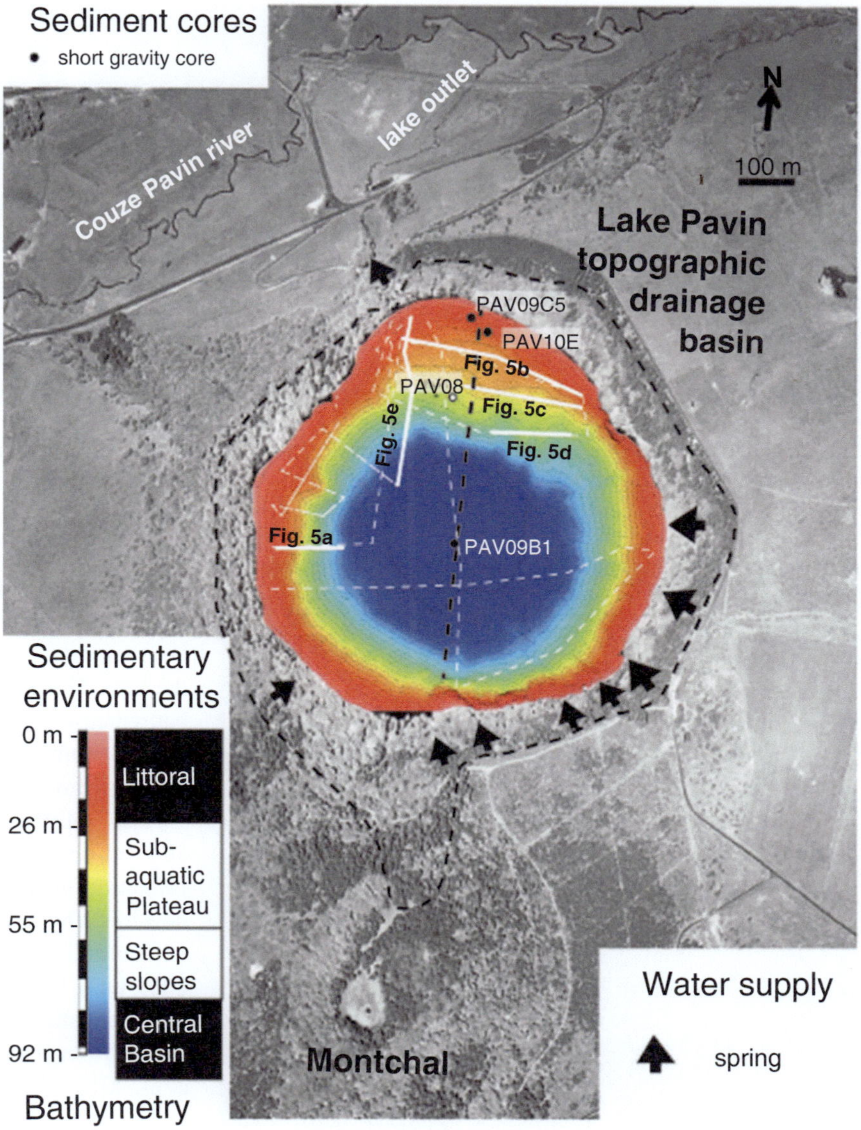

Fig. 22.2 Illustration of Lake Pavin multibeam bathymetry and location of sub-bottom profiles and short sediment cores. The grid of 12 kHz sub-bottom profiles (*white dotted lines*) and the 3.5 kHz sub-bottom profile (*black dotted lines*) were used together with short gravity cores to identify four main sedimentary environments from the littoral to the deep central basin. The locations of springs within the crater rim (*black arrows*) and of 12 kHz sub-bottom profiles illustrated in Fig. 22.5 (*thick white lines*) are also given

22 m deep and consist in two coalescing small **maars** damned by the Montcineyre **scoria cone** and **lava flow** which developed shortly before the Montchal **scoria cone** and **lavas** (Chapron et al. 2012). The Montchal **lavas** have then been partly destroyed by the Pavin **phreatomagmatic** eruption ca. 7000 years ago (Gewelt and Juvigné 1988; Chap. 6, this volume). This recent volcanic event formed Lake Pavin: a almost circular (750 m diameter) small but 92 m deep **maar** lake, today located at an altitude of 1197 m above sea level (a.s.l.) and draining a steep and well preserved **crater rim** reaching an altitude of 1253 m a.s.l.. As shown in Fig. 22.2, the edge of the Pavin **crater ring** matches the limit of the topographic drainage basin of Lake Pavin. It is however important to keep in mind that this **topographic drainage basin** is smaller than the still poorly defined **watershed** of Lake Pavin draining several subaerial and subaquatic springs (Jezequel et al. 2011).

22.2 Limnogeological Data Bases from Lakes Pavin and Chauvet

During the last decade advanced acoustic mapping techniques were applied in Lake Pavin in order to document subaquatic slope stabilities and to optimize the location of sediment cores (Figs. 22.2 and 22.3) to further understand the history and the evolution of this recent volcanic lake.

22.2.1 Acoustic Mapping Technics

In 2008, a Reason Sebat 8101 multibeam echosounder used with differential GPS positioning and an inertial navigation system allowed to precisely map the lake floor morphology (Figs. 22.2 and 22.4). This recent map presented in Chapron et al. (2010) significantly improved our understanding of

Fig. 22.3 Recently available geological data from in Lake Pavin and Lake Chauvet discussed in this chapter include short gravity coring performed from an UWITEC platform in 2008 (**a**); high resolution seismic reflection mapping survey from an inflatable boat in 2009 (**b**) and short gravity coring from a Limnoraft in 2009, 2010 and 2013 (**c**)

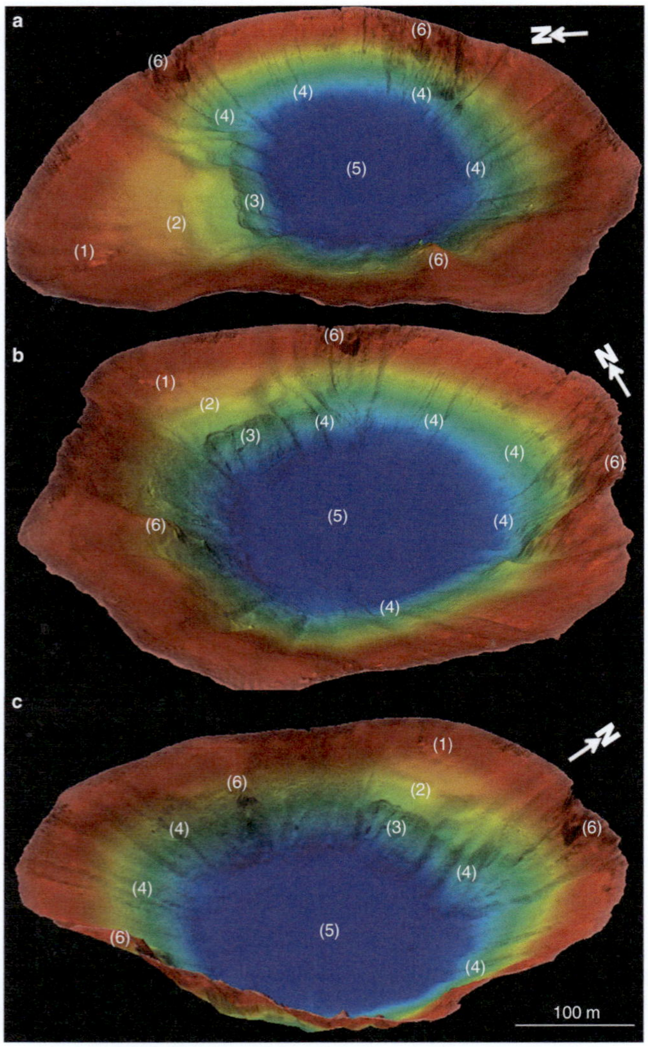

Fig. 22.4 3D views of Lake Pavin multibeam bathymetry illustrating a possible subaquatic outlet (*1*), the subaquatic plateau (*2*), the ca. AD 1300 slide scar (*3*), multiple active canyons along steep slopes (*4*), a deep and flat central basin (*5*) and outcropping volcanic rocks (*6*). These geomorphological features are further presented and discussed in the text

Lake Pavin sedimentary environments, in particular because it clearly illustrates the development of a subaquatic plateau between ca. 26 and 55 m water depths in the northern part of the lake. This key feature was not identified on the previous bathymetric map made by Delbecque (1898) on basis of manual water depth measurements along only two perpendiculars transects.

The multibeam bathymetric map of Lake Pavin also allowed optimizing the location of high-resolution **seismic reflection profiles** documenting the sub-bottom geometries developed by lacustrine sediments below the lake floor (Figs. 22.2 and 22.5). The sub-bottom profiling survey realized in 2009 used GPS positioning and a very high frequency (12 kHz) acoustic source in order to provide very high resolution **sub-bottom profiles** (ca. 8 cm vertical resolution) (Chapron et al. 2012). This strategy has been established based on the results of a previous seismic reflection survey performed in 2002 using a lower frequency seismic source (3.5 kHz) and a dense grid of profiles showing (i) that the maximum sediment thickness accumulated above the bedrock on the plateau is limited (ca. 5 m) and (ii) that the acoustic signal is very quickly absorbed by gas rich sediments in the deep flat basin of the lake (Chapron et al. 2010).

An exploratory sub-bottom profiling survey using the same 12 kHz seismic source and GPS positioning in Lake Chauvet (Fig. 22.6) was also conducted in 2009 based on available bathymetric data from Delbecque (1898). This survey allowed mapping lacustrine sediment thicknesses and geometries above the **bedrock** (along the lake shore) or subaquatic **moraine** deposits in the deep central basin were Late Glacial and Holocene sediments are clearly imaged (Chapron et al. 2012). In 2011, further bathymetric data has been collected in Lake Chauvet (i.e. along the shore line and several cross sections in between available seismic reflection profiles) with a single beam echo sounder using a 200 kHz acoustic source. This strategy allowed mapping

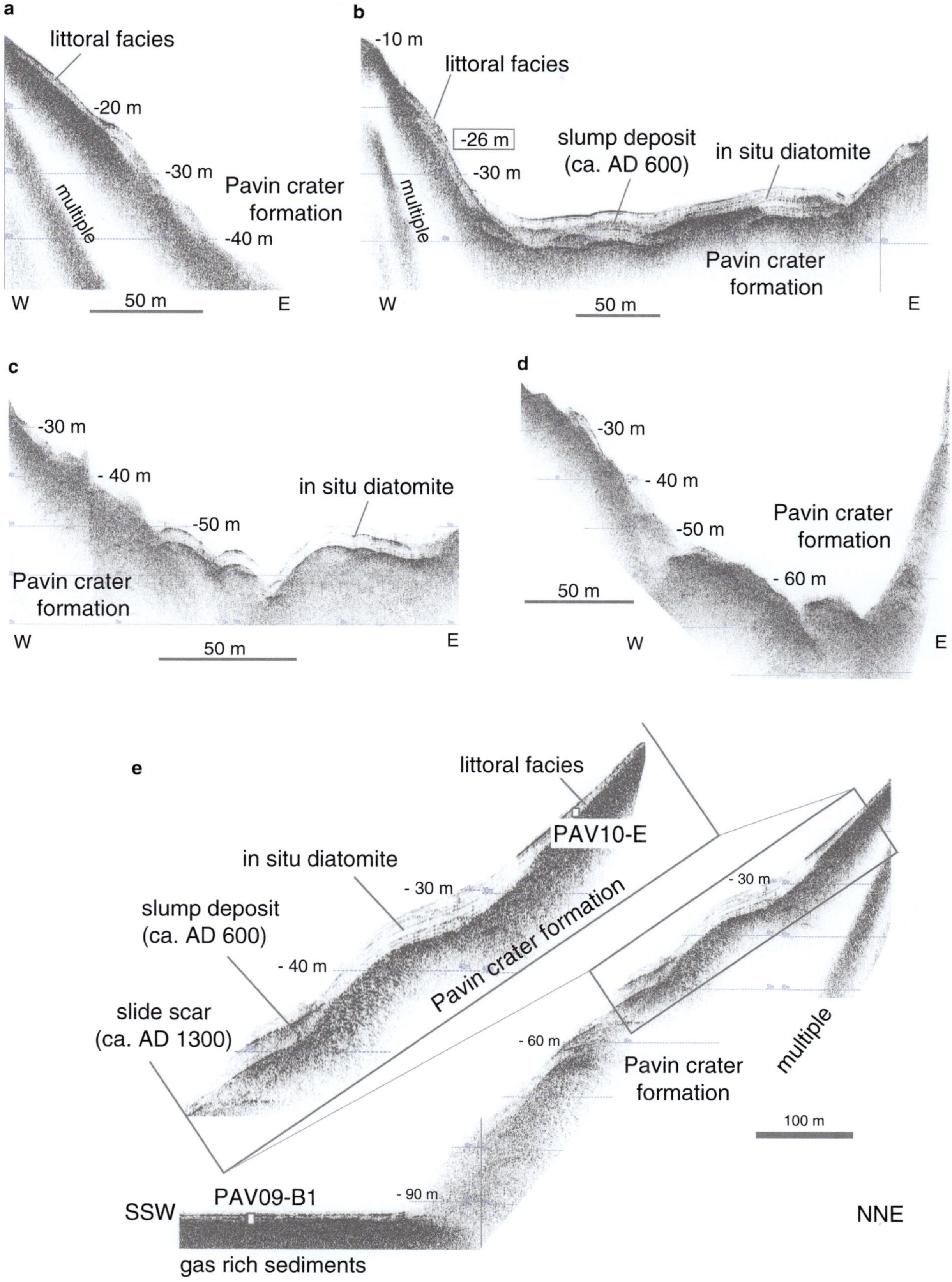

Fig. 22.5 Lake Pavin sub-bottom profiles illustrating the acoustic facies developed in littoral environments or along the subaquatic plateau (**a**, **b**, **c**, **d**, **e**) and down to the deep central basin (**e**). Note that all these sedimentary environments develop contrasted acoustic facies above the subaquatic plateau formed by the Pavin crater formation. The location of short gravity cores PAV09B1 and PAV10E is also indicated. As detailed in this chapter and in Chap. 23, in situ deposits (diatomites) on the plateau are clearly distinguished from reworked sediments (ca.AD 600 slump deposit and ca. AD1300 slide scar). The location of each seismic section is given in Fig. 22.2

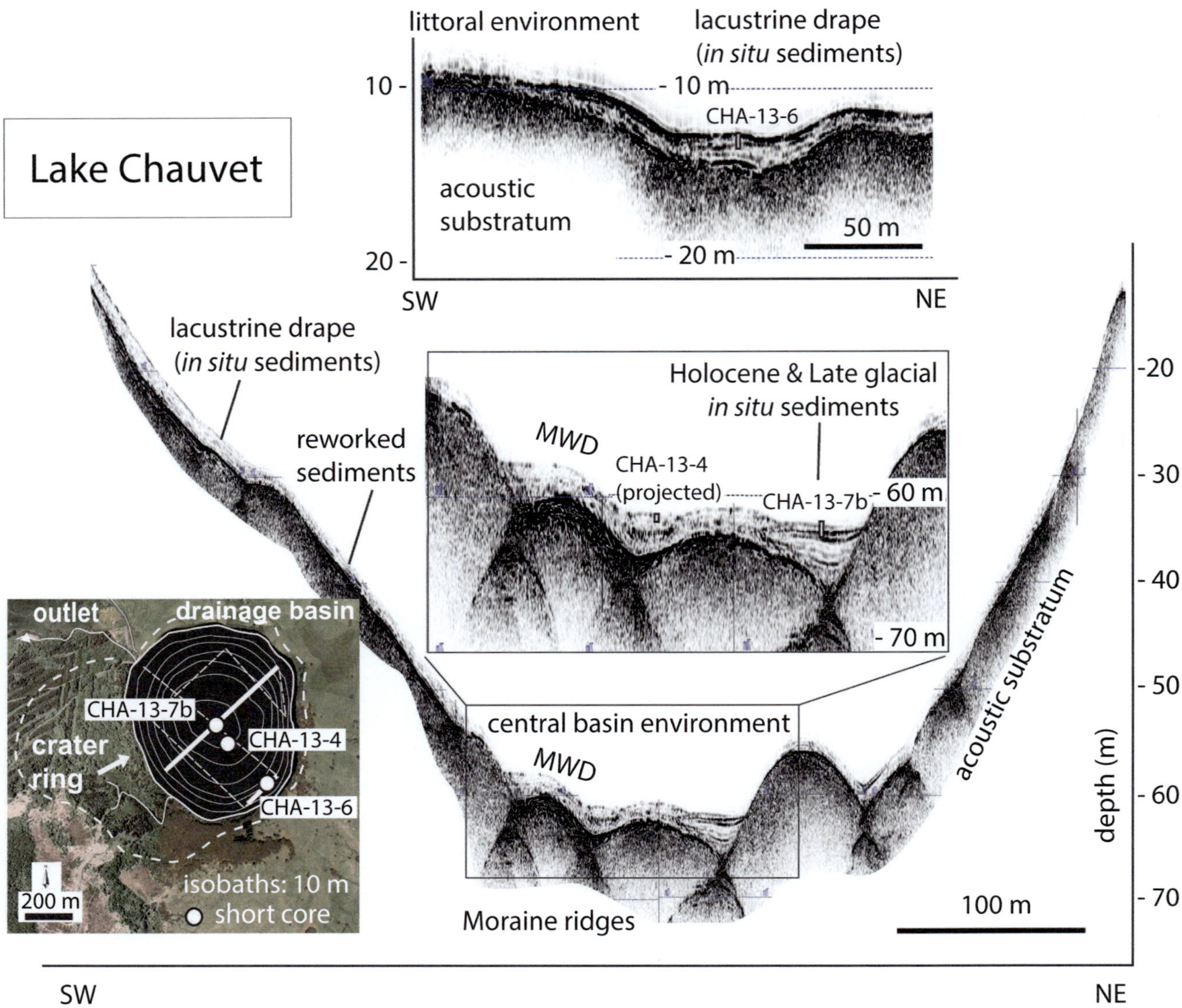

Fig. 22.6 Lake Chauvet sub-bottom profiles (12 kHz) of a littoral environment and of the central basin. The southwestern slopes of this maar lake are unstable and a recent mass wasting deposit (MWD) can be identified within the deep central basin were moraine ridges are ponded by Late glacial and Holocene lacustrine sediments. The bathymetric map, the location of the crater ring, the outlet, the extension of Lake Chauvet drainage basin, available short coring sites (white circle), seismic reflection profiles (*white dashed lines*) and illustrated seismic profiles (*thick white lines*) are also indicated

more precisely the subaqueous geomorphology of this maar lake.

22.2.2 Short Sediment Cores

Based on Lake Pavin multibeam bathymetry and sub bottom profiling, several short gravity cores were retrieved by the ISTO laboratory (i) in 2008 at 46 m water depth on the plateau (PAV08-P1, 33 cm long, 45°29.86′N/2°53.24′E), (ii) in 2009 from a littoral environment at 20 m water depth (PAV09-C5, 51 cm long, 45°29.93′N/2°53.32′E) and in the deep anoxic basin at 92 m water depth (PAV09-B1, 120 cm long, 45°29.74′N/2°53.28′E) and (iii) in 2010 (PAV10-E, 80 cm long, 45°29.92′N/2°53.35′E) at 17 m water depth (Fig. 22.2).

As shown in Fig. 22.6, three short gravity cores were also retrieved in 2013 by the ISTO laboratory in Lake Chauvet within littoral environments (CHA 13–6, 90 cm long) and in the deep basin; one within in situ deposits (CHA 13-7B, 95 cm long) and one within reworked deposits (CHA 13–4, 55 cm long) according to the seismic profiles.

A multi-proxy study of the lacustrine sediments was then conducted once the cores from both lakes were split in two halves. Hand-held measurements of sediment **magnetic susceptibility** (MS) with a Bartington MS2E point sensor and of sediment **diffuse spectral reflectance** with a Minolta 2600D spectrophotometer were both performed following the procedure described in Debret et al. (2010), at a sampling interval of 1 cm. Organic matter content and quality from lacustrine sediments were in addition punctually documented by Rock-Eval (RE) pyrolysis following the proce-

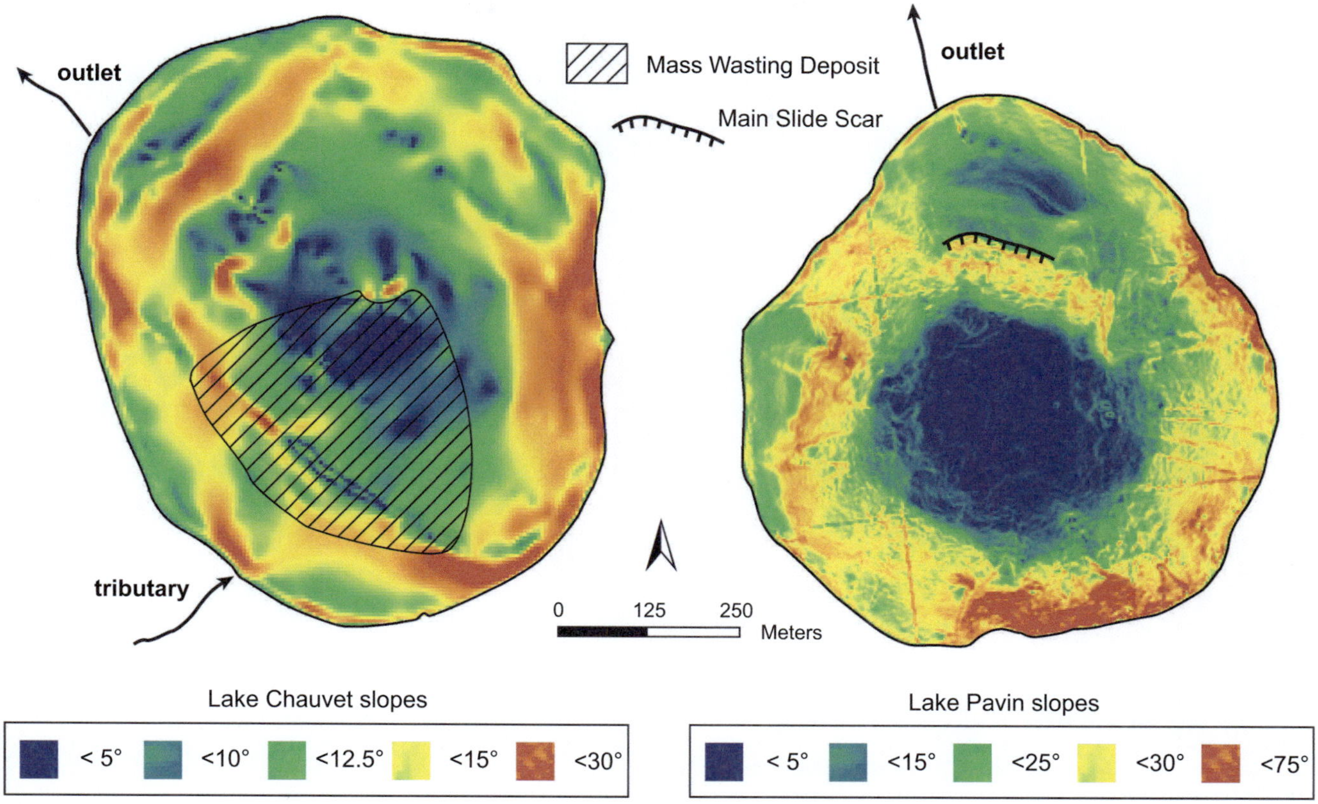

Fig. 22.7 Slope maps of maar lakes Chauvet (*left*) and Pavin (*right*)

dure described in Behar et al. (2001) and Simonneau et al. (2014).

22.3 Signatures of Littoral Environments in Lakes Pavin and Chauvet

22.3.1 Lake Pavin

Between Lake Pavin shore line and 26 m water depth, when slopes angles are below 30° (Fig. 22.7) a specific **acoustic facies** is observed on **sub-bottom profiles** (Fig. 22.5a, b, and e). This littoral facies is characterized by a transparent acoustic facies developed above the acoustic substratum and capped by a high amplitude **reflection**. This littoral facies is getting thinner towards the shoreline and can reach a maximum thickness of ca. 4 m in the northern part of the lake (Fig. 22.5e), while it is limited to ca. 3 m and 1 m in the northwestern (Fig. 22.5b) and western (Fig. 22.5a) parts of the lake, respectively.

Sediment cores PAV09-C5 and PAV10-E were retrieved along the northern part of Lake Pavin at 20 m and 17 m water depth, respectively, (Fig. 22.2) and allow characterizing the composition of the littoral acoustic facies.

Lake Pavin littoral sediments are made of massive brownish sandy silts with frequent sandy layers and numerous leaves and leave debris (Fig. 22.8). Sediment **magnetic susceptibility** (MS) is variable and oscillating between 35 and 3 10^{-6} SI. Two distinct **erosional events** producing sandy layers rich in leaves (E5 and E6) in PAV09-C5 were dated by **AMS radiocarbon** (Chapron et al. 2012) and are further discussed in Chap. 23. Only one of such outstanding sandy layer is identified in PAV10-E and characterized by high values of sediment MS. As shown in Fig. 22.8, this **erosive** sedimentary event (labeled E6?) matches an abrupt transition from laminated brownish sandy silts (with low MS values) towards a littoral sedimentary facies in the upper 20 cm of the core. Unfortunately, too little terrestrial organic matter (leave debris) did not allow **AMS radiocarbon** dating of this major change in sedimentation (Chapron et al. 2012). As further discussed in Chap. 23, the depth of this erosive layer in PAV10-E matches the depth of E6 in PAV09-C5 and might thus be contemporaneous.

22.3.2 Lake Chauvet

Between Lake Chauvet's shore line and ca. 30 m water depth, when slopes angles are below 30° (Fig. 22.7) a thin and transparent acoustic facies with a draping geometry is observed on **sub-bottom profiles** above the **acoustic substratum** (Fig. 22.6). This littoral facies is getting thinner towards the shore and can reach a maximum thickness of ca. 2.5 m in the southern part of the lake.

Up to five contrasted **sedimentary facies** are, however, identified within this littoral facies in sediment core CHA13-6 (Fig. 22.9). The upper Unit A consists of brown massive sediments with relatively high and fluctuating MS values and is interrupted by three distinct coarse sandy layers developing peaks in MS of a few cm wide with very high values (up to 350

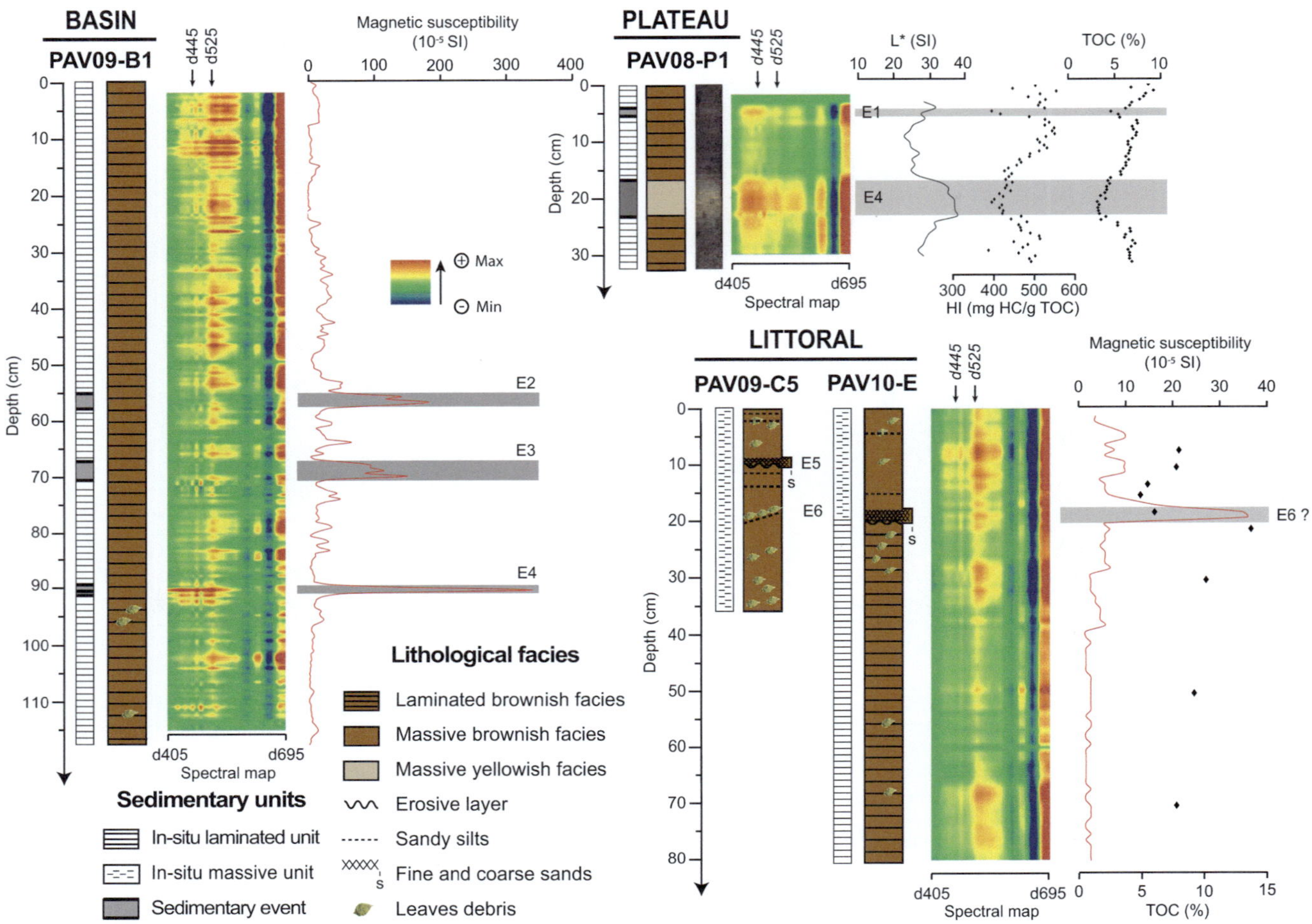

Fig. 22.8 Multidisciplinary characterization of Lake Pavin sediments retrieved by short gravity cores, in the deep central basin (PAV09-B1), on the plateau (PAV8-P1) and in littoral environments (PAV09-C5 and PAV10-E). Visual descriptions of sedimentary facies are further defined (i) by sediment diffuse spectral reflectance (here plotted on a 3D diagram where the X axis represent the wavelengths, Y is the depth in core and Z the first derivative value for the corresponding wavelength (in nm) expressed by a code of color); (ii) by sediment magnetic susceptibility (for PAV09-B1 and PAV10-E); (iii) core PAV08-P1 is in addition detailed by sediment reflectance (L*), sediment digital picture and organic carbon geochemistry (*TOC* total organic carbon, *HI* hydrogen index). The locations of these cores in Lake Pavin are also given in Fig. 22.2

10^{-5} SI) and labeled E2, E3 and E4. Below, Unit B is, on the contrary, a dark brownish massif unit with lower MS values and is locally interrupted by Unit C. Unit C is a brownish massive unit bearing much higher MS values than Unit B, and also the brownish Unit D identified at the base of core CHA13-6. Several leaves and leave debris founded in Units B and D are suitable for **AMS radiocarbon** dating and on-going analysis will provide some chronological controls on such contrasted sedimentation patterns in this Late Glacial **maar** lake.

22.4 Sedimentation Patterns Along the Slopes of Lakes Pavin and Chauvet

22.4.1 Lake Pavin Subaquatic Plateau

Between ca. 26 and 55 m water depths in the northern part of Lake Pavin, gentle slopes (below 15°, Fig. 22.7) are draped by an up to 5 m thick sedimentary sequence made of two contrasting **acoustic facies** (Chapron et al. 2010, 2012) as shown in Fig. 22.5b, e: a stratified unit bearing few low amplitude and continuous **reflections** is mainly identified downslope from the littoral facies and on top of a chaotic to transparent lens-shaped body. This chaotic deposit occurring above the Pavin crater formation has the typical signature of a **mass wasting deposit** (MWD). It has been sampled in PAV08 long piston core, interpreted as a **slump deposit** and dated to ca. **AD** 600 as detailed by Chapron et al. (2010) and further discussed in Chap. 23.

The stratified unit has also been sampled in short cores PAV08-P1 and in the lower part of PAV10-E (Fig. 22.2). These sediments are characterized by well-defined sandy silts developing brownish and greenish laminas rich in diatom bloom that are bearing low values of MS. At site PAV08, short core P1 highlight in addition high values of Total Organic Carbon (TOC) oscillating around 8 % and relatively high values of Hydrogen Index HI) between 450 and 550 mg HC/g TOC (Fig. 22.8). Following Schettler et al. (2007), Chapron et al. (2010) and Albéric et al. (2013), these sediments rich in

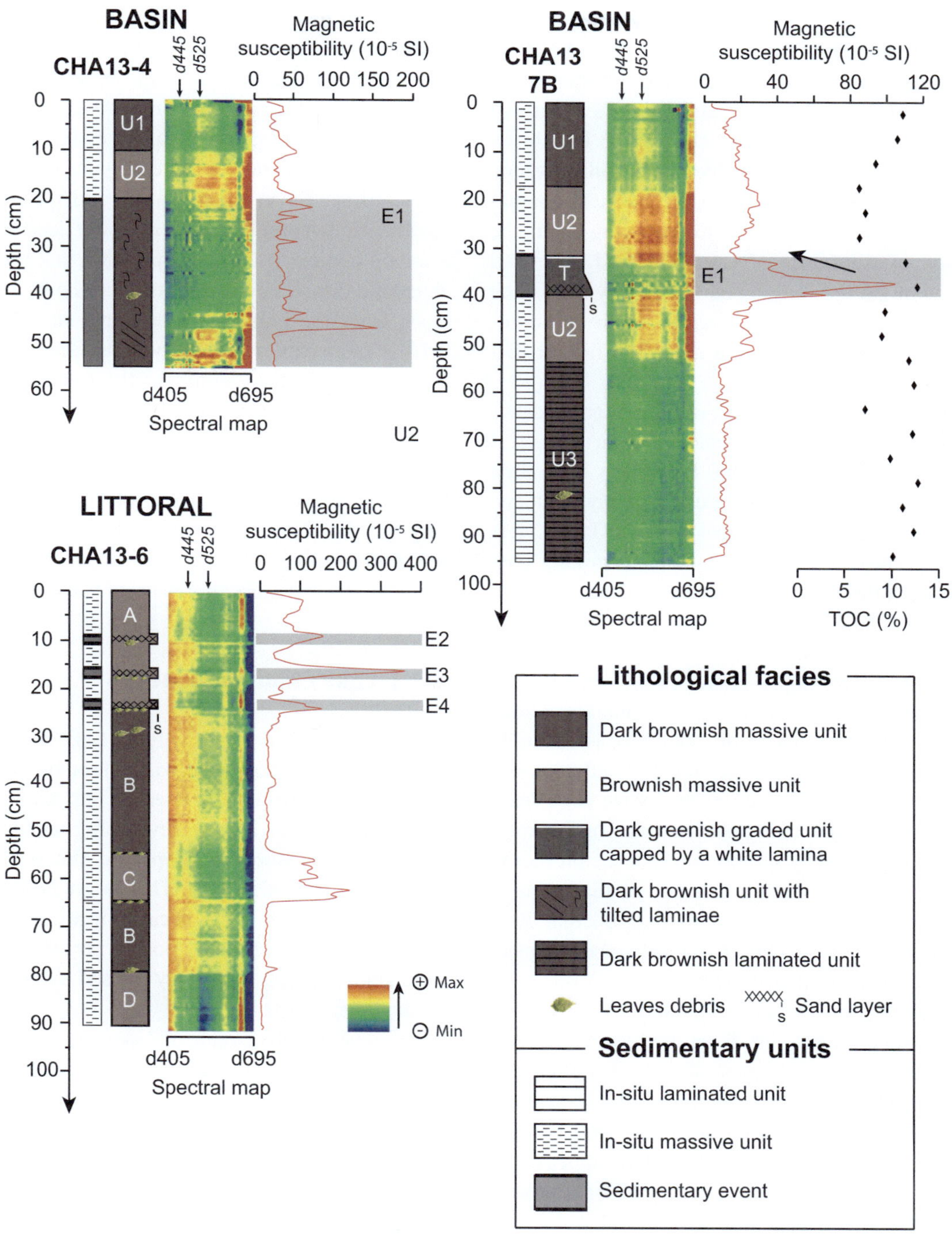

Fig. 22.9 Multidisciplinary characterization of Lake Chauvet sediments retrieved by short gravity cores in the deep central basin (CHA13-4 and CHA13-7B) and in a littoral environment (CHA13-6). Visual descriptions of sedimentary facies are further defined (i) by sediment diffuse spectral reflectance (here plotted on a 3D diagram where the X axis represent the wavelengths, Y is the depth in core and Z the first derivative value for the corresponding wavelength (in nm) expressed by a code of color); (ii) by sediment magnetic susceptibility and (iii) core CHA13-7B is in addition detailed by measurements of Total Organic Carbon (TOC). The locations of these cores in Lake Chauvet are also given in Fig. 22.6

organic matter essentially of algal origin developing a stratified sedimentary unit along the gentle slopes of Lake Pavin can be interpreted as in situ annually laminated **diatomite**.

Two **sedimentary events** (E1 and E4) are identified within the laminated **diatomite** at coring site PAV08 by light colored layers (i.e. higher values in **L* parameter**), and lower values of TOC and HI (Fig. 22.8). These characteristics suggest that E1 and E4 are resulting from the remobilization of a mixture of lacustrine and terrestrial material. They might thus correspond to gravity reworking phenomena initiated near the lake shore. Their chronologies and possible sediment source areas are further discussed in Chap. 23.

A fresh **slide scar** ca. 350 m long and 4 m high is identified at the southern edge of the subaquatic plateau (Figs. 22.4 and 22.5e) around 55 m water depth. Below this large **slide scar**, steep slopes (>30°, Fig. 22.7) are free of any sediment. This suggests that these steep slopes of Lake Pavin were unstable and recently submitted to gravity reworking phenomena that reach the deep central basin. This recent event has been dated around **AD** 1300 (Chapron et al. 2010) and is further discussed in Chap. 23.

Several smaller steep slopes breaks (>30°) identified at the lake floor on Figs. 22.4 and 22.7 at the southeastern edge of the subaquatic plateau (where the lake floor is generally characterized by slopes ranging between 15° and 25°) also suggest the development of recurent regressive **slide scars** and small scale gravity reworking phenomena. This interpretation is further supported by sub-bottom acoustic profiles in this area (Fig. 22.5c) illustrating that in situ **diatomite** is locally incised by several canyons. This suggests that these active canyons may have recently bypassed some sediment from the plateau to the deep central basin.

Finally, a bathymetric anomaly identified between ca. 12 and 26 m water depth just south form the lake outlet (Fig. 22.10) is suggesting the development of a deep but elongated depression at the lake floor. Such a geomorphological feature could highlight the occurrence of a subaquatic outlet in this part of the plateau (Chapron et al. 2010). This outlet could further explain the occurrence of a spring of water downstream from Lake Pavin into the canyon developed by its outlet at ca. 1180 m altitude a.s.l. (Jézéquel et al. 2011).

22.4.2 Lake Pavin Slopes

When slope angles in Lake Pavin are above 30° (Fig. 22.7) they are free of any sediment and characterized by the development of numerous steep **canyons** clearly visible on multibeam bathymetric data (Figs. 22.4, 22.7 and 22.10). **Sub-bottom profiles** along these steep slopes are thus only illustrating the morphology of the **acoustic substratum** (Fig. 22.5a, d and e). This acoustic facies has been sampled at the base on a long **piston core** (PAV08) and attributed to the Pavin crater formation (Chapron et al. 2010). It is therefore very probable that all these canyons draining the steep slopes of Lake Pavin crater are still active canyons and are sporadically bypassing sediment to the deep central basin. In such a context it is highly possible that sediment from subaquatic littoral environments, lake shores and sub aerial slopes from the crater ring draining into the lake (Fig. 22.10), can be exported directly to the deep central basin.

Locally very steep slopes at the eastern and southern edges of Lake Pavin are produced by **outcropping lavas** (Chapron et al. 2010). Some of these volcanic rocks are also locally **outcropping** within the inner slopes of the **crater ring** where they develop unstable cliffs (Fig. 22.10). Boulders along the shore lines and steep slopes of Lake Pavin near these **outcropping** volcanic rocks highlight the occurrence of relatively small scale but recurrent **rock falls**.

22.4.3 Lake Chauvet Slopes

Lake Chauvet is comparatively to Lake Pavin characterized by a shallower central basin, less steep slopes (Fig. 22.7) and no subaquatic plateaus, but several **moraines ridges** are however locally developing small topographic steps along the northern slopes of the basin (Juvignié 1992; Chapron et al. 2012). This **maar** lake is also quite different from Lake Pavin because it has a small but permanent tributary and a very poorly preserved **carter ring** where gentle slopes are locally incised by gullies draining into the lake.

Consequently, the slopes of Lake Chauvet are generally covered by a thin layer of sediments with a transparent acoustic facies. This facies is thus similar to the littoral facies, but generally only 1–2 m thick, except near the lake outlet, where a well-developed **moraine** ridge favored the accumulation of up to 2.8 m of sediments (Juvignié 1992; Chapron et al. 2012).

Offshore its tributary, a recent and relatively large **mass wasting deposit (MWD)** is in addition clearly identified along the southwestern slopes of the basin and down to the deep central basin (Fig. 22.7). Along the slopes this **MWD** is producing a slightly hummocky and transparent seismic facies (Fig. 22.6). It is thus very likely that this subaqueous slope failure reworked most of the delta deposits that were accumulated offshore the tributary (Fig. 22.11).

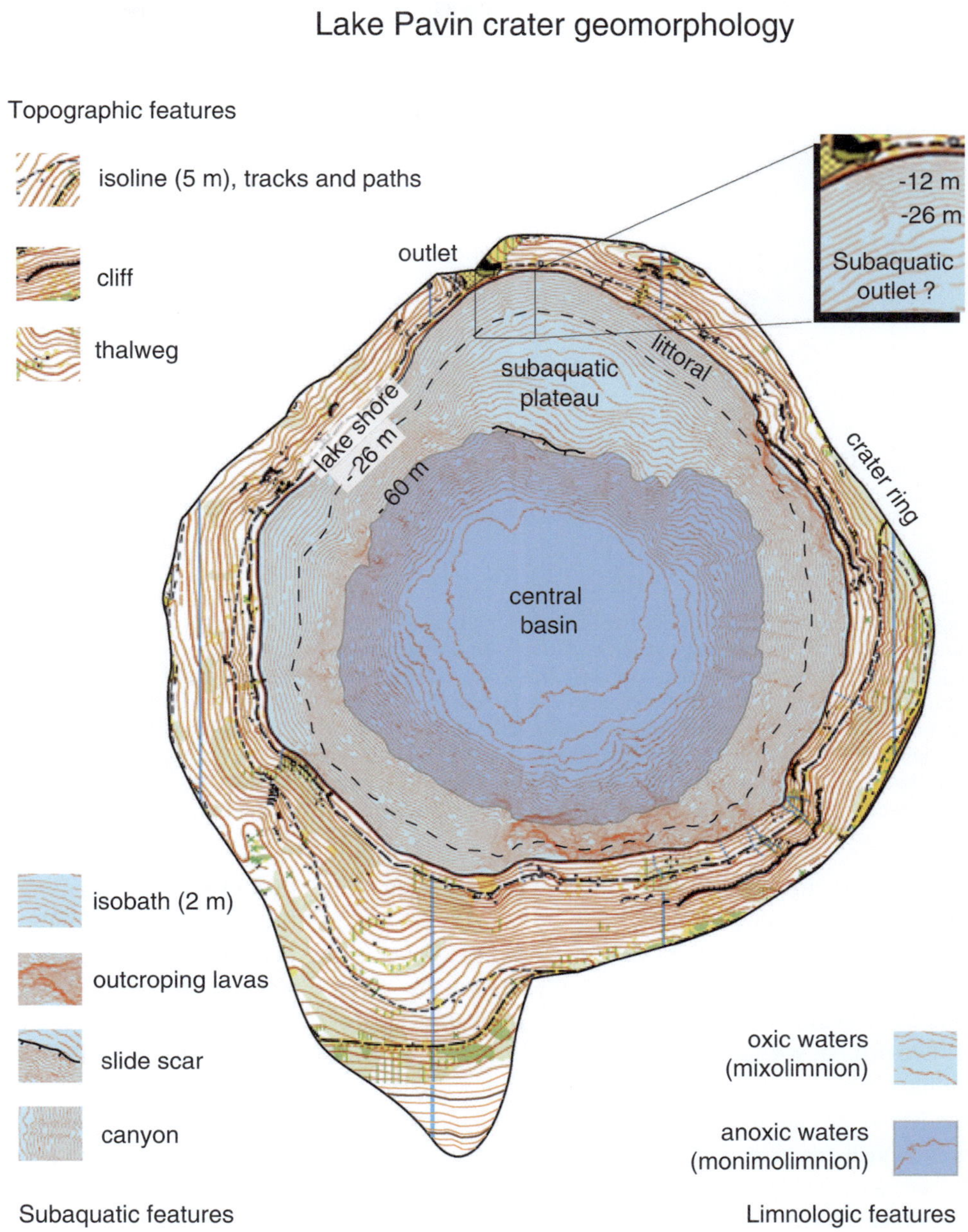

Fig. 22.10 Lake Pavin crater geomorphology from the crater ring to the deep central basin illustrating key topographic, subaquatic and limnologic features of this young and deep meromictic crater lake

22.5 Sedimentary Environments in the Deep Central Basins of Lakes Chauvet and Pavin

22.5.1 Lake Chauvet Basin

The deep central basin of Lake Chauvet is clearly imaged by 12 kHz sub-bottom acoustic profiles (Fig. 22.6). It is filled by up to 4.6 m thick well-stratified deposits were few continuous **reflections** are terminate by **onlap** against the edges of the basin and the **moraine ridges**. Locally, **moraine** ridges can also be partly draped by a thin layer of sediment developing a transparent acoustic facies.

Down slopes from the lake tributary, the **MWD** reworking slope sediments forms a lens-shaped body at the edge of the basin where it is characterized by a transparent to chaotic facies. This body is laterally thinning toward the central basin into a distinct high amplitude **reflection** that has been sampled in short core CHA13-7B (Fig. 22.6). Such a succession of acoustic facies is typical of a mass wasting phenomena evolving downslope from a mass movement into a

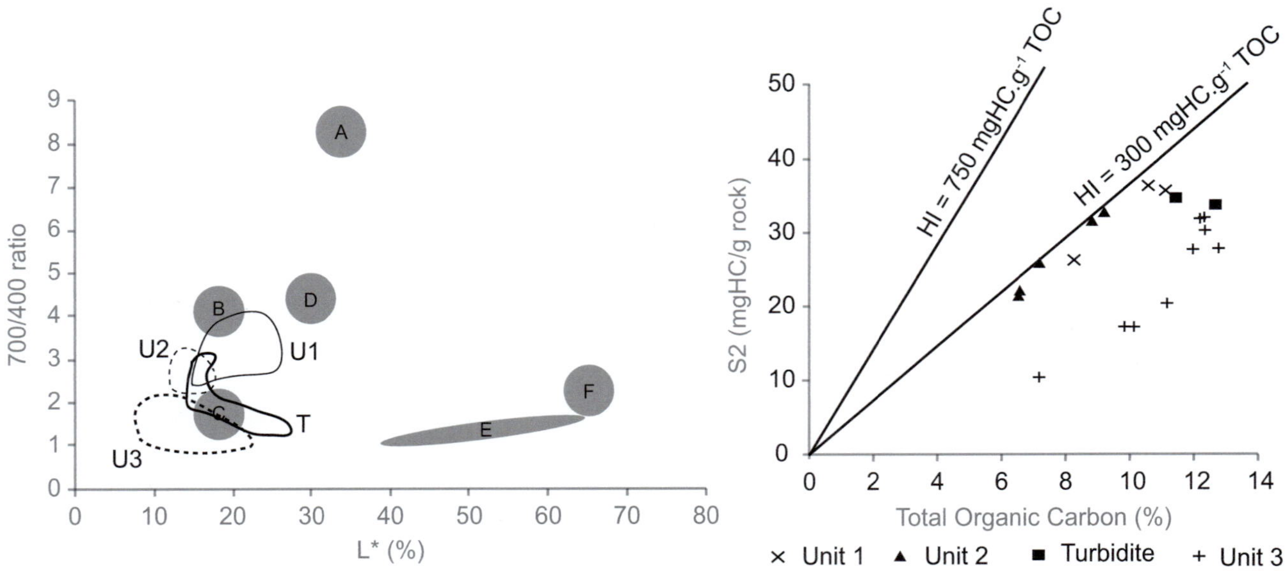

Fig. 22.11 Characterization of Chauvet sediments origin and sediment sources of sedimentary event E1 based on CHA13-7B sediment spectral diffuse reflectance Q7/4 diagram (*left*) and Rock-Eval pyrolysis results represented by S2 vs. TOC diagram (*right*) where Total Organic Carbon (TOC), together with Hydrogen Index (HI) and S2 (thermal cracking of the hydrocarbon compounds) are illustrated. The two linear domains of the hydrogen index (HI=750 and HI=300 mgHCg^{-1}TOC) corresponding to algal and terrestrial poles, respectively, are also represented. In the Q7/4 diagram, the diffuse spectral reflectance signature of Lake Chauvet sedimentary units 1, 2 and 3 (U1, U2, U3) together with the turbidite deposit associated with E1 (T) are compared to the five distinct poles of sediments defined by Debret et al .(2011): Iron-Rich deposits (**a**); Organic-rich deposits dominated by Melanoidine type (**b**); Organic-rich deposits dominated by altered organic matter (**c**); Organic-rich deposits dominated by Chlorophyll and by-products (**d**); Clayey deposits (**e**) and Carbonate deposits (**f**)

turbidity current (Mulder and Cochonat 1996; Chapron et al. 2006; St-Onge et al. 2012).

Sedimentary facies retrieved in cores CHA13-4 and CHA13-7B are in addition clearly illustrating the occurrence of a **sedimentary event** (E1) covered by a succession of two distinct **sedimentary units** (Unit 1 and 2, Fig. 22.9): While Unit 1 is a dark brownish massive unit with decreasing MS values but increasing TOC toward the top, Unit 2 is a brownish massive unit with fluctuating MS values and lower values of TOC. In core CHA13-4, E1 is on the contrary characterized by a dark brownish massive facies with locally some tilted laminas associated with strongly contrasting MS values. Following the classification of **mass wasting deposits** of Mulder and Cochonat (1996), this deposit is interpreted as a **slump** deposit. E1 is downslope evolving into a typical **turbidite** deposit in core CH13-7b where MS values and visual descriptions are clearly showing a fining upward structure above a sharp sandy base. Higher TOC content in E1 (Fig. 22.9) at this site are similar to the organic rich Unit 3 retrieved in the lower part of core CHA13-7B. But this dark brown laminated Unit 3 is in addition characterized by lower values of MS than the **turbidite** deposit.

Based on the seismic facies and on the **sedimentary facies** of cores CHA13-4 and CHA13-7B, it is thus possible to explain the formation of a lense-shaped body and the high-amplitude reflection by the deposition of a single **sedimentary event** (E1) that formed a **slump** deposit at the edge of the basin and a **turbidite** in the deep central basin. In addition it seems very likely that E1 remolded older sediments (such as Unit 3 and part of Unit 2) previously accumulated along the slopes and the basin edges of Lake Chauvet. This interpretation is further supported by the spectral signature and the organic geochemistry of Lake Chauvet sediments from core CHA13-7B (Fig. 22.10).

When plotting the spectral diffuse reflectance measurements of Lake Chauvet sediments in a Q 7/4 diagram as defined by Debret et al. (2011), it clearly appears that these maar sediments are organic-rich deposits essentially dominated by altered organic matter (Fig. 22.10). This is here further supported by Rock-Eval data indicating that Chauvet sediments are essentially within the terrestrial pole (especially Unit 3) based on the previous studies of Ariztegui et al. (2001) and Simonneau et al. (2013).

According to these two diagrams given in Fig. 22.10 it is thus possible to precise that Lake Chauvet sediments deposited in the central basin are mainly originating from the remobilization of altered terrestrial organic matter from its **catchment area** (i.e. organic rich soils material). This is particularly the case for Unit 3, but it seems that unit 1 and 2 are also containing some organic matter of algal origin. This is thus suggesting that the trophic state of Lake Chauvet may have recently changed and favored algal production. Based on these two diagrams and on the available seismic reflection profiles illustrated in Fig. 22.6, it is also possible to identify that the **turbidite** associated with E1 essentially remolded a mixture of lacustrine sediments rich in organic matter of terrestrial and algal origins typical from deltaic lacustrine environments.

22.5.2 Lake Pavin Basin

The deep central basin of Lake Pavin is poorly documented by seismic refection data because the acoustic signal is very quickly absorbed by gas-rich sediments (Chapron et al. 2010, Fig. 22.5e). Bathymetric data (Figs. 22.4 and 22.7) indicate that only the central part of the basin is very flat (2–5°), while its edges are locally affected by numerous small-scales steep slope breaks (<25°). This specific morphology suggests that the edges of Lake Pavin central basin are significantly shaped by sediment (or water) supply originating from the canyons developed along the steep slopes of the crater. Some of these canyons are in the continuity of gullies incised within the inner slopes of the **crater ring** (Fig. 22.10) and some of them are in addition linked with springs in the topographic **drainage basin** (Fig. 22.2).

Gravity cores from the deep central basin of Lake Pavin are characterized by organic rich in situ **diatomite** (Fig. 22.8) showing low values of MS but also several abrupt peaks with very high values of MS in core PAV09-B1. As further discussed in Chap. 23, the main peak in MS at 90 cm below the lake floor in the basin has been dated and correlated with **sedimentary event** E4 identified on the plateau at PAV08 coring site. This **sedimentary event** E4 was thus probably large enough to cross the plateau and reach the deep central basin. The two others outstanding peaks in MS identified on core PAV09-B1 can also be dated (see Chap. 23) but were not identified on the plateau at site PAV08. This suggest that these two MS peaks (labeled E2 and E3) might be more local **sedimentary events** supplied by some of the canyons draining Lake Pavin steep slopes and/or the edge of the plateau.

22.6 Implications for Natural Hazards in Lake Pavin

Maar lakes from the study area are filled with organic-rich lacustrine sediments and are exposed to subaqueous slope failures. It is today well-established that subaqueous slope instabilities in lakes (or in ocean realms) are either due (i) to changes (natural or human-induced) in sedimentation rates favoring sediment overloading; (ii) to changes in lake (sea) level controlling the weight of the water column (and thus loading of underlying sediments); (iii) to earthquake shaking producing an abrupt acceleration of gravity and cyclic loading when the site is impacted by seismic waves; (iv) to cyclic loading by waves; and/or (v) to gas hydrate destabilization within older sediments buried along margins of sedimentary basins (Mulder and Cochonat 1996; Van Rensbergen et al. 2002; Chapron et al. 2006; Girardclos et al. 2007; St-Onge et al. 2012; Phrampus and Hornbach 2012). All these factors may in addition combine with complex interactions in sedimentary basins to increase stresses or lower sediment strength and lead to sediment instability.

The **slump** and associate **turbidite** identified ca. 30 cm below the lake floor in Lake Chauvet are resulting from a recent subaquatic slope failure that affected its deltaic environment along relatively steep slopes (Fig. 22.7). Changes in sedimentation rates in lacustrine deltaic environments can either be due to climate changes or human impact (land use in the drainage basin for example) and could favor slope failure in Lake Chauvet. According to Juvignié (1992), Lake Chauvet has been affected by a significant and abrupt lake level drop, but during the last deglaciation, when glaciers from the Puy de Sancy were retreating outside the Chauvet **crater rim**, out of the lake's drainage area. The outlet of Lake Chauvet is today rather stable since its altitude is controlled by a **moraine** ridge (Juvignié 1992). Lake level change as a trigger for Lake Chauvet **MWD** is thus unlikely. Cyclic loading related with waves seems as well unlikely to explain this **MWD**, since Lake Chauvet is very small and not especially exposed to strong winds. Cyclic loading associated with earthquake shacking seems rather unlikely, but possible, since this volcanic area has a moderate regional seismicity (Boivin et al. 2004).

Ongoing **AMS radiocarbon** dating on core CHA13-7B should allow dating the formation of this **turbidite** in Lake Chauvet and this will be crucial to pinpoint earthquake triggering if this sedimentary event can be related to an historical earthquake and/or to a (prehistoric) period of contemporaneous **MWD** in lakes at a regional scale (Chapron et al. 2006; St-Onge et al. 2012).

Several generations of **MWDs** can be identified within Lake Pavin basin (Chapron et al. 2010, this study). Some small scale **sedimentary events** are identified either in littoral environments, on the plateau or in the basin. **Sedimentary events** affecting several sedimentary environments may however reflect an abrupt environmental change. The establishment of an **event stratigraphy** in a lake basin over longer time scales than historical chronicles or instrumental data can thus provide key elements to evaluate Natural Hazards in a given area.

Over the last millennium, a larger **slump deposit** dated on the plateau of Lake Pavin (at site PAV08) to ca. **AD** 600 (Chapron et al. 2010) can be related to **sedimentary event** E6 dated at site PAV09-C5 (Chapron et al. 2012). The occurrence of such an erosive sandy layer in shallow waters contemporaneous to the **slump deposit** on the plateau suggest that this **slump** may have been triggered by an abrupt lake level change or may have induced a destructive wave along the lake shore (Chapron et al. 2012). Similarly, as further discussed in Chap. 23, the correlation of **sedimentary event** E5 at site PAV09-C5, with the formation of the large **slide scar** at the edge of the plateau, suggest that this significant slope failure may also have been triggered by a change in lake level or may have formed and propagated a destructive wave in the lake.

As shown in Fig. 22.10, the large **slide scar** in Lake Pavin occurs at 55 m water depth just next to the upper limit of the **monimolimnion**. It is thus likely that such a slope failure remolding gas rich sediments from the steep slopes and the deep central basin, can significantly impact gas content within these permanently anoxic waters and may favor the generation of a **limnic eruption** (Chapron et al. 2010; Chap. 3 and 23, this volume).

22.7 Conclusions and Perspectives

Lakes Pavin and Chauvet are two nearby relatively similar maar lakes, since they are both filled with relatively little amount of organic rich sediments highlighting different **acoustic facies** and **sedimentary facies** within littoral environments and in their deep basins. Both lakes were also relatively recently exposed to subaqueous slope failures.

The different ages of these two nearby **maar lakes** offer the possibility to track in their sedimentary archives regional environmental changes and to better define the exposure of this volcanic area of Western Europe to natural hazards related with subaquatic slides.

Lake Pavin is in addition characterized by the absence of any tributary and the presence of a wide subaquatic plateau located above its **monimolimion**. Such a specific geomorphologic and limnologic charachteristics allows the accumulation of **diatomites** either within the **mixolimnion** and the **monimolimnion**. A challenging perspective in this young volcanic lake is thus to reconstruct is the timing of meromictic conditions in the deep basin and to document how much former subaqueous slope failures from the edges of its plateau may have impacted its limnologic condition. This might as well help to better understand the exposure of volcanic areas to **limnic eruptions**, i.e. one of the less well known natural hazards but potentially very dangerous in a touristic area such the Pavin and the Puy de Sancy area of the French Massif Central.

Future investigations in these deep and steep maar lakes should also concern numerical modeling of subaquatic slope failures in order to better understand the development of **turbidites** and the generation of waves. The identification of a subaquatic outlet in Lake Pavin should also be confirmed by in situ measurements in order to better identify its impact on Pavin limnology and geochemistry.

Acknowledgements We wish to thanks Philippe Rocher (BRGM-Auvergne, Clermont-Ferrand) for providing us the multibeam bathymetric data collected within the frame of the MEEDDAAT project on the stability of Lake Pavin. Many thanks also to François and Hélène Joubert from "Le Lac Pavin" for accommodation, logistical support and fruitful discussions on Lake Pavin. Didier Jézéquel (Université Paris-Diderot) is also greatly acknowledged for logistical support, scientific discussions and the collection of gravity core PAV09-B1 from the middle of Lake Pavin. We also thanks master students from Orléans University for their contribution to the collection of gravity cores in 2009, 2010, 2011 and 2013 in lakes Pavin and Chauvet. Investigations in Lake Chauvet were possible in 2009, 2011 and 2013 thanks to the support of Christian Froissart, Thierry Amblar and the fishing club from Lake Chauvet. Pieter Van Rensbergen kindly accepted to revise this chapter and we wish to thank him for all his constructive comments on this study.

References

Albéric P, Jézéquel D, Bergonzini L, Chapron E, Viollier E, Massault M, Michard G (2013) Carbon cycling and organic radiocarbon reservoir effect in a meromictic crater lake (lac Pavin, Puy-de-Dôme, France). Radiocarbon 55(2–3):1029–1043

Ariztegui D, Chondrogianni C, Lami A, Guilizzoni P, Lafargue E (2001) Lacustrine organic matter and the Holocene paleoenvironmental record of Lake Albano (central Italy). J Paleolimnol 26:283–292

Behar F, Beaumont V, De BPenteado HL (2001) Rock-Eval 6 technology: performances and developments. Oil Gas Sci Technol – Rev IFP 56(2):111–134

Boivin P, Besson J-C, Briot D, Camus G, de Goër de Herve A, Gougaud A, Labazuy P, de Larouzière F-D, Livet M, Mergoil J, Mialler D, Morel J-M, Vernet G, Vincent P-M (2004) Volcanologie de la Chaîne de Puys, 4ème édition, Editions du Parc Régional des Volcans d'Auvergne, Aydat, 179 pp

Chapron E, Ariztegui D, Mulsow S, Villarosa G, Pino M, Outes V, Juvignié E, Crivelli E (2006) Impact of the 1960 major subduction earthquake in Northern Patagonia (Chile, Argentina). Quat Int 158:58–71

Chapron E, Albéric P, Jézéquel D, Versteeg W, Bourdier J-L, Sitbon J (2010) Multidisciplinary characterization of sedimentary processes in a recent maar lake (Lake Pavin, French Massif Central) and implication for natural hazards. Nat Hazards Earth Syst Sci 10:1–13

Chapron E, Ledoux G, Simonneau A, Albéric P, St-Onge G, Lajeunesse P, Desmet M (2012) New evidence of Holocene mass wasting events in recent volcanic lakes from the French Massif Central (Lakes Pavin, Montcineyre and Chauvet) and implications for natural hazards. In: Yamada Y et al (eds) Submarine mass movements and their consequences, vol 31, Advances in natural and technological hazards research. Springer, Dordrecht, pp 255–264

Debret M, Chapron E, Desmet M, Rolland-Revel M, Magand O, Trentesaux A, Bout-Roumazeille V, Nomade J, Arnaud F (2010) North western Alps Holocene paleohydrology recorded by flooding activity in Lake Le Bourget, France. Quat Sci Rev 29:2185–2200

Debret M, Sebag D, Desmet M, Balsam W, Copard Y, Mourier M, Susperrigui A-S, Arnaud F, Bentaleb I, Chapron E, Lallier-Vergès E, Winiarski T (2011) Spectrocolorimetric interpretation of sedimentary dynamics: the new "Q7/4 diagram". Earth Sci Rev 109:1–19

Delbecque A (1898) Les lacs français. Chamerot & Renouard, Paris, 436 p

Gewelt M, Juvigné E (1988) Téphrochronologie du Tardiglaciaire et de l'Holocène dans le Cantal, le Cézallier et les Monts Dore (Massif Central, France): résultats nouveaux et synthèse. Bulletin de l'Association française pour l'étude du quaternaire 25:25–34

Girardclos S, Schmidt OT, Sturm M, Ariztegui D, Pugin A, Anselmetti FS (2007) The 1996 AD delta collapse andlarge turbidite in Lake Birenz. Mar Geol 241:137–154

Jézéquel D, Sarazin G, Prévot F, Viollier E, Groleau A, Michard G (2011) Bilan hydrique du lac Pavin. Revue des Sciences Naturelles d'Auvergne 74–75:67–90

Juvigné E (1992) Approche de l'âge de deux cratères volcaniques lacustres d'Auvergne (France). C.R. Acad Sci Paris 314:404–412

Mulder T, Cochonat P (1996) Classification of offshore mass movements. J Sediment Res 66:43–57

Phrampus BJ, Hornbach MJ (2012) Recent changes to the Gulf Stream causing widespread gas hydrate destabilization. Nature 490:527–530

Schettler G, Schwab MJ, Stebich M (2007) A 700-year record of climate change based on geochemical and palynological data from varved sediments (Lac Pavin, France). Chem Geol 240:11–35

Simonneau A, Chapron E, Vannière B, Wirth SB, Gilli A, Di Giovanni C, Anselmetti FS, Desmet M, Magny M (2013) Mass-movement and flood-induced deposits in Lake Ledro, southern Alps, Italy: implications for Holocene paleohydrology and natural hazards. Clim Past 9:825–840

Simonneau A, Chapron E, Garçon M, Winiarski T, Graz Y, Chauvel C, Debret M, Motelica-Heino M, Desmet M, Di Giovanni C (2014) Tracking Holocene glacial and high-altitude alpine environments fluctuations from minerogenic and organic markers in proglacial lake sediments (Lake Blanc Huez, Western Frenc Alps). Quat Sci Rev 89:27–43

St-Onge G, Chapron E, Mulsow S, Salas M, Viel M, Debret M, Foucher A, Mulder T, Winiarski T, Desmet M, Costa PJM, Ghaleb B, Jaouen A, Locat J (2012) Comparison of earthquake-triggered turbidites from the Saguenay (Eastern Canada) and Reloncavi (Chilean margin) fjords: implications for paleoseismicity and sedimentology. Sediment Geol 243–244:89–107

Van Rensbergen P, De Batist M, Klerkx J, Hus R, Poort J, Vanneste M, Gramin N, Khlystov O, Krinitsky P (2002) Sublacustrine mud volcanoes and methane seeps caused by dissociation of gas hydrates in Lake Baikal. Geology 30:631–634

Lake Pavin Paleolimnology and Event Stratigraphy

23

Léo Chassiot, Emmanuel Chapron, Yannick Miras, Markus J. Schwab, Patrick Albéric, Aude Beauger, Anne-Lise Develle, Fabien Arnaud, Patrick Lajeunesse, Renata Zocatelli, Sylvain Bernard, Anne-Catherine Lehours, and Didier Jézéquel

Abstract

In this chapter we present an up-to-date database of sedimentary sequences retrieved from Lake Pavin during the last 50 years in both oxic and anoxic waters. The detailed history of this mid Holocene crater lake can be reconstructed from the correlation of radiocarbon dated sedimentary sequences retrieved from the deep central basin, a subaquatic plateau and littoral environments. High-resolution measurements of sediment composition (diffuse spectral reflectance, XRF core scanning) combined with the analysis of organic matter composition and preliminary pollen and diatom assemblages investigations on selected sediment cores are used to reconstruct (i) the evolution since ca. 7000 cal BP of Lake Pavin limnology together with its radiocarbon reservoir effect and (ii) the impact of a wide range of subaquatic slope failure events. Such a multidisciplinary approach of Lake Pavin basin fill revealed contrasted sedimentation patterns just after the volcanic eruption and following the development of a dense vegetation cover along the slopes of the crater. Pavin sedimentation is rapidly and largely dominated by organic rich and finely laminated diatomite formation, but several short periods of enhanced mineral inputs might reflect the influence of wetter periods, such as the Little Ice Age. Over the last millennium two large subaquatic mass wasting events are also identified and may have significantly impacted its limnology.

L. Chassiot • P. Albéric • R. Zocatelli
ISTO Institut des Sciences de la Terre d'Orléans, Observatoire des Sciences de l'Univers en région Centre, UMR 7327 (CNRS-Université d'Orléans-BRGM), Campus Géosciences, 1A rue de la Férolerie, 45071 Orléans Cedex 2, France

E. Chapron (✉)
Present Address: ISTO Institut des Sciences de la Terre d'Orléans, Observatoire des Sciences de l'Univers en région Centre, UMR 7327 (CNRS-Université d'Orléans-BRGM), Campus Géosciences, 1A rue de la Férolerie, 45071 Orléans Cedex 2, France

Present Affiliation: GEODE UMR 5602 CNRS, Université de Toulouse 2 Jean Jaurès, 5 Allée A. Machado, 31058 Toulouse Cedex 9, France
e-mail: emmanuel.chapron@univ-tlse2.fr

Y. Miras • A. Beauger
GEOLAB UMR 6042 CNRS-Université Blaise Pascal-Maison des Sciences de l'Homme, 4 rue Ledru, 63057 Clermont-Ferrand Cedex 1, France

M.J. Schwab
GFZ, Helmholtz Zentrum Potsdam, Telegrafenberg, 14473 Potsdam, Germany

A.-L. Develle • F. Arnaud
EDYTEM Environnement, Dynamiques et Territoire de la Montagne, UMR 5204 CNRS Université de Savoie, Bâtiment Pôle Montagne, 73376 Le Bourget du Lac Cedex, France

P. Lajeunesse
CEN Centre d'études nordiques, Département de Géographie, Université Laval, Québec G1VOA6, Canada

S. Bernard
LMCM Laboratoire de Minéralogie et Cosmochimie du Museum, UMR 7202 CNRS-Museum National d'Histoire Naturelle (MNHN) USM 0205, Case Postale 52, 61 Rue Buffon, 75005 Paris, France

A.-C. Lehours
LMGE Laboratoire Micro organismes: Génome et Environnement, UMR 6023 CNRS-Université Blaise Pascale, Bâtiment de Biologie A, Les Cézeaux, Avenue des Landais, BP 80026, 63171 Aubière Cedex, France

D. Jézéquel
IPGP (Institut de Physique du Globe de Paris), Sorbonne Paris Cité, Univ. Paris Diderot, UMR 7154 CNRS, 1 rue Jussieu, 75005 Paris, France

T. Sime-Ngando et al. (eds.), *Lake Pavin*, DOI 10.1007/978-3-319-39961-4_23

Four smaller scale sedimentary events related with more limited subaquatic slope failures are in addition identified, dated and correlated with regional historical earthquakes. One slope failure event may be eventually associated with a "moderate" limnic eruption in AD 1783. Since the end of the eighteenth century, enhanced subaquatic slope instabilities (and thus a higher sensitivity to regional seismicity) may have resulted from the perturbation of subaqueous sediment pore pressure after the artificial lake level drop by ca. 4 m.

Keywords
Paleolimnology • Slope failures • Sedimentary event • Natural hazard • Crater lake

23.1 Introduction

While Lake Pavin limnology has been intensively studied in the past (Parts I and II, this volume), its evolution since the lake formation ca. 7000 years ago, is still poorly known. Such a reconstruction of past environmental changes in this young crater lake can be achieved through the reconstruction of its **paleolimnology** based on a multidisciplinary approach of its sedimentary archives. This chapter aims thus (i) at presenting an up to date synthesis of available long sedimentary sequences, their chronologies and sediment proxies of environmental changes retrieved in the basin of Lake Pavin and (ii) at illustrating how the integration of different aspects of earth sciences (geomorphology, sedimentology, geophysics, geochemistry and geochronology), ecology and historical archivess can provide instructive understanding of sedimentary processes and sedimentary record of environmental changes associated with climate changes, geological hazards and human activities in a lake of volcanic origin.

Maar lakes basin fills are frequently considered as key environments for paleoclimate reconstructions (Sifeddine et al. 1996; Thouveny et al. 1994; Oldfield 1996; Ariztegui et al. 2001; Brauer et al. 1999; Caballero et al. 2006; Augustinus et al. 2012; Zolitschka et al. 2013) but little is known about the triggering factors of gravity reworking phenomena and related natural hazards (Giresse et al. 1991; Truze and Kelts 1993; Bacon et al. 2002; Bani et al. 2009; Zolitschka et al. 2013). There is particularly a need to improve our understanding of subaquatic slope stabilities, since **maar** basins are frequently characterized by steep slopes, a conical morphology and no large inflows (Chap. 22, this volume). Maar bassins thus constitute peculiar environments to investigate the impact of sub-aquatic landslide(s) and the possible generation of violent waves or crater outburst(s). In a **meromictic maar** lake such as Lake Pavin (or Lake Nyos, Cameroon) where the development of a permanent anoxic deep water body (i.e. **monimolimnion**) can favor high concentrations of biogenic and mantle-derived gases (such as CO_2 and CH_4, Camus et al. 1993; Aeschbach-Hertig et al. 1999; Albéric et al. 2013), an additional natural hazard may also be related to deep water degassing (i.e. **limnic eruption**, cf. Sigurdsson et al. 1987; Schmid et al. 2005; Caracausi et al. 2009; Mott and Woods 2010).

Radiocarbon dated long sediment cores recently collected in Lake Pavin (Fig. 23.1) provide new insights on dominating sedimentary processes in a **maar** lake and environmental history of this mid latitude volcanic region of Western Europe largely exposed to the climatic influence of the Atlantic Ocean (Stebich et al. 2005; Schettler et al. 2007).

23.2 Specific Setting of Lake Pavin

The **outlet** of the lake is deeply incised into the northern walls of the crater rim (Chapron et al. 2010) and connected to the Couze Pavin, a tributary of the Allier River in the drainage basin of the Loire River (Figs. 22.1 and 23.1). The topographic drainage basin of Lake Pavin is presently densely covered by mixed deciduous/coniferous forest (Fig. 23.1) compared to nearby environments in this mid altitude region where the vegetation cover has been deeply affected by human activities (and particularly agropastoralism) since the Roman period and the Middle Age (Stebich et al. 2005; Miras et al. 2004; Lavrieux et al. 2013).

This Lake Pavin region is characterized by an oceanic-montane climate (Stebich et al. 2005; Schettler et al. 2007) with significant annual thermal amplitude (between −5 and 20 °C, mean annual temperature of 6.5 °C) and precipitations (mean annual value between 1600 and 1700 mm). Because of the morphology of its **crater rim**, Lake Pavin is protected from regional winds, and is poorly exposed to sunlight. It is thus usually frozen during winter months while its drainage basin is frequently snow covered.

Another key feature of Lake Pavin is the occurrence of underwater springs that provide oxygenated waters at around 60 m water depth within the lake (Bonhomme et al. 2011; Jézéquel et al. 2011; Albéric et al. 2013, Fig. 22.2), matching the boundary between the **monimolimnion** and the **mixolimnion** (Fig. 22.10). Not far from the subaerial outlet (occurring at 1197 m altitude), a subaquatic outlet has also been identified between 12 and 26 m water depths on multibeam bathymetric data. As summarized in Fig. 23.1 and further detailed in Chap. 22 (this volume), the isobath −26 m also matches the boundary between a littoral sedimentary environment contrasting with the accumulation of in situ **diatomite** on the gentle slopes of a subaquatic plateau

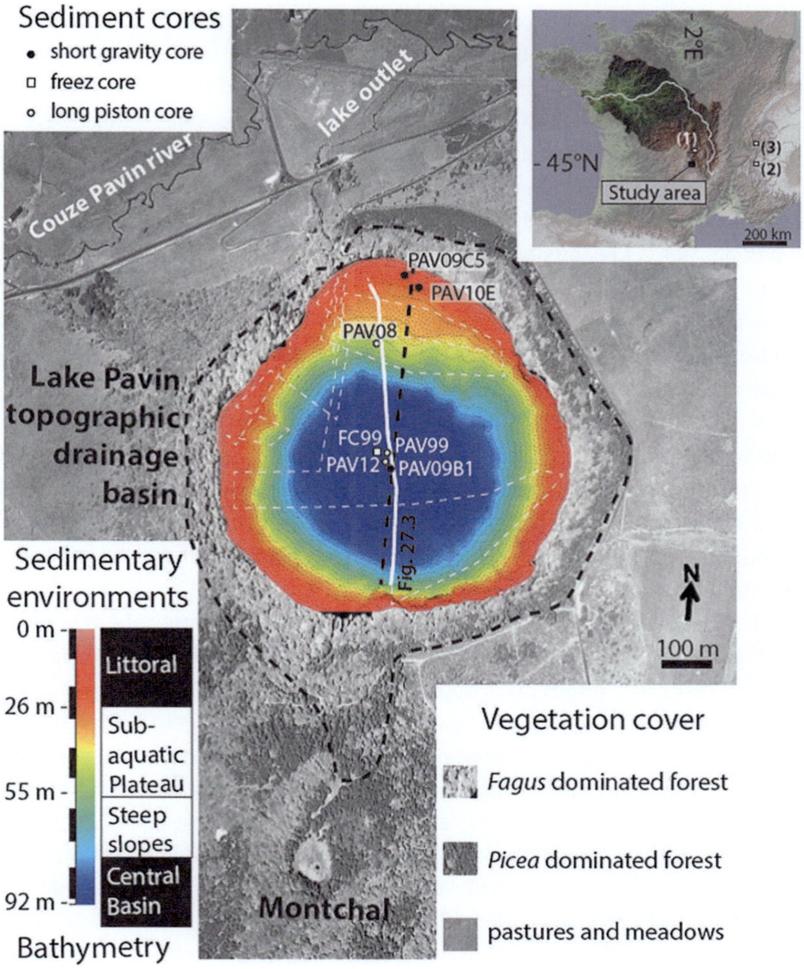

Fig. 23.1 General location of Lake Pavin in the French Massif Central (*upper right panel*) and detailed location of Lake Pavin key sediment cores discussed in the text (*central panel*). Present day vegetation cover of Lake Pavin drainage basin is also indicated. The multibeam bathymetric map and the grid of 12 kHz seismic reflection profiles (*white dotted lines*) and the 3.5 kHz seismic reflection profile (*black dotted lines*) used together with gravity, piston and freeze cores to identify four main sedimentary environments from the littoral to the deep central basin are also indicated an further presented in Chap. 22 (this volume). The location of the 12 kHz seismic profile illustrated in Fig. 23.3 is also given (*thick white line*). The general location of Lake Aydat (1), Lake Blanc Huez (2) and Lake Le Bourget (3) discussed in the text is also given

extending down to −55 m water depth in the northern part of Lake Pavin. The southern edge of this subaquatic plateau is in addition characterized by a fresh **slide scar** clearly identified on bathymetric and seismic reflection data (Figs. 22.4, 22.5 and 22.10). Finally, a 92 m deep flat central basin characterized by gas-rich sediments dominated by **diatomites** is draining steep slopes and numerous active canyons.

Such a specific hydrological and geomorphological context of Lake Pavin has to be taken into consideration when reconstructing the evolution of its limnology and surrounding environments over several millennia.

23.3 Lake Pavin Sedimentary Sequences

23.3.1 Sediment Cores and Sedimentation Rates

First short cores (<1 m long) collected in Lake Pavin were realized at the end of the ‘60’ by divers in littoral environments and these sediment cores were sampled at site. Based on bulk sediment radiocarbon dates, Delibrias et al. (1972) discussed the age of Lake Pavin and suggested very low sedimentation rates. In the early ‘90’, Martin et al. (1992) published, however, much higher recent accumulation rates on the plateau by 48 m water depths (between 1 and 4 mm/year) and in the deep central basin (between 0.8 and 7 mm/year) of Lake Pavin based on short gravity cores sampled at site and coupling radionuclide dating (^{210}Pb, ^{226}Ra) with cosmogenic isotope measurements (^{32}Si) on sediments.

High sedimentation rates (ranging between 1 and 3.4 mm/year over the last 700 years) were confirmed latter by Stebich et al. (2005) and Schettler et al. (2007) in the deep central basin of Lake Pavin based on sediment annual laminations (i.e. **varves**) counting with a microscope from sediment microscopic **thin-section** sub sampled on a 182 cm long gravity core (PAV 1–3) retrieved in 1999 and on a 198 cm long freeze-core (FC1) retrieved in 2001 (Fig. 23.1). Freeze-core technology (Kulbe and Niederreiter 2003) was here particularly adapted to retrieve a well-preserved sedimentary sequence in such fine-grained and gas rich lacustrine deposits accumulated in the deep and cold waters. These conditions in Lake Pavin are indeed favoring gas expansion when a sediment core is taken out of the lake.

The first long piston core PAV99 (Figs. 23.1 and 23.2) was collected in 1999 by the GFZ Potsdam in the deep central basin of Lake Pavin from an UWITEC coring platform using a 3 m long UWITEC piston corer. An 11 m long synthetic core lithology has then been established based on the correlation of overlapping split core sections retrieved at nearby locations. The study of this first long core from Lake Pavin has not yet been published, because (i) of the bad quality of sediment piston cores (due to sediment expansion by degasing) precluding the generation of good quality **thin sections** and **varve** counting, (ii) a complex succession of contrasted lithologies and (iii) a limited number of organic macro-remains found on split core sections suitable for **AMS radiocarbon** dating technics (Table 23.1).

The apparent contradiction between low sedimentation rates estimated by Delibrias et al. (1972) and higher ones identified by Martin et al. (1992) also supported by **varve** counting over several centuries (Stebich et al. 2005; Schettler et al. 2007), has only recently been explained by Albéric et al. (2013). This recent study identified an organic **radiocarbon reservoir effect** (ca. 2500 yrs) in Lake Pavin linked to its meromicticity comparing **AMS radiocarbon ages** from different organic carbon pools in the lake waters, together with **AMS radiocarbon ages** from bulk samples of organic rich lacustrine sediments and organic macro remains (leaves) from PAV08 piston core (Table 23.1). Based on this study, a **radiocarbon reservoir effect** is clearly identified over the last 1300 years at least, but is only suspected and modeled earlier because no organic macro remains were found at the base of PAV08 and could be compared to radiocarbon ages of bulk sediment samples.

PAV08 coring site is including short core PAV08-P1 presented in Chap. 22 (this volume) and is up to ca. 5 m long (Fig. 23.2). It was collected in 2008 by EDYTEM and ISTO laboratories on the subaquatic plateau by 46 m water depth (Figs. 23.1 and 23.3) from an UWITEC coring platform using either a 3 m or a 2 m long UWITEC piston corer as detailed in Chapron et al. (2010). This coring site (45°29.86'N/2°53.24'E) was selected within the **mixolimnion** of Lake Pavin based on multibeam bathymetric and seismic reflection data (Fig. 23.3 and Chap. 22, this volume).

Short gravity cores PAV10-E, PAV09-C5, and PAV09-B1 were in addition retrieved by 17.5 m, 20 m and 92 m water depth, respectively (Chap. 22, this volume). As shown in Fig. 23.2 and Table 23.1 two leave debris were dated by **AMS radiocarbon** from core PAV09-C5 (Chapron et al. 2012), and also both leave debris and bulk sediment samples from three different horizons in core PAV09-B1 (Albéric et al. 2013). These radiocarbon ages allowed dating two major **sedimentary events** in core PAV09-C5 retrieved in a littoral environment (see Chap. 22) and to document **radiocarbon reservoir effect** within the **monimolimnion** in core PAV09-B1.

During summer 2012, a 14 m long piston core (PAV12, Figs. 23.1 and 23.2) was finally retrieved by EDYTEM and LMGE laboratories from the center of the lake at the same location than core PAV09-B1 (45°29.74'N/2°53.28'E). Up to eleven samples of leaves debris and two samples of bulk sediment were recently dated by **AMS radiocarbon** (Table 23.1) in order to establish an age-depth model and to conduct a multi proxy study of contrasted sedimentary facies and successive **sedimentary events** as detailed below.

23.3.2 Sedimentary Records Within the Mixolimnion

As shown in Fig. 23.1 and further described in Chap. 22 (this issue), two main sedimentary environments are identified below the lake floor on **seismic reflection profiles** and sediment cores in Lake Pavin within its **mixolimnion**:

(i) a littoral environment (extending from the shore lines to the 26 m isobath) characterized by a transparent **acoustic facies** and a massive brownish **sedimentary facies** with some sandy layers and frequent leave debris as illustrated in cores PAV09-C5 and in the upper part of PAV10-E in Fig. 23.2 and,
(ii) *in situ* **diatomite** deposits between ca. 26 m and 55 m water depth on the subaquatic plateau developed in the northern part of the lake (Fig. 23.1). These lacustrine sediments are characterized by a faintly stratified **acoustic facies** with few low amplitude continuous **reflections** (Fig. 23.3) and a finely laminated **sedimentary facies** developing brownish and greenish laminas rich in diatoms as illustrated in PAV08 piston core (Chapron et al. 2010) and in the lower part of PAV10-E short core (Fig. 22.8).

The signature of these littoral sediments (Fig. 22.8) and diatomite sediments (Fig. 23.4), measured by **spectral diffuse reflectance** (SDR), **magnetic susceptibility** (MS) and Rock-Eval (RE) **pyrolysis** allows further characterizing the different **sedimentary units** documented by Chapron et al. (2010) in core PAV08. While SDR and MS measurements are considered as good indicators of sediment composition (Debret et al. 2010), RE **pyrolisis** is documenting organic matter geochemistry by the quantification of total organic carbon (TOC), **hydrogen index** (HI) and **oxygen index** (OI) as detailed in Behar et al. (2001) and used to precise the origin (terrestrial or lacustrine) of sedimentary organic matter in lacustrine environments as illustrated by Simonneau et al. (2013a, 2014) and Schettler and Albéric (2008).

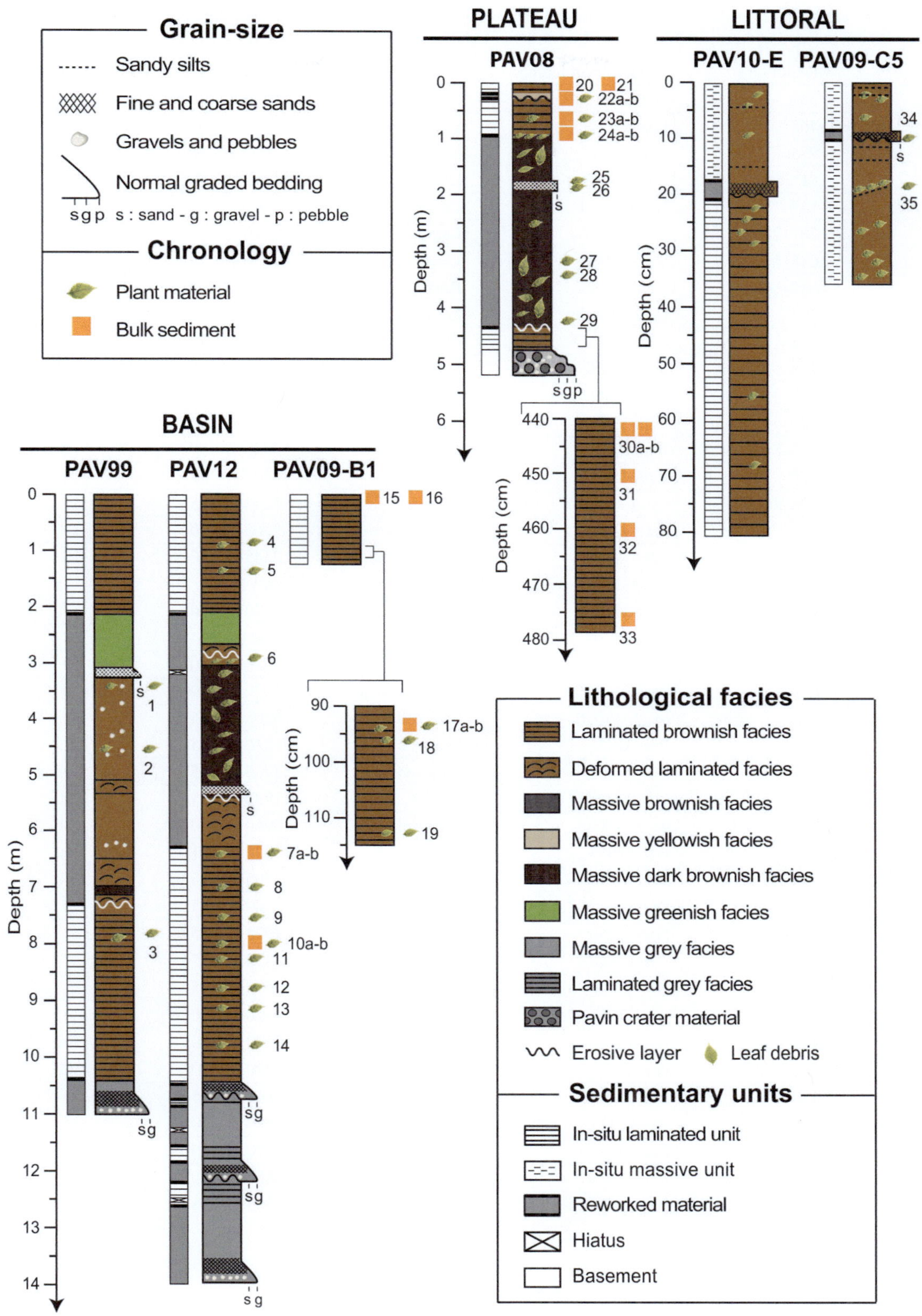

Fig. 23.2 Synthetic illustration of lithological logs of key sediment cores from Lake Pavin retrieved in the deep central basin (PAV99, PAV12, PAV09-B1), the plateau (PAV08) and littoral environments (PAV10-E and PAV09C5). The location and code of radiocarbon samples (bulk sediment or plant material) detailed in Table 23.1 are also indicated

Table 23.1 Radiocarbon dates obtained from terrestrial organic remains (leaves) or bulk organic sediment samples retrieved in sediment cores from the deep central basin, the plateau and littoral environments of Lake Pavin

Core	Depth (cm)	Laboratory reference	Material	Radiocarbon age (BP)	Reference number
PAV99	335	Poz-656	Leaves	1490 +/– 30	1
PAV99	465	Poz-657	Leaves	1700 +/– 35	2
PAV99	795	Poz-655	Leaves	2180 +/– 30	3
PAV12	81	Beta-336274	Leaves	1190 +/– 30	4
PAV12	137	Lyon-10961	Leaves	220 +/– 30	5
PAV12	287–289	Beta-336272	Leaves	2210 +/– 30	6
PAV12	645–646	SacA34984	Leaves	1730 +/– 30	7a
PAV12	645–646	SacA34985	Bulk sediment	5513 +/– 30	7b
PAV12	701	Lyon-10963	Leaves	2195 +/– 35	8
PAV12	755	Beta-336273	Leaves	2400 +/– 30	9
PAV12	798	SacA34983	Leaves	4170 +/– 30	10a
PAV12	798	SacA34997	Bulk sediment	5535 +/– 30	10b
PAV12	827	Beta-335372	Leaves	3400 +/– 30	11
PAV12	880–881	Beta-335371	Leaves	3940 +/– 30	12
PAV12	919	Beta-335370	Leaves	4400 +/–40	13
PAV12	978.5	Lyon-10962	Leaves	5250 +/– 35	14
PAV09-B1	0.5–1	SacA-28952	Bulk sediment	2965 +/– 30	15
PAV09-B1	1–1.5	SacA-28953	Bulk sediment	2445 +/– 30	16
PAV09-B1	94.5	SacA-19661	Leaves	440 +/– 35	17a
PAV09-B1	94.5	SacA-19657	Bulk sediment	6170 +/– 50	17b
PAV09-B1	96.5	Poz-33126	Leaves	150 +/– 30	18
PAV09-B1	113	Poz-33125	Leaves	1010 +/– 30	19
PAV08	0.2–0.7	SacA-19655	Bulk sediment	2070 +/– 40	20
PAV08	3–4	SacA-19656	Bulk sediment	5205 +/– 45	21
PAV08	23	Poz-31851	Leaves	1210 +/– 65	22a
PAV08	22.5–23.5	Poz-31852	Bulk sediment	3765 +/– 35	22b
PAV08	71	Poz-27046	Leaves	1290 +/– 35	23a
PAV08	70.5–71.5	SacA-19658	Bulk sediment	5960 +/– 50	23b
PAV08	97	Poz-27047	Leaves	1430 +/– 21	24a
PAV08	96.5–97.5	SacA-19659	Bulk sediment	6795 +/– 45	24b
PAV08	178.5	Poz-31849	Leaves	3180 +/– 35	25
PAV08	181	Poz-31850	Leaves	1975 +/– 35	26
PAV08	320	Poz-31848	Leaves	4350 +/– 35	27
PAV08	344	Poz-27050	Leaves	4995 +/– 35	28
PAV08	421	Poz-27051	Leaves	4820 +/– 40	29
PAV08	441–442	Poz-45411	Bulk sediment	7620 +/– 50	30a
PAV08	441–442	Poz-48070	Bulk sediment	8370 +/– 50	30b
PAV08	450–451	Poz-45413	Bulk sediment	5865 +/– 35	31
PAV08	460–461	Poz-45414	Bulk sediment	6580 +/– 40	32
PAV08	476–479	Poz-27052	Bulk sediment	6090 +/– 40	33
PAV09-C5	9	ULA-2376	Leaves	855 +/– 15	34
PAV09-C5	18	Poz-33125	Leaves	1355 +/– 35	35

Reference numbers of each date are also given in Fig. 23.2

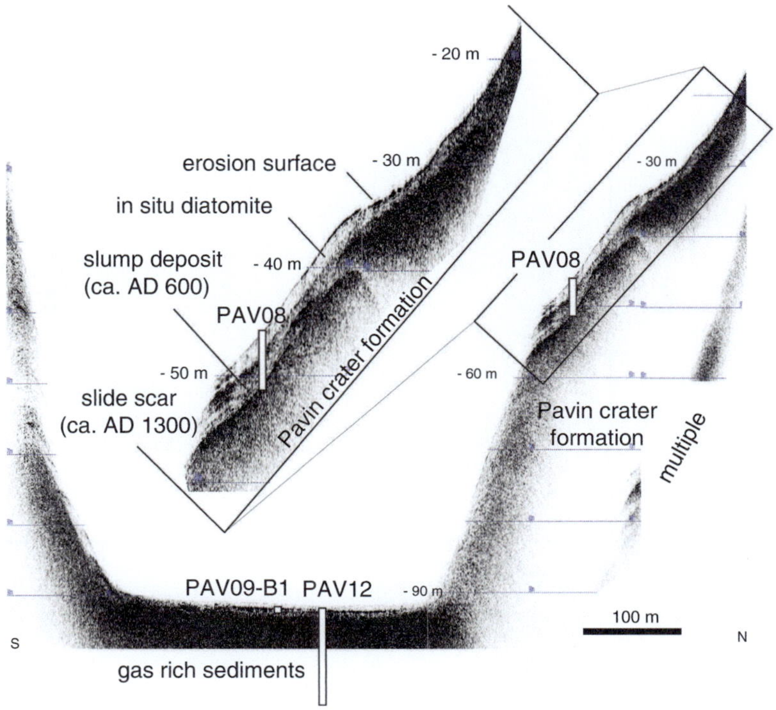

Fig. 23.3 Lake Pavin high-resolution seismic reflection profiles illustrating the development of contrasted acoustic facies above the subaquatic plateau formed by the Pavin crater formation. In the central basin, gas content in the sediment is preventing any penetration of the acoustic signal. In this part of the lake, the basin fill is thus only documented by sediment cores (PAV09-B1 and PAV12). The location of shallower core PAV08 is also indicated and allow a detail calibration of several acoustic facies associated with in situ deposits (diatomites) and reworked sediments (ca.AD 600 slump deposit and ca. AD1300 slide scar). The location of each seismic section is given in Fig. 23.1

Based on these new measurements, littoral sediments and **diatomites** are highlighting similar SDR spectrums, but littoral sediments are generally characterized by higher MS values than the **diatomites**. SDR spectrums from Lake Pavin sediments are in particular highlighting typical first derivative wavelength values of iron oxides such as Goethite characterized by higher values at 445 and 525 nm (Debret et al. 2011; Simonneau et al. 2014). TOC values between 5 and 8 % in littoral sediments are in addition generally slightly lower than the ones from the **diatomites** oscillating between 8 and 10 %.

In a former study, Chapron et al. (2010) identified two distinct **diatomite** units in core PAV08 based on their **sedimentary facies**. This distinction is further confirmed here by sediment measurements on core PAV08 clearly showing the occurrence of (Fig. 23.4):

- an upper **diatomite** unit (from the lake floor interface to 97 cm core depth) characterized by increasing TOC but decreasing HI values, and higher values of L* and b* (i.e. CIELab values of sediment reflectance, with L* parameter measuring sediment brightness, and b* parameter measuring sediment color variations from blue to yellow, cf. Debret et al. 2011);
- a lower **diatomite** unit (from 439 cm to 476 cm core depth) characterized by lower TOC values (around 8 %), higher HI (around 400 mg HC/g TOC) and darker sediments (lower L* and b* values).

Based on these high resolution measurements of sediment color and organic matter composition in PAV08, it is in addition possible to distinguish the intercalation of several contrasted layers within these **diatomites** accumulated on the plateau (Table 23.2). **Sedimentary event** 1 (E1 in Fig. 22.8) is a 2 cm thick light colored (higher L* values) layer identified at 5 cm core depth associated with an abrupt drop in TOC and HI values. **Sedimentary event** 4 (E4 in Figs. 22.8 and 23.4) is a thicker (up to 6.5 cm thick) and very similar layer (with lower TOC and HI values) easily visible on digital pictures between 17 cm and 23.5 cm core depth (Fig. 22.8). **Sedimentary event** 6 (E6 in Fig. 23.4) is a 340 cm thick dark brown layer (lower L* and b* values) rich in leave debris identified between 97 cm and 439 cm core depth. Unlike E1 and E4, E6 was previously documented in PAV08 and on seismic profiles (Chapron et al. 2010; 2012). This thick sedimentary event is developing a transparent to chaotic **acoustic facies** (Fig. 23.3) and is identified above most of the plateau (Fig. 22.5). The remarkable variability of L*, b*, TOC and HI values within E6, together with fluctuating sediment density, available **AMS radiocarbon** dates (Table 23.1) and its acoustic facies are indicating that this layer is a **slump deposit** following the classification of Mulder and Cochonat (1996). This subaquatic **slump deposit** is remolding a mixture of lacustrine and terrestrial material and is capped by a remarkable accumulation of numerous leaves and leave debris that were dated to 1290 +/− 35 ^{14}C Before Present (BP) by **AMS radiocarbon** (Table 23.1).

Two **sedimentary events** (E 5 and E6) are also identified within a littoral environment in core PAV09-C5 (Fig. 22.8): E

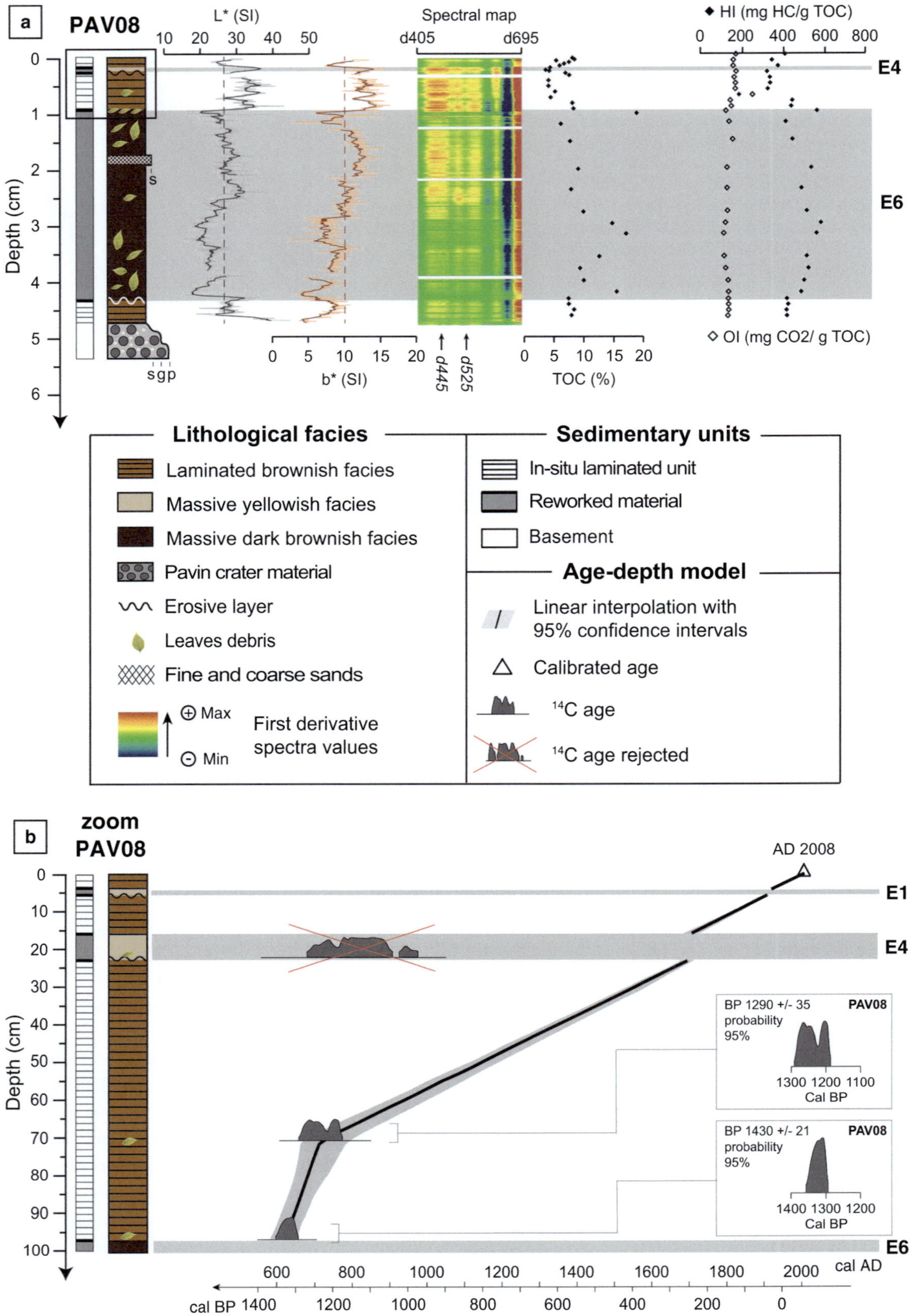

Fig. 23.4 Multi-proxies analyses on core PAV08 (**a**) and associated age-depth model of the upper unit (**b**) with calibration results with IntCal13 (Reimer et al. 2013). For continuous measurements, *thick* and *dashed lines* represent respectively a moving average and the mean. Sedimentary events E1, E4 and E6 discussed in the text are highlighted by *grey bars*

Table 23.2 Ages and thicknesses of sedimentary events identified in Lake Pavin sediment cores and possible associated historical earthquakes in the study area reported by Sisfrance data base from the French Geological Survey (BRGM)

Lake Pavin				Regional Seismicity				
Event	Core	Thickness	Age Cal AD	Historical earthquake	MSK scale intensity	Distance from Pavin	Type of event recorded in lakes	Likelihood a seismic triggering
E1	PAV08-P1	2 cm	1915 +/– 5	La Bourboule 1921	4.5	15 km		High
E2	PAV09-B1	2 cm	1880 +/– 70	Mont-Dore 1863	5	11 km	Slump in Lake Guéry (1)	Very high
				Issoire 1892	7	30 km		
E3	PAV09-B1	2 cm	1840 +/– 80	Chambon-sur-lac 1844	5.5	9 km		High
				Blesle 1833	7	30 km		
				Issoire 1833	6	30 km		
E4	PAV09-B1	1 cm	1775 +/– 90	Lepaud 1783	??	??	Limnic eruption in Lake Pavin ?	High
	PAV08	6.5 cm	1700 +/– 15	Mainsat 1783	??	??		
E5	PAV12	420 cm	1300 +/– 50	Uzerche 1348	??	??	Slumps in Lakes Guéry and Montcineyre (1,2)	High
	PAV99	510 cm	??					
	PAV09-C5	1 cm	1190 +/– 30	Limousin plateau 1233	??	??		
E6	PAV08	340 cm	620 +/– 20	No information				Moderate
	PAV09-C5	1 cm	660 +/– 50					

Also indicated is the type of contemporaneous mass wasting deposits documented in nearby lakes and further detailed (1) in Chassiot et al (in prep.) and (2) in Chapron et al (2012), as discussed in the text

5 is a 1 cm thick erosive sand layer rich in leave debris at 9 cm core depth, while E6 is a 1 cm thick layer at 18 cm core depth, very rich in leaves and also containing few pebbles and some fine sands (Chapron et al. 2012). As shown in Table 23.1, leave debris from E5 and E6 were dated by **AMS radiocarbon** to 855 +/– 15 ^{14}C BP and 1355 +/– 35 ^{14}C BP, respectively.

Using the IntCal13 radiocarbon calibration curve from Reimer et al. (2013), **AMS radiocarbon** ages ^{14}C BP can be accurately calibrated and converted in year cal BP or cal AD (Anno Domini, i.e. calendar year). This exercise is giving similar calibrated ages for E6 in core PAV08 (cal AD 717 +/– 39) and in core PAV09-C5 (cal AD 663 +/– 18). Based on their similar age clustering around AD 680, striking sediment layers identified within Lake Pavin littoral environment (at 18 cm in core PAV09-C5) and between **diatomite** deposits (from 97 cm to 439 cm in core PAV08) can be correlated and linked to a single **sedimentary event** (E6). This event E6 has different sedimentary signatures in shallower and deeper parts of the plateau, suggesting that such subaquatic sediment slumping along the plateau was probably sufficiently large and fast to generate erosive waves along the shore lines of Lake Pavin and to develop at site PAV09-C5 a coarse grained layer rich in terrestrial organic macro remains by 20 m water depth (Chapron et al. 2012).

No organic macro remains were found on top of E1 and E4 in core PAV08, but it is possible to estimate their age, based on the establishment of an age-depth model using calibrated **AMS radiocarbon** date following Reimer et al. (2013) and the depth of radiocarbon samples. Considering that these **sedimentary events** were rapidly deposited, in such age-depth model, the core depths should be, however, corrected from the thicknesses of each **sedimentary events**. Figure 23.4 B illustrates such PAV08 age-depth model based on the linear interpolation of two available dates in the upper diatomite unit using the CLAM software (Blaauw 2010). One radiocarbon age being too old was found within E4. Because this sample was likely reworked from the catchment area it can't be used to construct the age-depth model. Figure 23.4, illustrate thus the new age-depth model of the upper **diatomite** unit in PAV08 and allows dating E1 to cal AD 1915 +/–5 and E4 to cal AD 1700 +/–15 (Tables 23.1 and 23.2).

At the base of PAV08, between 476 cm and 509 cm core depth, a light grey to brownish gravels and pebbles in a coarse sand and silty matrix were sampled and analyzed (Chapron et al. 2010). Within this coarse-grained sedimentary facies, ca. 40 % in clast number consists of crystalline rocks, ca. 30 % of various trachyandesitic **lavas**, while the rest is made of basaltic looking **lavas** and of variously vesicular yellowish **pumices**. This **sedimentary facies** has been related to the Pavin crater material developing its **crater rim** and to the **acoustic substratum** on seismic reflection profiles. A sample of bulk organic rich **diatomite** sediments at 476–479 cm core depth just above the Pavin crater material has been dated by **AMS radiocarbon** and gave an age of

6090 +/−40 ^{14}C BP corresponding to 6971 +/−61 cal BP (Table 23.1). Because this age of lacustrine sediment is very close to the age of Pavin **tephra** layers found in regional outcrops and peat deposits (Chap. 6, this volume), it suggest that the onset of lacustrine sedimentation in Pavin started very quickly after the crater formation at site PAV08, without any significant **radiocarbon reservoir effect** (Albéric et al. 2013).

23.3.3 Sedimentary Records Within the Monimolimnion

All the sediment cores within the **monimolimnion** in Lake Pavin were collected in the deep central basin (Fig. 23.1) and are dominated by finely laminated **diatomites** (Fig. 23.2). Within both long piston cores PAV99 and PAV12, two distinct **diatomite** units can be identified within the two first meters below the lake floor (upper **diatomite** unit), and above ca. 1050 cm core depth were a coarse grained greyish unit characterizes the base of these piston cores (Fig. 23.2). This lower **diatomite** unit is ca. 3 m thick in PAV99 (between 730 and 1045 cm core depth), but up to ca. 4 m thick in PAV12 (between 626 and 1046 cm core depth).

The **sedimentary facies** and SDR measurements within these **diatomites** from the deep central basin are roughly similar to the ones from the plateau (Figs. 22.8, 23.4 and 23.5). RE analyses are however more variable, and generally showing lower TOC (up to 6 %) and IH (down to 200 mg HC/g TOC) values within the upper meters, while the lower unit has higher TOC (up to 18 %) and IH (oscillating around 500 mg HC/g TOC) values. Figure 23.6 is illustrating the stratigraphic correlation of available sediment cores from the central basin of Lake Pavin and the age-depth model established for PAV12 using the CLAM software (Blaauw 2010) based (i) on calibrated **AMS radiocarbon** dates on terrestrial organic macro remains found in PAV99, PAV09-B1 and PAV12 (see Table 23.1 and applying the calibration curve from Reimer et al. 2013) and (ii) on varve counting in the upper **diatomite** unit by Schettler et al. (2007) preformed on freeze core FC1. This figure shows that the upper **diatomite** unit contains several reworked organic material witch were not used for the age-depth model, while only one radiocarbon date has been removed in the lower **diatomite** unit to build up its chronology.

RE analysis (TOC and HI, Fig. 23.5) from PAV12 sediments are very different from the values obtained in core CHA13-7B retrieved in nearby Lake Chauvet central basin (Fig. 22.11). Based on the studies of Ariztegui et al. (2001), Behar et al. (2001) and Simonneau et al. (2013a), the organic matter in PAV12 samples are effectively essentially made of algae, and only few macro remains of terrestrial organic matter (leaves and leaves debris) seems therefore to characterize the organic fraction of Lake Pavin **diatomites** in the deep central basin. This is further supported by Scanning Electron Microscopy (SEM) images of selected samples from PAV12 (Fig. 23.7) illustrating the predominance of diatoms frustules and cysts more or less well-preserved within Lake Pavin sediments.

In order to further document PAV12 sediment geochemistry and provenance, X-ray fluorescence (XRF) measurements were performed on an Avaatech core scanner at EDYTEM laboratory every 5 millimeters with a Rhodium tube source. The settings were adjusted to 10 kV and 0.75 mA with an acquisition time of 20 seconds in order to measure Al to Fe relative intensities. Titanium (Ti) is for example an element relatively easy to measure and frequently used to track the evolution of clastic sediment supply in lacustrine basins (cf. Arnaud et al. 2012). Figures 23.5 and 23.8 generally illustrate low Ti content in PAV12 laminated **diatomites** and thus low clastic supply from Pavin catchment area, but several Ti peaks are however punctually identified, although maximum values in Ti are found in the light grey basal facies further detailed below.

Locally, several **sedimentary events** characterized by very high values of MS are also contrasting in core PAV09-B1 with the low MS values associated with **diatomites** in the upper unit (see E2, E3 and E4 in Fig. 22.8). These **sedimentary events** are, however, probably very punctual because they were not documented within the upper meters of other cores PAV99 (Stebich et al. 2005), FC1 (Schettler et al. 2007) and PAV12 (this study).

A major **sedimentary event** (E5) is on the contrary clearly identified in between the upper and lower **diatomite** units in both piston cores PAV99 and PAV12 (Figs. 23.2 and 23.5). In both cores this **sedimentary event** is characterized by a different thickness (up to ca. 510 cm thick in PAV99, but ca. 420 cm thick in PAV12) and a complex succession of contrasted lithological units.

In PAV99, the base of E5 consist in highly deformed **diatomites** were laminations are still visible (between ca. 730 and 650 cm core depth and between ca. 535 and 510 core depth). This deformed laminated facies is locally interrupted (i) by a massive dark brownish facies (ca. 18 cm thick) showing a sharp base around 715 cm core depth and (ii) by a massive brownish facies (between ca. 650 and 535 cm core depth) were laminations are generally destroyed or mixed and locally associated with some gravels and pebbles. Between 510 and 325 cm core depth, a similar massive brownish facies with destroyed laminations and some gravels and pebbles is again identified together with few organic macro remains (leave debris) suitable for **AMS radiocarbon** dating (Table 23.1). Above 325 cm core depth, a sharp based sandy layer ca. 4 cm thick, quickly evolves into a massive greenish facies ending at around 220 cm core depth, just below the upper diatomite unit (Fig. 23.2).

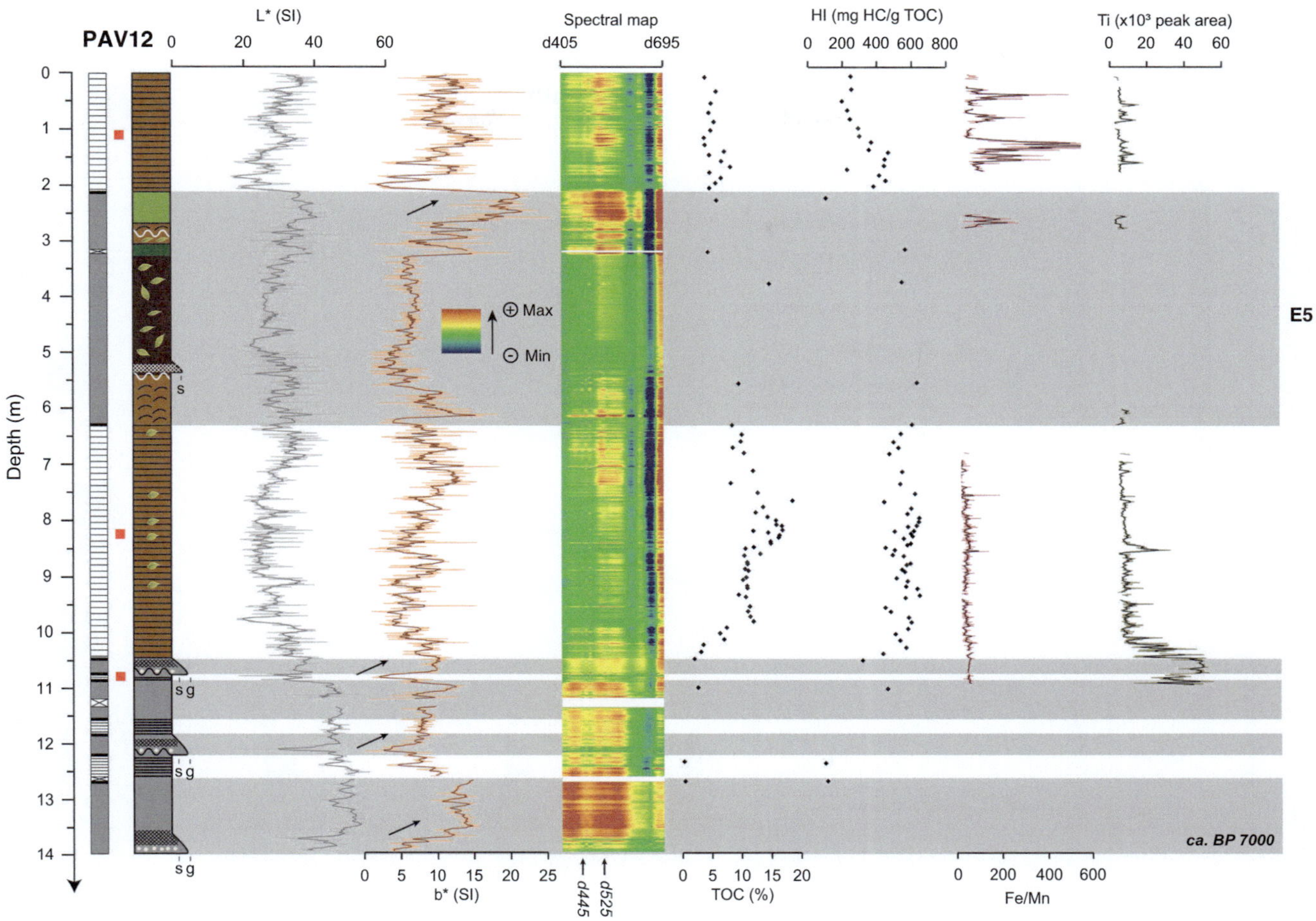

Fig. 23.5 Detailed multi-proxies analyses performed on core PAV12. *Grey squares* represent reworked material. L* and b* are parameters derived from spectrophotometric analyses. The spectral map shows first derivatives relative intensities with d445 and d525 used as proxies of iron oxides. Percent TOC and Hydrogen Index (HI) are provided by Rock-Eval 6 analyses. Ti content and Fe/Mn where measured by X-Ray Fluorescence. *Red squares* report samples for SEM images presented in Fig. 23.8. *Black arrows* are highlighting turbidites intercalated within laminated sediments (basal unit) or above a major Mass-Wasting Deposit (E5)

In PAV12, the base of E5 is characterized by a sharp transition from the lower diatomite unit towards highly deformed **diatomites** were laminations are still visible (between ca. 625 and 543 cm core depth). This facies is nicely reflected by sharply fluctuating values of b* parameter (Fig. 23.5). Between ca. 543 and 309 cm core depth, a massive dark brownish facies rich in organic macro remains (leave debris) occurs and is associated with fluctuating and lower values of L* and b* parameters. A sharp based sandy layer is in addition observed between 540 and 536 cm core depth. Between ca. 308 and 265 cm core depth another facies made of deformed **diatomites** (were some laminations are still visible) is locally interrupted by an erosive horizon rich in leaves debris. Above this contrasted facies, a massive greenish facies occurs between 265 and 207 cm core depth just below the upper **diatomite** unit (Fig. 23.5) and is characterized by a gradual increase of b* and higher content in goethite as reflected by 445 and 525 nm first derivative values.

As shown in Fig. 23.6 and detailed in Chapron et al. (2010), the top of E5 is dated to ca. 700 cal BP (ca. AD 1300) based on the extrapolation of **varve** counting from FC1 performed by Schettler et al. (2007). This chronology is now further supported by the age of two leaves debris from PAV09-B1 and PAV12 collected within the upper **diatomite** facies. Organic macro remains found within E5 either in PAV99 or PAV12 (Table 23.1) are, in addition, systematically older and not in stratigraphic order. Together with the above mentioned contrasted lithological descriptions within E5 and the occurrence of erosional surfaces within E5, these specificities imply that E5 is a major reworked deposit. According to the youngest age found at the top of the lower **diatomite** unit below E5 in PAV12 (ca. AD 300), it is also clear that this large event E5 has been erosive and probably remolded some of the previously deposited **diatomites** in the deep central basin. This is further supported by the different thickness of preserved **diatomites** below E5 within PAV99 and PAV12. Based on these arguments and following the classification of

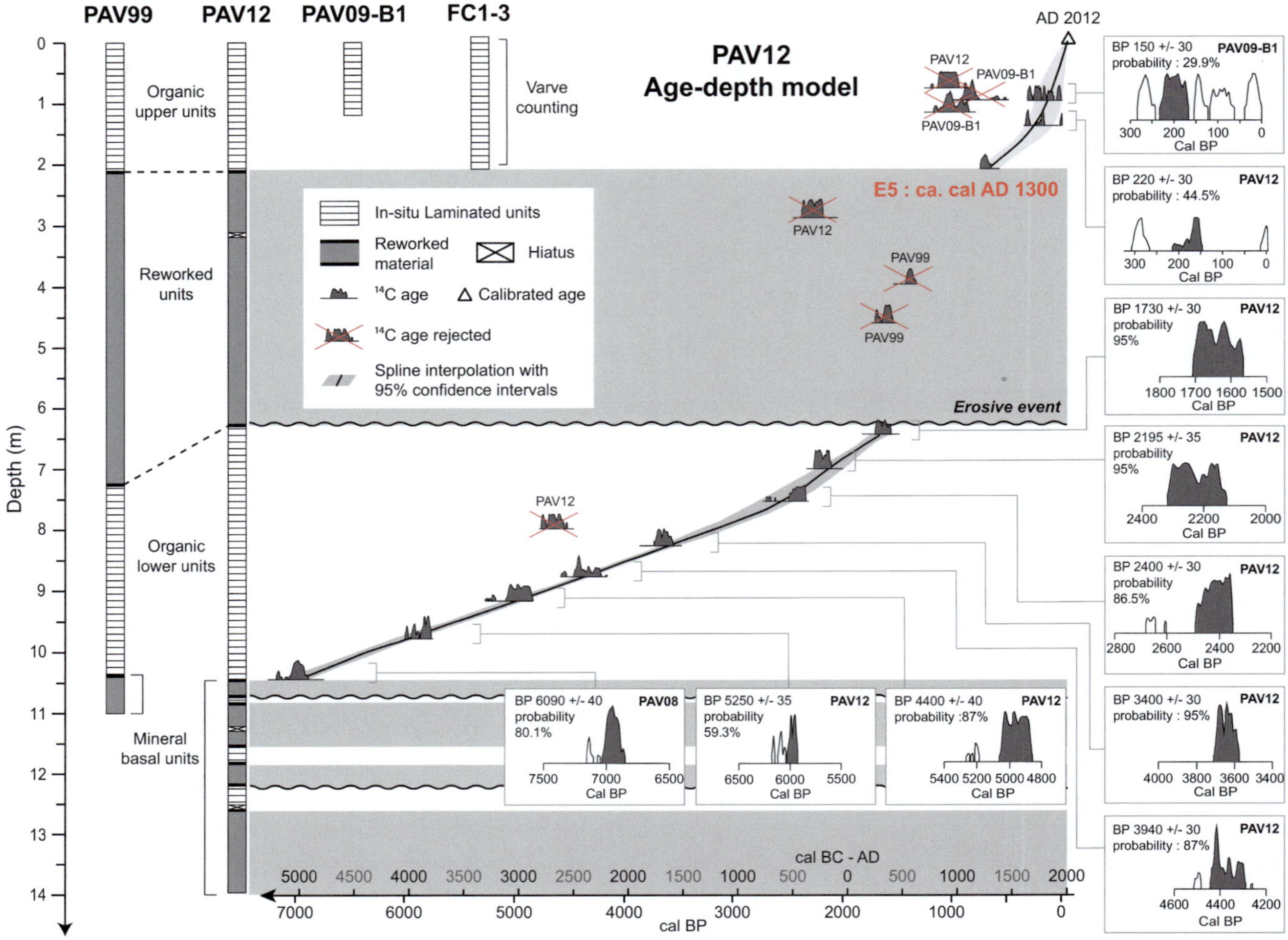

Fig. 23.6 Age depth model of core PAV12 based on AMS radiocarbon ages from organic macro remains using the Intcal09 calibration curve from Reimer et al. (2013), the identification of reworked deposits intercalated within the central basin fill and the application of a spline interpolation between the dating points using the CLAM software (Blaauw (2010). Varves counting established on a twin core retrieved from deep basin of Lake Pavin (Schettler et al. (2007) is supported by two radiocarbon dates (one from PAV09-B1 and one from PAV12). *Grey squares* correspond to reworked material associated with sedimentary events as discussed in the text

Mulder and Cochonat (1996), E5 can be subdivided in two successive mass wasting deposits: (i) first, a pluri-metric and erosive **slump deposit** remolding more or less deformed lacustrine sediments, and (ii) secondly, a pluri-decimetric greenish and fine grained **turbidite.** This **turbidite** is bearing a well-preserved and normally graded sandy base at 325 cm below the lake floor in PAV99, but is essentially made of a similar greenish and massive fine-grained sequence on both piston cores. The **slump** deposit and this fine-grained **turbidite** sequence are also clearly thinner in PAV12 than in PAV99, suggesting that the latter piston core is localized in a more proximal position than PAV12 from this mass wasting event E5. A classical fining upward pattern within the fine-grained turbiditic sequence retrieved in PAV12 is nicely illustrated by increasing b* values (Fig. 23.5). This is suggesting that this large and erosive mass wasting event remobilized and re-suspended a significant volume of lacustrine sediments within the water column, before massive settling occurred in the deep central basin of this maar lake.

Coarser but similar **turbidite** deposits are also identified in the basal grey unit retrieved in both piston cores from Lake Pavin central basin (Figs 23.2 and 23.5). These light-greyish **turbidites** are characterized by pluri-centrimetric thick sandy bases bearing some few gravel particles composed of similar volcanic materials than the Pavin crater formation retrieved at the base of PAV08. While PAV99 only contains one single **turbidite** between ca. 1050 and 1100 cm core depth, up to three similar **turbidites** are found at the base of PAV12. Their coarse-grained sandy bases are capped by a fining upward sequence that is centimetric to decimetric in thickness, fine-grained and systematically characterized by progressively increasing b* values. In PAV12, these **turbidites** are intercalated in between finely laminated and light-grey colored sediments (high L* and b* values) that are bearing low TOC and high Ti values (Fig. 23.5). It is also

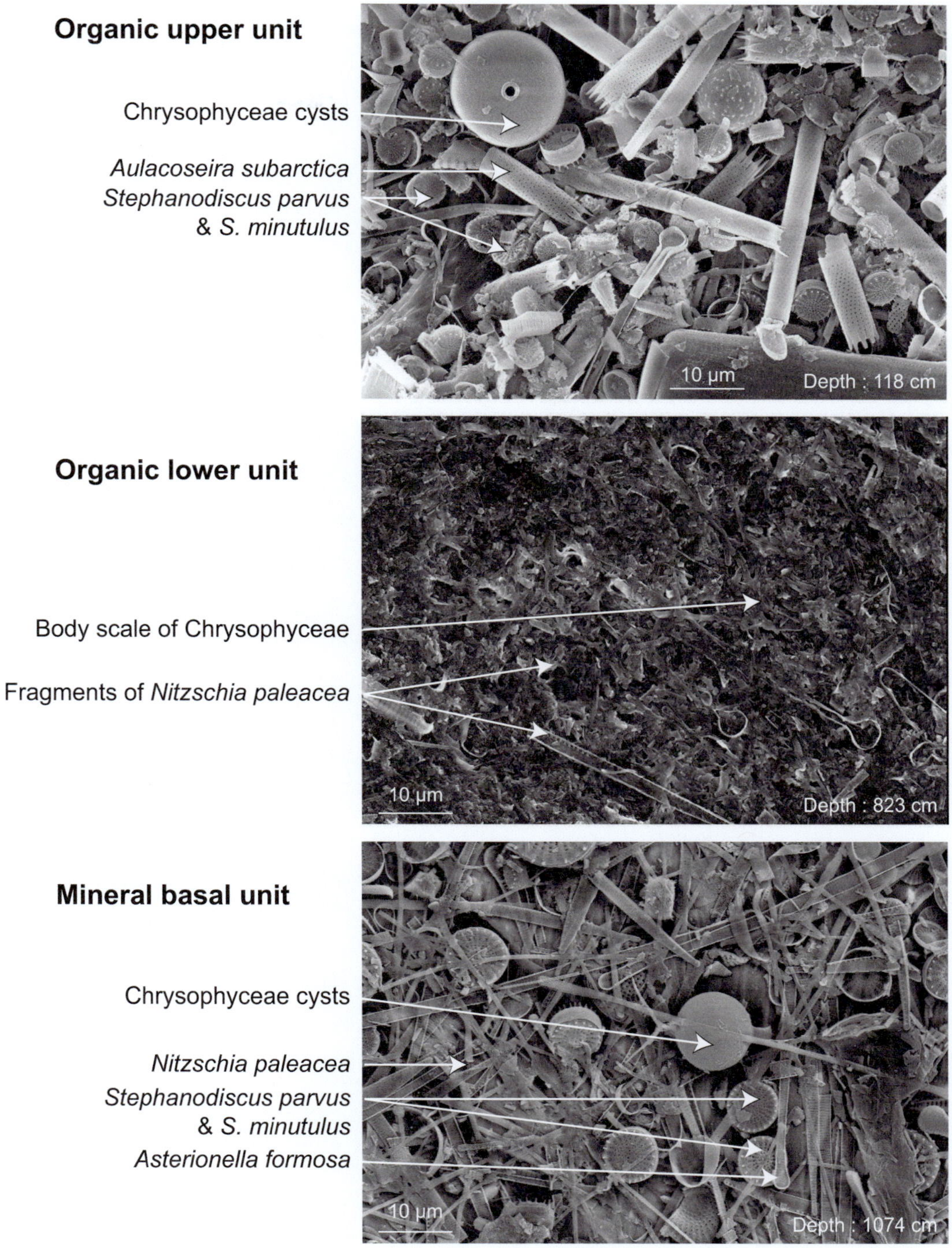

Fig. 23.7 Scanning Electron Microscopy images for PAV12 samples (*red squares* in Fig. 23.5) reflecting changes in diatoms assemblages inside the organic upper and lower units and the mineral basal unit

interesting to note that this basal mineral unit of Lake Pavin is rich in diatoms but made of different assemblages than the upper and lower **diatomite** units (Fig. 23.7) as further detailed below.

The chronology at the base of PAV12 is still poorly constrained because no organic macro remains were found in-between these light grey **turbidites** or at the base of the lower **diatomite** unit. Assuming that the onset of organic rich **diatomite** accumulation is synchronous throughout the lake and, like in PAV08, occurring around 7000 cal BP, it seems very likely that this mineral unit developed shortly after the Pavin eruption, in a recently formed maar lake. As a working hypothesis, to establish the lower boundary of the age-depth model in PAV12 with the CLAM software, one may thus use the calibrated **radiocarbon age** of bulk sediment retrieved just above the Pavin crater formation in

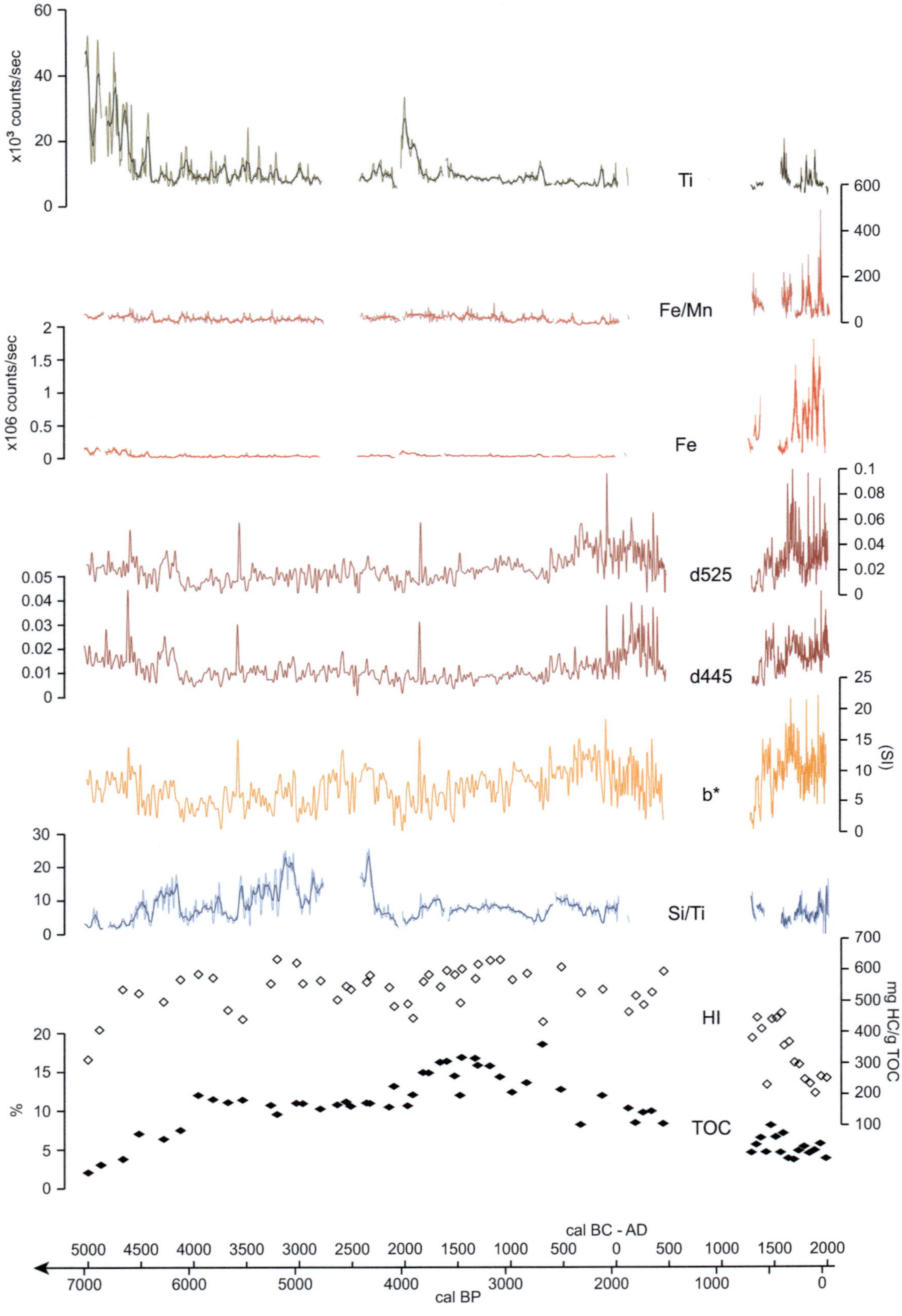

Fig. 23.8 Evolution through the last 7000 years of organic and mineral sedimentation in PAV12, as documented by geochemical proxies (Ti; Fe/Mn; Fe; Si/Ti curves form XRF core scanning), organic matter geochemistry (Hydrogen Index, HI and Total Organic Carbon, TOC) and spectrophotometric data (first derivative d445 and d525 and b* parameter)

PAV08 (Figs. 23.6, 23.2, Table 23.1). This basal turbiditic sequence seems here to have been deposited when the recently formed crater presented steep slopes surrounded by post-eruptive unsteady material, which have been quickly reworked into the deepest part of the crater.

In summary, Lake Pavin sedimentation in the central basin has been largely dominated by organic rich and finely laminated **diatomite** formation over the last ca. 7000 years. Roughly 700 years ago (ca. AD 1300), a major mass wasting event (E5) took place eroding and reworking approximately one millennia of **diatomite** deposition in the deep central basin. More recently, two smaller clastic layers (E3 and E2) only intercalated in-between the upper **diatomite** unit in a single core (PAV09-B1), can be dated to AD 1840 +/− 80 and AD 1880+/− 70, respectively, using an age-depth model with the CLAM software combining radiocarbon dating (one age from PAV09-B1 and two from PAV12) together with **varve** counting chronology established by Schettler et al. (2007) in a nearby freeze core (FC1).

23.4 Lake Pavin Stratigraphic Record of Environmental Changes

23.4.1 Lake Level Evolution in Lake Pavin

The level of Lake Pavin is controlled (i) by the altitude of its sub-aerial outlet, (ii) by the flow of its subaquatic outlet and (iii) by climate (precipitation regimes, lake water evaporation). The V-shape geomorphology of its sub-aerial outlet deeply incised into the walls of Pavin **crater rim** suggest that this maar lake has been exposed to a lake level lowering. Such morphology may result either from a progressive incision of the lake outlet into a heterogeneous and relatively poorly consolidated volcanic phreato-magmatic formation (Delbecque 1898; Chap. 5). It may also result from a relatively recent and abrupt collapse of this sector of the **carter rim** (i.e., a crater outburst) triggered either spontaneously (and induced by the weight of the lake water column) or favored (i) by a lake level rise, (ii) by the propagation of a violent wave into the lake outlet or (iii) by earthquake shaking (Chapron et al. 2010). The flow of the subaquatic outlet of Lake Pavin is poorly documented, but available data suggest it is relatively limited compared to the one of the sub-aerial outlet (Jézéquel et al. 2011). A recent synthesis by Magny et al. (2013) discussed the impact of climate on well-dated synchronous phases of lake level changes during the Holocene across Western European mountain ranges or around the Mediterranean Sea, but little is still known about the amplitudes of these lake level changes. Because such lake level reconstructions in Western Europe are typically performed in carbonated lake systems, none of these phases of lake level changes were ever documented in the volcanic area of the French Massif Central during the present interglacial period (Truze and Kelts 1993; Chapron et al. 2012; Lavrieux et al. 2013). Finally, the outlet of Pavin has been stabilized at an altitude of 1197 m only recently, by the building of human infrastructures at the end of the eighteenth century (Chap. 1, this issue).

One abrupt change of sedimentation pattern in core PAV10E occurring at ca. 20 cm below the lake floor on the subaquatic plateau (Fig. 23.2) has been related to a significant and rapid lake level drop (Chapron et al. 2012). The transition from in situ **diatomite** formation in the lower part of core PAV10E retrieved by 17 m water depth, into the deposition of a littoral facies just above an erosive sandy layer bearing some small sized organic macro remains (leaves debris) could not be dated by **AMS radiocarbon**. This change is interpreted as resulting from a lake level drop of ca. 9 m, considering that the deposition of the littoral facies in Lake Pavin occurs between the isobaths −26 m and the lake shore (Chap. 22, Fig. 23.1). Because excavation of the natural aerial outlet artificially dropped the level of Lake Pavin by ca. 4 m in the late eighteenth century (Chap. 1, this issue), these sedimentation changes in PAV10E can be related to an abrupt lake level drop of roughly 13 m.

Figure 23.9 compares DSR and RE pyrolysis measurements on cores PAV10E, PAV08 and PAV12. Following Debret et al. (2011), DSR data on a Q7/4 diagram suggest that either sediments from the littoral facies and in situ **diatomites** from these cores are organic-rich deposits dominated by Melonine type (B pole), by altered organic matter (C pole) and Chlorophyll and by-products (D pole). RE data represented by **S2** vs. TOC diagrams support DSR data and shows that Pavin sediments organic matter is always clustering within the algal pole. The organic matter from the littoral facies are, however, characterized by significantly lower values than in situ **diatomites** retrieved either in a littoral environment (PAV10E), on the plateau (PAV08) or in the basin (PAV12) of Lake Pavin. Interestingly, these diagrams also highlight that the upper **diatomite** unit from cores PAV08 and PAV12 are in addition characterized by lower values than the lower **diatomite** unit. This change in organic matter composition in Lake Pavin occurred just after the formation of the ca. AD 600 **slump** deposit on the subaquatic plateau (Figs. 23.4, 23.5 and 23.9) and can be explained by the progressive erosion and remobilization of oxidized littoral organic matter into the lake waters and its incorporation into the organic matter composition of the upper **diatomite** unit accumulated within the **mixolimnion** and the **monimolimnion** since this large mass wasting event. Such a progressive remobilization and incorporation of littoral organic matter within the upper **diatomite** after **sedimentary event** E6, suggest that this large **slump** deposit was contemporaneous to a major lake level drop of ca. 13 m at ca. AD 600.

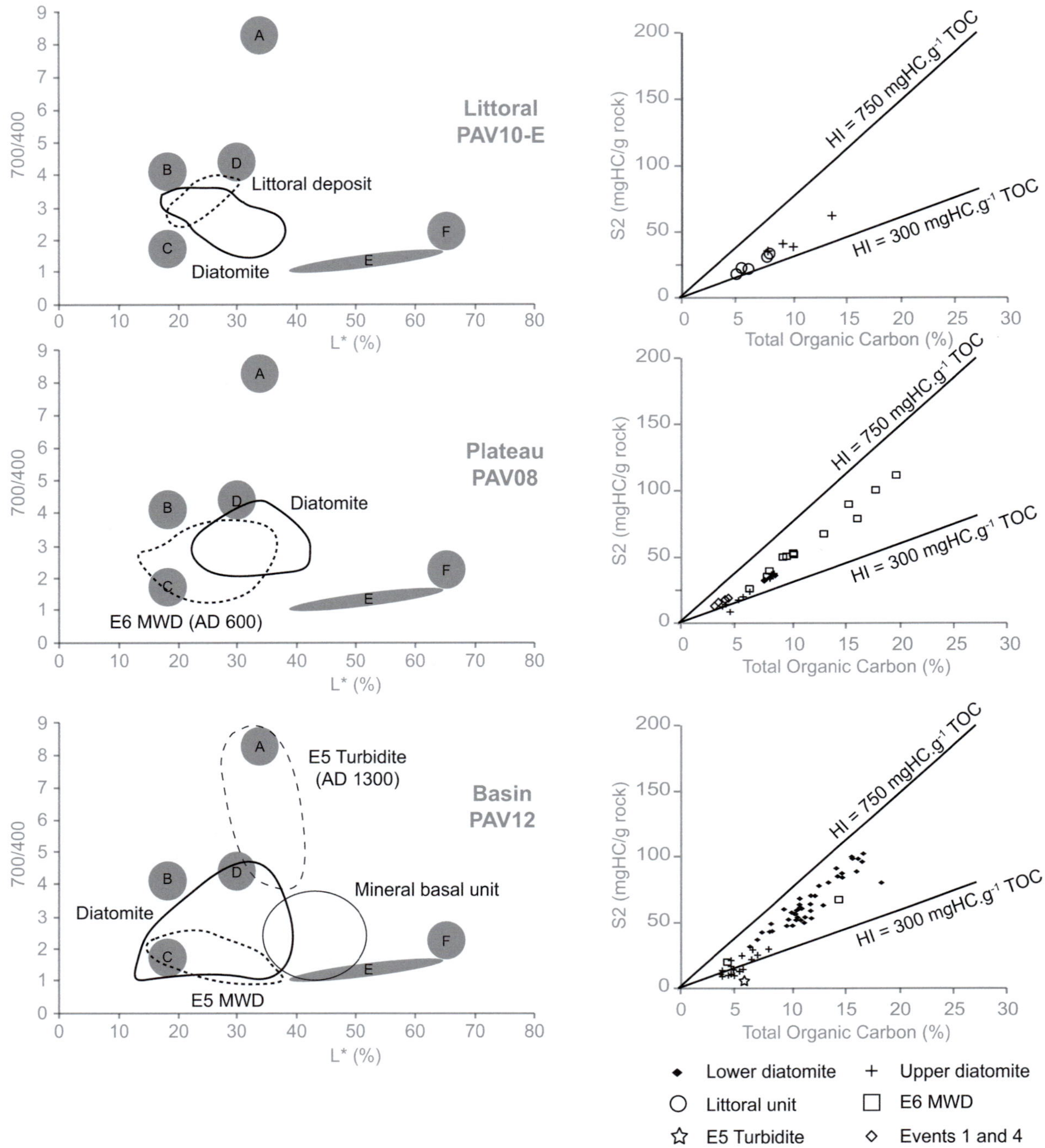

Fig. 23.9 Characterization of Pavin sediments origin and sediment source based on diffuse spectral reflectance Q7/4 diagram (*left*) and Rock-Eval pyrolysis represented by **S2** vs. TOC diagram (*right*) for cores PAV10-E (littoral), PAV08 (plateau) and PAV12 (basin). In the Q7/4 diagram, the diffuse spectral reflectance signature of littoral deposit, diatomite, Mass-Wasting Deposit (MWD), Turbidite and Mineral Basal unit are compared to the five distinct poles of sediments defined by Debret et al (2011): Iron-Rich deposits (**a**); Organic-rich deposits dominated Melanoidine type (**b**); Organic-rich deposits dominated by altered organic matter (**c**); Organic-rich deposits dominated by Chlorophyll and by-products (**d**); Clayey deposits (**e**) and Carbonate deposits (**f**). In the **S2** vs. TOC diagram, Total Organic Carbon (TOC), together with Hydrogen Index (HI) and **S2** (thermal cracking of the hydrocarbon compounds) are illustrated. The two linear domains of the hydrogen index (HI = 750 and HI = 300 mgHCg^{-1}TOC) corresponding to algal and terrestrial poles, respectively, are also represented

23.4.2 Evolution of Pavin Limnology

In some anoxic subaquatic environments, b* parameter measured on sediment cores has been used to track the evolution of diatoms content (Debret et al. 2006, 2011) and used as a proxy to document the evolution of the productivity of aquatic environments. As shown in Fig. 23.8, in Lake Pavin, b* parameter is, however, clearly different from the evolution of Si/Ti ratio measured by XRF core scanning and used here as an indicator of diatom production. It suggests that b* parameter can't be used here to document the evolution of its productivity. Lake Pavin sequence highlights a progressive and fluctuating increase in Si/Ti since 7000 cal BP culminating around 5000 cal BP, but this ratio is later on slightly reduced since 4200 cal BP and remained much more constant afterwards. This trend suggests that Lake Pavin productivity has been more variable and intense in the early Holocene and then stabilized during the mid-Holocene. Interestingly, TOC in PAV12 sediments shows a slightly different evolution through time, with a progressive increase since 7000 cal BP culminating around 3500 cal BP, and a decreasing trend afterwards. This bimodal evolution is, however, not observed in HI values that are quickly rising within the basal mineral unit and then remaining relatively constant within the lower **diatomite** unit, before sharply dropping in the upper **diatomite** unit. Such a complex evolution of organic sedimentation in Lake Pavin suggests that this crater lake underwent several steps and important changes in its productivity and in the preservation of organic matter on its floor.

Following Albéric et al. (2013), the confrontation between **AMS radiocarbon** dates obtained from bulk sediments and either leaves debris sampled at similar depths or corresponding model ages in the lower **diatomite** unit from core PAV12 (Table 23.1, Fig. 23.6), is, in addition, suggesting the development of a significant **radiocarbon reservoir effect** ranging around 2500 yrs at ca. 8 m core depth (i.e. near 3000 cal BP) and up to ca. 3800 yrs near 6.5 m core depth (i.e. around 1700 cal BP). These new data have been added in Fig. 23.10, adapted from Albéric et al. (2013), where radiocarbon reservoir effect (either in ^{14}C age scale or $\Delta^{14}C$ scale) is plotted versus calibrated model ages of Lake Pavin sediments. Although the link between the meromicticity and the existence of large reservoir effects is not univocal (Albéric et al. 2013), this suggests that the **meromicticity** of Lake Pavin started early in the lake history and may have been variable through time.

Lake Pavin **meromicticity** is also known to impact the geochemistry and biogeochemical cycles in the deep central basin (Part III, this issue) and favoring the storage of dissolved Fe (by the so-called iron-well process (Martin 1985)) and Mn in sediments from the **monimolimnion** (Viollier et al. 1995; Jezequel et al. 2011; Schettler et al 2007). A clear increase in Fe/Mn ratio within the upper **diatomite** unit in

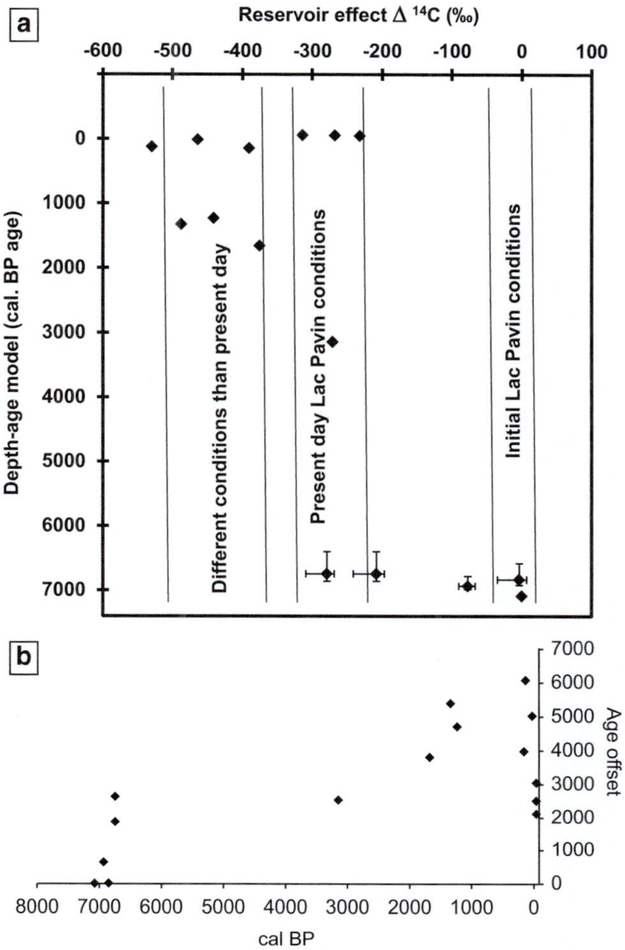

Fig. 23.10 Radiocarbon reservoir effect (**a**) and age offset (**b**) evolution since Lake Pavin formation (Modified after Albéric et al. (2013)

PAV12 (Figs. 23.5 and 23.8) may thus partly reflect an intensification of the **meromicticity** after sedimentary event E5. These XRF measurements were, however, performed several weeks after core sections opening, and it seems also very likely that such a significant increase in Fe/Mn ratio in the upper **diatomite** unit is partly an artefact link with an intense oxidation of PAV12 sediment interstitial water. SDR measurements on PAV12 were, on the contrary, performed just after core sections opening on "fresh" sediments and may thus be more reliable to track the evolution of iron oxides in Pavin sediments. Interestingly, the evolution of the first derivative values at 525 and 445 nm, corresponding to the occurrence of goethite in the sediments (Debret et al. 2011; Simonneau et al. 2013b), is highlighting a major increase in the lower **diatomite** unit ca. 2700 yrs ago (Fig. 23.8). As a working hypothesis, it seems therefore likely that the **meromicticity** of Lake Pavin developed during the late Holocene period, rather than just after the crater formation.

Ongoing investigations on the evolutions of diatoms assemblages in PAV12 sediments can bring important additional arguments to reconstruct changes in lacustrine

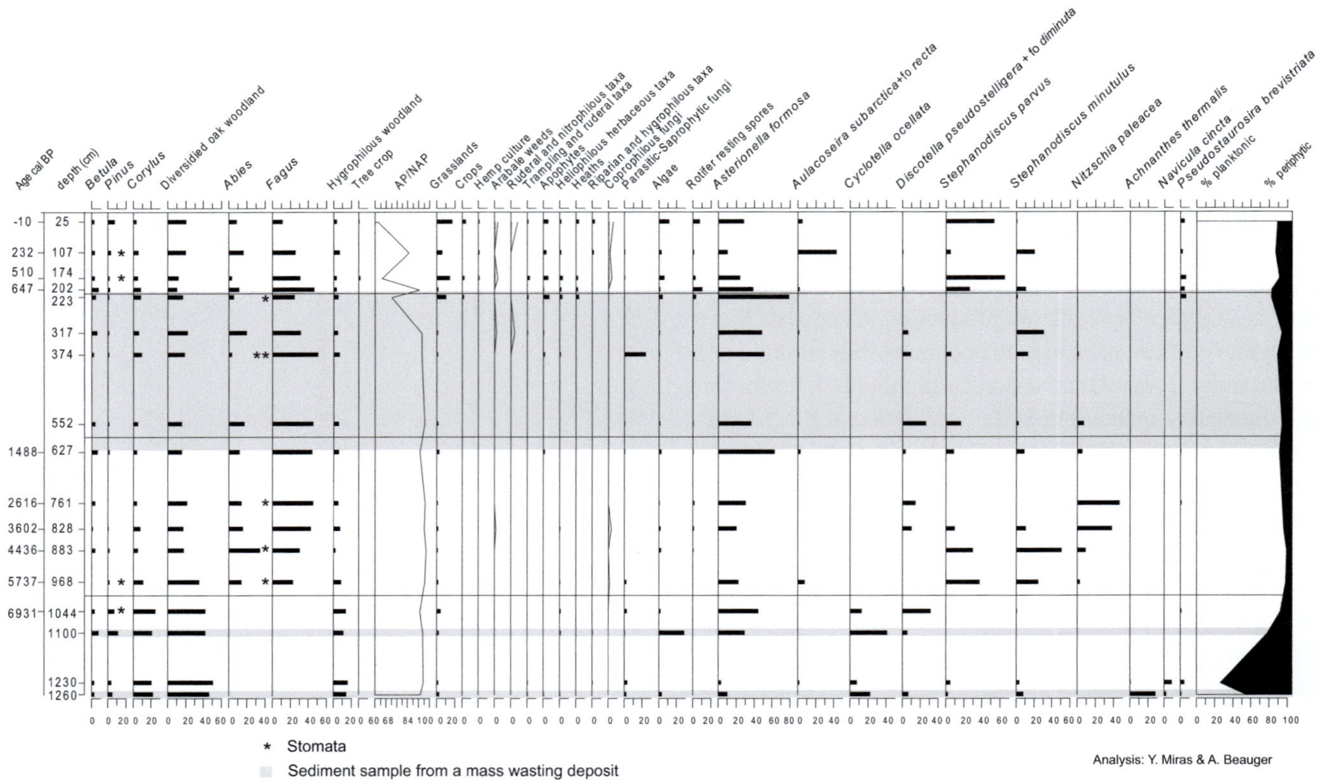

Fig. 23.11 Preliminary pollen (*left*) and diatom (*right*) diagram illustrating the evolution of species assemblages in sediment from PAV12 sampled either in reworked deposits (i.e. turbidite or mass wasting deposit; grey levels) and *in situ* deposits. The ratio from arboreal and none-arboreal pollen (AP/NAP) is also given

dynamic. Figure 23.11 shows preliminary results of analysis performed either within *in situ* laminated lacustrine sediments or within mass wasting deposits. Such a strategy can provide insights on environmental changes and on sediment source areas in remolded deposits, as discussed below. In the basal unit around 7000 cal BP, the diatom community during this period are either typical from alkaline and electrolyte rich waters (*Achnanthes thermalis* (Rabenhorst), Schoenfled var. *thermalis, Fragilaria famelica* (Kützing) Lange-Bertalot var. *fameilica* and *Navicula cincta* (Her.) Ralfs), littoral taxa (*Staurosirella pinnata* Ehrenberg, *Staurosira venter* (Ehr.) Cleve & Möller) or epithytic taxa (*Epithemia* spp, *Cocconeis* spp, *Rhopalodia gibba* (Her.) O. Muller) associated with the development of macrophytes (Walker and Paterson 1986; Van Dam 1994; Lange-Bertalot 2001; Hindakova 2009; Gutowski et al 2011). Opportunistic diatom species (*C. ocellata*) together with *Botryococcus* algal spores are in addition typical from deep and oligotrophic lakes (Blomqvist et al. 2001; Rioual et al. 2007). These assemblages suggest that Lake Pavin became a deep and oligotrophic lacustrine system quickly after the Pavin eruption. The presence of littoral diatom taxa in the deep central basin within laminated clastic sediments and within **turbidites** can also be explained by the development of numerous active canyons incised into the Pavin crater formation: these features are still visible in the morphology along the steep slopes of the lake (Fig. 22.10) and could easily bypassed material in the past from surrounding sub aerial and littoral environments to the deep water basin. In the lower **diatomite** unit, non-pollen palynomorphs assemblages mainly dominated by algal spores (*Debarya* and *Spyrogyra*), together with the development of spring diatom bloomers (different species of *Stephanodiscus*) and a change in rotifer resting eggs (*Conochilus hippocrepis*-type is progressively replaced by *Keratella*-type, *Filinia longiseta*-type and *Brachionus*-type) are suggesting an increasing trend in the trophic state of the lake (Lotter and Bigler 2000; Barbiero and Warren 2011). After 2600 cal BP, there is a replacement of *Stephanodiscus* by different planktonic genuses, with the dominance of *Nitzschia paleacea* (*Grunow*) indicating higher concentration of NH4 in the lake (Voigt et al. 2008). Interestingly, a peak in Ti (Figs. 23.5, 23.8 and 23.12) is observed during this time window and such trophic change could be explained by sediment inputs in the lake, but further studies (and higher resolution samples of diatoms assemblages) are needed to confirm this assumption. The disappearance of littoral taxa within this lower **diatomite** unit, suggest a limited influence of canyons. This is apparently also true for the upper **diatomite** unit. In the latter unit, dominant taxa (*A.formosa* and different species of *Stephanodiscus*) are typical from nutrient-rich waters. The

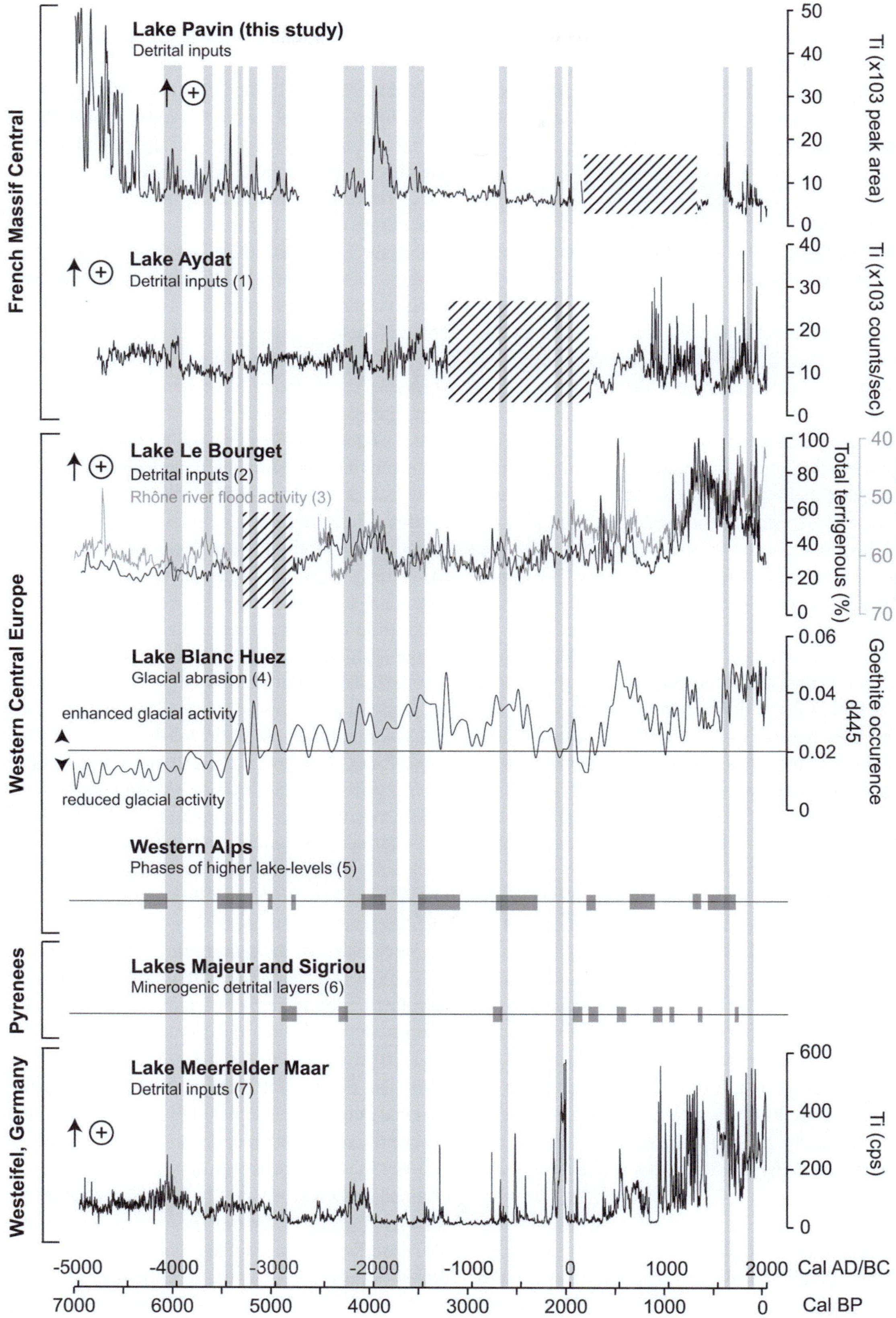

Fig. 23.12 Comparison of detrital inputs in Lake Pavin with other records in Western Europe (1) Lavrieux et al. (2013); (2) Arnaud et al. (2012); (3) Debret et al. (2010); (4) Simonneau et al. (2014); (5) Magny (2006); (6) Simonneau et al. (2013b); (7) Martin-Puertas et al. (2012). *Grey squares* represent phases of enhanced terrestrial supplies in Lake Pavin (this study). Hatched areas correspond to unrecorded periods due to Mass-Wasting Deposits

dominance of *Aulacoseira subarctica* (O. Muller) Haworth ca. 200 years ago in association with higher frequencies of small periphytic *Fragilaria*, *Staurosira* together with the decrease in diatoms and chrysophyceae cysts densities suggest, however, the development of a colder period and a change in the trophic status (Van Dam 1994).

23.5 Impact of Climate, Human and Geological Hazards on Lake Pavin Sedimentation

Ongoing investigations on the evolutions of pollen assemblages in PAV12 sediments are also given in Fig. 23.11 and are reflecting either local or regional changes in the vegetation cover. Together with independent indicators of terrestrial inputs to Lake Pavin, pollen data may help to disentangle environmental changes induced by climate and human activities. As discussed below, pollen and diatom assemblages from mass wasting deposits in core PAV12, can in addition provide insights on sediment source areas and further support stratigraphic and chronological reconstructions to precise the impact of geological hazards on Lake Pavin sedimentation.

In between the **turbidites** from the basal unit around 7000 cal BP, pollen data clearly indicate the occurrence of a diversified deciduous forest dominated by oak (*Quercus*) and also filled by lime tree (*Tilia*), elm (*Ulmus*), maple (*Acer*) and ash (*Fraxinus*). This forest dominated the regional landscape during the Atlantic period (Reille et al. 1992) and the pollen frequencies reached in PAV12 samples suggest that this vegetation was very close to Lake Pavin. The identification of alder (*Alnus*), birch (*Betula*), Hazel (*Corylus*) and pine (*Pinus*) and also of pine stomata further indicate that the slopes of the carter rim of Lake Pavin were quickly colonized by plants. Maximum values of Ti in this basal unit (Figs. 23.5 and 23.12) and the occurrence of coarse-grained **turbidites** suggest, however, a limited vegetation cover within the inner slopes of the **crater rim** and a significant erodability of (subaerial and subaquatic) slopes draining into the lake. It seems thus likely that plant colonization along the inner slopes of the **crater rim** came from the crest of the rim. In addition, in this early stage of Lake Pavin, it is likely that the lake surface was much higher than today and closer to crest of the rim.

In the lower **diatomite** unit, the development and maturity of beech (*Fagus*) and fir (*Abies*) woodlands matches the regional climatic variation of the Mid-Holocene toward wetter and cooler conditions (Magny and Hass 2004; Lavrieux et al. 2013). Such conditions also favored the development of vegetation on the shore of Lake Pavin dominated by alder. Regular occurrences of fir stomata and parasitic and saprophytic fungi of tree (Cugny et al. 2010) underline the presence of fir and other caducifolious tree species. These reconstructions indicate that the inner slopes of the crater were densely covered by vegetation and are in agreement with the observed limited clastic sediment supply reflected (i) by low Ti content within this lower **diatomite** unit and (ii) by increasing concentration of algal organic matter. Since the onset of this organic rich sedimentary unit, the erodability of subaerial slopes draining into the lake was thus much probably reduced. The identification of several peaks in Ti within the lower **diatomite** unit (Figs. 23.5, 23.8 and 23.12) suggest, however, that short periods of enhanced erosion occurred within the drainage basin of Lake Pavin. Because some of these erosive periods at Lake Pavin are also matching periods of enhanced clastic inputs in nearby Lake Aydat (Fig. 23.1), but also in more remote lakes from the western Alps (lakes Bourget and Blanc Huez), the northern Pyrenees (lakes Majeur and Sigriou) and eventually in the Eiffel volcanic province in Germany (maar lake Meerfelder) as shown in Fig. 23.12, they might reflect larger scale climate shifts. Phases of enhanced precipitation at the onset of the Neoglacial period (between 6000 and 5000 cal BP), during the Bronze Age (between 4300 and 3500 cal BP), at the beginning of the Iron Age (between 2800 and 2600 cal BP) and during the Roman period (around 2000 cal BP) in Lake Pavin might for example match periods of enhanced soil erosion in Lake Aydat (Lavrieux et al. 2013), higher Rhone River flooding activity in Lake Le Bourget (Debret et al. 2010; Arnaud et al. 2012), increasing glacier activity in Lake Banc Huez (Simonneau et al. 2014), phases of higher lake-levels in the Western Europe (Magny 2006; Magny et al. 2013), reactivation of canyons draining into lakes Majeur and Sigriou (Simonneau et al. 2013b) and soil erosion in maar lake Meerfelder (Martin-Puertas et al. 2012). These clastic input peaks in Lake Pavin could either result from the direct impact of periods of heavy rainfalls on runoff along the steep slopes of the **crater rim** characterized by numerous gullies and thalwegs (Fig. 22.10) or from enhanced soil erosion favored by the effect of snow amount on runoff during periods of snowmelt (Tanasienko et al. 2011; Simonneau et al. 2013b).

In the upper **diatomite** unit, a radical change in the regional landscape is depicted by pollen data. The woodland cover present a significant decreasing trend while grasslands and heathlands increase as illustrated by the ratio between arboreal and none arboreal pollen concentrations (AP/NAP) shown in Fig. 23.11. During this period, Stebich et al. (2005) also describe a more open landscape and pollen assemblages typically resulting from the development of grazing, crops and hemp cultures. In core PAV12, apophytes, trampling and ruderal pollen indicators together dung-related fungal spores (e.g. *Sporormiella* and *Coniochaeta lignaria*) indicate grazing activities throughout the zone and as well as the predominance of the rotifer assemblage (*Conochilus natans*) suggest a change in the lake water conditions. The vegetation of the lake shore constits of a sedge community and both lower arboreal pollen frequencies and the disappearance of tree stomata could suggest that woodlands are not so extended in the lake shore than previously. It is, however, important to keep in mind that Lake Pavin watershed is larger than its topographic drainage basin due to the occurrence of numerous subaerial and subaquatic springs (Fig. 22.1, Chap. 22). The precise limits of Pavin watershed are still poorly defined, but it has been shown that nowadays the lake water trophic level is highly sensitive to agricultural practices outside the crater rim (Chap. 1). It seems thus very likely that this has been the same in the past. Accordingly, the development over the last seven centuries of agricultural practices outside the **crater rim** probably impacted land use (and thus pollen rain at a regional scale) but also the trophic level of Lake Pavin (and thus diatom assemblages or non-pollen palynomorphs such as rotifer resting eggs). This interpretation is in agreement with old paintings from Lecoq realized in AD 1867 discussed in Chapron et al. (2010) and illustrating a former Lake Pavin panorama dominated by grasslands outside the **crater rim**, while its inner slopes (i.e., the topographic drainage basin of Lake Pavin) were still largely forested. Only two peaks in Ti are in addition observed in the upper **diatomite** unit in PAV12, while lakes Aydat, Bourget and Meerfelder are clearly characterized by increasing trends in detrital inputs over the last two millennia (Fig. 23.12). These trends of detrital inputs in these contrasted European lacustrine systems were related to increasing soil erosion due to development of land use and agriculture in their watersheds (Lavrieux et al. 2013; Debret et al. 2010; Arnaud et al. 2012 and Martin-Puertas et al. 2012). Because these two peaks in Ti are occurring during the Little Ice Age (LIA) in PAV12 and are contemporaneous with periods of enhanced flooding activity in lakes Aydat and Bourget (Lavrieux et al. 2013; Arnaud et al. 2012; Chapron et al. 2005), but also glacial activity in the Alps (Arnaud et al. 2012; Simonneau et al. 2014) and soil erosion in maar lake Meerfeler (Martin-Puertas et al. 2012), they are probably more reflecting a climatic signal (shits toward colder and wetter conditions) than anthropic activities. Following these authors (and numerous studies in Europe), it is, however, very likely that human practices during the LIA, had impacts on the watersheds vegetation cover and favored detrital inputs to these lakes, including in Lake Pavin.

Clear traces of recent human activities in the Pavin topographic drainage basin are sparse and concern essentially the building of successive infrastructures to stabilize its outlet (Chap. 1) and the development of some tracks and paths along the inner slopes of the **crater rim** (Fig. 22.10). To some extent, it is very likely that woodlands around Lake Pavin were used for domestic needs over the last centuries, but this maar lake might be in fact considered as one of the few preserved natural site in the French Massif Central. As further discussed in Chaps. 1, 2 and 3 this specificity of Lake Pavin in the region may be more related to its "dangerous" reputation than its relatively limited accessibility. In this paper and in Chap. 22, several stratigraphic and geomorphologic evidences are also highlighting the development of a wide range of subaquatic slope instabilities and at least two large events associated with the generation of violent waves since the lake formed ca. 7000 years ago (Chapron et al. 2012). It seems therefore possible that geological hazards in this maar lake may have contributed to its preservation from growing human activities over the last millennia.

Sedimentary event labeled E1 presented in Chap. 22 is only identified on the plateau at site PAV08 (Fig. 22.8). This 2 cm thick light colored layer (higher L* values) is characterized by lower TOC and HI values bearing a similar organic matter signature on a **S2** vs. TOC diagram than littoral sediments as shown in Fig. 23.9. According to the new age-depth model established for PAV08-P1 (Fig. 23.4), this **sedimentary event** E1 is dated to AD 1915 +/− 5 and can be correlated with the historical earthquake that stroke the town of La Bourboule in AD 1921 located at only 15 km from Lake Pavin (Fig. 23.13; Table 23.2). This earthquake reached an MSK intensity of 4.5 at the epicenter (Sisfrance database, BRGM, Lambert 1997). Compared to other studies documenting the impact of historical earthquakes on lacustrine sediments from European regions characterized by a moderate seismicity (Chapron et al. 1999; Monecke et al. 2004; Nomade et al. 2005; Strasser and Anselmetti 2008), it seems that Lake Pavin organic rich sedimentation in littoral areas is unstable and easily remobilized when submitted to gravity accelerations associated with seismic waves propagation. Following the conclusions of Strasser and Anselmetti (2008) it is, however, also very likely that this limited remobilization of littoral sediments down to PAV08 coring site has been favored by the perturbation of subaqueous sediment pore pressure after the artificial lake level drop by ca. 4 m in the late eighteenth century (Chap. 1, this issue). **Sedimentary event** labeled E2 described in Chap. 22 is only identified at a single site from the deep central basin in core PAV09-B1 (Fig. 22.8). This 2 cm thick layer is essentially characterized by very high values in MS and is dated to AD 1880 +/− 70. Given the dating uncertainties, this event interpreted has a fine grained **turbidite** originating from a canyon, is potentially contemporaneous with two nearby historical earthquakes that stroke the study area in AD 1863 (the Mont-Dore MSK intensity 5 event) and in AD 1892 (the Issoire MSK intensity 7 event) located at 11 km and 29 km from Lake Pavin, respectively (Fig. 23.13; Table 23.2). According to former studies on lacustrine sediments sensitivities to instrumental earthquakes (Monecke et al. 2004; Nomade et al. 2005) both of these historical earthquakes could be recorded in nearby Lake Pavin. Another similar study combining

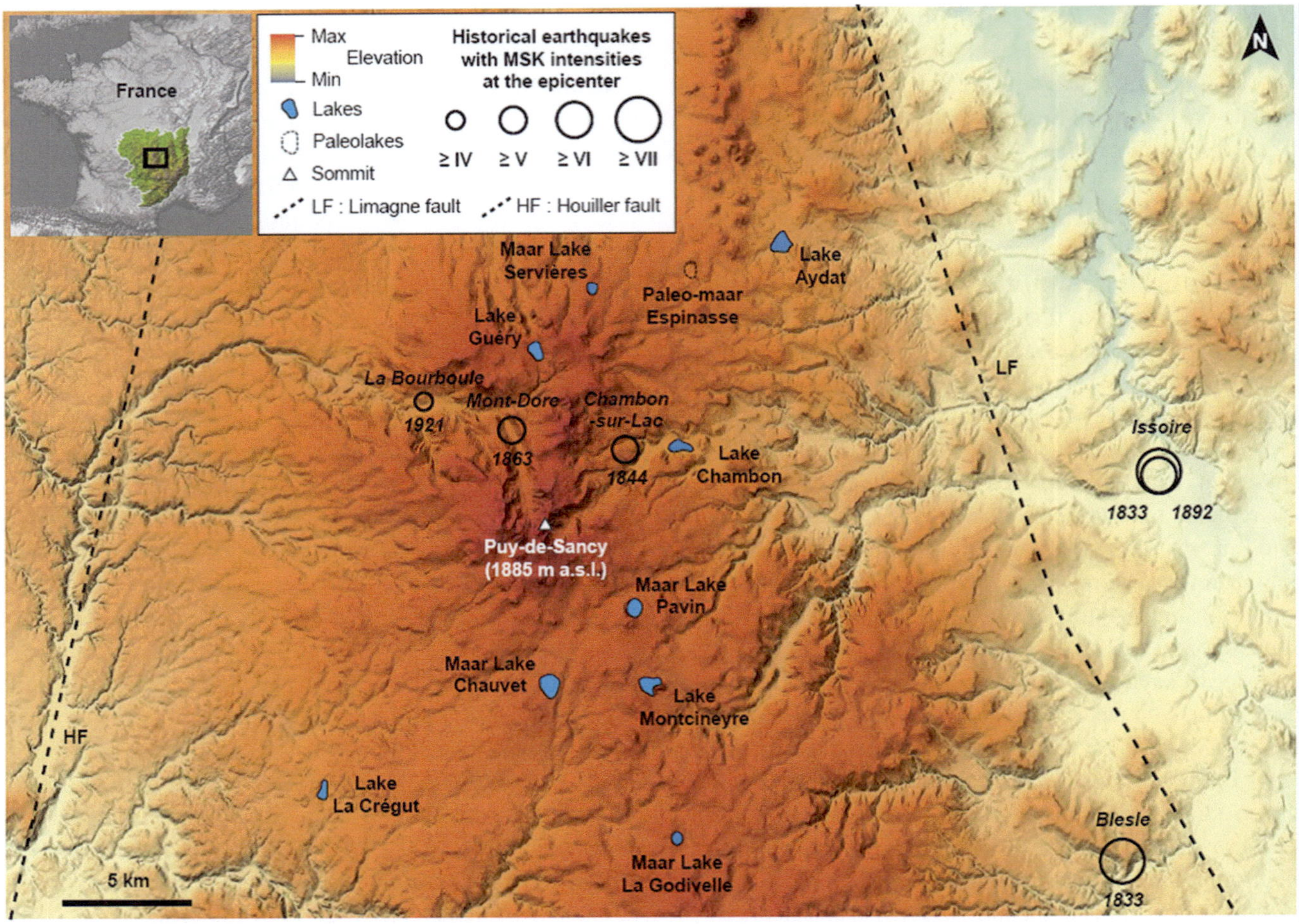

Fig. 23.13 Digital elevation model of the Northern part of the French Massif Central illustrating the location of lacustrine systems and MSK intensities of historical earthquake epicenters that stroke the study area. Regional faults (the Limagne Fault, LF and the Houiller Fault, HF) are also indicated by *dashed bold* lines

radiocarbon dated sediment cores from nearby Lake Guéry (Chassiot et al. 2016) also identified a pluri-decimetric **slump** deposit along its delta that is contemporaneous with the AD 1863 Mont-Dore earthquake (Fig. 23.13). As shown in the above mentioned former studies, regional contemporaneous slope instabilities in lacustrine environments are one of the stronger arguments to link a mass wasting deposit (MWD) with earthquake shaking. Another crucial criteria being that the age of the MWD fits with an historical earthquake. It seems thus likely that E2 in the Pavin central basin resulted from earthquake shaking in AD 1863. To support this interpretation, ongoing studies on this **sedimentary event** involve grain size and RE measurements, in order to document possible sediment source areas and to explain high magnetic susceptibility values in E2. Similarly, **sedimentary event** labeled E3 only documented in core PAV09B1 by MS measurements (Fig. 22.8) and dated to AD 1840 +/− 80, is potentially contemporaneous with two others nearby historical earthquakes that stroke the study area in AD 1844 (the Chambon-sur-lac MSK intensity 5.5 event) and in AD 1833 (the Issoire MSK intensity 6 event) located at 9 km and 29 km from Lake Pavin, respectively (Fig. 23.13; Table 23.2).

As shown in Fig. 22.8 and Table 23.2, **sedimentary event** labeled E4 is identified both on the plateau (PAV08-P1) and in the deep central basin (PAV09-B1). Clustering calculated ages for these two contrasted layers allow correlating them with a single event that occurred likely between AD 1685 and 1715, and eventually later (i.e. between AD 1685 and 1865 according to dating uncertainties on core PAV09-B1), since E4 on the plateau has an erosive base, and may thus be complicated to date as shown in Fig. 23.4. On the plateau, this 6.5 cm thick light colored layer, is like E1, containing organic matter originating from the littoral environments (Fig. 23.9). One **AMS radiocarbon age** within E4 on the plateau suggest, in addition, that remolded littoral sediments are significantly older than the event. In the deep central basin, E4 is thinner than sedimentary events E2 and E3, but characterized by a maximum value in MS and a clear change is SDR values (compared to the host sediment) that is similar to the one found for E4 on the plateau (Fig. 22.8). It seems thus here very likely that sedimentary E4 consist in reworked former littoral sediments developing a thin mass wasting deposit that froze on the plateau at PAV08 coring site, but evolved down slope into a thin and fine grained **turbidite** at PAV09B1 coring site. Due to age uncertainties, this sedimentary event may be related to perturbation of subaqueous littoral sediment pore pressure after the artificial lake level drop by ca. 4 m in the late eighteenth century (Chap. 1, this issue) and thus be human induced. It may also be linked to a former **limnic eruption** described back to AD 1783 in the so-called Godivel IV manuscript detailed by Michel Meybeck in Chap. 2 (this issue). The translation of this old manuscript suggest that the August 21, 1783 event was a "moderate degassing [event] due to lake rollover" eventually in relation with internal slumping according to Meybeck and potentially associated with two historical earthquakes (Table 23.2) of unknown MSK intensities and epicenter locations but documented in the Limousin region in the French historical earthquake catalogue Sisfrance (i) the same day (the Lepaud event) and (ii) in July (unknown precise date) that same year in a similar location (the Mainsat event). Further palaeoseismological studies are however required to better document the potential impact of these earthquakes in the study area.

Sedimentary event labeled E5 is a major **slump** deposit capped by a muddy **turbidite** found both in cores PAV99 (510 cm thick) and PAV12 (420 cm thick) in the deep central basin of Lake Pavin (Fig. 23.2; Table 23.2) and dated to AD 1282 +/− 20 according to the PAV12 age-depth model (Fig. 23.6). This major mass wasting deposit originates from a fresh **slide scar** at the edge of the plateau (Fig. 23.3) and is correlated with a slightly older but outstanding erosive sandy layer bearing organic macro remains dated to AD 1190 +/− 30 at coring site PAV09-C5 in the littoral environment identified north of the plateau (Fig. 23.2). This interpretation suggesting that this large **slide** was associated with violent and erosive waves along the lake shore is in agreement with available sedimentary facies, stratigraphy, SDR and RE data (Fig. 23.9). Firstly, only rare large waves could form an outstanding erosive sandy layer in littoral core PAV09-C5 and rework coarse littoral particles with numerous leaves debris. Secondly, the leave debris dated by **AMS radiocarbon** in this coarse layer can be older than the event. Thirdly, in a Q7/4 diagram, the E5 MWD in PAV12 is mainly made of **diatomite** material accumulated on the plateau (i.e. similar to the **diatomite** found in the lower unit of PAV10-E), while E5 **turbidite** as a different signature. Finally, in a **S2** vs. TOC diagram, the E5 MWD has a similar organic composition than the lower **diatomite** unit, but E5 **turbidite** contains mainly terrestrial organic matter. It seems thus likely that large and erosive waves triggered by the slide of **diatomite** could export down to the central basin some of the fine grained material deposited within the E5 **turbidite**, while E5 MWD was already frozen at the basin floor. Pollen data are showing variable percentages of assemblages dominated by arboreal pollen (mainly beech, oak, fir and hazel) within the base of E5 MWD, with a sudden drop of beech percentages synchronous with first record of tree crops like chestnut (*Castanea*) and walnut (*Juglans*) in the upper part of E5 MWD. Such results are in agreement with available **radiocarbon ages** in PAV99 within this deposit and are both suggesting sediment remolding and loss of chronological order within the **slump**. In the E5 **turbidite**, numerous parasitic and saprophytic fungal spore of hydrophilous vegetation and the noticeable presence of fir stomata are originating from the lakeshore and this pattern may have been favored by wave propagation along the lakeshore. Diatom assemblages in E5 MWD consist in a mix of different planktonic species from the pelagic zone (*A.formosa*, different species of *Stephanodiscus*), associated with diverse Fragilariaceae-benthic assemblage, as *Staurosira construens* known to live in the littoral zone. Once again, there is a change in E5 **turbidite** where different indicators (rotifer like *Brachionus* or cyanobacteria such as *Aphanizomenon* and diatom as *Eolimna minima*) suggest higher trophic level (Barbiero and Warren 2011). Consequently, it is thus likely that E5 **turbidite** had a slightly different sediment source than E5 MWD, due to the generation of violent waves along the lakeshore.

In a previousstudy, Chapron et al. (2012) suspected earthquake triggering for this large event E5 in Pavin, since it seems contemporaneous to a remarkable **slide** in nearby Lake Montcineyre (Fig. 23.13). This is further supported now by available **radiocarbon** dating in Lake Montcineyre dating this **slide** around AD 1320 +/− 10) and by the dating of another contemporaneous MWD back to cal. AD 1310 +/− 100 in Lake Guéry located in the Puy de Sancy area, ca. 7 km north from Pavin (Chassiot et al. 2016). Ongoing radiocarbon dating will tell if a third regional MWD in Lake La Crégut (at ca. 20 km from Pavin) and the outstanding slump and turbidite deposits in maar Lake Chauvet described in Chap. 22 (this issue), are also contemporaneous. As shown in Table 23.2, **sedimentary event** E5 may also be contemporaneous to a reported earthquake in the Limousin region in June, 13 AD 1348 (the Uzerche event of unknown both intensity and precise epicenter location). Since this event occurred few decades after the calculated age of E5 based on PAV12 chronology, it means that additional dating uncertainties in PAV12 age-depth model within the upper **diatomite** unit maybe related with unidentified sediment erosion episodes associated with the development of **turbididty currents** that deposited sedimentary events E2, E3 and E4 at site PAV09-B1.

A set of new data from **sedimentary event** E6 presented in this study, allows to further precise the sediment source areas and the consequences of this other major event that deeply impacted Lake Pavin (Chapron et al. 2010, 2012).

The changes in organic matter geochemistry observed just after this event (enrichment of organic matter of terrestrial origin or oxidized organic matter in the upper **diatomite** unit in both PAV8 and PAV12 coring sites) and related to the progressive erosion and remobilization of exposed former littoral lacustrine sediments following an abrupt lake level drop of ca. 13 m, suggest (i) that E6 MWD was much probably favored by the perturbation of subaqueous sediment pore pressure during the lake level drop and (ii) that this large MWD on the plateau was therefore associated with a major **lake outburst** and a catastrophic debris flow downstream in the Couze Pavin valley. As shown in Table 23.2, a similar age obtained from the second outstanding erosive sandy layer at site PAV09-C5, further points toward the propagation of violent waves during this event. The origin of the **lake outburst** is, however, still unknown. It may either result from a wetter period since there is a general increase in detrital sediment supply to regional lacustrine systems around AD 600 in Western Europe (Fig. 23.12), but it may also result from earthquake shaking, since several younger seismic events were apparently recorded in this maar lake by slope failures. Finally, as discussed in Chap. 3 (this issue), this major environmental crisis at Lake Pavin and downstream in the Pavin valley ca. 1400 years ago, may have cause the end of a Roman query at the shore of Lake Pavin documented in a text from Gregorius of Tours, one of the first historian of Gaul relating pagan lake cults and regular catastrophic events in Auvergne (Chap 2, this issue). It is however beyond the scope of this paper to further discuss the possible evolution of human activities within the topographic drainage basin of Lake Pavin since the Roman period, because this time window is not documented in core PAV12 due to sediment erosion at the base of sedimentary event E6.

23.6 Conclusions and Perspectives

The multidisciplinary study of Lake Pavin sedimentary infill combining acoustic soundings, and well-dated sediment cores together with the integration of regional and historical data sets on climate, human activities and natural hazards in the study area highlight that:

(i) Lake Pavin sedimentation was essentially dominated by mineral inputs from the drainage basin following the formation of Pavin crater;
(ii) Afterwards, the rapid development of vegetation cover along the inner slopes of the crater limited strongly mineral inputs toward the lake and favored the formation of **diatomite**, an organic rich sedimentation dominated by the productivity in the water column over the last ca. 7000 years;
(iii) The identification of several short periods of enhanced mineral inputs within the diatomite deposits contemporaneous with similar trends observed previously in contrasted lacustrine systems from western Europe, might reflect the influence of climate change (i.e. wetter periods) on Pavin sedimentation;
(iv) The development of Pavin **meromicticity** may have been favored by the occurrence of at least two major subaqueous slope failures dated at ca. AD 600 and ca. AD 1300. These two events were associated with unusual erosive waves at the lake shore and related with a catastrophic **lake outburst** and a lake level drop of ca. 13 m some 1400 years ago and regional earthquake shaking some 700 years ago, respectively;
(v) Apparently limited human impact on Pavin sedimentation since the Roman period is unusual in this region and may result from the catastrophic consequences of large subaquatic slides around AD 600 and AD 1300, but also recurrent more limited slopes failures since the Little Ice Age possibly triggered by regional historical earthquakes (in AD 1783; in either AD 1833 or AD 1844; in AD 1863 and in AD 1921) and eventually associated with a **limnic eruption** in AD 1783. Further studies integrating archeology, paleo ecology and lacustrine sedimentology are still needed to better document the impact of former societies on the environment in the study area. Enhanced slope instabilities since the end of the eighteenth century may, however, resulted from the perturbation of subaqueous sediment pore pressure after the artificial lake level drop by ca. 4 m.

Future and ongoing studies on Lake Pavin area should confirm the timing, causes and consequences of this event stratigraphy based on similar limnogeological approaches in several contrasted regional lakes, in order to further pinpoint the respective influences of climate changes and land use evolution on lacustrine sedimentation. Finally, this reconstruction of Lake Pavin paleolimnology and event stratigraphy should help reconstructing the history of its **meromicticity** and the evolution of its remarkable biodiversity.

Acknowledgements This study benefited from several projects funded by (i) the MEEDDAT project of the DDEA Puy-de- Dôme, (ii) the Agence de l'Eau Loire-Bretagne (AELB) project EDIFIS, (iii) the INSU project DICENTIM and (iv) the ARTEMIS facilities from INSU and INSHS for radiocarbon dating. L. Chassiot PhD grant is in addition funded by the Région Centre. We wish to thank Anaëlle Simonneau (ISTO, Orléans) for field work support and scientific discussions, Alice Recanati (LMCM, Paris) for MEB pictures and Michel Meybeck for scientific discussions. We wish to thank Prof. Marc Desmet for his constructive review of this chapter.

References

Aeschbach-Hertig W, Hofer M, Kipfer R, Imboden DM, Wieler R (1999) Accumulation of mantle gases in a permanently stratified volcanic lake (Lac Pavin, France). Geochim Cosmochim Acta 63(19–20):3357–3372

Albéric P, Jézéquel D, Bergonzini L, Chapron E, Viollier E, Massault M, Michard G (2013) Carbon cycling and organic radiocarbon reservoir effect in a meromictic crater lake (lac Pavin, Puy-de-Dôme, France). Radiocarbon 55(2–3):1029–1043

Ariztegui D, Chondrogianni C, Lami A, Guilizzoni P, Lafargue E (2001) Lacustrine organic matter and the Holocene paleoenvironmental record of Lake Albano (central Italy). J Paleolimnol 26:283–292

Arnaud F, Révillon S, Debret M, Revel M, Chapron E, Jacob J, Giguet-Covex C, Poulenard J, Magny M (2012) Lake Bourget regional erosion patterns reconstruction reveals Holocene NW European Alps soil evolution and paleohydrology. Quat Sci Rev 51:81–92

Augustinus P, Cochran U, Kattel G, D'Costa D, Shane P (2012) Late Quaternary paleolimnology of Onepoto maar, Auckland, New Zealand: implications for the drivers of regional paleoclimate. Quat Int 253:18–31

Bacon C, Gardner J, Mayer L, Buktenica M, Dartnell P, Ramsey D, Robinson J (2002) Morphology, volcanism and mass wasting in Crater Lake, Oregon. Geol Soc Am Bull 114:675–692

Bani P, Join J-L, Cronin SJ, Lardy M, Rouet I, Garaebiti E (2009) Characteristics of the summit lakes of Ambae volcano and their potential for generating lahars. Nat Hazards Earth Syst Sci 9:1471–1478

Barbierro RP, Warren GJ (2011) Rotifer communities in the Laurentian Great Lakes, 1983–2006 and factors affecting their composition. J Great Lakes Res 37:528–540

Behar F, Beaumont V, Penteado HL, De B (2001) Rock-Eval 6 technology: performances and developments. Oil Gas Sci Technol – Rev IFP 56(2):111–134

Blaauw M (2010) Methods and code for 'classical' age-modelling of radiocarbon sequences. Quat Geochronol 5:512–518

Blomqvist P (2001) Phytoplankton responses to biomanipulated grazing pressure and nutrient additions – enclosure studies in unlimed and limed Lake Njupfatet, central Sweden. Environ Pollut 111:333–348

Bonhomme C, Poulin M, Vinçon-Leite B, Saad M, Groleau A, Jézéquel D, Tassin B (2011) Maintaining meromixis in Lake Pavin (Auvergne, France): the key role of sublacustrine spring. Compt Rendus Geosci 343:749–759

Brauer A, Endres C, Günter C, Litt T, Stebich M, Negendank JFW (1999) High resolution sediment and vegetation responses to Younger Dryas climate change in varved lake sediments from Meerfelder Maar, Germany. Quat Sci Rev 18(3):321–329

Caballero M, Vazquez G, Lozano-Garcia S, Rodriguez A, Sosa-Najera S, Ruiz-Fernandez AC, Ortega B (2006) Present limnological conditions and recents (ca. 340 yr) paleolimnology of a tropical lake in the Sierra de Los Tuxtlas, eastern Mexico. J Paleolimnol 35:83–97

Camus G, Michard G, Olive P, Boivin P, Desgranges P, Jezequel D, Meybeck M, Peyrus J-C, Vinson J-M, Viollier E, Kornprobst J (1993) Risques d'éruption gazeuse carbonique en Auvergne. Bull Soc Geol Fr 164(6):767–781

Caracausi A, Mario Nuccio P, Favara R, Nicolosi M, Paternoster M (2009) Gas hazard assessment at the Monticchio crater lakes of Mt. Vulture, a volcano in Southern Italy. Terra Nova 21:83–87

Chapron E, Beck C, Pourchet M, Deconinck JF (1999) 1822 earthquake-triggered homogenite in Lake Le Bourget (NW Alps). Terra Nova 11:86–92

Chapron E, Arnaud F, Noël H, Revel M, Desmet M, Perdereau L (2005) Rhône-river flood deposits in Lake Le Bourget: a proxy for Holocene environmental changes in the NW Alps, France. Boreas 34(4):404–416

Chapron E, Albéric P, Jézéquel D, Versteeg W, Bourdier J-L, Sitbon J (2010) Multidisciplinary characterization of sedimentary processes in a recent maar lake (Lake Pavin, French Massif Central) and implication for natural hazards. Nat Hazards Earth Syst Sci 10:1–13

Chapron E, Ledoux G, Simonneau A, Albéric P, St-Onge G, Lajeunesse P, Boivin P, Desmet M (2012) New evidence of Holocene Mass-Wasting events in recent volcanic lakes from the French Massif Central (Lakes Pavin, Montcineyre and Chauvet) and implications for natural hazards. In: Yamada Y (ed) Submarine mass movements and their consequences, vol 31, Advances in natural and technological hazards research. Springer, Dordrecht, pp 255–264

Chassiot L, Chapron E, Di Giovanni C, Lajeunesse P, Tachikawa K, Garcia M, Bard E (2016) Historical seismicity of the Mont Dore volcanic province (Auvergne, France) unraveled by a regional lacustrine investigation: new insights about lake sensitivity to earthquakes. Sediment Geol 339:134–150

Cugny C, Mazier F, Galop D (2010) Modern and fossil non-pollen palynomorphs from the Basque mountains (Western Pyrenees, France): the use of coprophilous fongi to reconstruct pastoral activity. Veg Hist Archaeobotany 19:391–408

Debret M, Desmet M, Balsam W, Copard Y, Francus P, Laj C (2006) Spectrophotometer analysis of Holocene sediments from an anoxic fjord: Saanich Inlet, British Columbia, Canada. Mar Geol 229:15–28

Debret M, Chapron E, Desmet M, Rolland-Revel M, Magand O, Trentesaux A, Bout-Roumazeille V, Nomade J, Arnaud F (2010) North western Alps Holocene paleohydrology recorded by flooding activity in Lake Le Bourget, France. Quat Sci Rev 29:2185–2200

Debret M, Sebag D, Desmet M, Balsam W, Copard Y, Mourier M, Susperrigui A-S, Arnaud F, Bentaleb I, Chapron E, Lallier-Vergès E, Winiarski T (2011) Spectrocolorimetric interpretation of sedimentary dynamics: the new "Q7/4 diagram". Earth Sci Rev 109:1–19

Delbecque A (1898) Les lacs français. Chamerot & Renouard, Paris, 436 p

Delibrias G, Guillier MT, Labeyrie J (1972) Gif natural radiocarbon measurements VII. Radiocarbon 14:280–320

Giresse P, Maley J, Kelts K (1991) Sedimentation and paleoenvironment in crater lake Barombi Mbo, Cameroon, during the last 25 000 years. Sediment Geol 71:151–175

Gutowski A, van de Weyer K, Hofmann G, Doege A (eds) (2011) Makrophyten und Phytobenthos Indikatoren für den ökologischen Gewässerzustand. Druckschrift des Sächsischen Landesamts für Umwelt, Landwirtschaft und Geologie des Freistaates Sachsen, Dresden, Mai 2012, 188 pp

Hindáková A (2009) On the occurrence of Achnanthes thermalis var. rumrichorum (Bacillariophyceae) in Slovakia. Fottea 9:193–198

Jézéquel D, Sarazin G, Prévot F, Viollier E, Groleau A, Michard G, Agrinier P, Albéric P, Binet S, Bergonzini L (2011) Bilan hydrique du lac Pavin. Revue des Sciences Naturelles d'Auvergne 74–75:67–90

Kulbe T, Niederreiter RJ (2003) Freeze coring of soft surface sediments at a water depth of several hundred meters. J Paleolimnol 29:257–263

Lambert J, Levret-Arbaret A, Czitrom G, Dubie J-Y, Godefroy P (1997) Les tremblements de terre en France. Editions BRGM, Orléans, 127 p

Lange-Bertalot H (2001) Navicula sensu stricto, 10 Genera separated from Navicula sensu lato Frustulia. Diatoms of Europe 2:526

Lavrieux M, Disnar J-R, Chapron E, Bréheret J-G, Jacob J, Miras Y, Reyss J-L, Andrieu-Ponel V, Arnaud F (2013) 6700 yr sedimentary record of climatic and anthropogenic signals in Lake Aydat (French Massif Central). The Holocene 23(9):1317–1328

Lotter AF, Bigler C (2000) Do diatoms in the Swiss Alps reflect the lenght of ice-cover? Aquat Sci 62:125–141
Magny M (2006) Holocene fluctuations of lake levels in west-central Europe: methods of reconstruction, regional patterns, palaeoclimatic significance and forcing factors. Encycl Quat Geol 3:1389–1399
Magny M, Haas JN (2004) A major widespread climatic change around 5300 cal. Yr BP ago at the time of the Alpine Iceman. J Quat Sci 19:423–430
Magny M, Combrieu Nebout N, de Beaulieu J-L, Bout-Roumazeilles V, Colombaroli D, Desprat S, Francke A, Joannin S, Peyron O, Revel M, Sadori L, Siani G, Sicre M, Samartin S, Simonneau A, Tinner W, Vannière B, Wagner B, Zanchetta G, Anselmetti F, Brugiapaglia E, Chapron E, Debret M, Desmet M, Didier J, Essallami L, Galop D, Gilli A, Haas JN, Kallel N, Millet L, Stock A, Turon J, Wirth S (2013) North-south paleohydrological contrasts in the central Mediterraenean during the Holocene: tentative synthesis and working hypotheses. Clim Past 9:1901–1967
Martin J-M (1985) The Pavin crater lake. In: Stumm Y (ed) Chemical processes in lakes. Wiley, New York, pp 169–188
Martin J-M, Meybeck M, Nijampurkar VN, Somayajulu BLK (1992) ^{210}Pb, ^{226}Ra and ^{32}Si in Pavin lake (Massif Central, France). Chem Geol 94:173–181
Martin-Puertas C, Brauer A, Dulski P, Brademann B (2012) Testing climate-proxy stationarity throughout the Holocene: an example from the varved sediments of Lake Meerfelder Maar (Germany). Quat Sci Rev 58:56–65
Miras Y, Laggoun-Defarge F, Guenet P, Richard H (2004) Multi-disciplinary approach to changes in agro-pastoral activities since the Sub-boreal in the surroundings of the "narse d'Espinasse" (Puy-de-Dôme, French Massif Central). Veg Hist Archaeobot 13:91–103
Monecke K, Anselmetti F, Becker A, Sturm M, Giardini D (2004) The record of historic earthquakes in lakes sediments of Central Switzerland. Tectonophysics 394:21–40
Mott RW, Woods AW (2010) A model of overturn of CO2 laden lakes triggered by bottom mixing. J Volcanol Geotherm Res 192(3–4):151–158
Mulder T, Cochonat P (1996) Classification of offshore mass movements. J Sediment Res 66:43–57
Nomade J, Chapron E, Desmet M, Reyss JL, Arnaud F, Lignier V (2005) Reconstructing historical seismicity from lake sediments (Lake Laffrey, Western Alps, France). Terra Nova 17:350–357
Oldfield F (1996) The PASICLAS project: synthesis and overview. Memoire dell'instituto italiano di Idrobiologia 55:329–357
Reille M, Pons A, de Beaulieu J-L (1992) Late and postglacial vegetation, climate and human action in the French Massif Central. Laboratoire de Botanique Historique et Palynologie, Marseille
Reimer PJ, Bard E, Bayliss A, Warren Beck J, Blackwell PG, Bronk Ramsey C, Buck CE, Cheng H, Lawrence Edwards E, Friedrich M, Grootes PM, Guilderson TP, Haflidason H, Hajdas I, Hatté C, Heaton TJ, Hoffmann DL, Hogg AG, Hughen KA, Felix Kaiser K, Kromer B, Manning SW, Niu M, Reimer RW, Richards DA, Scott EM, Southon JR, Staff RA, Turney CSM, van der Plicht J (2013) IntCal13 and Marine13 radiocarbon age calibration curves 0–50,000 years cal BP Radiocarbon 55:1869-1887
Rioual P, Andrieu-Ponel V, de Beaulieu J-L, Reille M, Svobodovac H, Battarbee RW (2007) Diatom responses to limnological and climatic changes at Ribains Maar (French Massif Central) during the Eemian and Early Würm. Quat Sci Rev 26:1557–1609
Schettler G, Albéric P (2008) Laghi di Monticchio (Southern Italy, Region Basilicata): genesis of sediments - a geochemical study. J Paleolimnol 40:529–556
Schettler G, Schwab MJ, Stebich M (2007) A 700-year record of climate change based on geochemical and palynological data from varved sediments (Lac Pavin, France). Chem Geol 240:11–35
Schmid M, Halbwachs M, Wehrli B, Wüest A (2005) Weak mixing in Lake Kivu: new insights indicate increasing risk of uncontrolled gas eruption. Geochem Geophys Geosyst 6(7):1–11
Sifeddine A, Bertrand P, Lallier–Vergès E, Patience AJ (1996) Lacustrine organic fluxes and palaeoclimatic variations during the last 15 ka: lac du Bouchet (Massif Central, France). Quat Sci Rev 15:203–211
Sigurdsson H, Devine JD, Tchua FM, Presser FM, Pringle MKW, Evans WC (1987) Origin of lethal gas burst from Lake Monoun, Cameroun. J Volcanol Geotherm Res 31:1–16
Simonneau A, Chapron E, Vannière B, Wirth SB, Gilli A, Di Giovanni C, Anselmetti FS, Desmet M, Magny M (2013a) Mass-movement and flood-induced deposits in Lake Ledro, Southern Alps, Italy: implications for Holocene paleohydrology and natural hazards. Clim Past 9:825–840
Simonneau A, Chapron E, Courp T, Tachikawa K, Le Roux G, Baron S, Galop D, Garcia M, Di Giovanni C, Motellica-Heino M, Mazier F, Foucher A, Houet T, Desmet M, Bard E (2013b) Recent climatic and anthropogenic imprints on lacustrine systems in the Pyrenean Mountains inferred from minerogenic and organic clastic supply (Vicdessos valley, Pyrenees, France). The Holocene 23:1764–1777
Simonneau A, Chapron E, Garçon M, Winiarski T, Graz Y, Chauvel C, Debret M, Motelica-Heino M, Desmet M, Di Giovanni C (2014) Tracking Holocene glacial and high-altitude alpine environments fluctuations from minerogenic and organic markers in proglacial lake sediments (Lake Blanc Huez, Western Frenc Alps). Quat Sci Rev 89:27–43
Stebich M, Brüchmann C, Kulbe T, Negendank JFW (2005) Vegetation history, human impact and climate change during the last 700 years recorded in annually laminated sediments of Lac Pavin, France. Rev Palaeobot Palynol 133:115–133
Strasser M, Anselmetti F (2008) Mass-movement events stratigraphy in Lake Zurich: a record of varying seismic and environmental impacts. Beiträge zur Geologie der Schweiz, Geotechnische Serie 95:23–41
Tanasienko A, Yakutina O, Chumbaev A (2011) Effect of snow amount on runoff, soil los and suspended sediment during periods of snowmelt in Southern West Siberia. Catena 87:45–51
Thouveny N, de Beaulieu JL, Bonifay E, Creer KM, Guiot J, Icole M, Johnsen S, Jouzel J, Reille M, Williams T, Williamson D (1994) Climate variations in Europe over the past 140-Kyr deduced from rock magnetism. Nature 371:503–506
Truze E, Kelts K (1993) Sedimentology and Paleoenvironment from the maar lac du Bouchet for the last climatic cycle, 0–120,000 years (Massif Central, France). In: Negendank JFW, Zolitschka B (eds) Paleolimnology of European Maar Lakes, vol 49, Lecture notes in earth sciences., pp 237–375
Van Dam H, Mertens A, Sinkeldam J (1994) A coded checklist and ecological indicator values of freshwater diatoms from the Netherlands. Neth J Aquat Ecol 28:117–133
Viollier E, Albéric P, Jezequel D, Michard G, Pèpe M, Sarazin G (1995) Geochemical study of a crater lake: the Pavin Lake, France-trace element behavior in the monimolimnion. Chem Geol 125:1–72
Voigt R, Grüger E, Baier J, Meischner D (2008) Seasonal variability of Holocene climate: a palaeolimnological study on varved sediments in Lake Jues (Harz Mountains, Germany). J Paleolimnol 40:1021–1052
Walker IR, Paterson CG (1986) Associations of diatoms in the surficial sediments of lakes and peat pools in Atlantic Canada. Hydrobiologia 134:265–272
Zolitschka B, Anselmetti F, Ariztegui D, Corbella H, Francus P, Lücke A, Maidana NI, Ohlendorf C, Schäbitz F, Wastegard S (2013) Environment and climate of the last 51000 years – New insights from the Potrok Aike maar lake Sediment Archive Drilling prOject (PASADO). Quat Sci Rev 71:1–12

Glossary

$\delta^{13}C$ this parameter corresponds to the isotopic carbon composition (so-called isotopic signature) of a sample, expressed as $^{13}C/^{12}C$ ratio which is compared to a reference material:

$$\delta^{13}C = \left(\frac{\left(\frac{^{13}C}{^{12}C} \right)_{sample}}{\left(\frac{^{13}C}{^{12}C} \right)_{reference}} - 1 \right) \times 1000‰$$

Both isotopes have similar physicochemical properties but slight behavior differences, mainly due to the mass difference, lead to isotope fractionation. For example, during photosynthesis, faster diffusion of ^{12}C relative ^{13}C in leaf tissues and other chemical processes results in a greater uptake of ^{12}C versus ^{13}C in the organic matter. The $\delta^{13}C$ of the organic matter is consequently lower (typically −25‰ for C3 plants) than to that of the dissolved CO_2 or atmospheric CO_2 (≈−8‰). By consequence of photosynthesis, in water, the residual dissolved CO_2 (or DIC) is enriched in heavy isotope. $\delta^{13}C$ is then a powerful tool to infer biogeochemical reactions, as well as carbon sources (as isotope carbon composition depends also on the producer organisms).

^{14}C ages dating of materials containing carbon is now a days performed through the determination of the ratio of ^{14}C radioactive isotope *versus* ^{12}C stable isotope in a sample that initially incorporated ^{14}C from atmospheric CO_2 (piece of wood, leaf, charcoal, particulate organic matter (POC), carbonate precipitate from water DIC content…). The principle is based on the radioactive decay of ^{14}C (half-life ⊕5730 yrs). There are several ways to express the result: (1) age in years BP (before present = before 1950); (2) however, due to some variation in the $^{14}CO_2$ production rate in the atmosphere (^{14}C is continuously produced in the upper atmosphere due to solar radiation), the raw age value has to be corrected: age is expressed in years cal. BP (calibrated BP); (3) alternatively the result may be expressed in percent of modern carbon (PMC): 100 % PMC means that the sample is "modern" – actual – whereas 0 % PMC means that the sample is old (contains "dead" carbon, *i.e.* more than tenfold the half-life).

18S rDNA (18S rRNA) Abbreviation of 18 S ribosomal DNA (or RNA). 18S rDNA is a part of the rribosomal DNA. The small subunit (SSU) 18S rDNA is one of the most frequently used genes in phylogenetic studies.

18S rDNA Gene encoding the RNA component of the small ribosomal subunit, and widely used to identify and classify organisms. clone library Heterogeneous collection of cloned sequences (often 18S rDNA) derived from a complex assemblage of organisms.

AD Latin abbreviation for Anno Domini used in Julian and Gregorian calendars.

AMS radiocarbon radiometric method using an accelerator mass spectrometry (AMS) to measure isotopic record of ^{14}C carbon in order to date an organic sample. The isotopic composition of C can be used for stratigraphic correlations.

ASCONIT engineering consulting firm

ATHOS engineering consulting firm

Abiotic Not associated with or derived from living organisms.

Acoustic facies acoustic signature of sedimentary material according to the presence, frequency, amplitude, continuity and geometry of reflections on seismic profiles.

Acoustic substratum horizon or formation absorbing acoustic waves on sub-bottom profiles generally associated with bedrock.

Aerobic Relating to, involving, or requiring free oxygen.

Alcian Blue a dye used to stain acidic polysaccharides.

Aliphatic Compound having an open chain structure, not aromatic.

Alkalinity by definition this chemical quantity is:

$$alk = [HCO_3^-] + 2[CO_3^{2-}] + [OH^-] + [bases] - [H^+] - [acids]$$

where [bases] corresponds to basic species engaged in acid-base couple with pKa>4 and [acids] corresponds to acid species engaged in acid-base couple with pKa<4 (*e.g.* $[NH_3]$, $[HPO_4^{2-}]$, $2\times[PO_4^{3-}]$… or $[H_3PO_4]$ respectively). Alkalinity is modified by dissolution or precipi-

T. Sime-Ngando et al. (eds.), *Lake Pavin*, DOI 10.1007/978-3-319-39961-4 tation of carbonate minerals, by basifying or acidifying reactions (*e.g.* alk is increased by photosynthesis and decreased by oxidation of Fe^{2+} to $Fe(OH)_3$), but alk is not modified by CO_2 input or escape from the water. Alk is generally determined by HCl titration. Knowing alk and pH allows calculating the partial pressure of CO_2(pCO_2) and DIC (more generally the determination of two quantities among pH, alk, DIC or pCO_2 allows the calculation of the two other ones).

Anaerobic Relating to, involving, or requiring an absence of free oxygen.

Anoxia state of waters without dissolved oxygen, as in Pavin monimolimnion.

Anthropogenic Caused or produced by human.

Aromatic Unsaturated compound containing one or more ring, not aliphatic.

Bacterivores (bacterivorous) Microorganisms that ingest bacteria as a source of energy.

Bedrock a general term for the solid rock underlying unconsolidated material such as lacustrine sediments.

Beine a narrow terrace, immersed and/or emerged, at the lake shore.

Biodegradation Process by which organic substances are decomposed by living-organisms into simpler substances.

Biogeochemical cycle The flow of chemical elements and compounds between living organisms and the physical environment.

Biogeography is the study of the distribution of species and ecosystems in geographic space and time.

Bioremediation The use of biologically mediated processes to remove or degrade pollutants from specific environments.

Biosphere The part of the earth and its atmosphere in which living organisms exist or that is capable of supporting life.

Biotic Associated with or derived from living organisms.

Burst size The number of new phages liberated by spontaneous lysis of an infected host cell.

CPUE Catch Per Unit Effort is an indirect measure of the abundance of target species. It corresponds of the number of catching fish by the number of UFE. It's useful to track evolution of catches at constant effort.

Canyons a deep and narrow depression between escarpments.

Carboxyl Group (symbolized as COOH) has both a carbonyl and a hydroxyl group attached to the same carbon atom.

Catchment area extent of the area around the lake collecting and draining rainwater.

Caudovirales group of viruses also known as the tailed bacteriophages.

cauliflower bombs peculiar aspect of magmatic bombs that are quenched during a phreatomagmatic explosion.

Chemocline a cline (see cline) caused by a strong, vertical chemistry gradient within a body of water.

Chlorination Introduction of chlorine into a chemical compound.

Chondrites meteorites of which chemical composition is used as a reference for bulk earth composition (except volatile elements).

Chytridiomycosis Infection of organisms caused by fungal pathogens (e. g. chytrids).

Clearance rate Equivalent to filtration rate for unicellular eukaryotes.

Climax the final state of ecological succession and the most stable state in the existing conditions.

Cline horizontal layer within a fluid, in which a property of the fluid varies greatly over a relatively short vertical distance.

Cloning/sequencing Method used in molecular biology for sequencing long DNA strands.

Co-metabolism The transformation of an organic compound by a microorganism incapable of using the substrate as a source of energy or of one of its constituent elements.

Coenzyme organic small molecule of non-protein nature promoting the activity of the enzyme, and are often even essential.

Cofactor Any non-protein substance required by a protein for biological activity.

Consortium, consortia association between different partners of microorganisms.

Consortium A physical association of at least two different organisms, usually beneficial to both organisms.

Conjugation Transfer of genetic material (e.g. plasmid) between bacterial cells by direct cell-to-cell contact or by a bridge-like connection between two cells.

Cosmographiae early geographic encyclopedia covering the known world (sixteenth and seventeenth century).

Crater of elevation Theory proposed in 1820 by the German geologist L. von Buch, which will be taken and firmly defended by A. Dufrénoy and L. Élie de Beaumont who used Cantal and the Monts-Dore volcanoes as examples: "*Le cratère de soulèvement est une portion de l'écorce terrestre qui poussée par une force quelconque dont l'action a été dirigée de bas en haut a été ainsi soulevée de manière à prendre une forme plus ou moins circulaire/The crater of elevation is a portion of the earth's crust which was pushed by some strength, the action of which was driven bottom up and was raised so as to take a more or less circular shape*".

Crater rim (or ring) a low-relief rim of fragmental material surrounding a maar lake.

Cyanobacteria Major group of photosynthetic eubacteria that have two photosystems, and realise oxygenic photosynthesis, previously called *blue-green algae*.

Cyanophage Virus that infects cyanobacteria.

Cytochrome intermediate coenzyme of the respiratory chain.

DAPI 4′,6′-diamidino-2-phénylindole, a fluorescent stain that binds strongly to A-T rich regions in DNA.

DIC (Dissolved Inorganic Carbon) also named ΣCO_2, this parameter is the sum of dissolved inorganic carbon species concentrations: $DIC=[H_2CO_3]+[HCO_3^-]+[CO_3^{2-}]$ The $[H_2CO_3]$ term contains the true dissolved $CO_{2(aq.)}$ (aqueous CO_2) and the hydrated form H_2CO_3 i.e. the carbonic acid. Various bio- and physicochemical processes affect DIC: exchange with atmosphere (CO_2), phytoplankton uptake or CO_2 release from several respiration or fermentation pathways, precipitation or dissolution of carbonate minerals… Proportion of the three species is strongly related with the pH of water.

Dechlorination The process of removing the chlorine residue from chlorinated compounds.

Defortunate abandoned places left to the forces of nature, where religious settlements were built. Vassivière mountain, close to Pavin (1300 m a.s.l.), could be one of them.

Degassing controversy scientific controversy about the occurrence, the origin or the mechanism of lake degassing.

Degassing grid see sensory grid.

Degree days It is the product of the daily temperature of the water by the number of days. For Arctic char eggs hatch into fish larvae after 450° days.

Dehalogenase One of a group of enzymes that dehalogenate organohalide substrates.

Diatom bloom A rapid increase or accumulation in the population of diatoms in an aquatic ecosystem.

Diatom a major group of algae, among the most common types of phytoplankton.

Diatomite soft friable siliceous sedimentary rock or unit made of opaline frustules of the diatom, a unicellular aquatic plant related to the algae.

Diffuse spectral reflectance ratio of energy reflected by a material at a specific wavelength.

Dimictic attribute of lakes mixed twice a year at homothermy

Dioxygenase One of a group of enzyme that catalyses the incorporation of both atoms of molecular oxygen into substrates.

Disproportionation specific type of redox reaction in which a species is simultaneously reduced and oxidized to form two different products.

Dissipation is the part of turbulence that is not expressed in terms of movement but in the form of heat (related to the viscosity of the fluid).

Dragon fantastic creature present in ancient maar-lake stories as Pavin and possibly Albano); presented as living in the lake, it is very powerful, aerial, aquatic and/or telluric creature with a poisonous breath and can be deadly (e.g. Silvester dragon). These dragons, worshipped during pagan times, often with human sacrifices, are often defeated in early Christian times.

Ecosystem a biotic community and its surroundings, part inorganic (abiotic) and part organic (biotic), the latter including producers, consumers, and decomposers.

Electron acceptor A chemical entity that accepts electrons transferred to it from another compound.

Electron donor A chemical entity that donates electrons to another compound.

Enzyme A protein causing and accelerating chemical reactions.

Epifluorescence microscopy Technique that allows observation by fluorescence of very small cells retained on a filter.

Epilimnetic relating to epilimnion (see epilimnion).

Epilimnion the layer of water above the thermocline in stratified lakes (see thermocline).

Epiphytic An organism that grows on a plant for support.

Equilibrium constants the value of the reaction quotient (see reaction quotient) when the reaction has reached equilibrium. At any given temperature, the equilibrium constant has a value which is independent of the initial and actual concentrations of the reactant and product species.

Erosive layer coarse grained sediment deposit deposited on top of an erosion surface that cuts into and eroded part of the underlying sediments. This results in a gap in time recording by sediment archives at the base of the layer.

Erratic formation A term, now abandoned, that was used to identify the deposits formed by the action of glaciers such as moraines. Nopwadays, we would use till or moraine.

Euryarchaeota phylum within the *Archaea* including methanogens.

Event stratigraphy chronologic arrangement (stratigraphy) that record geological rare events at any scale.

Exergonic process process in which there is a positive flow of energy from the system to the surroundings.

FDPPMA 63 Departmental Federation of Fisheries and Protection of the Aquatic Environment of the Puy-de-Dôme.

Fairies fantastic creatures found in ancient Pavin legends. They are essentially aerial and lure the passers by. They are interpreted as $C0_2$ emissions.

Fantastic creatures representations of maar-lakes misbehaviors in ancient times stories (dragon, fairies) or in Antique and medieval iconography (Minotaurus, Medusa, Hydra, Mephytes).

Ferruginous enriched in dissolved ferrous iron (Fe^{2+}).

Flow cytometry Laser-based, biophysical technology for counting, examining and sorting microscopic particles (after staining with fluorescent dye) by suspending them in a stream of fluid and passing them through an optical and/or electronic detection apparatus. It allows simultaneous multiparametric analysis of the physical and/or chemical characteristics of up to thousands of particles per second.

Fluorescent *in situ* hybridization (FISH) Technique used to detect and localize the presence or absence of specific DNA sequences on chromosomes. FISH uses fluorescent probes that bind to only those parts of the chromosome with which they show a high degree of sequence complementarity.

Fluorochrome Any of a group of fluorescent dyes (chemical entity, such as a molecule or group) used to label biological material.

Frequency of Visibly Infected Cells (FVIC) Proportion of prokaryotic cells containing viral particles.

Frustule the siliceous skeleton of diatoms.

Fungi, true fungi, fungi True fungi (or fungi) are eukaryotic organisms presenting a chitin cell wall. All these organisms are classified as a kingdom, Fungi.

GSI Gonado Somatic Index: In percent. It is the report of the weight of gonads on the weight of the fish.

Gas lake outburst or limnic eruption massive release of gases (mainly CO_2, CH_4) from the monimolimnion (deep part of a permanently stratified lake), which was recognized as a type natural hazard since the Lake Monoun (1984) and Lake Nyos (1986) disasters. The catastrophic gas outburst may be provoked by a sublacustrine or aerial landslide, which itself may be triggered by an earthquake. Another way to destabilize the water column leading to the massive gas escape is when the waters are supersaturated in gases: when the ebullition process is initiated, the resulting destabilization of waters is irreversible and lead to the massive escape.

Geomythology perception of geological hazards by local people, as transmitted across generations through myths, legends, rites or representations.

Geothermal process Process relating to the Earth's interior heat.

Granitic basement This is the basement of the French Massif central, comprising a widespread, flattened erosional surface, which is the result of the long-lasting erosion of the hercynian mountain range. The majority of rocks are granites and gneisses. The volcanic massifs lie directly on the erosional surface.

Gravity cores sedimentary cores acquired by dropping a core barrel with a weight on the lake floor.

Gregorius of Tours Saint Gregorius (c 538 – 594) is a Gallo-Roman historian and Bishop of Tours, born in Urbs Arverna –now Clermont. In his chronicles he mentioned several extra-ordinary events as the Tauredunum catastrophic event which destroyed Geneva (proven in 2012 to be a lake tsunami). It is here hypothesized that two of his stories concern past pagan lake cult at Pavin and a catastrophic overspill near 590 (Hermit Caluppa'dragons encounters).

Halide ion A singlet halogen atom, which is an anion with a charge of −1.

Halogen A family of chemical elements that comprises astatine, fluorine, chlorine, bromine and iodine.

Halogenase One of a group of enzymes that halogenate organic substrates using oxygen and a halide ion (Cl−, Br−, or I−).

Haloperoxidase One of a group of enzymes that halogenates organic substrates using hydrogen peroxide and a halide ion (Cl−, Br−, or I−). Haloperoxidases contain either heme or vanadium.

Heteroaromatic Having the characteristics of an aromatic compound whilst having at least one non-carbon atom in the ring.

Heterotrophic flagellates free-living protozoa, their movement and feeding mainly through flagella, heterotrophic way to get food or energy.

Hotspot microenvironment a microscale spot with high microbial activity.

Human sacrifices mentioned at maar-lakes in some pre-christian lake cults legends and stories.

Hydrogen Index (HI) amount of hydrocarbons released during pyrolysis in Rock-Eval analyzer, normalized to TOC (unit expressed in mg $HC.g^{-1}$ TOC).

Hydrological cycle The natural sequence through which water passes into the atmosphere as water vapor, precipitates to earth in liquid or solid form, and ultimately returns to the atmosphere through evaporation.

Hydrostatic pressure The pressure exerted by a fluid at equilibrium at a given point within the fluid, due to the force of gravity. Hydrostatic pressure increases in proportion to depth.

Hyperhalophiles type of extremophiles which includes prokaryotes and some eukaryotes (algae and fungi) that live in environments characterized by very high salt concentrations.

Hyperthermophiles subset of extremophile microorganisms that thrive in extremely hot environments (80 °C and above) such as hot springs.

Hypolimnetic relating to hypolimnion (see hypolimnion).

Hypolimnion is the dense, bottom layer of water in a thermally-stratified lake. It is the layer that lies below the thermocline.

IRSTEA National institute of research in sciences and technologies for the environment and the agriculture.

Iron isotope composition although iron has four stable isotopes (^{54}Fe, ^{56}Fe, ^{57}Fe, ^{58}Fe), only two isotopes are generally used to define the isotope composition of a specific sample. The isotope composition is expressed in the δ notation as:

$$\delta^{56}Fe = [(^{56}Fe/^{54}Fe)_{sample}/(^{56}Fe/^{54}Fe)_{standard} - 1] \times 1000$$

where the international standard is a synthetic metal from the Institute for Reference Material and Measurement (*i.e.* IRMM-014). It provides the deviation of the isotope ratio of a sample relative to the standard, with a permil unit (‰).

Iron isotope fractionation variations of the iron isotope distribution between two different chemical species; for instance dissolved ferrous Fe and a solid precipitate.

Isopach map map of the equal thickness of the deposits. It is necessary to determine the energy of the eruption, its volume and the plume height.

Jointed basalt Jointing is frequently shown by lava flows when they are cut by erosion or in a quarry as a series of parallel columns several meters high and a few decimeters in diameter. The jointing results from shrinkage (decrease in volume) at the end of lava cooling, when lava changes eventually from a semi-liquid state to a totally solid state.

Juvenile pyroclast in volcanic deposits, they consist of particles made of the magma erupting.

k-strategist species (k-strategist) In ecology, r/K selection theory relates to the selection of combinations of traits in an organism that trade off between quantity and quality of offspring. K-selected species are described as "equilibrium" whereas r-selected species are referred to as "opportunistic".

L* parameter a specific spectrophotometric parameter providing the brightness of sediment.

LMGE Laboratoire Microorganismes: Génome et Environnement (LMGE, CNRS/Université Blaise Pascal Clermont-Ferrand 2).

Lahar mudflow of volcanic origin.

Lake cult most maar-lakes had specific lake cults in pre or early Christian times, generally with peculiar pagan (Mephites, Diana, Jupiter) or Christian (StMichael, Holy Virgin) deities.

Lake degassing emission of dissolved gases contained in lake deep waters, generally CO_2, sometimes with H_2S and CH_4. Intensity range from light bubbling to water fountain, limnic erution, overspills. It can be a permanent state or sudden events. Many degassing events described since 2400 years in maar-lakes remained unknown to modern scientists until 2000s.

Lake overspill rise of lake water levels due to increasing dissolved gas content, then overspill above the lake rim. The Albano Lake catastrophic event (398 BC) that threatened Rome has been the first to be demonstrated by Earth scientists in the 2000s; it was well known by ancient Rome historians and described as a major *prodigium*.

Lava a general term for a molten extrusive volcanic formation and also for the rock that is solidified from it.

Leaching The action to drain away from soil, ash, or similar material by the action of percolating liquid, especially rainwater.

Limnic eruption sudden violent explosion of lake surface waters due to deep water degassing, often catastrophic to nearby populations. Although Lake Nyos event (1986, 1700 victims) is the first being analysed by modern scientists, the Lake Monoun explosion occurred before in 1984.

Limnogeology the branch of geology that deals with lakes aiming at reconstructing continental environmental changes through geological times, based on a multidisciplinary study of fossil or recent lacustrine sedimentary systems.

Limnology the scientific study of the physical, chemical, meteorological, biological and ecological conditions and characteristics of all inland waters.

Lineament a linear, topographic feature of regional extent that is believed to reflect a tectonic feature due to crust deformation.

Lorica Shell-like protective outer covering, often reinforced with sand grains and other particles that some protozoans and loricifera metazoans secrete. Usually it is tubular or conical in shape.

Lysogeny Replication cycle in which temperate bacteriophage integrates its nucleic acid into a host genome. Thus, the newly integrated genetic material may be transmitted to daughter cells at each subsequent cell division without viral particles production and host cell lysis.

Maar-lake lake located in an explosive volcanic crater (maar), often with steep clifs and generally without river input or outlet.

Maar German word, which initially points to the circular lakes in the Eifel area located in Rhineland (Rhénanie). Crater of explosion, generally showing a large diameter, occupied by a lake or filled by sediment, and that has cut out the pre-existing basement. The fragments extracted from the basement always represent a noticeable, often prevailing percentage of expelled products. The maar contains two elements:

- The crater, a circular depression with steep walls, the diameter of which varies between 100 m and 2 km, and profoundly rooted in the basement (generally approximately 1000 m). It can be occupied by a lake, a wet zone, an accumulation of lava, a cone of scoriae, or by a lava flow stemming from a neighbouring volcano.
- The projections form a ring or low-relief crescent. The stratified deposits contain a large proportion of angular fragments scoured from the basement. The magma of the eruption (juvenile magma) is represented by dense clasts and heavy bombs termed "cauliflower bombs" due to their varicose surface.

A maar forms when the ascending magma meets en route surfacial water, groundwater or a stream. The instantaneous evaporation of this water fuels a very explosive activity. A part of the ejected products is launched vertically and dispersed by a plume, and another part moves as ground-hugging flows due to the lateral expansion of gases, which wreack very destructive effects. Simultaneously, the maar enlarges as a result of circular collapses of the bedrock.

Magnetic susceptibility measurement of volumic induced magnetization in a sediment linked with its grain size and content in magnetic material.

Marvelous extra-ordinary natural lake event or behavior, difficult to explain with contemporary knowledge and worth to be reported.

Mass wasting deposit a general term for a deposit resulting from the dislodgement and downslope transport of material under the direct application of gravitational body stresses.

Meromictic a lake that undergoes incomplete mixing of its waters during periods of circulation. Most often consists of a free circulating upper fresh water layer (*mixolimnion*) and a stagnant, often anoxic, bottom water layer (*monolimnion*).

Meromixis A lake or a pond is considered as meromictic when it is composed of a monimolimnion.

Marvelous extra-ordinary natural lake event or behavior, difficult to explain and worth to be reported.

Mesological in this case mean objective and measurable environmental data.

Metagenomics culture-independent methods where the genetic material (metagenomes) recovered directly from environmental samples is used to characterize microbial communities thriving in a particular ecosystem.

Methanogenesis The production of methane by bacteria or other living organisms.

Methylotrophy The consumption of C1 compounds (*i.e.*, methanol, methylamine, methane).

Microbial biogeography study of the distribution of microbial species with a special focus on the existence of geographic barriers and endemic species.

Microbial food web refers to the combined trophic interactions among microbes which included dissolved organic matter, prokaryotes, and both photosynthetic and heterotrophic protists (such as ciliates and flagellates) in aquatic environments.

Microbial loop Trophic pathway in the microbial food web where dissolved organic carbon (DOC) is returned to higher trophic levels *via* its incorporation into bacterial biomass, and then coupled with the classic food chain formed by phytoplankton-zooplankton-nekton.

Microcosm An experimental tool that brings a small part of the natural environment under controlled conditions.

Misbehavior deviation of a normal lake functioning perceived by local witness. Applicable to diverse state of lake degassing.

Misericords inferior part of wooden credences in church choir, carved with figurative or non-figurative representations. In Besse church the exceptional misericords are interpreted to be related to Pavin misbehavior and its damages on populations.

Mitomycin C Antibiotic produced by the soil actinomycete *Streptomyces caespitosu*. It is a DNA cross-linker and a DNA-alkylating agent leading to mispairing of bases, DNA strand breakage, and cross-linking of complimentary strands which prevents DNA synthesis. In aquatic virology, mitomycin C is used to activate prophages.

Mixolimnion The upper layer of a meromictic lake, characterized by mixing events.

Mixotrophic An organism capable of existing as either/both an autotroph or/and heterotroph.

Molecular diffusion Molecular diffusion is caused by the agitation of water molecules in the water column due to temperature and not by a mean advective transport or turbulence.

Molecular diversity Diversity based on the composition of one or several molecular markers. These markers could be nucleic acids, proteins, carbohydrates, metabolites… The molecular diversity is studied by sequencing.

Monimolimnion the lower, dense stratum of a meromictic lake that does not mix with the waters above.

Monooxygenase One of a group of enzymes that catalyses the incorporation of one hydroxyl group into substrates.

Moraine a mount, ridge, or other distinct accumulation of unsorted, unstratified glacial drift, predominantly till, deposited chiefly by direct action of glacier ice.

Multibeam bathymetric measurement measurement of the lake's topography (bottom and slope) using depth sounding from an echosounder (sonar) mounted beneath a boat.

Nanoalgivore Algivorous organisms with a size between 2 and 20 μm.

Nanophytoplankton A phytoplanktonic organism with a size between 2 and 20 μm.

Nanoplanktonic A planktonic organism with a size between 2 and 20 μm.

Nutrients dissolved nitrogen, phosphorus and silica necessary for algal production in waters.

ONEMA French National Agency for Water and Aquatic Environments.

Oligonucleotide probes Short sequence of nucleotides that are synthesized to match a specific region of DNA or RNA then used as a molecular probe to detect the specific DNA or RNA sequence.

Oligotrophic state of a lake in which algal production is low

Onlap description of the termination of a stratigraphic horizon (or seismic reflection) whereby the horizon terminates against the margins of a depositional basin.

Operational Taxonomic Units (OTU) Operational definition of a species or group of species often used when only DNA sequence data is available.

Operational taxonomic unit (OTU) a group of environmental sequences sharing a given similarity level required for quantitative and comparative sequence analyses.

Organohalide/Organohalogen Organic compounds containing a C-halogen bond.

Osmotrophy Uptake of dissolved organic compounds by transmission across the cell membrane.

Outcropping part of a geologic formation appearing at the surface of the Earth.

Outlet a stream flowing out of a lake, pond, or other body of standing water.

Oxidation The loss of one or more electrons by an atom, molecule, or ion.

Oxycline a sharp gradient on oxygen concentrations.

Oxygen Index (OI) amount of oxygen content in the OM released during pyrolysis in Rock-Eval analyzer, normalized to TOC (expressed in mg $CO_2.g^{-1}$ TOC).

Oxygen chemocline (so-called oxycline) strong dissolved oxygen gradient in the water column, typically representing a redox boundary.

PCR Tag-sequencing NGS (next generation sequencing) technique. Before sequencing, a PCR step of a hypervariable sequence is realized. A small oligonucleotide sequence (tag sequence) specific to sample is added to the primers used for PCRs, similarly to the technique of barcoding.

PEG-Model Plankton Ecology Group (PEG) Model is a standard template to describe factors driving the seasonal wax and wane of phyto- and zooplankton in lakes.

Paleolimnology the study of ancient lakes from their sediments and fossils aiming at reconstructing the paleoenvironments of inland waters. Such studies are based on analyses of sediment cores including the physical, chemical and mineralogical properties of sediments and divers biological records.

Palynostratigraphy Determination of pollen grains is commonly used as a stratigraphical tool in fossil terrains ant more specially in peat-bog and lake deposits to identify various pollen spectrums (palynozones) corresponding with palaeoenvironmental changes. Regarding the recent volcanic activity in the Massif Central, the following palynozones are concerned from −14.000 y.BP: Lateglacial; Allerød/Young Dryas/-11.500/Holocene: Préboréal/Boréal/Atlantic/Subboreal/Subatlantic/Present. Tephra layers, when clearly identified in various pollen diagrams are used a linkage tool of precise time markers.

Parasitism (Parasites) Non-mutual symbiotic relationship between species, where one species, the parasite, benefits at the expense of the other, the host. Parasites show a high degree of specialization, and reproduce at a faster rate than their hosts.

Partition (or distribution coefficients): concentration ratio of an element in the solid over the melt at equilibrium.

Passive tracer A passive tracer is a chemical molecule or a microorganism that is transported in the water body and does not react or is transformed during its transport in the water body.

Pavin event well dated and contextualized extra-ordinary event of Pavin, diverging from a normal non-degassing lake behavior. They are reported in Vassivière archives, (e.g. in 1547, 1551), as a fantastic story, (1632) in an official publication (1783), by a witness account (1936) and possibly in a Gregorius text (Caluppa's dragon encounter, around 590 AD). Events archived in Pavin sediments complement these (Chap. 23).

Pavin experiments they include throwing a stone into the lake to check its assumed immediate thunder, lightning and storm, and shooting a gun to generate a thunder noise. They were performed at Pavin since de Frétat's visit in 1672 until the mid-nineteenth century.

Pavin legends ancient legends, describing pre or early Christian representations of Pavin its surrounding and Pavin events. So far few of them are found: *Fairies garden*, the *Fairies dance*.

Pavin stories stories told in the eighteenth and nineteenth century to Pavin visitors by their local guides, as the *Thrown stone* and the *Whirl and storm*. They describe traditional Pavin (mis) behavior and are generally accompanied with Pavin experiments.

Permafrost soil permanently frozen (or for at least 2 years) and therefore impermeable.

Phagotrophy The process by which unicellular organisms derive their food from engulfing and digesting other cells.

Phreatomagmatic a volcanic explosion that extrudes both magmatic gases and steam caused by the contact of magma with groundwater or shallow surface water.

Phylogenetic clade Set of related sequences that originate from a single common ancestor.

Phylotype Observed similarity that classifies a group of organisms by their phenetic relationship. This phenetic similarity may reflect the evolutionary relationships.

Phytoplankton bloom a rapid increase or accumulation in the population of algae in an aquatic ecosystem.

Picoplankton Heterogeneous assemblage of organisms 1–5 μm in size that live suspended in the water column.

Pierre au Trésor (Treasure stone) an enigmatic 250 kg carved stone retrieved from Pavin waters in 1909. Its antique origin has been confirmed as a Pompeian meta millstone.

Piston cores sedimentary cores retrieved from a basin fill by using a piston inside a cylinder which reduces friction by creating suction.

Plague a possible representation of mass deaths due to CO_2 degassing; also related to buboes found on CO_2 intoxicated victims.

Primer Strand of nucleic acid that serves as a starting point for DNA synthesis.

Prodigium unexplicable terrifying events in ancient Rome. Interpreted as a disruption of the cosmic harmony or of the anger of gods, they were reported to special authorities who decided the necessary public rites to reset the harmony. They have been listed by Julius Obsequens. Many prodigii are related to the Colli Albani volcano near Rome and its maar-lakes, including Lake Albano overspill in 398 BC.

Prophage a stable, inherited form of bacteriophage in which the viral genetic material is integrated into the bacterial host's genome and is potentially replicated and expressed.

Proteome all the proteins in a cell, a tissue, an organ or an organism.

Proteomic science that studies the proteome (see proteome) at a given time and under given conditions.

Protists (Protista syn. Protoctista) eukaryotic organisms that are unicellular and sometimes colonial or less often multicellular and that typically include the protozoans, most unicellular algae, and often some fungi (as slime molds).

Pseudolysogeny defined as the stage of stalled development of a bacteriophage in a host cell without any multiplication of the phage genome (as in lytic development) nor its synchronized replication with the cellular cycle (as in lysogenic cycle) without viral genome degradation and allowing the subsequent restart of virus development.

Pulse field gel electrophoresis technique used for the separation of large DNA molecules into a gel matrix by applying an electric field that periodically changes direction in order to minimize overlap of the spots due to diffusion

Pumice a pyroclast (tephra) that consists of highly vesicular, rough textured volcanic glass, which may or may not contain crystals.

Pumices volcanic rock consisting of highly vesicular rough textured volcanic glass, created when super-heated and high-pressurized rock is violently ejected from a volcano.

Pyrite iron sulfide mineral (Fe(II)S(0)S(−II)) in the cubic crystal system.

Pyrolysis thermochemical decomposition of organic material at elevated temperatures in the absence of oxygen.

Pyrosequencing Pyrosequencing involves sequencing of DNA by synthesizing the complementary strand of a single base at a time, while determining the specific nucleotide being incorporated during the synthesis reaction. The reaction occurs on immobilized single stranded template DNA where the four deoxyribonucleotides (dNTP) are added sequentially and the unincorporated dNTPs are enzymatically degraded before addition of the next dNTP to the synthesis reaction.

Quantitative real-time PCR (qPCR) Technique based on the polymerase chain reaction (PCR) which is used to amplify and simultaneously detect or quantify a targeted DNA molecule.

Quorum Sensing set of regulatory mechanisms that control the coordinated expression of some bacterial genes within a bacterial population.

REE REE, the 17 chemical elements in the periodic table of the La(57) to Lu(71). LREE: light REE from La to Eu (63); HREE: from Gd (64) to Lu (71). Sometimes Sc(21) and Y(39) are also included as their behaviours are respectively close to those of La and Lu.

Radiative Forcing difference of radiant energy received by the Earth and energy radiated back to space.

Radiocarbon reservoir effect impact of carbon cycling hydrodynamic conditions in the environment on radiocarbon age determinations. Radiocarbon ages of materials are calculated by comparison to normal atmospheric CO_2 isotopic ratio. Dating materials not issued from terrestrial plant remains but made of more carbon sources than atmospheric CO_2 requires correcting radiocarbon ages by specific age offsets depending of locally particular conditions of the reservoir effect.

Rare biosphere collection of a huge number of low abundant taxa that characterize microbial communities when studied by molecular surveys.

Rarefaction curves Plot of the number of species as a function of the number of samples. In ecology, rarefaction is a technique to assess species richness from the results of sampling. Rarefaction allows the calculation of species richness for a given number of individual samples, based on the construction of so-called rarefaction curves.

Reaction quotient a function of the activities or concentrations of the chemical species involved in a chemical reaction.

Recombination Molecular processes by which parts of two different DNA molecules join and form a unique hybrid molecule.

Reduction The gain of one or more electrons by an atom, molecule, or ion.

Reflection acoustic horizon on seismic reflection or subbottom profiles linked to a change in acoustic impedance that is the product of sediment bulk density and sound wave velocity.

Rhizoid (Rhizoidal system) Short, thin filament found in fungi and in certain plants and sponge that anchors the growing (vegetative) body of the organism to a substratum and that is capable of absorbing nutrients. In fungi, the rhizoid is found in the thallus and resembles a root. It may serve either as a feeding organ (*Rhizopus*) or to anchor the thallus to its substratum (*Chytridium*).

Rock fall the free falling of a newly detached segment of bedrock of any size from a cliff or other very steep slopes.

Runoff Water which is not absorbed by the soil and flows to lower ground, eventually draining into a stream, river, or other bodies of water.

r-strategist species (r-strategists) *Cf* K-strategist species

S2 amount of hydrocarbon that escapes from the sample during the thermal cracking in Rock-Eval analyzer, expressed in mg HC/g sediment.

SSU rDNA Small SubUnit of Ribosomal DNA.

Saprophytism Ability of certain organisms to use dead or decaying organic matter as source of energy.

Sauroctonous dragon killer, mostly used in early Christian times (e.g. St Georges).

Scoria cone a volcanic cone formed by lava and or pyroclastics build up around a volcanic vent.

Seasonal overturn During winter, the lake water column becomes unstable when the uppest layers are low dense than the deepest ones. Therefore, the lake is affected by a convective movement of the layers of the water column that is called seasonal overturn.

Sedimentary event individual sedimentary bed recording rare geological events at any scale.

Sedimentary facies lithological signature of sediments including the color and the occurrence of sedimentary structures, bedding and grain size distribution.

Sedimentary unit sedimentary layer composed by similar sedimentary facies or any of the many characters, properties or attributes that sediments may possess.

Seismic reflection profile a cross section through the subsurface produced by the propagation and reflection of an acoustic wave into a sedimentary basin. It represents the layering geometry in the subsurface tens of meters to kilometers below the surface, depending of the frequency.

Sensory grid of degassing a collection of sensory perception of lake misbehavior (sight, hearing, taste, smell, touch) assembled from scientist's records of degassing lakes in Cameroun and Italy (Table 1.2). It includes observations at the lake or its surroundings, their far away effects on the environment and damages to populations and living beings. Used to re-interprete historical or legendary accounts.

Sestonic particulate matter suspended in the water column.

Siderite iron carbonate mineral ($Fe(II)CO_3$) in the trigonal crystal system.

Slide scar a scratch left on a sea floor/lake floor by subaqueous sliding or slumping.

Slump the sliding-down of a mass of sediment shortly after its deposition on an underwater slope characterized by a shearing and rotary movement along a curved basal slip surface.

Social responses to degassing The social responses generated by lake disorders are very diverse. They depend on the perioid and culture. They range from purely technical (tunnels and canals) to lake cults with specific deities and Christian communities on lake shores or in their vicinity, lake-related legends and fantastic stories featuring fantastic animals, as dragons, and/or marvelous events, including the damages caused to populations (e.g. a "plague") and their iconographic representations.

Solifluction flowage of oversaturated soil and colluvium on steep slope. Slow flowage produces smooth, gentle, concave and convex landforms on slopes.

Sporogenesis Production of spores

Sub-bottom profile a very high resolution reflection profile detailing the first few meters of the subsurface (m to tens of meters).

Sunken city legends common legend found for many lakes featuring a misbehaving city sunk into the lake; interpreted as a transcription of the biblical Sodom and Gomorrha myth. It appears in Pavin in the late XIXth and is thought by guidebooks authors to be its original legend since then

Surge or pyroclastic surge One of the pyroclastic density currents (PDCs), a diluted current interpreted as moving with a turbulent regime. Surges usually form fine grained, stratified and cross-bedded deposits.

Syntrophic association A biological relationship in which microorganisms of different species or strains are mutually dependent on one another for nutritional requirements.

T-RFLP Terminal Fragment Length Polymorphism is a culture- independent, rapid, sensitive and reproducible method of assessing diversity of complex communities without the need for any genomic sequence information. The technique provides information on a collection of microorganisms that may be present in a given sample.

Tephra The word 'tephra' is derived from the Greek *tephra* (ϕ) meaning ash, and is a collective term for all the unconsolidated, primary pyroclastic products of a volcanic eruption. It is inclusive of all grain sizes, including fine ash (dust) (<0.06 mm), coarse ash (0.06–2 mm), lapillus/i (2–64 mm), and block/s or bomb/s (>64 mm). Strictly, 'tephra' is singular (it does not occur in Greek in the plural form) and hence it is used as a collective with singular verb forms. Note that in compound words in English based on Greek roots, the original final vowel ('a') is always replaced with 'o' to form derivatives such as tephrostratigraphy.

Tephrochronology (*Sensu Stricto*) 'Tephrochronology' (*sensu stricto*) was defined as a dating method based on the identification, correlation and dating of tephra layers that is, essentially using tephra deposits as time-stratigraphic marker beds to establish numerical or relative ages.

Plinian volcanic eruption are of peculiar interest for tephrochronology as far as huge amount of tephra are ejected into the stratosphere (mushroom shape), and consequently widespread giving rise to ash-fall sometimes around the world.

Tephropalynology This is essentially the joint application of tephrochronology and palynology, which involves the correlation and dating of pollen-based palaeoenvironmental reconstructions using tephra layers as the linkage. It also encompasses pollen-based studies of the impacts of tephra fall in various environments.

Tephrostratigraphy 'Tephrostratigraphy' is the study of sequences of tephra layers and related deposits and their relative ages. It involves definition, description and the identification and characterization ('fingerprinting') of tephras or tephra sequences using their physical, mineralogical, or geochemical properties.

Thalli (Thallus) Entire body of a multicellular non-moving organism in which there is no organization of the tissues into organs.

Thermal stratification In the water column of a lake, a condition that may develop during the summer in which the warmer, less dense layer (epilimnion) overlies a colder denser layer (hypolimnion) over a short vertical distance and prevents mixing between them.

Thermocline (sometimes metalimnion in lakes) is a thin but distinct layer in a large body of fluid (e.g., water, such as an ocean or lake, or air, such as an atmosphere) in which temperature changes more rapidly with depth than it does in the layers above or below.

Thermokarst Lake formed by meltwater from the thawing permafrost.

Thin-section laboratory preparation of rocks, minerals, sediments or organic compounds for use with a microscope.

Topographic drainage basin a surface area occupied by a drainage system bounded by drainage divide and draining into a water body.

Tracer Anything that allows us to follow or 'trace' a process. A **conservative** tracer is one that can only be altered by mixing in the ocean interior.

Transduction Transfer of genetic material or characteristics from one microorganism to another by a viral agent (as a bacteriophage)

Transformation The genetic alteration of a cell resulting from the direct uptake and incorporation of exogenous genetic material (exogenous DNA) from its surroundings and taken up through the cell membrane(s).

Transporter proteins are specific proteins involved in facilitated diffusion, where it binds to larger molecule needed within the cell and guide it into the cell, moving particles against the concentration gradient.

Trophic network (food web) Representation of all food chains existing in a single ecosystem.

Trophic status the degree of biological production within a aquatic ecosystem.

Trophodynamics (trophic dynamics) The dynamics of nutrition or metabolism

Tunnel technical means to limit maar-lake surge and overspill (Albano, Nemi, Averno, Laacher See lakes). They were excavated across the-lake rims as early as 394 BC by Romans in Lake Albano and most were still in use in the XXth century. Canals dug out had the same function (Monticchio, Eifel maars).

Turbidite a sediment deposited from a turbidity current and characterized by the development of graded bedding and moderate sorting of particles.

Turbulent kinetic energy For a water parcel of a water body, the turbulent kinetic energy (TKE) represents the part of the kinetic energy that is due to turbulence in the water column. This part of kinetic energy is not related to any mean advective current (average of speeds is zero)

UFE Units of Fishing Effort. The surface of the nets used is multiplied by how long they are set in the water, which determines a certain number of UFE, a fishing effort corresponding to 1 m^2 of net set for 1 h.

Ultracentrifugation technique using a high-velocity centrifuge in order to separate colloidal or submicroscopic particles

Varves annual layer of sediment.

Vertical diffusivity (or K_z) K_z is the vertical coefficient of diffusion in the water body, considering both molecular and turbulent diffusion. Modelling of turbulence in the form of a diffusion law is an approximation, this model being one of the most simple. Other models of turbulence for water bodies exist.

Viral loop (or shunt) The viral-mediated movement of nutrients from organisms to pools of dissolved and non-living particulate organic matter in aquatic systems.

Viral lysis (or lytic cyle) lysis of host cell associated to the production of new viral particules by diverting the host cellular machinery

Virome The genomes of all the viruses that inhabit a particular organism or environment

Vivianite hydrated iron phosphate mineral ($Fe(II)_3(PO_4)_2.8(H_2O)$) in the monoclinic crystal system.

Volcanology the branch of geology that deals with volcanism, its causes and phenomena.

WFD Water Framework Directive. European directive to improve the quality of surface and subterranean water.

Water fountain kinds of natural geysers in degassing lakes. They can last long (Lake Monticchio, 5 m high) or be related to a single violent event (Lake Nyos, 50–80 m high). They are suspected at Pavin (1632), and also in Lake Ulmen (Eifel).

Watershed The sum of all the lands and smaller bodies of water which drains into a particular stream, lake or river.

Wheathering The physical, chemical and biological processes that decompose rock at and below the surface of the Earth through low pressures and temperatures and the presence of air and water.

Zoosporic organisms, zoospores Zoosporic organisms are organisms that produce motile asexual spore that uses a flagellum for locomotion/dissemination: zoospores.

Index

T. Sime-Ngando et al. (eds.), *Lake Pavin*, DOI 10.1007/978-3-319-39961-4

MIX
Papier aus verantwortungsvollen Quellen
Paper from responsible sources
FSC® C105338

If you have any concerns about our products,
you can contact us on
ProductSafety@springernature.com

In case Publisher is established outside the EU,
the EU authorized representative is:
Springer Nature Customer Service Center GmbH
Europaplatz 3, 69115 Heidelberg, Germany

Printed by Libri Plureos GmbH
in Hamburg, Germany